The Genetics and Biology of Sexual Conflict

A subject collection from *Cold Spring Harbor Perspectives in Biology*

OTHER SUBJECT COLLECTIONS FROM *COLD SPRING HARBOR PERSPECTIVES IN BIOLOGY*

The Origin and Evolution of Eukaryotes
Endocytosis
Mitochondria
Signaling by Receptor Tyrosine Kinases
DNA Repair, Mutagenesis, and Other Responses to DNA Damage
Cell Survival and Cell Death
Immune Tolerance
DNA Replication
Endoplasmic Reticulum
Wnt Signaling
Protein Synthesis and Translational Control
The Synapse
Extracellular Matrix Biology
Protein Homeostasis
Calcium Signaling
The Golgi
Germ Cells
The Mammary Gland as an Experimental Model
The Biology of Lipids: Trafficking, Regulation, and Function
Auxin Signaling: From Synthesis to Systems Biology
The Nucleus

SUBJECT COLLECTIONS FROM *COLD SPRING HARBOR PERSPECTIVES IN MEDICINE*

The Skin and Its Diseases
MYC *and the Pathway to Cancer*
Bacterial Pathogenesis
Transplantation
Cystic Fibrosis: A Trilogy of Biochemistry, Physiology, and Therapy
Hemoglobin and Its Diseases
Addiction
Parkinson's Disease
Type 1 Diabetes
Angiogenesis: Biology and Pathology
HIV: From Biology to Prevention and Treatment
The Biology of Alzheimer Disease

The Genetics and Biology of Sexual Conflict

A subject collection from *Cold Spring Harbor Perspectives in Biology*

Edited by

William R. Rice
University of California, Santa Barbara

Sergey Gavrilets
University of Tennessee

COLD SPRING HARBOR LABORATORY PRESS
Cold Spring Harbor, New York • www.cshlpress.org

The Genetics and Biology of Sexual Conflict
A Subject Collection from *Cold Spring Harbor Perspectives in Biology*
Articles online at www.cshperspectives.org

Printed in the United States of America

Executive Editor	Richard Sever
Managing Editor	Maria Smit
Senior Project Manager	Barbara Acosta
Permissions Administrator	Carol Brown
Production Manager	Denise Weiss
Production Editor	Diane Schubach
Cover Designer	Denise Weiss
Publisher	John Inglis

Front cover artwork: After losing one antler in fights with other males over premier breeding grounds in Arizona's Sonoran Desert, a male mule deer (*Odocoileus hemionus*) meets local females. (Photo provided by Alex Badyaev/tenbestphotos.com.)

Library of Congress Cataloging-in-Publication Data

The genetics and biology of sexual conflict / edited by William R. Rice, University of California, Santa Barbara, and Sergey Gavrilets, University of Tennessee.
pages cm
"A subject collection from Cold Spring Harbor Perspectives in Biology."
Includes bibliographical references and index.
ISBN 978-1-62182-059-8 (hardcover : alk. paper)
1. Sexual dimorphism (Animals) 2. Evolutionary genetics. 3. Sexual selection in animals. I. Rice, William R. (William Richard), 1951- II. Gavrilets, Sergey.

QP81.5.G465 2014
591.4'6--dc23

2014019587

10 9 8 7 6 5 4 3 2 1

All World Wide Web addresses are accurate to the best of our knowledge at the time of printing.

For a complete catalog of all Cold Spring Harbor Laboratory Press publications, visit our website at www.cshlpress.org.

Contents

Preface

THE FIELD OF SEXUAL CONFLICT BEGAN in the late 1970s with a seminal book chapter written by Geoff Parker entitled "Sexual Selection and Sexual Conflict."[1] The field experienced a major burst of activity in the mid-1990s as sexual conflict became more fully integrated into the field of sexual selection. The next decade of studies was then expertly summarized, analyzed, and integrated with earlier work in the highly cited book entitled *Sexual Conflict,* by Göran Arnqvist and Locke Rowe published in 2005.[2] Now, almost a decade later, we are bringing together the expertise of scientists from a broad cross section of the disciplines that collectively cover the field of genetic sexual conflict. In the Foreword, we describe the breadth of the field of sexual conflict and how essentially all aspects of the discipline derive from a simple reproductive dichotomy: microgametes (sperm) and macrogametes (eggs) that fuse to produce sexual offspring.

We would like to express our appreciation to two people at Cold Spring Harbor Laboratory Press: Richard Sever for inviting us to edit this book and Barbara Acosta for extensive help in the editing and production process. We also are thankful for the welcome response we received from the diverse community of scientists studying sexual conflict—nearly no one declined our request to contribute to this volume. Last, we dedicate this book to our wives, Kathryn Schoenrock and Galia Gavrilets, who have been a constant source of encouragement despite our frequent preoccupation with the development of this book.

BILL RICE
SERGEY GAVRILETS

[1] Parker GA. 1979. Sexual selection and sexual conflict. In *Sexual selection and reproductive competition in insects* (ed. Blum MS, Blum NA), pp. 123–166. Academic, New York.

[2] Arnqvist G, Rowe L. 2005. *Sexual conflict.* Princeton University Press, Princeton, NJ.

Foreword

SEXUAL GENETIC CONFLICT OCCURS WHEN genetic variation increases the fitness of one sex at the expense of the other sex. It can occur at many levels—from alleles at a single locus that are selected divergently between the sexes, to different genomic components that are transmitted asymmetrically between the sexes, to alleles at different gene loci that favor opposite sexes during sexual interactions. Some forms of sexual conflict are special cases of intragenomic conflict. The opportunity for sexual genetic conflict originates with the production of two types of interdependent gametes.

Anisogamy. Nearly all multicellular organisms reproduce sexually through anisogamy (aka heterogamy), in which a microgamete (sperm) fuses with a macrogamete (egg) to form an offspring. This fundamental sexual asymmetry in gamete size generates a cascade of secondary asymmetries (like a fundamental tone with its harmonic overtones), which in turn contribute to the diverse forms of sexual genetic conflict that are collectively covered in this volume. Below we describe the major asymmetries and their contribution to sexual genetic conflict.

Unequal meiotic cell division in egg production. In many species, eggs are produced by a pair of unequal meiotic cell divisions that result in one large and functional ovum and three minuscule, dead-end polar bodies. This asymmetry in size and function produces strong selection on heterozygous alleles to cheat on Mendel's law of random segregation and increase their probability of ending up in the functional egg—a process called female meiotic drive. A broad spectrum of both plants and animals has independently evolved female meiotic drive, and indirect evidence indicates that vast stretches of the genome (i.e., the pericentric heterochromatin that can encompass up to 50% of a genome) evolved primarily to influence this process. Female meiotic drive may generate sexual conflict because many lines of evidence indicate that centomeric regions that produce successful drivers in females during oogenesis commonly cause infertility in males during spermatogenesis.

Asymmetrical propagation of genomic components. Meiotic cell division is symmetrical during spermatogenesis, but sperm have evolved to expel most of their cytoplasm during their ontogeny, including nonnuclear genes such as those of the mitochondria and endosymbiotic bacteria. The resulting asymmetry, in which mitochondria and endosymbionts are transmitted only (or predominantly) via the egg, has led (in diverse species) to cytoplasmic genes evolving to kill sons when they compete with their sisters for limiting resources and/or when sibling mating harms daughters via inbreeding depression. Egg-limited genetic transmission has also led to mitochondria in hundreds of species of plants that have evolved to kill developing pollen. Sperm and pollen production leads to sexual conflict (son or pollen killing by female-propagated cytoplasmic genes) because of the asymmetries in transmission that it generates.

Dimorphic sex chromosomes. Separate sexes have commonly led to the evolution of dimorphic sex chromosomes that do not recombine along all or much of their length. The asymmetries in transmission of the X and Y sex chromosomes (X to daughters and Y to sons), and in their genetic relatedness (compared with homologous pairs of autosomes, the X and Y have markedly reduced relatedness), generate many additional opportunities for genomic conflict. In XY males, each sex chromosome is selected to kill or incapacitate the sperm carrying the other type of sex chromosome in order to gain a transmission advantage. Such selection has led to widespread X- and Y-linked segregation distortion (aka male meiotic drive or selective sperm killing) in organisms as diverse as flies, beetles, and mammals. The X and Y are also selected to kill or debilitate the noncarrier sex of offspring when there is sibling competition and/or when brothers harm their sisters via sibling mating (SA-zygotic drive). This phenomenon has recently been documented in flies. Both sex-linked segregation distortion and zygotic drive are forms of sexual conflict because they increase production of one sex at the expense of the other sex.

Another sex chromosome asymmetry occurs because, unlike the autosomes that spend half of their time across generations in each sex, the X spends 67% of its time in females and is always hemizygous when in males. These transmission and expression patterns generate sexual conflict because they facilitate, on the X, the accumulation of dominant female-benefit/male-detriment alleles and recessive male-benefit/female-detriment alleles. This conflict can strongly influence genomic structure in the context of the location of the sex-determining switch. Tight linkage of a new, sex-determining mutation to a sexually antagonistic allele segregating on an autosome can produce a selectively favored supergene that can, in theory, supplant an extant sex-determining locus. This process may explain why it is not uncommon for sex-determining chromosomes to differ between sister taxa.

Intralocus sexual conflict. Male and female function, whether expressed in separate sexes or as separate functions within hermaphrodites, is a fundamental biological dichotomy that generates intrinsically different selective environments and hence sex-specific optimal alleles at many loci scattered across the genome. This selective asymmetry between the sexes generates an inevitable tension within the genome—called intralocus sexual conflict—because different alleles will commonly be favored by male versus female function. As a consequence, adaptive evolution by one sex will be at the expense of the other sex unless gene duplication or sex-specific gene regulation permits each sex to evolve independently. Intralocus sexual conflict can be so widespread that the genome-wide genetic correlation for adult fitness between the sexes was found to be strongly negative in a laboratory populations of flies, and recent studies indicate that gene expression levels underlie much of the observed intralocus sexual conflict.

Interlocus sexual conflict. With the exception of obligately self-fertilizing hermaphrodites, sexual reproduction requires coordinated interactions between different individuals, which we will assume here to be males and females, but the same logic can be applied to male and female function in outcrossing hermaphrodites. These reproductive interactions have both costs and benefits, many of which can differ between the sexes—another form of secondary sexual asymmetry. When there is lifelong monogamy—with no extra-pair reproduction—the costs and benefits of reproductive interactions are symmetrically shared by the pair because they act as a single reproductive unit, and conflict between the sexes is precluded. But most plants and animals are promiscuous to at least some degree, and this deviation from monogamy reduces the correlation in lifetime fitness between mating partners and opens up manifest opportunities for male versus female sexual conflict. The conflict operates through a shared interaction trait, like a prospective pair's decisions: to copulate or not, how many offspring to produce after mating, and how much to invest in each resulting offspring. The genes mediating the shared interaction trait will generally be different in males and females, creating the opportunity for interlocus sexual conflict—for example, a gene in males producing a seminal fluid protein that acts as a pheromone that binds a female receptor and thereby increases her egg-laying rate, and a different gene expressed in females coding for the seminal fluid protein's receptor or other genes influencing its downstream effects on her egg-laying rate. When the optimal egg-laying rate differs between the sexes, this difference sets the stage for interlocus sexual conflict and antagonistic coevolution between the loci in males evolving to increase female fecundity and different loci in females to resist this influence.

There is a surprisingly diverse array of ways for interlocus sexual conflict to be manifest. Some examples in this volume include:

- spiders in which females kill and consume their mates after mating and males that break off their penises during mating to produce plugs that prevent their mates from remating with other males
- many invertebrates in which males have evolved injection-like penises that enable them to fertilize a resisting female through almost any part of her body, and females that have evolved to digest sperm delivered in this way
- outcrossing hermaphrodites that castrate their mates in order to increase the damaged mate's investment in their shared offspring
- females of many bird species that regularly mate with males besides their pair-bonded social mate, causing males of some species to provision and protect a substantial proportion of unrelated offspring
- members of both sexes that have evolved to desert their mates while rearing a family, thereby forcing the deserted partner to rear their shared offspring alone
- males of many mammal species that routinely kill all extant offspring belonging to their newly acquired harem of females
- seminal fluid proteins that collectively increase a male's performance in sperm competition but that are also toxic to their mates
- paternally imprinted genes that modify gene expression in the fetus and thereby increase his mate's resource allocation to fetuses during gestation, and maternally imprinted genes that counterbalance these male-induced changes in gene expression
- spines on the penis that produce a fertilization advantage to males but cause substantial tissue damage to the reproductive tract of females
- genes coding for sperm binding proteins or their receptors on eggs that rapidly and antagonistically coevolve
- homosexuality as a consequence of sexually antagonistic alleles
- chase-away antagonistic coevolution between the sexes that can contribute to speciation and sex chromosome evolution

This volume will explore the diverse kinds and consequences of sexual conflict on the genetics and biology of sexual multicellular organisms. Although we cannot address all of the forms of sexual conflict here, this book brings together perspectives from a broad range of disciplines.

The Genetics and Biology of Sexual Conflict

A subject collection from *Cold Spring Harbor Perspectives in Biology*

The Sexual Cascade and the Rise of Pre-Ejaculatory (Darwinian) Sexual Selection, Sex Roles, and Sexual Conflict

Geoff A. Parker

Department of Evolution, Ecology and Behaviour, Institute of Integrative Biology, University of Liverpool, Liverpool L69 7ZB, United Kingdom

Correspondence: gap@liverpool.ac.uk

After brief historic overviews of sexual selection and sexual conflict, I argue that pre-ejaculatory sexual selection (the form of sexual selection discussed by Darwin) arose at a late stage in an inevitable succession of transitions flowing from the early evolution of syngamy to the evolution of copulation and sex roles. If certain conditions were met, this "sexual cascade" progressed inevitably, if not, sexual strategy remained fixed at a given stage. Prolonged evolutionary history of intense sperm competition/selection under external fertilization preceded the rise of advanced mobility, which generated pre-ejaculatory sexual selection, followed on land by internal fertilization and reduced sperm competition in the form of postcopulatory sexual selection. I develop a prospective model of the early evolution of mobility, which, as Darwin realized, was the catalyst for pre-ejaculatory sexual selection. Stages in the cascade should be regarded as consequential rather than separate phenomena and, as such, invalidate much current opposition to Darwin–Bateman sex roles. Potential for sexual conflict occurs throughout, greatly increasing later in the cascade, reaching its peak under precopulatory sexual selection when sex roles become highly differentiated.

Sexual selection and sexual conflict are vast fields in evolutionary biology; when possible, here, I refer to reviews. I begin with brief general historic overviews of sexual selection and sexual conflict; more detail can be found in Andersson (1994), Simmons (2001), Chapman et al. (2003), and Arnqvist and Rowe (2005). Much of the current state of the field of sexual conflict is covered in this collection.

My principal aim, however, is to outline how sexual selection and sexual conflict have changed through evolutionary time, from mostly gamete competition in early unicellular eukaryotes, intense sperm competition in ancestral sessile and relatively immobile organisms, to both pre-ejaculatory (Darwinian) and postejaculatory sexual selection. These transitions in the evolution of sexual strategy arise as logical consequences whenever certain successive conditions are met, and together form what may be termed the "sexual cascade."

Cite this article as *Cold Spring Harb Perspect Biol* doi: 10.1101/cshperspect.a017509

HISTORIC OVERVIEW: SEXUAL SELECTION

Precopulatory Sexual Selection

Darwin (1871, p. 256) defined sexual selection as depending "on the advantage which certain individuals have over other individuals of the same sex and species, in exclusive relation to reproduction." Although his definition clearly embraces all possible aspects of sexual selection, Darwin's evidence and discussions related exclusively to competition over mating and ejaculation. The past 40–50 years have seen massive research efforts devoted to this subject. Darwin's two major categories of pre-ejaculatory sexual selection (male combat and female choice) are now described in all evolutionary texts. The monograph by Andersson (1994) gives an excellent survey of all but the last two decades.

Interest and controversy surrounded sexual selection for half a century after Darwin, generating both detailed surveys (e.g., Richards 1927) and notable conceptual advances (Fisher 1930), but attention faded after Huxley's influence in the 1930s (Huxley 1938a,b). Despite the occasional inspirational study (e.g., Jacobs 1955), the subject languished for almost three decades, probably because of the (implicit) group/species selection interpretations of adaptation prevailing throughout that time. The notable exception was Bateman's (1948) now classic study of *Drosophila melanogaster*, which he used to support his argument that sexual selection arises because male fitness typically increases more steeply with number of matings with different females than female fitness mating with different males, a fact that (following Darwin) he attributed ultimately to anisogamy. He used this to explain why males show "undiscriminating eagerness" and females "discriminating passivity," behavioral characteristics of sexual selection stressed by Darwin (1871) and considered to be supported by general observation. The relation between fitness and number of matings is now termed the "Bateman gradient," and the predictions surrounding this classic view of sexual selection have been termed the Darwin–Bateman paradigm or DBP (Dewsbury 2005).

Despite the major early conceptual advances after Darwin made by Fisher (1930), an explosion of interest, including attention to Bateman's ideas, gained impetus much later during the behavioral ecology revolution of the 1970s (Parker 2006a), although the first signs came earlier. Maynard Smith's (1958) essay helped to clarify the different effects of natural and sexual selection, which were becoming confused, and Fisher's (1930) work was republished in 1958. For sexual selection by female choice, Fisher had proposed (1) that females showing a preference for males with traits indicating their biological fitness ("indicators"; see Andersson 1994) would be favored, which (2) leads to positive feedback between the gene for the preferred male trait and the gene for the female preference ("Fisher's sons effect"), accelerating fixation of both genes, and (3) female choice may exaggerate the development of the male trait beyond its natural selection optimum (the "runaway" process) until natural selection finally prevents further development. Fisher's last research student, Peter O'Donald, pioneered population genetic models of female choice (O'Donald 1962, 1980), concluding that conditions for runaway were restrictive. However, notably, more optimistic advances concerning Fisher processes were made by Lande (1981; see also Kirkpatrick 1982). "Indicators" include cues of "good genes" or immediate female benefits such as good male territories or parenting (e.g., Williams 1966; Orians 1969; Trivers 1972), "handicaps" that signal "good genes" (Zahavi 1975; Grafen 1990), and brightness of male coloration correlating with parasite resistance (Hamilton and Zuk 1982). Andersson (1994) gives an excellent review of the historical development of these ideas.

Male–male competition for matings has attracted less theoretical attention than female choice, possibly because it is simpler and less interesting conceptually. Trivers' (1972) proposal that relative parental investment (PI) determines the intensity of sexual competition undoubtedly catalyzed the sudden interest in sexual selection that began in the 1970s. It was followed by suggestions that the operational sex ratio (OSR) (Emlen and Oring 1977) or relative potential rates of reproduction (PRR) (Clutton-Brock and Vincent 1991) represented the key index. More recently, emphasis has been placed on the difference in variance of reproductive success of the sexes (Shuster and Wade 2003), an index expressing the opportunity for sexual selection and related to Bateman's (1948) original predictions. All measures (PI, OSR, PRR, and relative variance in reproductive success) are likely to covary with the relative "times in" and "times out" of the mating pool of the two sexes (Clutton-Brock and Parker 1992; see Parker and Birkhead 2013) and have good reasons to correlate with sexual selection intensity (Jennions and Kokko 2010; Kokko et al. 2012; Parker and Birkhead 2013), although debate still continues over exactly what measure best reflects the intensity of sexual selection (reviewed in Parker and Birkhead 2013).

Qualitative evidence for male–male competition has always been very strong compared with female choice. My PhD studies of dung flies, published mostly in 1970, may represent the first attempt to validate this aspect of sexual selection quantitatively (reviewed in Parker 1978a,b, 2006a). Field observations fitted well with predictions based on competitive optimization models, suggesting that sexual selection could indeed be the dominant selective force shaping male reproductive behavior. Males behaved so as to maximize fertilization rate given that other males do the same, that is, their behavior patterns appeared to be evolutionarily stable strategies (ESSs) (sensu Maynard Smith and Price 1973; Maynard Smith 1982) set by the selective pressure of intramale competition (reviewed in Parker and Pizzari, in press). There can be little doubt that intrasexual selection has played a major part in determining male mate searching, contest, and mate-guarding strategies, something that has been rather overshadowed in recent decades by the focus on female choice.

Postcopulatory Sexual Selection

It is interesting that Darwin dismissed lower invertebrates: "The lowest classes will detain us for a very short time. . . ." (Darwin 1871, p. 300), considering them to be unaffected by sexual selection: "In the lowest classes the two sexes are not rarely united in the same individual, and therefore secondary sexual characters cannot be developed. In many cases in which the two sexes are separate, both are permanently attached to some support, and the one cannot search or struggle for the other. . . . Hence in these classes, such as the Protozoa, Coelenterata, Echinodermata, Scolecida, true secondary sexual characters do not occur; and this fact agrees with the belief that such characters in the higher classes have been acquired through sexual selection, . . ." (Darwin 1871, p. 321). Darwin debated why bright coloration occasionally occurs in some sessile or relatively immobile invertebrates, concluding this to be the product of natural rather than sexual selection. Although he rejected the relatively mobile annelids and cephalopod mollusks as candidates for sexual selection, he devoted much discussion to the various arthropod taxa, particularly in the insects, in which copulation is often well developed and male fighting apparatus and bright coloration sometimes apparent.

Arthropods aside, Darwin thus gave virtually all his attention to vertebrates, particularly those with copulation and internal fertilization. He did not mention competition between ejaculates or massive male gametic expenditure in his "lowly organized" sessile or relatively immobile animals. I argue here that such stages played a major role, with their high testes expenditure because of competition between ejaculates, in the cascade of events in the evolution of sexual strategy that culminated in precopulatory sexual selection that Darwin so brilliantly perceived and described in 1871. That he so positively dismisses these taxa suggests that he did not suppress insights into ejaculate competition simply because of the sensitivities of Victorian culture. Remarkably, "Darwin probably never made the intellectual leap that would have allowed him to identify the possibility of postcopulatory sexual selection" (Birkhead 2010) or competition between ejaculates in general.

It is now well appreciated that sexual selection does not end when mating begins. Birkhead (2010) gives an excellent account of the historic aspects of postcopulatory sexual selection. My review of sperm competition in insects (Parker 1970) is claimed to have stimulated development of the field of postcopulatory sexual selection (e.g., Birkhead 2000, 2010; Simmons 2001; Eberhard 2009). Since 1970, enormous amounts of research, both empirical (e.g., Smith 1984; Birkhead and Møller 1992, 1998; Simmons 2001) and theoretical (reviewed in Parker and Pizzari 2010) have been devoted to the topic. Later, attention also focused on female control of fertilizations, that is, "cryptic female choice" in the form of sperm selection or ejaculate manipulation (Thornhill 1983; Eberhard 1985, 1996, 2009). Sperm competition is now seen as the postejaculatory analog of Darwinian male–male competition for mating and cryptic female choice, the analog of Darwinian female choice of males before mating.

Postcopulatory sexual selection, in fact, encompasses adaptations both during and after copulation (Eberhard 2009). Although categorization into pre- and postcopulatory sexual selection serves a useful purpose, it presents problems for the many externally fertilizing taxa in which copulation is lacking; "pre- and postejaculatory" are more general terms. Competition among ejaculates will normally be a strong selective force in external fertilizers, including those that are sessile and weakly mobile (Levitan 2010). Failure to appreciate this not only was an oversight by Darwin, but has also impeded a full understanding of the succession of evolutionary events (the "sexual cascade") that generated internal fertilization and precopulatory sexual selection.

HISTORIC OVERVIEW: SEXUAL CONFLICT

A detailed history of the field of sexual conflict can be found in Arnqvist and Rowe (2005). "Battle of the sexes" concepts (the notion that males and females have conflicting interests) go back into history and usually relate to the immediate behavior associated with conflicts between the sexes. In evolutionary biology, sexual conflict concerns instances in which the evolutionary interests of males and females are different (Parker 1979) and arises very commonly out of sexual selection when a mutation causes strategic changes that would increase fitness in one sex, but decrease fitness in the other sex (a sexually antagonistic mutation). In fact, sexual conflict concepts can be applied to all anisogamous sexual systems (Schärer et al. 2014) Hermaphrodites combine male and female in one individual, and sexual conflict operates differently in sequential and simultaneous hermaphrodites (Schärer et al. 2014). At the level of nuclear genes, conflict can be interlocus if it concerns traits determined by different loci or intralocus if it concerns a trait determined by alleles at a single locus (Rice and Holland 1997; Parker and Partridge 1998). Cytoplasmic genetic elements are maternally inherited and require a different logic (e.g., Zeh 2004; Hurst and Frost 2014). For interlocus conflict, a sexually antagonistic mutation will spread unless a counterstrategy evolves to prevent it at a locus expressed in the other sex. Various definitions have been proposed (see Arnqvist and Rowe 2005; Kokko and Jennions 2014), but it is important to note that sexual conflict in evolutionary biology relates to conflict in evolutionary time, which may or may not be manifest in the behavior observed in male–female interactions.

Although Darwin (1871) appears not to have discussed sexual conflict directly, he was well aware that sex-limited inheritance could relate to different selective forces acting on males and females. For example, he envisaged that adult female size was often increased by selection to produce more ova and cryptic female coloration could reflect camouflage against predators (i.e., because of natural selection), whereas increased male size, weaponry, and coloration often reflected the result of sexual selection. Darwin–Bateman notions of indiscriminate eagerness in males and discriminating passivity in females also hint at mating conflict.

The concept that a gene may be beneficial in one sex but costly in the other was appreciated for many decades by population geneticists (e.g., Fisher 1931, p. 363), but little detailed discussion of the opposing interests of the sexes can be found before 1970 (Arnqvist and Rowe 2005); with rare exceptions, the prevailing view was that reproduction is an exclusively cooperative venture between the sexes. The modern rise of the concept of sexual conflict, as with sexual selection, is linked to the behavioral ecology revolution that had its roots in the late 1960s and exploded in the 1970s (Parker 2006a). During this period, the concept became quite widespread among those involved. Williams (1966) discussed an "evolutionary battle of the sexes" in which males may be adapted to overcome female resistance to mating, which would generate counteradaptation in females. (Williams' index includes the term "sexual conflict," but in the sense of intrasexual competition between males.) Clear expositions in the biological literature appear to be those by Trivers (1972) and Dawkins (1976) (see also Dawkins and Krebs 1979).

Arnqvist and Rowe (2005) have claimed that an important early foundation was my paper (Parker 1979) on sexual conflict and sexual selection, which was delayed in press (for historic accounts, see Arnqvist and Rowe 2005; Parker 2006a, 2010, 2013). It had been stimulated by my dung fly research in the late 1960s in which I analyzed male and female interest separately and appreciated that they could often be in conflict. Realizing that this had general implications, the final paragraph of my sperm competition review concluded with (Parker 1970, p. 559):

> The female cannot be regarded as an inert environment in and around which this form of adaptation (male adaptation to sperm competition) evolves. Supposing, for example, a mating plug reduces a female's reproductive rate, this natural selection disadvantage will affect both male and female. Provided that the plug confers a sexual selective advantage on the male, which outweighs its natural selective disadvantage, it should evolve or be maintained. Resultant modifications within the female to prevent or reduce the disadvantageous effects of the plug might be expected; these adaptations may conflict with the line of adaptation in the male sex. The female, however, is not isolated from intramale selection because a female mating with a male possessing a character of sexual selective advantage will gain if the character is present in her male offspring.

This foreshadowed my 1979 analysis, which began by investigating the parameter space (the "conflict zone" or "battleground") within which sexual conflict will apply, given that the male and female have shared offspring (see also Andrés and Morrow 2003; Parker 2006b). Consider a mutant gene that increases a male's fitness through sexual selection, but has a collateral cost via the female affecting their shared offspring. If the cost is too high, there is no conflict, and the gene will not spread. If the cost is sufficiently low and the benefit high, the gene will spread, and so will a gene that causes females to prefer to mate with a mutant male. But for a large parameter space, there will be conflict; the male trait is favored, but so is female resistance to males with the trait, that is, the character is in male, but not female, interests. The analysis went on to define and analyze biological cases and conflict zones in which sexual conflict was likely, and two simulation models to investigate conflict resolution, one of which resulted in continual cycling of coevolutionary strategies ("unresolvable evolutionary chases").

This paper attracted little interest and, as a subject area, sexual conflict remained inconspicuous until the mid-1990s. Although exceptions during this period were few (see Arnqvist and Rowe 2005), some were notable. For example, laboratory studies showed significant costs to *D. melanogaster* females inflicted by males during mating (Partridge et al. 1986); these result from male accessory gland proteins, many of which appear to benefit males, but conflict with female interests (reviewed by Simmons 2001 and Chapman et al. 2003). Field studies also occasionally provided strong evidence for sexual conflict (a notable example is Davies 1992).

A surge of activity, however, occurred in the mid-1990s. Key players were Bill Rice and his student Brett Holland in America, Göran Arnqvist in Sweden, and Locke Rowe in Canada, although many others contributed. Arnqvist began his study of sexual conflict by working on mating systems of water striders, *Gerris* (Arnqvist 1989), as did Locke Rowe (Rowe 1992). They were quick to collaborate (Rowe et al. 1994) and produced strong empirical evidence of a coevolutionary arms race between male-clasping devices to overcome female resistance, and female adaptations to aid in male rejection (Arnqvist and Rowe 1995). In addition to their independent contributions to sexual conflict, their collaborations, culminating in their major monograph (Arnqvist and Rowe 2005), have both notably advanced and consolidated the field.

Bill Rice came to the subject through evolutionary genetics; his early theoretical work showed that sex chromosomes assist sexual dimorphism evolution by enabling genes coding for sexually dimorphic traits to accumulate on the X (or Z) chromosome (Rice 1984). He used genetic markers and experimental evolution to create a new sex-determining locus in *D. melanogaster* and elegantly showed that, as predicted, genes detrimental to the homogametic sex accumulated close to this locus (Rice 1992). Appreciating that male and female interests often could not be satisfied simultaneously and that this may generate sexually antagonistic coevolution (a form of intraspecific Red Queen) (Rice and Holland 1997; Holland and Rice 1998), Rice (1996) showed by experimental evolution that when female *D. melanogaster* were prevented from coevolving with males, males quickly adapt, generating high costs to females. Holland and Rice (1999) showed that in *D. melanogaster* populations in which monogamy was enforced during experimental evolution, males evolved to be less harmful to females, which evolved to be less resistant to male-induced harm. Gavrilets (2000) devised an important formal model of "chase away" coevolution similar to that envisaged by Rice and Holland (1997) and Holland and Rice (1998), showing that this has the potential to drive speciation (although this depends on how conflict is resolved) (Parker 1979; Parker and Partridge 1998).

In the past decade or so, opposition to sexual conflict interpretations has been spearheaded by Bill Eberhard (e.g., Eberhard 2005), who proposed that apparent behavioral conflict is often in female interests and represents a mechanism by which the female filters out (rejects) lower-quality males and "gains by losing" to persistent males of high mating advantage because they generate high fitness progeny. This catalyzed debate on the importance of sexual conflict (see Hosken and Snook 2005 and articles in the same supplement; Tregenza et al. 2006 and articles in the same theme issue), the prevailing view supporting the ubiquity of sexual conflict. Sexual conflict has, by now, become an important and currently vigorous field in evolutionary biology.

EVOLUTION OF SEXUAL STRATEGIES: THE SEXUAL CASCADE

The various evolutionary adaptations surrounding sexuality have presented a formidable challenge for evolutionary biologists and many, from meiosis to sexual selection and sexual conflict, continue to generate debate. Here and elsewhere (Parker and Pizzari, in press), a deductive approach is developed that seeks to explain sexuality as a sequence of events within a causal framework. It is important to distinguish between irreversible evolutionary transitions that are ubiquitous and fixed in most extant advanced animal taxa and more labile transitions that can be more readily reversed by ecological changes. The primary sexual step of meiosis and syngamy leads predictably to gamete competition and (especially as organismal complexity increases) to the evolution of two sexes (anisogamy), which generates unitary sex ratios. Ancestral forms of sexual selection (gamete competition and selection) would have been prevalent in early external fertilizers. With the evolution of mobility and behavioral complexity, copulation and "traditional" sex roles emerge inevitably in most anisogamous systems lacking parental care (i.e., most animal species). However, parental care of the zygote is a much more labile transition that allows deviations from traditional sex roles through the evolution of paternal investment.

During this evolutionary succession (the "sexual cascade"), the opportunity for different episodes of sexual selection changes dynamically, and alternative stable states occur at many steps. After the ancestral phase of gamete and sperm competition under external fertilization, precopulatory sexual selection eventually follows as a result of such features as multicellularity, the unity sex ratio, enhanced mobility and increased behavioral capacity for male–

male competition, and the evolution of internal fertilization and is associated with a reduction in the strength of postcopulatory sexual selection and an associated increase in precopulatory sexual selection, with a consequent higher differentiation of male and female sexual strategies. There is a notable logical compulsion in this sequence of events that flows inevitably from the early evolution of meiosis and gamete formation and eventually results in sex-role differentiation and its associated conflicts (Parker and Pizzari, in press), supporting the view of Schärer et al. (2012) that sex-specific selection, arising ultimately from anisogamy, drives the evolution sexual dimorphism, rather than chance differences in male–female life history and mating traits.

TRANSITIONS IN THE SEXUAL CASCADE

The main transitions in the cascade in Figure 1 are outlined below as a series of discrete steps. Although, in general, the evolution of one step precedes and creates the selective pressure for the next, some synchrony in adaptation across adjacent steps seems likely. The cascade is driven initially by gamete competition and later (after the evolution of anisogamy) by sperm competition. As Darwin (1871) surmised, it is the rise of mobility and complex behavior that catalyzes the development of pre-ejaculatory (Darwinian) sexual selection; what Darwin missed was the long preceding history of sexual selection on gametes and gonads. As the importance of pre-ejaculatory sexual selection increases, the optimal economic balance between pre- and post-ejaculatory expenditures shifts toward reduced testes mass and sperm production (i.e., toward lower precopulatory expenditure) (Parker et al. 2013).

The sexual cascade (Fig. 1) provides a logical imperative for pre-ejaculatory (Darwinian) sexual selection and for Darwin's (1871) and Bateman's (1948) claims about the link between the primary sexual differentiation (anisogamy) and the evolution of traditional sex roles, which has recently been challenged (in my view, unjustifiably) (see Parker and Birkhead 2013; Parker and Pizzari, in press). A more detailed account of some of the transitions is given in a companion paper to the present article (Parker and Pizzari, in press).

Transition A: Events Leading to an Ancestral Isogamous Eukaryote

A suite of transitions in evolution, ranging from the origin of life up to meiosis, recombination, gametic fusion (syngamy), mating types, and the haploid–diploid cycle in eukaryotes (see Maynard Smith and Szathmáry 1995), are here all grouped together (transition A, Fig. 1). Much of the evolution of these early stages remains shrouded in mystery, and interest in the selective advantage of sexuality over asexuality, which expanded in the 1970s (e.g., Williams 1975; Maynard Smith 1978), continues to preoccupy evolutionary biologists (Hartfield and Keightley 2012).

Transition B: Evolution of Gamete Size Dimorphism (Two Sexes)

The sexual cascade in eukaryotes begins with an isogamous unicell (Fig. 1). Transition B, from isogamy to anisogamy, is the key transition in the cascade because it marks the origin of two sexes (in my view, sexes should be defined in terms of gamete size produced and not confused with mating types) (Parker 2011; but see Hoekstra 2011 for an opposing opinion). Anisogamy is often associated with an increase in cellular complexity and is universal in Metazoa (see reviews of Lessells et al. 2009; Parker 2011).

Lessells et al. (2009) review the three main theories for the evolution of anisogamy from an ancestral isogamous unicellular eukaryote. Beginning with Cosmides and Tooby (1981), one theory proposes that anisogamy evolved in response to conflict between nuclear genes and genes in cytoplasmic elements such as mitochondria, chloroplasts, and intracellular parasites (reviewed by Hoekstra 2011). Lessells et al. (2009) conclude that cytoplasmic conflict models are unlikely to be an adequate sole explanation of anisogamy, although they may help to maintain it once evolved.

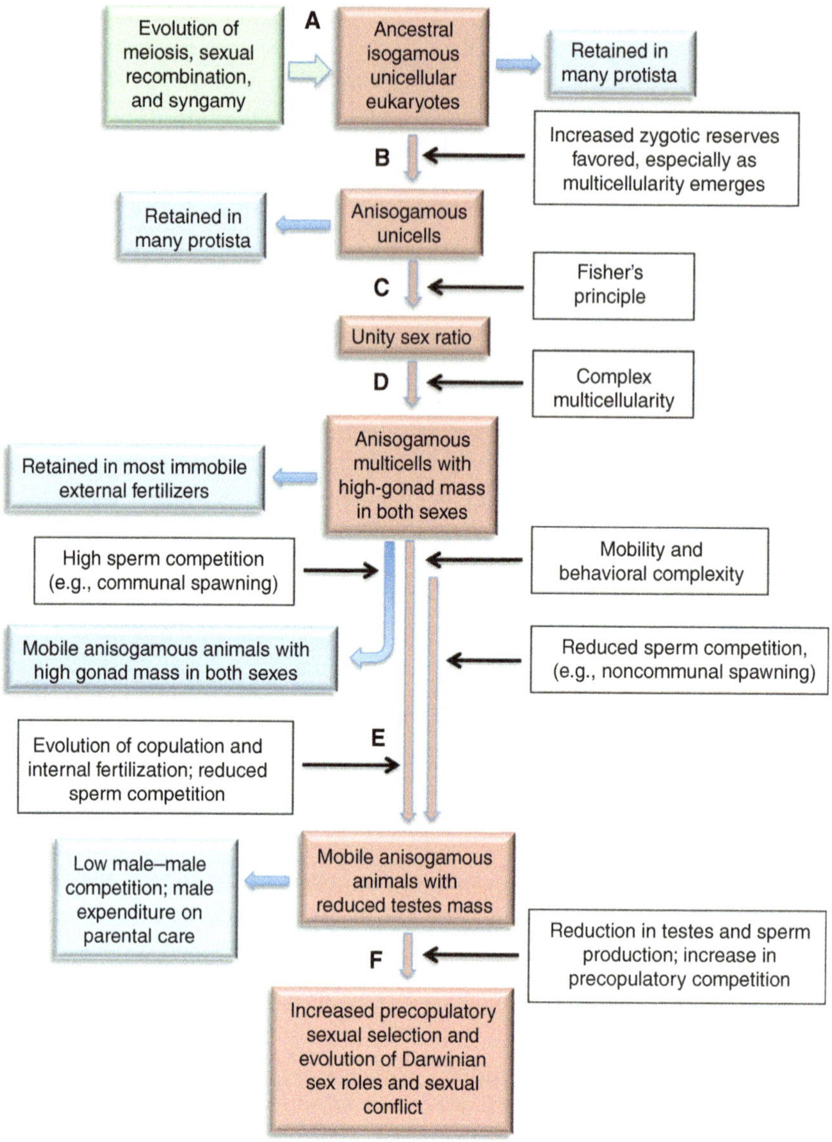

Figure 1. The sexual cascade (succession of evolutionary events leading to Darwinian sexual selection) (pink boxes and arrows) showing main transitions and selective forces (white boxes and black arrows) and alternative stable states (blue boxes and arrows). The transitions (*A–F*) are explained in the text.

Two other theories appear more plausible because both are based only on a trade-off between size and number of gametes (inevitable in an ancestral unicell) and the fact that, over some range, the fitness (i.e., success or viability) of a zygote must increase with its size. The "gamete limitation" theory (e.g., Levitan 1996; beginning with a population-level proposal by Kalmus 1932) relates to the fact that anisogamy could yield more surviving zygotes than isogamy. If some parents produce vast numbers of tiny gametes, this increases the probability of fertilization of the relatively low number of large gametes that survive well as zygotes. The "gamete competition" theory (Parker et al. 1972) relates to the fact that gamete competition can generate disruptive selection on an isogamous population leading to anisogamy (reviewed by Parker 2011). Again, large gametes survive well as zygotes, but intermediate and small gametes compete to gain most fusions with the large gametes. The ESS does not depend on whether the model assumes random fusion between gametes or begins with mating types. If selection favors increased complexity (e.g., toward multicellularity), then higher zygote size is favored, which increases the likelihood of anisogamy (Parker et al. 1972; Bulmer and Parker 2002).

An elegant recent analysis (Lehtonen and Kokko 2011) shows the relation between these two theories. Previously, both had been modeled as a large population of adults releasing gametes into an external medium (sea water). In Lehtonen and Kokko's model, (1) the number of adults in the local mating group could vary from two (i.e., no gamete competition) to infinity (maximum gamete competition), and (2) the rate of fusions of gametes could also vary, either by reducing gamete production rates or encounter rates or by increasing gamete mortality rates (see also Scudo 1967). They showed that anisogamy could evolve from isogamy through either gamete limitation or competition, depending on these two factors. When gametes can fuse fairly readily, even low levels of gamete competition (three or more adults per breeding group) generate anisogamy. However, gamete limitation (low gamete fusion rates) can also generate anisogamy, provided the number of adults per group remains low (i.e., low gamete competition). Which effect (gamete competition or limitation) has had the larger influence on the origin of anisogamy (and hence two sexes) depends on conditions experienced in ancestral isogamous unicells, but gamete competition appears the more robust selective force (Lessells et al. 2009) and one that leads naturally to sperm competition under external fertilization once anisogamy (and hence the two gamete-producing morphs, males and females) become established. Although this question may be difficult to resolve, gamete (sperm) competition certainly offers a plausible solution for the maintenance of anisogamy in most current populations (Parker 1982, 2011).

Early gamete competition models assumed simplistically that a gamete could survive successfully provided it was above a critical minimum size (Parker et al. 1972; Maynard Smith 1982); whether selection favors isogamy or anisogamy then depends critically on the importance of zygote size. More recent models (e.g., Bulmer and Parker 2002) include a function relating gamete survival to gamete size, as well as a function relating zygote fitness to zygote size. Which ESS (isogamy or anisogamy) is attained depends on how similar these two functions are. If similar, the ESS is isogamy, but if the zygote size-to-fitness function shifts away from the gamete size-to-fitness function, requiring much greater size to achieve the same fitness, the ESS is anisogamy (Bulmer and Parker 2002; Lehtonen and Kokko 2011). Exactly this sort of change toward larger zygotes is likely to have accompanied increasing organismal complexity during the evolution of multicellularity, and there is also empirical evidence that this is accompanied by a trend toward anisogamy (see review of Parker 2011). However, if selective pressures do not favor multicellularity, isogamy should remain fixed if the two functions remain rather similar, as seems likely in many unicells (although many show primitive forms of anisogamy).

Once sexual reproduction and syngamy have evolved, transition B (anisogamy) is thus likely to accompany the evolution of multicellularity and increased organismal complexity. It is associated with an immediate and primordial sexual conflict. If gametes can fuse readily, protofemales do better when their ova are fertilized by other ova rather than by parasitic, noncontributing sperm (Parker et al. 1972; Parker 1978c). Parker et al. (1972) proposed two reasons why sperm producers were likely to win: (1) Because the available mutations would be proportional to the number of gametes produced, sperm may have achieved a higher adaptation rate in coevolutionary battles. (2) If sperm–sperm fusions produced nonviable zygotes, sperm producers may have been under stronger selection to avoid sperm–sperm fusions and gain sperm–ovum fusions than ovum producers were to avoid sperm–ovum fusions. Parker (1978c, 2011) gives further reasons why an ovum producer carrying a mutation that prevented sperm–ovum fusions is likely to have been unsuccessful. However, if gametes cannot fuse readily and there is little gamete competition, anisogamy may evolve without conflict as a cooperative solution, as an individual selection extension of Kalmus' (1932) original proposal (Lehtonen and Kokko 2011).

Transition C: Evolution of the Unity Sex Ratio

The evolution of anisogamy may involve mutations that change gamete size by altering the number of cell divisions per unit of reproductive resource. Such a mutation would similarly affect all gametes produced by a given mutant individual, ultimately resulting in separate sexes (gonochorism) arising from two mating type lineages. This argument suggests that separate sexes would typically be ancestral for anisogamy arising in unicells. However, some primitive colonial forms such as *Volvox* include both hermaphrodite and gonochoristic species, so that it is not immediately clear whether the ancestral states would have been separate sexes or hermaphroditism for anisogamy arising in multicells. Because a hermaphrodite individual requires two types of gonad, economy suggests that specialism into separate sexes generally would be favored (Heath 1977), with hermaphroditism (or more complex hermaphroditic forms such as gynodioecy or androdioecy) being favored only under certain special conditions (Charnov et al. 1976).

Under gonochorism, Fisher's (1930) principle immediately operates to favor and maintain equal numbers of males and females. Essentially, the rarer sex has a fitness advantage that is lost only when equality is attained. The unity sex ratio (transition C) is thus an immediate consequence for populations with two sexes (anisogamy) and arose in the original gamete competition simulations (Parker et al. 1972). Unless special circumstances apply (see Hamilton 1967; Charnov 1982; West 2009), equal production of the two sexes is maintained in a tight equilibrium. This selective constraint forms a vital part of sexual selection and conflict by ensuring that (1) in sessile or weakly mobile external fertilizers, numerous sperm must compete for rare ova, and (2) in internal fertilizers, if males have less "time out" than females, males must compete for matings with females.

Transition D: Sperm Competition and Evolution of Complex Organisms with High Gonad Mass

The evolution of multicellularity has been much discussed; it has evolved at least 25 times including several times in eukaryotes, but only once in animals (e.g., Grosberg and Strathmann 2007; Bonner 1998), and involves many conflicts (reviewed by Grosberg and Strathmann 2007). Complex multicellularity may be difficult to acquire, and its origin in plants and animals can be traced back to ancestral marine forms. Marine plants and immobile or weakly mobile invertebrates spend most of their reproductive effort on gametes (see Parker and Pizzari, in press); other expenditures include pheromones and various adaptations to enhance fertilization such as spawning synchrony. Because they have few other means of achieving reproductive success, massive gametic expenditure in both sexes results in a high proportion of the total body mass (the gonado-somatic index or GSI) being expended on gonads and gamete production (Parker and Pizzari, in press). Broadcast spawning, often synchronous, results in intensive sperm competition, which is predicted to maintain male gonad expenditure at very high levels (Parker and Pizzari 2010). It is interesting that male GSI in sessile or relatively immobile broadcast spawning marine animals can considerably exceed female GSI in some taxa (Parker and Pizzari, in press). Why this should be so is unclear; one might imagine that GSI, here, would be equal, with males showing reduced GSI when expenditure increases on traits associated with precopulatory sexual selection. In their vertebrate-dominated study, Hayward and Gillooley (2011) found similar GSI in both sexes, but much higher gamete production rates in females.

At this ancestral stage of immobile (or relatively mobile) marine broadcast spawners, sexual selection is essentially postejaculatory and sexual conflict relates mainly to fertilization conflicts.

Transition E: Mobility/Behavioral Complexity and Evolution of Reduced Sperm Competition

Darwin (1871, p. 274) suggested that the reason male gametes are "brought to" female gametes, rather than the reverse, relates to the relative difficulty of transporting ova. Because of the risks of "at least a short transit through the waters of the sea," he plausibly suggested that as soon as organisms evolved some mobility, males would be favored if they "were to acquire the habit of approaching the female as closely as possible" to reduce risk to unprotected gametes. A transition from external to internal fertilization (usually by copulation) is associated with the colonization of the land (e.g., Dawkins and Carlisle 1976).

There are probably many reasons why mobility and behavioral complexity has arisen in animals, including more efficient feeding and escape from predators. However, once low motility was possible in certain animal taxa, sperm competition may have been a potent selective force favoring the evolution of increased mobility and "female-targeted" sperm release by males, leading ultimately to copulation with internal fertilization (Parker 1970, 1984), which all probably arose through combined selection to increase (1) fertilization gains in the intense sperm competition conditions characteristic of marine broadcast spawning and (2) the proportion of eggs fertilized, either through sperm limitation or high gamete mortality (e.g., after broadcast spawning). The dawn of mobility and behavioral complexity is, thus, the catalyst that permitted the rise of pre-ejaculatory sexual selection out of sexual selection by ejaculate competition. Mobile external fertilizers, such as fishes and many amphibians, typically show high degrees of female targeting, and whereas sperm competition remains intense, male GSI scores remain relatively high. Thus, mobility and behavioral complexity in external fertilizers appears inevitably to be linked with female targeting (e.g., amplexus behavior in anurans), although when sperm competition is intense, GSI often remains high, reducing only as sperm competition reduces (e.g., for fishes, see Stockley et al. 1997; for amphibians, see Byrne et al. 2002).

An immobile broadcast spawner has little option other than to maximize gamete production (in both sexes); males are thus committed to expend all their reproductive effort on sperm. The evolutionary steps by which increased mobility and behavioral complexity may, in some taxa, lead to reduced sperm competition and testis expenditure are interesting to contemplate. A possible sequence is as follows:

1. The weakly mobile ancestor has broadcast spawning with high GSI in both sexes. All reproductive effort is diverted to gametes, there is little secondary sexual dimorphism, and sperm limitation is relatively common.
2. Weak mobility permits and favors aggregation for synchronous spawning to increase fertilization probability for both sexes. The sex ratio in spawning aggregations would be close to unity because both sexes expend similarly on gametes.
3. Mutant males showing primitive female targeting are favored because they achieve sperm competition advantages (see below).
4. As female targeting spreads and becomes more advanced, then, because the sex ratio on spawning grounds is close to unity, sperm competition decreases as many of the spawnings occur in pairs.
5. Close association of the sexes at spawning decreases sperm limitation, reducing the advantage of spawning aggregation and synchrony to females.
6. Female targeting and pre-ejaculatory male competition advance further and costs to females of aggregation and male harassment favor female dispersion.
7. This further decreases sperm competition, reducing male GSI more and increasing

male expenditure on pre-ejaculatory traits such as mate searching and other forms of male–male competition.

In support, many of the above steps appear to be "fossilized" in modern species. For example, following step 1, echinoderms are typically marine broadcast spawners with weak mobility, equal sex ratios, and GSIs; percentage fertilization is very variable, so that sometimes there is considerable sperm limitation (Levitan and Petersen 1995). Obvious secondary sexual dimorphism is very rare and restricted to special ecological conditions (Levitan 2005). As in step 2, many species show synchronous spawning in aggregations (e.g., Himmelman et al. 2008), in which fertility can then be high (Levitan 2010). Remarkably, steps 3–5 appear to be mimicked in the asteroid, *Archaster typicus* (Run et al. 1988). Here, especially, the males display increased mobility as the breeding season begins and show female targeting; they seek out and mount females and begin to release sperm immediately after the female spawns. Some 85% of spawnings occur in this way at peak breeding season. Fertilization efficiency is very high (95% fertility occurred after pair spawnings). Male GSI is much less than that of females. The female/male GSI ratio = 4.8, which is aberrant and the highest found in the literature for 18 asteroid species (Table 3 in Run et al. 1988), otherwise ranges from 2.1 to 0.4. Run et al. (1988), suggest that this may relate to the trade-off of gonad against the energetic demands of higher movement rates and female mounting in males. Himmelman et al. (2008) commonly witnessed similar "pseudocopulation" behavior during spawning in the ophiuroids *Ophiopholis aculeata* and *Ophiura robusta*, and (with much lower frequency, and in a less clearly defined manner) in the sea star *Asterias vulgaris* (which had the second most reduced male GSI in Table 1 in Run et al. 1988). The Antarctic sea star *Neosmilaster georgianus*, a brooding asteroid that probably reproduces continuously, may even have advanced to steps 6 and 7. Here the female targeting behavior involves one or several males mounting a reproductively active female for a period of minutes to hours before spawning (Slattery and Bosch 1993), indicating that female harassment may be present.

Reduction in sperm competition is predicted theoretically to reduce male gonad expenditure (see models reviewed in Parker and Pizzari 2010). There is some empirical evidence for this in external fertilizers and abundant evidence across many internally fertilizing taxa (reviewed in Simmons and Fitzpatrick 2012), so that relative testes size is now commonly used as an index of the level of sperm competition. However, the key to reduced sperm competition in the above sequence of steps relates to female targeting coupled with spatial dispersion of females. Ejaculate and testis economy arise because expenditure devoted to finding and targeting sperm release toward females can yield higher fertilization gains than sheer numerical sperm productivity alone. But paradoxically, as pre-ejaculatory expenditure increases at the expense of postejaculatory expenditure, the OSR is likely to become increasingly male-biased, which can allow some restoration of sperm competition level, depending on female propensity for multiple mating.

Male GSI in broadcast spawners is typically much higher than in taxa with copulation and internal fertilization (Table 1 in Parker and Pizzari, in press). But is sperm competition generally reduced by internal fertilization? One argument might be that, in internally fertilizing species, sperm is often provisioned within the female sperm stores, giving greater longevity and more opportunity for sperm competition than external fertilizers, where the typical sperm life is short. However, many external fertilizers have high spawning synchrony, sometimes in mass aggregations, whereas internal fertilizers are often more dispersed. Furthermore, internal fertilization gives females the opportunity to resist copulations, and varying degrees of female unreceptivity (often after an initial mating) are common in many species, probably because the costs of polyandry to females are often significant, whereas the benefits are often (but not always) marginal. Thus, if females move to areas where males occur in high density, they often mate with one male and then leave. Very low sperm competition risk tends to occur in internal fertilizers that typically mate just once per reproductive cycle. When there is multiple mating, male GSI in internal fertilizers tends to be rather comparable with that of aggregating mobile external fertilizers, but less than (and, often, much less than) that of immobile or weakly mobile broadcast spawners (Table 1 in Parker and Pizzari, in press). Thus, in general, sperm competition appears to be lower under internal fertilization, although there is much overlap depending on the mating system.

Is male GSI high in broadcast spawners because of sperm limitation or competition? Note that sperm limitation per se does not affect the level of sperm competition, which increases with the number of non-self sperm in competition with self-sperm; even if only a small proportion of eggs are fertilized, with broadcast spawning, a given ejaculate may compete with many other ejaculates for fertilizations. Whether species are subject to sperm limitation or competition, then all else being equal, it pays for males to increase sperm numbers. Only when males face neither sperm competition nor limitation can it pay to decrease sperm numbers. Immobile species have little option other than to spend their reproductive effort on gametes, but mobility can generate an alternative, more profitable expenditure on pre-ejaculatory traits associated with female targeting.

Transition F: The Rise of Pre-Ejaculatory Sexual Selection, Sex Roles, and Sexual Conflict

The above key transition E, that is, from broadcast spawning to searching for and targeting sperm release toward females, leading ultimately to the ubiquitous occurrence of copulation in most terrestrial animals, thus marks the rise of pre-ejaculatory (Darwinian) sexual selection. As copulation and internal fertilization evolve, relative potential gains through precopulatory sexual competition can become more favorable than sperm expenditure, especially if sperm competition reduces. Thus, precopulatory expenditure rises at the expense of testis and ejaculate expenditure. In contrast, female gametic expenditure typically remains high. Ultimately, it is the transition to extensive male precopulatory expenditure that results in a high differentiation of sex roles as Darwin envisaged (i.e., the DBP) and high levels of sexual conflict. Sperm economy results in a different balance between time spent on gametic investment for males and females (Williams 1966), which translates into the sexual asymmetry in parental investment (Trivers 1972), OSR (Emlen and Oring 1977), and PRR (Clutton-Brock and Vincent 1991), all of which can be related to times in or out of the mating pool (Clutton-Brock and Parker 1992), and generate divergence in sex roles envisaged by Darwin (see Parker and Birkhead 2013).

It is important to separate the evolution of parental care, which is a labile, secondary phenomenon that is actually relatively infrequent, viewed across the animal kingdom as a whole, from the primary, prezygotic parental investment in gametes. It commonly involves sexual conflict and sometimes also cooperation (see review of Székely 2014). Gametic parental investment occurs symmetrically for parents in isogamy, but for sound theoretical reasons (Parker 1982; Lehtonen and Kokko 2011), it is confined to female investment as ovum provisioning in species with advanced anisogamy (the selective forces underlying sperm and ovum size are reviewed in Parker 2011). In contrast, patterns of parental care of zygotes are highly diverse and have evolved many times independently (Clutton-Brock 1991), but maternal care is much more common than paternal care. The selective mechanisms underlying parental care patterns are less clearly understood than those underlying gametic investment, and exactly what shapes the form of parental care and how it links to anisogamy (which, empirically, it appears to) remains controversial. The ecological lability of parental care allows, not infrequently in certain taxa, deviations from traditional DBP sex roles through the evolution of paternal care. Although Trivers (1972) linked the predominance of female care directly to anisogamy, this has been disputed many times since. Dawkins and Carlisle (1976) pointed out that previous investment does not a priori affect future investment, and argued plausibly that other biological circumstances, such as external versus internal fertilization, could affect the pattern of care. For instance, males of internally fertilizing species are typically not present when their young are released, so male care is, here, harder to evolve than in externally fertilizing species (see also Maynard Smith 1977). Since then, the literature has expanded extensively (reviewed by Klug et al. 2013a), and a direct effect of anisogamy in weighting parental care toward females has been increasingly questioned (see, e.g., Queller 1997; Kokko and Jennions 2008; Klug et al. 2013a). Most recently, Klug et al. (2013a,b) have examined life history conditions favoring transitions from existing states. When the ancestral state is no care, although females initially invest more into each zygote than males, paternal, maternal, and biparental care are equally likely to evolve if males and females are otherwise similar (Klug et al. 2013a). However, once some form of parental care has evolved, then as male and female gametic investment becomes more disparate, with females investing heavily in eggs in a way that decreases their future reproductive success, transitions to increased maternal care (paternal → maternal, paternal → biparental, biparental → maternal) are more likely to be favored (Klug et al. 2013b) as the stable state. This link with anisogamy may well help to explain the predominance of female care in nature.

When ecological or life history conditions do reverse sex roles, selection cannot produce a drive back to isogamy by allowing males to increase sperm size to assist in provisioning zygotes, but males can instead evolve paternal investment. An obvious reason for this is that although paternal care can be focused on zygotes, extra investment per sperm for zygote provisioning would largely be wasted on the vast numbers of sperm that do not fertilize ova (Parker 1982, 2011; Lehtonen and Kokko 2011). Once attained and specialized, the two-sex state becomes highly stable.

THE EVOLUTION OF MOBILITY AND FEMALE-TARGETED SPERM RELEASE

The key event leading to the rise of pre-ejaculatory (Darwinian) sexual selection is the evolution of advanced mobility, transition E (Fig. 1). Although the initial impetus for mobility could have been related to pressures such as improved food gathering and predator avoidance, these are likely to have been greatly enhanced by intrasexual selection on males to release gametes closer to females, and females to move closer to males to reduce sperm limitation (see above). Because of the importance, as Darwin (1871) recognized, of the evolution of mobility and behavioral complexity, I have investigated a simple prospective model to explain (1) the conditions important for transition E through sexual selection alone, and (2) why some invertebrate groups remain fixed as sessile or weakly mobile external spawners with vast expenditure on testes and sperm production.

The model (for details, see Box 1) considers a hypothetical marine invertebrate with weak mobility in which males and females gather in large groups to undergo broadcast spawning (Fig. 2). It calculates the initial "boost" to increased mobility given by sexual selection on males to reduce sperm competition by examining the fitness of a rare mutant male that moves to "target" his sperm release toward spawning females. A biologically realistic interpretation of the model could be that both sexes use their weak mobility to aggregate for spawning—they move closer together and then become stationary, awaiting the start of gamete release. The mutation we consider causes males to enhance this movement further; instead of remaining stationary, the mutant moves toward spawning females while releasing sperm. Although the model considers only the first step in female targeting, it is easy to see how continued drive for such a process could lead to females conserving energy by reducing their aggregation behavior, because as female targeting by males becomes more advanced, sperm limitation reduces (see above). This, in turn, increases the benefits of mate searching by males, driving ultimately toward highly mobile mate searching and highly female-targeted sperm release, and often, ultimately, copulation and internal fertilization. It is not inconceivable that autosomal mutations for enhanced mobility that were initially favored in males through this form of sexual selection were

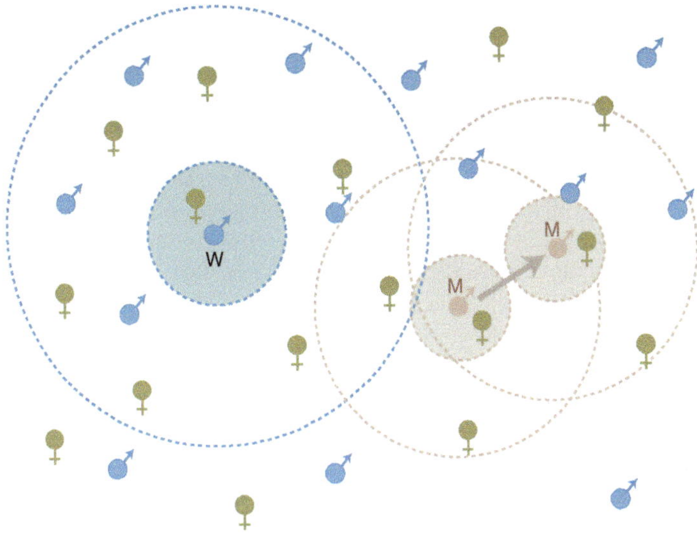

Figure 2. Schematic representation of the model for the evolution of female-targeted sperm release. Wild-type males (blue male symbols) and females (brown female symbols) have low mobility and broadcast gametes. Males releasing sperm have a fertilization zone, β, depicted by the outer broken circles: In this zone, they compete with males with overlapping zones. They gain higher fertilization success with a proportion, α, of females ("nearby" females, represented as those within the blue shaded area around the focal wild-type male), but lower fertilization success with a proportion $(1-\alpha)$ of females ("distant" females, represented as the white area between the concentric broken blue circles). A mutant male (pink male symbol) that moves to target females pays for this by having a reduced total sperm release, which changes zone β, but increases the proportion of nearby females.

also of benefit to both sexes for more efficient food foraging or predator avoidance.

To return to the model: A male's total sperm production is assumed to be proportional to his expenditure G on gonad, and his sperm can fertilize eggs shed by spawning females in a zone β around him (Fig. 2). Within this zone, he has higher fertilization success with a proportion α of females closest to him, and lower success with the proportion $(1-\alpha)$ of females in β that are further away. The male can also expend M on limited movement during the spawning period. This trades off with gonad expenditure within a fixed reproductive budget, R (thus, $R = M + G$). Suppose, now, that a mutant male arises that increases M so as to target females, that is, by moving toward spawning females, its proportion α of "nearby" fertilizations increases. The model, which is prospective only, seeks the ESS expenditure (M^*) on this primitive form of mate searching.

Despite the fact that the mutant's fertilization zone β will deviate from the rest of the population because of the change in M and G, the ESS (at least in this simple model) does not depend on β. Varying β changes the number of females, but similarly changes the number of competing males in the fertilization zone such that the two effects cancel each other out. Thus, the ESS expenditure, M^* (see Box 1, Equation 3) depends only on (1) the total resources for reproduction, R, (2) ϕ, an index of the fertilization skew toward nearby females ($\phi = 1$ if there is no fertilization difference between nearby and distant females, and rises above 1.0 as fertilization skew toward nearby females increases), and (3) on $\alpha(M)$, which defines how mobility spent on female targeting increases the proportion of nearby females in the fertilization zone β.

Some implications of this ESS were investigated using exponentially diminishing returns, that is, $\alpha(M) = 1 - \exp(-\mu M)$, in which increasing μ increases the rate at which $\alpha(M)$ approaches its asymptotic value of 1.0 (Fig. 3A). Figure 3B shows how the ESS, M^*, changes with the relative benefit of nearby females

BOX 1. THE EARLY EVOLUTION OF FEMALE-TARGETED SPERM RELEASE

The following model examines the conditions that would favor the early evolution of female-targeted sperm release in an ancestral broadcast spawner; it makes a number of simplifying assumptions in the interest of mathematical tractability and is heuristic only. A schematic representation is given in Figure 2.

Imagine a population of relatively immobile animals that shed gametes into the sea for fertilization. Each female releases E ova during spawning, and each male sheds s sperm. Adults have an amount of resource, R, to spend on reproductive effort, most of which (G units) is spent on gonad (and, hence, gametes) and rather little (M units) on (limited) mobility. Thus, $R = M + G$, and for males, the number of sperm produced, s, increases linearly with the gonad investment as $G = gs$. In this population, each wild-type (nonmutant) male releases $\hat{s}$ sperm and has allocations $\hat{M}$, $\hat{G}$. When such a male releases sperm, he is closer to a proportion $\alpha(\hat{M}) = \hat{\alpha}$ of "nearby" females, and hence (per female) fertilizes relatively more of their ova, and relatively less (per female) with the proportion $(1 - \hat{\alpha})$ of "distant" females (Fig. 2). Suppose that there is a local spawning population density of m males and f females. Males gain fertilizations in a zone β around them and experience sperm competition depending on the number of other males with overlapping zones. The zone around each wild-type male increases with his sperm expenditure, $\hat{G}$. We use the function $\beta(G,M)$ to define the limits of the fertilization zone. For wild-type males, $\beta(\hat{G}, \hat{M}) = \hat{\beta}$, so that the zone overlaps with $\hat{\beta}f$ spawning females and $\hat{\beta}m$ spawning males (Fig. 2).

We calculate fertilization gains using a loaded raffle model (Parker and Pizzari 2010). The sperm competition advantage in fertilization with nearby females is modeled by giving a focal male's sperm a higher "competitive weight" with nearby females and a lower competitive weight with distant females by applying different loading factors to sperm in nearby and distant conditions (respectively, h_n and h_d, where $h_n > h_d$). Thus, a wild-type focal male competing with $(\hat{\beta}m - 1)$ similar males gains $w_n(\hat{s}, \hat{\alpha}) = \hat{\beta}fEg\hat{s}\hat{\alpha}h_n/\{g\hat{s}[\hat{\alpha}h_n + (1 - \hat{\alpha})]h_d(\hat{\beta}m - 1)]\}$ fertilizations with nearby females, and $w_d(\hat{s}, \hat{\alpha}) = \hat{\beta}fEg\hat{s}(1 - \hat{\alpha})]h_d/\{g\hat{s}[\hat{\alpha}h_n + (1 - \hat{\alpha})]h_d(\hat{\beta}m - 1)]\}$ with distant females, so that his total fertilizations are $w(\hat{s}, \hat{\alpha}) = w_n(\hat{s}, \hat{\alpha}) + w_d(\hat{s}, \hat{\alpha}) = fE/m = E$ if the sex ratio is unity ($f = m$).

Now, consider the fertilization success of a mutant focal male with increased mobility ($M > \hat{M}$) that releases $s < \hat{s}$ sperm (because $G < \hat{G}$). He experiences two effects as a result of his increased mobility (see Fig. 2): (1) He targets his sperm release toward nearby spawning females by moving toward them and releasing sperm, thus increasing his proportion of nearby spawning females to $\alpha = \alpha(M) > \hat{\alpha}$, that is, a primitive form of mate searching, and (2) he alters his fertilization zone β, so that the population of individuals (of both) sexes overlapping with his zone changes to $\beta = \beta(G, M) \neq \hat{\beta}$ compared with wild-type males.

The nature of the change in β is harder to predict than that in α. If a male increases his mobility from $\hat{M}$ to M, he reduces the number of sperm he releases because of the trade-off, $G = R - M$, so that $s < \hat{s}$. This suggests that $\beta < \hat{\beta}$ because the mutant has a lower total sperm release, but because the mutant moves more through the spawning population, this assumption could be questioned. However, if we assume that the mutant experiences the same proportionate change in the number of male competitors and spawning females, he competes with $(\beta m - 1)$ males, each releasing $\hat{s}$ sperm, for fertilizations of βfE eggs, and we shall see later that the form of $\beta(G,M)$ does not influence the ESS.

The mutant's total fertilizations are

$$w(s, \hat{s}) = \frac{\beta fEgs(\alpha h_n + [1 - \alpha]h_d)}{\{sg[\alpha h_n + (1 - \alpha)h_d] + \hat{s}g[\hat{\alpha}h_n + (1 - \hat{\alpha})]h_d(\beta m - 1)\}}.$$

Dividing both numerator and denominator by h_d and simplifying, we can write this as

$$w(s, \hat{s}) = \frac{\beta fEsg[\alpha(\phi - 1) + 1]}{\{sg[\alpha(\phi - 1) + 1] + \hat{s}g[\hat{\alpha}(\phi - 1) + 1](\beta m - 1)\}}, \quad (1)$$

where $\phi = h_n/h_d > 1$, an index of the fertilization skew toward nearby females. Note that when the focal mutant plays the population strategy, that is, $s = \hat{s}$ so that $\alpha = \hat{\alpha}$, we see that again $w(\hat{s}, \hat{s}) = fE/m$.

To find the ESS value for s, that is, s^*, we set

$$\left.\frac{dw}{ds}\right|_{s=\hat{s}=s^*} = 0, \quad \text{subject to} \left.\frac{d^2 w}{ds^2}\right|_{s=\hat{s}=s^*} < 0,$$

for a maximum, remembering that both α and β are functions of s. This differentiation of Equation 1 involves terms β and $d\beta/ds$, but we eventually obtain

$$-s^* \frac{d\alpha(M^*)}{ds} = \alpha(M^*) + \frac{1}{\phi - 1}, \tag{2}$$

where the asterisks denote ESS values. Note that the β terms disappear. Effectively, decreasing β changes the number of females in the fertilization zone, but also similarly changes the number of males in competition, and the two effects cancel out each other. A more advanced model may examine this result in more detail, but we believe it to be a reasonable first approximation for this heuristic model.

Allocations M and G trade off within the male's reproductive budget, R. For simplicity, we have assumed that the amount of sperm released, s, is a linear function of the total mass of the gonad, that is, $G(s) = gs$ and $M(s) = R - gs$. Thus,

$$d\alpha(M^*)/ds = -g \cdot d\alpha(M^*)/dM.$$

Equation 2 then gives

$$(R - M^*)\frac{d\alpha(M^*)}{dM} = \alpha(M^*) + \frac{1}{\phi - 1}. \tag{3}$$

We now have a solution solely in terms of (i) the function $\alpha(M)$, which defines how female targeting increases the proportion of "nearby" females, (ii) ϕ, which defines the fertilization skew toward nearby females, and (iii) R, the male's reproductive budget.

A plausible form for $\alpha(M)$ in this ancestral state would be an increasing function with diminishing returns. Setting $\alpha(M) = 1 - \exp(-\mu M)$, μ defines the rate of approach to the asymptotic value of $\alpha(M) = 1.0$ (Fig. 3A). We cannot obtain an explicit solution for M^*, but Equation 3 gives

$$\exp(-\mu M^*)[1 + \mu(R - M^*)] = \phi/(\phi - 1), \tag{4}$$

which can be used to iterate solutions for M^* (see Fig. 3B). As $(\phi - 1)$ increases, the ESS mobility level increases toward an asymptote determined by μ. Note that highest ESS values for mobility and female targeting (M^*) are obtained when μ is low and $(\phi - 1)$ high, although the range over which at least some degree of female targeting can be favored increases with μ.

We can ask what conditions permit the initial evolution of female targeting, that is, for a relatively immobile broadcast spawner to begin to move toward spawning females (the condition for $M^* > 0$). This depends only on the starting gradient of $\alpha(M)$, that is, on $d\alpha(0)/dM$. Applying either the exponential function above or the linear form $\alpha(M) = \mu M$ to Equation 3, we find that the condition for female targeting to begin to evolve is simply that

$$\mu(\phi - 1) > R^{-1}. \tag{5}$$

The product of the marginal gain from increasing nearby females (μ) and relative fertilization skew $(\phi - 1)$, must exceed the reciprocal of the male's resource available for reproduction; if this can occur, the transition from sheer numerical sperm productivity toward sperm economy and female targeting begins.

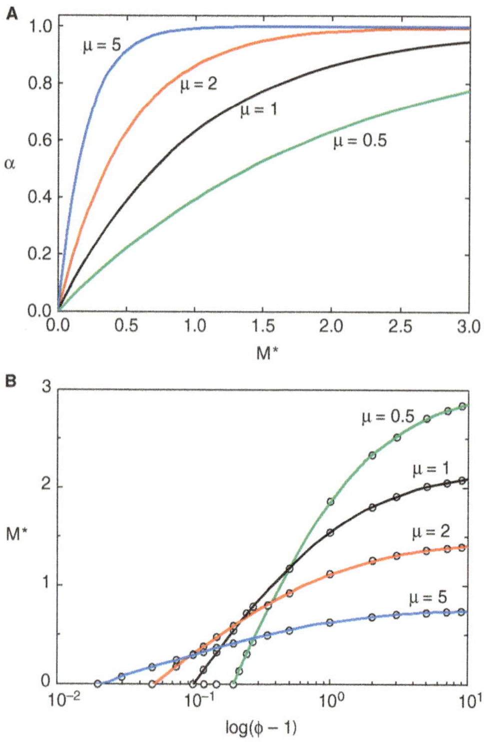

Figure 3. (*A*) Relation between α, the proportion of "nearby" females and a male's expenditure on monility and female targeting, *M*. (*B*) Relation between the ESS mobility, M^*, and the log index of fertilization skew toward nearby females, log $(\phi - 1)$, at different values for μ, which defines the rate at which $\alpha(M)$ increases to its asymptote.

$[\log(\phi - 1)]$. If the relative fertilization benefit of nearby females is small ($\phi \rightarrow 1$), female targeting will not evolve and the ESS is to remain as an immobile broadcast spawner, and if μ is also low, even moderate benefit of nearby females will not permit mobility to evolve. However, high benefits of nearby females ($\phi \gg 1$) result in significant mobility, generating very high M^* when μ is low. When μ is low, $\alpha(M)$ is slow to approach its asymptote (Fig. 3A), so that the marginal benefits of continued mobility investment are high. When μ is high, large amounts of female targeting can be achieved with low *M*, so high gonad investment can be maintained to aid in sperm competition.

So, what biological conditions allow immobility and high gonad mass to be maintained in broadcast spawners? Many multicellular plants are sessile and fixed to the substrate, and hence cannot move, as is the case for many marine invertebrates. Marine invertebrates with weak or moderate mobility often have highly synchronous spawning, which may reduce the benefits of female targeting because gametes are shed very quickly and synchronously. Similarly, if individuals are much dispersed, costs of moving to and targeting spawning females may be high. Empirical studies of the relevant parameters in weakly mobile marine invertebrates would repay detailed investigation.

This simplistic model thus shows the sort of conditions under which female-targeted sperm release may have begun to evolve, and also why many taxa retain broadcast spawning. It predicts the initial rise of mobility for female targeting if $\mu(\phi - 1) > R^{-1}$ (Box 1, Equation 5). The three conditions favoring this crucial transition for pre-ejaculatory sexual selection are (1) high initial increases in the proportion of nearby females by diverting resources from gonad to mobility (high μ), (2) high relative fertilization benefits with nearby females (high ϕ), and (3) high resources for reproduction (high *R*). The first two requirements would seem more likely to be satisfied as behavioral complexity and high mobility advance. A possibility is that an initial drive in certain taxa to increase mobility for food foraging purposes raised resources (increased *R*) and made female targeting more effective (raised μ), causing a sudden acceleration in female targeting. Selection would then act to increase its effectiveness, resulting in many taxa in a drive toward mate searching by males, copulation and internal fertilization, and greatly reduced sperm competition. This would be accompanied by greater potential for expenditure on more intense premating competition (*M*) with consequently reduced gonad expenditure (*G*) (Parker and Pizzari 2010; Parker et al. 2013). Effectively, this transition signifies the rise of pre-ejaculatory sexual selection, the form of selection discussed by Darwin.

SEXUAL CONFLICT IN THE SEXUAL CASCADE

Although pre-ejaculatory (Darwinian) sexual selection appears only at late stages in the cascade (Fig. 1), sexual selection acting on gametes begins after the evolution of syngamy and is likely to have been fundamental in generating two sexes (anisogamy). At this initial stage, sexual conflict concerns the fact that protofemales share the high investment they put into protoova with protomales, whose gametes (protosperm) contribute little, and as anisogamy advances, nothing. This is essentially the twofold cost of sex (Maynard Smith 1978). The reasons why males were likely to win this primordial sexual conflict, and why anisogamy is maintained, are reviewed by Parker (2011). As complex multicellularity advanced, potent sexual selection persists as sperm/pollen competition and cryptic female choice in the two major kingdoms, with sperm/pollen competition maintaining high relative male gonad investment unless (as in many animal taxa) mobility evolves. In sessile animals, sexual conflict will involve fertilization conflicts relating to, for example, selection of favorable sperm, genetic compatibility, and prevention of polyspermy. It will also concern parental care, which has evolved mainly (but not exclusively) as maternal care in several sessile or weakly mobile invertebrates, but sometimes also as paternal care in mobile external fertilizers such as several amphibians and many fish.

With the evolution in many animal taxa of advanced mobility and female targeting, often involving copulation, testes reduction accompanies the rise of pre-ejaculatory (Darwinian) sexual selection and evolution of sex roles. Testes reduction and the typical absence of male parental care result in male time in the mating pool greatly exceeding female time and relates to high intensity of sexual selection on males (e.g., Clutton-Brock and Parker 1992; Kokko et al. 2012; Parker and Birkhead 2013). Precopulatory sexual selection and the highly differentiated sex roles that often result from it, now generate a second dimension for sexual conflict: mating conflict. Conflicts over mating decisions will most typically be of the DBP type with males selected to persist and females to resist and may accompany mating decisions over inbreeding, mate quality, and crosses between ecotypes, and, hence, the probability of speciation (Parker 1979, 2006b; Parker and Partridge 1998). However, conflict will also arise from adaptations to sperm competition and may, too, be extensive (reviewed by Stockley 1997; Edward et al. 2014).

Recent critiques that DBP sex roles are unrelated to the primary sexual differentiation (anisogamy) are therefore unfounded (see also Schärer et al. 2012); sex roles flow logically from anisogamy as a consequence of the sexual cascade and, as such, present no fundamental inconsistency (Parker and Birkhead 2013; Parker and Pizzari, in press).

ACKNOWLEDGMENTS

I am grateful to Bill Rice and Sergey Gavrilets for the invitation to write this review, which has allowed me to elaborate on my ideas surrounding the sexual cascade that I developed for a monograph in the 1970s, but failed to complete, and ultimately abandoned. I am greatly indebted to Hanna Kokko, Lucas Schärer, and an anonymous reviewer for some very helpful suggestions that have much improved the manuscript. I also thank Tom Pizzari for several earlier discussions on this topic.

REFERENCES

**Reference is also in this collection.*

Andersson MB. 1994. *Sexual selection.* Princeton University Press, Princeton, NJ.

Andrés JA, Morrow EH. 2003, The origin of interlocus conflict: Is sex linkage important? *J Evol Biol* **16:** 219–223.

Arnqvist G. 1989. Multiple mating in a water strider: Mutual benefits or intersexual conflict? *Anim Behav* **38:** 749–756.

Arnqvist G, Rowe L. 1995. Sexual conflict and arms races between the sexes: A morphological adaptation for control of mating in a female insect. *Proc R Soc Lond B* **261:** 123–127.

Arnqvist G, Rowe L. 2005. *Sexual conflict.* Princeton University Press, Princeton, NJ.

Bateman AJ. 1948. Intra-sexual selection in *Drosophila. Heredity* **2:** 349–368.

Birkhead TR. 2000. *Promiscuity: An evolutionary history of sperm competition.* Harvard University Press, Cambridge, MA.

Birkhead TR. 2010. How stupid not to have thought of that: Post-copulatory sexual selection. *J Zool* **281:** 78–93.

Birkhead TR, Møller AP. 1992. *Sperm competition in birds: Evolutionary causes and consequences.* Academic, London.

Birkhead TR, Møller AP. 1998. *Sperm competition and sexual selection*. Academic, London.

Bonner JT. 1998. The origins of multicellularity. *Integr Biol* **1:** 27–36.

Bulmer MG, Parker GA. 2002. The evolution of anisogamy: A game-theoretic approach. *Proc R Soc Lond B* **269:** 2381–2388.

Byrne PG, Roberts JD, Simmons LW. 2002. Sperm competition selects for increased testes mass in Australian frogs. *J Evol Biol* **15:** 347–355.

Chapman T, Arnqvist G, Bangham J, Rowe L. 2003. Sexual conflict. *Trends Ecol Evol* **18:** 41–47.

Charnov EL. 1982. *The theory of sex allocation*. Princeton University Press, Princeton, NJ.

Charnov EL, Maynard Smith J, Bull JJ. 1976. Why be an hermaphrodite? *Nature* **263:** 125–126.

Clutton-Brock TH. 1991. *The evolution of parental care*. Princeton University Press, Princeton, NJ.

Clutton-Brock TH, Vincent ACJ. 1991. Sexual selection and the potential reproductive rates of males and females. *Nature* **351:** 58–60.

Clutton-Brock TH, Parker GA. 1992. Potential reproductive rates and the operation of sexual selection. *Q Rev Biol* **67:** 437–456.

Cosmides LM, Tooby J. 1981. Cytoplasmic inheritance and intragenomic conflict. *J Theor Biol* **202:** 74–101.

Darwin C. 1871. *The descent of man selection in relation to sex*. John Murray, London.

Davies NB. 1992. *Dunnock behaviour and social evolution*. Oxford University Press, Oxford.

Dawkins R. 1976. *The selfish gene*. Oxford University Press, Oxford.

Dawkins R, Carlisle TR. 1976. Parental investment, mate desertion and a fallacy. *Nature* **262:** 131–133.

Dawkins R, Krebs JR. 1979. Arms races between and within species. *Proc R Soc Lond B* **205:** 489–511.

Dewsbury DA. 2005. The Darwin-Bateman paradigm in historical context. *Integr Comp Biol* **45:** 831–837.

Eberhard WG. 1985. *Sexual selection and animal genitalia*. Harvard University Press, Cambridge, MA.

Eberhard WG. 1996. *Female control: Sexual selection by cryptic female choice*. Princeton University Press, Princeton, NJ.

Eberhard WG. 2005. Evolutionary conflicts of interests: Are female sexual decisions different? *Am Nat* **165:** S19–S25.

Eberhard WG. 2009. Postcopulatory sexual selection: Darwin's omission and its consequences. *Proc Natl Acad Sci* **106:** 10025–10032.

* Edward DA, Stockley P, Hosken DJ. 2014. Sexual conflict and sperm competition. *Cold Spring Harb Perspect Biol* doi: 10.1101/cshperspect.a017707.

Emlen ST, Oring LW. 1977. Ecology, sexual selection, and the evolution of mating systems. *Science* **197:** 215–223.

Fisher RA. 1930. *The genetical theory of natural selection*. Oxford University Press, Oxford.

Fisher RA. 1931. The evolution of dominance. *Biol Rev* **6:** 345–368.

Gavrilets S. 2000 Rapid evolution of reproductive barriers driven by sexual conflict. *Nature* **403:** 886–889.

Grafen A. 1990. Biological signals as handicaps. *J Theor Biol* **144:** 517–546.

Grosberg RK, Strathmann RR. 2007. The evolution of multicellularity: A major minor transition? *Ann Rev Ecol Evol Syst* **38:** 621–654.

Hamilton WD. 1967. Extraordinary sex ratios. *Science* **156:** 477–488.

Hamilton WD, Zuk M. 1982. Heritable true fitness and bright birds: A role for parasites? *Science* **213:** 384–387.

Hartfield M, Keightley PD. 2012. Current hypotheses for the evolution of sex and recombination. *Integr Zool* **7:** 192–209.

Hayward A, Gillooley JF. 2011. The cost of sex: Quantifying energetic investment in gamete production by males and females. *PLoS ONE* **6:** e16557.

Heath DJ. 1977. Simultaneous hermaphroditism: Cost and benefit. *J Theor Biol* **64:** 363–373.

Himmelman JH, Dumont CP, Gaymer CF, Vallières C, Drolet D. 2008. Spawning synchrony and aggregative behaviour of cold-water echinoderms during multi-species mass spawnings. *Mar Ecol Prog Ser* **361:** 161–168.

Hoekstra RF. 2011. Nucleo-cytoplasmic conflict and the evolution of gamete dimorphism. In *The evolution of anisogamy: A fundamental phenomenon underlying sexual selection* (ed. Togashi T, Cox PA), pp. 111–130. Cambridge University Press, Cambridge.

Holland B, Rice WR. 1998. Chase-away sexual selection: Antagonistic seduction versus resistance. *Evolution* **52:** 1–7.

Holland B, Rice WR. 1999. Experimental removal of sexual selection reverses intersexual antagonistic coevolution and removes a reproductive load. *Proc Natl Acad Sci* **96:** 5083–5088.

Hosken D, Snook R. 2005. How important is sexual conflict? *Am Nat* **165:** S1–S4.

* Hurst GDD, Frost CL. 2014. Reproductive parasitism: Maternally inherited symbionts in a biparental world. *Cold Spring Harb Perspect Biol* doi: 10.1101/cshperspect.a017699.

Huxley J. 1938a. Darwin's theory of sexual selection and the data subsumed by it, in the light of recent research. *Am Nat* **72:** 416–433.

Huxley JS. 1938b. The present standing of the theory of sexual selection. In *Evolution: Essays on aspects of evolutionary biology* (ed. de Beer GR), pp. 11–42. Oxford University Press, Oxford.

Jacobs ME. 1955. Studies on territorialism and sexual selection in dragonflies. *Ecology* **36:** 566–86.

Jennions MD, Kokko H. 2010. Sexual selection. In *Evolutionary behavioral ecology* (ed. Westneat DF, Fox CW), pp. 343–364. Oxford University Press, Oxford.

Kalmus H. 1932. On the adaptive value of phenotypic (morphological) anisogamy and the evolution of the first sex differences. *Biol Zentral* **52:** 716–726.

Kirkpatrick M. 1982. Sexual selection and the evolution of female choice. *Evolution* **36:** 1–12.

Kokko H, Jennions MD. 2008. Parental investment, sexual selection, and sex ratios. *J Evol Biol* **21:** 919–948.

* Kokko H, Jennions MD. 2014. The relationship between sexual selection and sexual conflict. *Cold Spring Harb Perspect Biol* doi: 10.1101/acshperspect.a017517.

Kokko H, Klug H, Jennions MD. 2012. Unifying cornerstones of sexual selection: Operational sex ratio, Bateman gradient and the scope for competitive investment. *Ecol Lett* **15:** 1340–1351.

Klug H, Bonsall MB, Alonzo SH. 2013a. The origin of parental care in relation to male and female life history. *Ecol Evol* **3:** 779–791.

Klug H, Bonsall MB, Alonzo SH. 2013b. Sex differences in life history drive evolutionary transitions among maternal, paternal, and bi-parental care. *Ecol Evol* **3:** 792–806.

Lande R. 1981. Models of speciation by sexual selection on polygenic traits. *Proc Natl Acad Sci* **78:** 3721–3725.

Lehtonen J, Kokko H. 2011. Two roads to two sexes: Unifying gamete competition and gamete limitation in a single model of anisogamy evolution. *Behav Ecol Sociobiol* **65:** 445–459.

Lessells CM, Snook RR, Hosken DJ. 2009. The evolutionary origin and maintenance of sperm: Selection for a small, motile gamete mating type. In *Sperm biology: An evolutionary perspective* (ed. Birkhead TR, Hosken D, Pitnick S), pp. 43–67. Academic, London.

Levitan DR. 1996. Effects of gamete traits on fertilization in the sea and the evolution of sexual dimorphism. *Nature* **382:** 153–155.

Levitan DR. 2005. The distribution of male and female reproductive success in a broadcast spawning marine invertebrate. *Integr Comp Biol* **45:** 848–855.

Levitan DR. 2010. Sexual selection in external fertilizers. In *Evolutionary behavioral ecology* (ed. Westneat DF, Fox CW), pp. 365–378. Oxford University Press, Oxford.

Levitan DR, Petersen C. 1995. Sperm limitation in the sea. *Trends Ecol Evol* **10:** 228–231.

Maynard Smith J. 1958. Sexual selection. In *A century of Darwin* (ed. Barnet SA), pp. 230–244. Heineman, London.

Maynard Smith J. 1977. Parental investment: A prospective analysis. *Anim Behav* **25:** 1–9.

Maynard Smith J. 1978. *The evolution of sex.* Cambridge University Press, Cambridge.

Maynard Smith J. 1982. *Evolution and the theory of games.* Cambridge University Press, Cambridge.

Maynard Smith J, Price GR. 1973. The logic of animal conflicts. *Nature* **246:** 15–18.

Maynard Smith J, Szathmáry E. 1995. *The major transitions in evolution.* Freeman Spektrum, Oxford.

O'Donald P. 1962. The theory of sexual selection. *Heredity* **17:** 541–552.

O'Donald P. 1980. *Genetic models of sexual selection.* Cambridge University Press, Cambridge.

Orians GH. 1969. On the evolution of mating systems in birds and mammals. *Am Nat* **103:** 589–603.

Parker GA. 1970. Sperm competition and its evolutionary consequences in the insects. *Biol Rev* **45:** 525–567.

Parker GA. 1978a. Searching for mates. In *Behavioural ecology: An evolutionary approach* (ed. Krebs JR, Davies NB), pp. 214–244. Blackwell, Oxford.

Parker GA. 1978b. Selfish genes, evolutionary games, and the adaptiveness of behaviour. *Nature* **274:** 849–855.

Parker GA. 1978c. Selection on non-random fusion of gametes during the evolution of anisogamy. *J Theor Biol* **73:** 1–28.

Parker GA. 1979. Sexual selection and sexual conflict. In *Sexual selection and reproductive competition in insects* (ed. Blum MS, Blum NA), pp. 123–166. Academic, New York.

Parker GA. 1982. Why are there so many tiny sperm? Sperm competition and the maintenance of two sexes. *J Theor Biol* **96:** 281–294.

Parker GA. 1984. Sperm competition and the evolution of animal mating strategies. In *Sperm competition and the evolution of animal mating systems* (ed. Smith RL), pp. 1–60. Academic, Orlando, FL.

Parker GA. 2006a. Behavioural ecology: The science of natural history. In *Essays on animal behaviour: Celebrating 50 years of animal behaviour* (ed. Lucas JR, Simmons LW), pp. 23–56. Elsevier, Burlington, MA.

Parker GA. 2006b. Sexual conflict over mating and fertilization: An overview. *Phil Trans R Soc B* **361:** 235–259.

Parker GA. 2010. Reflections at dusk. In *Leaders in animal behavior: The second generation* (ed. Drickamer L, Dewsbury DA), pp. 429–464. Cambridge University Press, Cambridge.

Parker GA. 2011. The origin and maintenance of two sexes (anisogamy), and their gamete sizes by gamete competition. In *The evolution of anisogamy: A fundamental phenomenon underlying sexual selection* (ed. Togashi T, Cox PA), pp. 17–74. Cambridge University Press, Cambridge.

Parker GA. 2013. A personal history of the development of animal contest theory and its role in the 1970s. In *Animal contests* (ed. Hardy ICW, Briffa M), pp. xi–xx. Cambridge University Press, Cambridge.

Parker GA, Birkhead TR. 2013. Polyandry: The history of a revolution. *Phil Trans R Soc B* **368:** 20120335.

Parker GA, Partridge L. 1998. Sexual conflict and speciation. *Phil Trans R Soc B* **353:** 261–274.

Parker GA, Pizzari T. 2010. Sperm competition and ejaculate economics. *Biol Rev* **85:** 897–934.

Parker GA, Pizzari T. Sexual selection: The logical imperative. In *Current perspectives on sexual selection: What's left after Darwin* (ed. Houquet T). Springer, Berlin (in press).

Parker GA, Baker RR, Smith VGF. 1972. The origin and evolution of gamete dimorphism and the male–female phenomenon. *J Theor Biol* **36:** 529–553.

Parker GA, Lessells CM, Simmons LW. 2013. Sperm competition games: A general model for pre-copulatory male–male competition. *Evolution* **67:** 95–109.

Partridge L, Fowler K, Trevitt S, Sharp W. 1986. An examination of the effects of males on the survival and egg-production rates of female *Drosophila melanogaster. J Insect Physiol* **11:** 925–929.

Queller DC. 1997. Why do females care more than males? *Proc R Soc Lond B* **264:** 1555–1557.

Rice WR. 1984. Sex chromosomes and the evolution of sexual dimorphism. *Evolution* **38:** 735–742.

Rice WR. 1992. Sexually antagonistic genes: Experimental evidence. *Science* **256:** 1436–1439.

Rice WR. 1996. Sexually antagonistic male adaptation triggered by experimental arrest of female evolution. *Nature* **361:** 232–234.

Rice WR, Holland B. 1997. The enemies within: Intergenomic conflict, interlocus contest evolution (ICE), and the intraspecific Red Queen. *Behav Ecol Sociobiol* **41:** 1–10.

Richards OW. 1927. Sexual selection and allied problems in the insects. *Biol Rev* **2:** 298–364.

Rowe L. 1992. Convenience polyandry in a water strider: Foraging conflicts and female control of copulation frequency and guarding duration. *Anim Behav* **44:** 189–202.

Rowe L, Arnqvist G, Krupa J, Sih A. 1994. Sexual conflict and the evolutionary ecology of mating patterns: Water striders as a model system. *Trends Ecol Evol* **9:** 289–293.

Run J-Q, Chen C-P, Chang K-H, Chia F-S. 1988. Mating behaviour and reproductive cycle of *Archaster typicus* (Echinodermata: Asteroidea). *Mar Biol* **99:** 247–253.

Schärer L, Rowe L, Arnqvist G. 2012. Anisogamy, chance and the evolution of sex roles. *Trends Ecol Evol* **27:** 260–264.

* Schärer L, Janicke T, Ramm SA. 2014. Sexual conflict in hermaphrodites. *Cold Spring Harb Perspect Biol* doi: 10.1101/cshperspect.a017673.

Scudo FM. 1967. The adaptive value of sexual dimorphism: I. Anisogamy. *Evolution* **21:** 285–291.

Shuster SM, Wade MJ. 2003. *Mating systems and strategies.* Princeton University Press, Princeton, NJ.

Simmons LW. 2001. *Sperm competition and its evolutionary consequences in the insects.* Princeton University Press, Princeton, NJ.

Simmons LW, Fitzpatrick JL. 2012. Sperm wars and the evolution of male fertility. *Reproduction* **144**, 519–534.

Slattery M, Bosch I. 1993. Mating behavior of a brooding antarctic asteroid, *Neosmilaster georgianus. Invertebr Reprod Dev* **24:** 97–102.

Smith RL, ed. 1984. *Sperm competition and the evolution of animal mating systems.* Academic, New York.

Stockley P. 1997. Sexual conflict resulting from adaptations to sperm competition. *Trends Ecol Evol* **12:** 154–159.

Stockley P, Gage MJG, Parker GA, Møller AP. 1997. Sperm competition in fish: The evolution of testis size and ejaculate characteristics. *Am Nat* **149:** 933–954.

* Székely T. 2014. Sexual conflict between parents: Offspring desertion and asymmetrical parental care. *Cold Spring Harb Perspect Biol* doi: 10.1101/cshperspect.a017665.

Thornhill R. 1983. Cryptic female choice and its implications in the scorpionfly *Harpobittacus nigriceps. Am Nat* **122:** 765–788.

Tregenza T, Wedell N, Chapman T. 2006. Introduction. Sexual conflict: A new paradigm? *Phil Trans R Soc B* **361:** 229–234.

Trivers R. 1972. Parental investment and sexual selection. In *Sexual selection and the descent of man 1871–1971* (ed. Campbell B), pp. 139–179. Aldine, Chicago.

West SA. 2009. *Sex allocation.* Princeton University Press, Princeton, NJ.

Williams GC. 1966. *Adaptation and natural selection.* Princeton University Press, Princeton, NJ.

Williams GC. 1975. *Sex and evolution.* Princeton University Press, Princeton, NJ.

Zahavi A. 1975. Mate selection—A selection for a handicap. *J Theor Biol* **53:** 205–214.

Zeh JA. 2004. Sexy sons: A dead end for cytoplasmic genes. *Proc R Soc B* **271:** S306–S309.

The Relationship between Sexual Selection and Sexual Conflict

Hanna Kokko and Michael D. Jennions

Center of Excellence in Biological Interactions, Ecology, Evolution & Genetics, Research School of Biology, The Australian National University, Canberra ACT 0200, Australia

Correspondence: hanna.kokko@anu.edu.au, Michael.jennions@anu.edu.au

Evolutionary conflicts of interest arise whenever genetically different individuals interact and their routes to fitness maximization differ. Sexual selection favors traits that increase an individual's competitiveness to acquire mates and fertilizations. Sexual conflict occurs if an individual of sex A's relative fitness would increase if it had a "tool" that could alter what an individual of sex B does (including the parental genes transferred), at a cost to B's fitness. This definition clarifies several issues: Conflict is very common and, although it extends outside traits under sexual selection, sexual selection is a ready source of sexual conflict. Sexual conflict and sexual selection should not be presented as alternative explanations for trait evolution. Conflict is closely linked to the concept of a lag load, which is context-dependent and sex-specific. This makes it possible to ask if one sex can "win." We expect higher population fitness if females win.

Many published studies ask if sexual selection or sexual conflict drives the evolution of key reproductive traits (e.g., mate choice). Here we argue that this is an inappropriate question. By analogy, G. Evelyn Hutchinson (1965) coined the phrase "the ecological theatre and the evolutionary play" to capture how factors that influence the birth, death, and reproduction of individuals (studied by ecologists) determine which individuals reproduce, and "sets the stage" for the selective forces that drive evolutionary trajectories (studied by evolutionary biologists). The more modern concept of "eco-evolutionary feedback" (Schoener 2011) emphasizes that selection changes the character of the actors over time, altering their ecological interactions. No one would sensibly ask whether one or the other shapes the natural world, when obviously both interact to determine the outcome.

So why have sexual conflict and sexual selection sometimes been elevated to alternate explanations? This approach is often associated with an assumption that sexual conflict affects traits under direct selection, favoring traits that alter the likelihood of a potential mate agreeing or refusing to mate because it affects the bearer's immediate reproductive output, whereas "traditional" sexual selection is assumed to favor traits that are under indirect selection because they increase offspring fitness. These "traditional" models are sometimes described as "mutualistic" (e.g., Pizzari and Snook 2003; Rice et al. 2006), although this term appears to be used

Cite this article as *Cold Spring Harb Perspect Biol* doi: 10.1101/cshperspect.a017517

only when contrasting them with sexual conflict models. The investigators of the original models never describe them as "mutualistic," which is hardly surprising given that some males are rejected by females.

In this review, we first define sexual conflict and sexual selection. We then describe how the notion of a "lag load" can reveal which sex currently has greater "power" in a sexual conflict over a specific resource. Next, we discuss why sexual conflict and sexual selection are sometimes implicitly (or explicitly) presented as alternative explanations for sexual traits (usually female mate choice/resistance). To illustrate the problems with the assumptions made to take this stance, we present a "toy model" of snake mating behavior based on a study by Shine et al. (2005). We show that empirical predictions about the mating behavior that will be observed if females seek to minimize direct cost of mating or to obtain indirect genetic benefits were overly simplistic. This allows us to make the wider point that whom a female is willing to mate with and how often she mates are often related questions. Finally, we discuss the effect of sexual conflict on population fitness.

DEFINING SEXUAL SELECTION AND SEXUAL CONFLICT

With the exception of artificial experimental breeding designs (see below), sexual conflict is likely to be present whenever there is sexual reproduction. This is true regardless of whether there is anisogamy with males and females or isogamy with two or more mating types (Matsuda and Abrams 1999). We therefore argue that little is gained by asking how sexual selection will differ in the presence and in the absence of sexual conflict. It makes more sense to ask what kinds of traits are favored in each sex, and why, given the inevitable existence of sexual conflict over at least some aspects of reproduction. Our argument relies on the fact that once sexual conflict and sexual selection are appropriately defined, it is near impossible to envisage a biologically plausible situation in which sexual selection exists without sexual conflict, or vice versa.

There are many definitions of sexual selection, extending Darwin's original proposal that sexual selection "depends on the advantage that certain individuals have over other individuals of the same sex and species, in exclusive relation to reproduction" (Darwin 1871, p. 256). One widely accepted definition is that "sexual selection is the differences in reproduction that arise from variation among individuals in traits that affect success in competition over mates and fertilizations" (Andersson 1994, p. 31). There can be practical difficulties in applying this definition to identify sexually selected traits. For example, is a male stickleback that builds a nest selected to do so because it improves offspring survival (natural selection) or because he needs a nest to attract mates? Such "gray zones" aside, sexual selection is a straightforward notion.

Sexual conflict has been defined in many subtly different ways, but these definitions all rely on the fact that evolutionary conflicts of interest can arise whenever two genetically different individuals ("actors") interact (Dawkins 1976; Dawkins and Krebs 1979; Parker 1979). Sexual conflict is comparable to parent–offspring conflict (Parker 2006). In both cases there are shared interests between the two actors (parents "want" their offspring to succeed; males and females "want" to breed), but this does not preclude disagreements over how this common goal is achieved.

Sexual conflict can occur over every facet of breeding. It starts with who will mate search (Hammerstein and Parker 1987), proceeds to whether to reject or accept a potential mate (Parker 1979), and then encompasses how mating and gamete transfer occur. After mating, there might be conflict over whether or not additional matings with other individuals are favored (Baer et al. 2001; Fromhage 2012). Finally, there is conflict over how many offspring are produced, when they are produced, and how much each parent invests into these offspring (e.g., disputes over parental care [Lessells 2006]). Sexual conflict extends beyond traits under sexual selection in one sex and responses by the other sex to their effects on its fitness. It is a much broader concept.

Despite consensus that sexual conflict arises because of genetic differences between potential mates, its definitions are a confusing mixture of statements about costs imposed on a single individual during a mating encounter, differences in mean fitness between the sexes (or temporal changes in mean fitness), or lowered population fitness. To avoid some easy pitfalls, and to make conceptual points, we propose an unorthodox definition that is, hopefully, thought-provoking.

There is sexual conflict if a hypothetical cost-free "tool" allowed some individuals of sex A to alter what individuals of sex B do at a cost to B, such that sex A individuals with the "tool" are then selectively favored over those without it.

The noun "tool" and the verb "to do" should be interpreted broadly. Most obviously, individuals of sex B could change their behavior or morphology when interacting with A individuals who have the tool. For example, instead of B incessantly attempting to mate with A, it now leaves A in peace; or B no longer includes chemicals in its ejaculate that reduce A's lifespan. Our definition is therefore readily applied to interlocus sexual conflict. It can also be applied to intralocus sexual conflict (i.e., genes that elevate fitness in an individual of sex B but reduce it in one of sex A [van Doorn 2009; also see work by Haig et al. 2014]). Here, the tool would be one that allowed an individual of sex A to alter its genes (or their expression) so that individuals with the tool are selectively favored over those without it, but these genetic modifications then impose a cost when expressed in an individual of sex B.

We are not assuming that the tool exists. Sexual conflict only requires that the current situation is suboptimal relative to a hypothetical alternative. It is the potential for selection should such a tool arise. It should also be noted that we could imagine scenarios in which several potential tools could improve fitness. It is then sufficient that one of them satisfies the definition for there to be conflict. For example, sex-specific gene expression might be a tool that benefits both sexes. This does not satisfy the part of the definition in which we require that tool use is costly to the other sex. Do we then erroneously conclude that sexual conflict is absent when sex-specific expression has not yet evolved but would be selected for if it arose? No, because the potential for sex-specific expression to resolve a conflict does not prevent another tool from existing (e.g., alleles that simply lead to the optimal female phenotype at the expense of male fitness). Our definition as a whole resembles that of Hosken et al. (2009) who asked what would happen if one sex had complete control of trait expression in the other sex.

We have specified that possession of the hypothetical tool is cost-free because the presence of costs might be the very factor preventing an individual from reducing sexual conflict (from its perspective). This, of course, would not mean that sexual conflict had vanished. For example, resisting mating attempts might be so costly for a female turtle that her best option is to acquiesce (convenience polyandry [Lee and Hays 2004]). If she could do something to prevent males from mating, however, and this was cost-free, then the actual behaviors of males—that is, how often they actually mated with her—would change. If additional matings were costly for females, then females with the cost-free tool would be favored by selection. This thought experiment reveals that an underlying sexual conflict exists despite females showing no overt resistance to male mating attempts.

In reality, evolutionary innovations can equip individuals with "tools" that, although rarely cost-free, are still selected for in situations with sexual conflict. It is reasonable to assume that the resultant changes will reduce the relative fitness of the affected sex B individuals. If these changes were beneficial for both A and B, they are likely to have already arisen. Sexual conflict therefore generates sexually antagonistic selection (Lessells 2006; van Doorn 2009).

Our definition emphasizes changes in relative fitness within a single sex in a single generation. This avoids the complications that ensure when trying to interpret statements such as "evolutionary changes in male behavior impose costs on females" or "on absolute fitness," given that female fitness might remain unchanged when measured in a population at demographic equilibrium (see Frank and Slatkin 1992; Frank

2012). For example, if female lifetime fecundity declines from one generation to the next because of male-imposed costs, density-dependent juvenile survival might compensate so that each female still recruits on average one daughter to form the next generation (the life history stage of compensatory density-dependent mechanisms need not be the juvenile stage, but if no equivalent mechanisms were in place, populations would generally not persist). Also, we avoid making statements that inappropriately contrast the effect of sexual conflict on the mean fitness of each sex (e.g., Pizzari and Snook 2003; see Arnqvist 2004). In diploid species each offspring has a mother and father so the total number of offspring sired by males cannot differ from the number produced by females ("Fisher condition" [Houston and McNamara 2005]).

LAG LOADS AND WHAT WE CAN LEARN FROM THEM

Our definition captures the idea that sexual conflict involves each sex imposing a "lag load" on the other sex (Rice et al. 2006). The concept of a lag load was used by Maynard Smith (1976) to describe cases in which the average trait value in a population is below its optimum because the selective environment deviates from an earlier state. For sexual conflict, this can be envisaged as the fitness improvement that a sex A individual could achieve over same-sex competitors if it made sex B individuals act to maximize A's reproductive value. This is the deviation of the current environment from an earlier, hypothetical best-case scenario for A.

Sexual conflict is a broad concept, so our definition is intentionally imprecise about the exact nature of the lag load. For example, if A is male and B is female, is "she will mate with you" the scenario for which lag load is measured, or is it "she will mate with you and then never mate again with anyone else"? Lag loads are context-dependent, and realizing this highlights why sexual conflict is so ubiquitous. For example, consider a peahen that is freely able to choose the best peacock with whom to mate. There is still a lag load for her regarding parenting. A mutant female with the power to turn peacocks into ostrich-like males who care for her chicks could benefit by more rapidly producing an additional egg clutch.

Even if we momentarily ignore conflicts over parental care, sexual conflict is still present in even the most "traditional" sexual selection scenario of displaying males and choosy females. Why? A male who is never rejected by females will have greater fitness. This does not happen because choosy females reject many males. Sexual conflict is therefore ongoing. It is even more instructive to recognize the difference between a signaling system that makes it easiest for females to rank males and one that evolves because males try to outcompete each other (van Doorn and Weissing 2006). The fact that no matings occur in the "conflict zone" (*sensu* Parker 1979, 2006), in which only one of the two individuals benefits from a mating, does not make the system fully mutualistic. Male–male competition introduces inefficiencies into mate assessment that are detrimental to females.

Who Has Power?

This leads us to another topic: power, and whether it is possible for one sex to be on the winning side of sexual conflict. If we focus on conflicts over whether mating occurs after encountering a potential mate, it is useful to ask whether one sex currently "controls" or has "power" in this zone of conflict. Consider an example with major fitness consequences: There is a zone of conflict over the values of inbreeding depression that can make sibling mating beneficial for each sex. For a male who suffers no opportunity cost to mating, it is adaptive to mate with his sister up to an inbreeding depression δ value as high as $2/3$. For a female, who cannot have the same egg fertilized by another (unrelated) male if her brother fertilizes it, the threshold is lower at $\delta = 1/3$ (Parker 1979, 2006).

The lag load for females is clearly lower if they have the power to decide whether mating with a sibling occurs (in which case there is no female lag load) than if males can overcome their sisters' resistance to mating when $1/3 < \delta < 2/3$. In this sense, statements that it is meaningless to ask which sex "wins" during a specific sexual conflict are overly strong (e.g., Arnqvist 2004). Sex A "has its own way" more than sex B if there are no conceivable ways in which B could behave better for the fitness of an individual of sex A in the context of the specified conflict. In this sense, A is the winner. This makes it valid to ask, although not necessarily easy to answer, questions such as whether certain adaptive traits, such as flight, have shifted sexual power toward females in birds when compared to the situation in less mobile mammals in which females cannot escape harassment as readily (Pradhan and van Schaik 2009). This might have consequences for the relative prevalence of ornaments versus armaments, if these trait types have different effects on female fitness.

The comparisons necessary to make statements about lag loads force us to deal with hypothetical scenarios. Pragmatically speaking, sensible questions about lag load are constrained by what occurs naturally or can be experimentally mimicked. The latter can involve clever experiments. For example, Rice et al. (2006) investigated male lag load when already mated females resist male courting effort. Males provided with a limitless supply of virgin females had a sevenfold potential reproductive output compared with a more natural setting that takes into account nonvirgin females' reluctance to remate (Rice et al. 2006). This difference can be interpreted as a measure of the male lag load—namely, the increase in reproductive success if he could find a cost-free way to ensure that every mated female encountered would mate with him (leaving aside issues about the share of paternity).

Removing Sexual Conflict Also Removes Sexual Selection

In nature, it is very rare to find monogamous pairs that form randomly, mate for life, and never remate if widowed (Hosken et al. 2009). Nevertheless, this hypothetical scenario provides a useful benchmark because it favors traits that remove sexual conflict and is experimentally tractable (see the review by Edward et al. 2010). The evolutionary interests of every interacting male and female (i.e., each pair) become identical, and selection acts to maximize their joint lifetime reproductive output. One can then quantify by how much lifetime reproductive success increases when wild-type individuals encounter opposite sex individuals from the experimental rather than their own population. The greater the increase in reproductive output, the larger the lag load due to sexual conflict.

Monogamous experimental populations, with random assignment of mates, are equally well described as ones in which sexual selection has been removed. The equivalency between sexual selection and conflict in this setting is unsurprising. Whenever sexual selection occurs, we can always conceive of ways in which, for example, a female would benefit if her mate invested less into trying to acquire other females or gain paternity assurance and more into her own survival or fecundity. To remove sexual conflict, it is therefore necessary to remove sexual selection.

Eliminating sexual selection, in contrast, does not always remove sexual conflict, because the conflict extends beyond mating and gaining access to opposite sex gametes to, for example, parental investment decisions. Nonetheless, sexual selection is clearly the most important source of sexual conflict. To see that it is more ubiquitous than, for example, conflicts over care, one only has to recall that neither sex offers parental care in most animal species (and offspring therefore do not "expect" to receive it).

The pervasiveness of sexual selection and the associated sexual conflict (and its context specificity) is well illustrated by considering a seemingly highly cooperative scenario in which the sexes always agree to mate when they meet: finding mates in a low-density population in which both sexes' reproductive rates are limited by mate encounters. There is no conflict over mating per se. Strong sexual selection on mate-location ability, likewise, seems mutually beneficial because individuals of neither sex can be certain that they will encounter mates at a sufficiently high rate. The Bateman gradient, a measure of whether selection favors an increase in the mating rate, is positive for both sexes when it is mainly measured between zero or one mate (Kokko et al. 2012).

Even so, sexual conflict persists. The problem is precisely analogous to parental care, in which individuals of sex A benefit if sex B cares more. This allows sex A to reduce care and invest in other, more self-serving, fitness components. If increased mate searching is energetically costly or dangerous, or if signals that reveal an individual's location are costly, then a female would benefit if males "worked harder" to find her and vice versa. Clearly, one of these activities must happen, or individuals will not find each other. Models indeed show that if one sex searches more efficiently, the other sex then adaptively reduces its own search effort (Hammerstein and Parker 1987; Kokko and Wong 2007).

ONCE WE REALIZE THERE IS ALWAYS SEXUAL CONFLICT, WHAT HAPPENS TO GENETIC BENEFITS?

There are three main types of traits favored by sexual selection because they increase access to mates. Traits can (a) make it easier to locate mates (e.g., locomotor efficiency, olfactory powers); (b) confer an advantage during direct physical contests between rivals (i.e., weapons); and (c) increase attractiveness to the opposite sex (e.g., ornaments). These categories appear to differ in the extent to which we need to consider the coevolution of the two sexes. Specifically, traits that increase attractiveness are associated with mate choice and mating preferences in the opposite sex. This can generate genetic correlations between ornaments, preferences, and other components of fitness (i.e., viability) (Kokko et al. 2006). Much of the debate about the relationship between sexual conflict and sexual selection over the last decade has been about mate choice. For brevity, we follow convention and describe cases in which females choose males.

Given sexual conflict, it is easy to see that females might increase their relative fitness by being choosier (i.e., choosiness is a "tool"). For example, females could prefer males that offer more nutritious nuptial gifts (Lewis and South 2012) or are less likely to inflict damage while mating (if this can be assessed before copulating). It is uncontroversial that sexual conflict can promote the evolution of mating preferences that involve a female actively sampling males and choosing those that confer greater direct benefits or, phrased differently, impose smaller costs. Both yield a direct benefit because choice elevates a female's lifetime fecundity. Here, we can explain the evolution of female behavior without considering genetic (indirect) benefits. Adaptive mate choice need not elevate offspring fitness because of the genes they inherit.

Slightly differently, it has been argued that female choice (which includes behavioral resistance to mating because this affects which types of males become mates) is favored not because it increases the likelihood that certain types of males mate, but rather because it lowers a female's mating rate (i.e., choosiness reduces the number of potential mates; Gavrilets et al. 2001). The evolution of choice is then seen to be the result of selection due to sexual conflict over mating. Conflict occurs because males usually gain fitness from each successive mating (males almost always have a positive Bateman gradient), whereas females that mate multiply show little gain, and usually a decline, in fitness after a certain number of matings (e.g., Arnqvist et al. 2004). This decline can reflect time and predation costs or is itself the product of sexual conflict that favors males with traits that increase their share of paternity and/or the female's rate of offspring production but, in so doing, damage females (Morrow et al. 2003). Again, it is argued that choosiness is selected for because of direct benefits that arise because of sexual conflict, not genetic benefits (Gavrilets et al. 2001).

There is much debate as to whether sexual conflict over mating rates or benefits of mating with specific males best explain the evolution of female mating behavior (for empirical examples, see Friberg and Arnqvist 2003; Forstmeier 2004; Shine et al. 2005). We believe this dichotomy is an unhelpful way to frame the question. The real issue is more nuanced, and it was extensively discussed before the explosion of interest in sexual conflict in the early 2000s (e.g., Kirkpatrick and Barton 1997). How much do genetic and direct benefits respectively contribute to selection that maintains costly mating preferences?

It is crucial to recognize that preference or resistance traits (for their different connotations, see Gavrilets et al. 2001) that affect the likely identity of the average mate will usually also change the number of mates per breeding cycle. For example, if a female has a strong preference for unusually large males, the rarity of such males in the population will automatically reduce her mating rate if she follows a rule of only mating with those males that exceed a size threshold (Kokko and Mappes 2013). Mate number and identity can be decoupled if females show strict monandry, but this is extremely rare (Jennions and Petrie 2000). Completely indiscriminate mating can also, in principle, eliminate this relationship, but in reality it often leads to indirect mate choice *sensu* Wiley and Poston (1996). If, for example, females arriving on a lek mated with all the males present (an unlikely scenario) or, alternatively, with just one randomly chosen male, their behavior would cause strong selection for males with stamina. Indeed, lek attendance is a strong correlate of male mating success (Hill 1991; Focardi and Tinelli 1996; Rintamäki et al. 2001; Friedl and Klump 2005). Mating truly randomly can be surprisingly difficult.

It is understandable why the correlation between mate identity and mating rate is easily forgotten: The history of the field encourages us to ignore it. The literature is largely split into "sexual conflict models" that emphasize selection for an optimal mating rate (e.g., Gavrilets et al. 2001; Rowe et al. 2005; Hoyle and Gilburn 2010; Kazancioglu and Alonzo 2012) and "mate choice models" that emphasize the identity of mates and the associated gains (direct or genetic), but do not explicitly address the number of mates or matings (for reviews, see Kokko et al. 2006; Kuijper et al. 2012). Consequently, tests of "sexual conflict models" as the driver of female mating behavior tend to consider the perceived alternative (i.e., the behavior predicted by mate choice models) in overly simplistic terms. Reality is more complex. For example, Rice et al. (2006) describe a fascinating set of experiments in *Drosophila* that removed the possibility of "trading up" on an earlier mating by preventing females from remating. Females mating less often is then equated with the absence of genetic benefits. In many cases, however, mating less often is indicative of stronger, not weaker, choosiness (Kokko and Mappes 2013; see also the debate on resistance as choice in Cordero and Eberhard [2003] and Rowe et al. [2005]). In general, it is challenging to predict what type of female behavior indicates increased choosiness. To highlight the complexity of the issue, we draw inspiration from a representative study on garter snakes (*Thamnophis sirtalis*) by Shine et al. (2005).

How to Be Choosy in a World of Coercive Mating

Garter snakes overwinter in communal dens. In summer, females rapidly disperse from the den, where male density is very high, to outlying areas with fewer males. Dispersal is a form of evasive, resistance-like behavior that reduces encounters with males that attempt to forcibly mate. Shine et al. (2005) asked whether female evasion acts to minimize cost imposed by male sexual harassment (labeled "sexual conflict") or is a form of indirect mate choice that generates a selective filter favoring certain males (labeled "traditional or mutualistic sexual selection," and even described as "cooperative" by the investigators). They explicitly equate choosiness with less effort spent evading males: "The hypothesis of indirect mate choice generates specific predictions... females will tolerate prolonged courtship rather than attempt to evade their suitors (because prolonged courtship would allow the female to assess male phenotype and thus select an appropriate mate)." We present a toy mathematical model inspired by mating in garter snakes to show that predicting female behavior from the mate choice hypothesis is more difficult than it first appears.

To investigate a hypothesis about indirect mate choice, we need to specify how female behavior affects the distribution of both mate number and identity. To allow for a conceptually simple possibility of indirect genetic benefits, we assume two types of males: those that are either efficient or inefficient at harassing females into mating when other males also compete for females. Producing efficient harassers will elevate a sons' mating success, regardless of whether or not other females "prefer" to be harassed. This creates the potential for genetic benefits of nonrandom mating. Let the proportion of efficient harassers be q and that of inefficient harassers $1 - q$.

Females move away from the den area at speed s. To provide a minimal model, we ignore any cost to moving faster, although the model could be easily modified such that reaching a large s is only possible if the female possesses physiological adaptations that impose costs on other life history components. In the den area, the number of males that successfully harass a female into mating is a declining function of s: Faster females leave the area sooner. The function is weighted so that efficient harassers obtain more matings when females move at a given speed, and the performance difference between efficient and inefficient harassers is more pronounced for faster females. Our examples are derived assuming that the number of efficient harassers that achieve a mating with the female is $q5e^{-s}$, whereas the number of inefficient harassers doing so is $(1 - q)4e^{-1.2s}$ (Fig. 1). These are the means of a Poisson distribution of mate numbers. So, for example, if all males are good harassers and a female moves very slowly ($s \approx 0$), she will, on average, mate five times before leaving the den area if all males are efficient harassers or four times if all males are inefficient.

We assume that female reproductive success (direct fitness) is maximized if she mates exactly once in the den area ($w_1 = 1$) and is lower if she mates more often: generally $w_n = e^{-k(n-1)}$, in which $n \geq 1$ is the number of matings and k is a parameter. A mathematical prediction also borne out by data (Shine et al. 2005) is that a fast female might not mate in the den area. Little is known about the fitness consequences of this, so we vary it using the factor w_0 (here 0 refers to zero matings in the den). It is clearly possible for a female to mate outside the den; efficient harassers might still outperform inefficient ones here, but for simplicity we assume their performance is now equally efficient, so the probabilities that a female mates with each type of male is q and $1 - q$, respectively, when outside the den. If finding a male outside the den requires effort, or if matings there are relatively more costly, then $w_0 < 1$. There are two questions:

1. To maximize direct benefits, how fast should a female move given the negative effect of the number of mates on fitness (this favors higher speeds) and a decline in fitness outside the den area (i.e., $w_0 < 1$; this favors lower speeds)?
2. What speed maximizes genetic benefits?

We assume $w_0 < 1$ to be able to provide a thought-provoking answer to these questions, and we leave it to the reader to visualize why the same set of results would apply for $w_0 > 1$ if we also introduce physiological costs that prevent a female from moving infinitely fast.

The answer to (1) is relatively straightforward (Fig. 2A,B). For (2), there are modeling

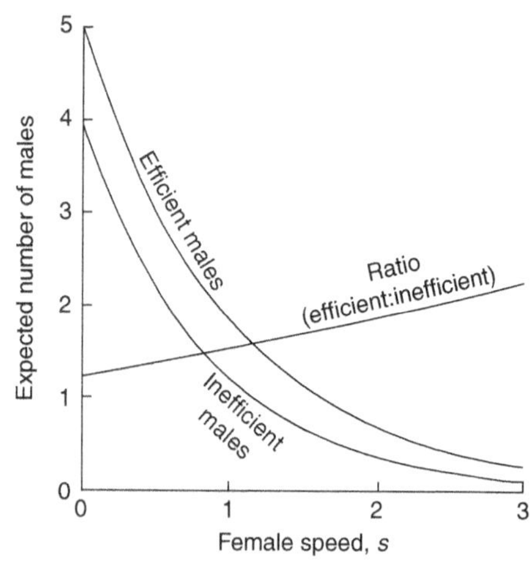

Figure 1. The assumptions of the toy model. The relationship between female speed, s, and the expected number of efficient harassers that mate with her (upper declining curve) and inefficient harassers (lower declining curve), each plotted for the scenario in which this male type predominates in the population. The increasing curve shows that the proportion of eggs fertilized by the former increases with s, here plotted assuming that both types are equally common ($q = 0.5$).

 Cite this article as *Cold Spring Harb Perspect Biol* doi: 10.1101/cshperspect.a017517

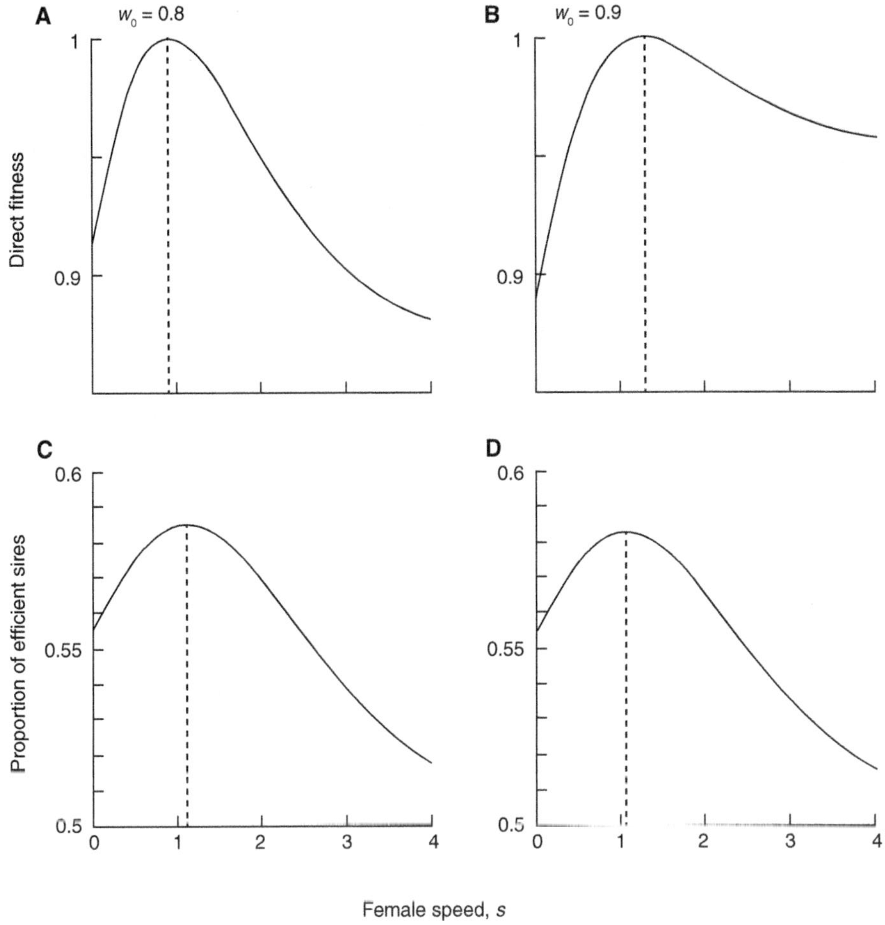

Figure 2. Two scenarios (*A,C*) $w_0 = 0.8$ and (*B,D*) $w_0 = 0.9$, where we plot (*A,B*) the consequent direct fitness, measured as

$$w_0\, e^{-\lambda} + \sum_{n=1}^{\infty} e^{-k(n-1)} \frac{\lambda^n}{n!} e^{-\lambda},$$

where

$$\lambda = q\, 5e^{-s} + (1-q)5e^{-1.2s};$$

and (*C,D*) the proportion of males that mate with a female that are efficient harassers, computed as $a/(a+b)$, where

$$a = q\, w_0\, e^{-\lambda} + \sum_{n=1}^{\infty} e^{-k(n-1)} \frac{\lambda^n}{n!} e^{-\lambda} \frac{q\, 5e^{-s}}{q\, 5e^{-s} + (1-q)5e^{-1.2s}},$$

and

$$b = (1-q)w_0\, e^{-\lambda} + \sum_{n=1}^{\infty} e^{-k(n-1)} \frac{\lambda^n}{n!} e^{-\lambda} \frac{(1-q)\, 5e^{-1.2s}}{q\, 5e^{-s} + (1-q)5e^{-1.2s}}.$$

Note that $e^{-\lambda}$ is the probability that the female does not mate in the den and that the speed that maximizes direct benefits can be either lower (*A,C*) or higher (*B,D*) than the one that is best for finding an efficient male to sire the offspring. The difference is, however, slight. The consequences of this discrepancy are minor for selection acting on males (see text).

techniques to analyze coevolution between resistance and male mating effort, and whether equilibria depends on genetic covariance between female and male traits (e.g., Kokko 2005). Here, to make a conceptual point, we simply assume that efficiency is heritable, so that indirect selection favors mating with efficient harassers to increase sons' mating success.

The fitness of sons is an increasing function of the proportion of mates who are efficient harassers. This proportion depends on female speed in a nonlinear way (Fig. 2C,D). Consider the extremes: Very slow females gain offspring sired by the two male types according to the ratio $5q{:}4(1-q)$ (efficient:inefficient harassers). Very fast females escape the den without mating, and the ratio is $q{:}1-q$ when mating outside the den. Intermediate speeds are the best for screening purposes. At a relatively high speed, females mate in the den, but most can still escape the mating attempts of inefficient harassers.

These nonlinearities per se are less interesting than realizing that the female speed s that maximizes genetic benefits can be faster or slower than the speed that maximizes direct fitness (Fig. 2A vs. 2C, and 2B vs. 2D). Given how little is known about paternity and fitness outside the den (Shine et al. 2005), nor about other important parameters such as heritabilities and antagonistic (or not) genetic effects on sons' versus daughters' fitness, we do not want to make statements about whether these differences produce a significant shift in female speeds compared with a scenario involving maximizing direct fitness. The point we want to make here is far simpler: It is inappropriate to assume that choosier females will move more slowly (so that faster speeds are automatically indicative of maximization of direct fitness) simply because models of mate choice traditionally use a framework in which sampling more males increases the eventual mate's quality either explicitly (Janetos 1980; Real 1990; Johnstone 1997) or implicitly (i.e., stronger preferences are assumed to be costlier [e.g., Pomiankowski et al. 1991]). In reality, choosiness for genetic benefits can require that more time be spent with males (e.g., when actively sampling displaying males; Byers and Waits 2006) but, as our model shows, it can equally be associated with a reduction in the time spent with males relative to that which minimizes costs. The lesson for empiricists is that all models make assumptions about an organism's natural history. If the biology of a species differs from these assumptions, then generic concepts such as choosiness have to be interpreted carefully.

In the garter snake scenario above, we simply assumed that the genetic trait of interest is the same male trait that is imposing costs on females. In reality the situation can be far more complicated. For example, what if females seek genetic benefits that are uncorrelated with a male's ability to acquire mates? As an illustrative example, female Chinook salmon are more aggressive toward major histocompatibility complex (MHC)-similar males (potentially indicating choice for MHC dissimilarity), but males appear to be able to overcome this preference (Garner et al. 2010). Intriguingly, this study provides evidence that females are better able to control the decision to mate when the sex ratio becomes more female-biased, as offspring MHC diversity was found to increase under such conditions.

Are There True Differences between Models? We Can Think of an Important One

It is often a challenge to determine exactly what females will do if there has been selection for genetic benefits. In the model above there is a sexual asymmetry, however, because selection on male behavior is largely independent of whether these benefits modify female behavior. Being efficient at harassing is strongly selected for in all cases other than when females move at extremely high speed s (and this is only because we arbitrarily assumed that harassment efficiency has no effect outside the den).

Although we have primarily argued for the need to consider sexual selection in the light of sexual conflict, the above sexual asymmetry highlights a more relevant distinction from "traditional" female choice scenarios than the (in reality near identical) degree to which different models involve sexual conflict. In the garter snakes, selection on male behavior shifts only relatively slightly with changes in the magnitude of female escape effort (s). In contrast, in classic female preference models it is difficult to explain the evolution of male traits without considering the nuances of female behavior and the selective forces that act on them. In the former case, female cost-minimization ("resist most matings") permits strong selection on males, whereas in the traditional models, the matter is somewhat less clear. Although models often assume that mating is random if females lack preferences, in reality, scenarios in which females lack costly preferences might still generate nonrandom mating. Males can still be selected to succeed in scramble competition or bear other traits that increase their mate encounter rate ("passive attraction" models [Wiley and Poston 1996]).

CONCLUDING REMARKS: WHAT HAPPENS TO POPULATIONS?

We have argued that there is a sense in which one sex can "win" a conflict. We can ask which sex carries a larger lag load (relative to a sensibly chosen hypothetical alternative state, such as "which sex has greater ability to decide on whether a mating occurs" or "which sex provides parental care"). Understanding such sexual asymmetries is important because females tend to drive demography. Simply put, a population consisting of a majority of males is expected to yield fewer offspring than a population consisting of mostly females (Rankin and Kokko 2007). If females have a high lag load (females "lose"), population fitness will tend to suffer more than when males have a high lag load.

An intriguing example is provided by the cichlid fish *Lamprologus callipterus*, in which breeding sites in empty snail shells are a limited resource (Maan and Taborsky 2008). Males can recognize that they have not spawned with a female and take over her shell so that she loses her current reproductive attempt. The expelled female is typically not yet ready to spawn again and must find another empty shell. Sexual conflict is clear (an individual of sex A would have higher fitness if it could somehow prevent its site from being stolen by individuals of sex B). Males take over nests from females, but not the reverse, indicating a greater female than male lag load. The population-level consequence is that many nests fail. The population would conceivably be more productive if females "won" the conflict.

The most fundamental detrimental population-level consequence of sexual conflict is, of course, the twofold cost of sex. The evolution of anisogamy (Parker et al. 1972; Parker 2014) leads to the great majority of resources used for offspring production only being channeled through females (hence, the expected halving of population growth in sexual populations compared with equivalent asexual counterparts [Lehtonen et al. 2012]). If males always contributed all their energy to improving offspring number or survival and none to the zero-sum game of outcompeting other males, population growth would theoretically double. This statement is, however, accompanied by several caveats. If there is biparental care the cost is less than twofold as males are then also investing into offspring. Even here, however, sexual conflict persists: Two caring parents do not necessarily raise twice as many offspring as a single parent, especially as egg production is still a female-only task. For example, in a data set of 980 bird species, removal of male care led to a 20%, rather than 50%, reduction in female productivity (Sibly et al. 2012). Conversely, if sexual conflict not only results in males failing to provide care but additionally selects for traits that reduce the ability of females to produce offspring, then the costs of sex can be more than twofold (Lehtonen et al. 2012). Population collapse is even theoretically possible (Le Galliard et al. 2005; Rankin et al. 2011).

Short-term demography is, of course, not the final word on the topic of population fitness. We need to consider long-term processes. There are few, if any, species with truly random pair formation and lifelong monogamy to look at the evolutionary consequences of the permanent removal of sexual conflict. We should, however, remind ourselves that there is another way to reproduce that achieves the same end: Asexuality removes both sexual conflict and sexual selection. It is difficult to reconcile the view that sex, with all the attendant conflict, is ultimately harmful for population fitness, with evidence that asexuality tends not to persist over evolutionary time (Neiman et al. 2005; but see Janko 2014). This could be interpreted as support for the view that sexual selection is ultimately "good" for populations (e.g., Sharp and Agrawal 2013). The difficulty, of course, is that comparisons in which "all else is equal" are a major challenge. Asexual species may retain maladaptive vestigial traits that relate to their previous sexual life cycles (Bengtsson 2009). In addition, asexuality removes not only conflict but also genetic recombination (Neher et al. 2010; Hartfield and Keightley 2012). The costs and benefits of sex and the costs of sexual conflict are measured on different timescales; how to reconcile them is a fascinating topic that deserves greater attention.

ACKNOWLEDGMENTS

We thank Geoff Parker, Sergey Gavrilets, and an anonymous reviewer for very helpful comments, and the Australian Research Council and the Academy of Finland for funding.

REFERENCES

Reference is also in this collection.

Andersson M. 1994. *Sexual selection.* Princeton University Press, Princeton, NJ.

Arnqvist G. 2004. Sexual conflict and sexual selection: Lost in the chase. *Evolution* **58:** 1383–1388.

Arnqvist G, Nilsson T, Katvala M. 2004. Mating rate and fitness in bean weevils. *Behav Ecol* **16:** 123–127.

Baer B, Morgan ED, Schmid-Hempel P. 2001. A nonspecific fatty acid within the bumblebee mating plug prevents females from remating. *Proc Natl Acad Sci* **98:** 3926–3928.

Bengtsson BO. 2009. Asex and evolution: A very large-scale overview. In *Lost sex: The evolutionary biology of parthenogenesis* (ed. Schön I, Martens K, van Dijk P), pp. 1–19. Springer, Berlin.

Byers JA, Waits L. 2006. Good genes sexual selection in nature. *Proc Natl Acad Sci* **103:** 16343–16345.

Cordero C, Eberhard WG. 2003. Female choice of sexually antagonistic male adaptations: A critical review of some current research. *J Evol Biol* **16:** 1–6.

Darwin C. 1871. *The descent of man, and selection in relation to sex.* Murray, London.

Dawkins R. 1976. *The selfish gene.* Oxford University Press, Oxford.

Dawkins R, Krebs JR. 1979. Arms races between and within species. *Proc R Soc B* **205:** 489–511.

Edward DA, Fricke C, Chapman T. 2010. Adaptations to sexual selection and sexual conflict: Insights from experimental evolution and artificial selection. *Proc R Soc B* **365:** 2541–2548.

Focardi S, Tinelli A. 1996. A structural-equations model for the mating behaviour of bucks in a lek of fallow deer. *Ethol Ecol Evol* **8:** 413–426.

Forstmeier W. 2004. Female resistance to male seduction in zebra finches. *Anim Behav* **68:** 1005–1015.

Frank SA. 2012. Wright's adaptive landscape versus Fisher's fundamental theorem. In *The adaptive landscape in evolutionary biology* (ed. Svensson E, Calsbeek R), pp. 41–57. Oxford University Press, Oxford.

Frank SA, Slatkin M. 1992. Fisher's fundamental theory of natural selection. *Trends Ecol Evol* **7:** 92–95.

Friberg U, Arnqvist G. 2003. Fitness effects of female mate choice: Preferred males are detrimental for *Drosophila melanogaster* females. *J Evol Biol* **16:** 797–811.

Friedl TWP, Klump GM. 2005. Sexual selection in the lek-breeding European treefrog: Body size, chorus attendance, random mating and good genes. *Anim Behav* **70:** 1141–1154.

Fromhage L. 2012. Mating unplugged: A model for the evolution of mating plug (dis-)placement. *Evolution* **66:** 31–39.

Garner SR, Bortoluzzi RN, Heath DD, Neff BD. 2010. Sexual conflict inhibits female mate choice for major histocompatibility complex dissimilarity in Chinook salmon. *Proc R Soc B* **277:** 885–894.

Gavrilets S, Arnquist G, Friberg U. 2001. The evolution of female mate choice by sexual conflict. *Proc R Soc B* **268:** 531–539.

* Haig D, Úbeda F, Patten MM. 2014. Specialists and generalists: The sexual ecology of the genome. *Cold Spring Harb Perspect Biol* doi: 10.1101/cshperspect.a017525.

Hammerstein P, Parker GA. 1987. Sexual selection: Games between the sexes. In *Sexual selection: Testing the alternatives* (ed. Bradbury JW, Andersson MB), pp. 119–142. Wiley, Hoboken, NJ.

Hartfield M, Keightley PD. 2012. Current hypotheses for the evolution of sex and recombination. *Integr Zool* **7:** 192–209.

Hill WL. 1991. Correlates of male mating success in the ruff *Philomachus pugnax*, a lekking shorebird. *Behav Ecol Sociobiol* **29:** 367–372.

Hosken DJ, Stockley P, Tregenza T, Wedell N. 2009. Monogamy and the battles of the sexes. *Ann Rev Entomol* **54:** 361–378.

Houston AI, McNamara JM. 2005. John Maynard Smith and the importance of consistency in evolutionary game theory. *Biol Philos* **20:** 933–950.

Hoyle A, Gilburn AS. 2010. Sexually antagonistic co-evolution: A model and an empirical test. *J Evol Biol* **23:** 166–174.

Hutchinson GE. 1965. *The ecological theatre and the evolutionary play.* Yale University Press, New Haven, CT.

Janetos AC. 1980. Strategies of female choice: A theoretical analysis. *Behav Ecol Sociobiol* **7:** 107–112.

Janko K. 2014. Let us not be unfair to asexuals: Their ephemerality may be explained by neutral models without invoking any evolutionary constraints of asexuality. *Evolution* **68:** 569–576.

Jennions MD, Petrie M. 2000. Why do females mate multiply? A review of the genetic benefits. *Biol Rev* **75:** 21–64.

Johnstone RA. 1997. The tactics of mutual mate choice and competitive search. *Behav Ecol Sociobiol* **40:** 51–59.

Kazancioglu E, Alonzo SH. 2012. The evolution of optimal female mating rate changes the coevolutionary dynamics of female resistance and male persistence. *Phil Trans R Soc B* **367:** 2339–2347.

Kirkpatrick M, Barton NH. 1997. The strength of indirect selection on female mating preferences. *Proc Natl Acad Sci* **94:** 1282–1286.

Kokko H. 2005. Treat 'em mean, keep 'em (sometimes) keen: Evolution of female preferences for dominant and coercive males. *Evol Ecol* **19:** 123–135.

Kokko H, Mappes J. 2013. Multiple mating by females is a natural outcome of a null model of mate encounters. *Entomol Exp Appl* **146:** 26–37.

Kokko H, Wong BBM. 2007. What determines sex roles in mate searching? *Evolution* **61:** 1162–1175.

Kokko H, Jennions MD, Brooks R. 2006. Unifying and testing models of sexual selection. *Annu Rev Ecol Evol Syst* **37:** 43–66.

Kokko H, Klug H, Jennions MD. 2012. Unifying cornerstones of sexual selection: Operational sex ratio, Bateman gradient, and the scope for competitive investment. *Ecol Lett* **15:** 1340–1351.

Kuijper B, Pen I, Weissing FJ. 2012. A guide to sexual selection theory. *Annu Rev Ecol Evol Syst* **43:** 287–311.

Le Galliard J-F, Fitze PS, Ferrière R, Clobert J. 2005. Sex ratio bias, male aggression, and population collapse in lizards. *Proc Natl Acad Sci* **102:** 18231–18236.

Lee PLM, Hays GC. 2004. Polyandry in a marine turtle: Females make the best of a bad job. *Proc Natl Acad Sci* **101:** 6530–6535.

Lehtonen J, Jennions MD, Kokko H. 2012. The many costs of sex. *Trends Ecol Evol* **27:** 172–178.

Lessells CM. 2006. The evolutionary outcome of sexual conflict. *Phil Trans R Soc B* **361:** 301–317.

Lewis S, South A. 2012. The evolution of animal nuptial gifts. *Adv Study Behav* **44:** 53–97.

Maan ME, Taborsky M. 2008. Sexual conflict over breeding substrate causes female expulsion and offspring loss in a cichlid fish. *Behav Ecol* **19:** 302–308.

Matsuda H, Abrams PA. 1999. Why are equally sized gametes so rare? The instability of isogamy and the cost of anisogamy. *Evol Ecol Res* **1:** 769–784.

Maynard Smith J. 1976. What determines the rate of evolution? *Am Nat* **110:** 331–338.

Morrow EH, Arnqvist G, Pitnick S. 2003. Adaptation versus pleiotropy: Why do males harm their mates? *Behav Ecol* **14:** 802–806

Neher RA, Shraiman BI, Fisher DS. 2010. Rate of adaptation in large sexual populations. *Genetics* **184:** 467–481.

Neiman M, Jokela J, Lively CM. 2005. Variation in asexual lineage age in *Potamopyrgus antipodarum*, a New Zealand snail. *Evolution* **59:** 1945–1952.

Parker GA. 1979. Sexual selection and sexual conflict. In *Sexual selection and reproductive competition in insects* (ed. Blum MS, Blum NA), pp. 123–166. Academic, New York.

Parker GA. 2006. Sexual conflict over mating and fertilization: An overview. *Phil Trans R Soc B* **361:** 235–259.

* Parker GA. 2014. The sexual cascade and the rise of pre-ejaculatory (Darwinian) sexual selection, sex roles, and sexual conflict. *Cold Spring Harb Perspect Biol* doi: 10.1101/cshperspect.a017509.

Parker GA, Baker RR, Smith VGF. 1972. The origin and evolution of gamete dimorphism and the male–female phenomenon. *J Theor Biol* **36:** 181–198.

Pizzari T, Snook RR. 2003. Sexual conflict and sexual selection: Chasing away paradigm shifts. *Evolution* **57:** 1223–1236.

Pomiankowski A, Iwasa Y, Nee S. 1991. The evolution of costly mate preferences: I. Fisher and biased mutation. *Evolution* **45:** 1422–1430.

Pradhan GR, van Schaik CP. 2009. Why do females find ornaments attractive? The coercion-avoidance hypothesis. *Biol J Linn Soc* **92:** 372–382.

Rankin DJ, Kokko H. 2007. Do males matter? The role of males in population dynamics. *Oikos* **116:** 335–348.

Rankin DJ, Dieckmann U, Kokko H. 2011. Sexual conflict and the tragedy of the commons. *Am Nat* **177:** 780–791.

Real L. 1990. Search theory and mate choice: I. Models of single-sex discrimination. *Am Nat* **136:** 376–405.

Rice WR, Stewart AD, Morrow EH, Linder JE, Orteiza N, Byrne PG. 2006. Assessing sexual conflict in the *Drosophila melanogaster* laboratory model system. *Phil Trans R Soc B* **361:** 287–299.

Rintamäki PT, Höglund J, Alatalo RV, Lundberg A. 2001. Correlates of male mating success on black grouse (*Tetrao tetrix* L.) leks. *Ann Zool Fennici* **38:** 99–109.

Rowe L, Cameron E, Day T. 2005. Escalation, retreat, and female indifference as alternative outcomes of sexually antagonistic coevolution. *Am Nat* **165:** S5–S18.

Schoener TW. 2011. The newest synthesis: Understanding the interplay of evolutionary and ecological dynamics. *Science* **331:** 426–429.

Sharp NP, Agrawal AF. 2013. Male-biased fitness effects of spontaneous mutations in *Drosophila melanogaster*. *Evolution* **67:** 1189–1195.

Shine R, Wall M, Langkilde T, Mason RT. 2005. Do female garter snakes evade males to avoid harassment or to enhance male quality? *Am Nat* **165:** 660–668.

Sibly RM, Witt CC, Wright NA, Venditti C, Jetz W, Brown JH. 2012. Energetics, lifestyle, and reproduction in birds. *Proc Natl Acad Sci* **109:** 10937–10941.

van Doorn GS. 2009. Intralocus sexual conflict. *Ann NY Acad Sci* **1168:** 52–71.

van Doorn GS, Weissing FJ. 2006. Sexual conflict and the evolution of female preferences for indicators of male quality. *Am Nat* **168:** 742–757.

Wiley RH, Poston J. 1996. Indirect mate choice, competition for mates, and coevolution of the sexes. *Evolution* **50:** 1371–1381.

Specialists and Generalists: The Sexual Ecology of the Genome

David Haig[1], Francisco Úbeda[2], and Manus M. Patten[3]

[1]Department of Organismic and Evolutionary Biology, Harvard University, Cambridge, Massachusetts 02138

[2]School of Biological Sciences, Royal Holloway, University of London, Egham, Surrey TW20 0EX, United Kingdom

[3]Department of Biology, Georgetown University, Washington, D.C. 20057

Correspondence: dhaig@oeb.harvard.edu

Sexual antagonism occurs when an allele is beneficial in one sex but costly in the other. Parental antagonism occurs when an allele is beneficial when inherited from one sex but costly when inherited from the other because of fitness interactions among kin. Sexual and parental antagonisms together define four genetic niches within the genome that favor different patterns of gene expression. Natural selection generates linkage disequilibrium among sexually and parentally antagonistic loci with male-beneficial alleles coupled to alleles that are beneficial when inherited from males and female-beneficial alleles coupled to alleles that are beneficial when inherited from females. Linkage disequilibrium also develops between sexually and parentally antagonistic loci and loci that influence sex determination. Genes evolve sex-specific expression to resolve sexual antagonism and evolve imprinted expression to resolve parental antagonism. Sex-specific chromosomes allow a gene to specialize in a single niche.

Every diploid individual of a sexually reproducing population is derived from an egg fertilized by a sperm. Therefore, males considered collectively have the same reproductive value as females because half the genes of the next generation will be derived from males and half from females (Kokko et al. 2006). This fundamental symmetry is independent of sex ratio and mating system and applies also to hermaphrodites in male and female roles. Despite their equal stakes in posterity, males and females have evolved distinct morphologies and reproductive strategies as indirect consequences, often very indirect, of the ancient dichotomy between production of larger gametes by one sex and smaller gametes by the other (Queller 1997). Vive la différence!

Not only are half the genes of the next generation present in females of the current generation, but half the genes of the current generation were inherited from females. Genes of maternal and paternal origin (hereafter matrigenes and patrigenes) are each transmitted to 50% of the next generation of gametes and resulting offspring. Therefore, matrigenic and patrigenic alleles benefit equally from an individ-

Cite this article as *Cold Spring Harb Perspect Biol* doi: 10.1101/cshperspect.a017525

ual's survival and reproduction. This symmetry is broken when organisms interact with kin to whom they are unequally related via their mother and father (Haig 1997; Úbeda and Gardner 2010). Adaptive responses to such asymmetries of relatedness can favor gene expression that is conditional on parental origin. "Imprinted" expression can cause matrigenes and patrigenes to have opposing effects on disputed phenotypes (Haig 2000a; Holman and Kokko 2014).

Anatomical, physiological, and behavioral sex differences are thus associated with two selective asymmetries, one obvious (natural selection acts differently on genes in female and male bodies) and the other less obvious (natural selection acts differently on genes of maternal and paternal origin). These asymmetries define orthogonal partitions of the gene pool into pairs of "environments" in which there is niche-specific selection (Fig. 1A). Differential selection in male and female niches reinforces sexual dimorphism by processes of sex-specific adaptation. Differential selection in matrigenic and patrigenic niches can result in imprinted gene expression and molecular adaptations acting at cross-purposes within organisms.

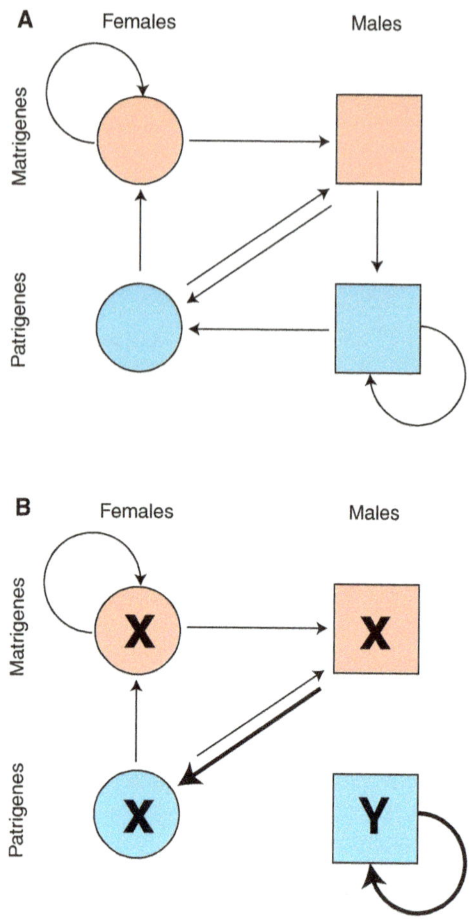

Figure 1. Sex-related partitions of the gene pool. The gene pool is independently subdivided into female (circles) and male (squares) niches and into matrigenic (pink) and patrigenic (blue) niches. Gene flow between the environments is represented by arrows. Thin arrows represent 50% gene flow in each generation. Thick arrows represent 100% gene flow. (*A*) Autosomal alleles experience all four environments. (*B*) Y-linked genes experience only the patrigenic male environment. X-linked genes are excluded from this environment.

PHENOTYPIC CONFLICT AND ALLELIC COMPETITION

"Sexual conflict," as the term is currently used in the literature, encompasses a set of loosely connected phenomena. It sometimes refers to conflictual relations between males and females and sometimes to antagonistic pleiotropy of a gene's effects in male and female bodies. Rather than review how terms have been used and misused, this section will define how terms will be used in this article. We will use conflict to refer to interactions in which different phenotypic agents have contradictory goals; competition to refer to processes by which alleles at a locus compete for representation in the gene pool; and antagonism to refer to fitness trade-offs in competition among alleles (for similar distinctions between conflict and competition, see Cosmides and Tooby 1981; Haig 1997).

Sexual conflict will refer to phenotypic interactions in which males and females "prefer" different outcomes. The classic example is when a male attempts to mate with a female who attempts to avoid his mating attempt (Parker 1979). Parental conflict will refer to phenotypic interactions in which matrigenes and patrigenes "prefer" different outcomes. The classic example is when patrigenes and matrigenes have opposing effects on fetal growth (Haig and Graham 1991). Sexual antagonism occurs when an allele has a relative advantage in one sex but relative disadvantage in the other (Rice 1987). By analogy, parental antagonism occurs when an allele has a relative advantage when inherited from one sex but relative disadvantage when inherited from the other (Haig 1997).

Sexual and parental conflict both occur within a generation: the first between male and female organisms and the second between patrigenic and matrigenic suborganisms. In contrast, sexual and parental antagonism involve pleiotropic effects of a single gene in different generations. Sexual antagonism occurs because an allele finds itself sometimes in males and sometimes in females; parental antagonism occurs because an allele is inherited sometimes from males and sometimes from females.

"Parental" is not the ideal label for conflict between the haploid contributions of parents because parental conflict is easily misinterpreted as referring to conflict between the diploid parents (unfortunately, we have found no better adjective). Genes of mothers and matrigenes of offspring have distinct evolutionary "interests," as do genes of fathers and patrigenes of offspring (Haig 1992; Burt and Trivers 1998; Wilkins and Haig 2002; Queller 2003). Parental conflict between matrigenes and patrigenes is not the same as sexual conflict between mothers and fathers.

Sexual conflict and sexual antagonism are distinct. Neither implies the other. Sexual conflict involves antagonistic adaptations that are mutually reinforcing. Males possess adaptations to overcome resistance because females resist; females possess adaptations to resist coercion because males coerce. The genes responsible for resistance are maintained because of the genes responsible for coercion. They justify each other's existence. Sexual antagonism, in contrast, is a relation between competing alleles in which one allele's gain is another allele's loss. It may involve traits that do not involve any direct interaction between the sexes. Our distinction, between sexual conflict and sexual antagonism, corresponds to the distinction of Chapman et al. (2003) between interlocus and intralocus sexual conflict. We prefer our terminology because lumping together these disparate notions as "sexual conflict" is a recipe for conceptual confusion.

Consider two illustrative hypotheses: (1) Seminal proteins benefit a male's sperm in competition with sperm of other males but reduce the longevity of females with whom he mates (Chapman et al. 1995). Sexual conflict is present because a trait that increases male fitness reduces the fitness of females, but sexual antagonism is absent because genes that encode male-specific proteins are never expressed in females. (2) Tyrosine at position 282 of the hemachromatosis protein (HFE) causes greater retention of iron and is costly in men because of iron overload but beneficial in women because of periodic loss of blood at menstruation and childbirth (Little 1996). Sexual antagonism is present because the *HFE* allele that encodes Tyr282 is beneficial in females but costly in males. Sexual conflict is absent because the allele's expression in females does not reduce the fitness of males.

Parental conflict and parental antagonism are also distinct. Consider an extended example. *Igf2* encodes a growth enhancer (IGF-II) and *Igf2r* a growth inhibitor (IGF2R). In the absence of imprinting, patrigenes at both loci benefit from greater fetal growth than matrigenes, but allelic competition favors a compromise between levels of expression that are optimal in matrigenic and patrigenic niches (because unimprinted alleles are constrained to the same expression in both niches). At the optimal compromise, an *Igf2* allele with increased expression would have a fitness benefit in the patrigenic niche outweighed by a fitness cost in the matrigenic niche, whereas an *Igf2r* allele with increased expression would have a fitness cost in the patrigenic niche outweighed by a fitness benefit in the matrigenic niche. Thus, expression at both loci is subject to parental antagonism. However, matrigenes and patrigenes have identical expression and there is no phenotypic conflict. IGF-II and IGF2R cooperate to produce the level of growth that is optimal for unimprinted genes.

Imprinted alleles are expressed differently in matrigenic and patrigenic niches and thereby create the possibility of phenotypic conflict. An

Igf2 allele with decreased expression as a matrigene but unchanged expression as a patrigene can outcompete an unimprinted allele, as can an *Igf2r* allele with decreased expression as a patrigene but unchanged expression as a matrigene. Patrigenes and matrigenes belong to different suborganisms with different optimal levels of growth. Parental conflict between the suborganisms is expressed by IGF-II pushing growth toward the patrigenic optimum and IGF2R pushing growth toward the matrigenic optimum (Haig and Graham 1991).

In a model of Wilkins and Haig (2001), parentally antagonistic selection favored progressively lower matrigenic expression of *Igf2* and progressively higher patrigenic expression, with the opposite at *Igf2r*. At both loci, imprinted alleles outcompeted unimprinted alleles and highly expressed imprinted alleles outcompeted more lowly expressed imprinted alleles. In the domain of allelic competition, imprinted alleles with increased expression of *Igf2* and *Igf2r* were evolutionary allies because each justified the other's existence in competition with alternative imprinted and unimprinted alleles at their respective loci. At the evolutionary equilibrium, there was all or none parent-specific expression at both loci. Parental antagonism was resolved because successful alleles could do no better in either matrigenic or patrigenic niches. However, in the domain of parental conflict, patrigenic expression of *Igf2* and matrigenic expression of *Igf2r* had opposing effects on growth.

SEXUAL ANTAGONISM

Sexual antagonism occurs when natural selection favors different alleles in different environments, in this case male and female bodies. Each generation, alleles that have survived selection in one sex are mixed with alleles that have survived selection in the other sex and are randomly assigned to male and female bodies of the next generation. Single-locus models of antagonistic selection with frequent gene mixing can maintain polymorphism but only under restrictive conditions. The usual outcome is fixation of whichever allele does best, on average, across the ensemble of selective environments (Hedrick 1999; Prout 2000).

Stable polymorphism requires that heterozygotes have higher average fitness than both homozygotes. In the case of sexual antagonism, selection in males and females occurs in parallel with changes in zygotic allele frequencies determined by the harmonic means of sex-specific fitnesses (Levene 1953). For the same arithmetic mean, the harmonic mean decreases with larger differences between the sex-specific fitnesses. Therefore, conditions for stable polymorphism become less restrictive for stronger sexual antagonism under the assumption that fitness differences between the sexes are greater for homozygotes than for heterozygotes (Prout 2000).

Sexual antagonism can occur with intragenomic conflict. Meiotic drive is usually sex-specific because male and female meiosis are mechanistically distinct (Úbeda and Haig 2004, 2005). If a driving allele is to be maintained at a polymorphic equilibrium, then its "unfair" advantage in spermatogenesis or oogenesis must be balanced by viability or fertility costs elsewhere in the life cycle. If these costs are shared by the sexes, rather than restricted to the sex in which drive occurs, then the polymorphism will be sexually antagonistic. In other words, the driving allele will be more fit than the alternative allele in the sex in which it drives with the relative allelic fitnesses reversed in the other sex.

PARENTAL ANTAGONISM

Kin are significant actors in the lives of many organisms. Sex differences of survival, dispersal, mating behavior, or reproductive role cause organisms to interact differently with their mothers and fathers or with their mothers' and fathers' kin. As a result, matrigenes and patrigenes experience different relatedness to social partners and are subject to antagonistic selection when phenotypic outcomes that increase matrigenic inclusive fitness reduce patrigenic inclusive fitness, or vice versa (Haig 2000b; Brandvain 2010; Úbeda and Gardner 2010, 2011, 2012; Van Cleve et al. 2010; Brandvain et al. 2011). Matrigenic and patrigenic niches may be further subdivided by sex if males and females experience different kinship environments (Fig. 1A).

Optimal outcomes differ for an allele in matrigenic and patrigenic niches when its locus influences a trade-off between the fitness of self (ego) and of "asymmetric kin." An individual is asymmetric kin of ego if the individual has different probabilities of carrying copies of ego's matrigenes and patrigenes (Haig 1997). Thus, mothers are asymmetric kin of their offspring but offspring are symmetric kin of their mothers. Paternal half-siblings (patrisibs) and maternal half-siblings (matrisibs) are asymmetric kin but full-siblings are symmetric kin. Aunts, uncles, and cousins are usually asymmetric kin.

A necessary condition for parental antagonism is that genotypes with the same pair of alleles inherited from opposite parents make unequal contributions to the next generation either because the reciprocal heterozygotes have unequal fitness or are produced in unequal numbers. We will adopt the convention that *Aa* heterozygotes inherit *A* from their mother, whereas *aA* heterozygotes inherit *A* from their father. Two simple scenarios will illustrate how interactions with asymmetric kin can cause *Aa* and *aA* to experience different social environments.

1. Individual males of the greater spear-nosed bat, *Phyllostomus hastatus*, defend harems of unrelated females and father most of a harem's offspring. Females of a harem give birth synchronously, each to a single pup. Pups huddle together in their mothers' absence (McCracken and Bradbury 1981; Porter and Wilkinson 2001; Haig 2010). Suppose that *A* is a rare allele causing pups to divert a portion of their milk intake from individual growth to thermogenesis. An *Aa* pup will huddle solely with *aa* patrisibs because only its mother carries *A*, but an *aA* pup will huddle with *aa* and *aA* patrisibs because their mutual father is heterozygous for *A*. Therefore, *aA* pups will be warmer than *Aa* pups for the same expenditure of fuel and will have higher survival than *Aa* pups.

2. A longer interval until the birth of a younger sibling increases a child's chances of survival but reduces its mother's fecundity, such that the child has fewer younger siblings. Suppose that each of a mother's children has a different father and *A* is a rare allele that increases the interbirth interval, perhaps by promoting more intense suckling and prolonging lactational amenorrhea of the mother. *Aa* and *aA* sucklings cause the same prolongation of the subsequent birth interval, relative to *aa*, and therefore enjoy the same increment in fitness, but *aA* sucklings occur in families in which *aA* patrisibs are each cared for by different mothers whereas *Aa* sucklings occur in families in which *Aa* matrisibs are cared for by the same mother. As a consequence of *A*'s effects on maternal fecundity, there will be more *aA* patrisibs than *Aa* matrisibs and *A* will make a greater contribution to the next generation as a patrigene than as a matrigene.

The balance of selective forces shifts with allele frequencies and the changing genetic composition of families. The above scenarios considered the simple situation when *A* was rare. As *A* increases in frequency, it begins to occur in families in which both parents are *Aa* or *aA* and then in families in which one or both parents are *AA*. Thus, the genetic environments of *A* as a matrigene and as a patrigene will depend on allelic frequency, as will the genotypic fitnesses of *aa*, *Aa*, *aA*, and *AA*.

MULTILOCUS INTERACTIONS

Consider a gene pool divided between two habitats, marsh (M) and fen (F), with limited migration between M and F. Habitat-specific selection would result in marsh-adapted alleles attaining high frequency in M and fen-adapted alleles attaining high frequency in F. Marsh-adapted alleles would be associated with marsh-adapted alleles, and fen-adapted alleles with fen-adapted alleles, at all loci subject to habitat-specific selection. Now suppose that M and F represent males and females. In this case, there is complete hybridization between sexual "habitats" in every generation, followed by random assignment of progeny to habitats in the next generation. Can nonrandom allelic associations be maintained under these conditions?

Linkage disequilibrium (LD) is generated between two sexually antagonistic loci with independent effects on sex-specific fitness (Patten et al. 2010; Úbeda et al. 2011). This LD arises from recurrent admixture of gene pools subject to selection in male and female niches. Sperm are enriched for male-beneficial alleles at both loci and eggs for female-beneficial alleles. Therefore, the next generation of zygotes is enriched for haplotypes that couple male-beneficial alleles and for haplotypes that couple female-beneficial alleles. LD is also generated between two parentally antagonistic loci with independent effects on parent-specific fitness (Patten et al. 2013). Favored haplotypes couple patrigenically beneficial alleles or couple matrigenically beneficial alleles. Nonrandom associations can be intuitively understood as arising because alleles that do well in the same niche, and occur on the same chromosome, tend to be transmitted together to the next generation. The strength of these associations increases with the tightness of linkage between the loci.

Sexually antagonistic alleles have a relative advantage in one sex and are thus preferentially inherited from that sex. Therefore, alleles at a parentally antagonistic locus in the patrigenic niche will tend to occur on haplotypes with male-beneficial alleles but on haplotypes with female-beneficial alleles in the matrigenic niche. Therefore, selection favors haplotypes that couple male-beneficial alleles with patrigenically beneficial alleles and haplotypes that couple female-beneficial alleles with matrigenically beneficial alleles (Patten et al. 2013). The process resembles mutual hitchhiking in which alleles at the two loci are carried along in each other's draft.

Local adaptation can create habitat-specialist islands within a habitat-generalist genome (Nosil et al. 2009) and select for chromosomal inversions that bind together coadapted clusters of alleles (Kirkpatrick and Barton 2006). Sex-specific selection can be seen as a special case of divergent selection in different habitats. By analogy to models of ecological divergence, sexual antagonism can create complementary haplotypes with sexually antagonistic effects and favor reduced recombination between the haplotypes.

SEX DETERMINATION

Sexually antagonistic haplotypes can be viewed as differentially adapted for distinct sexual "habitats." From this perspective, sex determination is the process by which alleles occupy habitats, and sex-biasing alleles influence habitat choice. Sex-biasing alleles also influence the choice of matrigenic or patrigenic habitats in the generation after they bias sex determination. Therefore, a male-beneficial/patrigenically beneficial haplotype would "prefer" to be present in males whereas a female-beneficial/matrigenically beneficial haplotype would "prefer" to be present in females. Thus, the former class of haplotypes will become preferentially associated with male-biasing alleles and the latter with female-biasing alleles. The ability to match genotype to habitat facilitates ecological specialization in models of sympatric speciation (Ravigné et al. 2004, 2009). By analogy, associations of sexually and parentally antagonistic loci with sex-biasing loci facilitate sexual and parental specialization in the genome. For simplicity of exposition, the discussion that follows will focus on the interaction between sexually antagonistic loci and sex-biasing loci.

Consider a baseline scenario in which sex chromosomes and sex-specific meiotic drive are absent. All chromosome segments benefit from an individual developing as whichever sex has higher expected fitness, given environmental conditions and the individual's genotype at sexually antagonistic loci. Different segments will have spent different amounts of time in male and female bodies because of their associations with sexually antagonistic and sex-biasing alleles.

Natural selection favors haplotypes that couple male-beneficial alleles and complementary haplotypes that couple female-beneficial alleles, with the resulting LD accentuating the local strength of sexually antagonistic selection (Patten et al. 2010). Natural selection will also generate LD that associates male-beneficial haplotypes with male-biasing alleles and female-beneficial haplotypes with female-biasing alleles because such associations ensure that haplotypes are preferentially present in the sex in which they have a relative advantage. Sex-biasing alleles, in effect, influence the choice between male and female habitats and become associated with genes that benefit from the choice. They create genomic neighborhoods that spend more than the average time in one sex and therefore accumulate sexually antagonistic polymorphisms (Jordan and Charlesworth 2011).

A system in which sex-biasing haplotypes are scattered throughout the genome is inherently unstable (Rice 1986) because local conditions for sexually antagonistic polymorphism are progressively relaxed as a genomic region gains influence in sex determination, whereas conditions for sexually antagonistic polymorphism become more restrictive in regions that lose influence. The "victor" of competition between unlinked sex-biasing loci for control of sex determination will generally be the locus that is associated with greater sexually antagonistic variance in fitness (van Doorn and Kirkpatrick 2007). This dynamic process tends to favor control of sex determination being concentrated in a single genomic region with a high hurdle of sexual antagonism to be overcome for the region's replacement by a new sex-determining locus. The generation of complementary haplotypes—one male-biasing and male-beneficial, the other female-biasing and female-beneficial—favors the suppression of recombination between the haplotypes which, in turn, favors further accumulation of sexually antagonistic differences between the nascent sex chromosomes (Rice 1984; Charlesworth 1991).

Sex-specific meiotic drive sows discord within the genome with respect to sex determination because driving haplotypes will accumulate alleles that promote development as the sex in which drive occurs, even if this is not the sex that maximizes fitness for loci segregating independently of drive (Úbeda and Haig 2004, 2005; Burt and Trivers 2006). At the same time, haplotypes that are driven against will accumulate sex-biasing alleles that promote development of the nondriving sex. Genomic segments that segregate independently of drive should also accumulate sex-biasing alleles to counter drive-associated departures from the optimal sex ratio.

Sex-specific drive could thus explain some cases of replacement of polygenic sex determination by sex chromosomes or of one system of sex chromosomes by another (van Doorn and Kirkpatrick 2007; Yoshida and Kitano 2012). Although distortions of sex ratio are expected during transitions of one system of sex determination to another, the genome is expected to eventually return to consensus sex ratios via the parliament of genes and the suppression of drive. In some cases, the driving chromosome that precipitated a change will be lost in the process, and, in others, it will be retained but with its drive suppressed (Kozielska et al. 2010). Systems of meiotic drive are usually associated with inversions (Charlesworth and Hartl 1978; Burt and Trivers 2006) and these inversions would facilitate the recruitment of sex-biasing and sexually antagonistic alleles onto driving haplotypes.

SEX CHROMOSOMES

Autosomal genes experience all four niches defined by partition of the gene pool by sex and sex of origin (Fig. 1A), whereas sex-linked genes experience some niches but not others (Fig. 1B). Specifically, Y-linked genes are restricted to the patrigenic–male niche and X-linked genes are excluded from it. As a consequence, sexual and parental antagonism are experienced differently by autosomal, X-linked, and Y-linked loci.

Y-linked genes are always patrigenes in males and are therefore not subject to parentally or sexually antagonistic selection. They should be strongly biased to promote male fitness and patrigenic interests (Hurst 1994).

The situation for X-linked genes is more complex for two reasons. First, an X-linked allele is present in female bodies twice as often as in male bodies. Second, it shares female bodies with another X-linked allele but is on its own in male bodies. These factors cancel for X-linked alleles with additive effects on a sexually antagonistic phenotype because each allele has twice the influence on phenotype in males as in females. Dominant effects of rare X-linked alleles are expressed twice as often in females as in males, whereas rare recessive effects are experienced predominantly by males. Therefore, X chromosomes will tend to accumulate polymorphisms in which female-beneficial alleles have dominant effects and male-beneficial alleles have recessive effects (Rice 1984). A population genetic model found a slight bias toward fixing male-beneficial alleles on the X chromosome (Patten and Haig 2009).

With respect to parental antagonism, an X-linked gene is a patrigene only in females but a matrigene in both sexes. Matrigenic effects are unopposed by a patrigenic allele in males whereas X-linked matrigenes and patrigenes have equal selective weight in females. Therefore, parentally antagonistic selection is predicted to be biased toward matrigenic interests at X-linked loci relative to autosomal loci (Patten and Haig 2009).

"RESOLUTION" OF PARENTAL AND SEXUAL ANTAGONISM

Sexual antagonism is resolved (with respect to standing allelic variation) by the loss of sexually antagonistic polymorphism. It is resolved (with respect to mutation) by the fixation of an allele that is unbeatable in both sexes. Similarly, parental antagonism is resolved in the short term by loss of polymorphism and in the long term by the fixation of an allele that is unbeatable in both matrigenic and patrigenic niches.

Sexual antagonism could be resolved by a gene duplication in which the two paralogs become specialized for different sexes (Connallon and Clarke 2011, 2013; Gallach and Betrán 2011). In theory, one could imagine duplications at each and every locus subject to sexual antagonism with the genome composed of specialist teams, one on the field in male bodies but on the bench in female bodies, the other with roles reversed. This arrangement would resemble the game called football in the United States in which different teams take the field on offense and defense. With respect to sexual specialization, the genome resembles more the game called football in the rest of the world in which compromises between offensive and defensive prowess are ubiquitous because players must perform in both roles.

The formal possibility of male-specific and female-specific subgenomes poses the question why the sexes are so similar. Gonochorism (separate sexes) has ancient origins on the tree of life, yet males and females are often immediately recognizable as members of the same species, presumably because of constraints on sexual divergence imposed by a shared genome. Extreme sexual dimorphism, as shown by scale insects or barnacles with dwarf males, is the exception rather than the rule, perhaps analogous to the limited global appeal of American football.

Two consequences of duplication and sexual subfunctionalization may contribute to the long-term maintenance of sexual resemblance. First, substitutions at one locus that would be beneficial in both sexes are not readily transferrable between loci. Taken in isolation, this effect is cancelled by duplicate genes providing twice the target for mutation. Second, each duplicate will be subject to selection in only one sex. Therefore the strength of selection relative to drift and mutation is weakened for substitutions that would have benefited both sexes. Species in which most genes are sexual specialists may have a long-term evolutionary disadvantage because the two sexes must evolve separately to meet shared environmental challenges.

Sexual and parental antagonism can occur not only with respect to the amino acid sequence of proteins but also with respect to the strength and tissue specificity of promoters. Sexual and parental antagonism over levels of expression are resolved in different ways because expression in male and female niches occurs in different bodies but expression in matrigenic and patrigenic niches occurs in shared bodies, indeed in shared nuclei. If optimal expression differs between the sexes, then sexual antagonism can be resolved by the evolution of sex-specific expression with the female optimum achieved in female bodies and the male optimum in male bodies (Fig. 2A).

In contrast, the level of expression within a nucleus is determined by the sum of matrigenic

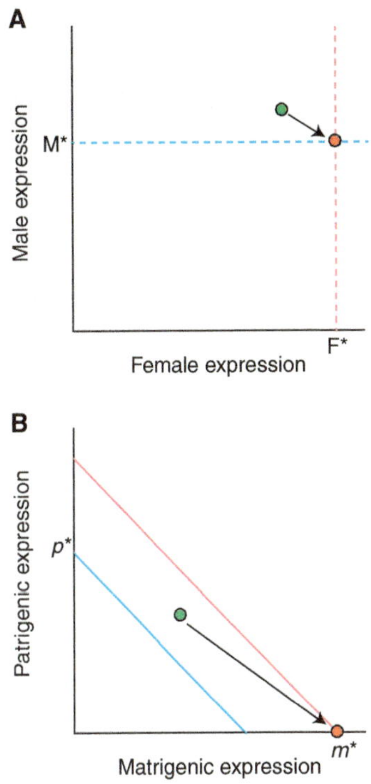

Figure 2. Resolution of sexual and parental antagonism. (*A*) The optimal level of expression in males is M* and in females F*. Sex expression is initially the same in both sexes at a compromise between their respective optima (green dot). Sex-specific expression allows M* to be achieved in males and F* in females (red dot). (*B*) The optimal level of expression for patrigenes is p^* and for matrigenes m^* (in this example, $m^* > p^*$). The blue line represents all combinations of matrigenic and patrigenic expression that sum to p^* and the reddish line represents all combinations that sum to m^*. Gene expression is initially unimprinted at a compromise between m^* and p^* (green dot). Imprinted expression allows silencing of the patrigenic allele and expression of m^* by the matrigenic allele (red dot).

and patrigenic expression, and this sum cannot simultaneously maximize matrigenic and patrigenic inclusive fitness when there is parental antagonism (Fig. 2B). Silencing of an allele when it occupies the niche that favors the lesser sum, but its expression at the optimal level in the other niche, is the evolutionarily stable outcome of conflict between matrigenes and patrigenes over level of transcription (Haig and Westoby 1989; Úbeda and Haig 2003). All-or-none imprinting is evolutionarily stable because the amount of mRNA is optimal in the niche in which an allele is expressed but the allele cannot reduce its expression below zero in the other niche. This resolution of parental antagonism has been called "the loudest voice prevails" (Wilkins and Haig 2003).

Quantitative biases of parent-specific expression, as distinct from all-or-none imprinting, have been reported (Khatib 2007). Expression at these loci is most easily explained as not being subject to antagonistic selection because biased expression is not predicted to be evolutionarily stable in the presence of parental antagonism (although see Greenwood-Lee et al. 2001). Alleles with biased expression could be selectively neutral variants at dosage-insensitive loci or could represent asymmetric "divisions of labor" between matrigenic and patrigenic roles at dosage-sensitive loci. The optimal level of expression at a dosage-sensitive locus not subject to parental antagonism is the same for matrigenes and patrigenes. Any combination of matrigenic and patrigenic expression that sums to this optimal amount is evolutionarily stable (Haig 1997, 2000a). Unbiased expression, with each allele producing half the optimal amount, has been considered the most plausible way of achieving optimal expression because matrigenes and patrigenes have been presumed to be equivalent in the absence of reasons to believe otherwise. This default assumption can be questioned within gene clusters that contain loci subject to all-or-none imprinting because matrigenic and patrigenic haplotypes are demonstrably nonequivalent.

CONCLUDING REMARKS

A gene is typically uncertain of whether it will occupy a male or female habitat in the next generation and, by extension, of whether it will occupy a matrigenic or patrigenic habitat in the generation after that. Differential selection in each pair of habitats gives rise to sexual and parental antagonism. Two kinds of resolutions are possible. In the first, a gene evolves different expression in different habitats: sexually dimorphic expression in response to sexual antagonism or imprinted expression in response to parental antagonism. Second, a gene might "choose" its habitat—either by influencing sex determination directly or by linkage to a sex-determining locus. A gene that occupies only one habitat can specialize. Not every antagonism can be resolved by tampering with expression levels nor can every gene choose its sex. Therefore, the sexual ecology of the genome will involve interactions among habitat specialists and habitat generalists.

REFERENCES

Brandvain Y. 2010. Matrisibs, patrisibs, and the evolution of genomic imprinting on autosomes and sex chromosomes. *Am Nat* **176:** 511–521.

Brandvain Y, Van Cleve J, Úbeda F, Wilkin JF. 2011. Demography, kinship, and the evolving theory of genomic imprinting. *Trends Genet* **27:** 251–257.

Burt A, Trivers R. 1998. Genetic conflicts in genomic imprinting. *Proc R Soc B* **265:** 2393–2397.

Burt A, Trivers R. 2006. *Genes in conflict.* Harvard University Press, Cambridge, MA.

Chapman T, Liddle LF, Kalb MF, Partridge L. 1995. Cost of mating in *Drosophila melanogaster* females is mediated by male accessory gland products. *Nature* **373:** 241–244.

Chapman T, Arnqvist G, Bangham J, Rowe L. 2003. Sexual conflict. *Trends Ecol Evol* **18:** 41–47.

Charlesworth B. 1991. The evolution of sex chromosomes. *Science* **251:** 1030–1033.

Charlesworth B, Hartl DL. 1978. Population dynamics of the segregation distorter polymorphism of *Drosophila melanogaster. Genetics* **89:** 171–192.

Connallon T, Clarke AG. 2011. The resolution of sexual antagonism by gene duplication. *Genetics* **187:** 919–937.

Connallon T, Clarke AG. 2013. Antagonistic versus non-antagonistic models of balancing selection: Characterizing the relative timescales and hitchhiking effects of partial selective sweeps. *Evolution* **67:** 908–917.

Cosmides LM, Tooby J. 1981. Cytoplasmic inheritance and intragenomic conflict. *J Theor Biol* **89:** 83–129.

Gallach M, Betrán E. 2011. Intralocus sexual conflict resolved through gene duplication. *Trends Ecol Evol* **26:** 222–228.

Greenwood-Lee JM, Taylor PD, Haig D. 2001. The inclusive fitness dynamics of genomic imprinting. *Selection* **2:** 101–116.

Haig D. 1992. Genomic imprinting and the theory of parent–offspring conflict. *Sem Devel Biol* **3:** 153–160.

Haig D. 1997. Parental antagonism, relatedness asymmetries, and genomic imprinting. *Proc R Soc B* **264:** 1657–1662.

Haig D. 2000a. The kinship theory of genomic imprinting. *Annu Rev Ecol Syst* **31:** 9–32.

Haig D. 2000b. Genomic imprinting, sex-biased dispersal, and social behavior. *Ann NY Acad Sci* **907:** 149–163.

Haig D. 2010. The huddler's dilemma: A cold shoulder or a warm inner glow. In *Social behaviour: Genes, ecology and evolution* (ed. Székely T, Moore AJ, Komdeur J), pp. 107–109. Cambridge University Press, Cambridge.

Haig D, Graham C. 1991. Genomic imprinting and the strange case of the insulin-like growth factor-II receptor. *Cell* **64:** 1045–1046.

Haig D, Westoby M. 1989. Parent-specific gene expression and the triploid endosperm. *Am Nat* **134:** 147–155.

Hedrick PW. 1999. Antagonistic pleiotropy and genetic polymorphism: A perspective. *Heredity* **82:** 126–133.

Holman L, Kokko H. 2014. The evolution of genomic imprinting: Costs, benefits and long-term consequences. *Biol Rev Camb Philos Soc* doi: 10.1111/brv.12069.

Hurst LD. 1994. Embryonic growth and the evolution of the mammalian Y chromosome: I. The Y as an attractor for selfish growth factors. *Heredity* **73:** 223–232.

Jordan CY, Charlesworth D. 2011. The potential for sexually antagonistic polymorphism in different genome regions. *Evolution* **66:** 505–516.

Khatib H. 2007. Is it genomic imprinting or preferential expression? *Bioessays* **29:** 1022–1028.

Kirkpatrick M, Barton N. 2006. Chromosome inversions, local adaptation and speciation. *Genetics* **173:** 419–434.

Kokko H, Jennions MD, Brooks R. 2006. Unifying and testing models of sexual selection. *Annu Rev Ecol Syst* **37:** 43–66.

Kozielska M, Weissing FJ, Beukeboom LW, Pen I. 2010. Segregation distortion and the evolution of sex-determining mechanisms. *Heredity* **104:** 100–112.

Levene H. 1953. Genetic equilibrium when more than one ecological niche is available. *Am Nat* **87:** 331–333.

Little P. 1996. Woman's meat, a man's poison. *Nature* **382:** 494–495.

McCracken GF, Bradbury JW. 1981. Social organization and kinship in the polygynous bat *Phyllostomus hastatus. Behav Ecol Sociobiol* **8:** 11–34.

Nosil P, Funk DJ, Ortiz-Barrientos D. 2009. Divergent selection and heterogenous genomic divergence. *Mol Ecol* **18:** 375–402.

Parker GA. 1979. Sexual selection and sexual conflict. In *Sexual selection and reproductive competition in insects* (ed. Blum MS, Blum NA), pp. 123–166. Academic, New York.

Patten MM, Haig D. 2009. Maintenance or loss of genetic variation under sexual and parental antagonism at a sex-linked locus. *Evolution* **63:** 2888–2895.

Patten MM, Haig D, Úbeda F. 2010. Fitness variation due to sexual antagonism and linkage disequilibrium. *Evolution* **64:** 3638–3642.

Patten MM, Úbeda F, Haig D. 2013. Sexual and parental antagonism shape genomic architecture. *Proc R Soc B* **280:** 20131795.

Porter TA, Wilkinson GS. 2001. Birth synchrony in greater spear-nosed bats, *Phyllostomus hastatus. J Zool* **253:** 383–390.

Prout T. 2000. How well does opposing selection maintain variation? In: *Evolutionary genetics: From molecules to morphology* (ed. Singh RS, Krimbas CB), pp. 157–181. Cambridge University Press, Cambridge.

Queller DC. 1997. Why do females care more than males? *Proc R Soc B* **264:** 1555–1557.

Queller DC. 2003. Theory of genomic imprinting conflict in social insects. *BMC Evol Biol* **3:** 15.

Ravigné V, Olivieri I, Dieckmann U. 2004. Implications of habitat choice for protected polymorphisms. *Evol Ecol Res* **6:** 125–145.

Ravigné V, Dieckmann U, Olivieri I. 2009. Live where you thrive: Joint evolution of habitat choice and local adaptation facilitates specialization and promotes diversity. *Am Nat* **174:** E141–E169.

Rice WR. 1984. Sex chromosomes and the evolution of sexual dimorphism. *Evolution* **38:** 735–742.

Rice WR. 1986. On the instability of polygenic sex determination: The effect of sex-specific selection. *Evolution* **40:** 633–639.

Rice WR. 1987. The accumulation of sexually antagonistic genes as a selective agent promoting the evolution of reduced recombination between primitive sex chromosomes. *Evolution* **41:** 911–914.

Úbeda F, Gardner A. 2010. A model for genomic imprinting in the social brain: Juveniles. *Evolution* **64:** 2587–2600.

Úbeda F, Gardner A. 2011. A model for genomic imprinting in the social brain: Adults. *Evolution* **65:** 462–475.

Úbeda F, Gardner A. 2012. A model for genomic imprinting in the social brain: Elders. *Evolution* **66:** 1567–1581.

Úbeda F, Haig D. 2003. Dividing the child: Genomic imprinting and evolutionary games. *Genetica* **117:** 103–110.

Úbeda F, Haig D. 2004. Sex-specific meiotic drive and selection at an imprinted locus. *Genetics* **167:** 2083–2095.

Úbeda F, Haig D. 2005. On the evolutionary stability of Mendelian segregation. *Genetics* **170:** 1345–1357.

Úbeda F, Haig D, Patten MM. 2011. Stable linkage disequilibrium owing to sexual antagonism. *Proc R Soc B* **278:** 855–862.

Van Cleve J, Feldman MW, Lehmann L. 2010. How demography, life history, and kinship shape the evolution of genomic imprinting. *Am Nat* **176:** 440–455.

van Doorn GS, Kirkpatrick M. 2007. Turnover of sex chromosomes induced by sexual conflict. *Nature* **449:** 909–912.

Wilkins JF, Haig D. 2001. Genomic imprinting of two antagonistic loci. *Proc R Soc B* **268:** 1861–1867.

Wilkins JF, Haig D. 2002. Parental modifiers, antisense transcripts and loss of imprinting. *Proc R Soc B* **269:** 1841–1846.

Wilkins JF, Haig D. 2003. What good is genomic imprinting: The function of parent-specific gene expression. *Nat Rev Genet* **4:** 359–368.

Yoshida K, Kitano J. 2012. The contribution of female meiotic drive to the evolution of neo-sex chromosomes. *Evolution* **66:** 3198–3208.

Sexual Conflict and Seminal Fluid Proteins: A Dynamic Landscape of Sexual Interactions

Laura K. Sirot[1], Alex Wong[2], Tracey Chapman[3], and Mariana F. Wolfner[4]

[1]Department of Biology, College of Wooster, Wooster, Ohio 44691

[2]Department of Biology, Carleton University, Ottawa, Ontario K1S 5B6, Canada

[3]School of Biological Sciences, University of East Anglia, Norwich Research Park, Norwich NR4 7TJ, United Kingdom

[4]Department of Molecular Biology and Genetics, Cornell University, Ithaca, New York 14853

Correspondence: mfw5@cornell.edu; tracey.chapman@uea.ac.uk

Sexual reproduction requires coordinated contributions from both sexes to proceed efficiently. However, the reproductive strategies that the sexes adopt often have the potential to give rise to sexual conflict because they can result in divergent, sex-specific costs and benefits. These conflicts can occur at many levels, from molecular to behavioral. Here, we consider sexual conflict mediated through the actions of seminal fluid proteins. These proteins provide many excellent examples in which to trace the operation of sexual conflict from molecules through to behavior. Seminal fluid proteins are made by males and provided to females during mating. As agents that can modulate egg production at several steps, as well as reproductive behavior, sperm "management," and female feeding, activity, and longevity, the actions of seminal proteins are prime targets for sexual conflict. We review these actions in the context of sexual conflict. We discuss genomic signatures in seminal protein (and related) genes that are consistent with current or previous sexual conflict. Finally, we note promising areas for future study and highlight real-world practical situations that will benefit from understanding the nature of sexual conflicts mediated by seminal proteins.

Both sexes benefit from successful reproduction, but the different reproductive strategies adopted by males and females may result in differential costs and benefits. This can result in sexual conflict before, during, and after mating. Conflict in the more familiar form of competition can also occur between females and between males, with the latter situation including interejaculate competition. Of the many "weapons" in these conflicts and competitions, this article focuses on the seminal fluid proteins (SFPs) that are made by males and transferred to females during mating. These proteins represent a crucial interface of functional activity between male and female. Transfer of SFPs can affect physiology and, in some animals, the behavior and life span of mated females (reviewed in Chapman 2001; Gillott 2003; Poiani 2006; Avila et al. 2011; Rodríguez-Martínez et al. 2011). Because SFPs have important effects on the most intimate of interactions between the sexes, they are prime candidates to become subject to sex-

Cite this article as *Cold Spring Harb Perspect Biol* doi: 101101/cshperspect.a017533

ually antagonistic selection (Arnqvist and Rowe 2005). With increasing knowledge of the functions of SFPs, their roles in inter- and intrasexual conflict and their evolutionary responses to conflict are becoming ever more apparent. Here, we explore the roles, evolution, and significance of these male-derived players in sexual conflict. We refer the reader to previous reviews for much of the detailed functional information on SFPs (e.g., Chapman 2001; Gillott 2003; Kubli 2003; Arnqvist and Rowe 2005; Poiani 2006; Sirot et al. 2009; Avila et al. 2011; Rodríguez-Martínez et al. 2011) and focus here instead on selected examples, drawn largely from the study of insects.

THE BATTLEGROUND

The opportunity for postmating conflict is played out through the behavior and physiology of the mated female and the fate of sperm in the female reproductive tract. After mating has begun, male-derived molecules and sperm interact with female-derived molecules, cells, and tissues both within the female reproductive tract (e.g., Yapici et al. 2008; Hasemeyer et al. 2009; Yang et al. 2009; Dean 2013; Rubinstein and Wolfner 2013; Bromfield et al. 2014) and, in the case of some male-derived molecules, elsewhere in the female (e.g., Lung and Wolfner 1999). These interactions result in changes in female gene expression, behavior, physiology, life span, and morphology (reviewed in Gillott 2003; Poiani 2006; Robertson 2007; Avila et al. 2011). These changes, in turn, affect both male and female reproductive success. As a result, the battleground between males and females over postmating responses occurs in several parts of the female body including the reproductive tract and the nervous system.

The Potential Weapons

The proteins that accompany sperm into the female—once dismissed as simply a supportive medium for sperm—are now known to be potent modulators of female reproductive biology; in some cases, they even have effects on offspring (reviewed in Martan and Shepherd 1976; Chapman 2001; Gillott 2003; Kubli 2003; Poiani 2006; Sirot et al. 2009; Avila et al. 2011; Rodríguez-Martínez et al. 2011; Ratto et al. 2012; Bromfield et al. 2014). Most SFPs are the products of secretory glands in the male reproductive tract; these include the prostate glands, epididymi, and seminal vesicles of mammals (Poiani 2006; Rodríguez-Martínez et al. 2011) and the male accessory glands and ejaculatory ducts/bulb of arthropods (Gillott 2003). These tissues produce, modify, and store secreted proteins that are then transferred to females during mating, along with sperm. For some species there are comprehensive data on SFPs from large-scale transcriptomic and/or proteomic analyses (e.g., Collins et al. 2006; Pilch and Mann 2006; Dottorini et al. 2007; Findlay et al. 2008, 2009; Sirot et al. 2008, 2011; Walters and Harrsion 2008, 2010; Baer et al. 2009; Dean et al. 2009; Ramm et al. 2009; Rogers et al. 2009; Claw 2013). In others, only a partial complement of SFPs has been identified so far (e.g., Andrés et al. 2006, 2008; Davies and Chapman 2006; South et al. 2011; Simmons et al. 2013).

In organisms in which SFPs have been globally characterized, an amazing complexity is apparent. First, they are numerous; latest estimates are just more than 200 different SFPs in *Drosophila melanogaster* (Swanson et al. 2001; Ravi Ram and Wolfner 2007; Findlay et al. 2008; Avila et al. 2009); several hundred in mice (Dean et al. 2009) and humans (Pilch and Mann 2006); and between 50 and 100 in honeybees (Baer et al. 2009) and in mosquitoes (Rogers et al. 2009; Sirot et al. 2011). Yet, even these SFP inventories are likely incomplete. Second, the molecular characteristics of SFPs are fascinating. On the one hand, many of the biochemical classes of molecules in seminal fluids are similar across all animals (Mueller et al. 2004; Baer et al. 2009). For example, seminal fluids contain many proteases and protease inhibitors, sugar-binding lectins, cysteine-rich secretory proteins (CRISPs), antimicrobial and antioxidant proteins, and coagulation proteins such as transglutaminases, small peptides, and larger, prohormone-like, molecules (Avila et al. 2011). On the other hand, although these conserved types of proteins are observed in the seminal fluid of all animals examined to date, an unusually high fraction of SFPs show rapid sequence evolution (e.g., Swanson and Vacquier 2002; Clark and Swanson 2005; Mueller et al. 2005; Clark et al. 2006; Haerty et al. 2007; Walters et al. 2010). SFPs of the same classes in different species are often not orthologous. For example, although the seminal fluid of two mosquito species and of *D. melanogaster* contains trypsin-class proteases that are predicted to have biochemically similar activities, those proteins are not orthologs (Findlay et al. 2008; Mancini et al. 2011; Sirot et al. 2011). Moreover, there is much redundancy in the types of protein present in the seminal fluid. For example, there are more than 15 trypsin-class proteases in *D. melanogaster* seminal fluid, and multiple serine proteases in mammalian seminal fluid (see LaFlamme and Wolfner 2012 for review). Although each SFP may have particular target proteins, it seems likely that different SFPs can also compensate for one another. These latter features (rapid evolution and nonorthology, as well as the frequent presence of redundancy) require explanation. One possibility is that this evolutionary exuberance is caused by coevolutionary arms races between SFPs in males and their receptors in females driven by sexual conflict. The evidence for this hypothesis is discussed in a section further below (see Does Sexual Conflict Shape SFP Evolution?).

Another likely contributor to complexity is that multiple seminal proteins could be needed to accomplish a single functional goal. For example, a network of at least eight *Drosophila* SFPs is required to bind a ninth SFP (the sex peptide, SP) to sperm. This binding to sperm prolongs the phenotypic effect of SP in females (Peng et al. 2005; Ravi Ram et al. 2009; Findlay et al. 2014). In another example from *Drosophila*, at least two SFPs (ovulin and Acp36DE) are proteolytically processed once inside the female (Fig. 1), and ectopic expression studies of

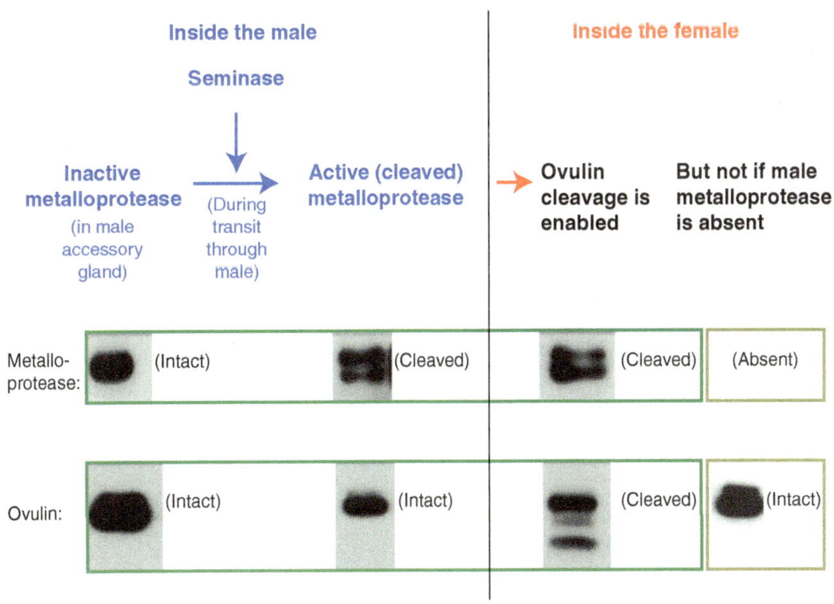

Figure 1. The molecular cascade that governs the cleavage of the seminal protein ovulin. Biochemical studies show that ovulin cleavage requires activation of a male metalloprotease during transit through the male during mating (Park and Wolfner 1995; Ravi Ram et al. 2006; LaFlamme et al. 2012, 2014). Ovulin and the two proteases that are needed to cleave it (seminase and the metalloprotease CG11864) are all made in the male's accessory gland but are stored uncleaved in this tissue. During transit through the male, the metalloprotease is activated by cleavage that is initiated by the serine protease "seminase." Although this metalloprotease is essential for ovulin cleavage within mated females, even after being activated, it does not cleave ovulin until all the proteins have entered females. (Western blot panels from Ravi Ram et al. 2006; reprinted, with permission, from the authors.)

ovulin suggest that the processing generates a more active portion of ovulin (Heifetz et al. 2005). The proteolytic processing only occurs once the proteins are within the female, and is mediated by at least two seminal proteases. These proteases are made in the same tissue as their targets (i.e., the male accessory glands) and are kept inactive during storage. They are activated through sequential proteolysis as they and their targets move through the male and into and through the female reproductive tract (Park and Wolfner 1995; Heifetz et al. 2005; Ravi Ram et al. 2006; LaFlamme et al. 2012, 2014). A proteolytic cascade involving multiple seminal proteins is also observed in mammals: Liquefaction of the seminal clot, a process that is thought to facilitate sperm movement, is regulated by a cascade of multiple seminal proteases and protease inhibitors (Pampalakis and Sotiropoulau 2007). Evolutionary pressures derived from sexual conflict could affect various members of the multiple proteins that regulate a specific effect on females, so interpreting data on the evolutionary dynamics of any one molecule is complex.

A further reason for complexity of SFPs is that different members of a given biochemical family could potentially be co-opted in different lineages to carry out a particular function. This might reflect pressures to overcome female resistance to more ancestral versions of SFPs. In this sense, the apparent complexity of seminal fluid may reflect previous conflicts. Finally, an aspect of the complexity of the action of SFPs is that they can, and often do, involve obligate participation of female molecules. Three examples, from *Drosophila*, are that (1) ovulin processing cannot be completed without female contributions (Park and Wolfner 1995; Heifetz et al. 2005; Ravi Ram et al. 2006; LaFlamme et al. 2012, 2014); (2) the effects of a sperm-bound seminal protein (sex peptide [SP]) require at least four female-expressed genes (Yapici et al. 2008; Findlay et al. 2014); and (3) SFPs such as ovulin act by turning up (or on) physiological pathways within females (Rubinstein and Wolfner 2013). That certain SFPs interact in molecular pathways with female proteins may constrain the evolution of those SFPs. Moreover, the existence of such constraints could open the door for other SFPs to evolve to target the pathway in other ways but with the same overall effect on the female.

It is also important to note that there are other, nonproteinaceous molecules in seminal fluid that could be players in postmating sexual conflict (e.g., steroid hormones [Baldini et al. 2013], juvenile hormone in insects [Borovsky et al. 1994; Clifton et al. 2014], prostaglandins [Destephano et al. 1974; Loher et al. 1981; Robertson 2007], vesicles [Leiblech et al. 2012; Aalberts et al. 2013], and noncoding RNAs such as miRNAs [e.g., Li et al. 2012]). Much less is known about the evolution, identity, and function of these other molecules in comparison to SFPs, but their study promises a fruitful avenue for future research. It is possible that, in some species, these molecules play roles performed by SFPs in other species. It is tempting to speculate that such a situation might, in part, contribute to interspecific differences in the SFP composition and complexity.

DO SFPs CONTRIBUTE TO POSTMATING SEXUAL CONFLICT?

SFPs may serve as biochemical and physiological agents that manipulate female behavior and physiology in ways that benefit males. Yet, females may also be taking advantage of SFPs as reliable cues for initiating physiological reproductive processes to coordinate with the receipt of sperm (Chapman 2001; Ratto et al. 2012; Baldini et al. 2013). Below, we present the processes over which postmating conflict is predicted to occur and the evidence for roles of SFPs in each; these ideas are summarized in Table 1.

Remating

Females of many insect species show dramatic changes in postmating behavior (reviewed in Gillott 2003; Sirot et al. 2009; Avila et al. 2011). In some species, mated females do not remate, or they remate at very low levels. These changes in female mating propensity could be advantageous to both sexes by reducing the risk of further sperm competition to males and reducing the costs of mating to females. However,

Table 1. Summary of potential conflicts mediated by SFPs

Trait over which there is potential conflict	Nature of the potential conflict	Types of SFPs that are potentially responsible for the conflict
Remating	Conflict over remating may occur between a female and her previous mates and between a mated female and courting males. Remating may benefit the female but at a cost to her previous mates because there are potential benefits for females of topping up or receiving better-quality, fresh sperm. These benefits may be at odds with the interests of males whose sperm are already in storage. There are also potential conflicts arising from the deleterious effects of male–male sperm competition. Yet, extensive remating could be costly to a female but beneficial to the males attempting to mate with her.	Those that modulate female receptivity, attractiveness, activity levels, pheromone production, and emission
Sperm transfer, storage, retention, usage	Conflict may arise over how many sperm are transferred during mating, stored by the female, and released for fertilization. The interests of the sexes over the efficiency of individual sperm usage should be aligned. However, sperm usage may become inefficient as a result of elevated rates of egg production, over which there is separate potential conflict, see below.	Those that mediate sperm transfer, sperm storage, sperm competition, the musculature of the female reproductive tract, including the sperm storage organs
Egg production, ovulation, egg provisioning, oviposition	Males may often gain from elevating current egg-production or egg-provisioning rates more than is the case for females. Females may trade off current versus future investment and gain from a longer-term strategy. Males, on the other hand, may gain from increased current investment by the female, despite any future costs. Any divergence over the rate of ovulation may negatively impact the efficiency of fertilization, as noted above.	Those that mediate ovulation, release of reproductive hormones, egg production, and provisioning
Food intake	Females may need to increase or change their nutrient intake to support increased reproductive rate. This will be favored in an open-ended way by males, who may have little interest in any future costs that the female might incur. However, increased nutrient trafficking decreases female life span, thus representing potential sexual conflict in terms of a female's future reproductive capacity.	Those that mediate feeding behavior and nutrient balancing
Activity	The female's sleep/wake cycles can be altered by SFPs and could impact energy usage. Increased activity and reduced siesta sleep in females (e.g., in *Drosophila*) could incur long-term costs for females but not males, reflecting a potential conflict.	Those that mediate activity patterns and circadian rhythms
Immune activation	Striking changes to the immune system occur during reproduction, the significance of which is not yet globally clear. However, there is potential for conflict if there is a suboptimal under- or overexpression of immunity in the female, with immune activity traded off against future reproductive capacity.	SFPs that have antimicrobial activity or that cause the expression of immune genes in females
Life span	SFPs that reduce female life span can be selected because any cost is incurred in the future. Therefore, the effect of such costs is felt unequally by males and females.	SFPs that either directly or indirectly exert costs leading to reduced female life span

females may also experience costs from not remating, as this may prevent the acquisition from future mates of any direct (e.g., food or other resources) or indirect benefits (e.g., "better," more compatible, or diverse alleles for offspring). Furthermore, conflict over remating can also occur when males attempt to mate with previously mated females. Males that court and attempt to mate with unreceptive, recently mated females may experience the costs of courtship without the benefit of offspring production. Additional costs can also arise; for example, *D. melanogaster* males that are rejected when they attempt to mate with previously mated females are less likely to court the females that they subsequently encounter and regardless of whether those new females have mated (Siegel and Hall 1979).

SFPs influence female remating in a manner consistent with sexual conflict. In several insect species, receipt of SFPs decreases the probability of female remating (reviewed in Gillott 2003; Sirot et al. 2009; Avila et al. 2011). Initial decreases in remating may not involve SFPs and can result from processes such as transfer of cuticular pheromones (e.g., in *Drosophila* [Zawistowski and Richmond 1986]) or from physical mate guarding. Longer-term changes in female receptivity across a wide range of insects, however, derive from the actions of SFPs (reviewed in Gillott 2003; Arnqvist and Rowe 2005; Avila et al. 2011). In species that form mating plugs, specific mating plug proteins (e.g., PEB2 in *Drosophila*) can increase latency to remating in females (Bretman et al. 2010). In all cases in which SFPs influence female remating, they appear to decrease the probability or frequency of remating. Thus far, little is known about the role of female-derived molecules. One exception, however, is the sex peptide receptor (SPR) of *D. melanogaster* (Yapici et al. 2008). SPR is a G-protein-coupled receptor, and its activity in a set of reproductive tract neurons in the female is necessary for sex peptide (SP), a 36-amino-acid peptide with diverse phenotypic effects (Chen et al. 1988; Kubli 2003), to suppress receptivity through activation of SPR (Hasemeyer et al. 2009; Yang et al. 2009). The ortholog of SPR in *Helicoverpa armigera* is also essential for female postmating responses such as pheromone production (Fan et al. 1999; Hanin et al. 2012). Recent studies suggest that there may be additional receptors for SP, and that SPR may facilitate access of SP to the nervous system (Ja et al. 2009; Haussmann et al. 2013). Additional female-encoded molecules have been identified that are also essential for SP effects (including on receptivity [Findlay et al. 2014]). Furthermore, recent evidence from *Drosophila* suggests that there are female reproductive proteins that promote rapid remating (Sirot et al. 2014). This finding is consistent with the hypothesis that females are in conflict with their mates over the frequency of remating.

Sperm Transfer, Storage, Retention, and Usage

Females of all internally fertilizing animals store the sperm that they receive during mating (Neubaum and Wolfner 1999), for days (e.g., most mammals), weeks (e.g., *Drosophila* [Neubaum and Wolfner 1999]; some birds [Sasanami et al. 2013]), or even years (e.g., some Hymenoptera and reptiles [Gist and Congdon 1998]). Sperm are stored in females in specialized organs (e.g., spermathecae, seminal receptacles in insects [Neubaum and Wolfner 1999]) or in special regions of the reproductive tract (e.g., isthmus in mammals [Suarez 2008]) and their release must be coordinated with the opportunity to encounter eggs. Storage of sperm is potentially advantageous to both sexes by allowing for extended progeny production after mating. However, it also provides the opportunity for conflicts, such as those resulting from cryptic female choice and sperm competition. Moreover, there could be sexual conflict over which sex controls the rate of sperm release. For example, it might be beneficial for the female to replace older sperm with newer, fresher sperm or sperm from a better quality male (e.g., Lüpold et al. 2013). This would be at odds with the male's interest in maintaining his own sperm in storage so they can be used for fertilization. It might also be advantageous to the female to modulate the rate of sperm release to coordinate with the timing of egg production, whereas it

 Cite this article as *Cold Spring Harb Perspect Biol* doi: 10.1101/cshperspect.a017533

might benefit the male to have his sperm used in fertilizations as quickly as possible.

SFPs play important roles in the transit of sperm into storage. For example, transglutaminases—the cross-linking enzymes in the seminal fluid of mice (Dean 2013) and anopheline mosquitoes (Rogers et al. 2009)—are necessary for the coagulation of seminal fluids and the consequent efficient movement of sperm to storage sites. An SFP in *D. melanogaster* (Acp36DE) is necessary to regulate the uterine contractions that move sperm into storage (Avila and Wolfner 2009). SFPs in bovine seminal plasma bind to sperm (Gwathmey et al. 2006) and to annexins (Ignotz et al. 2007) on the oviductal epithelium. This binding allows sperm to be stored in the cows' sperm reservoirs. SFPs also are important in regulating the release of sperm from storage for fertilization. In *Drosophila*, the lectin Acp29AB is required for maintenance of sperm in the female's storage organs (Wong et al. 2008), and SP is required for efficient rate of release of sperm from storage in mated females (Avila et al. 2010). Furthermore, several other SFPs are necessary to transport SP to the necessary location to exert its effects. In cows, SFP-mediated release of sperm from storage allows the sperm to move to the fertilization site (Hung and Suarez 2010). During the process of sperm storage, SFPs have the potential to directly affect the sperm of a female's previous mates. In vitro studies in Hymenoptera (den Boer et al. 2010) and Diptera (Holman and Snook 2008) show that incubating sperm in SFPs affects their viability. In Hymenoptera, SFPs from one male can decrease the viability of rival males' sperm and thus the ability of those sperm to successfully fertilize eggs (den Boer et al. 2010). In sum, SFPs affect the ability of a male's sperm to survive and travel to the appropriate storage sites to be stored, retained, and released for fertilization (Xue and Noll 2000). They can also negatively impact the survival or success of competitor sperm in the female. In this capacity, SFPs can play important roles in sexual conflict over sperm use patterns (also see work by Edward et al. 2014).

A female may also benefit by influencing sperm use patterns (Eberhard 1996). This phenomenon may be mediated by female-derived molecules. Hymenopteran female spermathecal secretions can act to diminish male–male conflicts by countering the negative viability effect of seminal fluid on the sperm of different males (den Boer et al. 2009). This pattern could be consistent with an intense "conflict phase" between the males followed by a subsequent dampening down of conflict in order for the female to retain viable sperm for a long time, hence prolonging progeny production over an extended period. In addition, several lines of evidence suggest that females play a physically active role in influencing sperm use patterns. For example, genetic evidence and allelic correlations to sperm competition outcomes indicate that in *D. melanogaster*, a functioning female nervous system is necessary for proper sperm storage or to influence the outcome of sperm competition (Arthur et al. 1998; Chow et al. 2013). In another example, sperm storage is compromised in dead or anesthetized *Tribolium castaneum* females, again suggesting an active role for the female in sperm management (Bloch Qazi et al. 1998).

Egg Production

The reproductive success of both sexes depends largely on the quantity and quality of fertilized eggs. The female has to make eggs that are provisioned for future embryonic development, release them from the ovaries (ovulation), and move them through the reproductive tract to the correct location for further development (either inside or outside of the female, depending on the species). We consider these events together as the "egg-production process" (see Heifetz et al. 2000). In species with little or no oogenesis, ovulation or (if appropriate) egg laying before mating, egg production must increase after mating.

Increased egg production should benefit both female and male if more progeny are produced. In an additional benefit to the female, by coupling her level of egg production to mating, she does not "waste" resources making eggs when she lacks sperm to fertilize them. Once she has mated, however, it pays to increase egg production (all steps) because those eggs can be fertilized. Yet, even these apparently mutually beneficial changes may engender potential conflicts: Mating may induce a higher short-term level of egg production in the female than that which maximizes her lifetime reproductive success. Females risk trading off resources that they might otherwise have used for defense against pathogens or for other survival traits. Furthermore, in species that lay eggs, oviposition can be costly to females in terms of time allocation, energy expenditure, and exposure to predators and parasites. Thus, the optimal rate of egg production is a source of sexual conflict.

Ovulation frequency is perhaps a less obvious arena than overall egg production for sexual conflict (particularly in mammals without induced ovulation, in which ovulation does not increase after mating). Ovulation could, however, be a source of sexual conflict in insects (and potentially in mammals who are induced ovulators; e.g., camelids [Adams and Ratto 2013]). For example, in *Drosophila*, the buildup of many mature (but unovulated) oocytes in the ovaries of unmated females feeds back to arrest oogenesis (Chapman et al. 2001; Drummond-Barbosa and Spradling 2001). So, males could benefit by increasing ovulation rates, thereby relieving this feedback inhibition and thus accelerating the increase in oogenesis. The resulting increased egg production can exert a physiological cost on females and thus, as noted above, represents a source of sexual conflict.

In insects, SFPs influence the egg-production process, at several stages (reviewed in Gillott 2003; Arnqvist and Rowe 2005; Avila et al. 2011). For example, *D. melanogaster* SP increases the number of eggs made, consistent with its effects in raising juvenile hormone titers (Moshitzky et al. 1996), and also through effects on specific neurons that innervate the reproductive tract (Hasemeyer et al. 2009; Yang et al. 2009). *D. melanogaster* SP can also increase egg production in the moth *H. armigera* on injection (Fan et al. 2000), and knockdown of the *H. armigera* ortholog of the sex peptide receptor blocks postmating increases in egg production (Hanin et al. 2012). In *D. melanogaster*, an SFP has been identified (ovulin) that increases ovulation rate specifically (Herndon and Wolfner 1995; Heifetz et al. 2000). Increased ovulation triggered by SFPs is not, however, restricted to insects. For example, although many mammals are spontaneous ovulators (i.e., independent of mating or SFPs), camelid ovulation is induced by the nerve growth factor in seminal fluid (Kershaw et al. 2012; Ratto et al. 2012), analogous in outline to the situation found in *Drosophila*. Therefore, through their effects on egg production and release, SFPs may mediate conflict over these steps in the egg-production process.

Recent evidence from dipteran insects has provided insights into how seminal fluid molecules interact with female molecules to influence specific steps in egg production. For example, *D. melanogaster* ovulin stimulates ovulation by increasing the amount or activity of a female-derived neuromodulator (octopamine) that modulates muscle contractions that accompany ovulation (Rubinstein and Wolfner 2013). In the mosquito, *Anopheles gambiae*, a seminal fluid derived hormone induces increased production of a female reproductive tract protein that is necessary for mating-induced changes in egg development (Baldini et al. 2013). Such molecular interactions between male- and female-derived molecules provide opportunities for both partners to adjust egg-production rate toward their own optima and, therefore, create an opportunity for sexually antagonistic coevolution.

Food Intake and Processing

The nutritional needs of females change once they are reproductively active. For example, in arthropods, the increased production of eggs requires increased food resources to provide and sustain energy. In mammals, the physiological demands of pregnancy also impact the amount of food needed by the female. Consistent with the idea that increased egg production requires additional resources, mating can alter food preferences in arthropods. Mated *D. melanogaster* females eat more protein (Ribeiro and Dickson 2010), mated *Aedes aegypti* mosquito females take larger blood meals (Houseman and Downe 1986), and a male-derived factor stimulates blood feeding and engorgement in ticks

(Weiss and Kaufman 2004; Donohue et al. 2009).

Conflicts may occur in this arena as well. For example, it could be beneficial to the male to increase food consumption by his mate if it allowed her to produce more eggs and hence more progeny. But increased food consumption is associated with a decrease in longevity (in flies, nematodes, and mice [Chapman and Partridge 1996; Partridge and Gems 2002; Gems and Partridge 2013]) and thus may be costly to the female. In addition, the increased time spent eating, or finding different food sources, can potentially expose females to predation and pathogens, providing additional potential sources of conflict.

The role of SFPs in feeding related processes is clear in some insects. For example, receipt of SP increases the amount of food consumed by *D. melanogaster* females (Ribeiro and Dickson 2010) and changes the female's food preferences. SP also affects the rate of intestinal transit, with mated *D. melanogaster* females having slower intestinal transit than virgins, producing more concentrated excreta, presumably because they have absorbed more water (and nutrients) from their food (Cognigni et al. 2011; Apger McGlaughon and Wolfner 2013). In contrast, mated female *Ae. aegypti* mosquitoes digest their blood meals more rapidly than virgins, an effect that has also been attributed to SFPs (Downe 1975).

Activity Level

Mated *D. melanogaster* females sleep less and move around more than do unmated females (Isaac et al. 2010). This postmating response, dependent on SP, also could result in conflict. The additional movement could be detrimental for females by decreasing their regenerative sleep time, taking energy that could be used for somatic maintenance or to produce eggs, and potentially exposing them to predation. In the tephritid fruit fly (*Ceratitis capitata*), mating (or injection with seminal fluids) changes the attraction of females from a preference for male odors to a preference for fruit used for oviposition (Jang 1995). Similarly, in the mosquito *Ae. aegypti*, mated females are less likely than unmated females to fly toward a host. Injections of male accessory glands into unmated females decrease their response to host cues, suggesting this change is mediated, in part, by seminal fluid molecules (Fernandez and Klowden 1995). Consistent with this hypothesis, male *Ae. aegypti* transfer an SFP that can change host-seeking behavior when applied topically to females (Naccarati et al. 2012).

Immune Activation

Mating causes changes in immune capacity in a number of animals, although the ultimate reasons are as yet poorly understood. Immune changes could protect the female from microbes introduced during mating, could reflect trade-offs occasioned by the need to produce eggs (insects), or could lessen the probability of rejecting the fetus (mammals). These changes provide another arena for conflict; whatever benefits changes in immunity might have for male (or female) fertility might be accompanied by negative effects on the female's immune capacity or by trade-offs with other physiological processes. For example, a *Drosophila* female's immunity to systemic infection drops after mating (Fedorka et al. 2007; Short and Lazzaro 2010; but see also Zhong et al. 2013) and up-regulation of female immunity has been proposed to generate a hostile environment for sperm. That seminal fluid plays a role in mating related immunity is known in *Drosophila*, mice, and humans (Robertson et al. 2009; Guerin et al. 2009, 2011; Sharkey et al. 2012; Short et al. 2012), and seminal fluid of all animals tested to date includes predicted antimicrobial compounds (e.g., see Poiani 2006; Avila et al. 2011). The effect of SFPs on immunity is complex, at least in *Drosophila*. Although numerous studies have indicated that mating, and SFPs in particular, induce an increase in expression of antimicrobial peptide (AMP) genes, or in AMP production, mated female *Drosophila* show an SFP-induced decrease in systemic immunity after mating (Short et al. 2012). It is possible that there are tissue-specific differences in immune response to mating—perhaps the AMPs protect the reproductive tract at the expense of systemic immunity.

Life Span

Reductions in life span associated with elevated reproduction are common in both sexes, but the nature of those costs often differs substantially between males and females. Sexual conflict can occur because adaptations that increase the reproductive success of males can be selected even if they also cause decreased life span in females, and vice versa. Females are expected to, and indeed do, have the capacity to evolve to ameliorate these costs. This leads to adaptation and counter adaptation (i.e., sexually antagonistic coevolution).

Laboratory experimental evolution in *D. melanogaster* provides some support for the idea that sexual conflict can be manipulated experimentally, over evolutionary time, with predictable outcomes for life span. For example, the level of sexual conflict can be manipulated in the laboratory by altering sex ratios. A high ratio of males to females is expected to increase the opportunity for sexual conflict over the optimum value of reproductive traits such as the frequency of mating in this species. In populations of *D. melanogaster* evolving under conditions with elevated potential for sexual conflict (arising from increased mating frequency), females evolved to live longer in the presence of males (Wigby and Chapman 2004). This suggests that females evolved resistance to the harmful (i.e., life-span-shortening) effects of elevated mating frequency. The evidence that females can evolve to minimize male-imposed costs suggests the strong potential for sexual conflict, in particular over optimum female life span. In terms of corresponding male adaptations, there are conflicting reports, with either no differences in male-induced harm (Wigby and Chapman 2004) or reduced harm under low-conflict conditions (Nandy et al. 2013). Experiments in which selection is limited to males, which predict the evolution of male benefit phenotypes, have also given various results. Rice (1998) reported the evolution of increased male harm under male limited selection, whereas Jiang et al. (2011) found no changes, using the same approach and source population. However, to date no study has explicitly examined the effects of SFPs on the evolution of male harm (or female resistance). That such effects might be mediated by SFPs is suggested by the finding that experimental evolution under altered adult sex ratios (and hence opportunities for sexual conflict) can result in divergent ejaculate allocation strategies (Linklater et al. 2007).

SFPs are thought to be involved in the conflict between adaptations that increase male reproductive success and decrease female life span because receipt of high levels of SFPs can be costly to females, causing a decrease in female longevity and reproductive success (Chapman et al. 1995). The shortening of female life span in response to SFPs could be a side effect of SFP function (e.g., of increased male permating fitness or a direct "toxic" effect that is selected to reduce the likelihood of female remating and/or to increase current investment in reproduction (Johnstone and Keller 2000; Lessells 2005). Support for the "side effect" idea comes from data that some SFP-mediated fitness traits are directly linked to reductions in female life span. Such a relationship is observed between increased male sperm defense (success of a first mating male after subsequent matings) and early female mortality in *D. melanogaster* (Civetta and Clark 2000). Furthermore, SFPs with identified or predicted reproductive functions that benefit males are implicated in causing female mating costs in *Drosophila* (SP [Wigby and Chapman 2005]) or have been suggested to play such a role (e.g., SP, CG8137, and CG10433 [Mueller et al. 2007]; Acp62F [Lung et al. 2002; Civetta et al. 2005]). However, the situation may be complex, as a recent study in *D. melanogaster* found no evidence that elevation of sexual conflict, via increased male to female sex ratio, affected the frequency of a null allele of Acp62F (Wong and Rundle 2013).

Evidence for the idea that SFPs are directly selected to be "toxic" is difficult to gather unambiguously, as the toxic effect must be linked to some male benefit (otherwise it would not be selected). One possible line of evidence for a direct toxic effect of SFPs would be if the toxicity occurred via a different pathway from the effect of the SFP on male reproductive success. To our knowledge, however, there is little evidence for this possibility. Although the toxicity of ectopic expression of SFPs (e.g., Lung et al. 2002; Mueller et al. 2007) could represent such evidence, it must be noted that that the ectopic expression was at very high levels in those experiments. Studies are needed in which potential toxicity of these SFPs is measured in situations in which their ectopic expression is at levels closer to those delivered during mating.

It should also not be ruled out that longevity "costs" may be lower than first perceived if there are beneficial effects of SFPs on females earlier in their lives. Although no such benefits have yet clearly been identified (e.g., Brommer et al. 2012), costs to females could be lowered if SFPs caused early life egg production to increase, minimizing the fitness effect of life span reduction (Edward et al. 2011). Males may also suffer longevity costs from ongoing SFP-related sexual conflict. For example, if females evolve decreased sensitivity to SFPs, selection may act on males for increased SFP production, which, in turn, could result in decreased male life span. So far, this prediction has not been tested. The finding that males that invest in longer matings throughout their lives can suffer reduced life span and late life mating capacity (Bretman et al. 2013) suggests that such effects could occur.

DOES SEXUAL CONFLICT SHAPE SFP EVOLUTION?

Sexual conflict can occur in two modes—intra- and interlocus; these terms define whether the conflict occurs between the same or different loci, respectively. Sexual conflict mediated by SFPs is likely to be exclusively in the interlocus mode, because SFPs generally interact with non-SFP proteins encoded by different loci (including female-encoded proteins) to influence the outcome of male–female encounters (e.g., timing or amount of remating or egg production). If SFPs participate in ongoing sexual conflicts described above (see the section Do SFPs Contribute to Postmating Sexual Conflict?), then the genes that encode them should show evidence of sexually antagonistic selection. This evidence could include changes in sequence or expression level between the interacting loci. These changes, in turn, could result in modifications in seminal fluid composition over time (Perry et al. 2013).

Molecular Signatures of Conflict

A general prediction (Parker 1979) is that sexual conflict can, under some conditions, lead to continual evolutionary chases between the adaptations in one sex and the counteradaptations in the other (Tregenza et al. 2006). Such chases can result in coevolution between the genes in males and females that encode these adaptations and counteradaptations. This coevolution could involve changes in protein sequence or in expression level of genes subject to sexual conflict. Theory predicts that this type of coevolution between males and females can lead to diversification in the sequences of the interacting loci involved. If this diversification occurs in reproductive genes in individuals from different populations of the same species, it can facilitate reproductive isolation between populations, and ultimately speciation, owing to breakdown in reproductive processes. Dynamic change in expression levels could also result in rapid diversification between different populations if it promoted regulatory incompatibilities between the reproductive genes of individuals from different populations. However, this possibility has not yet been considered in any detail.

Theory since Parker's original formulation (Parker 1979), from simple to multilocus models, quantitative to explicit genetics, predicts that sexual conflict can lead to five different types of dynamic outcomes between interacting loci in males and females (Parker 1979; Frank 2000; Rowe et al. 2003, 2005; Gavrilets and Hayashi 2005; Hayashi et al. 2007). These outcomes are continuous evolutionary chases, cyclic coevolution, evolution toward an equilibrium, differentiation in female traits, and differentiation in male and female traits (Fig. 2). In addition, females could vary in their sensitivity to SFPs (or in their activation threshold to SFPs). Theory predicts that when this occurs female

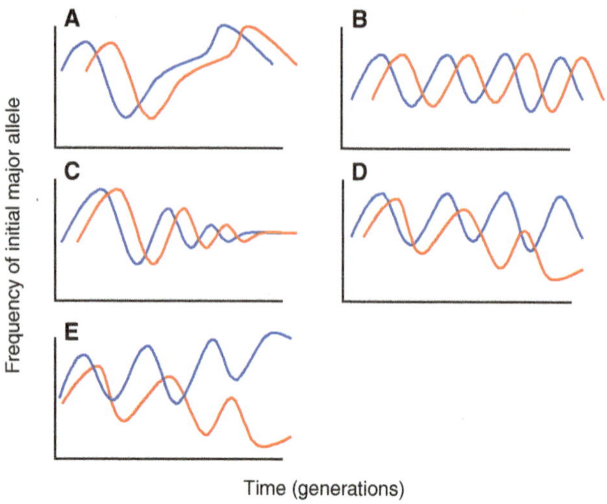

Figure 2. The simplified dynamic outcomes of sexual conflict between SFPs and their receptors in females, based on data adapted from the predictions of Parker (1979) and Hayashi et al. (2007). In this case, sexually antagonistic coevolution is occurring between different loci in males and females, each with two alleles. The frequency of the initial most frequent allele for each of the loci in males (blue) versus females (red) is shown. (*A*) A continuous coevolutionary chase, in which the frequency of the female allele tracks that of the male, with no underlying pattern, through time. (*B*) Cyclic coevolution, in which the female allele frequency tracks that of the male, with the coevolution having a cyclical pattern through time. (*C*) Evolution toward an equilibrium, in which the frequency of the female allele tracks that of the male toward a stable invariant frequency. (*D*) Differentiation in female, but not male allele frequency (an example of Buridan's ass). Here, the frequency of the major male allele continues to fluctuate through time, but that of the female, though initially tracking that of the male, diverges and in this case significantly decreases in frequency. Therefore, the coevolution has led to divergence in the frequencies of the two female alleles. (*E*) Differentiation in male and female allele frequencies. The frequency of the major male and female alleles initially show coevolutionary fluctuations; however, over time there is divergence in the frequency of both male and female alleles, with, in this case, the initial major allele in males becoming more frequent and that of the female becoming less so. Therefore, the initial coevolution has led to divergence in the frequencies of the two male and the two female alleles.

traits can evolve to "ignore" male adaptations to sexual conflict (Rowe et al. 2005). For example, females might become completely resistant to a male SFP that induces high levels of egg production. Expression level variation could mediate this type of regulatory change (e.g., through down-regulation of a female receptor).

To date, four kinds of evidence for signatures of sexual conflict have been described.

(i) Rapid Evolution, in Sequence or Regulation, in Some SFP Loci and Their Receptors in Females

Sexual conflict involving SFPs is expected to result in selection for male molecules with increasingly "manipulative" (male benefit/female detriment) functions, and for female SFP receptors to become more resistant to the effects of such SFPs (Rice 1996; Pitnick et al. 2001). We would therefore expect rapid sequence and/or expression changes in these molecules. There is substantial evidence for rapid SFP evolution, both at the sequence level and with respect to seminal fluid composition. Sequence-based studies in a wide range of species, including insects (e.g., Aguadé et al. 1992; Swanson et al. 2001; Begun et al. 2000) and mammals (Ramm et al. 2008, 2009; Karn et al. 2008), have identified SFP loci that diverge rapidly between species and/or whose patterns of sequence evolution are indicative of positive selection. Furthermore, in lineages for which genome-wide evolutionary comparisons have been performed, SFPs are typically found to evolve more rapidly, on average, than other functional classes of protein (e.g., Clark and Swanson 2005; Haerty et al. 2007; Dean et al. 2009; Findlay and Swanson 2010; Wong 2010). Similarly, SF composition appears to change particularly rapidly in a number of species. For example, many genes encoding known SFPs in *D. melanogaster* are undetectable in other members of the genus (Mueller et al. 2005; Wagstaff and Begun 2005; Haerty et al. 2007), and novel species-specific SFP genes have been described in *Drosophila* (Wagstaff and Begun 2007; Findlay et al. 2008, 2009; Almeida and Desalle 2009). Systematic characterization of seminal fluid composition in other groups of animals (e.g., mammals) will help to determine the generality of this trend.

It is worth noting that, along with sequence level changes, there can also be significant variation in the expression level of SFP loci and of SFP receptors within and between populations (Smith et al. 2009; Ayroles et al. 2011). Studies of this type of variation in the context of sexual conflict have not yet been reported. The extent to which expression level variation occurs in genes encoding female proteins that interact with SFPs is of significant interest. It could, for example, represent some of the "missing" variation that is evident in comparisons between the rapid evolution of male-expressed SFPs versus the typically lower levels of sequence variation seen within the female-expressed loci with which SFPs potentially interact (e.g., Swanson et al. 2004; Prokupek et al. 2009). Females might respond to new SFP sequence variants by changing the expression level or sensitivity of existing receptors and interacting molecules. This could subsequently select for ever-newer SFP sequence variants in the male, or "improved" versions of old SFPs. We suggest that further work into the extent and significance of expression level variation in genes encoding SFPs and the female proteins with which they interact will prove very informative.

(ii) Correlations between Rates of SFP Evolution and Mating Systems

If the intensity of sexual conflict or sexual selection is responsible for driving the rapid evolution of SFP loci, this would predict a correlation between measures of the intensity of postcopulatory selection (such as mating system) and rate of SFP evolution. Mating system can be a good proxy for tests of such correlations because the intensity of postcopulatory sexual selection is generally expected to increase with female promiscuity. A number of studies have sought to test these ideas, and specifically, the hypothesis that rates of SFP sequence evolution should correlate with measures of female promiscuity and/or sperm competition risk. The results of these studies are, however, mixed. Some SFP genes show clearly elevated rates of amino acid substitution in species in which females mate with many males (polyandry); however, many SFP loci do not show this pattern (reviewed in Wong 2011; see also Claw 2013; Schumacher et al. 2014). The reason for this inconsistency could be that the strength of selection may not be consistent for the same SFP across multiple species, perhaps because different signals and pathways may operate in different lineages. This would obscure correlations between mating system and SFP evolution in studies that focus on a few or single loci. This problem could be avoided if the average rates of evolution for multiple SFPs in monandrous versus polyandrous species were compared. Such analyses should include investigations of sequence variation across different members of the same biochemical gene family. This can reveal whether different family members are coopted for similar functions in different species. Studies using these approaches and including multigene comparisons report a higher average rate of evolution for testis-biased or testis-specific genes in chimpanzees (in which sperm competition is more intense) in comparison to humans (Wong 2010; see Turner and Hoekstra 2008 for a similar comparison in rodents).

(iii) Coevolution between the Genes Encoding Interacting Male and Female Proteins

In addition to affecting the rate of sequence evolution of SFPs, sexual conflict is expected to result in the coevolution of SFPs and receptors in females. This process can be characterized by successive waves of directional selection or fluctuating selection, leading to continual evolutionary chases (Fig. 2). In the simplest scenario, in which sexual conflict is mediated by interactions between a single SFP and a single female receptor, it is expected that these two molecules should coevolve over time. Such signatures should, in theory, be detectable, given knowledge of pairs of loci in males and females that are subject to sexually antagonistic selection, and robust methods for identifying coevolution (Clark and Aquadro 2010; Findlay et al. 2014). Unfortunately, it has been difficult to test this prediction, because female receptors have been identified for only a very few SFPs (Yapici et al. 2008; Haussmann et al. 2013). In addition, there is a lack of population genomic studies targeted at detecting potentially divergent evolutionary trajectories between different populations of the same species.

Even when a receptor has been identified, detecting coevolution is not straightforward. For example, SPR, the *D. melanogaster* SP receptor, has nonreproductive ligands in addition to SP (the myoinhibitory peptides, MIPs [Kim et al. 2010; Poels et al. 2010; Vandersmissen et al. 2013]). This may constrain its ability to coevolve with SP potentially independent of whether it is subject to sexual conflict. Another confounding situation would occur if the receptor and ligand interact through molecular features that are not directly reflected in the primary amino acid sequence of either protein. For example, in cows, annexins in the oviducts interact with BSP seminal proteins to mediate sperm storage—but this molecular interaction involves the annexins binding to sugar (fucose) modifications on the BSPs—modifications that are added posttranslationally (Ignotz et al. 2007). Although it has not yet been possible to test the hypothesis of coevolution with SFPs and their receptors, some ideas of what might be found come from results on coevolving egg (vitelline envelope) and sperm ligand/receptor pairs in abalone (VERL and lysin, respectively [Clark et al. 2009]). VERL and lysin are in linkage disequilibrium, even though their loci are not physically linked. They also show evidence for coevolution at the intrapopulation and between-species levels (Clark et al. 2009), including correlated rates of amino acid substitution.

Sexually antagonistic interactions are, however, unlikely to be limited to the simple scenario above, in which the interaction is only between one SFP and its female receptor. For example, sexual conflict could be mediated by SFPs that do not have a receptor, as in coevolution between SFPs that coagulate to form a mating plug and the female proteases that degrade the plug. Conflicts could also be mediated through interactions between SFPs and female morphology or behavioral traits. This type of conflict could be evident, for example, in coevolution between mating plug-forming SFPs and female plug removal behaviors, or female genital tract morphology changes. Such female traits may indeed be easier to identify than SFP receptors themselves and should be more widely considered. A second example is that antagonistic interactions between males and females could be impacted by interactions of either sex with pathogens that access the reproductive tract or cells. There is evidence that some immunity genes evolve more rapidly in promiscuous species, consistent with the idea that multiple mating is associated with increased immune challenge (Wlasiuk and Nachman 2010). In such a situation, it could be difficult to differentiate antagonistic coevolution between SFPs and their receptors from antagonistic coevolution between female receptors and the pathogens. This would be particularly difficult if the pathogen interacted directly with the SFP receptor in the female, because coevolution between the female receptor and the pathogen could also "drag along" SFP coevolution to retain SFP function through the female receptor. The primary driving interaction in this scenario would be the host–pathogen interaction, not the male–female conflict. However, without knowledge of host–pathogen interactions, the correct explanation for any coevolutionary patterns observed between SFPs and receptors could be hard to pin down.

One outcome of sexual conflict, predicted by theory when the intensity of sexual conflict between males and females is relatively weak, is the differentiation of reproductive traits in one sex but not in the other (the so-called Buridan's ass outcome, Fig. 2) (Gavrilets and Waxman 2002; Parker 2006; Hayashi et al. 2007). If this process occurs within a species, it could assist in the evolution of divergent reproductive forms, and hence reproductive isolation via the sympatric mode (i.e., in the absence of physical barriers to gene flow [Gavrilets and Waxman 2002]). Interestingly, there is empirical evidence for Buridan's ass in the apparently antagonistic interactions concerning the speed and efficiency of sperm entry into the egg; this evidence concerns the abalone sperm protein lysin, and the egg protein VERL to which it binds, as discussed above (Clark et al. 2009). The empirical observation is that male-derived lysin has low genetic variability and hence shows little phenotypic variation, whereas its female partner VERL shows abundant polymorphism and manifests multiple different phenotypes in females, associated in some cases with incipient speciation. These data have been interpreted as evidence for Buridan's ass in that the invariant male lysin is "stranded" between the divergent female VERLs (Panhuis et al. 2006). A Buridan's ass situation appears to be a unique signature of sexual conflict. Further work will show whether this phenomenon is seen for SFPs.

It is important to note that, although each of the evolutionary patterns described so far could arise as a result of sexually antagonistic coevolution, none is a unique signature of sexual conflict, with the possible exception of the Buridan's ass scenario, as discussed above. As such, the presence or absence of rapid SFP evolution is consistent with sexual conflict but does not rule out other selective regimes. Furthermore, even where sexual conflict over a particular outcome (e.g., egg-laying rate or remating) is expected to result in coevolution between an SFP and its receptor, coevolution may be impeded by pleiotropy (where the loci involved have additional, potentially nonreproductive, functions) or if the conflict is mediated by multiple interacting loci. Thus, care must be taken in interpreting patterns of SFP molecular evolution in the light of sexual conflict.

(iv) Chromosomal Distribution of SFP Loci

In organisms with chromosomal-based sex determination, there is an inevitable asymmetry of passage of sex chromosomes through the sexes (see work by Mank et al. 2014). In XX/female and XY/male systems, for example, X chromosomes are present more often in females than in males. In such a situation, there is a greater chance for selection for genes of benefit to females on the X chromosome and less chance to select for genes of benefit to males. This would be predicted to lead to a situation in which female benefit genes should be X-enriched (Rice 1984). Consistent with this idea, the gene for SPR is X-linked in *D. melanogaster*. As additional SFP receptors are discovered it will be interesting to see whether they, too, are X-linked.

Genes encoding *D. melanogaster* SFPs are predominantly located on the autosomes. This could reflect that alleles of genes that more often benefit males can accumulate more easily on the autosomes, or perhaps that sex limitation in males is easier to evolve on autosomes, because of additional requirements for dosage-compensation control of X-linked genes in male *Drosophila*.

Evidence for Conflict Resolution

In addition to evidence for ongoing conflict mediated by SFPs, it should be possible to detect signatures of previous, but resolved, conflicts. These resolutions might, of course, themselves open up alternative possibilities, or routes for, novel sexual conflicts. Signatures of previous sexual conflict in the genome are predicted to be detectable as genomic "baggage" such as pseudogenes, gene redundancy, duplication, and/or gross genome rearrangements between sibling species. Examples of all of these phenomena are observed among genes encoding SFPs.

One line of evidence for resolution of conflict could be the loss of function of SFP genes, as females overcome or avoid their effects. The "lost" genes could be replaced by others, including by co-option of other members of the same biochemical family, in response. Such SFP turnover appears to be particularly rapid in *Drosophila* species subgroups and in the primate copulatory plug protein SEMG2 (Kingan et al. 2003; Carnahan and Jensen-Seaman 2008). Female gorillas, unlike their chimpanzee cousins, are usually monandrous during each reproductive bout, removing the opportunity for postcopulatory sexual conflict. Because primate copulatory plug proteins are thought to be maintained, at least in part, owing to the advantage they provide males in preventing sperm competition, in species with monandrous females, selection is expected to be relaxed on such proteins, and thus loss-of-function mutations within them would not be removed from the population (see also Claw 2013).

Resolution of one type of conflict can also provide conditions for new conflicts to emerge. For example, intralocus sexual conflict, which results from differing "demands" by males and females on a single gene, is often proposed to be resolved by the evolution of sex-limited expression of the relevant gene (see work by Ingleby et al. 2014). Sex limited expression for SFPs would have released SFPs from any functional constraint imposed by having to operate in females. It may also have allowed the effects of SFPs to be more easily tailored to coordinate reproductive processes. For example, an SFP signal from males received by females only once sperm transfer has been achieved would be an efficient way of activating reproductive processes in females at the correct time and in the correct sequence (Chapman 2001). This mechanism benefits both sexes. However, once present, sex limitation instantly opens up the possibility for relatively unconstrained direct manipulation of reproduction in the female by males, with the possibility of initiating a new interlocus sexual conflict. Hence, new routes for sexual conflict via the interlocus mode may emerge following resolution of intralocus sexual conflict via the evolution of sex limitation.

CONCLUDING REMARKS

SFPs provide an informative molecular system in which to study sexual conflict because they not only provide mechanisms that can underlie sexual conflict but also show signatures at the gene sequence and expression level of past and current conflict. In this section, we conclude by considering additional aspects of the interface between SFPs and sexual conflict and proposing future research directions.

Is It All Conflict?

The evidence provided in the sections above can be interpreted as indicating the presence of sexual conflict. However, we do not yet know to what extent conflict actually operates and how important it is in relation to processes such as sexual selection. As noted above (in the section The Potential Weapons), there are examples of biochemical pathways that include contribution by males and females (e.g. proteolytic processing of *Drosophila* ovulin, Fig. 1), males turning on molecular pathways in females (e.g., Rubinstein and Wolfner 2013), males activating already prepared pools of preexisting RNAs or proteins in females (e.g., Heifetz and Wolfner 2004; Lawniczak and Begun 2004; McGraw et al. 2004; Mack et al. 2006; Kapelnikov et al. 2008; Rubinstein and Wolfner 2013), and apparent male/female synergy in achieving sperm storage (e.g., Ignotz et al. 2007). These examples may reflect the requirement to coordinate the many complex processes needed in order for reproduction to be successful. It is worth considering, however, that sexual conflict can occur within even the most complex and apparently mutually beneficial of these interactions. SFPs have the potential to engage females in a "sensory trap" (West Eberhard 1979) through the use of SFP signals to which females "must" respond to reproduce successfully, but which can then be used to divert females from their optimal investment in reproduction. Females cannot evolve complete insensitivity; otherwise they achieve lower fitness (hence "trapped"). Only a combination of mechanistic knowledge with manipulative experiments can determine whether such a scenario exists.

Further, the intensity of sexual conflict mediated by SFPs may vary over time, both across and within generations. The intensity of conflict now may not indicate its relative importance in the past or future. Resource levels may also cause variability in the extent to which conflict is expressed (Fricke et al. 2010). For example, when food is scarce, females may be unable to exhibit increased egg laying following mating, no matter how much SF has been received. Additional experiments into the lifetime costs and benefits of SFP transfer from both the male and the female's perspective, made under a range of biologically relevant conditions, are now necessary to resolve the extent of sexual conflict mediated by SFPs.

Practical Applications

Understanding how SFPs shape and are shaped by sexual conflict has numerous potential practical applications, and also provides cautionary insights. For example, this knowledge could provide insights for the control of pest insects. One option for increased efficiency of insect control is to select for, or genetically engineer, males that have a greater effect on female reproductive physiology and/or behavior. This manipulation could be coupled with pest control strategies such as the sterile male technique or release of genetically modified strains that are altered in such a way to reduce damage (e.g., Fu et al. 2007). Proof of principle for the use of artificial selection in creating more "manipulative" males comes from studies in *D. melanogaster*. For example, female *D. melanogaster* mated to males that had experimentally evolved in a polygamous mating system took longer to remate and produced fewer offspring after mating than females mated to males that had evolved in a monogamous mating system (Holland and Rice 1999; Pitnick et al. 2001). Furthermore, *D. melanogaster* males selected for increased accessory gland size produced and transferred more SP and sired more offspring when in competition with other males than control males (Wigby et al. 2009). The results of these studies suggest that standard artificial selection could increase the ability of laboratory-reared males to induce postmating responses in their mates. Similar effects can be achieved simply by exposing males to rivals (Bretman et al. 2009). Selection or genetic engineering could be used to make SFPs more stable in females so that their effects persist, or to allow them to be effective even in the absence of sperm (in the case of SFPs, like SP, whose effects are sperm dependent). Another logical application of the use of SFPs in pest management would be to chemically interfere with the production of particular SFPs in males or their receptors in females. This approach may be successful for effects in females that are mediated by a single SFP–receptor pair, but as noted earlier there is significant redundancy in SFP functional classes and/or lack of specificity in SFP receptors that might make such interference with a single SFP or receptor less effective. Moreover, the rapid rate of SFP evolution suggests that pest species may evolve resistance to such strategies, as these strategies simulate what might happen under conflict scenarios. Rapid SFP evolution also makes it unlikely that SFP-based molecules used to target one species would negatively affect other species.

In addition to controlling pest populations, our knowledge of the role of sexual conflict in shaping SFPs could also inform strategies to promote successful reproduction outcomes in species of conservation concern, agricultural animals (e.g., honey bees and farm animals), and even humans. For example, for nonhuman animals, SFPs from males in populations in which females mate promiscuously may enhance the fertilizing ability of sperm used in assisted reproduction. Similarly, exposure of males to cues of sperm competition could increase the quantity or fertilizing ability of sperm in their ejaculates (Wedell et al. 2002; Killgallon and Simmons 2005; Bretman et al. 2009, 2011).

Future Prospects

SFPs are experimentally accessible molecules that exert precise and measurable effects on the female—yet are made by the male. These characteristics make them a fascinating and tractable system for dissecting molecular interactions that can participate in, and underlie, sexual conflict. In the sections above, we have highlighted some of these interactions and how they can contribute to or mediate sexual conflict. Here, we conclude by suggesting five areas of SFP research related to sexual conflict that are so far underexplored.

1. Identification of the SFP-interacting loci in females to conduct tests of coevolution and chromosomal distribution predictions.
2. Transcriptomic and proteomic studies of variation in SFP and receptor expression levels to identify previously unrecognized signatures of sexual conflict.
3. Genomic population studies in natural populations, analogous to those conducted for reproductive phenotypes (e.g., Andrés and Arnqvist 2001), to detect cycles of sexually antagonistic coevolution through time and space.
4. Studies of genome and expression level changes in response to experimental evolution under sexual conflict.
5. Cross-population and cross-species comparisons of SFP effectiveness in relation to the intensity of sexual conflict.

Future studies of the above areas, as well as continued dissection of how SFPs exert their effects on females, promise to reveal profound and fundamental insights into the evolutionary potential of conflict and cooperation between the sexes.

ACKNOWLEDGMENTS

We are grateful to Bill Rice, Larry Harshman, and two anonymous reviewers for valuable comments on this article and to Bill Rice and Sergei Gavrilets for the opportunity to write it. We thank the members of our laboratories for inspiration and helpful discussions. We appreciate support from National Institutes of Health (NIH) grants R01-HD038921 (to M.F.W.), R01-HD059060 (to M.F.W.), R01-AI095491 (to M.F.W.); Natural Environment Research Council (NERC) and Biotechnology and Biological Sciences Research Council (BBSRC) grants (to T.C.); and a Natural Sciences and Engineering Research Council (NSERC) Discovery Grant (to A.W.).

REFERENCES

*Reference is also in this collection.

Aalberts M, Stout TA, Stoorvogel W. 2013. Prostasomes: Extracellular vesicles from the prostate. *Reproduction* **147:** R1–R14.

Adams GP, Ratto MH. 2013. Ovulation-inducing factor in seminal plasma: A review. *Anim Reprod Sci* **136:** 148–156.

Aguadé M, Miyashita N, Langley CH. 1992. Polymorphism and divergence in the *Mst26A* male accessory gland gene region in *Drosophila*. *Genetics* **132:** 755–770.

Almeida FC, Desalle R. 2009. Orthology, function and evolution of accessory gland proteins in the *Drosophila repleta* group. *Genetics* **181:** 235–245.

Andrés JA, Arnqvist G. 2001. Genetic divergence of the seminal signal-receptor system in houseflies: The footprints of sexually antagonistic coevolution? *Proc Biol Sci* **268:** 399–405.

Andrés JA, Maroja LS, Bogdanowicz SM, Swanson WJ, Harrison RG. 2006. Molecular evolution of seminal proteins in field crickets. *Mol Biol Evol* **23:** 1574–1584.

Andrés JA, Maroja LS, Harrison RG. 2008. Searching for candidate speciation genes using a proteomic approach: Seminal proteins in field crickets. *Proc Biol Sci* **275:** 1975–1983.

Apger-McGlaughon J, Wolfner MF. 2013. Post-mating change in excretion by mated *Drosophila melanogaster* females is a long-term response that that depends on sex peptide and sperm. *J Insect Physiol* **59:** 1024–1030.

Arnqvist G, Rowe L. 2005. *Sexual conflict*. Princeton University Press, Princeton, NJ.

Arthur BI, Hauschteck-Jungen E, Naothiger R, Ward PI. 1998. A female nervous system is necessary for normal sperm storage in *Drosophila melanogaster*: A masculinized nervous system is as good as none. *Proc Biol Sci* **266:** 1749–1753.

Avila FW, Wolfner MF. 2009. Acp36DE is required for uterine conformational changes in mated *Drosophila* females. *Proc Natl Acad Sci* **106:** 15796–15800.

Avila FW, Ravi Ram K, Bloch Qazi MC, Wolfner MF. 2010. Sex peptide is required for the efficient release of stored sperm in mated *Drosophila* females. *Genetics* **186:** 595–600.

Avila FW, Sirot LK, LaFlamme BA, Rubinstein CD, Wolfner MF. 2011. Insect seminal fluid proteins: Identification and function. *Annu Rev Entomol* **56:** 21–40.

Ayroles JF, Laflamme BA, Stone EA, Wolfner MF, Mackay TF. 2011. Functional genome annotation of *Drosophila* seminal fluid proteins using transcriptional genetic networks. *Genet Res (Camb)* **93:** 387–395.

Baer B, Heazlewood JL, Taylor NL, Eubel H, Millar AH. 2009. The seminal fluid proteome of the honeybee *Apis mellifera*. *Proteomics* **9:** 2085–2097.

Baldini F, Gabrieli P, South A, Valim C, Mancini F, Catteruccia F. 2013. The interaction between a sexually transferred steroid hormone and a female protein regulates oogenesis in the malaria mosquito *Anopheles gambiae*. *PLoS Biol* **11:** e1001695.

Begun DJ, Whitley P, Todd BL, Waldrip-Dail HM, Clark AG. 2000. Molecular population genetics of male accessory gland proteins in *Drosophila*. *Genetics* **156:** 1879–1888.

Bloch Qazi M, Aprille JR, Lewis SM. 1998. Female role in sperm storage in the red flour beetle, *Tribolium castaneum*. *Comp Biochem Physiol A: Mol Integr Physiol* **120:** 641–647.

Borovsky D, Carlson DA, Hancock RG, Rembold H, van Handel E. 1994. De novo biosynthesis of juvenile hormone III and I by the accessory glands of the male mosquito. *Insect Biochem Mol Biol* **24:** 437–444.

Bretman A, Fricke C, Chapman T. 2009. Plastic responses of male *Drosophila melanogaster* to the level of sperm competition increase male reproductive fitness. *Proc Biol Sci* **276:** 1705–1711.

Bretman A, Lawniczak MK, Boone J, Chapman T. 2010. A mating plug protein reduces early female remating in *Drosophila melanogaster*. *J Insect Physiol* **56:** 107–113.

Bretman A, Gage MJ, Chapman T. 2011. Quick-change artists: Male plastic behavioural responses to rivals. *Trends Ecol Evol* **26:** 467–473.

Bretman A, Westmancoat JD, Gage MJ, Chapman T. 2013. Costs and benefits of lifetime exposure to mating rivals in male *Drosophila melanogaster*. *Evolution* **67:** 2413–2422.

Bromfield JJ, Schjenken JE, Chin PY, Care AS, Jasper MJ, Robertson SA. 2014. Maternal tract factors contribute to paternal seminal fluid impact on metabolic phenotype in offspring. *Proc Natl Acad Sci* **111:** 2200–2205.

Brommer JE, Fricke C, Edward DA, Chapman T. 2012. Interactions between genotype and sexual conflict environment influence transgenerational fitness in *Drosophila melanogaster*. *Evolution* **66:** 517–531.

Carnahan SJ, Jensen-Seaman MI. 2008. Hominoid seminal protein evolution and ancestral mating behavior. *Am J Primatol* **70:** 939–948.

Chapman T. 2001. Seminal fluid-mediated fitness traits in *Drosophila*. *Heredity* **87:** 511–521.

Chapman T, Partridge L. 1996. Female fitness in *Drosophila melanogaster*: An interaction between the effect of nutrition and of encounter rate with males. *Proc Biol Sci* **263:** 755–759.

Chapman T, Liddle LF, Kalb JM, Wolfner MF, Partridge L. 1995. Cost of mating in *Drosophila melanogaster* females is mediated by male accessory gland products. *Nature* **373:** 241–244.

Chapman T, Herndon LA, Heifetz Y, Partridge L, Wolfner MF. 2001. The Acp26Aa seminal fluid protein is a modulator of early egg hatchability in *Drosophila melanogaster*. *Proc Biol Sci* **268:** 1647–1654.

Chen PS, Stumm-Zollinger E, Aigaki T, Balmer J, Bienz M, Bohlen P. 1988. A male accessory gland peptide that regulates reproductive behavior of female *D. melanogaster*. *Cell* **54:** 291–298.

Chow CY, Wolfner MF, Clark AG. 2013. A large neurological component to genetic differences underlying biased sperm use in *Drosophila*. *Genetics* **193:** 177–185.

Civetta A, Clark AG. 2000. Correlated effects of sperm competition and postmating female mortality. *Proc Natl Acad Sci* **97:** 13162–13165.

Civetta A, Montooth KL, Mendelson M. 2005. Quantitative trait loci and interaction effects responsible for variation in female postmating mortality in *Drosophila simulans* and *D. sechellia* introgression lines. *Heredity* **94:** 94–100.

Clark NL, Aquadro CF. 2010. A novel method to detect proteins evolving at correlated rates: Identifying new functional relationships between coevolving proteins. *Mol Biol Evol* **27:** 1152–1161.

Clark NL, Swanson WJ. 2005. Pervasive adaptive evolution in primate seminal proteins. *PLoS Genet* **1:** e35.

Clark NL, Aagaard JE, Swanson WJ. 2006. Evolution of reproductive proteins from animals and plants. *Reproduction* **131:** 11–22.

Clark NL, Gasper J, Sekino M, Springer S, Aquadro CF, Swanson WJ. 2009. Coevolution of interacting fertilization proteins. *PLoS Genet* **5:** e1000570.

Claw KG. 2013. "Proteomic identification and evolutionary analysis of primate reproductive proteins," PhD thesis, University of Washington, Seattle.

Clifton ME, Correa S, Rivera-Perez C, Nouzova M, Noriega FG. 2014. Male *Aedes aegypti* mosquitoes use JH III transferred during copulation to influence previtellogenic ovary physiology and affect the reproductive output of female mosquitoes. *J Insect Physiol* **64:** 40–47.

Cognigni P, Bailey AP, Miguel-Aliaga I. 2011. Enteric neurons and systemic signals couple nutritional and reproductive status with intestinal homeostasis. *Cell Metab* **13:** 92–104.

Collins AM, Caperna TJ, Williams V, Garrett WM, Evans JD. 2006. Proteomic analyses of male contributions to honey bee sperm storage and mating. *Insect Mol Biol* **15:** 541–549.

Davies SJ, Chapman T. 2006. Identification of genes expressed in the accessory glands of male Mediterranean fruit flies (*Ceratitis capitata*). *Insect Biochem Mol Biol* **36:** 846–856.

Dean MD. 2013. Genetic disruption of the copulatory plug in mice leads to severely reduced fertility. *PLoS Genet* **9:** e1003185.

Dean MD, Clark NL, Findlay GD, Karn RC, Yi X, Swanson WJ, MacCoss MJ, Nachman MW. 2009. Proteomics and comparative genomic investigations reveal heterogeneity in evolutionary rate of male reproductive proteins in mice (*Mus domesticus*). *Mol Biol Evol* **26:** 1733–1743.

den Boer SP, Boomsma JJ, Baer B. 2009. Honey bee males and queens use glandular secretions to enhance sperm viability before and after storage. *J Insect Physiol* **55:** 538–543.

den Boer SP, Baer B, Boomsma JJ. 2010. Seminal fluid mediates ejaculate competition in social insects. *Science* **327:** 1506–1509.

Destephano DB, Brady UE, Lovins RE. 1974. Synthesis of prostaglandin by reproductive tissue of the male house cricket, *Acheta domesticus*. *Prostaglandins* **6:** 71–79.

Donohue KV, Khalil SM, Ross E, Mitchell RD, Roe RM, Sonenshine DE. 2009. Male engorgement factor: Role in stimulating engorgement to repletion in the ixodid tick, *Dermacentor variabilis*. *J Insect Physiol* **55:** 909–918.

Dottorini T, Nicolaides L, Ranson H, Rogers DW, Crisanti A, Catteruccia F. 2007. A genome-wide analysis in *Anopheles gambiae* mosquitoes reveals 46 male accessory gland genes, possible modulators of female behavior. *Proc Natl Acad Sci* **104:** 16215–16220.

Downe AE. 1975. Internal regulation of rate of digestion of blood meals in the mosquito, *Aedes aegypti*. *J Insect Physiol* **21:** 1835–1839.

Drummond-Barbosa D, Spradling AC. 2001. Stem cells and their progeny respond to nutritional changes during *Drosophila* oogenesis. *Dev Biol* **231:** 265–278.

Eberhard WG. 1996. *Female control: Sexual selection by cryptic female choice.* Princeton University Press, Princeton, NJ.

Edward DA, Fricke C, Gerrard DT, Chapman T. 2011. Quantifying the life-history response to increased male exposure in female *Drosophila melanogaster*. *Evolution* **65:** 564–573.

* Edward DA, Stockley P, Hosken DJ. 2014. Sexual conflict and sperm competition. *Cold Spring Harb Perspect Biol* doi: 10.1101/cshperspect.a017707.

Fan Y, Rafaeli A, Gileadi C, Kubli E, Applebaum SW. 1999. *Drosophila melanogaster* sex peptide stimulates juvenile hormone synthesis and depresses sex pheromone production in *Helicoverpa armigera*. *J Insect Physiol* **45:** 127–133.

Fan Y, Rafaeli A, Moshitzky P, Kubli E, Choffat Y, Applebaum SW. 2000. Common functional elements of *Drosophila melanogaster* seminal peptides involved in reproduction of *Drosophila melanogaster* and *Helicoverpa armigera* females. *Insect Biochem Mol Biol* **30:** 805–812.

Fedorka KM, Linder JE, Winterhalter W, Promislow D. 2007. Post-mating disparity between potential and realized immune response in *Drosophila melanogaster*. *Proc Biol Sci* **274:** 1211–1217.

Fernandez NM, Klowden MJ. 1995. Male accessory gland substances modify the host seeking behavior of gravid *Aedes aegypti* mosquitoes. *J Insect Physiol* **41:** 965–970.

Findlay GD, Swanson WJ. 2010. Proteomics enhances evolutionary and functional analysis of reproductive proteins. *BioEssays* **32:** 26–36.

Findlay GD, Yi X, Maccoss MJ, Swanson WJ. 2008. Proteomics reveals novel *Drosophila* seminal fluid proteins transferred at mating. *PLoS Biol* **6:** e178.

Findlay GD, MacCoss MJ, Swanson WJ. 2009. Proteomic discovery of previously unannotated, rapidly evolving seminal fluid genes in *Drosophila*. *Genome Res* **19:** 886–896.

Findlay GD, Sitnik JL, Aquadro CF, Clark NL, Wolfner MF. 2014. Evolutionary rate covariation identifies new members of a protein network required for *Drosophila* female post-mating responses. *PLoS Genet* **10:** e1004108.

Frank SA. 2000. Sperm competition and female avoidance of polyspermy mediated by sperm-egg biochemistry. *Evol Ecol Res* **2:** 613–625.

Fricke C, Bretman A, Chapman T. 2010. Female nutritional status determines the magnitude and sign of responses to a male ejaculate signal in *Drosophila melanogaster*. *J Evol Biol* **23:** 157–165.

Fu G, Condon KC, Epton MJ, Gong P, Jin L, Condon GC, Morrison NI, Dafa'alla TH, Alphey L. 2007. Female-specific insect lethality engineered using alternative splicing. *Nat Biotechnol* **25:** 353–357.

Gavrilets S, Waxman D. 2002. Sympatric speciation by sexual conflict. *Proc Natl Acad Sci* **99:** 10533–10538.

Gems D, Partridge L. 2013. Genetics of longevity in model organisms: Debates and paradigm shifts. *Annu Rev Physiol* **75:** 621–644.

Gillott C. 2003. Male accessory gland secretions: Modulators of female reproductive physiology and behavior. *Annu Rev Entomol* **48:** 163–184.

Gist DH, Congdon JD. 1998. Oviductal sperm storage as a reproductive tactic of turtles. *J Exp Zool* **282:** 526–534.

Guerin LR, Prins JR, Robertson SA. 2009. Regulatory T-cells and immune tolerance in pregnancy: A new target for infertility treatment? *Hum Reprod Update* **15:** 517–535.

Guerin LR, Moldenhauer LM, Prins JR, Bromfield JJ, Hayball JD, Robertson SA. 2011. Seminal fluid regulates accumulation of FOXP3$^+$ regulatory T cells in the preimplantation mouse uterus through expanding the FOXP3$^+$ cell pool and CCL19-mediated recruitment. *Biol Reprod* **85:** 397–408.

Gwathmey TM, Ignotz GG, Mueller JL, Manjunath P, Suarez SS. 2006. Bovine seminal plasma proteins PDC-109, BSP-A3, and BSP-30-kDa share functional roles in storing sperm in the oviduct. *Biol Reprod* **75:** 501–507.

Haerty W, Jagadeeshan S, Kulathinal RJ, Wong A, Ravi Ram K, Sirot LK, Levesque L, Artieri CG, Wolfner MF, Civetta A, et al. 2007. Evolution in the fast lane: Rapidly evolving sex-related genes in *Drosophila*. *Genetics* **177:** 1321–1335.

Hanin O, Azrielli A, Applebaum SW, Rafaeli A. 2012. Functional impact of silencing the *Helicoverpa armigera* sex-peptide receptor on female reproductive behaviour. *Insect Mol Biol* **21:** 161–167.

Hasemeyer M, Yapici N, Heberlein U, Dickson BJ. 2009. Sensory neurons in the *Drosophila* genital tract regulate female reproductive behavior. *Neuron* **61:** 511–518.

Haussmann IU, Hemani Y, Wijesekera T, Dauwalder B, Soller M. 2013. Multiple pathways mediate the sex-peptide-regulated switch in female *Drosophila* reproductive behaviours. *Proc Biol Sci* **280:** 20131938.

Hayashi TI, Vose M, Gavrilets S. 2007. Genetic differentiation by sexual conflict. *Evolution* **61:** 516–529.

Heifetz Y, Wolfner MF. 2004. Mating, seminal fluid components, and sperm cause changes in vesicle release in the *Drosophila* female reproductive tract. *Proc Natl Acad Sci* **101:** 6261–6266.

Heifetz Y, Lung O, Frongillo EA Jr, Wolfner MF. 2000. The *Drosophila* seminal fluid protein Acp26Aa stimulates release of oocytes by the ovary. *Curr Biol* **10:** 99–102.

Heifetz Y, Vandenberg LN, Cohn HI, Wolfner MF. 2005. Two cleavage products of the *Drosophila* accessory gland protein ovulin can independently induce ovulation. *Proc Natl Acad Sci* **102:** 743–748.

Herndon LA, Wolfner MF. 1995. A *Drosophila* seminal fluid protein, Acp26Aa, stimulates egg laying in females for one day after mating. *Proc Natl Acad Sci* **92:** 10114–10118.

Holland B, Rice WR. 1999. Experimental removal of sexual selection reverses intersexual antagonistic coevolution and removes a reproductive load. *Proc Natl Acad Sci* **96:** 5083–5088.

Holman L, Snook RR. 2008. A sterile sperm caste protects brother fertile sperm from female-mediated death in *Drosophila pseudoobscura*. *Curr Biol* **18:** 292–296.

Houseman JG, Downe AER. 1986. Methods of measuring blood meal size and proteinase activity for determining effects of mated state on digestive processes of female *Aedes aegypti* (L.) (Diptera: Culicidae). *Can Entomol* **118:** 241–248.

Hung PH, Suarez SS. 2010. Regulation of sperm storage and movement in the ruminant oviduct. *Soc Reprod Fertil Suppl* **67:** 257–266.

Ignotz GG, Cho MY, Suarez SS. 2007. Annexins are candidate oviductal receptors for bovine sperm surface proteins and thus may serve to hold bovine sperm in the oviductal reservoir. *Biol Reprod* **77:** 906–913.

* Ingleby FC, Flis I, Morrow EH. 2014. Sex-biased gene expression and sexual conflict throughout development. *Cold Spring Harb Perspect Biol* doi: 10.1101/cshperspect.a017632.

Isaac RE, Li C, Leedale AE, Shirras AD. 2010. *Drosophila* male sex peptide inhibits siesta sleep and promotes locomotor activity in the post-mated female. *Proc Biol Sci* **277:** 65–70.

Ja WW, Carvalho GB, Madrigal M, Roberts RW, Benzer S. 2009. The *Drosophila* G protein-coupled receptor, Methuselah, exhibits a promiscuous response to peptides. *Protein Sci* **18:** 2203–2208.

Jang EB. 1995. Effects of mating and accessory gland injections on olfactory-mediated behavior in the female Mediterranean fruit fly, *Ceratitis capitata*. *J Insect Physiol* **41:** 705–710.

Jiang PP, Bedhomme S, Prasad NG, Chippindale A. 2011. Sperm competition and mate harm unresponsive to male-limited selection in *Drosophila*: An evolving genetic architecture under domestication. *Evol* **65:** 2448–2460.

Johnstone RA, Keller L. 2000. How males can gain by harming their mates: Sexual conflict, seminal toxins, and the cost of mating. *Am Nat* **156:** 368–377.

Kapelnikov A, Zelinger E, Gottlieb Y, Rhrissorrakrai K, Gunsalus KC, Heifetz Y. 2008. Mating induces an immune response and developmental switch in the *Drosophila* oviduct. *Proc Natl Acad Sci* **105:** 13912–13917.

Karn RC, Clark NL, Nguyen ED, Swanson WJ. 2008. Adaptive evolution in rodent seminal vesicle secretion proteins. *Mol Biol Evol* **25:** 2301–2310.

Kershaw-Young CM, Druart X, Vaughan J, Maxwell WM. 2012. β-Nerve growth factor is a major component of alpaca seminal plasma and induces ovulation in female alpacas. *Reprod Fertil Dev* **24:** 1093–1097.

Kilgallon SJ, Simmons LW. 2005. Image content influences men's semen quality. *Biol Lett* **1:** 253–255.

Kim YJ, Bartalska K, Audsley N, Yamanaka N, Yapici N, Lee JY, Kim YC, Markovic M, Isaac E, Tanaka Y, et al. 2010. MIPs are ancestral ligands for the sex peptide receptor. *Proc Natl Acad Sci* **107:** 6520–6525.

Kingan SB, Tatar M, Rand DM. 2003. Reduced polymorphism in the chimpanzee semen coagulating protein, semenogelin I. *J Mol Evol* **57:** 159–169.

Kubli E. 2003. Sex-peptides: Seminal peptides of the *Drosophila* male. *Cell Mol Life Sci* **60:** 1689–1704.

LaFlamme BA, Wolfner MF. 2013. Identification and function of proteolysis regulators in seminal fluid. *Mol Reprod Dev* **80:** 80–101.

LaFlamme BA, Ravi Ram K, Wolfner MF. 2012. The *Drosophila melanogaster* seminal fluid protease "seminase" regulates proteolytic and post-mating reproductive processes. *PLoS Genet* **8:** e1002435.

LaFlamme BA, Avila FW, Ravi Ram K, Wolfner MF. 2014. A *Drosophila* protease cascade member, seminal metalloprotease-1, is activated stepwise by male factors and require female factors for full activity. *Genetics* **196:** 1117–1129.

Lawniczak MK, Begun DJ. 2004. A genome-wide analysis of courting and mating responses in *Drosophila melanogaster* females. *Genome* **47:** 900–910.

Leiblich A, Marsden L, Gandy C, Corrigan L, Jenkins R, Hamdy F, Wilson C. 2012. Bone morphogenetic protein- and mating-dependent secretory cell growth and migration in the *Drosophila* accessory gland. *Proc Natl Acad Sci* **109:** 19292–19297.

Lessells CM. 2005. Why are males bad for females? Models for the evolution of damaging male mating behavior. *Am Nat* **165:** S46–S63.

Li H, Huang S, Guo C, Guan H, Xiong C. 2012. Cell-free seminal mRNA and microRNA exist in different forms. *PLoS ONE* **7:** e34566.

Linklater JR, Wertheim B, Wigby S, Chapman T. 2007. Ejaculate depletion patterns evolve in response to experimental manipulation of sex ratio in *Drosophila melanogaster*. *Evol* **61:** 2027–2034.

Loher W, Ganjian I, Kubo I, Stanley-Samuelson D, Tobe SS. 1981. Prostaglandins: Their role in egg-laying of the cricket *Teleogryllus commodus*. *Proc Natl Acad Sci* **78:** 7835–7838.

Lung O, Wolfner MF. 1999. *Drosophila* seminal fluid proteins enter the circulatory system of the mated female fly by crossing the posterior vaginal wall. *Insect Biochem Molec* **12:** 1043–1052.

Lung O, Tram U, Finnerty CM, Eipper-Mains MA, Kalb JM, Wolfner MF. 2002. The *Drosophila melanogaster* seminal fluid protein Acp62F is a protease inhibitor that is toxic upon ectopic expression. *Genetics* **160:** 211–224.

Lüpold S, Pitnick S, Berben KS, Blengini CS, Belote JM, Manier MK. 2013. Female mediation of competitive fertilization success in *Drosophila melanogaster*. *Proc Natl Acad Sci* **110:** 10693–10698.

Mack PD, Kapelnikov A, Heifetz Y, Bender M. 2006. Mating-responsive genes in reproductive tissues of female *Drosophila melanogaster*. *Proc Natl Acad Sci* **103:** 10358–10363.

Mancini E, Tammaro F, Baldini F, Via A, Raimondo D, George P, Audisio P, Sharakhov IV, Tramontano A, Catteruccia F, et al. 2011. Molecular evolution of a gene cluster of serine proteases expressed in the *Anopheles gambiae* female reproductive tract. *BMC Evol Biol* **11**.

* Mank JE, Hosken DJ, Wedell N. 2014. Conflict on the sex chromosomes: Cause, effect, and complexity. *Cold Spring Harb Perspect Biol* doi: 10.1101/cshperspect.a017715.

Martan J, Shepherd BA. 1976. The role of the copulatory plug in reproduction of the guinea pig. *J Exp Zool* **196:** 79–84.

McGraw LA, Gibson G, Clark AG, Wolfner MF. 2004. Genes regulated by mating, sperm, or seminal proteins in mated female *Drosophila melanogaster*. *Curr Biol* **14:** 1509–1514.

Moshitzky P, Fleischmann I, Chaimov N, Saudan P, Klauser S, Kubli E, Applebaum SW. 1996. Sex-peptide activates juvenile hormone biosynthesis in the *Drosophila melanogaster* corpus allatum. *Arch Insect Biochem Physiol* **32:** 363–374.

Mueller JL, Ripoll DR, Aquadro CF, Wolfner MF. 2004. Comparative structural modeling and inference of conserved protein classes in *Drosophila* seminal fluid. *Proc Natl Acad Sci* **101:** 13542–13547.

Mueller JL, Ravi Ram K, McGraw LA, Bloch Qazi MC, Siggia ED, Clark AG, Aquadro CF, Wolfner MF. 2005. Cross-species comparison of *Drosophila* male accessory gland protein genes. *Genetics* **171:** 131–143.

Mueller JL, Page JL, Wolfner MF. 2007. An ectopic expression screen reveals the protective and toxic effects of *Drosophila* seminal fluid proteins. *Genetics* **175:** 777–783.

Naccarati C, Audsley N, Keen JN, Kim JH, Howell GJ, Kim YJ, Isaac RE. 2012. The host-seeking inhibitory peptide, Aea-HP-1, is made in the male accessory gland and transferred to the female during copulation. *Peptides* **34:** 150–157.

Nandy B, Gupta V, Sen S, Udaykumar N, Samant MA, Ali SZ, Prasad NG. 2013. Evolution of mate-harm, longevity and behavior in male fruit flies subjected to different levels of interlocus conflict. *BMC Evol Biol* **13:** 212.

Neubaum DM, Wolfner MF. 1999. Wise, winsome, or weird? Mechanisms of sperm storage in female animals. *Curr Top Dev Biol* **41:** 67–97.

Pampalakis G, Sotiropoulou G. 2007. Tissue kallikrein proteolytic cascade pathways in normal physiology and cancer. *Biochim Biophys Acta* **1776:** 22–31.

Panhuis TM, Clark NL, Swanson WJ. 2006. Rapid evolution of reproductive proteins in abalone and *Drosophila*. *Philos Trans R Soc B* **361:** 261–268.

Park M, Wolfner MF. 1995. Male and female cooperate in the prohormone-like processing of a *Drosophila melanogaster* seminal fluid protein. *Dev Biol* **171:** 694–702.

Parker GA. 1979. Sexual selection and sexual conflict. In *Sexual selection and reproductive competition in insects* (ed. Blum MS, Blum NA), pp. 123–166. Academic, New York.

Parker GA. 2006. Sexual conflict over mating and fertilization: An overview. *Philos Trans R Soc B* **361:** 235–259.

Partridge L, Gems D. 2002. Mechanisms of ageing: Public or private? *Nat Rev Genet* **3:** 165–175.

Peng J, Chen S, Busser S, Liu H, Honegger T, Kubli E. 2005. Gradual release of sperm bound sex-peptide controls female postmating behavior in *Drosophila*. *Curr Biol* **15:** 207–213.

Perry JC, Sirot L, Wigby S. 2013. The seminal symphony: How to compose an ejaculate. *Trends Ecol Evol* **28:** 414–422.

Pilch B, Mann M. 2006. Large-scale and high-confidence proteomic analysis of human seminal plasma. *Genome Biol* **7:** R40.

Pitnick S, Brown WD, Miller GT. 2001. Evolution of female remating behaviour following experimental removal of sexual selection. *Proc Biol Sci* **268:** 557–563.

Poels J, Van Loy T, Vandersmissen HP, Van Hiel B, Van Soest S, Nachman RJ, Vanden Broeck J. 2010. Myoinhibiting peptides are the ancestral ligands of the promiscuous *Drosophila* sex peptide receptor. *Cell Mol Life Sci* **67:** 3511–3522.

Poiani A. 2006. Complexity of seminal fluid: A review. *Behav Ecol Sociobiol* **60:** 289–310.

Prokupek AM, Kachman SD, Ladunga I, Harshman LG. 2009. Transcriptional profiling of the sperm storage organs of *Drosophila melanogaster*. *Insect Mol Biol* **18:** 465–475.

Ramm SA, Oliver PL, Ponting CP, Stockley P, Emes RD. 2008. Sexual selection and the adaptive evolution of mammalian ejaculate proteins. *Mol Biol Evol* **25:** 207–219.

Ramm SA, McDonald L, Hurst JL, Beynon RJ, Stockley P. 2009. Comparative proteomics reveals evidence for evolutionary diversification of rodent seminal fluid and its functional significance in sperm competition. *Mol Biol Evol* **26:** 189–198.

Ratto MH, Leduc YA, Valderrama XP, van Straaten KE, Delbaere LT, Pierson RA, Adams GP. 2012. The nerve of ovulation-inducing factor in semen. *Proc Natl Acad Sci* **109:** 15042–15047.

Ravi Ram K, Wolfner MF. 2007. Seminal influences: *Drosophila* Acps and the molecular interplay between males and females during reproduction. *Integr Comp Biol* **47:** 427–445.

Ravi Ram K, Wolfner MF. 2009. A network of interactions among seminal proteins underlies the long-term postmating response in *Drosophila*. *Proc Natl Acad Sci* **106:** 15384–15389.

Ravi Ram K, Sirot LK, Wolfner MF. 2006. Predicted seminal astacin-like protease is required for processing of reproductive proteins in *Drosophila melanogaster*. *Proc Natl Acad Sci* **103:** 18674–18679.

Ribeiro C, Dickson BJ. 2010. Sex peptide receptor and neuronal TOR/S6K signaling modulate nutrient balancing in *Drosophila*. *Curr Biol* **20:** 1000–1005.

Rice WR. 1984. Sex chromosomes and the evolution of sexual dimorphism. *Evolution* **38:** 735–742.

Rice WR. 1996. Sexually antagonistic male adaptation triggered by experimental arrest of female evolution. *Nature* **381:** 232–234.

Rice WR. 1998. Male fitness increases when females are eliminated from gene pool: Implications for the Y chromosome. *Proc Natl Acad Sci* **95:** 6217–6221.

Robertson SA. 2007. Seminal fluid signaling in the female reproductive tract: Lessons from rodents and pigs. *J Anim Sci* **85:** E36–E44.

Robertson SA, Guerin LR, Moldenhauer LM, Hayball JD. 2009. Activating T regulatory cells for tolerance in early pregnancy—The contribution of seminal fluid. *J Reprod Immunol* **83:** 109–116.

Rodriguez-Martinez H, Kvist U, Ernerudh J, Sanz L, Calvete JJ. 2011. Seminal plasma proteins: What role do they play? *Am J Reprod Immunol* **66:** 11–22.

Rogers DW, Baldini F, Battaglia F, Panico M, Dell A, Morris HR, Catteruccia F. 2009. Transglutaminase-mediated semen coagulation controls sperm storage in the malaria mosquito. *PLoS Biol* **7:** e1000272.

Rowe L, Cameron E, Day T. 2003. Detecting sexually antagonistic coevolution with population crosses. *Proc Biol Sci* **270:** 2009–2016.

Rowe L, Cameron E, Day T. 2005. Escalation, retreat, and female indifference as alternative outcomes of sexually antagonistic coevolution. *Am Nat* **165:** S5–S18.

Rubinstein CD, Wolfner MF. 2013. *Drosophila* seminal protein ovulin mediates ovulation through female octopamine neuronal signaling. *Proc Natl Acad Sci* **110:** 17420–17425.

Sasanami T, Matsuzaki M, Mizushima S, Hiyama G. 2013. Sperm storage in the female reproductive tract in birds. *J Reprod Dev* **59:** 334–338.

Schumacher J, Rosenkranz D, Herlyn H. 2014. Mating systems and protein–protein interactions determine evolutionary rates of primate sperm proteins. *Proc Biol Sci* **281:** 20132607.

Sharkey DJ, Tremellen KP, Jasper MJ, Gemzell-Danielsson K, Robertson SA. 2012. Seminal fluid induces leukocyte recruitment and cytokine and chemokine mRNA expression in the human cervix after coitus. *J Immunol* **188:** 2445–2454.

Short SM, Lazzaro BP. 2010. Female and male genetic contributions to post-mating immune defence in female *Drosophila melanogaster*. *Proc Biol Sci* **277:** 3649–3657.

Short SM, Wolfner MF, Lazzaro BP. 2012. Female *Drosophila melanogaster* suffer reduced defense against infection due to seminal fluid components. *J Insect Physiol* **58:** 1192–1201.

Siegel RW, Hall JC. 1979. Conditioned responses in courtship behavior of normal and mutant *Drosophila*. *Proc Natl Acad Sci* **76:** 3430–3434.

Simmons LW, Tan YF, Millar AH. 2013. Sperm and seminal fluid proteomes of the field cricket *Teleogryllus oceanicus*: Identification of novel proteins transferred to females at mating. *Insect Mol Biol* **22:** 115–130.

Sirot LK, Poulson RL, McKenna MC, Girnary H, Wolfner MF, Harrington LC. 2008. Identity and transfer of male reproductive gland proteins of the dengue vector mosquito, *Aedes aegypti*: Potential tools for control of female feeding and reproduction. *Insect Biochem Mol Biol* **38:** 176–189.

Sirot LK, LaFlamme BA, Sitnik JL, Rubinstein CD, Avila FW, Chow CY, Wolfner MF. 2009. Molecular social interactions: *Drosophila melanogaster* seminal fluid proteins as a case study. *Adv Genet* **68:** 23–56.

Sirot LK, Hardstone MC, Helinski ME, Ribeiro JM, Kimura M, Deewatthanawong P, Wolfner MF, Harrington LC. 2011. Towards a semen proteome of the dengue vector mosquito: Protein identification and potential functions. *PLoS Negl Trop Dis* **5:** e989.

Sirot LK, Findlay G, Sitnik J, Frasheri D, Avila F, Wolfner MF. 2014. Molecular characterization and evolution of a gene family encoding both female- and male-specific reproductive proteins in *Drosophila*. *Mol Biol Evol* **31:** 1554–1567.

Smith DT, Hosken DJ, Ffrench-Constant RH, Wedell N. 2009. Variation in sex peptide expression in *D. melanogaster*. *Genet Res (Camb)* **91:** 237–242.

South A, Sirot LK, Lewis SM. 2011. Identification of predicted seminal fluid proteins in *Tribolium castaneum*. *Insect Mol Biol* **20:** 447–456.

Suarez SS. 2008. Regulation of sperm storage and movement in the mammalian oviduct. *Int J Dev Biol* **52:** 455–462.

Swanson WJ, Vacquier VD. 2002. The rapid evolution of reproductive proteins. *Nat Rev Genet* **3:** 137–144.

Swanson WJ, Clark AG, Waldrip-Dail HM, Wolfner MF, Aquadro CF. 2001. Evolutionary EST analysis identifies rapidly evolving male reproductive proteins in *Drosophila*. *Proc Natl Acad Sci* **98:** 7375–7379.

Swanson WJ, Wong A, Wolfner MF, Aquadro CF. 2004. Evolutionary expressed sequence tag analysis of *Drosophila* female reproductive tracts identifies genes subjected to positive selection. *Genetics* **168:** 1457–1465.

Tregenza T, Wedell N, Chapman T. 2006. Introduction. Sexual conflict: A new paradigm? *Philos Trans R Soc B* **361:** 229–234.

Turner LM, Chuong EB, Hoekstra HE. 2008. Comparative analysis of testis protein evolution in rodents. *Genetics* **179:** 2075–2089.

Vandersmissen HP, Nachman RJ, Vanden Broeck J. 2013. Sex peptides and MIPs can activate the same G protein-coupled receptor. *Gen Comp Endocrinol* **188:** 137–143.

Wagstaff BJ, Begun DJ. 2005. Comparative genomics of accessory gland protein genes in *Drosophila melanogaster* and *D. pseudoobscura*. *Mol Biol Evol* **22:** 818–832.

Wagstaff BJ, Begun DJ. 2007. Adaptive evolution of recently duplicated accessory gland protein genes in desert *Drosophila*. *Genetics* **177:** 1023–1030.

Walters JR, Harrison RG. 2008. EST analysis of male accessory glands from *Heliconius* butterflies with divergent mating systems. *BMC Genomics* **9:** 592.

Walters JR, Harrison RG. 2010. Combined EST and proteomic analysis identifies rapidly evolving seminal fluid proteins in *Heliconius* butterflies. *Mol Biol Evol* **27:** 2000–2013.

Wedell N, Gage MJG, Parker GA. 2002. Sperm competition, male prudence and sperm-limited females. *Trends Ecol Evol* **17:** 313–320.

Weiss BL, Kaufman WR. 2004. Two feeding-induced proteins from the male gonad trigger engorgement of the female tick *Amblyomma hebraeum*. *Proc Natl Acad Sci* **101:** 5874–5879.

West Eberhard MJ. 1979. Sexual selection, social competition, and evolution. *Proc Am Phil Soc* **123:** 222–234.

Wigby S, Chapman T. 2004. Female resistance to male harm evolves in response to manipulation of sexual conflict. *Evol* **58:** 1028–1037.

Wigby S, Chapman T. 2005. Sex peptide causes mating costs in female *Drosophila melanogaster*. *Curr Biol* **15:** 316–321.

Wigby S, Sirot LK, Linklater JR, Buehner N, Calboli FC, Bretman A, Wolfner MF, Chapman T. 2009. Seminal fluid protein allocation and male reproductive success. *Curr Biol* **19:** 751–757.

Wlaskiuk G, Nachman MW. 2010. Promiscuity and the rate of molecular evolution at primate immunity genes. *Evolution* **64:** 2204–2220.

Wong A. 2010. Testing the effects of mating system variation on rates of molecular evolution in primates. *Evolution* **64:** 2779–2785.

Wong A. 2011. The molecular evolution of animal reproductive tract proteins: What have we learned from mating-system comparisons? *Int J Evol Biol* **2011:** 1–9.

Wong A, Rundle HD. 2013. Selection on the *Drosophila* seminal fluid protein Acp62F. *Ecol Evol* **3:** 1942–1950.

Wong A, Albright SN, Giebel JD, Ram KR, Ji S, Fiumera AC, Wolfner MF. 2008. A role for Acp29AB, a predicted seminal fluid lectin, in female sperm storage in *Drosophila melanogaster. Genetics* **180:** 921–931.

Xue L, Noll M. 2000. *Drosophila* female sexual behavior induced by sterile males showing copulation complementation. *Proc Natl Acad Sci* **97:** 3272–3275.

Yang CH, Rumpf S, Xiang Y, Gordon MD, Song W, Jan LY, Jan YN. 2009. Control of the postmating behavioral switch in *Drosophila* females by internal sensory neurons. *Neuron* **61:** 519–526.

Yapici N, Kim YJ, Ribeiro C, Dickson BJ. 2008. A receptor that mediates the post-mating switch in *Drosophila* reproductive behaviour. *Nature* **451:** 33–37.

Zawistowski S, Richmond RC. 1986. Inhibition of courtship and mating of *Drosophila* melanogaster by the male-produced lipid, *cis*-vaccenyl acetate. *J Insect Physiol* **32:** 189–192.

Zhong W, McClure CD, Evans CR, Mlynski DT, Immonen E, Ritchie MG, Priest NK. 2013. Immune anticipation of mating in *Drosophila*: *Turandot M* promotes immunity against sexually transmitted fungal infections. *Proc Biol Sci* **280:** 20132018.

Paradox of Mother's Curse and the Maternally Provisioned Offspring Microbiome

Michael J. Wade

Department of Biology, Indiana University, Bloomington, Indiana 47405-7005

Correspondence: mjwade@indiana.edu

Strict maternal transmission creates an "asymmetric sieve" favoring the spread of mutations in organelle genomes that increase female fitness, but diminish male fitness. This phenomenon, called "Mother's Curse," can be viewed as an asymmetrical case of intralocus sexual conflict. The evolutionary logic of Mother's Curse applies to each member of the offspring microbiome, the community of maternally provisioned microbes, believed to number in the hundreds, if not thousands, of species for host vertebrates, including humans. Taken together, these observations pose a compelling evolutionary paradox: How has maternal transmission of an offspring microbiome become a near universal characteristic of the animal kingdom when the genome of each member of that community poses a potential evolutionary threat to the fitness of host males? I review features that limit or reverse Mother's Curse and contribute to resolving this paradox. I suggest that the evolution of vertical symbiont transmission requires conditions that mitigate the evolutionary threat to host males.

The genomes of mitochondria, chloroplasts, and many symbiotic microbes are transmitted maternally by host females to their offspring. Maternal transmission can be transovariole (intracellular, within the egg) or contagious, during gestation, birth, or feeding (Sonneborn 1950; Smith and Dunn 1991; Gillham 1994; O'Neill et al. 1997). Vertically transmitted (VT) symbiont lineages tend to be genetically homogeneous within hosts (Birky et al. 1983, 1989; Funk et al. 2000). Maternal uniparental transmission creates an "asymmetric sieve" wherein mutations advantageous for females, but harmful for males, can spread through a population (Cosmides and Tooby 1981; Frank and Hurst 1996; Zeh and Zeh 2005; Burt and Trivers 2006). Such mutations spread because deleterious male-specific fitness effects do not affect the response to natural selection of the maternally transmitted entities. This adaptive process favoring the transmitting sex is called Mother's Curse (MC) (Gemmell et al. 2004) and it has been referred to as an irreconcilable instance of intralocus conflict: "... exclusively maternal transmission of cytoplasmic genes (e.g., in mitochondria) can result in suboptimal mitochondrial function in males ... a form of [intralocus sexual conflict] that apparently cannot be resolved, because selection on mitochondria in males cannot produce a re-

Cite this article as *Cold Spring Harb Perspect Biol* doi: 101101/cshperspect.a017541

sponse" (Bonduriansky and Chenoweth 2009, p. 285).

Mitochondria are ubiquitous in animals and despite the indisputable evolutionary logic of MC (Frank and Hurst 1996) there are no reported cases of sperm-killing or son-killing mitochondria (Burt and Trivers 2006). Moreover, many species of animals possess an offspring microbiome, a community of microbes transmitted uniparentally from mother to offspring at some point in development, whether prefertilization, postfertilization, or postnatal (Funkhouser and Bordenstein 2013). In some vertebrates, including humans, this community is believed to number in the hundreds of species (Funkhouser and Bordenstein 2013). Prolonged periods of maternal care, as in mammals and birds, as well as kin-structured sociality, afford many opportunities for maternal provisioning of microbes to developing offspring. The social insects, in particular, show obligate mutualisms with a microbiome that confers important nutritional benefits for its host (Baumann 2005; Engel and Moran 2013), the termites being a classic example (Ikeda-Ohtsubo and Brune 2009).

Together, the evolutionary logic of MC and the widespread existence of maternally transmitted hereditary symbioses pose a paradox for evolutionary biology. The maternally provisioned microbiome (MC) consists of tens to hundreds of genomes affording ample opportunity, along with mitochondrial and organelle genomes, for the occurrence of mutations that benefit females while harming host males. Assembling a VT community as a host nutritional or defensive adaptation requires evading MC not once, but from a continuous siege over evolutionary time. This is the Mother's Curse–microbiome (MC–MB) paradox. It conceptually affiliated with the "paradox of mutualism," the persistence of interspecific mutualisms despite the advantages of cheating by one or the other member of the mutualism (Heath and Stinchcombe 2014). Symbiont "cheating" on only half the members of a host species, the males, might offer marginal benefits relative to wholesale cheating on both host sexes. Nevertheless, the MC–MB paradox deserves research attention.

In this review, I discuss inbreeding, kin selection, compensatory evolution, and defensive advantages against more virulent pathogens (or predators and herbivores) as means for resolving the MC–MB paradox. First, I review the simple population genetics of MC. I discuss how host inbreeding and kin selection (Unckless and Herren 2009; Wade and Brandvain 2009), alone or in concert, allow for a response to selection on male fertility and viability fitness effects of maternally transmitted genomes. As a result, inbreeding and kin selection can limit or prevent the spread of mutations in a hereditary symbiosis (Cowles 1915) that are harmful to males. I will show that, for both inbreeding and kin selection, there exist conditions that "favor the spread of maternally transmitted mutations harmful to females"; a situation that is the reverse of MC. However, many outbreeding, asocial species harbor maternally provisioned microbiomes and these solutions cannot be applied to them.

I also consider the evolution of compensatory nuclear mutations that mitigate or eliminate the harm to males of organelles or symbionts, spreading via MC dynamics. However, I find that the relative rate of compensatory evolution is only 1/4 the rate of evolution of male-harming symbionts. Thus, an evolutionary rescue of host males via compensatory host nuclear mutations requires that there be fourfold or more opportunities for compensation offered by a larger host nuclear genome. The larger the number of species in a host microbiome, the more difficult it is to entertain host nuclear compensatory mutations as a resolution of the MC–MB paradox.

Next, I consider the situation in which a deleterious, VT symbiont harms its host but prevents host infection by a more severely deleterious contagiously transmitted pathogen (Lively et al. 2005; see also Clay 1988). This is a case in which absolute harm to a host by a maternally provisioned symbiont becomes a "relative" fitness advantage. This is a scenario that may be common in hosts with speciose microbial communities, especially if each microbial species increases host resistance or outright immunity to infectious, virulent pathogens.

Finally, I discuss models of symbiont domestication and capture via the evolution of vertical transmission from an ancestral state of horizontal transmission (Drown et al. 2013). I show that the evolution of vertical transmission requires conditions that tend to restrict the capacity for male harming by symbionts. Each of these scenarios significantly expands the range of evolutionary possibilities permitted for the coevolution of host–symbiont assemblages, especially those microbial communities that are maternally, uniparentally transmitted across host generations. Unfortunately, current data do not permit discriminating among these various evolutionary responses to MC, so none can be definitively considered a resolution of the MC–MB paradox.

MC

Viability Fitness

Consider a species with alternative, maternally inherited, cytoplasmic alleles, C_1 and C_2, in frequencies P and Q at birth before selection in both males and females. The alleles can be haploid genes of either organelles or symbionts. Let the C_1 allele have sex-specific fitness effects, $s_♂$ and $s_♀$, on the viability of males and females, respectively (Table 1). I assume there is no effect on fitness, positive or negative for either sex of allele C_2. In males, the change from birth to adult in P is $\Delta P_♂ = PQs_♂/W_♂$, in which mean male fitness $W_♂ = 1 + s_♂P$. Similarly, the change in C_1 allele frequency in females is $\Delta P_♀ = PQs_♀/W_♀$, in which $W_♀ = 1 + s_♀P$. Owing to maternal transmission, offspring receive the cytoplasmic allele of their mothers, so change in the frequency of C_1 across generations is simply, $\Delta P_♀ = PQs_♀/W_♀$. Thus, the allele spreads as long as $s_♀ > 0$. (Both sons and daughters receive mother's allele, justifying the assumption of equal frequencies in both sexes at birth.) The parameter $s_♂$ does not appear in the equation for allele frequency change, making cytoplasmic evolution blind to fitness effects on males.

Less than perfect maternal transmission or the paternal leakage of mitochondria has only a small effect on these equations. Let L be the probability that an offspring inherits the paternal mitochondria and $(1 - L)$ be the probability that it inherits the maternal mitochondria. Then, we have $P_♀' = (1 + s_♀)(P_♀ + LX)/W_♀$, in which $W_♀$ becomes $(1 + P_♀s_♀ + LXs_♀)$ and X is $(P_♂Q_♀ - P_♀Q_♂)$. Here, $\Delta P_♀ = \{P_♀Q_♀s_♀/W_♀\} + \{LX(1 + Qs_♀)/W_♀\}$. Note that, when $L = 0$, we return to the equations above. The second term, added to $\Delta P_♀$ by paternal leakage, will tend to be small as long as L is small because X tends to be much less than 1. Thus, small amounts of paternal leakage do not rescue MC.

Fertility Fitness

Again, consider a species with alternative, maternally inherited, cytoplasmic alleles, C_1 and C_2, in frequencies P and Q in both sexes at birth. The alleles can be genes of either organelles or symbionts. Let the C_1 allele have sex-specific fitness effects, $s_♂$ and $s_♀$, on the fertility of males and females, respectively (Table 1) and

Table 1. The fundamental viability and fertility fitness models of MC

Sire	Dam	Frequency	Sons	Daughters	Family
Family				**Viability fitness**	
C_1	C_1	$P_♂P_♀$	$1 + s_♂$	$1 + s_♀$	
C_1	C_2	$P_♂Q_♀$	1	1	
C_2	C_1	$Q_♂P_♀$	$1 + s_♂$	$1 + s_♀$	
C_2	C_2	$Q_♂Q_♀$	1	1	
				Fertility fitness	
C_1	C_1	$P_♂P_♀$	1	1	$1 + s_♂ + s_♀$
C_1	C_2	$P_♂Q_♀$	1	1	$1 + s_♂$
C_2	C_1	$Q_♂P_♀$	1	1	$1 + s_♀$
C_2	C_2	$Q_♂Q_♀$	1	1	1

assume, again, no effect of the C_2 allele. The change in C_1 frequency equals $\Delta P = PQs_{♀}/W$, in which mean family fitness, W, equals $(1 + P[s_{♂} + s_{♀}])$. As for the viability fitness model, the C_1 allele spreads as long as $s_{♀} > 0$ and evolution is blind to this allele's fertility effects on males. If mean fertility of the $C_1 \times C_1$ family was multiplicatively determined, that is, $(1 + s_{♂})(1 + s_{♀})$, then $\Delta P = PQs_{♀}(1 + Ps_{♂})/W$, in which $W = (1 + P[s_{♂} + s_{♀}] + P^2 s_{♂} s_{♀})$. Although $Ps_{♂}$ appears in the numerator of the ΔP expression, the term $[1 + Ps_{♂}]$ is always positive. Moreover, with $0 \leq (1 + Ps_{♂}) \leq 1$, multiplicative fertility effects impede, but do not prohibit, MC relative to the additive effects model.

For both models, MC is the case in which $s_{♂} < 0 < s_{♀}$, wherein mutations beneficial to female fitness ($0 < s_{♀}$) spread despite impairing male fitness ($s_{♂} < 0$).

MC WITH INBREEDING

Viability Fitness

For viability effects of C_1 (Table 2, upper), the change in C_1 allele frequency in females is $\Delta P_{♀} = PQs_{♀}/W_{♀}$, in which $W_{♀} = (1 + s_{♀}P)$ as before. Owing to maternal transmission, offspring receive the cytoplasmic allele of their mothers, so change in the frequency of C_1 across generations is also $\Delta P_{♀} = PQs_{♀}/W_{♀}$. Thus, the allele spreads as long as $s_{♀} > 0$, and, with effects only on viability, inbreeding has no effect at all on the rate of spread of C_1. Neither f nor $s_{♂}$ appears in the equation for allele frequency change, making cytoplasmic evolution blind to viability fitness effects on males and inbreeding.

Fertility Fitness

For this case (Table 2, lower), we find the change in C_1 frequency to be $\Delta P_{♀} = PQ(s_{♀} + fs_{♂})/W_{♀}$, in which $W_{♀} = 1 + (s_{♀} + s_{♂})P$. Whenever there is inbreeding ($f > 0$), the effect of the C_1 allele on sons' fertility "differentially" affects the fertility of their C_1 sisters. (Note that, when there is no inbreeding [$f = 0$], I recover the earlier equation.) Change in C_1 frequency will be positive as long as the quantity $(s_{♀} + fs_{♂})$ is positive. Earlier, Wade and Brandvain (2009) remarked on the relationship between this condition and Hamilton's rule (Hamilton 1967) when the effects on female fertility are positive, but those on males are negative. Recall that, MC is the case in which $s_{♂} < 0 < s_{♀}$. For gene frequency change to be positive in this case, the fitness benefit of C_1 to a female, $s_{♀}$, must exceed the product of the relatedness to her mate, f, times the fitness cost of C_1 to her mate, $s_{♂}$. Thus, inbreeding sets limits on MC. There are clear parallels between the restrictions on cytoplasmic male effects of MC with inbreeding and Hamilton's (1967) discovery that inbreeding limits X-linked or Y-linked selfish, sex ratio distorters.

In fact, there are conditions in which inbreeding favors maternally inherited cytoplasmic alleles with "deleterious" effects on females, but positive effects on males. Change in C_1 frequency is positive when $s_{♀} < 0 < s_{♂}$, as long as $fs_{♂}$ exceeds $s_{♀}$, which is the reverse of MC.

Table 2. Inbreeding and the viability and fertility fitness models of MC

Sire	Dam	Frequency	Sons	Daughters	Family
Family				**Viability fitness**	
C_1	C_1	$P_{♂}P_{♀} + Q_{♂}P_{♀}f$	$1 + s_{♂}$	$1 + s_{♀}$	
C_1	C_2	$P_{♂}Q_{♀}(1 - f)$	1	1	
C_2	C_1	$Q_{♂}P_{♀}(1 - f)$	$1 + s_{♂}$	$1 + s_{♀}$	
C_2	C_2	$Q_{♂}Q_{♀} + P_{♂}Q_{♀}f$	1	1	
				Fertility fitness	
C_1	C_1	$P_{♂}P_{♀} + Q_{♂}P_{♀}f$	1	1	$1 + s_{♂} + s_{♀}$
C_1	C_2	$P_{♂}Q_{♀}(1 - f)$	1	1	$1 + s_{♂}$
C_2	C_1	$Q_{♂}P_{♀}(1 - f)$	1	1	$1 + s_{♀}$
C_2	C_2	$Q_{♂}Q_{♀} + P_{♂}Q_{♀}f$	1	1	1

NUCLEAR COMPENSATORY MUTATIONS

Consider alternative host nuclear alleles, A_1 and A_2, such that the A_1 allele acts additively to restore the fitness of C_1 males from $(1 + s_♂)$ to 1, in which $s_♂ < 0$. More specifically, the $A_1A_1C_1$ homozygotes have fitness equal to 1, $A_1A_2C_1$ heterozygotes have fitness of $(1 + [s_♂/2])$, and $A_2A_2C_1$ homozygotes have fitness equal to $(1 + s_♂)$. The rate of evolution of the compensatory A_1 allele is $\Delta P_{A1} = P_{C1}P_{A1}\,P_{A2}\,s_♂/4W_♂$, in which $W_♂ = (1 + P_{A1}s_♂)$. The rate of compensatory evolution relative to the evolution of male-harming organelles or symbionts depends on the relative values of $s_♀$ and $s_♂$. If the benefit to females from C_1 equaled the harm to males, then the rate of compensatory evolution is only $(1/4P_{C1})$ that of the organelle or symbiont. The rate of evolution of the A_1 allele is diminished for three reasons. First, because its fitness advantage is only in the male sex, copies of A_1 in females are not screened by natural selection. All else being equal, this reduces the rate of evolution by a factor of 1/2 (Wade 1998; Demuth and Wade 2005; van Demuth and Wade 2007; Cruickshank and Wade 2008). Second, only those copies of the A_1 allele in males that occur on the C_1 background enjoy a fitness benefit. This reduces the rate of evolution by a factor of P_{C1}. And, third, the fitness difference between $A_1A_1C_1$ and $A_2A_2C_1$ homozygotes is $s_♂$, a scale set by the haploid-harming organelle or symbiont. Although the scale of the fitness difference between nuclear homozygotes can be either s or $2s$ without affecting ΔP, the scale is fixed by fitness effects caused by the other genome. This reduces the rate of evolution by another factor of 1/2.

Although nothing is known about the rate of mutation of organelles and symbionts to male-harming, female-beneficial states, it is unlikely that differential mutation rates, in which $\mu_{nuclear} > \mu_{symbiont}$, can offset the large reduction in rate $(1/4P_{C1})$ of host nuclear compensatory evolution. In most animals, the mutation-rate differential runs in the opposite direction (i.e., $\mu_{mitochondria} > \mu_{nuclear}$), and it is likely that the same is true for most symbionts. Moreover, if initially $\mu_{nuclear} = K(\mu_{symbiont})$, in which K is the number of symbionts in the microbiome, once the maternally provisioned microbiome reached $(K/4)$-species, the initial advantage of nuclear compensatory evolution over male-harming organelle and symbiont evolution would be eroded.

MC AND KIN SELECTION

Consider a species living in matrilineal family groups in which sons make essential contributions to the viability of their female relatives. Let C_1 and C_2 be alternative, maternally inherited, cytoplasmic alleles, in frequencies P and Q, and let f be the inbreeding parameter. The C_1 allele has two viability fitness effects, $s_♂$ and $s_♀$, one on the viability of C_1 males and the other on that of C_1 females, respectively. Additionally, let $s_{F♂}$ be the effect of viable C_1 males on the fitness of their sisters; thus, $s_{F♂}$ is a kin selection or family effect. We assume that only surviving males influence the fitness of their sisters, so that the effect on family fitness is modulated by the survival of C_1 males and becomes $\{1 + s_{F♂}(1 + s_♂)\}$ (see Table 3). We further assume that family and individual female viability fitnesses are multiplicative.

With these assumptions, the general equation describing the one-generation change in allele frequency caused by selection is

$$\Delta P_♀ = P_♀Q_♀(s_♀ + s_{KS_♂}[1 + s_♂] \times [1 + s_♀])/W_♀, \quad (1)$$

in which mean female fitness $W_♀ = 1 + P_♀[s_♀ + s_{KS♂}\{1 + s_♂\}\{1 + s_♀\}]$. This is also the change in frequency of C_1 between generations because the frequency of the C_1 allele in daughters of the next generation equals the frequency in the surviving females.

Here, inbreeding has no effect on the rate of evolution because the frequency of helpful sons born to C_1 mothers is not affected by the mating system. When $s_{KS♂}$ is 0, $\Delta P_♀$ reduces to the classic formula $P_♀Q_♀s_♀/W_♀$ (Wright 1969, p. 163, Equation [6.1]; or Hedrick 2000, p. 105). The direct effects of C_1 on male viability, $s_♂$, have no effect on the evolution of C_1 unless males are helping $(s_{KS♂} > 0)$ or harming $(s_{KS♂} < 0)$

Table 3. Model I: Kin selection in which sons assist sisters

Sire	Dam	Frequency	Family fitness	Sons	Daughters
Family					
C_1	C_1	$P_♂P_♀ + Q_♂P_♀ f$	$1 + s_{F♂}(1 + s_♂)$	$1 + s_♂$	$1 + s_♀$
C_1	C_2	$P_♂Q_♀(1 - f)$	1	1	1
C_2	C_1	$Q_♂P_♀(1 - f)$	$1 + s_{F♂}(1 + s_♂)$	$1 + s_♂$	$1 + s_♀$
C_2	C_2	$Q_♂Q_♀ + P_♂Q_♀ f$	1	1	1

their sisters. In the absence of the indirect effect on sisters ($s_{KS♂} = 0$), maternal transmission precludes an evolutionary response to male-specific viability selection (Frank and Hurst 1996; Burt and Trivers 2006).

Some taxa with mutualistic microbiomes have both inbreeding and male-helping behavior, for example, termites (Reilly 1987), but there are also taxa with microbiomes that are neither social nor inbreeding (Funkhouser and Bordenstein 2013). Resolving the MC–MB paradox in these taxa may be more difficult. I discuss two additional possibilities in the next two sections.

COMPETITIVE COEXISTENCE OF SYMBIONTS AND PATHOGENS

Members of the maternally provisioned microbiome are believed to be derived from free-living forms by a process sometimes referred to as "symbiont capture" or "symbiont domestication" (Maynard Smith and Szathmáry 1998). Importantly, vertical transmission favors decreased symbiont virulence because it aligns the evolutionary interests of host and symbiont (Smith 2007). For this reason, VT symbionts must enhance host fitness or they will be lost from the host population in the same manner as a deleterious mitochondrial gene (Fine 1975; Ewald 1987; Lipsitch et al. 1995, 1996). The nutritional benefits of the maternally provisioned microbiome have long been considered (reviewed in Funkhouser and Bordenstein 2013). Defensive benefits to the offspring provided by maternal symbiont transmission are also well known (Clay 1988; Clay and Schardl 2002). There is one such defensive benefit that increases host relative fitness that may be quite general, namely, a host fitness benefit that derives from possessing a VT symbiont, which prevents host infection by another free-living and more virulent member of the same or a similar taxon (Lively et al. 2005).

Free-living, virulent parasites often adapt to the most frequently encountered host genotypes, and they are believed to be one of the evolutionary forces driving the maintenance of sexual reproduction (Lively 2010). Although some such parasites encounter one host sex more than the other and, subsequently, specialize on that host sex (Duneau et al. 2010), many free-living, virulent parasites are host generalists and not host sex-specific in their incidence or fitness effects. The presence of infectious free-living virulent parasites can allow a VT symbiont, which prevents host infection by the free-living virulent form, to increase in frequency. Moreover, the rate of its spread increases as the fidelity of maternal transmission increases (Lively 2010). Importantly, in the models explored by Lively (2010), not only does the virulent form persist in the host population (albeit at very low frequency), but its virulence is also increased.

Once a symbiont species becomes a member of the maternally provisioned microbiome, the origin in its genome of male-specific harmful mutations benefiting host females becomes a possibility. However, the spread of such a male-harming mutation allows the coexisting, virulent species to specialize on host females (Duneau et al. 2010), which become the more abundant host sex as the male-harming mutation spreads. This results in a diminishing fitness advantage to host females of the male-harming symbiont. Once the fitness benefit to host females is lost, the male-harming mutation stops spreading. The enhanced virulence of the free-living pathogen (Lively 2010) may accelerate this process.

Overall, the harmful effects of the endosymbiont on host males are counterbalanced by the harmful effects of the free-living pathogen on host females. Thus, restriction of male-harming mutations, originating in the genomes of members of the maternally provisioned microbiome, may depend on the female-biased, sex-specific counteradaptations of the free-living virulent species, excluded from the host by its prior occupancy by the VT defensive symbiont.

This scenario accounts for both the high species diversity of the maternally provisioned microbiome and the restriction of the male-harming mutations arising from it. Like models for the evolution of sex, however, it depends on the virulence of a free-living pathogen relative to that of the VT symbiont. It requires a careful balance of sex-specific, deleterious effects on host fitness. The plausibility of establishing such an evolutionary balance for each member of the maternally provisioned microbiome remains to be explored in both theory and experiment.

EVOLUTION OF TRANSMISSION MODE IN OBLIGATE SYMBIONTS

Special conditions appear to be necessary to evolve from an ancestral state of horizontal transmission to a derived state of vertical transmission (Drown et al. 2013). These conditions are relevant to understanding the evolutionary origins of the maternally transmitted microbiome, and they may contribute to resolving the MC–MB paradox by reducing the capacity of a VT symbiont to generate male-harming, but female-beneficial mutations.

One necessary condition is transgenomic epistasis for fitness between host and symbiont genes, like that characteristic of matching allele models in host-pathogen models (Agrawal and Lively 2002) and additive-by-additive classical epistasis for fitness in population genetic models (Drown et al. 2013). Unless there are gene combinations that enhance the fitness of both host and symbiont, vertical transmission does not evolve from an ancestral state of horizontal transmission. It is selection favoring the mutually advantageous combinations that generates the indirect selection on mode of transmission.

A second necessary condition is repeated mutation in the genomes of both host and symbiont. Without repeated mutation, advantageous transgenomic combinations are quickly fixed and indirect selection favoring vertical transmission ceases. Repeated mutation is necessary for maintaining an influx of advantageous, transgenomic gene combinations in much the same way that rapid adaptation of virulent parasite genotypes to common host genotypes is required for the maintenance of sex. With a continuous influx of such mutations, vertical transmission is a stable evolutionary endpoint for a matching alleles model (Drown et al. 2013).

With regard to host-harming mutations, this mutational phase acts as an evolutionary sieve, removing deleterious transgenomic combinations and preserving favorable ones. Vertical transmission evolves only to the extent that mutational combinations arise and successfully pass through this sieve. This period in the evolution of maternally transmitted symbiont communities can be fairly long, so that many mutational combinations, in both host and symbiont, are screened. This sieve may represent such a significant barrier to admission of a species to the VT symbiont community of its host that very few species with a capacity for host male-harming mutations are admitted. Conversely, many species with a capacity for host-harming mutations are excluded.

This theory proposes a mutational sieve that occurs earlier in the process of the evolution of vertical transmission that leaves little, if any, genetic material for the "asymmetric sieve" of MC. The study of symbionts with both modes of transmission (Lipsitch et al. 1996) will be necessary to determine whether or not membership in the maternally provisioned biome is exclusive to only a few symbiont species. It will also help determine whether or not the reduction in genome size, often observed in such symbionts (Sloan and Moran 2012), restricts their capacity for generating host male-harming mutations.

CONCLUSIONS

The evolutionary logic of MC and the widespread existence of maternally provisioned symbioses pose a paradox for evolutionary biology. The high species diversity of the microbiome affords ample opportunity for the occurrence of mutations that benefit host females while harming host males. Host nuclear compensatory mutations that mitigate the male-specific fitness effects of VT symbionts may exist and spread, but their rate of evolution is less than 1/4 that of the male-harming symbiont mutations they are meant to counter.

Recent theory (Unckless and Herren 2009; Wade and Brandvain 2009) has shown that, whenever host males affect the fitness of female relatives or whenever there is inbreeding, there are evolutionary restrictions placed on MC. However, the existence of these effects alone cannot resolve the MC–MB paradox because many host species with a maternally provisioned microbiome are neither social nor inbreeding.

In this review, I have proposed two possible evolutionary scenarios that may contribute to the resolution of the MC–MB paradox: (1) the competitive coexistence of symbionts and pathogens, and (2) the mutational exclusivity of membership in the maternally provisioned microbiome. Although each is plausible, both proposals are speculative and require additional theoretical and empirical investigation. Each makes different empirical predictions. For example, the competitive coexistence hypothesis predicts that some members of the maternally provisioned community exclude infection by more virulent free-living pathogens. This could be tested by comparing the susceptibility of hosts with and without a specific VT symbiont. This is not sufficient, however, to establish the competitive coexistence hypothesis as true. A VT symbiont and one or more free-living pathogens must show sexually antagonistic effects on their host, with the VT symbiont differentially harming host males and the free-living pathogens differentially harming host females. The eradication of a free-living pathogen may result in the loss of a less virulent symbiont and, at least in some instances, should result in the spread of male-harming mutations caused by its VT counterpart.

Experimentally establishing the mutational exclusivity of membership in the microbiome may well be more difficult. But, the study of symbionts with both modes of transmission (Lipsitch et al. 1996) may offer the best opportunity. Those with higher levels of vertical transmission should be farther along in the evolutionary process than those with lower levels. The former should be capable of fewer host-harming mutations than the latter and such a difference might be revealed with replicated, mutation-accumulation experiments. On the other hand, if membership in the maternally provisioned biome is not exclusive, many member species will be of recent origin. Comparison of the communities of taxonomically related hosts may be useful in addressing this possibility.

ACKNOWLEDGMENTS

I thank the members of the Wade laboratory, Devin Drown, Amy Dapper, Doug Drury, and Jeff Adrion, and my colleagues Yaniv Brandvain, Curt Lively, and David E. McCauley for their thoughtful feedback on these ideas. I gratefully acknowledge the support of National Institutes of Health grant R01GM084238.

REFERENCES

Agrawal A, Lively CM. 2002. Infection genetics: Gene-for-gene versus matching-alleles models and all points in between. *Ecol Evol Res* **4:** 79–90.

Baumann P. 2005. Biology bacteriocyte-associated endosymbionts of plant sap-sucking insects. *Annu Rev Microbiol* **59:** 155–189.

Birky CW, Maruyama T, Fuerst. 1983. An approach to population and evolutionary genetic theory for genes in mitochondria and chloroplasts, and some results. *Genetics* **103:** 513–527.

Birky CW, Fuerst P, Maruyama T. 1989. Organelle diversity under migration, mutation, and drift: Equilibrium expectations, approach to equilibrium, effects of heteroplasmic cells, and comparison to nuclear genes. *Genetics* **121:** 613–627.

Bonduriansky R, Chenweth SF. 2009. Intralocus sexual conflict. *Trends Ecol Evol* **24:** 280–288.

Burt A, Trivers R. 2006. *Genes in conflict.* The Belknap Press of Harvard University Press, Cambridge, MA.

Clay K. 1988. Fungal endophytes of grasses: A defensive mutualism between plants and fungi. *Ecology* **69:** 10–16.

Clay K, Schardl CL. 2002. Evolutionary origins and ecological consequences of endophyte symbiosis with grasses. *Am Nat* **160:** S99–S127.

Cosmides LM, Tooby J. 1981. Cytoplasmic inheritance and intragenomic conflict. *J Theor Biol* **89:** 83–129.

Cowles HC. 1915. Hereditary symbiosis. *Bot Gaz* **59:** 61–63.

Cruickshank T, Wade MJ. 2008. Microevolutionary support for a developmental hourglass: Gene expression patterns shape sequence variation and divergence in *Drosophila*. *Evol Dev* **10:** 583–590.

Demuth J, Wade MJ. 2007. Maternal expression increases the rate of *bicoid* evolution by relaxing selective constraint. *Genetica* **129:** 37–43.

Dieter D. 2013. The epidemiology and evolution of symbionts with mixed-mode transmission. *Annu Rev Ecol Evol System* **44:** 623–643.

Drown DM, Zee PC, Brandvain Y, Wade MJ. 2013. Evolution of transmission mode in obligate symbionts. *Ecol Evol Res* **15:** 43–59.

Duneau D, Luijckx P, Ruder LF, Ebert D. 2010. Sex-specific effects of a parasite evolving in a female-biased host population. *BMC Biol* **10:** 104.

Engel P, Moran NA. 2013. The gut microbiota of insects—Diversity in structure and function. *Microbiol Rev* **37:** 699–735.

Ewald PW. 1987. Transmission modes and the evolution of the parasitism–mutualism continuum. *Ann NY Acad Sci* **503:** 295–306.

Fine PEM. 1975. Vectors and vertical transmission: An epidemiological perspective. *Ann NY Acad Sci* **266:** 173–194.

Frank SA, Hurst LD. 1996. Mitochondria and male disease. *Nature* **383:** 224.

Funk DJ, Helbling L, Wernegreen JJ, Moran NA. 2000. Intraspecific phylogenetic congruence among multiple symbiont genomes. *Proc R Soc Lond B* **267:** 2517–2521.

Funkhouser LJ, Bordenstein SR. 2013. Mom knows best: The universality of maternal microbiome transmission. *PLoS Biol* **11:** e1001631.

Gemmell NJ, Metcalf VJ, Allendorf FW. 2004. Mother's curse: The effect of mtDNA on individual fitness and population viability. *Trends Ecol Evol* **19:** 238–244.

Gillham MW. 1994. Organelle genes and genomes. Oxford University Press, Oxford.

Hamilton WD. 1967. Extraordinary sex ratios. A sex-ratio theory for sex linkage and inbreeding has new implications in cytogenetics and entomology. *Science* **156:** 477–488.

Heath KD, Stinchcombe JR. 2014. Explaining mutualism variation: A new evolutionary paradox? *Evolution* **68:** 309–317.

Hedrick PW. 2000. *Genetics of populations*. Jones and Bartlett, Sudbury, MA.

Ikeda-Ohtsubo W, Brune A. 2009. Cospeciation of termite gut flagellates and their bacterial endosymbionts: *Trichonympha* species and "*Candidatus* Endomicrobium trichonymphae." *Mol Ecol* **18:** 332–342.

Lipsitch M, Nowak MA, Ebert D, May R. 1995. The population dynamics of vertically and horizontal transmitted parasites. *Proc R Soc Lond B* **260:** 321–327.

Lipsitch M, Siller S, Nowak MA. 1996. The evolution of virulence in pathogens with vertical and horizontal transmission. *Evolution* **50:** 1729–1741.

Lively CM. 2010. A review of Red Queen models for the persistence of obligate sexual reproduction. *Heredity* **101:** S13–S20.

Lively CM, Clay K, Wade MJ, Fuqua C. 2005. Competitive coexistence of vertically and horizontally transmitted parasites. *Ecol Evol Res* **8:** 1183–1190.

Maynard Smith J, Szathmáry E. 1998. The major transitions in evolution. Oxford University Press, Oxford.

O'Neill SL, Hoffmann AA, Werren JH. 1997. Influential passengers: Inherited microorganisms and arthropod reproduction. Oxford University Press, New York.

Reilly LM. 1987. Measurements of inbreeding and average relatedness in a termite population. *Am Nat* **130:** 339–349.

Sloan JE, Moran NA. 2012. Genome reduction and co-evolution between the primary and secondary bacterial symbionts of psyllids. *Mol Biol Evol* **29:** 3781–3792.

Smith J. 2007. A gene's-eye view of symbiont transmission. *Am Nat* **170:** 542–550.

Smith JE, Dunn AM. 1991. Transovariol transmission. *Parasitol Today* **7:** 146–148.

Sonneborn TM. 1950. Partner of the genes. *Sci Am* **183:** 30–39.

Unckless RL, Herren JK. 2009. Population genetics of sexually antagonistic mitochondrial mutants under inbreeding. *J Theor Biol* **260:** 132–136.

van Dyken JP, Wade MJ. 2010. Quantifying the evolutionary consequences of conditional gene expression in time and space. *Genetics* **184:** 557–570.

Wade MJ. 1998. The evolutionary genetics of maternal effects. In *Maternal effects* (Mousseau J, Fox C, eds.), pp 5–21. Oxford University Press, Oxford.

Wade MJ, Brandvain Y. 2009. Reversing mother's curse: Selection on male mitochondrial fitness effects. *Evolution* **63:** 1084–1089.

Wright S. 1969. Evolution and genetics of populations. Vol. II. University of Chicago Press, Chicago.

Zeh JA, Zeh DW. 2005. Maternal inheritance, sexual conflict and the maladapted male. *Trends Genet* **21:** 281–286.

The Evolution of Sexually Antagonistic Phenotypes

Jennifer C. Perry[1,2] and Locke Rowe[3]

[1]Jesus College, University of Oxford, Oxford OX1 3DW, United Kingdom

[2]Edward Grey Institute, Department of Zoology, University of Oxford, Oxford OX1 3PS, United Kingdom

[3]Department of Ecology and Evolutionary Biology, University of Toronto, Toronto, Ontario M5S 3B2, Canada

Correspondence: jennifer.perry@zoo.ox.ac.uk

Sexual conflict occurs whenever there is sexually antagonistic selection on shared traits. When shared traits result from interactions (e.g., mating rate) and have a different genetic basis in each sex (i.e., interlocus conflict), then sex-specific traits that shift the value of these interaction traits toward the sex-specific optimum will be favored. Male traits can be favored that increase the fitness of their male bearers, but decrease the fitness of interacting females. Likewise, female traits that reduce the costs of interacting with harmful males may simultaneously impose costs on males. If the evolution of these antagonistic traits changes the nature of selection acting on the opposite sex, interesting coevolutionary dynamics will result. Here we examine three current issues in the study of sexually antagonistic interactions: the female side of sexual conflict, the ecological context of sexual conflict, and the strength of evidence for sexually antagonistic coevolution.

Conflict requires some interaction or common activity between males and females... which generates the con straint that the ideal optima for each sex cannot be achieved simultaneously.

—Geoff A. Parker 2006, p. 235

Many traits are shared between the sexes, either because they result from an interaction between the sexes or because they share an underlying genetic basis. Examples of the first set may include mating rate, copulation duration, reproductive schedule, fecundity, fertility, and remating rate (Fig. 1A). We can refer to this set of shared traits as interaction traits. They are likely to have different genetic bases in each sex, and therefore can be examples of interlocus conflict. Examples of the second include most traits that are expressed in both sexes with a common genetic architecture between the sexes, such as human height or pelvis shape. These traits can be examples of intralocus conflict. The evolutionary interests of the two sexes must often diverge because of their different life histories, beginning with anisogamy (Parker 1979; Holland and Rice 1998; Arnqvist and Rowe 2005; Schärer et al. 2012). Therefore, sexual conflict over the value of these shared traits is expected. More specifically, sex specific optima for shared traits yields sexually antagonistic selection on their values (Fig. 1A) (Rowe and Day 2006). Here we will focus on interlocus

Cite this article as *Cold Spring Harb Perspect Biol* doi: 101101/cshperspect.a017558

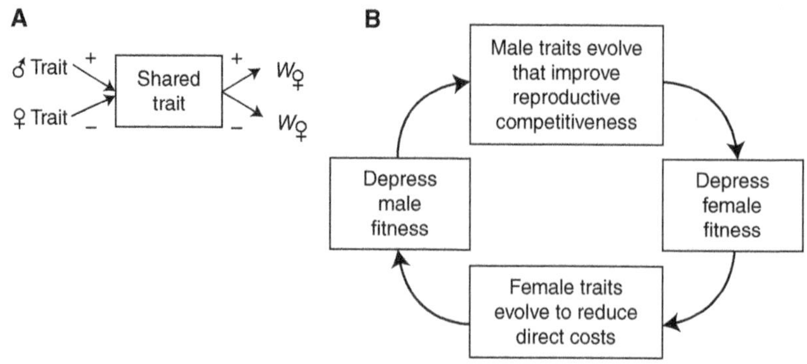

Figure 1. Sexual conflict and sexually antagonistic coevolution. (*A*) Sexual conflict arises over traits shared by the sexes, in this case traits that are the outcome of sexual interactions that impact fitness in both sexes, and is defined by opposite signs of selection on the trait in each sex. Conflict creates selection on sex-specific traits that shift the value of the interaction traits toward sex-specific optima. This conflict is interlocus because it is expected that the male and female traits involved will have a different genetic basis in each sex. (Adapted from Rowe and Day 2006.) (*B*) Sexually antagonistic coevolution arises as evolutionary change in each sex alters selection on the other, resulting in further evolutionary change. (Adapted from Arnqvist and Rowe 2005.)

sexual conflict over shared traits resulting from interactions between the sexes.

Interlocus sexual conflict is appealing to evolutionary biologists because it has the potential to be a powerful force driving phenotypic divergence of the sexes. Yet, at the same time, the genetic ties that bind the sexes may act to impede the evolution of each sex toward its optimum. These selective forces and genetic constraints have many additional interesting implications for population demography, reproductive isolation and speciation, and evolutionary diversification, including morphology, physiology, and behavior. The field has moved from an initial period when ideas were first developed mainly by Parker (1979) through a burst of interest sparked by Holland and Rice's (1998) description of chase-away selection, with further development and synthesis in Arnqvist and Rowe's (2005) monograph. By 2006, sexual conflict was declared a new paradigm (Tregenza et al. 2006).

This recent burst of research has not been without controversy, including early disagreements that often centered on defining the scope of the field and how to test its predictions (Chapman et al. 2003a; Cordero and Eberhard 2003; Eberhard and Cordero 2003; Pizzari and Snook 2003; Arnqvist 2004; Eberhard 2004a; Schärer et al. 2012; Ah-King 2013; Kokko et al. 2013), as well as recent critiques evaluating the potential for gender bias in sexual conflict theory and focus (Green and Madjidian 2011; Madjidian and Green 2012; Perry and Rowe 2012a; Dougherty et al. 2013). In recent years, a new era has begun as the field has matured, with studies considering implications for topics like aging (Promislow 2003; Bonduriansky et al. 2008), population structure (Eldakar et al. 2009b; Chang and Sih 2013), kin selection (Rankin 2011; Carazo et al. 2014), human health (Badcock and Crespi 2008; Morrow and Connallon 2013), and extinction (Kokko and Brooks 2003).

A key pathway from sexual conflict over some sexual interaction to diversity is the evolution of sexually antagonistic (SA) traits (Fig. 1A). When an interaction trait—such as mating rate or female remating—is under sexually antagonistic selection, then selection will favor the altered expression of existing traits or the evolution of entirely novel traits that effect a change in mating toward the bearer's optimal mating rate. For example, in a conflict over mating rate (the interaction), male grasping traits that increase mating, and female antigrasping traits that decrease mating (the SA traits) will be favored and may coevolve (Parker 1979; Arnqvist and Rowe 2005). Much progress has been made

 Cite this article as *Cold Spring Harb Perspect Biol* doi: 10.1101/cshperspect.a017558

in the study of these traits and the forces shaping their evolution. However, there are still major gaps in understanding some of the most basic aspects of sexual conflict and its influence on SA traits. In fact, many of the questions put forward in initial reviews remain pressing today.

Here, we review current understanding of male and female SA adaptations or armaments. We begin with a brief review of theory for the evolution of sexual armaments and SA coevolution, followed by a survey of male antagonistic adaptations (for a more detailed review, see Arnqvist and Rowe 2005). We then focus on recent advances in three important areas that we believe warrant further study: (a) the female side of sexual conflict; (b) the ecological signature in sexual conflict; and (c) sexually antagonistic coevolution. We discuss in detail four well-known case studies to highlight the diversity of recent advances and remaining questions.

SEXUALLY ANTAGONISTIC COEVOLUTION IN THEORY

At the heart of sexually antagonistic coevolution (SAC) is the potential for evolutionary change in one sex to modify selection acting on the opposite sex, such that the sexes continuously drive evolutionary change in each other (Fig. 1B). The theory for SAC was substantially worked out by Parker three and a half decades ago (Parker 1979). Subsequent verbal and mathematical models of this process focused primarily on one outcome: escalating arms races characterized by the investment in and exaggeration of sexual armaments in both sexes (Holland and Rice 1998; Gavrilets 2000; Gavrilets et al. 2001; Gavrilets and Waxman 2002; Gavrilets and Hayashi 2006). Further theoretical effort has made it clear that, although escalating arms races are one possible outcome, they are expected only in restricted circumstances or for transitory evolutionary periods (Härdling and Smith 2005; Rowe et al. 2005; Hayashi et al. 2007). Both earlier and later models show that a range of alternative forms of coevolutionary outcomes are possible, including stable equilibria, coevolutionary cycles, and de-escalation of armaments in both sexes.

These theoretical analyses show that the dynamics and outcome of SAC depend on (1) the form of female resistance, (2) the economics of mating and how selection acts on females in particular, and (3) population-specific factors, including local ecology and initial trait values.

The Form of Female Resistance

Models of SAC are often set up similarly to traditional models of sexual selection, in which there is a male trait and a female response (or preference), which interact to determine mating, with potential impacts for fitness in both sexes. In SAC models, there is an antagonistic sexually selected male trait (i.e., whose expression decreases female fitness) and a female antagonistic trait that determines female responses to that male trait. These interact to determine the value of the shared interaction trait (e.g., mating rate) over which sexual conflict occurs, and determine the extent of SA trait evolution (e.g., Gavrilets et al. 2001; Kokko et al. 2003; Rowe and Day 2006). Some models permit females to evolve resistance to harmful male traits only by escalating their defenses against males through an increased threshold of response to antagonistic male traits. In this scenario, escalating arms races are commonly expected, because males face strong selection favoring the exaggeration of SA traits that overcome female thresholds. However, females might alternatively evolve resistance through decreased sensitivity to harmful male traits, equivalent to a decreased slope for the relationship between female responses and male traits. When female sensitivity is free to evolve, models predict that arms races are much less likely, with other outcomes including the de-escalation of armaments in both sexes, and the evolution of female insensitivity to male traits (Rowe et al. 2005).

These models tell us that an important part of understanding SAC is characterizing the form of the female response function, as determined by female antagonistic traits, and variation in this function (Rowe et al. 2005). Specifically, we need to address the following questions. What is the genetic variation underlying female response functions to male persistence traits?

Are female responses free to evolve in any direction, or constrained—by a lack of genetic variation or by natural selection acting on female responses—to evolve in certain directions? Do different categories of female resistance traits differ in their evolvability?

The Economics of Mating

The economy of mating—the costs and benefits of mating for both sexes—determines the extent of conflict and therefore the dynamics of SAC (Arnqvist and Rowe 2005; Chapman 2006; Fricke et al. 2009). Of course, where increased mating is beneficial for both sexes, there will be no conflict over mating rate. But where there is conflict, recent models point to details about the economics of mating that should have a strong influence, and suggest that the economics of mating for females is particularly significant. For example, the relationship between the number of mates and mating costs is potentially important, especially whether every mating beyond the first mating is costly for females or there is an intermediate optimum number of matings (Arnqvist and Nilsson 2000; Kimura and Ihara 2009). If the latter, then the expected coevolutionary outcome varies with the initial state of the system, setting the stage for idiosyncratic outcomes. Furthermore, whether mating costs are fixed or increase with the degree of exaggeration of male sexually antagonistic traits matters, with the latter much less likely to lead to extreme exaggeration in harmful male traits (Parker 1979; see also Kazancioğlu and Alonzo 2012). One can imagine species in which both forms of mating costs occur: fixed per-mating costs are likely in water striders, in which mating costs are imposed externally by predators (Rowe 1994), whereas in seed beetles with spiny male genitalia, increasing spininess results in higher mating costs (Hotzy and Arnqvist 2009). In another example, Härdling and Smith (2005) showed that arms races are much less likely when costs depend on the interaction between the sexes (e.g., when costs arise from behavioral interactions) rather than when costs are fixed independent of interactions (e.g., when costs arise from investing in armament production).

These theoretical predictions prescribe a research program for sexual conflict that is based on measuring the economics of sexual interactions, particularly in females, a point we develop below.

Predictable or Stochastic?

Models have examined ways in which ecological conditions can determine the outcome of conflict. For example, when higher population density leads to more frequent sexual interactions, the intensity of sexual arms races is expected to increase (Kimura and Ihara 2009). Furthermore, ecological and population factors can promote assortative mating among males and females possessing different SA traits, and this kind of nonrandom mating can maintain intrapopulation variation in SA traits (Härdling and Bergsten 2006). However, this ecologically based predictability stands in contrast to other theoretical expectations that the outcome of SAC should vary with initial conditions and that stochastic effects, including genetic drift, should play a major role (Gavrilets 2000; Hayashi et al. 2007; Kimura and Ihara 2009). These expectations begin with classic models predicting that populations will move by genetic drift along a line of equilibrium on which SA traits are balanced between the sexes (Lande 1981; Gavrilets 2000). The potential importance of ecological variation means that it is vital to investigate the context-dependence of the economics of mating (Fricke et al. 2009; and below).

MALE SEXUAL ARMAMENTS

Male SA adaptations—"persistence" traits—are defined by their harmful effects on females that interact with males bearing the traits (Figs. 1 and 2). We therefore need measurements of costs in females to identify male harmful traits. In some cases, costs to females are so obvious that measuring them would be superfluous. But even when costs seem obvious—surely the hypodermic aedeagus of bed bugs must impose costs on females?—the evolution of female SA traits like the paragenitalia that mitigate costs can mean that male traits that were once harm-

Figure 2. Examples of putatively or well-supported SA traits in males and females. Clockwise from *top left*: female paragenitalia in the plant bug *Coridromius bulbopella*, close-up and whole-organism views; the female abdominal spine of the water strider *Gerris incognitus*; the furrowed antigrasping dorsal surface of a female diving beetle (*Dytiscus latissimus*); the claw of the male intromittent organ in guppies (*Poecilia reticulate*); the spines and jaw-like structure of the male aedeagus in the seed beetle *Callosobruchus subinnotatus*; the antennae of male water striders (*Rheumatobates rileyi*), colored to illustrate several grasping adaptations. (Photographs provided by N. Tatarnic, G. Arnqvist, J. Bergsten, L. Kwan, and A. Khila.)

ful may no longer be so, or the harm may be very much reduced.

These costs to females could arise either from their impact on interaction traits over which there is conflict (by shifting these traits away from the female optimum), or from direct harm to the females themselves. Harm to females can arise as a byproduct of selection on male traits in another context, such as male-male competition. As examples, longer genital spines in male seed beetles cause both increased male success in sperm competition and increased harm to females (Hotzy and Arnqvist 2009; Hotzy et al. 2012; and one allele of a male reproductive gene in *Drosophila melanogaster* is associated with both decreased female remating rate and decreased female longevity (Fiumera et al. 2006). Alternatively, harm to females can in theory be directly favored by selection (Johnstone and Keller 2000), although this prediction has yet to receive empirical support (e.g., Morrow et al. 2003).

Male antagonistic traits and their impacts are much better understood than those of females. This might be owing to the apparently greater frequency in males of morphological SA traits, which can be observed and measured with relative ease compared to behavioral and physiological traits. Classes of persistence traits include grasping appendages, ranging from the adhesive tarsae of diving beetles (Bergsten et al. 2001) to the enlarged and spined and hooked antennae of water striders (Rowe et al. 2006); spiny stabbing genitalia in seed beetles (Crudgington and Siva-Jothy 2000), garter snakes (Friesen et al. 2014), guppies (Fig. 2) (Kwan et al. 2013) and traumatically inseminating bugs (Stutt and Siva-Jothy 2001; Tatarnic et al. 2006); chemical traits, such as seminal proteins that influence female physiology and behavior (e.g., by boosting fecundity or delaying remating beyond the female optimum; Perry et al. 2013), and behavioral traits, such as the harassment of females by males observed in many species. Eberhard (2004a,b, 2010) has suggested that physiological and chemical male SA traits might result in arms races more commonly than morphological or behavioral traits, in part because physiological and chemical traits might be more complex and difficult for females to avoid. At this point, there are not enough data available to evaluate this hypothesis.

In many cases, costs for females are plausible but not straightforward. A prime example is the short-term boost in egg deposition or production after mating, caused by male seminal proteins in many insects (Vahed 2007; Chapman 2008; Perry et al. 2013). Such gonadotropic effects are often interpreted as harmful to females (e.g., Parker and Simmons 1989; Simmons and Parker 1989). But without careful measurement, it is not clear what fecundity stimulation means for female fitness. An increased reproductive rate is generally favored, but increased reproductive effort might entail costs to other life history components. A similar example is the chemical inhibition of female-remating rate by male seminal proteins in many arthropods. This effect is often taken as either male manipulation of female remating rate (e.g., Arnqvist and Nilsson 2000; Wolfner 2002; Arnqvist and Rowe 2005; Crudgington et al. 2005), or evidence of sexual cooperation (Dickson 2008). In fact, there is typically scant evidence for what optimal female remating rates are in natural conditions (Gwynne 2008), largely because measuring this optimum is empirically challenging (Maklakov and Arnqvist 2009).

SA selection and coevolution are expected to drive the rapid diversification of male SA traits. This prediction corresponds well with the substantial phenotypic diversity of male persistence traits, both among populations and species. As examples, both water striders and diving beetles show striking diversity across both populations and species in male grasping armaments (Bergsten et al. 2001; Bergsten and Miller 2007; Khila et al. 2012; Perry and Rowe 2012b), as do traumatically inseminating bug species in their stabbing aedeagus (Tatarnic and Cassis 2010), and there is population-level variation in the effects of male seminal substances on female fecundity and longevity in adzuki bean beetles (Yamane 2013). This phenotypic diversity is matched by the substantial standing genetic variation in male SA traits found in several experimental evolution studies. For example, the genital spines of *Callosobruchus* beetles respond to artificial selection for increased length within only a few generations (Hotzy et al. 2012). In *D. melanogaster*, the overall level of male harmfulness to females increases when females are prevented from coevolving with males (Rice 1996; but see Jiang et al. 2011), as well as in response to the experimental manipulation of polyandry (Holland and Rice 1999; Pitnick et al. 2001) and sex ratio (Nandy et al. 2013; but see Wigby and Chapman 2004).

Although there has been much progress, there is much to learn about harmful male traits. We know much more about condition-dependence in sexually selected ornaments and weapons than in SA armaments. We know little about the hypothesis that coevolution between the sexes should result in multiple harmful male traits, as females evolve resistance to each in succession (Arnqvist and Rowe 2005). It is intriguing to speculate that this kind of process can account for the diversity and redundancy of male seminal proteins in animal like fruit flies, but evidence is lacking (Arnqvist and Rowe 2005).

SEXUALLY ANTAGONISTIC TRAITS IN FEMALES

Numerous reviews have called for increased study of the female side of sexual conflict (Zeh and Zeh 2003; Arnqvist and Rowe 2005; Arnqvist 2006; Chapman 2006; Fricke et al. 2009). These reviews highlight the paucity of strong evidence for SA traits in females, and the very limited data available on forms of resistance, natural selection on female SA traits, and their genetic variability and evolvability. What progress has been made?

In short, there remains only limited evidence identifying female SA traits. To establish that a female trait is sexually antagonistic, we need to work out the relationships depicted in Figure 1A. We need to show SA selection acting on a trait that is the outcome of sexual interactions, and to show that a putative female SA trait impacts the interaction trait in a way that improves female fitness. The most convincing demonstrations of female resistance adaptations are from studies that identify an interaction trait that negatively impacts female fitness. The most common example is mating rate, which is often costly for females beyond some intermediate value. Studies that experimentally manipulate the interaction trait to evaluate the impact on female fitness are most valuable in establishing causality.

Measuring the relationship between interaction traits and female fitness is not trivial. A major difficulty is that we would like to have data from natural conditions, but for many organisms, natural mating behavior and the range of ecological settings is unknown or highly variable. A further difficulty is that for many species it is difficult to engineer phenotypes that are costly for either sex, because organisms have evolved strategies to avoid these phenotypes. Hence, it is easy to experimentally reduce mating rate, but difficult to increase it because females resist excess mating. Furthermore, we would ideally like to have data that describe variation in female lifetime reproductive success, rather than components of fitness (Maklakov and Arnqvist 2009). This is because the potential for trade-offs among life history components greatly complicates extrapolating fitness components to total fitness. For example, a negative effect of mating rate on female longevity is sometimes portrayed as a cost of mating, but may represent an adaptive shift in female life history. When this is the case, decreased longevity is not a cost imposed by males but by allocation tradeoffs between investment in reproductive effort and somatic maintenance. Complicating things further, it is clear that male SA traits can impact not just offspring number but also quality (Gasparini et al. 2012)—and possibly grandoffspring quality (Brommer et al. 2012).

Types and Forms of Female Resistance Traits

Comprehensive surveys in arthropods suggest that female SA traits might frequently be behavioral or physiological rather than morphological (e.g., Eberhard 1997; Eberhard 2004b). Alternatively, it may be that it is difficult to assess the evolutionary history and current function of female morphological traits from measurements made on static specimens. For example, grooves and slots in female genitalia that appear to be cooperative guides for male genitalia (Eberhard 2004b) might have evolved to minimize male-imposed costs. The spermalege of female bed bugs appears to have evolved to minimize costs from the hypodermic aedeagus of male bed bugs (Morrow and Arnqvist 2003). In any case, behavioral resistance seems likely when the economics of mating are very context-dependent, such that females need to fine-tune resistance to local conditions.

We have emphasized above the theoretical prediction that the form of female resistance—in terms of the threshold and sensitivity of female responses to male traits (Fig. 3)—should

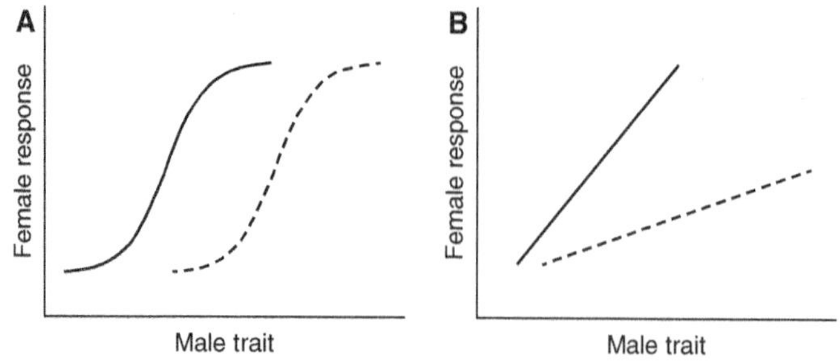

Figure 3. Female responses to male sexually antagonistic traits can take diverse forms. Increased female resistance to male traits is illustrated by a shift from the solid line to the dashed line within each panel, and can take the form of (*A*) an increased threshold of response, or (*B*) decreased sensitivity.

have significant impact on the outcome of SAC. We therefore need data that characterize which forms of female response are evolutionary labile. How does natural selection act on female SA traits? How much additive genetic variation do they display, and in which directions? This remains a significant gap, and there are almost no data available on this topic, although studies of genetic variation in female resistance are now being conducted.

Genetic Variation for Resistance to Harm

Several studies in *D. melanogaster* have detected additive genetic variation in female ability to resist harm from males (e.g., Wigby and Chapman 2004; Linder and Rice 2005; Lew et al. 2006; Brommer et al. 2012). However, functional understanding is often lacking, and in general it is not known what specific SA traits in females mediate this variation. Instead, the data generally come from studies of the genetic variation underlying female fecundity, sometimes along with longevity, in response to male harm. There are exceptions; for example, one study of *D. melanogaster* was able to link female reluctance to remate with increased resistance to harm from males (Lew et al. 2006). Another potential example comes from work on the nuptial feeding bushcricket *Gryllodes sigillatus*, in which there is genetic variation in the duration females feed on spermatophylax nuptial gifts from males (Gershman et al. 2013); these gifts are thought to manipulate female physiology in potentially costly ways for females.

THE CONDITIONAL ECONOMICS OF CONFLICT

We have previously argued that we need a much more ecological understanding of sexual conflict because ecological context determines the economics of conflict (Arnqvist and Rowe 2005; Fricke et al. 2009). The reasons for this are threefold. First, the ecological context of sexual interactions will often determine whether there is conflict or congruence, because costs and benefits to each sex depend on ecological context (Arnqvist and Rowe 2005). Moreover, the extent of conflict may vary among individuals within a population depending on their condition. For example, in *D. melanogaster,* males cause the greatest harm to females of highest fecundity (Lew et al. 2006; Long et al. 2009) and large males cause greater harm to females than small males (Friberg and Arnqvist 2003). Second, ecological context can determine the extent of natural selection on SA traits. For example, in water striders, antagonistic abdominal spines in females increase susceptibility to cannibalism and predation during molt (Arnqvist 1994). Therefore, natural selection against these spines will be a function of local density of conspecifics and predators. Third, both theoretical and empirical studies have shown that ecological context can determine the direction and extent of SAC (e.g., Martin and Hosken 2003; Rowe and Day 2006; Fricke et al. 2010a; Arbuthnott et al. 2014).

An excellent example of the importance of ecological context is the sex peptide of *D. melanogaster*, often considered a prime example of an SA male trait (Arnqvist and Rowe 2005; Lessells 2006; Rowe and Day 2006). Females experience strong postmating changes in behavior from receipt of sex peptide, including increased egg-laying in the short term and delayed remating, among a suite of other changes (Chapman et al. 2003b; Fricke et al. 2013). Sex peptide appears to decrease female fitness in some contexts. In response to receipt of sex peptide, females achieve lower lifetime reproductive success (Wigby and Chapman 2005). These and related results have sparked a great deal of interest in this protein. However, the magnitude and even the existence of costs for females depend strongly on nutrient levels (Rogina 2009; Fricke et al. 2010b). In fact, receipt of sex peptide permits females to lay more eggs under most nutrient levels, and it is only under extreme nutrient restriction that costs have been reported in two subsequent studies (Rogina 2009; Fricke et al. 2010b). Thus, depending on local food abundance, sex peptide could be considered a prime example of sexual conflict or sexual congruence.

A second intriguing example of the potential importance of ecological context for the evolution of antagonistic traits comes from a recent study of independently evolving *D. melanogaster* populations (Arbuthnott et al. 2014). Here, replicate populations evolved in one of two novel environments—food supplemented with cadmium or ethanol—and were then contrasted for male harm and female resistance. The investigators found suggestive evidence for ecologically mediated parallel evolution of both male harm and female resistance to that harm. Males evolved on ethanol-supplemented food depressed the lifespan of standard (i.e., not experimentally evolved) females more than did males evolved on cadmium-supplemented food; females evolved on ethanol-supplemented food had a smaller reduction in lifespan following high exposure to males compared with females evolved on cadmium-supplemented food. Depression of female lifespan might represent harm to females—and therefore sexual conflict—or a shift in life history that might have positive or neutral effects on female fitness. If the changes in lifespan do reflect harm, it would be intriguing to investigate the antagonistic traits that underlay harm and resistance to determine whether populations differed in suites of antagonistic traits or simply in the degree of expression of traits common to both sets of populations. However, the study does suggest that ecological conditions can substantially influence the course of evolution of traits involved in sexual conflict.

SEXUALLY ANTAGONISTIC COEVOLUTION

The limited evidence for female resistance traits inevitably means that there are few systems for which SAC has been well characterized. To our knowledge, these systems include water striders, diving beetles, traumatically inseminating bugs, and seed beetles, each of which is discussed in some detail below. Despite the paucity of examples, the potential for sexual coevolution is one of the most intriguing aspects of sexual conflict. For example, how often does conflict lead to coevolution, diversification, and speciation? There is suggestive and mixed evidence for all of these larger scale outcomes (Arnqvist et al. 2000; Martin and Hosken 2004; Rice et al. 2005; but see Plesnar-Bielak et al. 2013). Even smaller scale questions remain open. For example, as some have argued, does SAC necessarily lead to increased complexity in sexual armaments (Alexander et al. 1997; Eberhard 2004a, 2006; Rowe and Arnqvist 2012)? These questions must remain unanswered until we have more than the small handful of case studies now available. Below we review these case studies.

CASE STUDIES

Water Striders

Water striders remain among the best-characterized system for interlocus sexual conflict over sexual interactions. In this group, the sexes are in conflict over mating rate, which is mediated by social and ecological conditions. Several antagonistic traits in both sexes affect mating rate, and there is evidence of arms races between these traits (Arnqvist and Rowe 2005). There are quantitative data on the economics and ecology of conflict (Rowe et al. 1994), as well as the trajectory of SAC (Arnqvist and Rowe 2002a,b; Rowe and Arnqvist 2002). Since conflict in the group was last reviewed (Arnqvist and Rowe 2005), there have been several new insights into the behavioral context of conflict, antagonistic traits involved in the conflict, and their coevolution.

Most of the attention on female SA traits in water striders has been directed toward female resistance behavior and pregenital spines (Fig. 2), both of which vary among populations and species and affect mating outcomes (Arnqvist and Rowe 2005; Ronkainen et al. 2005; Perry and Rowe 2012b). Han and Jablonski (2009) have identified a new trait in females that may have a substantial effect on SAC. The basal state in *Gerris* spp. is an exposed vulvar opening, into which males may be able to insert their genitalia without female cooperation, although this has not been shown. In contrast, the pregenital segment 7 of *G. gracilicornis* largely covers the vulvar opening. Han and Jablonski (2009) have suggested that this precludes so-called forced copulation because females must inflate their genitalia to allow male access. An interesting hypothesis is that males have responded by making resistance (delay in inflating genitalia) costly to females, potentially through a "war of attrition"-type process. Han and Jablonski (2010) report that males in this species tap the water surface during the precopula stage, which attracts predators; females respond by inflating their genitalia.

One of the more dramatic grasping traits of male water striders is the enlarged, hooked, and spined male antennae in the genus *Rheumatobates* (Fig. 2) (Rowe et al. 2006). The fine-scale function and development of these SA traits in *Rheumatobates rileyi* have recently been reported by Khila et al. (2012). Of the several traits that make up the antennae, all seem finely crafted to match the grasping sites of resistant females during premating struggles. One interesting question is whether this pattern extends across the diverse range of antennal traits and shapes in the group. An apparent lack of morphological diversity in females corresponding to male morphology suggests that this may not be the case (Rowe et al. 2006). Alternatively, there may be many other finely crafted means to grasp females. In a second part of the study, Khila et al. (2012) used RNA interference to show a role for the gene distal-less (*dll*) in generating the grasping antennal traits of *R. rileyi*. A graded reduction of the grasping traits with knocked down *dll* expression resulted in a similarly graded reduction of mating success in males. These data suggest continuous selection for elaboration from a point close to the basal state for antennae in this genus.

In *Gerris* spp., male grasping traits include pregenital and genital segments, which are inflated and curled around the female genitalia. A recent comparative analysis of shape and complexity of these grasping segments suggests that they are evolving antagonistically with their counterparts in females (Rowe and Arnqvist 2012). Moreover, they are more complex and evolving more divergently than similar body parts that do not seem to be involved in sexual conflict. An exception to this is the terminal components of the genitalia, which are particularly complex and divergent, and do not appear to be involved in grasping, but do appear to be under sexual selection (Rowe and Arnqvist 2012). These data on both components of the genitalia provide quantitative evidence for the long-held hypothesis that genitalia are particularly complex and divergent in shape (Eberhard 1985).

The general pattern of SAC in the water strider genus *Gerris* is one of matched escalation and de-escalation of male grasping traits and female antigrasping traits (Arnqvist and Rowe 2002a,b). When these traits veer from matching they affect mating behaviors; when males are relatively armed mating rates go up, and when females become relatively armed, mating rate goes down (Arnqvist and Rowe 2002a). The connection of these coevolutionary dynamics at the species level to the well-known ecological effects on conflict is not well understood (Rowe and Arnqvist 2002). A recent study of SAC at the population level may be a step toward connecting these two scales. Perry and Rowe (2012b) report a pattern of coevolution in grasping and antigrasping traits, and mating behaviors at the population level that matches that reported at the species level. A similar study in a closely related species, *Gerris gillettei*, reported evidence of correlated evolution between the sexes in potentially antagonistic sexual traits (Gagnon and Turgeon 2011). However, in contrast to interspecific studies of *Gerris* (Rowe and Arnqvist 2002; Arnqvist and Rowe 2002b) and the interpopulation study of Perry and Rowe (2012b), the evolution of female spine size was not correlated to male genital traits. The next step in these interpopulation studies will be to directly examine local ecological (and social) conditions and attempt to connect them to the state of SA traits in the populations.

The social context of sexual conflict has received considerable recent attention in the stream water strider *Aquarius* spp. In the typical environment of pools connected by flowing water, individuals in different pools often experience different social environments because group composition varies (Sih and Watters 2005). Individual males tend to vary in their aggressiveness to females (and males), which may affect both their own mating fitness and the mating rate of the group (Eldakar et al. 2009a,b; Chang and Sih 2013). The multilevel nature of selection in this system sets up the potential for selection to work in opposition between levels. Eldakar et al. (2009a,b) has argued for a "tragedy of the commons" process in *Aquarius remigis*. Aggressive or hyperaggressive males tend to harass females at an extremely high frequency in an attempt to mate. They will also harass males and mating pairs. In some conditions, this results in increased relative mating success for these males, but reduced activity of other males (Eldakar et al. 2009a,b; Chang and Sih 2013). Females tend to disperse from pools containing very aggressive males, resulting in a reduction in average mating frequency in the original pool. A very interesting possibility is that male aggression may reduce female fecundity in addition to mating frequency, which would result in a tragedy of the commons over the resource (female ova) over which males compete. This hypothesis remains to be tested, but establishing a link between heightened male aggression and decreased female fecundity would provide strong evidence for a sexually antagonistic tragedy of the commons.

Traumatic Insemination in Bugs

Traumatic insemination—whereby males transfer ejaculates by piercing the female body cavity with syringe-like intromittent organs—has evolved independently in a variety of invertebrate taxa (reviewed by Lange et al. 2013; Tatarnic et al. 2014). The potential for sexual conflict in these systems is clear. Females face increased risk of infection and the costs of mounting an immune response, and these kinds of costs have been detected in a well-studied system of traumatic insemination, *Cimex* bed bugs (Stutt and Siva-Jothy 2001). Females in these and many other species have evolved remarkable morphological counteradaptations that mitigate the costs of hypodermic insemination, in the form of a unique structure called the spermalege (Fig. 2) (Reinhardt et al. 2003). The spermalege is, in many species, effectively a secondary female genitalic structure, typically consisting of a modified abdominal wall and an internal pocket where injected ejaculate is stored, with wide variation in the presence and form of these structures across bed bug species (Usinger 1966). Evidence that the spermalege is effective in reducing mating costs comes from functional studies in which artificial piercing and insemination within the spermalege region does not harm females, but the same action outside this region is very harmful and often fatal (Morrow and Arnqvist 2003; Reinhardt et al. 2003).

What has previously been lacking in understanding SAC for traumatic insemination and spermalege evolution is formal comparative analyses linking divergence in male and female morphology (Arnqvist and Rowe 2005). A recent study has done just this for traumatically inseminating plant bugs of the genus *Coridromius* (Tatarnic and Cassis 2010). In these species, male piercing genitalia varies extensively in size and shape; likewise females display a great deal of variation in spermalege morphology (Tatarnic et al. 2006). Across the genus, the diversification of male genital morphology and female spermalege morphology are closely correlated. Notably, there are exceptions. In some species, male morphology has diverged with little corresponding evolution in female traits. Functional studies in these species would be most interesting to investigate whether males have gained an advantage in increasing mating rate, overcoming female resistance, transferring more ejaculate, or whether female resistance has taken other forms (e.g., behavioral or physiological resistance).

Seed Beetles

Seed beetles, *Callosobruchus*, are renowned for their highly elaborated genitalia that damage female reproductive tracts (Fig. 2). Females resist mating by kicking to dislodge mating males. Crudgington and Siva-Jothy (2000) experimentally ablated female legs so that resistance was prevented. The effect of this was an increase in both mating duration and damage to females (see also Maklakov and Arnqvist 2009). Since these early studies, research has focused on the function of genital spines and on coevolution between male spines and the female reproductive tract.

In a comparative study of populations that varied in genital morphology, Hotzy and Arnqvist (2009) have shown a role for genital spines in sperm competition success (when males were last to mate with a female). Males from populations with longer spines had higher success in sperm competition and this effect remained after controlling for body size variation. In the same study, there was no correlation between harm to the female reproductive tract and male spine length. Two possible nonexclusive roles for the spines in sperm competition are that spines may act as a grasping device to prolong mating or aid in positioning (Edvardsson and Tregenza 2005), or alternatively, scarring of the female tract by the spines may facilitate uptake of accessory substances transferred by the male. Results of the Hotzy and Arnqvist (2009) study support the first hypothesis, as there was no independent effect of scarring on the outcome of sperm competition, and Hotzy and colleagues (2012) report increased female uptake of male ejaculate proteins following matings with males with longer spines.

One might expect an evolutionary response of females to the harm caused to them by spined male genitalia. In comparative study at the species level, Rönn et al. (2007) found a positive correlation across species between spine elaboration and the amount of connective tissue in the copulatory duct, where scarring occurs. These data provide a rare case in which there is strong support for the correlated evolution of a male persistence trait and a female response. It is argued that the male trait is antagonistic because its advantage to males comes at a cost to females. The thickening of the reproductive tract in females seems likely to act as a defensive trait in females; however, it is not yet clear how female reproductive tract thickness affects selection on the male trait. Male spine elaborations may be caused by some other aspect of female morphology or behavior, or caused entirely by environmental variables that differ between species. Therefore, it is not obvious that these data support a case of SA coevolution.

Diving Beetles

Diving beetles (Dytiscidae) provide one of the few cases in which we have strong evidence of sexual coevolution in nature that appears to be antagonistic. Males chase and grasp females in an attempt to mate and females attempt to dislodge males with repeated dives and erratic swimming. A cost of mating to females has not been quantified, but a lack of access to air (they are air breathers) has been suggested (Bergsten and Miller 2007). Males grasp females on the pronotum and elytra with tarsae modified to include a set of suction cups. The females of some species and populations have evolved rough dorsal surfaces that appear to impede grasping (Fig. 2). In an early study of *Graphoderus zonatus verrucifer*, Bergsten et al. (2001) provided evidence of a correlation between the expression of male grasping traits and the frequency of the rough dorsal surfaced female. This was the first evidence pointing to SAC at the population level.

Two key pieces of the story that were previously missing have now been addressed. First, the utility of the male suction cups in grasping females, and the obstruction of grasping by the rough females surface, had not been shown. A recent biomechanical study supports these hypotheses (Green et al. 2013). Moreover, the work indicated an interaction that suggests that males with tarsae suited to grasping rough surfaced females were unsuited to grasping smooth surfaced females. This interaction implies a trade-off, which would play into the correlated evolution of male and female features. Second, recent comparative analysis provides strong evidence for the correlated evolution of male and female structures at the species level (Bergsten and Miller 2007). The combination of known traits and correlated evolution at the both the population and species level makes this a very interesting group for intense study. A necessary step for further study of this group as a model for SAC is an understanding of the economics of mating.

CONCLUSIONS

Theoretical analysis points to the importance of the form of female resistance and the forces of selection acting on it. These analyses mirror those in more traditional sexual selection models (Kirkpatrick and Ryan 1991). Despite this, there are precious few identified SA traits in females—most of which are behavioral—and fewer still where their role in antagonistic coevolution has been studied. Therefore, we continue the call for further study of female SA traits and their evolutionary dynamics (Zeh and Zeh 2003; Rowe et al. 2005; Arnqvist 2006; Fricke et al. 2009; Findlay and Swanson 2010; Perry and Rowe 2012a).

Models also illustrate that the course of SAC is determined to a large degree by natural selection acting on both interaction traits and SA traits in the two sexes. There is very limited data here, particularly on SA traits. One reason to understand these forces of selection is to enable predictions about how the ecological context will determine both the extent of conflict and the course of SAC (Rice et al. 2006; Rowe and Day 2006). This means that we need to understand the economy of conflict in natural systems. Even in some of the better-known systems (e.g., water striders, diving beetles), the connection between the ecological context of conflict and the course of SAC is unknown.

Finally, the number of demonstrations of SAC remains a handful and it is dominated by studies in insects. More species of a broader taxonomic set (e.g., King et al. 2013; see works by Schärer et al. 2014 and Schneider, 2014) will allow us to assess the importance of SAC in driving the evolution of diversity, and will almost certainly reveal entirely novel processes.

ACKNOWLEDGMENTS

We thank S. Gavrilets and W. Rice for the invitation to contribute to this collection, and G. Arnqvist, W. Rice, and an anonymous reviewer for helpful comments that improved the manuscript. We thank those who kindly provided the images of Figure 2. J.C.P. is supported by Jesus College (Oxford). L.R. is supported by the Natural Sciences and Engineering Research Council and the Canada Research Chairs program.

REFERENCES

**Reference is also in this collection.*

Ah-King M. 2013. On anisogamy and the evolution of "sex roles." *Trends Ecol Evol* **28:** 1–2.

Alexander RD, Marshall DC, Cooley JR. 1997. Evolutionary perspectives on insect mating. In *The evolution of mating systems in insects and arachnids* (ed. Choe JC, Crespi BJ), pp. 4–31. Cambridge University Press, Cambridge.

Arbuthnott D, Dutton EM, Agrawal AF, Rundle HD. 2014. The ecology of sexual conflict: Ecologically dependent parallel evolution of male harm and female resistance in *Drosophila melanogaster*. *Ecol Lett* **17:** 221–228.

Arnqvist G. 1994. The cost of male secondary sexual traits: Developmental constraints during ontogeny in a sexually dimorphic water strider. *Am Nat* **144:** 119–132.

Arnqvist G. 2004. Sexual conflict and sexual selection: Lost in the chase. *Evolution (NY)* **58:** 1383–1388.

Arnqvist G. 2006. Sensory exploitation and sexual conflict. *Phil Trans R Soc* **361:** 375–386.

Arnqvist G, Nilsson T. 2000. The evolution of polyandry: Multiple mating and female fitness in insects. *Anim Behav* **60:** 145–164.

Arnqvist G, Rowe L. 2002a. Comparative analysis unveils antagonistic coevolution between the sexes in a group of insects. *Nature* **415:** 787–789.

Arnqvist G, Rowe L. 2002b. Correlated evolution of male and female morphologies in water striders. *Evolution (NY)* **56:** 936–947.

Arnqvist G, Rowe L. 2005. *Sexual conflict*, Princeton University Press, Princeton, NJ.

Arnqvist G, Edvardsson M, Friberg U, Nilsson T. 2000. Sexual conflict promotes speciation in insects. *Proc Natl Acad Sci* **97:** 10460–10464.

Badcock C, Crespi B. 2008. Battle of the sexes may set the brain. *Nature* **454:** 1054–1055.

Bergsten J, Miller KB. 2007. Phylogeny of diving beetles reveals a coevolutionary arms race between the sexes. *PLoS ONE* **2:** e522.

Bergsten J, Töyrä A, Nilsson AN. 2001. Intraspecific variation and intersexual correlation in secondary sexual characters of three diving beetles (Coleoptera: Dytiscidae). *Biol J Linn Soc Lond* **73:** 221–232.

Bonduriansky R, Maklakov A, Zajitschek F, Brooks R. 2008. Sexual selection, sexual conflict and the evolution of ageing and life span. *Func Ecol* **22:** 443–453.

Brommer JE, Fricke C, Edward DA, Chapman T. 2012. Interactions between genotype and sexual conflict environment influence transgenerational fitness in *Drosophila melanogaster*. *Evolution (NY)* **66:** 517–531.

Carazo P, Tan CKW, Allen F, Wigby S, Pizzari T. 2014. Within-group male relatedness reduces harm to females in *Drosophila*. *Nature* **505:** 672–675.

Chang AT, Sih A. 2013. Multilevel selection and effects of keystone hyperaggressive males on mating success and behavior in stream water striders. *Behav Ecol* **24:** 1166–1176.

Chapman TW. 2006. Evolutionary conflicts of interest between males and females. *Curr Biol* **16:** R744–R754.

Chapman T. 2008. The soup in my fly: Evolution, form and function of seminal fluid proteins. *PLoS Biol* **6:** e179.

Chapman T, Arnqvist G, Bangham J, Rowe L. 2003a. Response to Eberhard and Cordero, and Córdoba-Aguilar and Contreras-Garduño: Sexual conflict and female choice. *Trends Ecol Evol* **18:** 440–441.

Chapman T, Bangham J, Vinti G, Seifried B, Lung O, Wolfner MF, Smith HK, Partridge L. 2003b. The sex peptide of *Drosophila melanogaster*: Female post-mating responses analyzed by using RNA interference. *Proc Natl Acad Sci* **100:** 9923–9928.

Cordero C, Eberhard WG. 2003. Female choice of sexually antagonistic male adaptations: A critical review of some current research. *J Evol Biol* **16:** 1–6.

Crudgington HS, Siva-Jothy MT. 2000. Genital damage, kicking and early death. *Nature* **407:** 855–856.

Crudgington HS, Beckerman AP, Brüstle L, Green K, Snook RR. 2005. Experimental removal and elevation of sexual selection: Does sexual selection generate manipulative males and resistant females? *Am Nat* **165:** S72–S87.

Dickson BJ. 2008. Wired for sex: The neurobiology of *Drosophila* mating decisions. *Science* **322:** 904–909.

Dougherty LR, Burdfield-Steel ER, Shuker DM. 2013. Sexual stereotypes: The case of sexual cannibalism. *Anim Behav* **85:** 313–322.

Eberhard WG. 1985. *Sexual selection and animal genitalia*, Harvard University Press, Cambridge, MA.

Eberhard WG. 1997. Sexual selection by cryptic female choice in insects and arachnids. In *The evolution of mating systems in insects and arachnids* (ed. Choe JC, Crespi BJ), pp. 32–57. Cambridge University Press, Cambridge.

Eberhard WG. 2004a. Male–female conflict and genitalia: Failure to confirm predictions in insects and spiders. *Biol Rev* **79:** 121–186.

Eberhard WG. 2004b. Rapid divergent evolution of sexual morphology: Comparative tests of antagonistic coevolution and traditional female choice. *Evolution (NY)* **58:** 1947–1970.

Eberhard WG. 2006. Sexually antagonistic coevolution in insects is associated with only limited morphological diversity. *J Evol Biol* **19:** 657–681.

Eberhard WG. 2010. Evolution of genitalia: Theories, evidence, and new directions. *Genetica* **138:** 5–18.

Eberhard WG, Cordero C. 2003. Sexual conflict and female choice. *Trends Ecol Evol* **18:** 438–439.

Edvardsson M, Tregenza T. 2005. Why do male *Callosobruchus maculatus* harm their mates? *Behav Ecol* **16:** 788–793.

Eldakar OT, Dlugos MJ, Pepper JW, Wilson DS. 2009a. Aggressive mating as a tragedy of the commons in the water strider *Aquarius remigis*. *Behav Ecol Soc* **64:** 25–33.

Eldakar OT, Dlugos MJ, Pepper JW, Wilson DS. 2009b. Population structure mediates sexual conflict in water striders. *Science* **326:** 816.

Findlay GD, Swanson WJ. 2010. Proteomics enhances evolutionary and functional analysis of reproductive proteins. *BioEssays* **32:** 26–36.

Fiumera AC, Dumont BL, Clark AG. 2006. Natural variation in male-induced "cost-of-mating" and allele-specific association with male reproductive genes in *Drosophila melanogaster*. *Phil Trans R Soc* **361:** 355–361.

Friberg U, Arnqvist G. 2003. Fitness effects of female mate choice: Preferred males are detrimental for *Drosophila melanogaster* females. *J Evol Biol* **16:** 797–811.

Fricke C, Perry J, Chapman T, Rowe L. 2009. The conditional economics of sexual conflict. *Biol Lett* **5:** 671–674.

Fricke C, Andersson C, Arnqvist G. 2010a. Natural selection hampers divergence of reproductive traits in a seed beetle. *J Evol Biol* **23:** 1857–1867.

Fricke C, Bretman A, Chapman T. 2010b. Female nutritional status determines the magnitude and sign of responses to a male ejaculate signal in *Drosophila melanogaster*. *J Evol Biol* **23:** 157–165.

Fricke C, Green D, Mills WE, Chapman T. 2013. Age-dependent female responses to a male ejaculate signal alter demographic opportunities for selection. *Proc R Soc B* **280:** 20130428.

Friesen CR, Uhrig EJ, Squire MK, Mason RT, Brennan PL. 2014. Sexual conflict over mating in red-sided garter snakes (*Thamnophis sirtalis*) as indicated by experimental manipulation of genitalia. *Proc R Soc B* **281:** 20132694.

Gagnon M-C, Turgeon J. 2011. Sexual conflict in *Gerris gillettei* (Insecta: Hemiptera): Intraspecific intersexual correlated morphology and experimental assessment of behaviour and fitness. *J Evol Biol* **24:** 1505–1516.

Gasparini C, Devigili A, Pilastro A. 2012. Cross-generational effects of sexual harassment on female fitness in the guppy. *Evolution (NY)* **66:** 532–543.

Gavrilets S. 2000. Rapid evolution of reproductive barriers driven by sexual conflict. *Nature* **403:** 886–889.

Gavrilets S, Hayashi TI. 2006. The dynamics of two- and three-way sexual conflicts over mating. *Phil Trans R Soc* **361:** 345–354.

Gavrilets S, Waxman D. 2002. Sympatric speciation by sexual conflict. *Proc Natl Acad Sci* **99:** 10533–10538.

Gavrilets S, Arnqvist G, Friberg U. 2001. The evolution of female mate choice by sexual conflict. *Proc R Soc B* **268:** 531–539.

Gershman SN, Hunt J, Sakaluk SK. 2013. Food fight: Sexual conflict over free amino acids in the nuptial gifts of male decorated crickets. *J Evol Biol* **26:** 693–704.

Green KK, Madjidian JA. 2011. Active males, reactive females: Stereotypic sex roles in sexual conflict research? *Anim Behav* **81:** 901–907.

Green KK, Kovalev A, Svensson EI, Gorb SN. 2013. Male clasping ability, female polymorphism and sexual conflict: Fine-scale elytral morphology as a sexually antagonistic adaptation in female diving beetles. *J R Soc Interface* **10.**

Gwynne DT. 2008. Sexual conflict over nuptial gifts in insects. *Annu Rev Entomol* **53:** 83–101.

Han CS, Jablonski PG. 2009. Female genitalia concealment promotes intimate male courtship in a water strider. *PLoS ONE* **4:** e5793.

Han CS, Jablonski PG, Kim B, Park FC. 2010. Size-assortative mating and sexual size dimorphism are predictable from simple mechanics of mate-grasping behavior. *BMC Evol Biol* **10:** 359.

Härdling R, Bergsten J. 2006. Nonrandom mating preserves intrasexual polymorphism and stops population differentiation in sexual conflict. *Am Nat* **167:** 401–409.

Härdling R, Smith HG. 2005. Antagonistic coevolution under sexual conflict. *Evol Ecol* **19:** 137–150.

Hayashi TI, Vose M, Gavrilets S. 2007. Genetic differentiation by sexual conflict. *Evolution (NY)* **61:** 516–529.

Holland B, Rice WR. 1998. Perspective: Chase-away sexual selection: Antagonistic seduction versus resistance. *Evolution (NY)* **52:** 1–7.

Holland B, Rice WR. 1999. Experimental removal of sexual selection reverses intersexual antagonistic coevolution and removes a reproductive load. *Proc Natl Acad Sci* **96:** 5083–5088.

Hotzy C, Arnqvist G. 2009. Sperm competition favors harmful males in seed beetles. *Curr Biol* **19:** 404–407.

Hotzy C, Polak M, Ronn JL, Arnqvist G. 2012. Phenotypic engineering unveils the function of genital morphology. *Curr Biol* **22:** 2258–2261.

Jiang PP, Bedhomme S, Prasad NG, Chippindale AK. 2011. Sperm competition and mate harm unresponsive to male-limited selection in *Drosophila*: An evolving genetic architecture under domestication. *Evolution (NY)* **65:** 2448–2460.

Johnstone RA, Keller L. 2000. How males can gain by harming their mates: Sexual conflict, seminal toxins, and the cost of mating. *Am Nat* **156:** 368–377.

Kazancioğlu E, Alonzo SH. 2012. The evolution of optimal female mating rate changes the coevolutionary dynamics of female resistance and male persistence. *Phil Trans R Soc* **367:** 2339–2347.

Khila A, Abouheif E, Rowe L. 2012. Function, developmental genetics, and fitness consequences of a sexually antagonistic trait. *Science* **336:** 585–589.

Kimura M, Ihara Y. 2009. Replicator-dynamics models of sexual conflict. *J Theor Biol* **260:** 90–97.

King EDA, Banks PB, Brooks RC. 2013. Sexual conflict in mammals: Consequences for mating systems and life history. *Mamm Rev* **43:** 47–58.

Kirkpatrick M, Ryan MJ. 1991. The evolution of mating preferences and the paradox of the lek. *Nature* **350:** 33–38.

Kokko H, Brooks R. 2003. Sexy to die for? Sexual selection and the risk of extinction. *Ann Zool Fennici* **40:** 207–219.

Kokko H, Brooks R, Jennions M, Morley J. 2003. The evolution of mate choice and mating biases. *Proc R Soc B* **270:** 653–664.

Kokko H, Booksmythe I, Jennions MD. 2013. Causality and sex roles: Prejudice against patterns? A reply to Ah-King. *Trends Ecol Evol* **28:** 2–4.

Kwan L, Cheng YY, Rodd FH, Rowe L. 2013. Sexual conflict and the function of genitalic claws in guppies (*Poecilia reticulata*). *Biol Lett* **9:** 20130267.

Lande R. 1981. Models of speciation by sexual selection on plolygenic traits. *Proc Natl Acad Sci* **78:** 3721–3725.

Lange R, Reinhardt K, Michiels NK, Anthes N. 2013. Functions, diversity, and evolution of traumatic mating. *Biol Rev Camb Phil Soc* **88:** 585–601.

Lessells CM. 2006. The evolutionary outcome of sexual conflict. *Phil Trans R Soc* **361:** 301–317.

Lew TA, Morrow EH, Rice WR. 2006. Standing genetic variance for female resistance to harm from males and its relationship to intralocus sexual conflict. *Evolution (NY)* **60:** 97–105.

Linder JE, Rice WR. 2005. Natural selection and genetic variation for female resistance to harm from males. *J Evol Biol* **18:** 568–575.

Long TAF, Pischedda A, Stewart AD, Rice WR. 2009. A cost of sexual attractiveness to high-fitness females. *PLoS Biol* **7:** e1000254.

Madjidian JA, Green KK. 2012. A reply to Perry & Rowe: Costs in sexual conflict research. *Anim Behav* **83:** e14–e16.

Maklakov AA, Arnqvist G. 2009. Testing for direct and indirect effects of mate choice by manipulating female choosiness. *Curr Biol* **19:** 1903–1906.

Martin OY, Hosken DJ. 2003. The evolution of reproductive isolation through sexual conflict. *Nature* **423:** 979–982.

Martin OY, Hosken DJ. 2004. Reproductive consequences of population divergence through sexual conflict. *Curr Biol* **14:** 906–910.

Morrow EH, Arnqvist G. 2003. Costly traumatic insemination and a female counter-adaptation in bed bugs. *Proc R Soc B* **270:** 2377–2381.

Morrow EH, Connallon T. 2013. Implications of sex-specific selection for the genetic basis of disease. *Evol Appl* **6:** 1208–1217.

Morrow EH, Arnqvist G, Pitnick S. 2003. Adaptation versus pleiotropy: Why do males harm their mates? *Behav Ecol* **14:** 802–806.

Nandy B, Gupta V, Sen S, Udaykumar N, Samant MA, Ali SZ, Prasad NG. 2013. Evolution of mate-harm, longevity and behaviour in male fruit flies subjected to different levels of interlocus conflict. *BMC Evol Biol* **13:** 212.

Parker GA. 1979. Sexual selection and sexual conflict. In *Sexual selection and reproductive competition in insects* (ed. Blum MS, Blum NA), pp. 123–166. Academic, New York.

Parker GA. 2006. Sexual conflict over mating and fertilization: An overview. *Phil Trans R Soc* **361:** 235–259.

Parker GA, Simmons LW. 1989. Nuptial feeding in insects: Theoretical models of male and female interests. *Ethology* **82:** 3–26.

Perry JC, Rowe L. 2012a. Sex role stereotyping and sexual conflict theory. *Anim Behav* **83:** e10–e13.

Perry JC, Rowe L. 2012b. Sexual conflict and antagonistic coevolution across water strider populations. *Evolution* **66:** 544–557.

Perry JC, Sirot L, Wigby S. 2013. The seminal symphony: How to compose an ejaculate. *Trends Ecol Evol* **28:** 414–422.

Pitnick S, Miller GT, Reagan J, Holland B. 2001. Males' evolutionary responses to experimental removal of sexual selection. *Proc R Soc B* **268:** 1071–1080.

Pizzari T, Snook RR. 2003. Perspective: Sexual conflict and sexual selection: Chasing away paradigm shifts. *Evolution (NY)* **57:** 1223–1236.

Plesnar-Bielak A, Skrzynecka AM, Prokop ZM, Kolasa M, Dzialo M, Radwan J. 2013. No evidence for reproductive isolation through sexual conflict in the bulb mite *Rhizoglyphus robini. PLoS ONE* **8:** e74971.

Promislow DEL. 2003. Mate choice, sexual conflict and evolution of senescence. *Behav Genet* **33:** 191–201.

Rankin DJ. 2011. Kin selection and the evolution of sexual conflict. *J Evol Biol* **24:** 71–81.

Reinhardt K, Naylor R, Siva-Jothy MT. 2003. Reducing a cost of traumatic insemination: Female bedbugs evolve a unique organ. *Proc R Soc B* **270:** 2371–2375.

Rice WR. 1996. Sexually antagonistic male adaptation triggered by experimental arrest of female evolution. *Nature* **381:** 232–234.

Rice WR, Linder JE, Friberg U, Lew TA, Morrow EH, Stewart AD. 2005. Inter-locus antagonistic coevolution as an engine of speciation: Assessment with hemiclonal analysis. *Proc Natl Acad Sci* **102:** 6527–6534.

Rice WR, Stewart AD, Morrow EH, Linder JE, Orteiza N, Byrne PG. 2006. Assessing sexual conflict in the *Drosophila melanogaster* laboratory model system. *Phil Trans R Soc* **361:** 287–299.

Rogina B. 2009. The effect of sex peptide and calorie intake on fecundity in female *Drosophila melanogaster. Scientific-WorldJournal* **9:** 1178–1189.

Rönn J, Katvala M, Arnqvist G. 2007. Coevolution between harmful male genitalia and female resistance in seed beetles. *Proc Natl Acad Sci* **104:** 10921–10925.

Ronkainen K, Kaitala A, Huttunen R. 2005. The effect of abdominal spines on female mating frequency and fecundity in a water strider. *J Insect Behav* **18:** 619–631.

Rowe L. 1994. The cost of mating and mate choice in water striders. *Anim Behav* **48:** 1049–1056.

Rowe L, Arnqvist G. 2002. Sexually antagonistic coevolution in a mating system: Combining experimental and comparative approaches to address evolutionary processes. *Evolution (NY)* **56:** 754–767.

Rowe L, Arnqvist G. 2012. Sexual selection and the evolution of genital shape and complexity in water striders. *Evolution (NY)* **66:** 40–54.

Rowe L, Day T. 2006. Detecting sexual conflict and sexually antagonistic coevolution. *Phil Trans R Soc* **361:** 277–285.

Rowe L, Arnqvist G, Sih A, Krupa JJ. 1994. Sexual conflict and the evolutionary ecology of mating patterns: Water striders as a model system. *Trends Ecol Evol* **9:** 289–293.

Rowe L, Cameron E, Day T. 2005. Escalation, retreat, and female indifference as alternative outcomes of sexually antagonistic coevolution. *Am Nat* **165:** S5–S18.

Rowe L, Westlake KP, Currie DC. 2006. Functional significance of elaborate secondary sexual traits and their evolution in the water strider genus *Rheumatobates. Can Entomol* **138:** 568–577.

Schärer L, Rowe L, Arnqvist G. 2012. Anisogamy, chance and the evolution of sex roles. *Trends Ecol Evol* **27:** 260–264.

* Schärer L, Janicke T, Ramm SA. 2014. Sexual conflict in hermaphrodites. *Cold Spring Harb Perspet Biol* doi: 10.1101/cshperspect.a017673.

* Schneider JM. 2014. Sexual cannibalism as a manifestation of sexual conflict. *Cold Spring Harb Perspet Biol* doi: 10.1101/cshperspect.a017731.

Sih A, Watters JV. 2005. The mix matters: Behavioural types and group dynamics in water striders. *Behaviour* **142:** 1417–1431.

Simmons LW, Parker GA. 1989. Nuptial feeding in insects: mating effort versus paternal investment. *Ethology* **81:** 332–343.

Stutt AD, Siva-Jothy MT. 2001. Traumatic insemination and sexual conflict in the bed bug *Cimex lectularius. Proc Natl Acad Sci* **98:** 5683–5687.

Tatarnic NJ, Cassis G. 2010. Sexual coevolution in the traumatically inseminating plant bug genus *Coridromius. J Evol Biol* **23:** 1321–1326.

Tatarnic NJ, Cassis G, Hochuli DF. 2006. Traumatic insemination in the plant bug genus *Coridromius* Signoret (Heteroptera: Miridae). *Biol Lett* **2:** 58–61.

Tatarnic NJ, Cassis G, Siva-Jothy MT. 2014. Traumatic insemination in terrestrial arthropods. *Annu Rev Entomol* **59:** 245–261.

Tregenza T, Wedell N, Chapman TW. 2006. Introduction. Sexual conflict: A new paradigm? *Phil Trans R Soc* **361:** 229–234.

Usinger RL. 1966. *Monograph of cimicidae (Hemiptera, Heteroptera)*. Entomological Society of America, College Park, MD.

Vahed K. 2007. All that glisters is not gold: Sensory bias, sexual conflict and nuptial feeding in insects and spiders. *Ethology* **113:** 105–127.

Wigby S, Chapman TW. 2004. Female resistance to male harm evolves in response to manipulation of sexual conflict. *Evolution (NY)* **58:** 1028–1037.

Wigby S, Chapman T. 2005. Sex peptide causes mating costs in female *Drosophila melanogaster*. *Curr Biol* **15:** 316–321.

Wolfner MF. 2002. The gifts that keep on giving: physiological functions and evolutionary dynamics of male seminal proteins in *Drosophila*. *Heredity* **88:** 85–93.

Yamane T. 2013. Intra-specific variation in the effect of male seminal substances on female oviposition and longevity in *Callosobruchus chinensis*. *Evol Biol* **40:** 133–140.

Zeh JA, Zeh DW. 2003. Toward a new sexual selection paradigm: Polyandry, conflict and incompatibility. *Ethology* **109:** 929–950.

Sexual Conflict, Life Span, and Aging

Margo I. Adler and Russell Bonduriansky

Evolution & Ecology Research Centre and School of Biological, Earth and Environmental Sciences, University of New South Wales, Sydney 2052, New South Wales, Australia

Correspondence: margo.adler@gmail.com

The potential for sexual conflict to influence the evolution of life span and aging has been recognized for more than a decade, and recent work also suggests that variation in life span and aging can influence sexually antagonistic coevolution. However, empirical exploration of these ideas is only beginning. Here, we provide an overview of the ideas and evidence linking inter- and intralocus sexual conflicts with life span and aging. We aim to clarify the conceptual basis of this research program, examine the current state of knowledge, and suggest key questions for further investigation.

Sexual conflict arises because the sexes maximize their fitness via different, and often mutually incompatible, strategies, and its signature has been detected across a wide range of morphological, physiological, behavioral, and life-history traits in many species. A number of investigators have suggested that sexual conflict could play an important role in the evolution of two particularly interesting life-history traits: life span and aging (Svensson and Sheldon 1998; Promislow 2003; Bonduriansky et al. 2008; Maklakov and Lummaa 2013). Sexual conflict can affect life span and aging rate at both proximate (within-generation) and ultimate (evolutionary) scales. Sexually antagonistic behavioral or physiological interactions that increase mortality rate in one or both sexes (interlocus sexual conflict) could drive the evolution of faster life histories. Moreover, sex-specific optimization of reproductive strategies may often result in sex differences in life span and aging rates, and sexually antagonistic selection on shared genetic architecture can displace one or both sexes from their sex-specific optima for these traits (intralocus sexual conflict). Conversely, a change in life histories because of environmental fluctuations could affect the degree of sexual conflict in a population and influence sexual coevolution. Although evidence for sexual conflict is rapidly accumulating, our understanding of its relationship to life span and aging remains rudimentary. In this review, we provide a critical review of recent literature and highlight areas that require further investigation.

DOES INTERLOCUS SEXUAL CONFLICT AFFECT THE EXPRESSION AND EVOLUTION OF LONGEVITY AND AGING?

Proximate Effects of Sexually Antagonistic Interactions on Life Span and Aging

The near-ubiquity among species of sex differences in optimal mating rate has led to numerous examples of adaptations in one sex

Cite this article as *Cold Spring Harb Perspect Biol* doi: 10.1101/cshperspect.a017566

(typically males) that function to manipulate individuals of the opposite sex (usually females) into mating at a suboptimal rate (Parker 1979; Arnqvist and Rowe 2005). Such interactions may reduce life span or increase aging rate in one or both sexes and could thus influence the evolution of life histories. Male strategies that function to increase female mating rate or reproductive investment can increase immediate mortality in females through direct injury, resulting in reduced female life span. For example, in natural populations of many species of waterfowl, forced copulation of females by one or more extra-pair males appears to be a common phenomenon. Because females vigorously resist mating with extra-pair males, forced copulation attempts can result in severe injuries, and in extreme cases even death, for the female (Adler 2010). Similarly, in feral sheep, females are exposed to high levels of male harassment and injury during the breeding season, resulting in a female-biased mortality rate (Réale et al. 1996). Male-imposed injuries to females through coercive mating strategies are documented across taxa, sometimes resulting from bizarre adaptations, such as the "hypodermic genitalia" found in the true bug infraorder Cimicomorpha (Heteroptera), in which males inseminate females by stabbing them in the abdomen, often resulting in injury and infection (Stutt and Siva-Jothy 2001; Tatarnic et al. 2014).

Although mating-related injuries to females have generally been regarded as a side effect of male reproductive strategies, some investigators have suggested that such injuries may be adaptive for males. Injured females may be less likely to remate (to avoid further costs), thus reducing the risk of sperm competition for the males (Johnstone and Keller 2000). Injured females may also respond by increasing immediate reproductive output as a form of "terminal investment," an increased reproductive investment in individuals nearing the end of life (as conceived by Williams 1966), resulting in more or higher-quality offspring for the male (Michiels 1998). However, this idea remains controversial and requires direct empirical support (Morrow et al. 2003).

In addition to bodily injury, females may also suffer life span–reducing mating costs through exposure to sexually transmitted infections and parasites, as well as mating-related behaviors that reduce the ability to forage or increase vulnerability to predation (e.g., Jennions and Petrie 2000; Arnqvist and Rowe 2005). In water striders, premating struggles drastically increase the likelihood of predation, and females faced with such mortality risk often respond by engaging in "convenience polyandry" (Rowe 1994). In some species, males appear to intimidate females into mating by drumming on the water surface during struggles—a behavior that attracts predators, which are more likely to attack the female than the male (Han and Jablonski 2010).

In some species, sexually antagonistic interactions may result in latent costs that elevate aging rate. Latent costs may result from somatic wear and tear associated with courtship or premating struggles. For example, in the neriid fly *Telostylinus angusticollis*, both sexes suffered accelerated aging when the operational sex ratio was biased toward the opposite sex (Adler and Bonduriansky 2011). Males may use strategies that elevate female investment in the current reproductive bout. Such a strategy can maximize a male's fitness gain from the mating, but can result in latent costs for females as a result of the trade-off between reproduction and somatic maintenance (*sensu* Kirkwood 1977). A well-known example of a male adaptation to increase female immediate reproductive investment is the "toxic ejaculate" of male *Drosophila*, laden with accessory gland proteins (Acp), such as the "sex peptide," that can increase female oviposition rate but reduce life span (Chapman et al. 1995; Wigby and Chapman 2005), although this effect has been shown to vary with female nutritional status (Fricke et al. 2010). Similarly, males may be selected to elevate female resource allocation to offspring via epigenetic modification of genes in the sperm genome that are involved in resource extraction from the mother (Moore and Haig 1991; Haig 2000). Like Acp-mediated effects on female fecundity, such parent-of-origin (genomic imprinting) effects on female resource al-

 Cite this article as *Cold Spring Harb Perspect Biol* doi: 10.1101/cshperspect.a017566

location are thought to result in latent costs (i.e., accelerated aging) in females (Haig 2000). Females are expected to retaliate by silencing the corresponding genes in the egg—a parental tug-of-war that may account for the evolution of genomic imprinting at some loci (Moore and Haig 1991; Haig 2000).

Environmentally Mediated Changes in the Degree of Sexual Conflict May Affect Life Span and Aging Rate within Populations

There exist some interesting examples of environmental changes that alter the degree of sexual conflict within populations and thereby also bring about changes in life span and aging in females and/or males. Magurran and Seghers (1994) found that female guppies in high-predation sites reduced defense against unwanted males, whereas males at such sites exhibited more coercive behaviors. Because females in high-predation sites experienced an extremely high frequency of male harassment (approximately one sneaky mating attempt per minute), they tended to mate at higher than optimal rates and potentially experienced increased survival costs as a result. Hall et al. (2009) subjected a species of cricket with nuptial feeding behavior to experimental evolution under diets of varying quality. The investigators found that females from populations that evolved on a poor-quality diet exhibited increased nuptial feeding, which resulted in reduced male life span. Indeed, any ecological change that affects the intensity of sexual conflict could thereby affect life span and aging, at least in the short term.

Evolutionary Consequences of Interlocus Sexual Conflict for Life Span and Aging

Interlocus sexual conflict can result in a coevolutionary "arms race" between the sexes, in which the evolution of a male manipulative trait is countered by the coevolution of a female resistant trait, which then selects for increased manipulativeness in males, and so on. But investment in manipulative or defensive traits may come at the cost of survival. Promislow (2003) suggested that species experiencing pervasive sexual conflict are likely to have high mortality rates, because of both the harm inflicted on one sex by the other and the costs of developing traits designed to manipulate or resist the other sex. Although the mortality costs of sexually antagonistic interactions occur on a proximate level, standard evolutionary models predict that increased background mortality rate will result in the evolution of faster aging (Medawar 1946; Williams 1957). Moreover, selection may favor increased resource allocation to manipulative or defensive traits at the cost of investment in late-life survival, such that in populations with high rates of sexual conflict, faster life histories (and increased aging rates) may evolve.

It is important to note, however, that sexual conflict, even if it increases female mortality, might not have the effect of selecting for reduced life span or increased aging rate. Chen and Maklakov (2012) recently showed in the nematode *Caenorhabditis remanei* that high mortality rates resulted in the evolution of reduced life span only when mortality was random. In contrast, condition-dependent mortality resulted in the evolution of increased life span. Condition-dependent mortality may also account for the evolution of extended life spans in guppies from high-predation environments (Reznick et al. 2006). Insofar as sexually antagonistic interactions result in condition-dependent mortality, sexual conflict could result in strong selection on condition and thus drive the evolution of extended life span and reduced aging rate (Bonduriansky et al. 2008; Maklakov and Lummaa 2013). Interestingly, however, male harassment of females can also result in negative condition-dependent mortality, as has been shown for *Drosophila melanogaster*, in which males selectively harass larger (more fecund) females, perhaps thereby reducing the advantage for females of expressing a high-condition phenotype (Long et al. 2009). The consequences of sexually antagonistic interactions for mortality rate, and the relation between mortality and condition, require further research.

DOES INTRALOCUS SEXUAL CONFLICT AFFECT THE EVOLUTION OF LONGEVITY AND AGING?

Sex-Specific Selection Affecting Life Span and Aging

The distinct reproductive strategies pursued by the sexes may often generate very different patterns of selection on strategies that affect life span and aging. In particular, the sex (typically males) that competes more strongly for access to mates tends to be selected to pursue a "live fast, die young" strategy, sacrificing longevity for reproductive opportunity. This is because the sexes usually encounter different "opportunity landscapes," with the sex that experiences stronger sexual competition typically in a position to gain more fitness through risky or costly strategies with a high potential short-term payoff (Fig. 1). Consider a "typical" mating system involving males that compete for matings and females that compete for resources and invest heavily in offspring. For a female, high fitness can accrue only gradually, through successful accumulation of resources and production of young (Vinogradov 1998). Although female reproduction is resource-limited and females compete for resources, the fitness payoff from a high-quality resource patch is moderated by the amount of time required to exploit the resource and the many additional steps involved in the reproductive process. It rarely makes adaptive sense for a female to risk death or substantial somatic damage or to make a large resource investment, for a chance to acquire a food morsel or nesting site. In contrast, for a male, a single successful mating with a high-quality female can represent a considerable fitness gain. Thus, we suggest that males may often be selected to take greater risks than females or to invest a

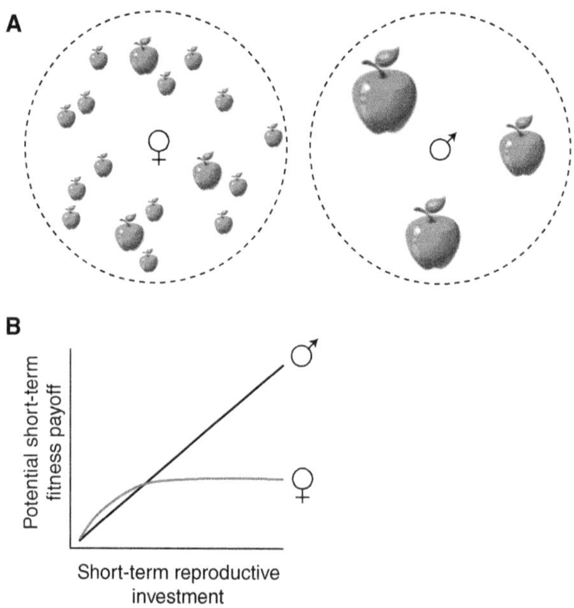

Figure 1. In a "typical" mating system involving stronger sexual competition among males and greater investment in offspring among females, the sexes tend to encounter different "opportunity landscapes" that generate sex-specific selection on longevity and aging rate. (*A*) Females can gain fitness only in relatively small increments (small apples), by accumulating resources and converting those resources into offspring. In contrast, males have greater opportunity for large, short-term fitness gain (large apples) through mating. (*B*) Thus, for males, large, short-term investments in reproduction (including strategies that involve high risk of injury or death) may be favored by selection because such investments may result in large fitness payoffs. In contrast, females cannot gain equally large short-term fitness payoffs and may therefore be selected to pursue relatively safe, long-term strategies involving greater risk avoidance and larger investment in somatic maintenance.

larger quantity of limiting resources for the sake of a reproductive opportunity, thereby sacrificing longevity for reproductive opportunity. This is because the potential short-term fitness payoffs available to males tend to be relatively large, such that males have a higher maximum potential reproductive rate than females (Vinogradov 1998).

The sex-specific optima for reproductive investment and life-history scheduling can be manifested in sexual dimorphism for a wide range of behavioral, physiological, and even morphological traits. For example, it is well established in many species that males are prone to engage in risky behavior, such as intrasexual combat over resources or sexual activity in the presence of predators (Andersson 1994; Magurran and Seghers 1994; Candolin 1998). Sex-differences in behavior are mediated hormonally in vertebrates (Adkins-Regan 2005; Juntti et al. 2010; but see Golinski et al. 2011), and there is evidence of a link between testosterone titer and risk-taking behavior (Sapienza et al. 2009; Stanton et al. 2011). Although the relationship between testosterone and sexual behavior appears to be complex (Marler and Moore 1988; Foerster and Kempenaers 2005; McGlothlin et al. 2008), there is evidence of sexually antagonistic selection on testosterone titer (Mills et al. 2012). In addition to stimulating risk-taking behaviors, testosterone also suppresses the immune system (Duffy et al. 2000; Schroderus et al. 2010). As a result of such phenotypic differences, females of many (but not all) species outlive males on average (Comfort 1979; Finch 1990), although sex differences in aging rate are less clear (Bonduriansky et al. 2008).

The Potential for Intralocus Sexual Conflict over Longevity and Aging

Like most other traits, the behavioral and physiological differences that contribute to differential survival and aging rate in the sexes are likely to share a common developmental-genetic architecture and can therefore be subject to intralocus sexual conflict. Because much of the genome is shared between sexes, sex-specific selection tends to favor different alleles at the same genetic loci in males and females, or to favor the same alleles to a different degree. Selection on one sex can thus displace the other sex from its sex-specific phenotypic optimum. Although selection will favor modifications that weaken genetic constraints on independent evolution of the sexes for such traits, a considerable body of evidence suggests that substantial sexually antagonistic fitness variation segregates in natural and captive populations (Bonduriansky and Chenoweth 2009; Cox and Calsbeek 2009; van Doorn 2009; Connalon et al. 2010).

Establishing intralocus sexual conflict in relation to longevity and aging requires two pieces of information. First, it is necessary to show that selection on these traits is sex-specific. Intralocus sexual conflict becomes increasingly likely with increasing divergence between male and female fitness surfaces, and the most intense conflict occurs when selection is sexually antagonistic (i.e., acts with different sign on sexually homologous traits in males and females). Second, it is necessary to show that genetic constraints impede independent evolution of the sexes. The potential for intralocus sexual conflict increases as the magnitude of the intersexual genetic correlation (r_{mf}) increases (Lande 1980; Bonduriansky and Chenoweth 2009). However, a more complete picture of intersexual genetic constraints requires the estimation of the between-sex genetic covariance matrix (**B**), which incorporates constraints arising from genetic covariances between multiple traits (Lande 1980; Lewis et al. 2011; Gosden et al. 2012). Because fitness and its life-history components are especially prone to environmental effects and genotype × environment interactions (Roff 1992; Stearns 1992), the most meaningful estimates of selection, as well as intersexual genetic constraints, will come from populations experiencing natural or seminatural environmental conditions.

Evidence of Intralocus Sexual Conflict over Longevity and Aging

Although the extent and intensity of intralocus sexual conflict in relation to longevity and aging rate remain poorly known, some intriguing evidence has come from studies documenting sexual dimorphism in longevity and/or aging, quantitative-genetic studies (sometimes combined with assays of sex-specific selection), and artificial evolution studies.

Evidence of Sexual Dimorphism in Longevity and Aging

The existence of sexual dimorphism in longevity and/or aging rate suggests that intralocus sexual conflict may have occurred during the evolution of these traits. However, in the absence of data on intersexual genetic constraints and sex-specific selection, such evidence is not sufficient to establish that intralocus sexual conflict is currently operating.

A few studies have compared longevity and aging in the sexes in natural or seminatural populations. In a natural population of black field crickets (*Teleogryllus commodus*), males experienced a higher mortality rate throughout the life span than females, but there was no evidence of faster aging in males (Zajitschek et al. 2009b). However, in a study on the same species in seminatural enclosures, females exhibited a later-starting but more rapid pattern of actuarial senescence (an increase in mortality rate with age) than males (Zajitschek et al. 2009a). In a natural population of the neriid fly *Telostylinus angusticollis*, extrinsic mortality rate was higher in females than in males, but actuarial aging was detected only in males (Kawasaki et al. 2008). In a wild population of red deer (*Cervus elaphus*), males experienced earlier and more rapid reproductive aging than females, although different components of performance declined at different rates in each sex (Nussey et al. 2009). In a natural population of blue-footed boobies (*Sula nebouxii*), the sexes exhibited similar patterns of reproductive senescence (Kim et al. 2011). In free-ranging populations of 24 species of large mammals (Bovidae and Cervidae), males experienced a higher mortality rate and more rapid actuarial senescence than females on average (Lemaitre and Gaillard 2013). A number of studies have also compared longevity and aging in the sexes in the laboratory (e.g., Kawasaki et al. 2008; Rodríguez-Graña et al. 2010; Ceballos and Kiørboe 2011; Archer et al. 2012b). However, data on laboratory populations must be interpreted with caution, given that relative mortality and aging rates of males and females may be environment-specific.

Some tentative, correlational evidence is also starting to emerge for modern human populations (reviewed in Maklakov and Lummaa 2013). Although women tend to outlive men in most contemporary human populations, sex differences in aging are much less clear, with some somatic indices of senescence showing more rapid deterioration with age in women. There is also evidence, from 18th- to 19th-century Finland, suggesting that the number of grandchildren produced increases with survival to old age for women but not for men (Lahdenpera et al. 2007, 2011). However, in recent human populations, such patterns are likely to be shaped by cultural and socioeconomic factors.

Evidence from Quantitative Genetic and Selection Analyses

A quantitative-genetic analysis of black field cricket (*Teleogryllus commodus*) life history provided no evidence of intralocus sexual conflict over longevity, because it was found that r_{mf} for this trait was not significantly different from zero (Zajitschek et al. 2007). Similarly, a very weak r_{mf} for longevity was observed in the dung fly *Sepsis cynipsea* (Mühlhäuser and Blanckenhorn 2004). A moderately strong and significant r_{mf} for longevity (although not aging rate) was observed in *D. melanogaster* (Lehtovaara et al. 2013). In contrast, in the cricket *Gryllodes sigillatus*, strong, positive r_{mf} was observed for both longevity and aging rate (Archer et al. 2012b). This study shows the potential for intralocus sexual conflict in *G. sigillatus*, but does not show that the sexes are actually displaced from their sex-specific phenotypic optima for these traits. Evidence suggesting intralocus sexual conflict over longevity also comes from a study on the moth *Plodia interpunctella* (Lewis et al. 2011). Selection was found to favor longer life span (with diminishing returns) in males, but shorter life span in females. Although r_{mf} for longevity was low, it was shown that multivariate genetic constraints, reflected in the male–female genetic covariance matrix (**B**), would tend to greatly weaken the potential for each sex (but especially females) to evolve toward its sex-specific longevity optimum. Substantial sexual dimorphism was observed for longevity and other life-history traits, but it remains unclear how far each sex is displaced from its sex-specific phenotypic optimum by intralocus sexual conflict.

Evidence from Artificial Selection Studies

More compelling evidence of intralocus sexual conflict over longevity was furnished by Berg and Maklakov's (2012) experiment on *Callosobruchus maculatus* beetles. In this study, artificial selection on males was applied over five generations. Selection on male longevity resulted in a response in females as well, demonstrating a positive intersexual genetic correlation ($r_{mf} > 0$) for longevity. Moreover, artificial selection on longevity had opposite effects on fitness in the sexes: In lines selected for longer life span, male fitness was (nonsignificantly) reduced, whereas female fitness was increased. These results suggest that, in the base *C. maculatus* population from which the experimental lines were derived, the sexes are displaced from their sex-specific longevity optima by sex-specific selection acting on a shared genetic architecture for longevity.

Where Is Intralocus Sexual Conflict Most Likely to Occur?

Intralocus sexual conflict in relation to longevity and aging rate can be predicted to occur in systems characterized by intense sexual selection on one sex. In such systems, the sexes are likely to exhibit highly divergent reproductive strategies and highly sex-specific patterns of selection on longevity and aging (Promislow 2003). More precisely, the potential for intralocus sexual conflict in relation to longevity and aging rate should relate to differences between the sexes in the maximum potential rate of reproduction (Vinogradov 1998). Although short-term reproductive payoff opportunities may be considerably greater for males in "typical" mating systems (Fig. 1), the "opportunity landscape" experienced by the sexes is likely to depend on mating system parameters, such as the degree of skew in the male mating success distribution, as well as ecological parameters, such as the spatial/temporal distribution and fitness value of food or nest-site resources for females. If matings are relatively easy to achieve, and an additional mating represents only a small fitness gain for a male, then males are unlikely to benefit from taking high risks for mating opportunities, unless they can assess female mate quality and identify particularly high-value mates. Conversely, if food or nest-site resources are sparse, females may be selected to compete and take risks to secure access to them (Clutton-Brock 2007, 2009). Likewise, reversed sex roles may select for reversed patterns of selection on longevity and aging (Maklakov and Lummaa 2013).

DOES VARIATION IN LIFE SPAN AND AGING AFFECT SEXUAL CONFLICT?

Above, we considered the potential for variation in the intensity of sexually antagonistic interactions (interlocus sexual conflict) or sex-specific selection on the life history (intralocus sexual conflict) to affect the expression and evolution of life span and aging. In the following section, we consider the converse possibility: the potential for variation in life span and aging rate to influence the intensity of sexual conflict.

Aging Can Generate Sexual Conflict

In age-structured populations, in which generations overlap, there is potential for conflict to occur over the age of mating partners. As fertility tends to increase early in life and then decline at older ages for most organisms, mating with a very young or very old individual has the potential to reduce fitness for the other mating partner. This can be true for both sexes, for numerous reasons, including reduced ability to provision or care for young and reduced fertility because of reproductive immaturity, low gamete number, or quality (Reinhardt 2007). In the case of very young mating partners, a number of studies have showed that the ability to mate successfully is no guarantee of full reproductive maturity. For example, in the male hide beetle *Dermestes maculatus* (Jones et al. 2007) and both sexes of the common lizard *Lacerta vivipara* (Richard et al. 2005), young individuals produce fewer and less viable offspring. Old mating partners, on the other hand, are less likely to be virgin, and thus perhaps more likely to be gamete-depleted or to carry sexually transmitted diseases. Moreover, mutations may accumulate with age in germ cells, reducing the viability, size, or other fitness-related traits of offspring (Johnson and Gemmell 2012). For example, in humans, advanced maternal age reduces offspring viability and survival and greatly increases risk of Down syndrome (reviewed in Gaulden 1992; Bewley et al. 2005), whereas advanced paternal age increases incidence of childhood cancers and neuropsychiatric conditions such as schizophrenia and, fascinatingly, is associated with reduced cognitive abilities (reviewed in Cannon 2009).

Across species, avoiding mating with suboptimal partners is likely to be more important for females, because females typically invest more in offspring and thus have more to lose by poor mate choice. Because surviving to old age might indicate high genetic quality, although somatic condition invariably declines in old age, there is considerable debate about whether females stand to lose or gain, on the whole, by mating with old males (reviewed in Johnson and Gemmell 2012). However, if old males represent suboptimal mating partners, females are expected to evolve preferences that reduce the likelihood of mating with them. Insofar as females can exert such preferences, there is no inherent conflict. However, sexual conflict may occur if males are able to deceive females about their age or if females are unable to choose. Female preferences for "honest" condition-dependent (i.e., costly) signals may be one mechanism by which females can discriminate against old males. For example, Velando et al. (2008) proposed that females may be able to avoid males with oxidatively damaged germline DNA by basing preference on oxidative-dependent sexual signals that decline with age and oxidative damage, such as foot color intensity in the blue-footed booby (Torres and Velando 2007). However, as Kokko (1997) points out, other costly sexual signals may be fixed or may increase with the age of the male, potentially causing sexual signals to become dishonest and resulting in sexual conflict. For example, Dean et al. (2010) showed that male mating rate declines sharply with age in feral fowl (*Gallus gallus domesticus*), but male dominance rank only declines with age in male-biased sex-ratio groups. Thus, females in a non-male-biased sex ratio group have the potential to suffer fitness costs from mating with low-fertility dominant males.

If selection tends to favor female avoidance of old males, this is likely to select in turn for increased male persistence or coercion with age. Old males may also elevate their mating effort because the opportunity costs of investing in current mating attempts should tend to decrease as a male ages and his chances of future mating decrease. In other words, selection may favor increased persistence as a form of "terminal investment." An example of this sort of exacerbated sexual conflict comes from the beetle *Tenebrio molitor*, in which females that mate with old males have smaller, lower-quality offspring (Carazo et al. 2011). As a result, females resist mating with old males and reduce guarding of spermatophores from old males, whereas old males increase guarding of their own spermatophores postcopulation (Carazo et al. 2011).

Males might also face costs by mating with old females, particularly in species in which males invest heavily in each mating. For example, in male copepods, mating entails high costs, and males are consequently limited in the number of females they can mate with during their short life span. Female fertility decreases with age, such that males waste reproductive investment by mating with subfertile females. Ceballos and Kiørboe (2011) suggest this scenario leads males to invest heavily in mate searching and to be particularly choosy. Even in systems in which males invest less per mating, there is potential for males to suffer fitness costs by mating with old females. Fricke et al. (2013) show that the increased egg-laying response of females to the male sex peptide is reduced in old females. As male ejaculate products have been shown to be costly for males to produce (Blay and Yuval 1997; Baker et al. 2003; Fricke et al. 2008), mating with old females is likely to result in lost fitness for males.

Variation in Life Expectancy Can Modulate Sexual Conflict

As noted above, males in polyandrous populations are selected to manipulate the reproductive effort of their mates via accessory-gland products that induce elevated fecundity (Chapman et al. 1995; Wigby and Chapman 2005) or epigenetic parent-of-origin effects that result in increased extraction of maternal resources by offspring (Moore and Haig 1991; Haig 2000). Such male strategies can confer benefits for males by increasing the number or quality of offspring produced from the mating, but females may suffer reduced fitness from overallocating resources to a single reproductive bout. These costs are likely to be latent—that is, manifested later in life as accelerated aging—because latent costs will have little or no effect on the fitness of the male that induces them. Thus, all else equal, the magnitude of such costs to females will depend on female longevity.

Interestingly, this suggests that female longevity may also determine the intensity of sexual conflict in such systems. Using a simple model, Bonduriansky (2014) showed that a change in background mortality rate (resulting, e.g., from increased predation pressure) can result in a short-term relaxation of sexual conflict: If female life expectancy is reduced, then male-induced elevation in female reproductive effort may become less harmful for females. Similar reasoning suggests that sexual conflict intensity within populations may depend on female condition: If female mortality risk is condition-dependent, then high-condition females (whose life expectancy is high) may experience more intense sexual conflict than low-condition females (whose life expectancy is low). This logic can be extended to female age as well: The reduced opportunity costs of increased investment in current reproduction late in life might also reduce the intensity of sexual conflict experienced by old females. Indeed, there are a number of reports of increased reproductive investment in old females. For example, brushtail possum females increase investment in lactation and offspring production later in life (Isaac and Johnson 2005), whereas old collared flycatcher females feed fledglings more and lose more weight, at the cost of survival to subsequent breeding seasons (Pärt et al. 1992). The hypothesized link between female life expectancy and sexual conflict intensity awaits empirical verification.

DIRECTIONS FOR FUTURE RESEARCH

Research on the link between sexual conflict, life span, and aging is still in its early stages, and much work remains to be done. Below, we highlight some gaps in our current understanding and suggest potentially fruitful avenues for future investigation.

Although potential links between sexual conflict, life span, and aging were first identified many years ago, and a number of investigators have discussed this topic (e.g., Svensson and Sheldon 1998; Promislow 2003; Bonduriansky et al. 2008; Maklakov and Lummaa 2013), there is a pressing need to explore these links more formally through analytical and simulation modeling. For example, it would be interesting to explore the consequences of the elevated mortality or accelerated aging that can result from sexually antagonistic interactions for the evolution of male and female life histories. It would also be interesting to model the effects of environmentally induced variation in aging rate on sexual antagonism. Likewise, although theory suggests that variation in extrinsic mortality rate can modulate the intensity of sexual conflict (Bonduriansky 2014), it would be interesting to examine the consequences of long-term changes in mortality rate for sexual coevolution.

Many questions also remain for empirical investigation. With regard to intralocus sexual conflict, comparative, experimental, and genomic studies are needed to establish when sex-specific selective pressures are most likely to generate strongly divergent selection on life span and aging in the sexes, and to what extent and by what means such conflicts are resolved. We also need a more complete picture of the phenotypic traits that mediate such conflict. Studies examining the ecological, behavioral, or physiological consequences of sex-specific reproductive strategies may illuminate this problem (Magurran and Seghers 1994; Long and Rice 2007; Archer et al. 2012a). The intensity and resolution of such conflicts are likely to be influenced by mating system parameters (Bonduriansky and Chenoweth 2009), but the relationship between mating system and selection on life span and aging may be complex. For example, in some species characterized by intense male–male competition, male fitness may depend largely on the ability to grow to a large body size, and males may therefore be selected for high investment in survival until the critical body size is attained (Graves 2007). Other factors, such as sexual conflict over diet choice (Maklakov et al. 2008; Adler et al. 2013; Reddiex et al. 2013) or the matrilineal inheritance of mitochondria, which limits the scope for evolutionary optimization of mitochondrial function in males (Dowling et al. 2008; Innocenti et al. 2011; Camus et al. 2013), may also influence the dynamics of such conflicts. With regard to interlocus sexual conflict, recent evidence shows that costs to females of male manipulative strategies are likely to be highly context-dependent (Fricke et al. 2009a,b, 2013; Edward et al. 2010). More research is thus needed to establish when and how male strategies impose harm, and how these strategies affect female fitness. To discern general patterns, we need studies under more realistic environmental conditions (see below), as well as on a wide variety of organisms.

A further problem, relating to all of the topics discussed in this paper, is that most of the available empirical evidence relates to life span (which is often used as a proxy for aging rate), although very little is known about the role of aging per se. As a general rule, longevity and aging rate may be expected to covary negatively among taxa because greater risk taking (e.g., by males in "typical" systems) may result in both elevated immediate mortality risk that reduces life expectancy and in greater somatic wear and tear that results in faster aging. Moreover, classic theory suggests that shorter life expectancy will lead to the evolution of faster aging (Williams 1957; Hamilton 1966). However, predictions in relation to aging are complicated by several factors. First, much somatic damage can, in principle, be repaired, so a high rate of wear and tear may be balanced by a large investment in somatic repair. Second, longevity and aging rate may be nonindependent. For example, an organism that survives to an advanced age as a result of low mortality risk in early and middle age may eventually suffer very rapid somatic deterioration (Zajitschek et al. 2009a). Third, when extrinsic mortality is nonrandom with respect to condition, elevated mortality can select for a more robust phenotype that ages less rapidly (Chen and Maklakov 2012), and, in principle, males may therefore evolve to age less rapidly despite having a shorter life expectancy than females (Maklakov and Lummaa 2013). Finally, the quantification of aging rate is a nontrivial problem because aging trajectories are often complex and nonlinear, and different components of performance may decline with age at different rates (Nussey et al. 2009). To understand the relationship between aging and sexual conflict, we need to know which features of the aging trajectory experience the strongest and most divergent patterns of selection in the sexes, and which are subject to the strongest intersexual genetic constraints.

Finally, empirical evidence on the relationship between sexual conflict, life span, and aging comes largely from laboratory studies. Although such studies allow for controlled experimentation, captive conditions may also affect (and often exacerbate) both inter- and intralocus sexual conflicts (Bonduriansky and Chenoweth 2009; Bonduriansky 2014). For example, if life span is extended in the laboratory relative to the wild source population (Kawasaki et al. 2008), then latent fitness costs of male manipulation may be more important for captive females than for their wild counterparts. Captive animals may also tend to experience higher population density than is typically the case for natural populations, and captive females may have little opportunity to escape from harassing males, potentially exacerbating the fitness effects of sexually antagonistic interactions. Likewise, because captivity tends to eliminate many sources of viability selection (most of which are shared by the sexes), while typically preserving sex-specific selection pressures associated with sexual competition, captive populations may experience more intense intralocus sexual conflict than wild populations. Even if model organisms such as *Callosobruchus maculatus* are well adapted to laboratory conditions, an environment in which predators and parasites are rare and resources are superabundant is probably not representative of the conditions experienced by most natural populations. To achieve a better understanding of the relationship between sexual conflict, life span, and aging, it will be necessary to carry out field studies, as well as laboratory studies that subject experimental animals to the kinds of challenges that are faced by natural populations.

ACKNOWLEDGMENTS

This work is supported by the Australian Research Council through a fellowship and research grant to R.B. We are grateful to Urban Friberg and an anonymous reviewer for helpful comments on a draft of this article.

REFERENCES

Adkins-Regan E. 2005. *Hormones and animal social behavior.* Princeton University Press, Princeton, NJ.

Adler M. 2010. Sexual conflict in waterfowl: Why do females resist extra-pair copulations? *Behav Ecol* **21:** 182–192.

Adler MI, Bonduriansky R. 2011. The dissimilar costs of love and war: Age-specific mortality as a function of the operational sex ratio. *J Evol Biol* **24:** 1169–1177.

Adler MI, Cassidy EJ, Fricke C, Bonduriansky R. 2013. The lifespan-reproduction trade-off under dietary restriction is sex-specific and context-dependent. *Exp Gerontol* **48:** 539–548.

Andersson M. 1994. *Sexual selection.* Princeton University Press, Princeton, NJ.

Archer CR, Sakaluk SK, Selman C, Royle NJ, Hunt J. 2012a. Oxidative stress and the evolution of sex differences in life span and ageing in the decorated cricket, *Gryllodes sigillatus. Evolution* **67:** 620–634.

Archer CR, Zajitschek F, Sakaluk SK, Royle NJ, Hunt J. 2012b. Sexual selection affects the evolution of lifespan and ageing in the decorated cricket *Gryllodes sigillatus. Evolution* **66:** 3088–3100.

Arnqvist G, Rowe L. 2005. *Sexual conflict.* Princeton University Press, Princeton, NJ.

Baker RH, Denniff M, Futerman P, Fowler K, Pomiankowski A, Chapman T. 2003. Accessory gland size influences time to sexual maturity and mating frequency in the stalk-eyed fly, *Cyrtodiopsis dalmanni. Behav Ecol* **14:** 607–611.

Berg EC, Maklakov A. 2012. Sexes suffer from suboptimal lifespan because of genetic conflict in a seed beetle. *Proc Biol Sci* **279:** 4296–4302.

Bewley S, Davies M, Braude P. 2005. Which career first? The most secure age for childbearing remains 20–35. *BMJ* **331:** 588–589.

Blay S, Yuval B. 1997. Nutritional correlates of reproductive success of male Mediterranean fruit flies (Diptera: Tephritidae). *Anim Behav* **54:** 59–66.

Bonduriansky R. 2014. The ecology of sexual conflict: Background mortality can modulate the effects of male manipulation on female fitness. *Evolution* **68:** 595–604.

Bonduriansky R, Chenoweth SF. 2009. Intralocus sexual conflict. *Trends Ecol Evol* **24:** 280–288.

Bonduriansky R, Maklakov A, Zajitschek F, Brooks R. 2008. Sexual selection, sexual conflict and the evolution of ageing and lifespan. *Funct Ecol* **22:** 443–453.

Camus MF, Clancy DJ, Dowling DK. 2013. Mitochondria, maternal inheritance, and male ageing. *Curr Biol* **22:** 1717–1721.

Candolin U. 1998. Reproduction under predation risk and the trade-off between current and future reproduction in the threespine stickleback. *Proc Biol Sci* **265:** 1171–1175.

Cannon M. 2009. Contrasting effects of maternal and paternal age on offspring intelligence. *PLoS Med* **6:** e1000042.

Carazo P, Molina-Vila P, Font E. 2011. Male reproductive senescence as a potential source of sexual conflict in a beetle. *Behav Ecol* **22.**

Ceballos S, Kiørboe T. 2011. Senescence and sexual selection in a pelagic copepod. *PLoS ONE* **6:** e18870.

Chapman T, Liddle LF, Kalb JM, Wolfner MF, Partridge L. 1995. Cost of mating in *Drosophila melanogaster* females is mediated by male accessory gland products. *Nature* **373:** 241–244.

Chen HY, Maklakov AA. 2012. Longer life span evolves under high rates of condition-dependent mortality. *Curr Biol* **22:** 2140–2143.

Clutton-Brock T. 2007. Sexual selection in males and females. *Science* **318:** 1882–1885.

Clutton-Brock T. 2009. Sexual selection in females. *Anim Behav* **77:** 3–11.

Comfort A. 1979. *The biology of senescence.* Elsevier, New York.

Connalon T, Cox RM, Calsbeek R. 2010. Fitness consequences of sex-specific selection. *Evolution* **64:** 1671–1682.

Cox RM, Calsbeek R. 2009. Sexually antagonistic selection, sexual dimorphism, and the resolution of intralocus sexual conflict. *Am Nat* **173:** 176–187.

Dean R, Cornwallis CK, Løvlie H, Worley K, Richardson DS, Pizarri T. 2010. Male reproductive senescence causes potential for sexual conflict over mating. *Curr Biol* **20:** 1192–1196.

Dowling DK, Friberg U, Lindell J. 2008. Evolutionary implications of non-neutral mitochondrial genetic variation. *Trends Ecol Evol* **23:** 546–554.

Duffy DL, Bentley GE, Drazen DL, Ball GF. 2000. Effects of testosterone on cell-mediated and humoral immunity in non-breeding adult European starlings. *Behav Ecol* **11:** 654–662.

Edward DA, Fricke C, Gerrard DT, Chapman T. 2010. Quantifying the life-history responses to increased male exposure in female *Drosophila melanogaster*. *Evolution* **65:** 564–573.

Finch CE. 1990. *Longevity, senescence, and the genome*. University of Chicago Press, Chicago.

Foerster K, Kempenaers B. 2005. Effects of testosterone on male–male competition and male–female interactions in blue tits. *Behav Ecol Sociobiol* **57:** 215–223.

Fricke C, Bretman A, Chapman T. 2008. Adult male nutrition and reproductive success in *Drosophila melanogaster*. *Evolution* **62:** 3170–3177.

Fricke C, Perry J, Chapman T, Rowe L. 2009a. The conditional economics of sexual conflict. *Biol Lett* **5:** 671–674.

Fricke C, Wigby S, Hobbs R, Chapman T. 2009b. The benefits of male ejaculate sex peptide transfer in *Drosophila melanogaster*. *J Evol Biol* **22:** 275–286.

Fricke C, Bretman A, Chapman T. 2010. Female nutritional status determines the magnitude and sign of responses to a male ejaculate signal in *Drosophila melanogaster*. *J Evol Biol* **23:** 157–165.

Fricke C, Green D, Mills WE, Chapman T. 2013. Age-dependent female responses to a male ejaculate signal alter demographic opportunities for selection. *Proc Biol Sci* **280:** 20130428.

Gaulden ME. 1992. Maternal age effect: The enigma of Down syndrome and other trisomic conditions. *Mutat Res* **296:** 69–88.

Golinski A, John-Adler H, Kratochvil L. 2011. Male sexual behavior does not require elevated testosterone in a lizard (*Coleonyx elegans*, Eublepharidae). *Horm Behav* **59:** 144–150.

Gosden TP, Shastri KL, Innocenti P, Chenoweth SF. 2012. The B-matrix harbors significant and sex-specific constraints on the evolution of multicharacter sexual dimorphism. *Evolution* **66:** 2106–2116.

Graves BM. 2007. Sexual selection effects on the evolution of senescence. *Evol Ecol* **21:** 663–668.

Haig D. 2000. The kinship theory of genomic imprinting. *Annu Rev Ecol Syst* **31:** 9–32.

Hall MD, Bussiere LF, Brooks R. 2009. Diet-dependent female evolution influences male lifespan in a nuptial feeding insect. *J Evol Biol* **22:** 873–881.

Hamilton WD. 1966. The moulding of senescence by natural selection. *J Theor Biol* **12:** 12–45.

Han CS, Jablonski PG. 2010. Male water striders attract predators to intimidate females into copulation. *Nat Commun* **1:** 1–6.

Innocenti P, Morrow EH, Dowling DK. 2011. Experimental evidence supports a sex-specific selective sieve in mitochondrial genome evolution. *Science* **332:** 845–848.

Isaac JL, Johnson CN. 2005. Terminal reproductive effort in a marsupial. *Biol Lett* **1:** 271–275.

Jennions MD, Petrie M. 2000. Why do females mate multiply? A review of the genetic benefits. *Biol Rev Camb Philos Soc* **75:** 21–64.

Johnson SL, Gemmell NJ. 2012. Are old males still good males and can females tell the difference? *Bioessays* **34:** 609–619.

Johnstone RA, Keller L. 2000. How males can gain by harming their mates: Sexual conflict, seminal toxins, and the cost of mating. *Am Nat* **156:** 368–377.

Jones TM, Featherson R, Paris DBBP, Elgar MA. 2007. Age-related sperm transfer and sperm competitive ability in the male hide beetle. *Behav Ecol* **18:** 251–258.

Juntti SA, Tollkuhn J, Wu MV, Fraser EJ, Soderborg T, Tan S, Honda SI, Harada N, Shah NM. 2010. The androgen receptor governs the execution, but not programming, of male sexual and territorial behaviors. *Neuron* **66:** 260–272.

Kawasaki N, Brassil CE, Brooks R, Bonduriansky R. 2008. Environmental effects on the expression of lifespan and aging: An extreme contrast between wild and captive cohorts of *Telostylinus angusticollis* (Diptera: Neriidae). *Am Nat* **172:** 346–357.

Kim SY, Velando A, Torres R, Drummond H. 2011. Effects of recruiting age on senescence, lifespan and lifetime reproductive success in a long-lived seabird. *Oecologia* **166:** 615–626.

Kirkwood TBL. 1977. Evolution of ageing. *Nature* **270:** 301–304.

Kokko H. 1997. Evolutionarily stable strategies of age-dependent sexual advertisement. *Behav Ecol Sociobiol* **41:** 99–107.

Lahdenpera M, Russell AF, Lummaa V. 2007. Selection for long lifespan in men: Benefits of grandfathering? *Proc Biol Sci* **274:** 2437–2444.

Lahdenpera M, Lummaa V, Russell AF. 2011. Selection on male longevity in a monogamous human population: Late-life survival brings no additional grandchildren. *J Evol Biol* **24:** 1053–1063.

Lande R. 1980. Sexual dimorphism, sexual selection, and adaptation in polygenic characters. *Evolution* **34:** 292–305.

Lehtovaara A, Schielzeth H, Flis I, Friberg U. 2013. Heritability of life span is largely sex limited in *Drosophila*. *Am Nat* **182:** 653–665.

Lemaitre JF, Gaillard JM. 2013. Male survival patterns do not depend on male allocation to sexual competition in large herbivores. *Behav Ecol* **24:** 421–428.

Lewis Z, Wedell N, Hunt J. 2011. Evidence for strong intralocus sexual conflict in the Indian meal moth, *Plodia interpunctella*. *Evolution* **65:** 2085–2097.

Long TAF, Rice WR. 2007. Adult locomotory activity mediates intralocus sexual conflict in a laboratory-adapted population of *Drosophila melanogaster*. *Proc Biol Sci* **274:** 3105–3112.

Long TAF, Pischedda A, Stewart AD, Rice WR. 2009. A cost of sexual attractiveness to high-fitness females. *PLoS Biol* **7:** e10000254.

Magurran AE, Seghers BH. 1994. Sexual conflict as a consequence of ecology: Evidence from guppy, *Poecilia reticulata*, populations in Trinidad. *Proc Biol Sci* **255:** 31–36.

Maklakov A, Lummaa V. 2013. Evolution of sex differences in lifespan and aging: Causes and constraints. *Bioessays* **35:** 717–724.

Maklakov AA, Simpson SJ, Zajitschek F, Hall MD, Dessmann J, Clissold F, Raubenheimer D, Bonduriansky R, Brooks RC. 2008. Sex-specific fitness effects of nutrient intake on reproduction and lifespan. *Curr Biol* **18:** 1062–1066.

Marler CA, Moore MC. 1988. Evolutionary costs of aggression revealed by testosterone manipulations in free-living male lizards. *Behav Ecol Sociobiol* **23:** 21–26.

McGlothlin JW, Jawor JM, Greives TJ, Phillips JL, Ketterson ED. 2008. Hormones and honest signals: Males with larger ornaments elevate testosterone more when challenged. *J Evol Biol* **21:** 39–48.

Medawar PB. 1946. Old age and natural death. *Mod Q* **2:** 30–49.

Michiels NK. 1998. Mating conflicts and sperm competition in simultaneous hermaphrodites. In *Sperm competition and sexual selection* (ed. Birkhead TR, Møller AP), pp. 219–254. Academic, London.

Mills SC, Koskela E, Mappes T. 2012. Intralocus sexual conflict for fitness: Sexually antagonistic alleles for testosterone. *Proc Biol Sci* **279:** 1889–1895.

Moore T, Haig D. 1991. Genomic imprinting in mammalian development: A parental tug-of-war. *Trends Genet* **7:** 45–49.

Morrow EH, Arnqvist G, Pitnick S. 2003. Adaptation versus pleiotropy: Why do males harm their mates? *Behav Ecol* **14:** 802–806.

Mühlhäuser C, Blanckenhorn WU. 2004. The quantitative genetics of sexual selection in the dung fly *Sepsis cynipsea*. *Behaviour* **141:** 327–341.

Nussey DH, Kruuk LEB, Morris A, Clements MN, Pemberton JM, Clutton-Brock TH. 2009. Inter- and intrasexual variation in aging patterns across reproductive traits in a wild red deer population. *Am Nat* **174:** 342–357.

Parker GA. 1979. Sexual selection and sexual conflict. In *Sexual selection and reproductive competition in insects* (ed. Blum MS, Blum NA), pp. 123–166. Academic, New York.

Pärt T, Gustafsson L, Moreno J. 1992. "Terminal investment" and a sexual conflict in the collared flycatcher (*Ficedula albicollis*). *Am Nat* **140:** 868–882.

Promislow D. 2003. Mate choice, sexual conflict, and evolution of senescence. *Behav Genet* **33:** 191–201.

Réale D, Boussès P, Chapuis JL. 1996. Female-biased mortality induced by male sexual harassment in a feral sheep population. *Can J Zool* **74:** 1812–1818.

Reddiex AJ, Gosden TP, Bonduriansky R, Chenoweth SF. 2013. Sex-specific fitness consequences of nutrient intake and the evolvability of diet preferences. *Am Nat* **182:** 91–102.

Reinhardt K. 2007. Evolutionary consequences of sperm cell aging. *Q Rev Biol* **82:** 375–393.

Reznick DA, Bryant M, Holmes D. 2006. The evolution of senescence and post-reproductive lifespan in guppies (*Poecilia reticulata*). *PLoS Biol* **4:** e7.

Richard M, Lecomte J, De Fraipont M, Clober J. 2005. Age-specific mating strategies and reproductive senescence. *Mol Ecol* **14:** 3147–3155.

Rodríguez-Graña L, Calliari D, Tiselius P, Hansen BW, Sköld HN. 2010. Gender-specific ageing and non-Mendelian inheritance of oxidative damage in marine copepods. *Mar Ecol Prog Ser* **401:** 1–13.

Roff DA. 1992. *Evolution of life histories: Theory and analysis*. Chapman & Hall, New York.

Rowe L. 1994. The costs of mating and mate choice in water striders. *Anim Behav* **48:** 1049–1056.

Sapienza P, Zaingales L, Maestripieri D. 2009. Gender differences in financial risk aversion and career choices are affected by testosterone. *Proc Natl Acad Sci* **106:** 15268–15273.

Schroderus E, Jokinen I, Koivula M, Koskela E, Mappes T, Mills SC, Oksanen TA, Poikonen T. 2010. Intra- and intersexual trade-offs between testosterone and immune system: Implications for sexual and sexually antagonistic selection. *Am Nat* **176:** E90–E97.

Stanton SJ, Liening SH, Schultheiss OC. 2011. Testosterone is positively associated with risk taking in the Iowa Gambling Task. *Horm Behav* **59:** 252–256.

Stearns SC. 1992. *The evolution of life histories*. Oxford University Press, Oxford.

Stutt AD, Siva-Jothy MT. 2001. Traumatic insemination and sexual conflict in the bed bug *Cimex lectularius*. *Proc Natl Acad Sci* **98:** 5683–5687.

Svensson E, Sheldon BC. 1998. The social context of life history evolution. *Oikos* **83:** 466–477.

Tatarnic NJ, Cassis G, Siva-Jothy MT. 2014. Traumatic insemination in terrestrial arthropods. *Ann Rev Entomol* **59:** 245–261.

Torres R, Velando A. 2007. Male reproductive senescence: The price of immune-induced oxidative damage on sexual attractiveness in the blue-footed booby. *J Anim Ecol* **76:** 1161–1168.

van Doorn GS. 2009. Intralocus sexual conflict. *Ann NY Acad Sci* **1168:** 52–71.

Velando A, Torres R, Alonso-Alvarez C. 2008. Avoiding bad genes: Oxidatively damaged DNA in germ line and mate choice. *BioEssays* **30:** 1212–1219.

Vinogradov AE. 1998. Male reproductive strategy and decreased longevity. *Acta Biotheoretica* **46:** 157–160.

Wigby S, Chapman T. 2005. Sex peptide causes mating costs in female *Drosophila melanogaster*. *Curr Biol* **15:** 316–321.

Williams GC. 1957. Pleiotropy, natural selection, and the evolution of senescence. *Evolution* **11:** 398–411.

Williams GC. 1966. Natural selection, the costs of reproduction, and a refinement of Lack's Principle. *Am Nat* **100:** 687–690.

Zajitschek F, Hunt J, Zajitschek SRK, Jennions MD, Brooks R. 2007. No intra-locus sexual conflict over reproductive fitness or ageing in field crickets. *PLoS ONE* **2:** e155.

Zajitschek F, Bonduriansky R, Zajitschek SRK, Brooks R. 2009a. Sexual dimorphism in life history: Age, survival, and reproduction in male and female field crickets *Teleogryllus commodus* under seminatural conditions. *Am Nat* **173:** 792–802.

Zajitschek F, Brassil CE, Bonduriansky R, Brooks R. 2009b. Sex effects on life span and senescence in the wild when dates of birth and death are unknown. *Ecology* **90:** 1698–1707.

Copulatory Wounding and Traumatic Insemination

Klaus Reinhardt[1], Nils Anthes[1], and Rolanda Lange[1,2]

[1]Animal Evolutionary Ecology, Institute of Evolution and Ecology, University of Tübingen, D-72076 Tübingen, Germany

[2]School of Biological Sciences, Monash University, Clayton 3800, Australia

Correspondence: k.reinhardt@uni-tuebingen.de

Copulatory wounding (CW) is widespread in the animal kingdom, but likely underreported because of its cryptic nature. We use four case studies (*Drosophila* flies, *Siphopteron* slugs, *Cimex* bugs, and *Callosobruchus* beetles) to show that CW entails physiological and life-history costs, but can evolve into a routine mating strategy that, in some species, involves insemination through the wound. Although interspecific variation in CW is documented, few data exist on intraspecific and none on individual differences. Although defensive mechanisms evolve in the wound recipient, our review also indicates that mating costs in species with CW are slightly higher than in other species. Whether such costs are dose- or frequency-dependent, and whether defense occurs as resistance or tolerance, decisively affects the evolutionary outcome. In addition to sexual conflict, CW may also become a model system for reproductive isolation. In this context, we put forward a number of predictions, including (1) occasional CW is more costly than routine CW, (2) CW is more costly in between- than within-population matings, and (3) in the presence of CW, selection may favor the transmission of sexually transmitted diseases if they induce resource allocation. Finally, we outline, and briefly discuss, several medical implications of CW in humans.

Males and females do not have the same fitness optima for all traits (Bateman 1948; Parker 1979; see contributions to this collection), which results in intraspecific, intragenomic sexual conflict (Rice and Chippendale 2001). Sex-limited expression of traits is one way in which the conflict over divergent trait optima can be temporarily resolved. Sex-limited trait expression may, however, result in intraspecific, intergenomic conflict in which traits in one sex can reduce the fitness of the other (Rice and Chippendale 2001). Traits related to mating may be particularly involved in this type of conflict. Whereas gamete transfer requires some degree of cooperation, the interaction between males and females during mating rarely results in perfect harmony. Some extreme cases have been reported in which mating attempts by waterfowl or toad males resulted in the immediate death of their mates (Holland Rice 1998). These cases have remained anecdotes, likely because such male behavior will not result in fitness gains and thus be selected against. In other species, mating routinely involves forceful and

Cite this article as *Cold Spring Harb Perspect Biol* doi: 101101/cshperspect.a017582

apparently aggressive physical contacts between mating partners. A well-studied system concerns the male clasping devices and behaviors to prevent females from escaping during copulation in water striders (Arnqvist and Rowe 2002a,b). This system also shows female evolution in response to male aggression (Arnqvist and Rowe 2002a,b). Other cases of sexual aggression are known as sexual punishment (Clutton-Brock and Parker 1995) or situation exploitation (Reinhardt et al. 2009b), in which members of one sex exploit circumstances or actively produce signals that reduce the ability of the other sex to defend themselves against mating attempts.

One form of seemingly aggressive mating interaction that has received systematic treatment only very recently (Lange et al. 2013a; Tatarnic et al. 2014) concerns cases in which the mating partner's genital tract is injured during copulation. These cases are defined as copulatory wounding (CW) (Box 1) (Lange et al. 2013a). Although CW may appear as an accidental exception and so easily dismissed as being anecdotal, a recent literature review revealing that CW independently evolved in at least 36 taxa (Fig. 1) (Lange et al. 2013a) instead identifies it as a phylogenetically widespread phenomenon.

CW occurs in various forms and can be differently categorized; here, we use the recently unified terminology and categorization of the different types of CW (Box 1) (Lange et al. 2013a). Before we deal with these technical aspects and definitions and discuss the origin of CW and evolutionary consequences of its costs and benefits, we introduce the topic by means of four illustrative case studies. These were selected among other well-studied animal model systems (reviewed by Lange et al. 2013a; Tatarnic et al. 2014) to represent the great diversity of CW mechanisms, while also providing a relatively advanced state of knowledge about their physiological underpinnings. We conclude by illustrating some of the medical and legal consequences of CW in humans.

BOX 1. DEFINITIONS OF TERMS RELATED TO REPRODUCTION AS USED IN THIS REVIEW

Copulation: Physical coupling of the sperm donor and recipient and intromission of the sperm donor's copulatory organ into the sperm recipient's body (Alexander et al. 1997).

Copulatory organ: Any device used to transfer or receive sperm, including genitalia, spermatophores, and structures that secondarily acquired the function of sperm transfer or receipt.

Copulatory wounding (CW): An infliction of a wound during copulation involving copulatory organs. The routine occurrence of CW (fixed trait) is classified as traumatic mating (TM).

Genitalia: In males, organs that initially in evolution function to transfer sperm and seminal fluid; in females, these are organs that initially served to accommodate male genitalia and/or receive sperm and seminal fluid.

Insemination: Sperm enter the recipient's body (Alexander et al. 1997).

Inverted copulation: A process by which the sperm recipient uses structures to intrude the body of the sperm donor in order to obtain sperm.

Mating: Generic term for animals interacting to breed; more generic than copulation.

Trauma: An externally inflicted physical wound.

Traumatic insemination (TI): The transfer of sperm through a wound (from Lange et al. 2013).

Traumatic mating (TM): Generic term to describe cases where, routinely, the partner's intraB or extragenital integument is wounded by specialized devices in the direct context of copulation (Lange et al. 2013). TM subsumes TI, TST, and TP.

Traumatic penetration (TP): TM without substances being transferred through the trauma caused (Lange et al. 2013).

Traumatic secretion transfer (TST): TM that includes the transfer of secretions, but not sperm through the trauma (from Lange et al. 2013).

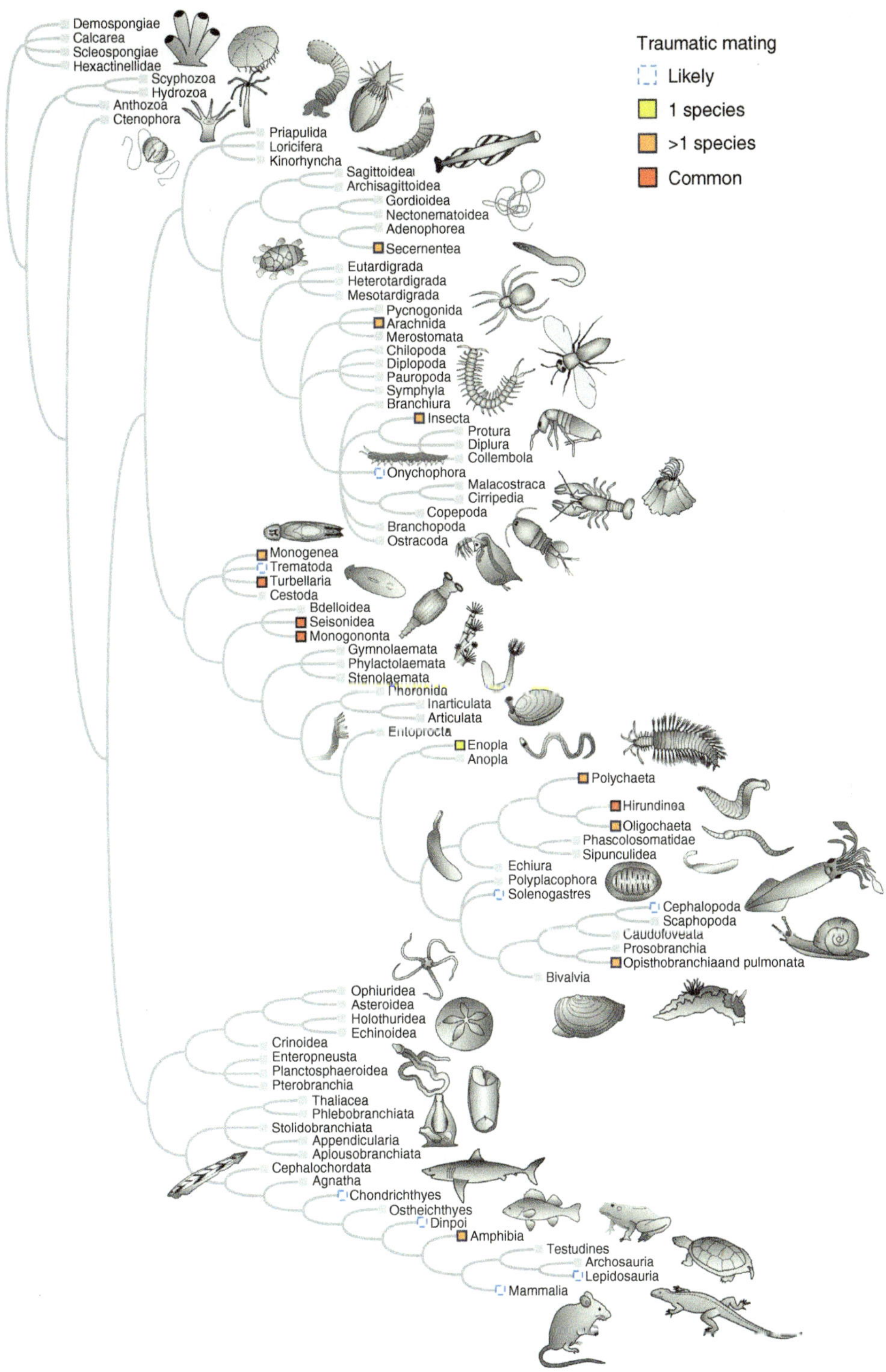

Figure 1. Phylogenetic distribution of TM in the animal kingdom. (Figure modified from Anthes 2010.)

CASE STUDIES OF CW

Drosophila: Fruit Flies

In 2005, CW in *Drosophila* fruit flies was independently observed by Yoshitaka Kamimura in Hokkaido (Kamimura 2007) and Helen Crudgington in Sheffield (*Drosophila subobscura*) (H Crudgington, unpubl.). Since then, fruit flies have become one of the best-studied taxa to reveal mechanisms and functions of CW. It occurs regularly and in the female is both intragenital, as well as extragenital (abdomen) (Kamimura 2007, 2010). Wounds are inflicted by appendages of male genitalia (Fig. 2) (Kamimura 2007; Kamimura and Mitsumoto 2011) in at least 23 species of the *Drosophila ananassae* and *melanogaster* subgroups, and a few other species (Kamimura 2010).

In the *Drosophila melanogaster* subgroup, two types of male structures puncture the female abdomen, both ventrally and dorsally (Kamimura and Mitsumoto 2011). The frequency by which copulations result in CW differs among species of this subgroup. In *D. melanogaster*, *Drosophila simulans*, and *Drosophila teisseri*, genital wounding occurred in all observed cases (Kamimura and Mitsumoto 2011, 2012) and represents "traumatic penetration" (TP; Box 1), in *Drosophila sechellia*, CW occurs in 66% of studied cases (Kamimura 2011).

Seminal fluid is not transferred through the wound in any of the species of the *melanogaster* group. In contrast, in *Drosophila eugracilis*, seminal fluid, but not sperm, is transferred through the genital wounds (Kamimura 2010), representing "traumatic secretion transfer" (TST; Box 1).

"Traumatic insemination" (TI; Box 1) has evolved from TP at least twice independently in *Drosophila*. In addition to the *melanogaster* group, species of the *D. bipectinata* species complex in the *ananassae* subgroup show TI through the genital wounds (Fig. 2) (Kamimura 2007).

The wound-causing genital spines in *D. ananassae* promote anchorage between partners and provide males a benefit in sperm competition (Grieshop and Polak 2012). Costs of this anchorage to females are unknown, but may arise from infection (Kamimura 2012). For instance, the female immune system in *D. melanogaster* responds to seminal fluids (Peng et al. 2005; Domanitskaya et al. 2007), but not to bacteria (Gendrin et al. 2009) introduced to the reproductive tract (Kamimura 2012). However, microbes experimentally deposited on male genital plates persisted for longer and induced stronger immune responses in the male than when deposited on other tissues (Gendrin et al. 2009).

Females in the *melanogaster* group show a pattern of sclerotized structures that are hypothesized to function as a resistance adaptation (Kamimura and Mitsumoto 2011; Yassin and Orgogozo 2013). Correspondingly, interspecific matings between the sister species *Drosophila santomea* and *Drosophila yakuba* resulted in more damaging CW than in intraspecific matings (Kamimura 2012).

Siphopteron: Sea Slugs

At least five species within the hermaphroditic sea slug genus *Siphopteron* show CW. All inject prostate fluids into the body of their mate during every mating using a genital appendage terminating in an injection stylet (Fig. 2) (Anthes and Michiels 2007a,b; Lange et al. 2012, 2013b, 2014). Wound infliction and ejaculate transfer through the wound involve different structures (Anthes and Michiels 2007a), providing the potential to isolate the costs of traumatic injections from general mating costs. Whereas high mating rates, in general, cause a decline in female fecundity beyond an intermediate fitness optimum (Lange et al. 2013a), these short-term fecundity costs could not be linked to the wound infliction per se (Lange et al. 2013b), making *Siphopteron* a candidate to test whether TST is maintained by a paternity benefit to the wound inflictor.

The various *Siphopteron* species have highly divergent and often species-specific traumatic injection sites (Lange et al. 2014). *Siphopteron* may, therefore, be a good model to test whether CW occurs localized because of constraints in the mating position or specific wounding sites produce different fitness outcomes. There

Figure 2. Well-studied model systems of CW. (*Top*) A mating pair of *Drosophila bipectinata*, as well as a close-up of a pair of basal processes of the aedeagus (highlighted in orange) and gonopods (highlighted in green) that make a pair of novel, harmful intromittent organs for sperm transfer. (Image from photographer and copyright holder, Y. Kamimura, Keio University, Yokohama, Japan; reprinted, with permission.) (*Bottom*) A mating pair of *Siphopteron* species 1 (sensu Lange et al. 2014), showing the position (dashed rectangle) and an enlarged view (*inset*) of the intromittent organ, a needle-like injection stylet. (Image from photographer and copyright holders, Johanna Werminghausen and Rolanda Lange, University of Tübingen, Germany; reprinted, with permission.)

is also substantial between- and within-species variation in the shape and length of penile spines and copulatory stylets (Reichel 2011), calling for studies to assess the relationship between variation in genitalia morphology and CW.

Cimicidae: Bedbugs

Bedbugs (Cimicidae) comprise approximately 100 species, which all show traumatic mating (Usinger 1966) that has been classified as TI (Lange et al. 2013a). Shape variation in male intromittent organs is relatively small between species (Usinger 1966), but length variation has been documented within several species (Usinger 1966; Reinhardt and Jacobs 2006). In the best-studied system, the common bedbug *Cimex lectularius* (Fig. 3), males harbor microbes on their intromittent organs and transfer them to females during mating (Reinhardt et al. 2005; Otti et al. 2013). Females cannot avoid mating by behavioral means (Reinhardt et al. 2009b) and, hence, can neither avoid CW nor infection. Females therefore rely on morphological or physiological defense to ameliorate the costs of CW. In most species, TI occurs localized where a secondary genital structure, the spermalege, has evolved (Fig. 3). In the common bedbug, the spermalege is a complex organ that activates sperm and digests seminal fluid (Davies 1966) and has experimentally been shown to reduce infection (Reinhardt et al. 2003), the effects of wounding (Morrow and Arnqvist 2003), and water loss (Benoit et al. 2012). Despite these defense functions, females do incur costs from an increased mating rate (Stutt and Siva-Jothy 2001; Morrow and Arnqvist 2003; Polanco et al. 2011). When the costs of the trauma and possible infection were kept constant experimentally, higher amounts of ejaculate substances accelerated female mortality, but overall provided benefits to females in the form of higher fecundity and delayed reproductive senescence (Reinhardt et al. 2009a). How precisely female aging is delayed is unknown, but it appears that immune-active ejaculate substances are involved (Otti et al. 2013).

The various cimicid species differ in the position of where the spermalege is situated on the body. In two species, the spermalege has evolved to having an actual opening that males inseminate into. Following Lange et al. (2013a), these cases are classified as representing TI. However, as is the case in other taxa of true bugs (Carayon 1966a; Tatarnic et al. 2014), mating is, strictly speaking, not traumatic anymore because no wounding is inflicted (see further discussion in the section Major Evolutionary Trends of CW).

Rather peculiarly, males in two cimicid species, *Afrocimex constrictus* and *Latrocimex spectans*, wound and inseminate other males (Carayon 1966b; Reinhardt et al. 2007; K Reinhardt and L Holman, unpubl.). Males of at least one of the species, *A. constrictus*, possess a spermalege, including those structures that have been shown to reduce copulatory infection in *C. lectularius* (Reinhardt et al. 2007). A certain proportion of females mimic this male-type of spermalege, and these females appear to experience a reduced mating rate (Reinhardt et al. 2007).

Callosobruchus: Bean Weevils

In several species of the bean weevil genus *Callosobruchus*, males show TP inflicted by spiny male genitalia (Fig. 3) (Crudgington and Siva-Jothy 2000; Rönn et al. 2007). Between-species variation is large and the genus has become an outstanding model for testing the functions of traumatic mating and its evolutionary maintenance.

Intraspecific variation in genital spine length exists, and longer genital spines were found to cause more harm to females (Hotzy et al. 2012) and lower female lifetime reproductive success (Rönn and Hotzy 2012). Males benefit from longer genital spines by an increased success in sperm competition (Hotzy and Arnqvist 2009; Hotzy et al. 2012), a success that was, however, not caused by an increased anchorage benefit of longer spines to males (Rönn and Hotzy 2012).

Research in *Callosobruchus* (Morrow et al. 2003; Edvardsson and Tregenza 2005) has also revealed that the wounding itself of the female is not adaptive for males under sperm competition as hypothesized by allocation theory (Lessels 2006). To the contrary, experimentally wounded

Figure 3. Additional well-studied model systems of CW. (*Top*) A mating pair of the common bed bug, *Cimex lectularius*, showing the normal mating position (*left*), as well as a ventral view of a mating pair. The position of an organ that evolved in response to male harm is indicated by a dashed rectangle. The *inset* shows the in situ insertion of the male intromittent organ, a hypodermic needle-like structure. (Image from photographer and copyright holder, Richard Naylor, CimexStore, Chepstow, UK; reprinted, with permission.) (*Bottom*) A mating pair of the seed beetle, *Callosobruchus maculatus*. The spiny structure is the male penis, which is inserted into the female genital tract regular, causing wounds that are visible as melanized scars (*inset*). (Image from photographer and copyright holders, PSmicrographs, UK and Mike Siva-Jothy, University of Sheffield, UK; reprinted, with permission.)

females remated sooner after CW and so promoted sperm competition (Morrow et al. 2003). Recent research further showed that a certain type of female behavior regarded as defensive (Crudgington and Siva-Jothy 2000) may actually carry costs itself and contribute to decreased female lifespan (Wilson and Tomkins 2014).

The following refers to all four previous examples.

With these examples, we have hopefully shown that CW, rather than being an obscure, anecdotal phenomenon, is real, widespread, and important. We will now consider its morphological, physiological, and evolutionary causes and consequences in more detail.

A MECHANICAL ORIGIN OF CW

CW can occur inside the genital tract of the wound recipient (intragenital wounding) or outside of it (extragenital wounding) (Carayon 1977; Lange et al. 2013a).

Intragenital CW

Intragenital CW likely derives from a mismatching of male and female genitalia or behavior. For example, reproductive tissue can be ruptured, overstretched, torn, or pierced during copulation if there is a mismatch between male and female reproductive morphology (Fig. 4), but also if behavioral elements in the sequence of copulation are exaggerated or mistimed (Fig. 4). In common bedbugs, *Cimex lectularius*, for example, in which every copulation is associated with CW (see the section Case Studies), females can make forceful jerking abdomen movements to shake off males that attempt to copulate. These sudden movements, however, stop as soon as the male pierces his intromittent organ through the female integument (K Reinhardt, unpubl.). A continuation of these movements during intromission would presumably increase tissue rupture and, thus, be mistimed from the female perspective. Perhaps, such female behavior can explain why, in seed beetles (see the section Case Studies), female behavioral defenses may contribute to decreased female lifespan (Wilson and Tomkins 2014).

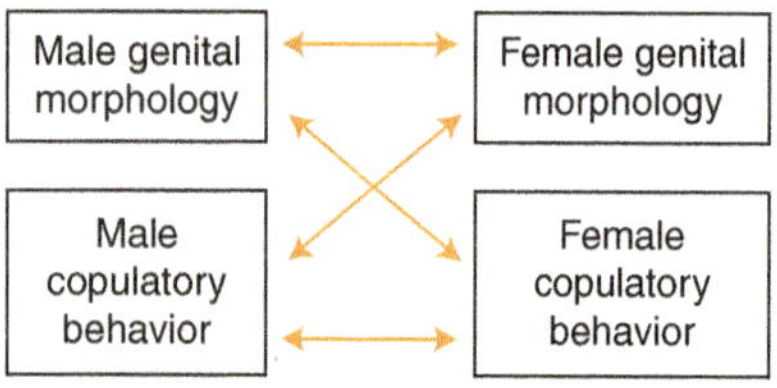

Figure 4. A scheme of mismatches leading to CW. During copulation, any mismatch between male and female genital morphology, or the occurrence of unusual copulatory behavior, can lead to wounding of the partner's, or their own, genital organs.

Theoretically, when coupled to a fitness benefit to either recipient or inflictor, such accidental, intragenital CW can evolve toward a routine mating strategy involving specialized wounding structures.

It is noteworthy, at this point, that similar mismatch mechanisms underlying copulatory genital injuries have been described in the medical literature using terms such as penovaginal disproportion, excessive force, or unusual behavior at coitus (Eke 2002) (for further discussion, see the section CW in Humans, with a Note on Its Medical and Legal Significance).

Extragenital CW

Explaining the origin of extragenital CW is more difficult and probably begins with a deviating male behavioral element, perhaps involving any of the anchorage or harming benefits discussed in Lange et al. (2013a). Before extragenital TI can evolve from accidental extragenital CW, the ability of sperm to survive outside the female sperm receptive organs must also have evolved.

Some species show large variation in the site of extragenital CW, whereas in others CW is strictly localized (Lange et al. 2013a, 2014). It is unclear why specific body sites are being targeted. Although there may be constraints on mating position, it has also been suggested that CW, at specific sites, may increase partner manipulation (Lange et al. 2014), accelerate partner survival (and, hence, increase its reproductive output), or sperm survival.

Not all copulatory mismatches will lead to CW. CW may also be affected by the material

 Cite this article as *Cold Spring Harb Perspect Biol* doi: 10.1101/cshperspect.a017582

properties of the reproductive tissues involved. For example, elastic female reproductive tissue will accommodate a larger range of male genital morphologies than sclerotized tissue without leading to CW.

WHICH SEX CAUSES CW?

Because of the different fitness optima for mating traits (Bateman 1948; Parker 1979; references in this collection), males and females experience different selection pressures and it will often be important to identify which sex inflicts the copulatory trauma. Although it is usually males that have specific wounding organs (Lange et al. 2013a), it may be worth pointing out a number of circumstances under which it is difficult to predict which sex inflicts the wound.

1. Consider a male whose copulatory organ does not inflict a genital wound in most female phenotypes, but does so in females that show unusual movement during copulation. Here, the mismatch causing CW arose from a behavioral deviation in the female. The term "wound inflictor" can therefore be ambiguous. In species in which CW is rare, such ambiguity can be circumvented by classifying whether it is the male or female that departs from the sex-specific morphological or behavioral population mean (cf. Fig. 4). However, even this approach is not free from anthropomorphic judgment because it may be the male that induces the behavioral deviation in the female. In species in which CW occurs routinely, the ancestral state of male and female morphology or behavior may be predicted from related species.

   The same considerations apply to cases in which the male intromittent organ is injured during copulation. This situation may be self-inflicted, but may also arise from female resistance behavior or morphology and, hence, not be "self"-inflicted.

2. In contrast to self-infliction caused by partner resistance, in several species, the intromittent organs routinely break and the whole or parts thereof remain in the female (nudibranchs: Sekizawa et al. 2013; spiders: Uhl et al. 2010; Izquierdio and Rubio 2011; earwigs: Kamimura 2003; bees: Franck et al. 2002). This self-inflicted CW can acquire adaptive function (such as preventing remating of the partner: Nessler et al. 2007; Uhl et al. 2010), in which case the term "wound inflictor" would, again, be ambiguous, even more so if specific break-off points evolved (e.g., Uhl et al. 2010; Sekizawa et al. 2013) that no longer result in wounding. In other species, injuries to the intromittent organs happen occasionally rather than routinely (moths: Cordero and Miller 2012; humans: the section Case Studies) and their adaptive function, if any, is unknown (but, see Cordero and Miller 2012).

3. Exceptional situations can emerge in hermaphrodites when an individual chews off the male intromittent organ of the mating partner after copulation (as in banana slugs: Leonard et al. 2002). Although such behavior fulfills our definition of CW (Box 1), it does not target the organ that receives the sperm.

In the following, we will use wound inflictor synonymously with male (or the male function in hermaphrodites) and wound recipient synonymously with female (or the female function in hermaphrodites). We do so to facilitate the presentation of our arguments, although it may oversimplify real-world situations, especially in hermaphrodites in which being wounded affects the entire organism of both sexual functions (see Schärer et al. 2014).

Regardless of whether CW is self-inflicted or inflicted by others, the resulting injuries will elicit a similar range of physiological responses.

PHYSIOLOGICAL EFFECTS OF CW

The Nature of the Physiological Costs

Once the copulatory trauma is inflicted, one or several immediate effects can occur in the wounded individual: (1) blood loss, (2) the entry of microbes, and (3) the entry of other nonself particles. These effects and subsequent defensive responses have been discussed earlier (Siva-Jothy 2009) and are only briefly considered here.

Preventing Blood Loss

The activation of repair cascades, the transport of cells to a wound, and wound closure are ubiquitous among vertebrates and common among invertebrates (Theopold et al. 2004). Although the process of wound healing after CW is almost unstudied, we, here, assume it is similar to general wound healing.

Preventing Microbial Infection

After CW, external male or female surface-dwelling or sexually transmitted microbes may enter the wound (Eke 2002; Mehta et al. 2009, 2010, 2012; Otti et al. 2009, 2013; Kamimura 2012; and references therein). This elicits a series of responses. For example, in insects, blood clots are formed and prevent systemic infection (Theopold et al. 2004; Haine et al. 2007). After clotting, the next line of defense is the activation of local and systemic immune cascades (Siva-Jothy et al. 2005). It is important to note that genital infections, at least in males, cause systemic responses and sepsis much more often than infection through other tissues (Gendrin et al. 2009).

Responses toward other Non-Self Particles

Other non-self objects, such as blood and sperm cells, seminal fluid molecules, or particles from the environment, may also elicit clotting or immune responses. Immune responses toward male ejaculate proteins occur widely and seem particularly energy intense (Peng et al. 2005; Domanitskaya et al. 2007; Morrow and Innocenti 2012).

CW Can Elicit Dose- or Frequency-Dependent Costs

Trauma effects, such as blood loss or ejaculate toxicity, may be dose-dependent such that energy costs increase when more blood is lost or when more seminal protein delivered. Other effects will be frequency-dependent. For example, infection may occur after one CW event regardless of how many microbes have entered. Wound healing and immune responses may also be activated after CW regardless of whether the trauma is a slight rupture, a mucus lesion, or a severe injury. The distinction of dose- and frequency-dependent costs of CW and its important evolutionary consequences (see the section Outlook and Future Research in CW) have not previously been recognized (but see Reinhardt et al. 2011 for a similar argument).

LIFE-HISTORY CONSEQUENCES OF CW

Physiological studies show that energy is required to reduce, prevent, or outweigh the consequences of trauma. Allocating resources to provide the energy necessary to compensate for blood loss, wound healing, or activation of the immune system is costly (Moret and Schmid-Hempel 2000). Energy into repair or immune function may be "directly" diverted from other activities, or obtained via increased feeding activity, which "indirectly" takes away resources from other body functions.

Resource allocation will depend on the amount of energy needed and wounded individual's expected lifespan, or more precisely, its residual reproductive value (Clutton-Brock 1984). If CW occurs early in life, more resources may be spent on immune function. If CW occurs late in life, or is very severe, the wounded individual may invest little into defense and, instead, allocate all remaining resources into maximizing reproductive output in the face of death. This process is known as terminal investment (Clutton-Brock 1984) and has important evolutionary consequences (Johnstone and Keller 2000; Lessels 2006): CW may benefit the wound inflictor by inducing resource allocation to reproduction in the recipient, and so persist evolutionarily. However, empirical studies testing whether CW induces resource allocation are lacking. Although females of three insect species that were injured just after females had mated showed no increased investment (Morrow et al. 2003), the females had been wounded at body sites unconnected to the reproductive tract. Any response in females that occurs specifically in response to CW, rather than normal wounding, would remain undetected. In contrast, infection is known to induce resource allocation (Gustafsson et al. 1994; Adamo 1999; Reaney and Knell 2010). Siva-Jothy (2009) argued that females are able to predict the time of mating and, therefore, the physiological and infection costs of CW. This predictability may lead to the evolution of CW-anticipating immune responses, the basic mechanism of which has recently been shown for a sexually transmitted disease (STD) (Zhong et al. 2014). Anticipated responses require energy and thus may alter the timing of resource allocation, but will not abolish it.

A speculative, previously unconsidered life-history consequence might be that immune responses after CW reduce the costs of later infections—akin to vaccination or immune priming. Because immune priming also occurs in invertebrates (e.g., Little and Kraaijeveld 2004), this mechanism may potentially be widespread. Although this possibility is not further discussed here, it illustrates the complexity of responses to CW (and may be tractable to a solution using mathematical modeling).

EVOLUTIONARY CONSEQUENCES OF CW

Sexual Conflict and the Evolution of Traumatic Mating

The physiological demands and their life-history consequences induced by CW will carry fitness costs and affect evolutionary change if CW varies among wound inflictors and recipients, and if characters affecting CW are heritable (Fig. 5). Similar to the success of any new mutation in a population, the initial occurrence of CW may be beneficial, neutral, or deleterious to either wound inflictor or recipient, and so will persist or disappear. Details of the costs and benefits of CW have extensively been discussed elsewhere (Lange et al. 2013a). Here, we only briefly consider scenarios that occur under sexual conflict, that is, when only one sex, but not the other, incurs fitness costs. This includes a new hypothesis on facilitated transfer of STDs.

Selection for Traumatic Penetration from Accidental CW

TP can evolve from accidental CW (Fig. 5) if the male fertilizes more eggs after CW because the recipient delays remating or lays more eggs (for details, see Lange et al. 2013a). For the latter possibility, altered resource allocation (see the section Life-History Consequences of CW) (Lessels 2006) is required. *D. sechellia* may offer a suitable system to study the transition from rare CW to regular TP. CW, in this species, is

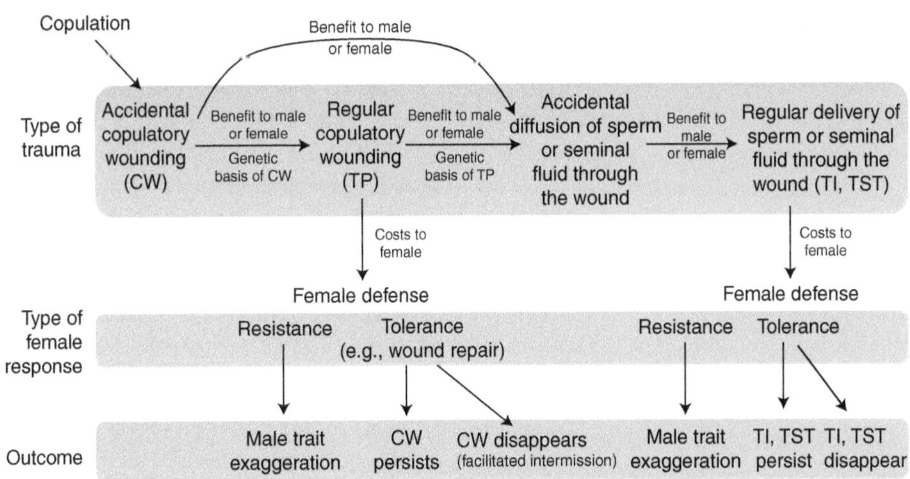

Figure 5. Flow diagram highlighting CW as a signature of sexual conflict. Several outcomes of sexual conflict are possible, including male and female genital exaggeration, male and genital female simplification, and, across populations, male and female genital diversity. The complete disappearance of CW is also possible. As not all genital characters under selection may respond in the same way, CW may also contribute to the evolution of genital complexity.

facultative and occurs in 66% of observed cases (Kamimura 2011).

Selection for TI and TST from Accidental Transfer of Sperm and Seminal Fluids

In several species, sperm or seminal fluids are exclusively transferred traumatically (TI, TST) (Lange et al. 2013a). TI or TST may evolve from accidental deliveries of sperm or seminal fluid if (1) the recipient's short-term reproductive output is increased compared with partners without traumatic delivery, or (2) if the paternity share of the wound inflictor is increased without necessarily increasing the wound recipient's reproductive output (for details, see Lange et al. 2013a). Paternity benefits that are exclusively caused by the traumatic delivery of sperm or seminal fluids are difficult to reveal experimentally, but have been predicted to occur if they bypass the often hostile reproductive tract of the wound recipient, particularly structures that resorb sperm or mix the sperm with that of other males. Such paternity benefits may also arise if the traumatic delivery of seminal fluids manipulates the recipient's physiology and reduces the resorption of currently received sperm.

Selection for the Transfer of STDs in the Wound Inflictor

If the short-term reproductive rate of the female increases because of resource reallocation (Lessels 2006) and if microbes are involved in eliciting resource allocation (see above), both arguments can be connected to form a novel, testable hypothesis detailed below.

Microbes are found on the intromittent organs of males in several species without necessarily infecting the male (see references in Reinhardt 2010). Such microbes can enter copulatory wounds (see references in Reinhardt et al. 2005; Otti et al. 2009; Siva-Jothy 2009), and elicit immune responses (Otti et al. 2009, 2013; Siva-Jothy 2009). Microbes can also increase an infected host's investment into current reproduction (Gustafsson et al. 1994; Adamo 1999; Reaney and Knell 2010), and sexually transmitted microbes particularly benefit from, and select for, increased mating activity of the host (albeit inducing sterility) (Lockhart et al. 1996). Under the condition that sterility does not occur immediately, copulatory infection, or the associated immune response, may induce the wound recipient to invest more into current reproductive output. This may then lead to selection acting on wound inflictors to harbor microbes on their genitalia or increase microbe transmission to the wound recipient. Under this hypothesis, sexual conflict increases, but aspects of the host–parasite conflict between males (wound inflictors) and their STDs would relax. Experimentally separating the effects of the microbes from the effects of CW will be challenging.

Defensive Adaptations and Variation in CW

Wound inflictors and recipients coevolve like other parties in evolutionary conflicts (this collection). This causes costs and benefits of CW, TI, and TST to vary over evolutionary time, and generates several predictions.

Low Costs of CW over Most of Evolutionary Time

Females will often carry large costs of CW and are expected to adapt very rapidly. As a consequence, the observable costs of CW may be low over much of evolutionary time (longitudinally) and, hence, in many study systems (corresponding to cross-sectional sampling). Therefore, the costs of mating in species with TM may not, generally, be expected to be higher than in species without TM or other forms of CW (Morrow and Arnqvist 2003). This can be tested by comparing female fitness in response to mating rates in species with CW and other species (Table 1, Fig. 6). Following the approach by Arnqvist and Nilsson (2000), we measured the response rate (defined in Table 1) of two fitness proxies to elevated mating rates for both species groups. Response ratios >1 indicate benefits, response ratios <1 indicate costs.

For female lifespan, the response rate of species with CW (overall 0.89) showed a modest

Table 1. The costs of mating in species in which CW routinely occurs

Species	Eggs (mean ± SE) at low mating rate	*N*	Eggs (mean ± SE) at high mating rate	*N*	RR	Lifespan (mean ± SE) at low mating rate (days)	*N*	Lifespan (mean ± SE) at high mating rate (days)	*N*	RR	References
Type 1											
Cimex lectularius	83.8 ± 4.50	20	61 ± 3.1	20	0.73						Polanco et al. 2011
Xylocoris flavipes	16.47 ± 1.87	21	18.1 ± 2.3	21	1.10	13.6 ± 0.9	21	11.9 ± 1.4	21	0.76	Backhouse et al. 2012
Callosobruchus maculatus	93.0 ± 25.30	20	99 ± 4.6	16	1.06	n.a.		n.a.			Credland and Wright 1989
	93.0 ± 12.60	10	93 ± 5.2	10	1.00	n.a.		n.a.			Credland and Wright 1989
	85.4 ± 12.60	98	93.1 ± 1.6	81	1.09	8.4	98	9.0 ± 0.1	81	1.07	Fox 1993
	127.7 ± 19.70	45	142.7 ± 3.4	37	1.12	23.5 ± 1.4	45	22.0 ± 1.5	37	0.94	Fox 1993
	52.4 ± 32.50	22	74.6 ± 8.3	16	1.42	n.a.		n.a.			Wilson et al. 1999
Callosobruchus analis	41.5 ± 16.40	35	59.3 ± 2.8	36	1.43	n.a.		n.a.			Wilson et al. 1999
Sepsis cynipsea	153.4 ± 18.51	49	151.7 ± 19.5	30	0.99	25.2 ± 2.0	41	21.9 ± 2.9	42	0.87	Blanckenhorn et al. 2002
Drosophila melanogaster	116.0 ± 30.9	54	169.7 ± 5.9	45	1.46	n.a.		n.a.			Gromko and Pyle 1978
	419.5 ± 129.2	4	368.5 ± 68.5	2	0.88	n.a.		n.a.			Lefevre and Jonsson 1962
	408.8 ± 65.6	6	323.2 ± 14.5	6	0.79	n.a.		n.a.			Lefevre and Jonsson 1962
	551.0 ± 225.0	26	959 ± 94.0	26	1.74	n.a.		n.a.			Pyle and Gromko 1978
Siphopteron quadrispinosum	1493.9 ± 224.95	29	1650.2 ± 170.7	29	1.10	n.a.		n.a.			Lange et al. 2012
Alderia modesta	1413.5 ± n.a.	21	1077.9 ± n.a.	14	0.76	n.a.		n.a.			Smolensky et al. 2009
C. maculatus	106.7 ± 29.90	40	118.5 ± 3.8	61	1.11	n.a.		n.a.			Wilson et al. 1999
Mean of species with CW					**1.14** (1.02–1.28)					**0.92** (0.71–1.14)	
Mean of all species					**1.23** (1.13–1.35)					**1.00** (0.88–1.50)	Arnqvist and Nilsson 2000
Type 2											
C. lectularius	294.4 ± 9.89	45	224.7 ± 16.1	45	0.76	147.2 ± 8.4	45	110.9 ± 5.0	45	0.75	Stutt and Siva-Jothy 2001
C. maculatus	67.9 ± 12.40	48	61.5 ± 2.2	53	0.91	n.a.		n.a.			Fox 1993
Mean of species with CW					**0.84** (0.52–1.15)					**0.75** (n.a.)	
Mean of all species					**1.19** (1.03–1.39)					**0.92** (0.78–1.10)	Arnqvist and Nilsson 2000

Continued

Table 1. *Continued*

Species	Eggs (mean ± SE) at low mating rate	N	Eggs (mean ± SE) at high mating rate	N	RR	Lifespan (mean ± SE) at low mating rate (days)	N	Lifespan (mean ± SE) at high mating rate (days)	N	RR	References
Type 3											
D. melanogaster	25.0 ± 16.90	35	20.7 ± 1.8	35	0.83	13.5 ± 0.6	35	14.5 ± 0.9	35	1.07	Chapman and Partridge 1996; T Chapman and L Partridge, unpubl.
	20.2 ± 8.9	35	24.1 ± 2.7	35	1.19	18.0 ± 1.8	35	21.1 ± 1.7	35	1.17	Chapman and Partridge 1996; T Chapman and L Partridge, unpubl.
	35.6 ± 14.6	35	33.9 ± 3.6	35	0.95	36.2 ± 1.9	35	27.4 ± 1.8	35	0.76	Chapman and Partridge 1996; T Chapman and L Partridge, unpubl.
	56.1 ± 14.6	35	58.4 ± 5.1	35	1.04	27.5 ± 2.4	35	23.3 ± 1.8	35	0.85	Chapman and Partridge 1996; T Chapman and L Partridge, unpubl.
	140.3 ± 14.6	35	90.2 ± 8.4	35	0.64	30.5 ± 1.5	35	19.5 ± 1.7	35	0.64	Chapman and Partridge 1996; T Chapman and L Partridge, unpubl.
S. quadrispinosum	1493.9 ± 224.95	29	1272.5 ± 149.2	29	0.85	n.a.		n.a.			Lange et al. 2012
Mean of species with CW					**0.91** (0.72–1.11)					**0.90** (0.70–1.09)	
Mean of all species					**0.88** (0.77–0.98)					**0.80** (0.73–0.90)	Arnqvist and Nilsson 2000
Overall mean in species with CW					**1.05** (0.94–1.16)					**0.89** (0.77–1.02)	
Overall mean of all species					**1.11** (1.04–1.18)					**0.88** (0.79–0.98)	Arnqvist and Nilsson 2000

All studies published before 2000 and treatment type definitions were taken from Arnqvist and Nilsson (2000). Type 1: Females mated once were compared with females that mated at a higher, controlled rate or were cohabited with a male, but the cohabitation was shorter than female lifetime. Type 2: Same as Type 1, but females were cohabitated with a male for the entire female's lifetime. Type 3: Experiments in which females mating at an intermediate rate were compared with females mating at a higher rate. Means ±95% CI of all species were taken from Figure 3 and Table 2 in Arnqvist and Nilsson (2000) for species showing no nuptial feeding. Response ratios (RR) were derived by dividing the mean of high mating rates by the mean at low mating rates. Means (in bold) and CI are weighted by the ln-transformed inverse of the sampling variance.

n.a., not available.

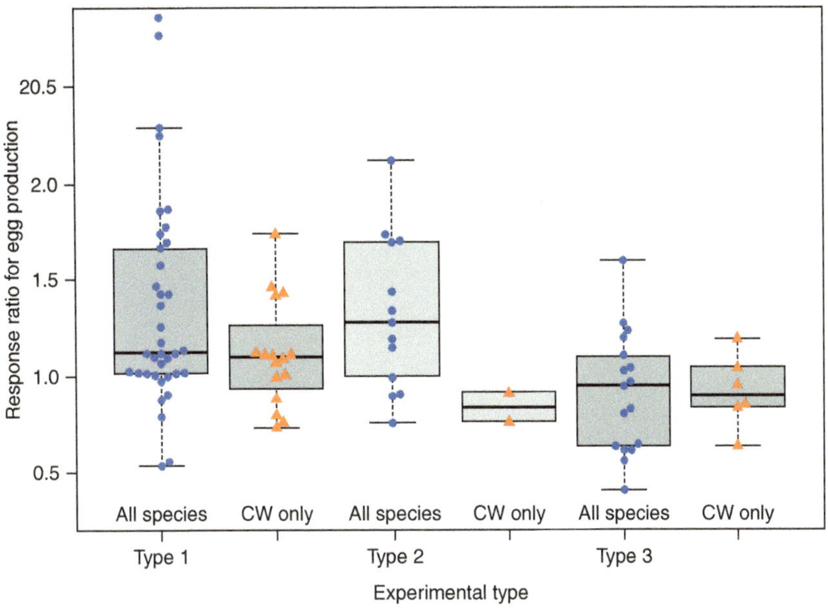

Figure 6. Distribution of response ratios (defined in Table 1) illustrating positive (values >1) or negative (values <1) effects of enhanced mating frequencies on female egg production. "All species" refers to all species listed by Arnqvist and Nilsson (2000) (blue circles). They are compared with species showing CW ("CW only," orange triangles). Original data for the CW species and definitions of the three experimental types are given in Table 1.

decline in females exposed to high, relative to low, mating rates, and this figure almost perfectly matches the mean response rate of all species (0.88) listed by Arnqvist and Nilsson (2000). Similar to all previously investigated species, species with CW show little variation in response rates among different types of measurement and experimental treatment (Types 1, 2, and 3 as defined in Table 1).

For egg production, a more immediate fitness measure than lifespan, the pattern is a little more differentiated (Fig. 6). The overall response rate in CW species (1.05) tended to be smaller than that reported for all species (1.11) by Arnqvist and Nilsson (2000) (Table 1). This pattern is largely caused by smaller benefits of females exposed to intermediate mating rates compared with low mating rates (Type 1) for CW species (response rate 1.23 vs. 1.14; Fig. 6), as well as the fact that high mating rates or lifelong exposure to males (Type 2) causes costs in females of CW species (0.84), but results in benefits to females of all other species (1.19) (Table 1; Fig. 6). No difference was visible when comparing intermediate mating rates with high mating rates (Type 3).

Taken together, the results indicate that female benefits of multiple mating (response ratios > 1) persist under routine CW, but these benefits are slightly reduced as a consequence of CW. It will be useful to (1) determine in how many of the species listed by Arnqvist and Nilsson (2000) CW occurs cryptically, but has not yet been revealed, and (2) whether this pattern persists once the current bias toward insect studies (Table 1) will be removed (see the section Outlook and Future Research in CW).

Intraspecific Variation in the Incidence of CW

If the costs of CW vary over evolutionary time as outlined above, we predict species to differ also in the rates at which CW is inflicted. However, this prediction cannot currently be tested because none of the few species in which CW is not fixed (i.e., CW rate < 100%) (Table 2) have been examined for the costs of CW (cf. Table 1).

Table 2. Selected examples of inter- and intraspecific variation in CW

Group	Species	Proportion of copulations with CW (*N* populations examined)	References
Earwigs (Pygidicranidae)	*Echinosoma denticulatum*	100% (1)	Kamimura and Lee 2014
Bedbugs (Cimicidae)	All species	100% (all)	Usinger 1966
Damselbugs	*Nabis pseudoferus*	3.5% (?)	Carayon 1966a
(Nabidae)	*Alleorhynchus furens*	~10% of males (?)	Carayon 1966a
Spiderweb bugs (Plokiophilidae)	*Embiophila species*	7% of males (?)	Carayon 1966a
Fruit flies (*Drosophila*)	*Drosophila sechellia*	66% (1)	Kamimura and Mitsumoto 2011
	Drosophila melanogaster	100% (1)	Kamimura 2010
	Drosophila simulans	100% (1)	Kamimura and Mitsumoto 2011
	Drosophila teisseri	100% (1)	Kamimura and Mitsumoto 2012
Ground beetles (Carabidae)	*Carabus iwakianus*	0% (1)	Sota and Kubota 1998
	Carabus maiyasanus	11% (1)	Sota and Kubota 1998
	C. iwakianus male × *C. maiyasanus* female	30% (1)	
	C. maiyasanus male × *C. iwakianus* female	47% (1)	Sota and Kubota 1998
Moths	*Nystalea aequipars*	One of two mated females (1)	Cordero and Miller 2012
	Scotura annulata	One of three mated females (1)	Cordero and Miller 2012
	Malacosoma	One of 68 mated females (1)	Bieman and Witter 1982
Land snails	*Arianta arbustorum*	23%–50% (3)	Baminger et al. 2000
	Cornu aspersum	73% (1)	Chase and Vaga 2006
Humans		10%–52% (several)	reviewed Astrup et al. 2012; Lincoln et al. 2013

The proportion of copulations resulting in CW differs across species. The table is based only on examples resulting from systematic anatomical dissections or observations.

The same prediction of variable rates of CW should hold for populations within a species that evolve independently, but no data exist to test this question. In fact, intraspecific variation in the incidence of CW has only been studied in bedbugs (no variation, 100%), humans, and dart-shooting snails (Table 2). In dart-shooting snails, Baminger et al. (2000) found that CW is attempted in 23%–50% of individuals. Whether intraspecific variation will evolve into interspecific variation depends on the genetic component of CW, whether CW is transient in populations or relatively fixed, and how strongly CW acts in reproductive isolation.

If the proportion of CW increases rapidly because wound recipients experiencing low costs are selected for, the costs of CW for individual wound recipients are predicted to be lower in populations in which CW is more common. No data exist that allow testing this prediction. If, in contrast, the proportion of CW is relatively stable, it may be the result of different individuals responding in different ways genetically to CW. This situation may lead to female-limited polymorphism or sympatric speciation (Gavrilets and Waxman 2002). Evidence consistent with that model comes from a study of *A. constrictus*, in which some female morphs attracted fewer wounds than others (Reinhardt et al. 2007).

Inter- and Intraindividual Differences

Intraspecific, interpopulation variation in the proportion of copulations resulting in CW may be caused by environmental or genetic differences between populations. Consistent genetic differences between populations require that interindividual differences in wounding, or in being wounded, are consistent and are smaller than between-population differences. No data on individual or genotypic differences exist so far. However, *Drosophila* mutants with genitalia differences (Masly et al. 2011; McNeil et al. 2011; House et al. 2013) may be used to uncover genotypic differences in CW.

CW and Reproductive Isolation

CW as a Barrier against Hybridization between Species

Almost universally, genitalia differ between species. Therefore, in cross-species mating attempts, mismatches between male and female morphology and copulatory behavior (Fig. 4) will be larger in inter- than in intraspecific pairings, and fitness costs will likely be higher. In this case, CW would represent a barrier against hybridization. Empirical examples have confirmed such barrier function. In ground beetles (Sota and Kubota 1998) and *Drosophila* (Kamimura 2012), interspecific matings resulted in more severe CW than intraspecific mating crosses, but in both cases, asymmetrically in only one direction. Remarkably, the risk of foreign, bacteria-sized particles to invade the copulatory wound was increased in both interspecific pairings (Kamimura 2012). This points toward the possibility that infection, in addition to CW itself, may generate a hybridization barrier. For traumatically inseminating plant bugs, Tatarnic and Cassis (2013) speculated that the interspecific morphological genitalia variation also has evolved by selection against hybridization (character displacement).

CW as an Isolating Factor in Reproductive Isolation

CW may not only represent a barrier against hybridization, but also cause reproductive isolation, that is, the emergence of new species. Between-population mismatches in genitalia morphology or copulatory behavior (Fig. 4) may be larger than within-population mismatches. If between-population matings result in more CW and reduced fitness compared with within-population matings, CW will represent an isolating factor (see Rice 1998 for general reasoning). This topic remains unexplored and itself rests on the unexplored assumption that between-individual differences in wounding or being wounded are consistent and have a genetic component.

Consistent interpopulation differences in CW may also drive evolutionary variation in wound healing, but no data exist on healing differences between populations.

Major Evolutionary Trends of CW

Three major evolutionary trends were identified to be associated with TM (Lange et al. 2013a): (1) increased male and female genital complexity, such as in dart-shooting mollusks (Koene and Schulenburg 2005), (2) convergent evolution to simple, needle-like organs (Eberhard 2006), and (3) thickened or sclerotized tissue at the sites of wounding, such as in bruchid beetles (Rönn et al. 2007) or some bug families (Carayon 1977). We wish to propose a fourth trend that is linked to female tolerance as a defense strategy: (4) the reduction of CW or TM (Fig. 5). In at least one strepsipteran and two bedbug species, as well as in the Lyctocoridae and Plokiophilidae (true bugs), female structures evolved that tolerate, or facilitate, rather than resist male intromission. At body sites in which sister species show copulatory wounds or thickened tissue to resist intromission, females in these groups evolved a novel opening in the integument (Usinger 1966; Carayon 1974, 1977; Beani et al. 2005). As a result, sperm delivery is not (any more) associated with trauma. Other families of true bugs contain males with needle-like copulatory organs, as well as females with secondary genital openings (Carayon 1977; Tatarnic and Cassis 2013; Tatarnic et al. 2014). This combination of traits makes it possible, even likely, that these families represent cases in which males initially performed CW, but female tolerance led to CW being lost, although no species exist in which CW currently is observed.

These cases concern the evolution of a secondary genital tract. However, the same arguments may be used to explain the evolution of a primary genital tract. For example, Lange et al. (2013a) highlight that TI occurs in some species within clades of otherwise externally fertilizing polychaete worms. If females in these species would be selected to evolve an opening to receive sperm rather than to experience TI, we would have the scenario that CW actually precedes the evolution of a female genital tract and, hence, copulation in the stricter sense (but, see our broader definition of copulation in Box 1).

CW IN HUMANS, WITH A NOTE ON ITS MEDICAL AND LEGAL SIGNIFICANCE

Epidemiology of Genital Wounds after Consensual Sexual Intercourse

Wounding of the female, and sometimes the male, genital organs regularly occurs during consensual sexual intercourse in humans. Including only the skin-breaching symptoms that obey our definition of CW (Box 1), such as mucus lesions, abrasions, or lacinations, CW was found in 10%–52% of the examined women (Astrup et al. 2012; Lincoln et al. 2013). However, a standard classification of genital injury after sexual intercourse was not available until recently (Kelly et al. 2013). Variation in CW between studies will therefore include actual variation between populations, but also variation as a result of different methodologies. This uncertainty is likely to also apply to animal studies.

Physiology

In humans, post-CW healing can be inferred from the fact that fewer genital wounds have been found 36 h after injury than after 0–24 h (Anderson et al. 2008; Berkowitz 2011; Astrup et al. 2012). The time course of wound healing seems unknown in animal studies, but may be easily considered in arthropods, in which repaired wounds are visible as black melanin scars.

Male Injuries during Consensual Intercourse

Penile fractures and prepuce lesions can occur during sexual intercourse (Yip et al. 1989; Lee et al. 2000; Szabo and Short 2000; Ozorak et al. 2014). Prepuce lesions are medically significant because they provide a route for systemic STD infections (Szabo and Short 2000; Mehta et al. 2009), and male circumcision during boyhood was associated with fewer injuries in adulthood during sexual intercourse (Mehta et al. 2010). It would be worthwhile to study whether efforts to reduce such STD infections in populations contributed to the cultural emergence of male circumcision.

Penile fractures appear to happen more frequently in stressful situations or when some degree of unusual copulatory behavior is involved (Kramer 2011), suggesting the general mismatch scheme (Fig. 4) does, in principle, also apply to humans.

Nonconsensual Sex, Forced Sex, and Rape

In cases of rape and nonconsensual sex, injuries to the victim are common (e.g., Riggs et al. 2000). Nonconsensual sex was found to result in much, moderately, or not significantly more female genital injuries compared with consensual sex (Edgardh et al. 1999; Jones et al. 2003; Anderson et al. 2009; Lincoln et al. 2013). However, the lack of a classification scheme of female genital injuries until recently (Kelly et al. 2013) and the occurrence of genital wounds after consensual sex made a recent review conclude that "the presence or absence of genital injury should not be used to render an opinion regarding consent to sexual intercourse" (Anderson and Sheridan 2012). We do not discuss this topic in more detail as such cases usually represent situations in which human males use sex as a weapon (Geist 1988). Although sperm transfer is involved in a large number of rape cases (e.g., Riggs et al. 2000) we, unlike some investigators, consider such cases as generically, rather than reproductively, aggressive male behavior (see also Coyne 2000; Coyne and Berry 2000). The limited relevance to applying sexual conflict theory is, among others, apparent from the fact that many victims are outside reproductive age (Riggs et al. 2000; Jones et al. 2009).

OUTLOOK AND FUTURE RESEARCH IN CW

The study of CW has gained momentum, especially in sexual conflict research. Below, we propose that CW is probably still grossly underreported. Getting a more complete picture of the taxonomic distribution and abundance of CW is an important task for future research. However, we also wish to emphasize that some of the existing data already indicate that the study of CW may provide exceptional contributions to some areas of evolutionary biology.

CW Is Likely Understudied and Underreported

The 36 cases of CW mentioned by Lange et al. (2013a) may be supplemented by at least six additional examples to include the cases of carabid beetles (Sota and Kubota 1998), two species of notodontid moths (Cordero and Miller 2012), an earwig (Kamimura and Lee 2014), as well as humans (see above), and one more case in fruit flies if CW in *D. teisseri* (Kamimura and Mitsumoto 2013) represents an independent event. Other recent studies confirmed the occurrence of TM in other taxa (e.g., flower bugs: Shapiro et al. 2010; Tatarnic et al. 2014). We propose that CW will be detected in many more taxa given that CW during consensual sexual intercourse in humans was described only relatively recently. More strikingly, CW was found to occur regularly in *D. melanogaster* only a few years ago (Kamimura 2007), although as one of biology's main model organisms, it had been under anatomical and physiological scrutiny for more than 100 years. There may be at least five additional reasons why CW is likely more common than currently appreciated:

1. The frequent evolution of copulation, and hence of intromittent organs, from spawning or spermatophore transfer has been shown for various groups, including hermaphrodites (Michiels 1998), insects (Proctor 1998; Heming 2003), and the Bilateria (Mann 1984). Such evolutionary volatility of copulation itself likely goes in hand with substantial variation in genital morphology and copulatory behavior, and so may cause CW by mismatch (Fig. 4).
2. Copulatory organs evolved from a diverse set of developmental structures, representing penises, ectodermal appendages (such as parameres in the insects: Carayon 1966b), mouthparts (such as the pedipalps in spiders: Uhl et al. 2010), and a range of structures collectively known as spermatophores of various origins (Davey 1960; Mann 1984). Such frequent developmental reorganizations may increase the likelihood of copulatory mismatches and, hence, CW.
3. It is a striking pattern that spines on male intromittent organs recurrently evolve. They are found in many taxa, ranging from primates (Stockley 2002), snakes (Zaher 1999), and insects (Rönn et al. 2007; Cordero and Miller 2012) to mollusks (Valdes 2004) or nematodes (Coomans et al. 1988). Some of them are known to cause CW (Lange et al. 2013a), but in most taxa, CW that has not been looked for.
4. In addition to the abundant sources of evolutionary novelties across species, evidence exists for intraspecific variation in genital morphology (Hotzy and Arnqvist 2009) and copulatory behavior in species without currently recognized CW (West-Eberhard 2003; Cordero et al. 2004; Oliveira et al. 2008). This variation may increase opportunities of mismatches (Fig. 4). Particularly

noteworthy is that intraspecific variation includes variation in male genital spines and other potential wounding devices in CW species (Hotzy and Arnqvist 2009; Cayetano et al. 2011; Masly et al. 2011). Increased potential for CW arises if variation in genital spines exists also in non-CW species.

5. Finally, not merely one, but several different processes contribute to variation in genital morphology and copulatory behavior, including sexual selection (reviews: Eberhard 1985, 1996; Arnqvist 1998; Hosken and Stockley 2004; experimental evidence: Cayetano et al. 2011; House et al. 2013), environmental variation and natural selection (Reinhardt 2010; experimental evidence: House et al. 2013), phenotypic plasticity (Neufeld and Palmer 2008), and pleiotropy (Müller 1962; Arnqvist and Thornhill 1998; Parzer and Moczek 2008). We suggest that this variety of sources of variation may also contribute to mismatches and CW.

Given the large number of sources of variation, we dare to speculate that no internally fertilizing species with intromittent organs exists in which copulation has not led, at least once, to CW.

Further Need of the Delineation of Costs and Benefits of CW

Throughout this contribution we have, at several places, suggested that sexual conflict and CW research would greatly benefit from quantifying the costs and benefits of CW. We still see this as a primary goal. Although some of the case studies, perhaps, come closest toward achieving this currently, we suggest it may be fruitful to concentrate on species in which CW is not obligatory. Perhaps, we will soon understand how important the rather accidental first step of a mismatch is in the evolution of TM (Fig. 4).

Are the Costs of CW Dose- or Frequency-Dependent?

The answer may decisively alter the predicted outcome of evolution. If CW results in frequency-dependent costs, traits affecting the mating rate will be under selection. In contrast, if costs are dose-dependent, genital morphology, physiology, or copulatory behavior will be subject to selection.

Understanding Intraspecific Genital Variation

If mismatches arise from genital variation (Fig. 4), solving the paradox of the extent and origin of intraspecific genital variation is important. On the one hand, variation between populations is so low that genitalia are routinely used to identify members of the same species or subspecies, even if they are from different populations. This low variation would make mismatches and CW rare. On the other hand, the evolutionary volatility and interspecific variation of copulation and copulatory organs suggests that some source generating genital variation must exist, in which case, CW will be more frequent.

Resistance and Tolerance as Separate Defense Strategies

Different outcomes of evolutionary conflicts occur depending on whether the offended party responds by resistance or tolerance (Fornoni 2011; Medzhitov et al. 2012). The same seems true for sexual conflict and CW research (see the section Major Evolutionary Trends of CW) (Fig. 5) and we explicitly recommend making this distinction (see also Lessels 2006 for an introduction of palliative traits). Resistance is a defense strategy that causes costs to the offending party and therefore represents a selection pressure that induces coevolution. Tolerance does not impose costs on the offending party and may instead, perhaps more strongly, select (or allow) for evolutionary novelties. As an example, consider the female response to piercing of the vaginal tissue. Females may respond by thickening the connective tissue, which may give males a reduced anchorage and so may reduce their fitness. Alternatively, females may instead accelerate resource allocation to rapid wound healing. Female wound healing does not impose costs on males (although variation in female healing may cause variation in male fitness) and so may not represent resistance.

CW and Sex-Role Reversal

A number of species exist in which females possess an intromittent organ and insert it into males (inverted copulations) (Box 1). It would be interesting to study whether CW occurs in these species as well, which include pipefish (Watanabe et al. 2000), planarians (Ullyot and Beauchamp 1931), but also insects (Jolivet 2008). Some of the latter species have normal sex roles and may be useful to separate whether wounding benefits males or the wound inflictor.

CONCLUSION

We have illustrated that CW is a common phenomenon with implications for the evolution of copulation, sexual conflict, and speciation. Although we have gathered some empirical evidence concerning the evolution, maintenance, and trajectories of CW, many suggestions remain speculative. Largely neglected for more than 100 years, the importance of CW has recently begun to be appreciated. We hope that this review stimulates further research by conceptually and empirically scrutinizing the hypotheses proposed herein, including the more speculative ones on the evolution of TM from accidental CW, the selection for STD transfer, or the proposed mechanisms of interaction in genital variation, behavioral variation, and speciation.

ACKNOWLEDGMENTS

We thank Ted Morrow, Bill Rice, and an anonymous reviewer for their constructive comments and suggestions, Steffen Roth for information on the Nabidae and, again, Ted Morrow for pointing out, and making available, the paper by Carayon (1966a), which we had overlooked for too long. We thank Yoshitaka Kamimura, Richard Naylor, and Mike Siva-Jothy for providing the images.

REFERENCES

*Reference is also in this collection.

Adamo SA. 1999. Evidence for adaptive changes in egg laying in crickets exposed to bacteria and parasites. *Anim Behav* **57:** 117–124.

Alexander RD, Marshall DC, Cooley JR. 1997. Evolutionary perspectives on insect mating. In *Mating systems in insects and arachnids* (ed. Choe JC, Crespi BJ), pp. 4–31. Cambridge University Press, Cambridge.

Anderson SL, Sheridan DJ. 2012. Female genital injury following consensual and nonconsensual sex: State of the science. *J Emerg Nurs* **38:** 518–522.

Anderson SL, Parker BJ, Bourguignon CM. 2008. Changes in genital injury patterns over time in women after consensual intercourse. *J Forensic Leg Med* **15:** 306–311.

Anderson SL, Parker BJ, Bourguignon CM. 2009. Predictors of genital injury after nonconsensual intercourse. *Adv Emerg Nurs J* **31:** 236–247.

Anthes N. 2010. Mate choice and reproductive conflict in simultaneous hermaphrodites. In *Animal behaviour: Evolution and mechanisms* (ed. Kappeler P), pp. 329–357. Springer, Berlin.

Anthes N, Michiels NK. 2007a. Precopulatory stabbing, hypodermic injections and unilateral matings in a hermaphroditic sea slug. *Biol Lett* **3:** 121–124.

Anthes N, Michiels NK. 2007b. Reproductive morphology, mating behaviour, and spawning ecology of cephalaspid sea slugs (Aglajidae and Gastropteridae). *Invert Biol* **126:** 335–365.

Arnqvist G. 1998. Comparative evidence for the evolution of genitalia by sexual selection. *Nature* **393:** 784–786.

Arnqvist G, Nilsson T. 2000. The evolution of polyandry: Multiple mating and female fitness in insects. *Anim Behav* **60:** 145–164.

Arnqvist G, Rowe L. 2002a. Antagonistic coevolution between the sexes in a group of insects. *Nature* **415:** 787–789.

Arnqvist G, Rowe L. 2002b. Correlated evolution of male and female morphologies in water striders. *Evolution* **56:** 936–947.

Arnqvist G, Thornhill R. 1998. Evolution of animal genitalia: Patterns of phenotypic and genotypic variation and condition dependence of genital and non-genital morphology in water striders (Heteroptera: Gerridae: Insecta). *Genet Res* **71:** 193–212.

Astrup BS, Ravn P, Lauritsen J, Thomsen JL. 2012. Nature, frequency and duration of genital lesions after consensual sexual intercourse—Implications for legal proceedings. *Forensic Sci Int* **219:** 50–56.

Backhouse A, Sait SM, Cameron TC. 2012. Multiple mating in the traumatically inseminating Warehouse pirate bug, *Xylocoris flavipes*: Effects on fecundity and longevity. *Biol Lett* **8:** 706–709.

Baminger H, Locher R, Baur B. 2000. Incidence of dart shooting, sperm delivery, and sperm storage in natural populations of the simultaneously hermaphroditic land snail *Arianta arbustorum*. *Can J Zool* **78:** 1767–1774.

Bateman AJ. 1948. Intra-sexual selection in *Drosophila*. *Heredity* **2:** 349–368.

Beani L, Giusti F, Mercati D, Lupetti P, Paccagnini E, Turillazzi S, Dallai R. 2005. Mating of *Xenos vesparum* (Rossi) (Strepsiptera, Insecta) revisited. *J Morphol* **265:** 291–303.

Benoit JB, Jajack AJ, Yoder JA. 2012. Multiple traumatic insemination events reduce the ability of bed bug females to maintain water balance. *J Comp Physiol B* **182:** 189–198.

Berkowitz CD. 2011. Healing of genital injuries. *J Child Sex Abuse* **20:** 537–547.

Bieman DN, Witter JA. 1982. Mating wounds in *Malacosoma*: An insight into bed bug mating-behavior. *Fla Entomol* **65:** 377–378.

Blanckenhorn WU, Hosken DJ, Martin OY, Reim C, Teuschl Y, Ward PI. 2002. The costs of copulating in the dung fly *Sepsis cynipsea*. *Behav Ecol* **13:** 353–358.

Carayon J. 1966a. Traumatic insemination and the paragenital system. In *Monograph of Cimicidae* (ed. Usinger RL), pp. 81–166. Thomas Say Foundation, Entomological Society of America, College Park, MD.

Carayon J. 1966b. Les inseminations traumatiques accidentelles chez certains Hémiptères Cimicoidea [Accidental traumatic insemination in some Hemiptera Cimicoidea]. *CR Acad Sc Paris* **262:** 2176–2179.

Carayon J. 1974. Étude sur les Hémiptères Plokiophilidae [Study on the Hemiptera Plokiophilidae]. *Ann Soc Fr (NS)* **10:** 499–525.

Carayon J. 1977. Insémination extra-génitale traumatique. In *Traite des zoologie–Anatomie, systématique, biologie—Tome VIII Fascicule V-A—Insectes: Gamétogenèses, fécondation, metamorphoses* [Extragenital traumatic insemination. In *Textbook of zoology–Anatomy, systematics, biology–Vol VIII Part V-A–Insects: Gametogenesis, fertilization, metamorphoses*] (ed. Grassé PP), pp. 351–390. Masson, Paris.

Cayetano L, Maklakov AA, Brooks RC, Bonduriansky R. 2011. Evolution of male and female genitalia following release from sexual selection. *Evolution* **65:** 2171–2183.

Chapman T, Partridge L. 1996. Female fitness in *Drosophila melanogaster*: An interaction between the effect of nutrition and of encounter rate with males. *Proc R Soc B* **263:** 755–759.

Chase R, Vaga K. 2006. Independence, not conflict, characterizes dart-shooting and sperm exchange in a hermaphroditic snail. *Behav Ecol Sociobiol* **59:** 732–739.

Clutton-Brock TH. 1984. Reproductive effort and terminal investment in iteroparous animals. *Am Nat* **123:** 219–229.

Clutton-Brock TH, Parker GA. 1995. Sexual coercion in animal societies. *Anim Behav* **49:** 1345–1365.

Coomans A, Verschuren D, Vanderhaeghen R. 1988. The demanian system, traumatic insemination and reproductive strategy in *Oncholaimus oxyuris* Ditlevsen (Nematoda, Oncholaimina). *Zool Scr* **17:** 15–23.

Cordero C, Miller JS. 2012. On the evolution and function of caltrop cornuti in Lepidoptera—Potentially damaging male genital structures transferred to females during copulation. *J Nat Hist* **46:** 701–715.

Cordero Rivera A, Andres JA, Cordoba-Aguilar A, Utzeri C. 2004. Postmating sexual selection: Allopatric evolution of sperm competition mechanisms and genital morphology in calopterygid damselflies (Insecta: Odonata). *Evolution* **58:** 349–359.

Coyne JA. 2000. Of vice and men. The fairy tales of evolutionary psychology. *New Republic* **3:** 27–34.

Coyne JA, Berry A. 2000. Rape as an adaptation. Is this contentious hypothesis advocacy, not science? *Nature* **404:** 121–122.

Credland PF, Wright AW. 1989. Factors affecting female fecundity in the cowpea seed beetle, *Callosobruchus maculatus* (Coleoptera: Bruchidae). *J Stored Prod Res* **25:** 125–136.

Crudgington HS, Siva-Jothy MT. 2000. Genital damage, kicking and early death. *Nature* **407:** 855–856.

Davey KG. 1960. The evolution of spermatophores in insects. *Proc R Entomol Soc Lond A Gen Entomol* **35:** 107–113.

Davies NT. 1966. Reproductive physiology. In *Monograph of Cimicidae (Hemiptera-Heteroptera)* (ed. Usinger RL). Thomas Say Foundation, Entomological Society of America, College Park, MD.

Domanitskaya EV, Liu H, Chen S, Kubli E. 2007. The hydroxyproline motif of male sex peptide elicits the innate immune response in *Drosophila* females. *FEBS J* **274:** 5659–5668.

Eberhard WG. 1985. *Sexual selection and animal genitalia*. Harvard University Press, Cambridge, MA.

Eberhard WG. 1996. *Female control: Sexual selection by cryptic female choice*. Princeton University Press, Princeton, NJ.

Eberhard WG. 2006. Sexually antagonistic coevolution in insects is associated with only limited morphological diversity. *J Evol Biol* **19:** 657–681.

Eberhard WG, Huber BA, Rodriguez RLS, Briceno RD, Salas I, Rodriguez V. 1998. One size fits all? Relationships between the size and degree of variation in genitalia and other body parts in twenty species of insects and spiders. *Evolution* **52:** 415–431.

Edgardh K, von Krogh G, Ormstad K. 1999. Adolescent girls investigated for sexual abuse: History, physical findings and legal outcome. *Forensic Sci Int* **104:** 1–15.

Edvardsson M, Tregenza T. 2005. Why do male *Callosobruchus maculatus* harm their mates? *Behav Ecol* **16:** 788–793.

Eke N. 2002. Urological complications of coitus. *BJU Int* **89:** 273–277.

Fornoni J. 2011. Ecological and evolutionary implications of plant tolerance to herbivory. *Funct Ecol* **25:** 399–407.

Fox CW. 1993. Multiple mating, lifetime fecundity and female mortality of the bruchid beetle, *Callosobruchus maculatus* (Coleoptera: Bruchidae). *Funct Ecol* **7:** 203–208.

Franck P, Solignac M, Vautrin D, Cornuet JM, Koeniger G, Koeniger N. 2002. Sperm competition and last-male precedence in the honeybee. *Anim Behav* **64:** 503–509.

Gavrilets S, Waxman D. 2002. Sympatric speciation by sexual conflict. *Proc Natl Acad Sci* **99:** 10533–10538.

Geist RF. 1988. Sexually related trauma. *Emerg Med Clin North Am* **6:** 439–466.

Gendrin M, Welchman DP, Poidevin M, Herve M, Lemaitre B. 2009. Long-range activation of systemic immunity through peptidoglycan diffusion in *Drosophila*. *PLoS Pathog* **5:** e1000694.

Grieshop K, Polak M. 2012. The precopulatory function of male genital spines in *Drosophila ananassae* [Doleschall] (Diptera: Drosophilidae) revealed by laser surgery. *Evolution* **66:** 2637–2645.

Gromko MH, Pyle DW. 1978. Sperm competition, male fitness, and repeated mating by female *Drosophila melanogaster*. *Evolution* **32:** 588–593.

Gustafsson L, Nordling D, Andersson MS, Sheldon BC, Qvarnström A. 1994. Infectious diseases, reproductive effort and the cost of reproduction in birds. *Phil Trans R Soc Lond B* **346:** 323–331.

Haine ER, Rolff J, Siva-Jothy MT. 2007. Functional consequences of blood clotting in insects. *Dev Comp Immunol* **31:** 456–464.

Heming BS. 2003. *Insect development and evolution*. Cornell University Press, Ithaca, NY.

Hosken DJ, Stockley P. 2004. Sexual selection and genital evolution. *Trends Ecol Evol* **19:** 87–93.

Hotzy C, Arnqvist G. 2009. Sperm competition favors harmful males in seed beetles. *Curr Biol* **19:** 404–407.

Hotzy C, Polak M, Rönn JL, Arnqvist A. 2012. Phenotypic engineering unveils the function of genital morphology. *Curr Biol* **22:** 2258–2261.

House CM, Simmons LW. 2005. The evolution of male genitalia: Patterns of genetic variation and covariation in the genital sclerites of the dung beetle *Onthophagus taurus*. *J Evol Biol* **18:** 1281–1292.

House CM, Lewis Z, Hodgson DJ, Wedell N, Sharma MD, Hunt J, Hosken DJ. 2013. Sexual and natural selection both influence male genital evolution. *PLoS ONE* **8:** e63807.

Izquierdo MA, Rubio GD. 2011. Male genital mutilation in the high-mountain goblin spider, *Unicorn catleyi*. *J Insect Sci* **11:** 118.

Johnstone RA, Keller L. 2000. How males can gain by harming their mates: Sexual conflict, seminal toxins, and the cost of mating. *Am Nat* **156:** 368–377.

Jolivet P. 2008. Inverted copulation. In *Encyclopedia of entomology* (ed. Capinera JL), pp. 2014–2044. Springer, New York.

Jones JS, Rossman L, Hartman M, Alexander CC. 2003. Anogenital injuries in adolescents after consensual sexual intercourse. *Acad Emerg Med* **10:** 1378–1383.

Jones JS, Rossman L, Diegel R, Van Order P, Wynn BN. 2009. Sexual assault in postmenopausal women: Epidemiology and patterns of genital injury. *Am J Emerg Med* **27:** 922–929.

Kamimura Y. 2003. Effects of broken male intromittent organs on the sperm storage capacity of female earwigs, *Euborellia plebeja*. *J Ethol* **21:** 29–35.

Kamimura Y. 2007. Twin intromittent organs of *Drosophila* for traumatic insemination. *Biol Lett* **3:** 401–404.

Kamimura Y. 2010. Copulation anatomy of *Drosophila melanogaster* (Diptera: Drosophilidae): Wound-making organs and their possible roles. *Zoomorphology* **129:** 163–174.

Kamimura Y. 2012. Correlated evolutionary changes in *Drosophila* female genitalia reduce the possible infection risk caused by male copulatory wounding. *Behav Ecol Sociobiol* **66:** 1107–1114.

Kamimura Y, Lee C-Y. 2014. Mating and genital coupling in the primitive earwig species *Echinosoma denticulatum* (Pygidicranidae): Implications for genital evolution in dermapteran phylogeny. *Arthrop Syst Phylog* **72:** 11–21.

Kamimura Y, Mitsumoto H. 2011. Comparative copulation anatomy of the *Drosophila melanogaster* species complex (Diptera: Drosophilidae). *Entomol Sci* **14:** 399–410.

Kamimura Y, Mitsumoto H. 2012. Genital coupling and copulatory wounding in *Drosophila teissieri* (Diptera: Drosophilidae). *Can J Zool* **90:** 1437–1440.

Kelly DL, Larkin HJ, Cosby CD, Paolinetti LA. 2013. Derivation of the Genital Injury Severity Scale (GISS): A concise instrument for description and measurement of external female genital injury after sexual intercourse. *J Forensic Leg Med* **20:** 724–731.

Koene JM, Schulenburg H. 2005. Shooting darts: Co-evolution and counter-adaptation in hermaphroditic snails. *BMC Evol Biol* **5:** 25.

Kramer AC. 2011. Penile fracture seems more likely during sex under stressful situations. *J Sex Med* **8:** 3414–3417.

Lange R, Gerlach T, Beninde J, Werminghausen J, Reichel V, Anthes N. 2012. Female fitness optimum at intermediate mating rates under traumatic mating. *PLoS ONE* **7:** e43234.

Lange R, Reinhardt K, Michiels NK, Anthes N. 2013a. Functions, mechanisms, and evolution of traumatic mating. *Biol Rev* **88:** 585–601.

Lange R, Werminghausen J, Anthes A. 2013b. Does traumatic secretion transfer manipulate mating roles or reproductive output in a hermaphroditic sea slug? *Behav Ecol Sociobiol* **67:** 1239–1247.

Lange R, Werminghausen J, Anthes N. 2014. Cephalo-traumatic secretion transfer in a hermaphrodite sea slug. *R Soc Lond B* **281:** 20132424.

Lee J, Singh B, Kravets FG, Trocchia A, Waltzer WC, Khan SA. 2000. Sexually acquired vascular injuries of the penis: A review. *J Trauma* **49:** 351–358.

Lefevre G Jr, Jonsson UB. 1962. Sperm transfer, storage, displacement, and utilization in *Drosophila melanogaster*. *Genetics* **47:** 1719–1736.

Leonard JL, Pearse JS, Harper AB. 2002. Comparative reproductive biology of *Ariolimax californicus* and *A. dolichophallus* (Gastropoda: Stylommatophora). *Invert Reprod Dev* **41:** 83–93.

Lessells CM. 2006. The evolutionary outcome of sexual conflict. *Phil Trans R Soc Lond B* **361:** 301–317.

Lincoln C, Perera R, Jacobs I, Ward A. 2013. Macroscopically detected female genital injury after consensual and non-consensual vaginal penetration: A prospective comparison study. *J Forensic Leg Med* **20:** 884–901.

Little TJ, Kraaijeveld AR. 2004. Ecological and evolutionary implications of immunological priming in invertebrates. *Trends Ecol Evol* **19:** 58–60.

Lockhart AB, Thrall PH, Antonovics J. 1996 Sexually transmitted diseases in animals: Ecological and evolutionary implications. *Biol Rev* **71:** 415–471.

Mann T. 1984. *Spermatophores: Development, structure, biochemical attributes, and role in the transfer of spermatozoa*. Springer, New York.

Masly JP, Dalton JE, Srivastava S, Chen L, Arbeitman MN. 2011. The genetic basis of rapidly evolving male genital morphology in *Drosophila*. *Genetics* **189:** 357–374.

McNeil CL, Bain CL, Macdonald SJ. 2011. Multiple quantitative trait loci influence the shape of a male-specific genital structure in *Drosophila melanogaster*. *G3 (Bethesda)* **1:** 343–351.

Medzhitov R, Schendier DS, Soares MP. 2012. Disease tolerance as a defense strategy. *Science* **335:** 936–941.

Mehta SD, Moses S, Agot K, Parker CB, Ndinya-Achola JO, Maclean I, Bailey RC. 2009. Adult male circumcision does not reduce risk of incident *Neisseria gonorrhoeae*, *Chlamydia trachomatis*, or *Trichomonas vaginalis* infection: Results from a randomized controlled trial in Kenya. *J Infect Dis* **200:** 370–378.

Mehta SD, Krieger JN, Agot K, Moses S, Ndinya-Achola JO, Parker C, Bailey RC. 2010. Circumcision and reduced risk of self-reported penile coital injuries: Results from a randomized controlled trial in Kisumu, Kenya. *J Urol* **184:** 203–209.

Mehta SD, Moses S, Parker CB, Agot K, Maclean I, Bailey RC. 2012. Circumcision status and incident herpes simplex virus type 2 infection, genital ulcer disease, and HIV infection. *AIDS* **26:** 1141–1149.

Michiels NK. 1998. Mating conflicts and sperm competition in simultaneous hermaphrodites. In *Sperm competition and sexual selection* (ed. Birkhead TR, Møller AP), pp. 219–254. Academic, San Diego.

Moret Y, Schmid-Hempel P. 2000. Survival for immunity: The price of immune system activation for bumblebee workers. *Science* **290:** 1166–1168.

Morrow EH, Arnqvist G. 2003. Costly traumatic insemination and a female counter-adaptation in bed bugs. *Proc R Soc Lond B* **270:** 2377–2381.

Morrow EH, Innocenti P. 2012. Female postmating immune responses, immune system evolution and immunogenic males. *Biol Rev* **87:** 631–638.

Morrow EH, Arnqvist G, Pitnick S. 2003. Adaptation versus pleiotropy: Why do males harm their mates? *Behav Ecol* **14:** 802–806.

Müller HJ. 1962. Die variabilität der genitalstrukturen bei zikaden und ihre biologische und taxonomische bedeutung [The variability of genitalic structures in planthoppers and its biological and taxonomic significance]. *Mittbl Insektenkd* **6:** 36–42.

Nessler SH, Uhl G, Schneider JM. 2007. Genital damage in the orb-web spider *Argiope bruennichi* (Araneae: Araneidae) increases paternity success. *Behav Ecol* **18:** 174–181.

Neufeld CJ, Palmer AR. 2008. Precisely proportioned: Intertidal barnacles alter penis form to suit coastal wave action. *Proc R Soc Lond B* **275:** 1081–1087.

Oliveira RF, Taborsky M, Brockmann HJ. 2008. *Alternative reproductive tactics: An integrative approach*. Cambridge University Press, Cambridge.

Otti O, Naylor R, Siva-Jothy MT, Reinhardt K. 2009. Bacteriolytic activity in the ejaculate of an insect. *Am Nat* **174:** 292–295.

Otti O, McTighe AP, Reinhardt K. 2013. In vitro antimicrobial sperm protection by an ejaculate-like substance. *Funct Ecol* **27:** 219–226.

Özorak A, Hoşcan MB, Oksay T, Güzel A, Koşar A. 2014. Management and outcomes of penile fracture: 10 years' experience from a tertiary care center. *Int Urol Nephrol* **46:** 519–522.

Parker GA. 1979. Sexual selection and sexual conflict. In *Sexual selection and reproductive competition in insects* (ed. Blum MS, Blum NA), pp. 123–163. Academic, New York.

Parzer HF, Moczek AP. 2008. Rapid antagonistic coevolution between primary and secondary sexual characters in horned beetles. *Evolution* **62:** 2423–2428.

Peng J, Zipperlen P, Kubli E. 2005. *Drosophila* sex-peptide stimulates female innate immune system after mating via Toll and Imd pathways. *Curr Biol* **15:** 1690–1694.

Polanco A, Miller D, Brewster C. 2011. Reproductive potential of field-collected populations of *Cimex lectularius* L. and the cost of traumatic insemination. *Insects* **2:** 326–335.

Proctor HC. 1998. Indirect sperm transfer in arthropods: Behavioural and evolutionary trends. *A Rev Entomol* **43:** 153–174.

Pyle DW, Gromko MH. 1978. Repeated mating by female *Drosophila melanogaster*: The adaptive importance. *Experientia* **34:** 449–450.

Reaney LT, Knell RJ. 2010. Immune activation but not male quality affects female current reproductive investment in a dung beetle. *Behav Ecol* **21:** 1367–1372.

Reichel V. 2011. "Penis morphology, mating strategies and fitness effects in simultaneous hermaphrodites." Diploma thesis, Eberhard Karls Universität, Tübingen, Germany, pp. 1–48.

Reinhardt K. 2010. Natural selection and genital variation: A role for the environment, parasites and sperm ageing? *Genetica* **138:** 119–127.

Reinhardt K, Jacobs DS. 2006. Abundance of *Cacodmus villosus* (Stål, 1855) (Heteroptera: Cimicidae) on its host, *Neoromica capensis* (Chiroptera: Verspertillionidae). *Afr Entomol* 14: 398–400.

Reinhardt K, Naylor RA, Siva-Jothy MT. 2003. Reducing a cost of traumatic insemination: Female bed bugs evolve a unique organ. *Proc R Soc Lond B* **270:** 2371–2375.

Reinhardt K, Naylor RA, Siva-Jothy MT. 2005. Potential sexual transmission of environmental microbes in a traumatically inseminating insect. *Ecol Entomol* **30:** 607–611.

Reinhardt K, Harney E, Naylor R, Gorb S, Siva-Jothy MT. 2007. Female-limited polymorphism in the copulatory organ of a traumatically inseminating insect. *Am Nat* **170:** 931–935.

Reinhardt K, Naylor RA, Siva-Jothy MT. 2009a. Ejaculate components delay reproductive senescence while elevating female reproductive rate in an insect. *Proc Natl Acad Sci* **106:** 21743–21747.

Reinhardt K, Naylor RA, Siva-Jothy MT. 2009b. Situation exploitation: Higher male mating success when female resistance is reduced by feeding. *Evolution* **63:** 29–39.

Reinhardt K, Naylor RA, Siva-Jothy MT. 2011. Male mating rate is constrained by seminal fluid availability in bedbugs, *Cimex lectularius*. *PLoS ONE* **6:** e22082.

Rice WR. 1998. Speciation by sexual conflict. In *Endless forms: Species and speciation* (ed. Howard DJ, Berlocher SH), pp. 261–270. Oxford University Press, Oxford.

Rice WR, Chippendale A. 2001. Intersexual ontogenetic conflict. *J Evol Biol* **14:** 685–693.

Riggs N, Houry D, Long G, Markovchick V, Feldhaus KM. 2000. Analysis of 1,076 cases of sexual assault. *Ann Emerg Med* **35:** 358–362.

Rönn JL, Hotzy C. 2012. Do longer genital spines in male seed beetles function as better anchors during mating? *Anim Behav* **83:** 75–79.

Rönn J, Katvala M, Arnqvist G. 2007. Coevolution between harmful male genitalia and female resistance in seed beetles. *Proc Natl Acad Sci* **104:** 10921–10925.

*Schärer L, Janick T, Ramm SA. 2014. Sexual conflict in hermaphrodites. *Cold Spring Harb Perspect Biol* doi: 10.1101/cshperspect.a017673.

Sekizawa A, Seki S, Tokuzato M, Shiga S, Nakashima Y. 2013. Disposable penis and its replenishment in a simultaneous hermaphrodite. *Biol Lett* **9:** 20121150.

Shapiro JP, Shirk PD, Kelley K., Lewis TM, Horton DR. 2010. Identity of two sympatric species of Orius (Hemiptera: Heteroptera: Anthocoridae). *J Insect Sci* **10:** 189.

Siva-Jothy MT. 2009. Reproductive immunity. In *Insect infection and immunity* (ed. Rolff J, Reynolds SE), pp. 241–250. Oxford University Press, Oxford.

Siva-Jothy MT, Moret Y, Rolff J. 2005. Insect Immunity: An evolutionary ecology perspective. *Adv Insect Physiol* **32:** 1–48.

Smolensky N, Romero MR, Krug PJ. 2009. Evidence for costs of mating and self-fertilization in a simultaneous hermaphrodite with hypodermic insemination, the Opisthobranch *Alderia willowi*. *Biol Bull* **216:** 188–199.

Sota T, Kubota K. 1998. Genital lock-and-key as a selective agent against hybridization. *Evolution* **52:** 1507–1513.

Stockley P. 2002. Sperm competition risk and male genital anatomy: Comparative evidence for reduced duration of female sexual receptivity in primates with penile spines. *Evol Ecol* **16:** 123–137.

Stutt AD, Siva-Jothy MT. 2001. Traumatic insemination and sexual conflict in the bed bug *Cimex lectularius*. *Proc Natl Acad Sci* **98:** 5683–5687.

Szabo R, Short RV. 2000. How does male circumcision protect against HIV infection? *Brit Med J* **320:** 1592–1594.

Tatarnic NJ, Cassis G. 2013. Surviving in sympatry: Paragenital divergence and sexual mimicry between a pair of traumatically inseminating plant bugs. *Am Nat* **182:** 542–551.

Tatarnic NJ, Cassis G, Siva-Jothy MT. 2014. Traumatic insemination in terrestrial arthropods. *A Rev Entomol* **59:** 245–261.

Theopold U, Schmidt O, Söderhall K, Dushay MS. 2004. Coagulation in arthropods: Defence, wound closure and healing. *Trends Immunol* **25:** 289–294.

Uhl G, Nessler SH, Schneider JM. 2010. Securing paternity in spiders: A review on occurrence and effects of mating plugs and male genital mutilation. *Genetica* **138:** 75–104.

Ullyot P, Beauchamp RSA. 1931. Mechanisms for the prevention of self-fertilization in some species of fresh-water triclads. *Q J Microsc Sci* **74:** 477–490.

Usinger RL. 1966. *Monograph of Cimicidae (Hemiptera-Heteroptera)*, Vol. 7. Thomas Say Foundation, Entomological Society of America, College Park, MD.

Valdes A. 2004. Morphology of the penial hooks and vaginal cuticular lining of some dorid nudibranchs (Mollusca, Opisthobranchia). *Am Malacol Bull* **18:** 49–53.

Watanabe S, Hara M, Watanabe Y. 2000. Male internal fertilization and introsperm-like sperm of the seaweed pipefish (*Syngnathus schlegeli*). *Zool Sci* **17:** 759–767.

West-Eberhard MJ. 2003. *Developmental plasticity and evolution*. Oxford University Press, Oxford.

Wilson CJ, Tomkins JL. 2014. Countering counteradaptations: Males hijack control of female kicking behavior. *Behav Ecol* **25:** 470–476.

Wilson N, Tufton TJ, Eady PE. 1999. The effect of single, double, and triple matings on the lifetime fecundity of *Callosobruchus analis* and *Callosobruchus maculatus* (Coleoptera: Bruchidae). *J Insect Behav* 12: 295–306.

Yassin A, Orgogozo V. 2013. Coevolution between male and female genitalia in the Drosophila melanogaster species subgroup. *PLoS ONE* **8:** e57158.

Yip A, Ng SK, Wong WC, Li MK, Lam KH. 1989. Injury to the prepuce. *Br J Urol* **63:** 535–538.

Zaher H. 1999. Hemipenial morphology of the South American xenodontine snakes: With a proposal for a monophyletic Xenodontinae and a reappraisal of colubroid hemipenes. *Bull Am Mus Nat Hist* **240:** 1–168.

Zhong W, McClure CD, Evans CR, Mlynski DT, Immonen E, Ritchie MG, Priest NK. 2014. Immune anticipation of mating in *Drosophila*: Turandot M promotes immunity against sexually transmitted fungal infections. *Proc R Soc B* **280:** 20132018.

Sexual Conflict Arising from Extrapair Matings in Birds

Alexis S. Chaine[1], Robert Montgomerie[2], and Bruce E. Lyon[3]

[1]Station d'Écologie Expérimentale du Centre National de la Recherche Scientifique (CNRS), 09200 Moulis, France

[2]Department of Biology, Queen's University, Kingston, Ontario K7L 3N6, Canada

[3]Department of Ecology and Evolutionary Biology, University of California, Santa Cruz, California 95064

Correspondence: lyon@biology.ucsc.edu

The discovery that extrapair copulation (EPC) and extrapair paternity (EPP) are common in birds led to a paradigm shift in our understanding of the evolution of mating systems. The prevalence of extrapair matings in pair-bonded species sets the stage for sexual conflict, and a recent focus has been to consider how this conflict can shape variation in extrapair mating rates. Here, we invert the causal arrow and consider the consequences of extrapair matings for sexual conflict. Extrapair matings shift sexual conflict from a simple two-player (male vs. female) game to a game with three or more players, the nature of which we illustrate with simple diagrams that highlight the net costs and benefits of extrapair matings to each player. This approach helps identify the sorts of traits that might be under selection because of sexual conflict. Whether EPP is driven primarily by the extrapair male or the within-pair female profoundly influences which players are in conflict, but the overall pattern of conflict varies little among different mating systems. Different aspects of conflict are manifest at different stages of the breeding cycle and can be profitably considered as distinct episodes of selection caused by conflict. This perspective is illuminating both because conflict between specific players can change across episodes and because the traits that evolve to mediate conflict likely differ between episodes. Although EPP clearly leads to sexual conflict, we suggest that the link between sexual conflict and multiple paternity might be usefully understood by examining how deviations from lifetime sexual monogamy influence sexual conflict.

The development of genetic tools for determining parentage fundamentally altered our understanding of animal mating systems (Jeffreys et al. 1985; Avise 1996; Reynolds 1996) and provided invaluable insights into the consequences and causes of females mating with more than one male. Particularly for the study of birds, these methods revealed that social pair bonds often fail to match the actual patterns of copulations that produced offspring (Gowaty and Karlin 1984; Birkhead and Møller 1992; Reynolds 1996; Petrie and Kempenaers 1998), revolutionizing the study of avian mating systems. Extensive research and two recent reviews point out the progress we have made in this field and show how little we still understand extrapair behavior (Griffith et al. 2002; Westneat and Stewart 2003).

Cite this article as *Cold Spring Harb Perspect Biol* doi: 10.1101/cshperspect.a017590

In the 1960s, David Lack compiled what was then known about mating systems in birds and concluded that >90% of species were monogamous, a pattern that provided an early framework for the development of mating system theory (Lack 1968; Orians 1969; Emlen and Oring 1977). When it was later discovered that sexual mating patterns did not match the social mating systems that Lack described, the field was turned on its head (Westneat et al. 1990; Avise 1996; Reynolds 1996; Zeh and Zeh 2001). In extreme cases, the mismatch between the social and sexual mating systems is nothing short of spectacular; in fairywrens (*Malurus* species) that are socially monogamous, cooperatively breeding species with helpers, the extrapair paternity (EPP) rate can exceed 75% of all offspring and 95% of all broods (Mulder et al. 1994). In socially monogamous birds, in general, the rate of EPP is typically on the order of 10% of offspring and 20% of broods (Griffith et al. 2002), but variation among species, and even populations within species, is extensive (Arnold and Owens 2002; Griffith et al. 2002). The occurrence of EPP has profound consequences for the evolution of social behavior, both because it alters the scope for the action of sexual selection (Webster et al. 1995; Sheldon and Ellegren 1999) and because it results in males often providing parental care to offspring they have not sired (Davies et al. 1992; Westneat and Sherman 1993).

Even in taxa with mating systems other than social monogamy, or in which there is no obvious pair bond, the ability to determine parentage genetically was revolutionary, allowing precise estimates of male reproductive success when females mate multiply. That focus on multiple mating also catalyzed an interest in sexual selection from the female's perspective, whereas previous attention had been strongly biased toward males and male traits. More specifically, it raised the questions as to why females would pursue and benefit from matings outside the social pair bond (Westneat et al. 1990; Petrie and Kempenaers 1998), and why a female would benefit from mating with more than one male for a given clutch or litter. This new focus on females brought attention to the issue of polyandry more generally (Jennions and Petrie 2000; Simmons 2005; Parker and Birkhead 2013; Pizzari and Wedell 2013).

Recently, the assumption that females control mating patterns, and thus that polyandry and EPP can be universally understood from the perspective of fitness benefits to females, has been questioned (Westneat and Stewart 2003). Focusing specifically on EPP, Westneat and Stewart (2003) suggested that, in some taxa, EPP could be driven entirely by benefits to the extrapair-seeking male. They also suggested that many aspects of EPP can be profitably explored from the perspective of sexual conflict, as had Petrie and Kempenaers (1998) before them. Previous interest in the relation between EPP and sexual conflict in birds was focused particularly on trying to explain the incidence and frequency of EPP within and among species. Westneat and Stewart (2003) recognized that that link was indirect. Instead, they suggested that sexual conflict theory might help us to identify traits that could arise from conflict and that those traits might inform the search for a general explanation of the huge variation in EPP rates both among and within bird species.

Sexual conflict, the conflicting fitness interests of males and females during mating (Parker 1979; Rice 1998; Arnqvist and Rowe 2005), can lead to antagonistic coevolution between the traits expressed in males and those expressed in females, traits that in some way influence mating outcomes. Traits in males and females are ultimately the drivers of conflict, and, reciprocally, conflict fuels further trait evolution. Sexual conflict theory is useful because it can potentially explain the evolution and maintenance of traits that are otherwise difficult to understand (Arnqvist and Rowe 2005). Thus, studies often examine the factors and traits that underlie different aspects of sexual conflict, as well as the types of morphological and behavioral traits that result from selection caused by the sexual conflict itself.

In this article, we build on the foundation provided by Westneat and Stewart (2003). They proposed that sexual conflict can help to explain variation in the occurrence of EPP among species and populations (Westneat and Stewart 2003). Here, we invert the focus and seek to understand the consequences that EPP can have for sexual conflict and the relation between EPP and other drivers of sexual conflict. Thus, we examine the players involved in the sexual conflict generated by EPP and the costs and benefits that underpin the conflicts among the different players. We point out that different conflicts are involved in the different stages of a single bout of reproduction, and we suggest that these represent sequential periods of conflict, each of which is a different episode of selection generated from that conflict. We then place these patterns of conflict into a broader context by contrasting how different patterns of fidelity and infidelity (including EPP) during a lifetime of mating can influence sexual conflict. Our review focuses on birds as examples, both because they have been extensively studied with respect to EPP and because they have been the subjects of most of our own research. Our goal, however, is to provide a framework for understanding trait evolution under the influence of sexual conflict caused by females mating multiply in any animal species.

THE PLAYERS

Sexual conflict is often seen as a conflict between males and females. However, when females mate with more than one male, sexual conflict necessarily becomes a multiplayer game. The players include at least the pair female, pair male, and extrapair male (Petrie and Kempenears 1998; Rice 1998; Westneat and Stewart 2003; Alonzo and Pizzari 2013). In some cases, the female paired with the extrapair male may also suffer fitness losses because of her mate's activities, in which case, the conflict becomes a four-player game (Petrie and Kempenaers 1998). Indeed, in specific cases, more than four players may also be involved if, for example, there are multiple extrapair males, multiple females are affected by the extrapair behavior in a polygynous system, or EPP affects the fitness of nonbreeding individuals, such as helpers in co operative breeders.

The complexities involved in the sorts of conflicts generated by EPP can be usefully illustrated with conflict-of-interest diagrams that show all of the players involved (Fig. 1; see also Petrie and Kempenaers 1998; Rice 1998). For example, the addition of an extrapair male to the simple conflict between the sexes in a mated pair potentially creates two sexual conflicts—one between the female and her mate and the other between the same female and the extrapair male—as well as intrasexual conflict between the two males (Fig. 1). The dynamics of the conflict between any two of these players can be affected by the other conflicts of interests in the triad (Westneat and Stewart 2003). Thus, to fully understand how sexual conflict affects trait evolution, we need to consider all conflict interactions, including how intrasexual conflict between males influences

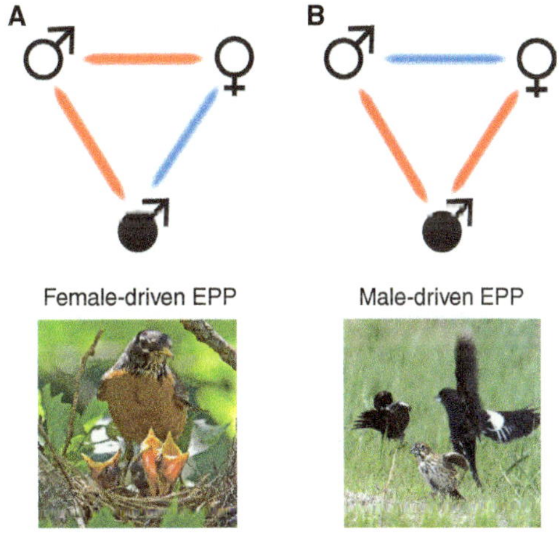

Figure 1. Conflicts of interests between males and females over the occurrence of extrapair matings in socially monogamous mating systems. Symbols for extrapair males are solid black. Red bars join individuals that are in conflict over the occurrence of extrapair matings, whereas blue bars join individuals that agree on the optimal level (zero for male-driven) of extrapair mating. Two different socially monogamous systems are shown here: (*A*) When females benefit from EPP (i.e., EPP is female driven), the female and EPP male both benefit from EPP and are in conflict with the within-pair male (e.g., American robin *Turdus migratorius*), and (*B*) when EPP is solely driven by the fitness interests of the extrapair male (i.e., EPP is male driven), both the female and within-pair male are in conflict with the extrapair male (e.g., lark bunting *Calamospiza melanocorys* shown here with a male attempting an EPC while the pair male tries to fend him off). (Both photos courtesy of B. Lyon.)

conflict between the sexes. It is important to recognize that extrapair matings generate different conflicts that manifest themselves at successive stages of reproduction (e.g., conflict over mating, fertilization, and postfertilization investment; see the section Episodes of Conflict over EPP: When and over What?). In this first section, we use conflict-of-interest diagrams to focus only on the conflict over the occurrence of extrapair copulations (EPCs) and not during the subsequent episodes of reproduction. We provide details on only one stage of reproduction because the basic nature of the conflict can change across episodes (see below), and hence a single diagram is often specific to one episode.

Our conflict-of-interest diagrams (Fig. 1) illustrate simply whether or not there is conflict between two players; we ignore the strength of the conflict and avoid assumptions about which player causes the conflict, hence the lack of arrows denoting a direction of conflict. Blue links indicate when two players both share the same optimal level (zero for male-driven) of extrapair mating, red links denote conflicts of interest over the occurrence of that mating, and, in Figure 2, thin black lines denote pairs of socially interacting individuals whose joint fitness inter-

Figure 2. Conflicts of interests between focal males and females in relation to breeding system. Symbols for extrapair males and their social mates in solid black. Red bars join individuals that are in conflict over the occurrence of the extrapair matings, whereas blue bars join individuals that share the same optimal level (zero for male-driven) or extrapair mating; thin black lines indicate no particular costs or benefits. All cases shown involve female-driven EPP, in which the female benefits from the extrapair mating. Photos show one example species of each mating system: (*A*) Tree swallow (*Tachycineta bicolor*), (*B*) purple martin (*Progne subis*), (*C*) red-winged blackbird (*Agelaius phoeniceus*), (*D*) Smith's longspur (*Calcarius pictus*), (*E*) red phalarope, and (*F*) splendid fairywren (*Malurus splendens*). (All photos courtesy of B. Lyon, except Smith's longspur, which is provided by Ian Davies.)

Cite this article as *Cold Spring Harb Perspect Biol* doi: 10.1101/cshperspect.a017590

ests are not affected by the extrapair mating. The two males will typically be in conflict over the occurrence of the EPCs because they compete to fertilize the same ova—usually a zero-sum game. Although male–male cooperation over matings with a single female is theoretically feasible (e.g., when courtship and copulation with different males greatly stimulate female fecundity [Alonzo and Pizzari 2013]), this idea remains to be tested empirically.

These conflict-of-interest diagrams help to illustrate how conflict changes under different social scenarios. A fundamental debate, for example, is whether or not EPP benefits females (Mulder et al. 1994; Westneat and Stewart 2003; Arnqvist and Kirkpatrick 2005; Griffith 2007; Dunn et al. 2013). The sex that promotes the occurrence of an extrapair mating also critically influences the pattern of sexual conflict over such matings: Male- and female-driven EPP result in opposite patterns of conflict over whether extrapair matings should occur (Petrie and Kempenears 1998) and hence opposite patterns of selection on traits (Fig. 1A,B). Importantly, sexual conflict over EPP occurs whether it is male or female driven; what changes is whether the conflict is within the pair or between the extrapair male and both members of the pair (Fig. 1A,B).

In socially monogamous species with female-driven EPP, the female and extrapair male share a benefit from EPP, whereas the female and her mate are in conflict over the occurrence of the EPC (Fig. 1A). Thus, we expect to see traits that enhance the extrapair male's ability to court the female, the pair male's ability to guard his mate, and the female's ability to evade mate guarding and thereby sneak away to copulate with an extrapair male. Mate guarding by the within-pair male appears to be particularly widespread (Møller and Birkhead 1991; Kokko and Morrell 2005; Pedersen et al. 2006), but there is also some evidence that the extrapair male and pair female use tactics to enhance their copulation success. Superb fairywren (*Malurus cyaneus*) females, for example, go on extrapair forays to seek EPCs before dawn (Double and Cockburn 2000), presumably to avoid detection by their social mate. Intriguingly, male fairywrens seek EPCs with a unique display behavior—a flower petal display—that they use only when courting females seeking EPCs (Mulder 1997; Karubian and Alvarado 2003). Alternatively, Dunnock (*Prunella modularis*) females fly quickly to try to "lose" their social mate when they are seeking extrapair mates, and copulations with those extrapair males are exceptionally rapid and cryptic (Davies 1992). Females may also adopt countertactics to incite male–male competition and thereby thwart tenacious mate guarding by their mates (Montgomerie and Thornhill 1989); female bearded reedlings (*Panurus biarmicus*), for example, produce a call that attracts the attention of multiple extrapair males and this has been interpreted as a means for the female to increase the frequency of EPCs (Hoi 1997).

When EPP is driven by extrapair males, the female and her mate might both benefit from preventing EPCs, and thus both would be in conflict with an extrapair male (Fig. 1B). In this situation, we expect evidence for forced or coerced copulations by the extrapair male and cooperation between the pair members to prevent them. Females, for example, might call to attract the attention of their mate when an extrapair male approaches in pursuit of copulations. There are no clear avian examples of such tactics, but female moose (*Alces alces*) call to attract the attention of the dominant male when subordinate males attempt to force copulations (Bowyer et al. 2011). Extrapair males, on the other hand, should evolve morphologies or behaviors to force matings that females are trying to avoid. Thus, cloacal protuberances in male birds may have been favored specifically because of benefits they provide in securing EPP, as they occur most commonly in species in which there is sperm competition because of extrapair matings (Briskie 1993; Tuttle et al. 1996). A few taxa of birds, particularly waterfowl and ratites, also have penises and these structures may have evolved particularly to help extrapair males overcome female defenses (Briskie and Montgomerie 1997; Brennan et al. 2007).

Variation in breeding system (i.e., in patterns of association between the sexes and parentage) might be expected to impact how

EPP influences sexual conflict because different breeding systems vary in the numbers of socially interacting individuals. It is therefore surprising that the overall patterns of conflict over the occurrence of extrapair matings would appear to be similar across breeding systems (Fig. 2A–F). On reflection, this makes sense because EPP is about the sexual (genetic) patterns of reproduction, and these are actually quite similar across social systems. The patterns of conflict only differ when EPP influences either the social patterns or parental care trade-offs among more than one male and female, such as in social polygynandry (Fig. 2D), when males provision offspring at more than one nest (Davies 1992; Briskie et al. 1998). Likewise, social polyandry can lead to mixed paternity in one or more clutches (Fig. 2E), either because females mate simultaneously with multiple males or because sperm from a first reproductive bout can be stored and sire offspring from a second clutch, a pattern documented for role-reversed species like spotted sandpipers (*Actitis macularius* [Oring et al. 1992] or red phalaropes (*Phalaropus fulicarius* [Dale et al. 1999]). Although these conflict patterns resemble the basic interactions in social monogamy, a key difference in polygnandrous and polyandrous breeding systems is that all of the males are socially paired with the same female, and the conflict thus comes from within the breeding association, not from outside.

In one other breeding system, cooperative breeding, the pattern of sexual conflict among breeding adults is similar to that in social monogamy, but the occurrence of EPP has a special influence on other social group members, the helpers (Fig. 2F). The presence of extrapair offspring in the brood reduces the relatedness between helpers and the offspring they are helping to raise, thus reducing the indirect benefits from helping (Charnov 1981). This creates conflict between helpers and their mother, which, in turn, should favor helpers that discourage the occurrence of extrapair matings. How cooperative breeding remains stable under the extremely high levels of EPP observed in some fairywrens (*Malurus* species) is puzzling. Possibly, females show very high fidelity to the same extrapair male across years, in which case, helpers would often be assisting their mother to raise full siblings.

BENEFITS AND COSTS OF EPP AS DRIVERS OF CONFLICT

Sexual conflict occurs because the fitness interests of the sexes can differ during the various stages of reproduction. Thus, to understand how and why EPP generates sexual conflict, and how this source of conflict compares with other forms of sexual conflict, we need to understand the costs and benefits of the extrapair matings to the different players.

The benefits that males gain from extrapair matings have long been clear (Trivers 1972): via EPP, males sire more offspring than would otherwise be possible. Any benefits gained by the extrapair male are thought to lead to direct fitness costs for the within-pair male because the within-pair male then sires fewer offspring than he would otherwise. But EPP may cause females to increase their investment, either through the stimulatory effects of ejaculates from multiple males (Alonzo and Pizzari 2013) or when females engage in EPP to upgrade the genetic quality of their offspring, resulting in differential allocation of maternal care (Burley 1988). Thus, changes in female investment should be considered when evaluating the fitness costs to the pair male and the effects on his contribution to parental care (Sheldon 2000, 2002).

For females, extrapair matings potentially bring a wide variety of direct and indirect fitness benefits, but the importance of indirect benefits remains controversial (Arnqvist and Kirkpatrick 2005; Griffith 2007). Direct benefits of EPC can accrue from an increase in the number of fertilized ova (e.g., when the pair male is infertile or has limited sperm available [Sheldon 1994]), access to more resources (e.g., extrapair male's territory or paternal care [Davies 1992; Briskie et al. 1998]), and a reduction in harassment by those males seeking EPCs (Rowe 1992; Gowaty and Buschhaus 1997). In species with sexually selected infanticide by males, females that mate multiply (via polyandry or EPP) might reduce the risk of infanticide of their eggs or offspring if infanticide is performed by

 Cite this article as *Cold Spring Harb Perspect Biol* doi: 10.1101/cshperspect.a017590

males that have not copulated with the female (Agrell et al. 1998; Soltis et al. 2000). Potential indirect benefits include the genetic benefits that accrue from an increase in the quality, rather than quantity, of offspring (good genes in a viability or sexual context [Slayter et al. 2012; Madsen et al 1992] or genetic complementarity [Zeh and Zeh 2006]), an increase in the genetic diversity of offspring (Slayter et al. 2012; Parker and Birkhead 2013), or a reduction of inbreeding depression when females are socially paired with a relative (Blomqvist et al. 2002; Cornell and Tregenza 2007). Note that these female benefits are identical to benefits from polyandry generally (i.e., when females mate with more than one male during a reproductive bout [Schlicht and Kempenaers 2011; Slayter et al. 2012; Parker and Birkhead 2013]).

There are also likely to be fitness costs to females from extrapair matings, especially when EPP is driven solely by the fitness interests of the extrapair male, such as when the copulations are forced or coerced (Westneat and Stewart 2003). In addition to the risk of injury or transfer of sexually transmitted diseases during mating, the female suffers a potential indirect loss of fitness by having some of her offspring sired by a nonpreferred male—a cost that is the inverse of the benefits a female gains when EPP is female driven. Females may also suffer a cost due to the loss of parental care from her social mate when he suspects EPP (Sheldon 2002; Griffin et al. 2013). Finally, there are also potential costs to the female mated to a male who seeks EPCs—his extrapair matings could increase her risk of contracting sexually transmitted diseases or could result in lower investment from her mate if he invests time seeking extrapair mates (Magrath and Elgar 1997; Magrath and Komdeur 2003). In theory, the extrapair male's mate could also suffer fitness losses if the extrapair male allows the other female to access his territorial resources or if he provides food to her offspring.

EPISODES OF CONFLICT OVER EPP—WHEN AND OVER WHAT?

In their broad survey of the natural history of sexual conflict, Arnqvist and Rowe (2005) noted that sexual conflict can occur during any interaction between a male and female, not just when deciding whom to mate with. They identified three main types of interactions as the "components" of conflict: (1) conflict before copulation, (2) conflict during the period from copulation to fertilization of the zygote, and (3) conflict after fertilization, over parental investment. We suggest that these three components delineate clear episodes of selection because they are temporally distinct bouts that connect in a temporal sequence to influence lifetime fitness (Arnold and Wade 1984; Moore 1990; McGlothlin 2010). Thinking about episodes of selection helps clarify how these different periods of conflict are linked through their effects on fitness, how selection driven by conflict might be resolved, and how different types of traits are often favored during the different episodes.

The different episodes of conflict differ in some critical ways with respect to EPP. First, the basic nature of the conflict can invert from one episode to the next. For example, with male-driven EPP, the female and her mate should cooperate to prevent extrapair mating attempts, but when an EPC does occur and is detected by the pair male, it can trigger conflict between the paired male and female over postfertilization investment. Males should not invest in offspring they have not sired whether that EPP was male or female driven. Second, the traits that mediate conflict (i.e., that evolve in response to conflict) differ across the episodes, both in the types of traits and in the degree to which they show canalization or plasticity among mating roles. The same traits that mediate conflict over the decision to mate are not likely to be the same as those that mediate fertilization success or parental investment.

Likewise, plastic traits are probably more useful in some episodes than in others. Traits shows plasticity when their expression depends on the mating role in which they are manifest (e.g., different expression in extrapair versus within-pair males) (Fig. 1) (Petrie and Kempenaers 1998; Ball and Parker 2003; Westneat and Stewart 2003; Parker 2006; Alonzo and Pizzari 2013). Role-specific traits must necessarily be flexible, and they also require an assessment mechanism that can link trait expression to its specific role. Behavioral traits provide the simplest way to achieve role flexibility. Thus, plasticity may be more likely in episodes in which behavior is important (e.g., premating decisions and parental care), whereas canalized traits may be more likely when physiology, rather than behavior, is the mediator of conflict (e.g., fertilization success).

Before copulation, there is sexual conflict that favors traits in the extrapair male that increase the success of his extrapair mating attempts, in the within-pair male that decrease the copulation success of the extrapair male, and in the females that enhance or decrease the success of extrapair matings, depending on whether or not the female benefits from those matings. Because the premating episode mainly involves behavioral interactions between individuals, there is much potential for role-specific trait expression and ample evidence for differences in the behaviors of the extrapair and within-pair males (e.g., territorial vs. sneaker males).

Once an EPC occurs, the conflict shifts to interactions that influence fertilization success within the female's reproductive tract. Sperm number and characteristics as well as ejaculate properties mediate that conflict for males, whereas sperm storage structures and reproductive tract secretions play a role in mediating the conflict over fertilization success for females (Birkhead et al. 1993; Briskie and Montgomerie 1993). Males thus have the opportunity to adjust sperm number or ejaculate size specifically in the context of pursuing EPP (Parker 1990; Birkhead and Fletcher 1992; Birkhead et al. 1995; Pizzari et al. 2003).

The third episode of conflict over EPP arises after fertilization and influences parental investment in offspring. Multiple paternity in a brood potentially alters sexual conflict over parental investment because it reduces the fitness the social male could have gained from the current reproductive bout. However, it is debatable whether this translates into a reduction in paternal care (e.g., Sheldon 2002; Griffin et al 2013), although there are clearly conflicts of interest over the fitness consequences of EPP. For example, theory suggests that there can be situations under which EPP does not affect the optimal parental investment of the male (Westneat and Sherman 1993; Griffin et al. 2013). On the other hand, there is some evidence that males, on average, invest less parental care in species with higher rates of EPP (reviewed in Griffin et al. 2013). Finally, extrapair males would be expected to benefit if they can cause an increased investment by both parents beyond their optimal levels, but it is unclear what mechanisms would enable them to do so facultatively.

VARIATION IN MATING PATTERNS DRIVES EPISODES OF SEXUAL CONFLICT

In this section, we focus particularly on how different mating patterns (including EPP) affect different episodes of conflict. To do this, we distinguish four basic mating patterns that differ in the degree of mating fidelity and the types of sexual conflict they generate (Fig. 3). At one extreme is lifetime sexual monogamy—one male and one female social partner for life, and they mate only with each other. This mating pattern is intriguing because, in theory, sexual conflict should be nonexistent because the lifetime fitness interests of a paired male and female are identical and entirely dependent once pairing has occurred (Rice and Holland 1997). Selection experiments with *Drosophila* support this theory—selection lines with experimentally imposed lifetime monogamy showed dramatic reductions in measures of sexual conflict relative to lines with the normal polyandrous pattern of mating (Holland and Rice 1999). Theory and comparative studies also suggest that lifetime sexual monogamy is crucial for the evolution of some strong forms of altruism, such as the helping behavior observed in social insects and some birds (Hughes et al. 2008; Cornwallis et al. 2010; Boomsma 2013).

Any deviations from lifetime sexual monogamy set the stage for sexual conflict (Holland and Rice 1997) in two different ways (Fig. 3): sexually monogamous matings within a reproductive bout, but males and females change mates between reproductive bouts, or females mate polyandrously within a given reproductive bout. Polyandry within a reproductive bout

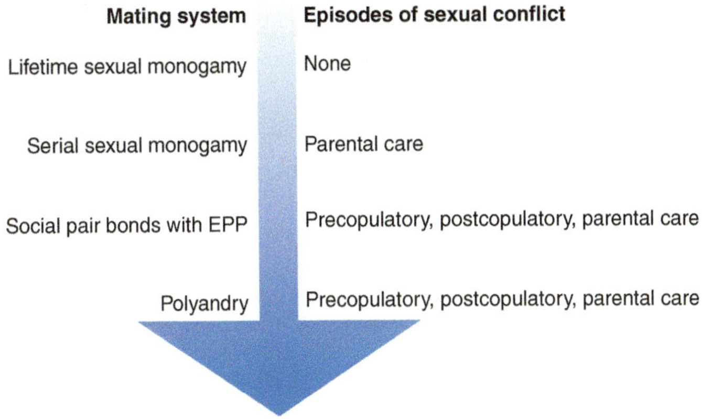

Figure 3. Deviation from lifetime monogamous mating creates sexual conflict, but different mating patterns create different patterns of conflict. All deviations from lifetime monogamy potentially create sexual conflict over how much a female invests in a given reproductive bout, but polyandrous matings (including EPP) are required for conflict over both matings and fertilization.

occurs in species both with and without pair bonds; the former is the special context referred to as extrapair matings, which gives rise to EPP. This sexual conflict framework based on mating patterns is useful because it allows us to partition the different factors that contribute to sexual conflict and identify which of them apply generally to all mating systems (except lifelong sexual monogamy) and which are unique to specific mating systems.

A couple of patterns emerge. First, sexual conflict over reproductive investment is ubiquitous in all systems in which lifetime monogamy does not occur (Fig. 3)—even without polyandrous matings, traits in males that enable them to extract levels of female parental investment above the female's optimum can be favored, including ejaculate components that manipulate females to increase their investment in the sire's brood (Chapman 2001; Wolfner 2002; Wedell et al. 2006), and imprinted genes that, when inherited from the male, cause an embryo to increase the extent of resource extraction from its mother (Haig and Westoby 1989; Haig 2000). Second, polyandrous matings and multiple paternity (including EPP) in a given clutch or litter generate two additional episodes of sexual conflict that are absent when females mate with only one male in a given breeding season—sexual conflict over both copulations and fertilization (Fig. 3). That general polyandry and EPP share these distinctive episodes of conflict supports the notion that EPP is simply a special case of polyandry in matings that involve pair bonds (Schlicht and Kempenaers 2011). For the conflict over copulations, which individuals are in conflict will depend on whether EPP is female or male driven, and this will affect the types of traits that are favored. For the conflict over fertilization, the scope for flexible role-specific traits in the competing males is probably limited with regard to sperm morphology, but it could be expressed as variation in copulation rate (Briskie 1993) or ejaculate size (Parker 1990; Birkhead and Fletcher 1992).

Finally, conflict that affects male parental care is clearly influenced by EPP. This is generally true because any deviation from lifetime sexual fidelity favors traits in both sexes that lead to increased investment in a given bout by the other sex. Mechanisms to achieve this increased investment include products in male ejaculates that alter female investment (Reinhardt et al. 2009) or genomic imprinting that alters offspring physiology or behavior to extract more resources from the female (and possibly the pair male [Haig and Westoby 1989; Haig 2000]). However, when sexual conflict is driven by EPP, a reduction in parental investment by the within-pair male is favored, and that would exacerbate the already present conflict over optimal parental investment by each parent. Of course, this only applies to pair-bonded species that show male parental care—a situation that is widespread in birds but not ubiquitous (Schwagmeyer et al. 1999; Cockburn 2006).

CONCLUSIONS

One of our goals in writing this review has been to give some structure to thinking about the kinds of traits that are expected to be favored in response to the sexual conflict that arises when there is extrapair mating in birds. Clearly, many of these traits are common to conflict when females mate multiply, whether or not there is a social mating system that includes extrapair matings. We see this as an important, and often neglected, point both because there is really not very much that is special about EPP in the context of sexual conflict and because those few traits that are special are all relevant to conflict over parental care. This is perhaps not so surprising because it is biparental care that underlies the structure of socially monogamous mating systems (or polygyny with male care), and the resolution of those conflicts might be an important source of selection on the evolution of fidelity in general.

Most research on the incidence of extrapair mating—in birds, at least—has focused on trying to explain why males and females might gain a fitness advantage by seeking EPCs. In part because promiscuity is probably the ancestral state (Corwallis et al. 2010), we suggest that it might be profitable to look, instead, at factors that influence the evolution of fidelity in birds and the reduction of sexual conflict caused by EPP. Although within-pair sexual fidelity might evolve for a variety of reasons (Black 1996), there would certainly be advantages from conflict reduction in the short term that might well outweigh the disadvantages of infidelity, some of which we have listed in this review. Thus, we suggest that, in species with biparental care in particular, models that consider the benefits of fidelity balanced against the costs of extrapair mating, including costs due to sexual conflict, might help us to understand the still bewildering incidence of EPP within and among bird species.

In a similar vein, it has been clear for a long time that demography could have a major influence on the value of extrapair mating in birds. As Arnold and Owens (2002) showed, about half of the variation in rates of EPP among bird species can be explained solely by the rates of adult mortality. Two factors might explain this pattern. First, a prerequisite for future cooperation between a breeding pair is that both the male and female of a pair survive to the following year, a probability that decreases exponentially with increased mortality rates (i.e., joint survival $\propto$ mortality^2). Second, longer-lived species simply have more opportunity to benefit from the advantages of sexual fidelity (i.e., more years of enjoying the benefits) and, as a consequence, have lower rates of EPP and EPP-driven sexual conflict—a notion that remains to be explored. This, again, suggests that an exploration of the factors that influence the evolution of fidelity could be a profitable way to understand the distribution of EPP both within and among species. As we have outlined above, traits that arise from sexual conflict vary during the breeding cycle, involving distinct episodes of selection before copulation, between copulation and fertilization, and during the period of parental care. Consideration of the costs and benefits of fidelity during each of these episodes is important because these will vary, directly influencing the benefits of sexual fidelity during each episode.

As has been frequently pointed out, there are many interesting parallels between the natural histories (social monogamy, biparental care) and modes of communication (vocal and color vision) in birds and humans (Birkhead 2012). As for birds, EPP in humans creates sexual conflict that is resolved to some extent by behaviors (mate guarding, kin recognition), physiologies (concealed ovulation), and the advantages of sexual fidelity, particularly with respect to parental care (Benshoof and Thornhill 1979; Burley 1979; Alvergne et al. 2009). We have focused this review on sexual conflict in birds, but we suggest that the perspectives we have taken here—on the players, changes in conflict across episodes of the breeding cycle, and the traits that evolve to reduce conflict—are generally applicable to the study of sexual conflict in any organism in which females mate multiply.

ACKNOWLEDGMENTS

Our research on the mating systems of birds is supported by grants from Agence Nationale de la Recherche (ANR)-LabEx "Tulip" and ANR-Programme Jeunes Chercheuses Jeunes Chercheurs (JCJC) "NetSelect" (A.S.C.), the Natural Sciences and Engineering Research Council of Canada and Queen's University (R.M.), and the National Science Foundation, National Geographic Society, and the University of California (A.S.C., B.E.L.).

REFERENCES

Agrell J, Wolff JO, Ylönen H. 1998. Infanticide in mammals: Strategies and counter-strategies. *Oikos* **84:** 507–517.

Alonzo SH, Pizzari T. 2013. Selection on female remating interval is influenced by male sperm competition strategies and ejaculate characteristics. *Phil Trans Roy Soc B* **368:** 20120044.

Alvergne A, Faurie C, Raymond M. 2009. Father–offspring resemblance predicts paternal investment in humans. *Anim Behav* **78:** 61–69.

Arnold KE, Owens IPF. 2002. Extra-pair paternity and egg dumping in birds: Life history, parental care and the risk of retaliation. *Proc Roy Soc B* **269:** 1263–1269.

Arnold SJ, Wade MJ. 1984. On the measurement of natural and sexual selection: Theory. *Evolution* **38:** 709–719.

Arnqvist G, Kirkpatrick M. 2005. The evolution of infidelity in socially monogamous passerines: The strength of direct and indirect selection on extrapair copulation behavior in females. *Am Nat* **165:** S26–S37.

Arnqvist G, Rowe L. 2005. *Sexual conflict.* Princeton University Press, Princeton, NJ.

Avise JC. 1996. Three fundamental contributions of molecular genetics to avian ecology and evolution. *Ibis* **138:** 16–25.

Ball MA, Parker GA. 2003. Sperm competition games: Sperm selection by females. *J Theor Biol* **224:** 27–42.

Benshoof L, Thornhill R. 1979. The evolution of monogamy and loss of estrus in humans. *J Soc Biol Str* **2:** 95–106.

Birkhead T. 2012. *Bird sense.* Bloomsbury, London.

Birkhead TR, Fletcher F. 1992. Sperm to spare? Sperm allocation by male zebra finches. *Anim Behav* **43:** 1053–1055.

Birkhead TR, Møller AP. 1992. *Sperm competition in birds: Evolutionary causes and consequences.* Academic, London.

Birkhead TR, Briskie JV, Møller AP. 1993. Male sperm reserves and copulation frequency in birds. *Behav Ecol Sociobiol* **32:** 85–93.

Birkhead TR, Fletcher F, Pellatt EJ, Staples A. 1995. Ejaculate quality and the success of extra-pair copulations in the zebra finch. *Nature* **377:** 422–423.

Black JM. 1996. *Partnerships in birds: The study of monogamy.* Oxford University Press, Oxford.

Blomqvist D, Andersson M, Küpper C, Cuthill IC, Kis J, Lanctot RB, Sandercock BK, Székely T, Wallander J, Kempenaers B. 2002. Genetic similarity between mates and extra-pair parentage in three species of shorebirds. *Nature* **419:** 613–615.

Boomsma JJ. 2013. Beyond promiscuity: Mate-choice commitments in social breeding. *Phil Trans Roy Soc B* **368:** 20120050.

Bowyer RT, Rachlow JL, Stewart KM, Van Ballenberghe V. 2011. Vocalizations by Alaskan moose: Female incitation of male aggression. *Behav Ecol Sociobiol* **65:** 2251–2260.

Brennan PLR, Prum RO, McCracken KG, Sorenson MD, Wilson A, Read AF, Birkhead TR. 2007. Coevolution of male and female genital morphology in waterfowl. *PLoS ONE* **2:** e418.

Briskie JV. 1993. Anatomical adaptations to sperm competition in Smith's Longspurs and other polygynandrous passerines. *Auk* **110:** 875–888.

Briskie JV, Montgomerie R. 1993. Patterns of sperm storage in relation to sperm competition in passerine birds. *Condor* **95:** 442–454.

Briskie JV, Montgomerie R. 1997. Sexual selection and the intromittent organ of birds. *J Avian Biol* **28:** 73–86.

Briskie JV, Montgomerie R, Poldmaa T, Boag PT. 1998. Paternity and paternal care in the polygynandrous Smith's longspur. *Behav Ecol Sociobiol* **43:** 181–190.

Burley N. 1979. The evolution of concealed ovulation. *Am Nat* **114:** 835–858.

Burley N. 1988. The differential allocation hypothesis: An experimental test. *Am Nat* **132:** 611–628.

Chapman T. 2001. Seminal fluid-mediated fitness traits in *Drosophila. Heredity* **87:** 511–521.

Charnov EL. 1981. Kin selection and helpers at the nest: Effects of paternity and biparental care. *Anim Behav* **29:** 631–632.

Cockburn A. 2006. Prevalence of different modes of parental care in birds. *Proc Roy Soc B* **273:** 1375–1383.

Cornell SJ, Tregenza T. 2007. A new theory for the evolution of polyandry as a means of inbreeding avoidance. *Proc Roy Soc B* **274:** 2873–2879.

Cornwallis CK, West SA, Davis KE, Griffin AS. 2010. Promiscuity and the evolutionary transition to complex societies. *Nature* **466:** 969–972.

Dale J, Montgomerie R, Michaud D, Boag PT. 1999. Frequency and timing of extrapair fertilisation in the polyandrous red phalarope (*Phalaropus fulicarius*). *Behav Ecol Sociobiol* **46:** 50–56.

Davies NB. 1992. *Dunnock behaviour and social evolution.* Oxford University Press, Oxford.

Davies NB, Hatchwell BJ, Robson T, Burke T. 1992. Paternity and paternal effort in dunnocks *Prunella modularis*: How good are male chick-feeding rules? *Anim Behav* **43:** 729–745.

Double M, Cockburn A. 2000. Pre-dawn infidelity: Females control extra-pair mating in superb fairy-wrens. *Proc Roy Soc B* **267:** 465–470.

Dunn PO, Bollmer JL, Freeman-Gallant CR, Whittingham LA. 2013. MHC variation is related to a sexually selected ornament, survival, and parasite resistance in common yellowthroats. *Evolution* **67:** 679–687.

Emlen ST, Oring LW. 1977. Ecology, sexual selection and the evolution of mating systems. *Science* **197:** 215–223.

Gowaty PA, Buschhaus N. 1997. Functions of aggressive and forced copulations in birds: Female resistance and the CODE hypothesis. *Am Zool* **38:** 207–225.

Gowaty PA, Karlin AA. 1984. Multiple maternity and paternity in single broods of apparently monogamous Eastern Bluebirds (*Sialia sialis*). *Behav Ecol Sociobiol* **15:** 91–95.

Griffin AS, Alonzo SH, Cornwallis CK. 2013. Why do cuckolded males provide paternal care? *PLoS Biol* **11:** e1001520.

Griffith SC. 2007. The evolution of infidelity in socially monogamous passerines: Neglected components of direct and indirect selection. *Am Nat* **169:** 274–281.

Griffith SC, Owens IPF, Thuman KA. 2002. Extra pair paternity in birds: A review of interspecific variation and adaptive function. *Mol Ecol* **11:** 2195–2212.

Haig D. 2000. The kinship theory of genomic imprinting. *Annu Rev Ecol Syst* **31:** 9–32.

Haig D, Westoby M. 1989. Selective forces in the emergence of the seed habit. *Biol J Linn Soc* **38:** 215–238.

Hoi H. 1997. Assessment of the quality of copulation partners in the monogamous bearded tit. *Anim Behav* **53:** 277–286.

Holland B, Rice WR. 1997. Cryptic sexual selection-more control issues. *Evolution* **51:** 321–324.

Holland B, Rice WR. 1999. Experimental removal of sexual selection reverses intersexual antagonistic coevolution and removes a reproductive load. *Proc Natl Acad Sci* **96:** 5083–5088.

Hughes WOH, Oldroyd BP, Beekman M, Ratnieks FLW. 2008. Ancestral monogamy shows kin selection is key to the evolution of eusociality. *Science* **320:** 1213–1216.

Jeffreys AJ, Wilson V, Thein SL. 1985. Hypervariable "minisatellite" regions in human DNA. *Nature* **314:** 67–73.

Jennions MD, Petrie M. 2000. Why do females mate multiply? A review of the genetic benefits. *Biol Rev* **75:** 21–64.

Karubian J, Alvarado A. 2003. Testing the function of petal-carrying in the Red-backed Fairy-wren (*Malurus melanocephalus*). *Emu* **1:** 87–92.

Kokko H, Morrell LJ. 2005. Mate guarding, male attractiveness, and paternity under social monogamy. *Behav Ecol* **16:** 724–731.

Lack D. 1968. *Ecological adaptations for breeding in birds.* Chapman and Hall, London.

Madsen T, Shine R, Loman J, Hakansson T. 1992. Why do female adders copulate so frequently? *Nature* **355:** 440–441.

Magrath MJL, Elgar MA. 1997. Paternal care declines with increased opportunity for extra-pair matings in fairy martins. *Proc Roy Soc B* **264:** 1731–1736.

Magrath MJL, Komdeur J. 2003. Is male care compromised by additional mating opportunity? *Trends Ecol Evol* **18:** 424–430.

McGlothlin JW. 2010. Combining selective episodes to estimate lifetime nonlinear selection. *Evolution* **64:** 1377–1385.

Møller AP, Birkhead TR. 1991. Frequent copulations and mate guarding as alternative paternity guards in birds: A comparative study. *Behaviour* **118:** 170–186.

Montgomerie R, Thornhill R. 1989. Fertility advertisement on female birds: A means of inciting male–male competition. *Ethology* **81:** 209–220.

Moore AJ. 1990. The evolution of sexual dimorphism by sexual selection: The separate effects of intrasexual selection and intersexual selection. *Evolution* **44:** 315–331.

Mulder RA. 1997. Extra-group courtship displays and other reproductive tactics of superb fairy-wrens. *Aust J Zool* **45:** 131–143.

Mulder RA, Dunn PO, Cockburn A, Lazenby-Cohen KA, Howell MJ. 1994. Helpers liberate female fairy-wrens from constraints on extra-pair mate choice. *Proc Roy Soc B* **255:** 223–229.

Orians GH. 1969. On the evolution of mating systems in birds and mammals. *Am Nat* **103:** 589–603.

Oring LW, Fleischer RC, Reed JM, Marsden KE. 1992. Cuckoldry through stored sperm in the sequentially polyandrous spotted sandpiper. *Nature* **359:** 631–633.

Parker GA. 1979. Sexual selection and sexual conflict. In *Sexual selection and reproductive competition in insects* (ed. Blum MS, Blum NA), pp. 123–166. Academic, New York.

Parker GA. 1990. Sperm competition games: Sneaks and extra-pair copulations. *Proc Roy Soc B* **242:** 127–133.

Parker GA. 2006. Sexual conflict over mating and fertilization: An overview. *Phil Trans Roy Soc B* **361:** 235–259.

Parker GA, Birkhead TR. 2013. Polyandry: The history of a revolution. *Phil Trans Roy Soc B* **368:** 20120335.

Pedersen MC, Dunn PO, Whittingham LA. 2006. Extraterritorial forays are related to a male ornamental trait in the common yellowthroat. *Anim Behav* **72:** 479–486.

Petrie M, Kempenaers B. 1998. Extra-pair paternity in birds: Explaining variation between species and populations. *Trends Ecol Evol* **13:** 52–58.

Pizzari T, Wedell N. 2013. The polyandry revolution. *Phil Trans Roy Soc B* **368:** 20120041.

Pizzari T, Cornwallis CK, Lovlie H, Jakobssen S, Birkhead TR. 2003. Sophisticated sperm allocation in male fowl. *Nature* **426:** 70–74.

Reinhardt K, Naylor RA, Siva-Jothy MT. 2009. Ejaculate components delay reproductive senescence while elevating female reproductive rate in an insect. *Proc Natl Acad Sci* **106:** 21743–21747.

Reynolds JD. 1996. Animal breeding systems. *Trends Ecol Evol* **11:** 68–72.

Rice WR. 1998. Intergenomic conflict, interlocus antagonistic coevolution, and the evolution of reproductive isolation. In *Endless forms: Species and speciation* (ed.

Howard DJ, Berlocher SH), pp. 261–270. Oxford University Press, New York.

Rice WR, Holland B. 1997. The enemies within: Intergenomic conflict, interlocus contest evolution (ICE), and the intraspecific red queen. *Behav Ecol Sociobiol* **41:** 1–10.

Rowe L. 1992. Convenience polyandry in a water strider: Foraging conflicts and female control of copulation frequency and guarding duration. *Anim Behav* **44:** 189–202.

Schlicht E, Kempenaers B. 2011 Extra-pair paternity and sexual selection. In *From genes to animal behavior* (ed. Inoue-Murayama M, Kawamura S, Weiss A), pp. 35–65. Springer, New York.

Schwagmeyer PL, St. Clair RC, Moodie JD, Lamey TC, Schnell GD, Moodie MN. 1999. Species differences in male parental care in birds: A reexamination of correlates with paternity. *Auk* **116:** 487–503.

Sheldon BC. 1994. Male phenotype, fertility, and the pursuit of extra-pair copulations by female birds. *Proc Roy Soc B* **257:** 25–30.

Sheldon BC. 2000. Differential allocation: Tests, mechanisms and implications. *Trends Ecol Evol* **15:** 397–402.

Sheldon BC. 2002. Relating paternity to paternal care. *Proc R Sco* **357:** 341–350.

Sheldon BC, Ellegren H. 1999. Sexual selection resulting from extrapair paternity in collared flycatchers. *Anim Behav* **57:** 285–298.

Simmons LW. 2005. The evolution of polyandry: Sperm competition, sperm selection, and offspring viability. *Ann Rev Ecol Evol Syst* **36:** 125–146.

Slatyer RA, Mautz BS, Backwell PRY, Jennions MD. 2012. Estimating genetic benefits of polyandry from experimental studies: A meta-analysis. *Biol Rev* **87:** 1–33.

Soltis J, Thomsen R, Matsubayashi K, Takenaka O. 2000. Infanticide by resident males and female counter-strategies in wild Japanese macaques *Macaca fuscata*. *Behav Ecol Sociobiol* **48:** 195–202.

Trivers RL. 1972. Parental investment and sexual selection. In *Sexual selection and descent of man* (ed. Campbell B), pp. 136–179. Aldine, Chicago, IL.

Tuttle EM, Pruett-Jones S, Webster M. 1996. Cloacal protuberances and extreme sperm production in Australian fairy-wrens. *Proc Roy Soc B* **263:** 1359–1364.

Webster MS, Pruett-Jones S, Westneat DF, Arnold SJ. 1995. Measuring the effects of pairing success, extra-pair copulations and mate quality on the opportunity for sexual selections. *Evolution* **49:** 1147–1157.

Wedell N, Kvarnemo C, Lessells CM, Tregenza T. 2006. Sexual conflict and life histories. *Anim Behav* **71:** 999–1011.

Westneat DF, Sherman PW. 1993. Parentage and the evolution of parental behavior. *Behav Ecol* **4:** 66–77.

Westneat DF, Stewart IRK. 2003. Extra-pair paternity in birds: Causes, correlates, and conflict. *Annu Rev Ecol Syst* **34:** 365–396.

Westneat DF, Sherman PW, Morton ML. 1990. The ecology and evolution of extra-pair copulations in birds. *Curr Ornithol* **7:** 331–369.

Wolfner MF. 2002. The gifts that keep on giving: Physiological functions and evolutionary dynamics of male seminal proteins in *Drosophila*. *Heredity* **88:** 85–93.

Zeh JA, Zeh DW. 2001. Reproductive mode and the genetic benefits of polyandry. *Anim Behav* **61:** 1051–1063.

Zeh JA, Zeh DW. 2006. Outbred embryos rescue inbred half-siblings in mixed-paternity broods of live bearing females. *Nature* **439:** 201–203.

Sexually Antagonistic Zygotic Drive: A New Form of Genetic Conflict between the Sex Chromosomes

Urban Friberg[1,2] and William R. Rice[3]

[1]Department of Evolutionary Biology, Uppsala University, 752 36 Uppsala, Sweden

[2]IFM Biology, Linköping University, 581 83 Linköping, Sweden

[3]Department of Ecology, Evolution and Marine Biology, University of California, Santa Barbara, California 93111

Correspondence: urban.friberg@liu.se

Sisters and brothers are completely unrelated with respect to the sex chromosomes they inherit from their heterogametic parent. This has the potential to result in a previously unappreciated form of genetic conflict between the sex chromosomes, called sexually antagonistic zygotic drive (SA-ZD). SA-ZD can arise whenever brothers and sisters compete over limited resources or there is brother–sister mating coupled with inbreeding depression. Although theory predicts that SA-ZD should be common and influence important evolutionary processes, there is little empirical evidence for its existence. Here we discuss the current understanding of SA-ZD, why it would be expected to elude empirical detection when present, and how it relates to other forms of genetic conflict.

When a diploid individual reproduces sexually, the two alleles at heterozygous loci are necessarily in competition because reproduction by one allele must be at the expense of the other. Such competition is an inescapable component of the organismal level of evolution that was originally advanced by Darwin and later integrated with the field of genetics during the modern synthesis of the early 20th century (Huxley 1942). If the competition is mediated by Mendelian segregation followed by (1) differences in the Darwinian fitness (i.e., survival and fecundity) that each allele produces in offspring, (2) random sampling (genetic drift), and/or (3) differences in the alleles' mutation or migration rates, then no genetic conflict exists and only canonical evolution at the organismal level occurs. But alleles can also compete outside the context of organismal evolution via diverse mechanisms of selection at the level of the gene that are collectively called genomic conflict (or selfish, ultraselfish, and parasitic DNA). These mechanisms can be divided into three general classes (Burt and Trivers 2006): (1) gonotaxis (in which the selfish elements bias Mendelian segregation by moving away from dead-end polar bodies and into the functional egg during oogenesis, i.e., meiotic or centromeric drive), (2) interference (in which the selfish element kills or debilitates noncarrier gam-

Cite this article as *Cold Spring Harb Perspect Biol* doi: 101101/cshperspect.a017608

etes or offspring, i.e., segregation distortion and zygotic drive), and (3) overreplication (in which the selfish element increases its copy number in the genome, e.g., biased gene conversion, transposable elements, and homing endonucleases). De novo mutations can also gain a transmission advantage by increasing the rate of stem cell division in the germ line (germline drive) (e.g., Yoon et al. 2013). All of these genomic conflict mechanisms have been described in detail in Burt and Trivers (2006).

Genomic conflict frequently leads to reduced fitness at the organismal level. Meiotic drive can harm the organism as a whole because the attributes that provide a segregation advantage in oogenesis (e.g., the structure of the centromere and neighboring heterochromatin) can be maladaptive during spermatogenesis and contribute to male sterility (see, for review, Elde et al. 2011). Segregation distorters and zygotic drivers can substantially reduce a carrier male's fitness because they kill up to half of his sperm (leading to reduced fertility) and offspring, respectively. Sex-linked, meiotic drivers in WZ females (like birds) and segregation distorters and zygotic drivers in XY males (like insects and mammals) cause biased sex ratios that reduce fitness with respect to Fisherian sex ratio selection and can also reduce population growth and potentially drive species to extinction. Biased gene conversion and germline drive (Yoon et al. 2013) reduce organismal fitness when harmful mutations accumulate to elevated levels (i.e., beyond the conventional values predicted by mutation-selection balance) because they have a molecular drive advantage over an allele that is more beneficial at the organismal level. Transposable elements insert at new places in the genome where they can disrupt gene function and thereby reduce their carrier's fitness.

Zygotic drive is an unusual form of genetic conflict because it directly reduces Darwinian fitness by killing or debilitating offspring. It is favored by gene-level selection when there is competition among siblings for limiting resources. By killing or weakening noncarrier competitor siblings, the gene(s) coding for zygotic drive gain a selective advantage because their survival is increased at the expense of siblings carrying alleles that are not identical by descent—despite any fitness loss to the parents, siblings, or other parts of the genome.

Zygotic drive of the autosomes has been observed in a wide diversity of model organisms (e.g., worms, beetles, and mice) (reviewed in Burt and Trivers 2006) in which it can be efficiently detected because of the availability of numerous genetic markers. In general, an autosomal zygotic driver must have both a driver allele at one locus and a protective allele at a responder locus. In worms (*Caenorhabditis elegans*), a molecular mechanism leading to zygotic drive was recently discovered. Here a zygotic driver is coded by a pair of tightly linked genes, in which an allele at one gene (*peel-1*) produces a toxin, the driver locus, which is packaged in the sperm and transmitted to the zygote, whereas an allele at another gene (*zeel-1*) produces an antidote (the protective allele, which is expressed very early in development) that rescues only those embryos that inherit *zeel-1* (and usually also the tightly linked driver, *peel-1*) (Seidel et al. 2011). Zygotic drive on the autosomes is expected to be difficult to evolve—and therefore to be relatively rare in genomes—because it requires an improbable phenotype (i.e., a functionally coupled driver gene product and a responder gene sequence or product) and genotype (i.e., very close linkage between the loci coding for the driver and responder).

SEXUALLY ANTAGONISTIC ZYGOTIC DRIVE

Here we focus on zygotic drive coded by the X and Y (or Z and W) sex chromosomes. This phenomenon is called sexually antagonistic zygotic drive (hereafter, SA-ZD) because any gain by one sex chromosome transmitted from a heterogametic parent is at the expense of the noncarrier sex of offspring (sons or daughters) (Rice et al. 2008). Like autosomal zygotic drive, SA-ZD is favored by gene-level selection when there is sib competition for limiting resources. In addition, SA-ZD is also favored when there is a nontrivial rate of brother–sister mating and inbreeding depression. This occurs because when sisters are harmed by producing inbred offspring from sib mating (but brothers benefit from sib mating owing to lower opportunity costs) (Haig 1999), the paternal X is selected to kill sons so that they do not fertilize their sisters, which inherit the paternal X.

Linkage to the sex chromosomes facilitates the evolution of a zygotic driver (SA-ZD) in four ways. For simplicity in the following, but without loss of generality, we will assume male heterogamety (X and Y sex chromosomes) and the terms X and Y will refer to their regions that do not recombine in males. First, all genes on the Y cosegregate across each generation, and as a consequence, the component genes coding for Y-linked SA-ZD (e.g., driver and responder genes, or polygenes coding for either trait) need not be positioned at nearby chromosomal loci—only located on the same-sex chromosome. Second, nonrecombining X and Y sex chromosomes are far more genetically diverged than homologous autosomes—because the latter are continually homogenized by recombination—making the number of potential responder loci (alleles not present on the same chromosome as the driver allele) far larger for the X and Y sex chromosomes than the autosomes. Third, there is no absolute requirement for a responder locus on a homologous chromosome because noncarrier offspring can be killed (or, more generally, harmed) when the driver gene product disrupts a sex-limited ontogenetic pathway (harming only the noncarrier sex, e.g., via epigenetic changes produced during spermatogenesis and transmitted to offspring) that is coded by any chromosome. Fourth, as mentioned above, SA-ZD is favored by gene-level selection in response to both sib competition (like autosomal zygotic drive) but also by nontrivial levels of sib–sib mating with any level of inbreeding depression.

The logic of the genetic conflict that underlies SA-ZD is identical to that of cytoplasmic genomes that are only transmitted across multiple generations via the matriline. In response to matrilineal transmission (or very low levels of patrilineal transmission in some species) (e.g., Nunes et al. 2013), mitochondria and cytoplasmically transmitted endosymbionts (like *Wolbachia* and *Spiroplasma)* have evolved to kill gametes or offspring that give rise to lineages that do not transmit them (pollen and sons) and compete for resources with the matrilineal offspring that do transmit them (reviewed in Burt and Trivers 2006). By the same logic, paternal X and Y sex chromosomes are selected to kill (or harm and make less competitive) the sex of offspring that does not reproduce them, but competes with the sex that does reproduce them. Given the widespread evolution of pollen killing in plants (coded by mitochondria in hundreds of species) (reviewed in Burt and Trivers 2006) and male killing and cytoplasmic incompatibility in animals (coded by bacterial endosymbionts in thousands of species of animals) (reviewed in Burt and Trivers 2006), it would be surprising if this same selection pressure did not lead to the widespread evolution of SA-ZD.

Collectively, the unique characteristics of sex chromosomes would be expected to cause SA-ZD to evolve via an unusually powerful evolutionary mechanism called a "green-beard effect" (Hamilton 1964; so named by Dawkins 1976) that was originally described in the context of animal behavior but later extended to molecular and cellular interactions as well (e.g., Haig 1996). A green-beard effect is a multifarious trait coded by a single gene that pleiotropically produces three distinct characteristics (or a group of tightly linked genes, each coding for a different part of the green-beard effect). First, it causes the carrier to produce a distinguishing phenotype (tag). Second, it causes the carrier to differentiate among other individuals—in the context of interactions—based on the presence or absence of the tag (tag differentiation). And third, the first two traits collectively lead to increased fitness of other individuals that express the tag (tag-directed aid). A green-beard effect is antagonistic when it increases the relative fitness of individuals carrying the gene(s) that codes for it by reducing the fitness of competitors that do not express the tag. Because green-beard effects require a complex and multifarious pleiotropy (or very tight linkage of a cluster of component genes), they have previously been presumed to be too rare in nature to be a major evolutionary force, although a few established examples are well documented (e.g., Keller and Ross 1998, Quellar et al. 2003, Sinervo et al. 2006). However, single sex-linked genes coding for SA-ZD can easily achieve all the characteristics of a green-beard effect because (1) the tag is any sexually dimorphic trait coded by any gene in the genome, (2) the tag differentiation is any phenotype coded by one of the sex chromosomes causing fathers (or siblings) to interact differently toward sons (brothers) and daughters (sisters), and (3) the tag-directed aid/harm is any trait included in (2) that increases the fitness of the sex of offspring that carries them. For example, an X-linked mutation with male-limited expression that caused a sire to invest more in daughters at the expense of sons would lead to SA-ZD, as would a Y-linked gene causing the father to invest more in sons. Similarly, an X-linked mutation expressed in sperm producing a gene product that epigenetically silenced a dosage-sensitive gene in a male-specific ontogenetic pathway (like dosage compensation in flies) would lead to SA-ZD if the epigenetic mark persisted across generations. A fuller account of how SA-ZD simplifies the operation of green-beard effects can be found in Rice et al. (2008).

Unlike segregation distortion, which only can operate via a very limited range of phenotypes that reduce the number of noncarrier functional sperm, SA-ZD can operate through four distinct mechanisms. First, through paternal effects (especially in the context of transgenerational epigenetic influences on gene expression) that kill or harm the noncarrier sex. As described above, dead or harmed noncarrier siblings free up shared resources among siblings and thereby increase the fitness of carrier siblings. Paternal effects have been historically assumed to be sufficiently rare and/or of small effect that they can be safely ignored in genetic analyses like those for estimating heritability (e.g., paternal half-sib analysis). However, recent evidence indicates that this assumption can be mistaken (reviewed in Rice et al. 2008). For example, a recent study in mice indicated that a father's Y chromosome influences a large suite of polygenic traits in his daughters (that do not carry it) just as much as his X chromosome (which they all carry). Both X- and Y-linked paternal effects on egg-to-adult survival have also recently been documented in fruit flies (Friberg et al. 2012).

Second, sib–sib interactions that favor same-sex sibs: All Y-linked genes are identical by descent among brothers (but not with sisters) sired by the same father, as are all paternal X-linked genes among sisters (but not with brothers). As a consequence, SA-ZD is predicted to operate via brothers helping each other more than their sisters and/or harming sisters to free up more shared resources for their brothers (and vice versa for sisters interacting with each other vs. brothers). Examples of sex-specific altruism and harm (including siblicide) are common in nature (see, for review, Rice et al. 2009), so this route to SA-ZD is clearly feasible. However, there are many potential adaptive scenarios that predict sex-specific sib–sib help and harm, so none of these examples provide unequivocal evidence for SA-ZD operating via this route. As a consequence, SA-ZD operating via sib–sib interactions would be ambiguous and easily overlooked.

Third, sex-specific parental investment and/or harm: SA-ZD is predicted to operate by X- and Y-coded phenotypes that cause fathers to help one sex of offspring at the expense of the other sex. This includes killing or directly harming the noncarrier sex of offspring. Sex-specific parental investment and harm are well documented in the behavioral literature (see review of evidence in Rice et al. 2008, 2009), so this route to SA-ZD is clearly feasible. However, like the data on sex-specific sib–sib interactions, it is not clear what selective agents were responsible for building these phenotypes. As a consequence, SA-ZD operating via parent–offspring interactions would be ambiguous and easily overlooked.

Fourth, sex-specific investment by paternal grandmothers in grandchildren: Because a father inherits his X only from his mother, paternal grandmothers are related through their X chromosomes only to paternal granddaughters and not to paternal grandsons. This asymmetry is absent between maternal grandmothers and their grandchildren. As a consequence, SA-ZD is predicted to (1) operate by grandmothers favoring their paternal granddaughters, including harming their paternal grandsons to free more shared resources, and (2) this pattern is predicted to be absent between grandmothers and their maternal grandchildren (Rice et al. 2010). Surprisingly, recent anthropological data appear to support this counterintuitive prediction in humans (Fox et al. 2010, 2011). Potential confounds, such as paternity uncertainty and genome-wide Hamiltonian relatedness, cannot explain the observed grand-maternal harm that occurs exclusively in the patriline (Fox et al. 2010, 2011; Rice et al. 2010). Paternal grandparental investment in grandoffspring is rare in nature (humans being an exception), so this route to SA-ZD is expected to have limited application.

WHY NO PRIOR EVIDENCE?

If SA-ZD is such a feasible mechanism for genetic conflict, then why has it not already been widely reported? We have already described in the above three paragraphs why SA-ZD via sib–sib, parent–offspring, and grandparent–offspring interactions would be expected to be easily confused with other causative agents, and therefore undetected. Here we focus on SA-ZD via transgenerational epigenetic effects that harm or kill the noncarrier sex. Unlike autosomal zygotic drive, SA-ZD leads to biased sex ratios, so Fisherian sex ratio selection (i.e., parents producing the rarer sex leave more descendent grand offspring) is expected to rapidly lead to repressors on the opposite sex chromosome and/or the autosomes. As a consequence, extant genes coding for SA-ZD will almost always be hidden because drive only operates episodically immediately after a de novo driver evolves or when extant but suppressed drivers temporarily escape suppression owing to new mutations or immigration of foreign drivers into a genetic background lacking the corresponding suppressors. However, this same constraint also applies to sex-linked segregation distorters and many examples of this phenomenon have been reported in the literature (Jaenike 2001). The difference with SA-ZD is that it is mediated by sex-specific mortality. Clear evidence for sex-linked segregation distortion is readily obtained (leading to detailed studies) when brood mortality is too low to account for the observed sex ratio distortion, making it relatively easy to identify this causative route to sex ratio distortions observed in nature. Brood mortality in the case of SA-ZD confounds this process with sex-specific mortality (coded anywhere in the genome) so no simple observations can unambiguously identify SA-ZD. For example, Noor and Coyne (1995) observed strong sex ratio distortion (female biased) in a laboratory stock of *Drosophila simulans* they named skew—leading them to suspect an X-linked segregation distorter. Many experiments were performed with the skew stock (and a control stock—named Florida City—with an even sex ratio) to characterize the genetics behind the sex ratio bias. However, when they checked the egg hatch rate, it was found to be elevated substantially when the sires were from the skew stock but not in the reciprocal cross with Florida City sires. When the extra mortality was applied to the rarer sex (males) there was a nearly perfect congruence with the observed sex ratio and Noor and Coyne abandoned their experiments concluding that the sex ratio bias was associated with some form of sex-specific mortality. However, the empirical pattern observed by Noor and Coyne was exactly what would be expected with SA-ZD. This example illustrates how SA-ZD could be prevalent but overlooked—even when its signature phenotype is observed (sex ratio bias associated with elevated mortality in broods), it is dismissed as mundane sex-specific mortality, i.e., as uninteresting. A few years ago we reanalyzed the skew stock studied by Noor and Coyne and found substantial—but not irrefutable—evidence that it does in fact express SA-ZD (Friberg et al. 2011).

In sum, we believe that SA-ZD is plausibly a relatively common but overlooked and unappreciated form of genetic conflict because (1) it is suppressed most of the time (like sex-linked segregation distortion) and when it is observed, it is too easily confused with sex-specific mortality (unlike sex-linked segregation distortion); (2) it can feasibly operate through a phenotype that debilitates but does not kill competing offspring of the noncarrier sex, which makes it harder to detect than segregation distorters that always skew the frequency of offspring toward those carrying the driver; and (3) moderately strong drivers lead to only modest sex ratio distortion and are therefore hidden by sampling error. To illustrate the last point, consider a driver that was present at a frequency of 50% and killed 25% of sons (a substantial level of male mortality). In this case, the expected population sex ratio (% females) would be 0.536. Even in families in which the sire carried the driver the sex ratio would be only 0.571. Limitations on sample sizes would cause these small deviations from an even sex ratio to be hidden by sampling error or confused with weak sex-specific mortality in most empirical studies. Now that a theoretical framework has been established for SA-ZD, future cases of sex ratio distortion associated with elevated mortality of the rarer sex will hopefully be tested for the operation of SA-ZD.

GAMETIC DRIVE AND SA-ZD—DISTINCT PHENOTYPES PRODUCED BY ONE MOLECULAR MECHANISM?

So far in this article, we have treated SA-ZD as separate and distinct from sex-linked segregation distortion. However, these two forms of genetic conflict may share a common origin in the context of SA-ZD mediated via paternal transgenerational epigenetic effects. In *Drosophila melanogaster*, the autosomal segregation distorter *Sd*, and in *D. simulans* the X-linked segregation distorters *Paris* and *Winters*, all drive by modifying Y-bearing sperm in such a way that they develop improperly during spermatogenesis (reviewed in Meiklejohn and Tao 2010; Larracuente and Presgraves 2012). In *D. melanogaster*, drive via *Sd* occurs because a pericentric satellite (Rsp^s, a repetitive DNA sequence in the heterochromatin neighboring the centromere) is present on sensitive chromosomes and responds to the presence of *Sd* (by an unknown molecular mechanism) to facilitate improper chromatin remodeling during spermatogenesis. This remodeling error occurs because the driver gene (now known to be a truncated duplication of the gene *RanGAP* and renamed *Sd–RanGAP*) blocks the histone-to-protamine replacement during spermatogenesis. When the responder satellite has a low copy number of repeats (Rsp^i), chromatin remodeling occurs properly and drive is suppressed. Defective chromatin remodeling during mitosis and/or spermatogenesis is also feasibly responsible for the sperm defects seen with the two sex-linked drivers in *D. simulans* (Cazemajor et al. 2000; Tao et al. 2007; Meiklejohn and Tao 2010). We will collectively refer to such improper chromatin remodeling as ectopic epigenetic marks (ectopic-epi-marks).

Suppressors of segregation distortion via defective chromatin remodeling can act in three ways: (1) suppress gene expression at the driver locus, (2) modify the responder locus so that ectopic-epi-marks are not formed, or (3) tolerate driver-induced ectopic-epi-marks during meiosis and spermatogenesis via some compensation mechanism. In the latter case, if the driver-induced ectopic-epi-mark(s) of Y-bearing sperm were transgenerationally transmitted to the zygote (e.g., the Y is known to be imprinted in *D. melanogaster*) (Maggert and Golic 2002) they could disrupt embryonic development in at least some sons—leading to SA-ZD. For example, if transgenerational ectopic-epi-marks on Y-bearing sperm slowed the transition between a condensed sperm nucleus into a decondensed pronucleus, then mitotic checkpoints—that operate independently in the male and female pronuclei in flies—would be expected to lead to asynchronous entrance into the first mitotic division in the zygote, causing (1) permanent arrest of subsequent cell division, or (2) aneuploid mitotic products that could induce death or malformations later in development (Landmann et al. 2009). This is the same cell-cycle asymmetry in timing between male and female pronuclei that leads to embryo death during cytoplasmic incompatibility (CI) in *Drosophila* (Landmann et al. 2009), but in this case only the males would be killed (as occurs with the CI-like mortality of sons produced by some male-killing *Wolbachia*) (e.g., Riparbelli et al. 2012) because only sons would have sperm-derived pronuclei with ectopic-epi-marks. In sum, suppressors of sex-linked segregation distortion can, in principle, convert segregation distortion into SA-ZD.

A SIMPLE MOLECULAR PATHWAY TO SA-ZD

To illustrate a potential molecular mechanism for SA-ZD via ectopic-epi-marks, again consider the phenomenon of CI caused by *Wolbachia* in flies. Here infected males produce sperm that cause embryonic mortality owing to defective chromatin remodeling of the sperm pronucleus before the first mitotic division of the zygote, but only when mated to uninfected females (Landmann et al. 2009). Like an X coding for son-killing SA-ZD via transgenerational ectopic-epi-marks, *Wolbachia* of infected sires are not transmitted to sons and must kill zygotes via modified sperm, which is accomplished via modified sperm chromatin that is not properly remodeled in *Wolbachia*-free female flies (Landmann et al. 2009). The molecular mechanisms underlying this chromatin remodeling defect are not fully resolved, but *Wolbachia*-infected sires are established to have reduced expression of the chromatin remodeling protein HIRA (Zheng et al. 2011). HIRA is important in remodeling a type of nucleosome (containing histone H3.3) that is far more common on the X compared with the Y sex chromosome. Sires that express a loss-of-function allele at the *Hira* locus produce male-biased sex ratios owing to higher early embryonic death of daughters compared with sons, presumably because X-bearing sperm carry a higher load of chromatin defects compared with Y-bearing sperm (owing to the higher frequency of HIRA-dependent nucleosomes on the X compared with the Y) (Zheng et al. 2011). Interestingly, dams that are infected with *Wolbachia* have elevated expression of HIRA in their eggs and this maternal effect is associated with complete rescue of daughters (from early embryonic death owing to sperm from HIRA-deficient sires) and restores an even brood sex ratio (Zheng et al. 2011). These HIRA phenotypes in males and females illustrate a simple pathway by which Y-linked SA-ZD could be produced in sires and suppressed via a maternal effect in dams.

SA-ZD AND HYBRID INFERTILITY

It is well established in flies that segregation distortion, especially when X and/or Y linked, has contributed importantly to the evolution of male hybrid infertility, at least in the F_2 or backcross generations (reviewed in Johnson 2010; Presgraves 2010). Like segregation distortion, SA-ZD could play an important role in the evolution of hybrid infertility and inviability during the speciation process. As described above, gene-level selection directly selects for the X to induce paternal-effect infertility in sons (to protect daughters from harmful sib mating with associated inbreeding depression), and for both the X and Y to code for paternal-effect inviability of sons and daughters, respectively (to reduce competition for shared limiting resources and to prevent sons from mating with their sisters). If SA-ZD and its suppressors evolved separately in sister species, and if such SA-ZDs were suppressed by maternal effects, then hybrid crosses in both directions would be expected to produce inviable and/or sterile offspring and contribute to postzygotic reproductive isolation in the F_1. If SA-ZDs were suppressed in the sires of each of the sister species, then the same phenotypes producing hybrid infertility and death would be expected, but in the progeny of the F_2 and backcrosses. The feasibility of SA-ZD contributing to postzygotic reproductive isolation is more fully discussed in Rice et al. (2008, 2009), as is its potential to contribute to the decay of Y and W sex chromosomes.

CONCLUSIONS

In closing we conclude that SA-ZD may be a prevalent form of genetic conflict that has the potential to have important ramifications in fundamental processes like speciation and the degeneration of Y and W sex chromosomes. It represents one of the few evolutionary phenomena that—like many cases in physics—was predicted to exist by the mathematically explicit theory alone, before any empirical evidence.

Nonetheless, SA-ZD is expected to be a largely unknown and unappreciated phenomenon that is "hidden in plain sight" because it is (1) only episodically expressed owing to repression by coevolved suppressors most of the time, (2) hidden by binomial sampling error when expressed and of weak or moderate strength, (3) confused with sex-specific viability factors when expressed at any strength, (4) confused with diverse alternative causations in the cases of sib–sib, parent–offspring, and grandparent–offspring behavioral interactions, and (5) very difficult to detect when its phenotype only reduces the competitive ability of noncarrier offspring instead of killing them. The first two factors are shared with sex-linked segregation distortion (which is well documented in nature), but the last three factors have feasibly caused this theoretically predicted form of genetic conflict to elude empirical detection when present, and consequently it would remain obscure and unappreciated even if prevalent.

ACKNOWLEDGMENTS

We thank Adam Chippindale for valuable comments on an earlier version of this review. This study is supported by grants from the Swedish Research Council and the Swedish Foundation for Strategic Research.

REFERENCES

Burt A, Trivers R. 2006. *Genes in conflict: The biology of selfish genetic elements.* Belknap Press of Harvard University Press, Cambridge, MA.

Cazemajor M, Joly D, Montchamp-Moreau C. 2000. Sex-ratio meiotic drive in *Drosophila simulans* is related to equational nondisjunction of the Y chromosome. *Genetics* **154:** 229–236.

Dawkins R. 1976. *The selfish gene.* Oxford University Press, Oxford.

Elde NC, Roach KC, Yao MC, Malik HS. 2011. Absence of positive selection on centromeric histones in *Tetrahymena* suggests unsuppressed centromere-drive in lineages lacking male meiosis. *J Mol Evol* **72:** 510–520.

Fox M, Sear R, Beise J, Ragsdale G, Voland E, Knapp LA. 2010. Grandma plays favourites: X-chromosome relatedness and sex-specific childhood mortality. *Proc R Soc B* **277:** 567–573.

Fox M, Johow J, Knapp LA. 2011. The selfish grandma gene: The roles of the X-chromosome and paternity uncertainty in the evolution of grandmothering behavior and longevity. *Int J Evol Biol* **2011:** 165919.

Friberg U, Stewart AD, Rice WR. 2011. Empirical evidence for son-killing X chromosomes and the operation of SA-zygotic drive. *PLoS ONE* **6:** e23508.

Friberg U, Stewart AD, Rice WR. 2012. X and Y chromosome linked paternal effects on a life history trait. *Biol Lett* **8:** 71–73.

Haig D. 1996. Gestational drive and the green-bearded placenta. *Proc Natl Acad Sci* **93:** 6547–6551.

Haig D. 1999. Asymmetric relations: Internal conflicts and the horror of incest. *Evol Hum Behav* **20:** 83–98.

Hamilton WD. 1964. Genetical evolution of social behaviour 2. *J Theor Biol* **7:** 17–52.

Huxley J. 1942. *Evolution: The modern synthesis.* Allen and Unwin, London.

Jaenike J. 2001. Sex chromosome meiotic drive. *Annu Rev Ecol Syst* **32:** 25–49.

Johnson NA. 2010. Hybrid incompatibility genes: Remnants of a genomic battlefield? *Trends Genet* **26:** 317–325.

Keller L, Ross KG. 1998. Selfish genes: A green beard in the red fire ant. *Nature* **394:** 573–575.

Landmann F, Orsi GA, Loppin B, Sullivan W. 2009. *Wolbachia*-mediated cytoplasmic incompatibility is associated with impaired histone deposition in the male pronucleus. *PLoS Pathog* **5:** e1000343.

Larracuente AM, Presgraves DC. 2012. The selfish segregation distorter gene complex of *Drosophila melanogaster. Genetics* **192:** 33–53.

Maggert KA, Golic KG. 2002. The Y chromosome of *Drosophila melanogaster* exhibits chromosome-wide imprinting. *Genetics* **3:** 1245–1258.

Meiklejohn CD, Tao Y. 2010. Genetic conflict and sex chromosome evolution. *Trends Ecol Evol* **25:** 215–223.

Noor MAF, Coyne JA. 1995. Research note. *Dros Inf Serv* **76:** 151–152.

Nunes MDS, Dolezal M, Schloetterer C. 2013. Extensive paternal mtDNA leakage in natural populations of *Drosophila melanogaster. Mol Ecol* **22:** 2106–2117.

Presgraves DC. 2010. The molecular evolutionary basis of species formation. *Nat Rev Genet* **11:** 175–180.

Queller DC, Ponte E, Bozzaro S, Strassmann JE. 2003. Single-gene greenbeard effects in the social amoeba, *Dictyostelium discoideum. Science* **299:** 105–106.

Rice WR, Gavrilets S, Friberg U. 2008. Sexually antagonistic "zygotic drive" of the sex chromosomes. *PLoS Genet* **4:** e1000313.

Rice WR, Gavrilets S, Friberg U. 2009. Sexually antagonistic chromosomal cuckoos. *Biol Lett* **5:** 686–688.

Rice WR, Gavrilets S, Friberg U. 2010. The evolution of sex-specific grandparental harm. *Proc R Soc B* **277:** 2727–2735.

Riparbelli MG, Giordano R, Ueyama M, Callaini G. 2012. *Wolbachia*-mediated male killing is associated with defective chromatin remodeling. *PLoS ONE* **7:** e30045.

Seidel HS, Ailion M, Li J, van Oudenaarden A, Rockman MV, Kruglyak L. 2011. A novel sperm-delivered toxin causes late-stage embryo lethality and transmission ratio distortion in *C. elegans. PLoS Biol* **9:** e1001115.

Sinervo B, Chaine A, Clobert J, Calsbeek R, Hazard L, Lancaster L, McAdam AG, Alonzo S, Corrigan G, Hochberg ME. 2006. Self-recognition, color signals, and cycles of greenbeard mutualism and altruism. *Proc Natl Acad Sci* **103:** 7372–7377.

Tao Y, Masly JP, Araripe L, Ke Y, Hartl DL. 2007. A sex-ratio meiotic drive system in *Drosophila simulans*: I. An autosomal suppressor. *PLoS Biol* **5:** e292.

Yoon SR, Choi SK, Eboreime J, Gelb BD, Calabrese P, Arnheim N. 2013. Age-dependent germline mosaicism of the most common noonan syndrome mutation shows the signature of germline selection. *Am J Hum Gene* **92:** 917–926.

Zheng Y, Ren P-P, Wang J-L, Wang Y-F. 2011. *Wolbachia*-induced cytoplasmic incompatibility is associated with decreased HIRA expression in male *Drosophila*. *PLoS ONE* **6:** e19512.

Sex Chromosome Drive

Quentin Helleu, Pierre R. Gérard, and Catherine Montchamp-Moreau

Laboratoire Évolution Génomes et Spéciation, CNRS UPR9034, Gif-sur-Yvette, France and Université Paris-Sud, Orsay, France

Correspondence: catherine.montchamp@legs.cnrs-gif.fr

Sex chromosome drivers are selfish elements that subvert Mendel's first law of segregation and therefore are overrepresented among the products of meiosis. The sex-biased progeny produced then fuels an extended genetic conflict between the driver and the rest of the genome. Many examples of sex chromosome drive are known, but the occurrence of this phenomenon is probably largely underestimated because of the difficulty to detect it. Remarkably, nearly all sex chromosome drivers are found in two clades, *Rodentia* and *Diptera*. Although very little is known about the molecular and cellular mechanisms of drive, epigenetic processes such as chromatin regulation could be involved in many instances. Yet, its evolutionary consequences are far-reaching, from the evolution of mating systems and sex determination to the emergence of new species.

Meiotic drivers are selfish genetic elements that subvert Mendelian segregation during gametogenesis for their own benefit. They are passed on to most, if not all, of the functional gametes produced by heterozygotes. Therefore, drivers can increase in frequency and invade populations even if they reduce individual fitness, which is usually the case. The drivers are typically expressed in one sex, of which fertility is impaired. This also has deleterious consequences for the opposite sex and is expected to promote adaptations to counteract drive through sexual selection and sexual conflict. Furthermore, sex-linked meiotic drivers expressed in the heterogametic sex typically lead to biased offspring sex ratios, which represents an additional cost and can exacerbate the sexual conflict.

Morgan et al. (1925) were the first to observe sex-biased offspring, which turned out to be caused by a sex-linked meiotic driver. Unfortunately, the *Drosophila affinis* strain was lost before any conclusive study could be performed. Later, Gershenson (1928) found that the offspring of some *Drosophila obscura* males were female biased. He showed that these males carried an X-linked genetic element (hereafter "*sex ratio*" or *SR*) responsible for the sex-ratio distortion, and showed that the *SR* did not affect the viability of the male offspring but acted as a gametic killer of Y-bearing sperm. Gamete killing or disabling is observed in males; in females, meiotic drive is usually a result of centromere competition for access to the egg.

In its original definition (from Gershenson's work and others), the term meiotic drive applies to the consequences of the mechanics of the meiotic divisions (Sandler and Novitski 1957). Here, under the term "sex chromosome drive," we will include more broadly any case of

Cite this article as *Cold Spring Harb Perspect Biol* doi: 10.1101/cshperspect.a017616

preferential transmission that results directly or indirectly from an event that took place before, during, or after meiosis. Sex chromosome drive is different from sex-ratio adjustment, in which the favored chromosome is not the actor of its drive (West and Sheldon 2002). As emphasized by Sandler and Novitski (1957), it is also different from selection in the haploid phase as a consequence of the gamete's intrinsic fitness.

Only a few dozen cases of sex chromosome drive have been described, mainly in *Drosophila* and other *Diptera* (reviewed in Jaenike 2001; Burt and Trivers 2006). One possible explanation for the rarity of reported cases is that a biased sex ratio is not evolutionarily stable. Fisher (1930) predicted that natural selection will favor a 1:1 sex ratio, and that any deviation will be counterselected. This means that variants with counteracting effects can be selected at unlinked loci. Consistent with this prediction, autosomal drive suppressors and resistant Y chromosomes have been found in several *Drosophila* species (De Carvalho and Klaczko, 1994; Carvalho et al. 1997; Cazemajor et al. 1997). Three different cryptic X-linked *SR* systems have been described in the same species (*Drosophila simulans*: Paris, Winters, and Durham systems), showing that they can evolve repeatedly and be completely neutralized in the wild, remaining undetectable unless appropriate genetic crosses are performed (Merçot et al. 1995; Tao et al. 2001, 2007a). *D. simulans* also teaches us that the time window leading up to neutralization can be very narrow (Bastide et al. 2013). However, variants that enhance distortion can be selected if they are linked to the distorter. Inversions should prevent recombination with nondriving X chromosomes and keep together the loci that interact to induce drive, as found in *D. pseudoobscura* (Wu and Beckenbach 1983). These examples illustrate the extended genetic conflict that can result from the evolution of sex chromosome drive.

Among the known cases of sex chromosome drive, X chromosome drive is much more common than Y chromosome drive. This may be because Y-linked drivers are always expressed, at each generation, unlike X-linked drivers. All else being equal, Y chromosome drive spreads faster and leads to a higher risk of extinction owing to the lack of females (Hamilton 1967). Furthermore, when the sex chromosomes are well differentiated, the Y chromosome usually has many fewer genes, which may provide fewer opportunities for a driver to evolve. On the other hand, heteromorphic sex chromosomes are expected to facilitate the evolution of meiotic drive. Indeed, the more divergent the sex chromosomes are, the less they recombine, reducing the risk of producing a suicide chromosome that carries both the driver and a sensitive allele at the target locus (Charlesworth and Hartl 1978; Frank 1991; Hurst and Pomiankowski 1991).

OCCURRENCE AND MECHANISMS OF SEX CHROMOSOME DRIVE

Nearly all of the well-established cases of sex chromosome drive occur in the distant *Diptera* and *Rodentia* taxa (reviewed in Jaenike 2001; Burt and Trivers 2006). Naturally occurring X-linked drivers have been reported in 13 *Drosophila* species with sporadic taxonomic distribution across the *genus*. A similar phenotype has been described in four species of stalk-eyed flies and in the tsetse fly. Y drive is much less common, but is known in two mosquito species: *Aedes aegyptii* and *Culex pipiens*.

Whereas the dipteran drivers usually operate during meiosis (Fig. 1), the sex chromosome drivers known in rodents first change the sex of their carriers (but see Cocquet et al. 2012). XY females have been described in lemmings, in voles, and in the African pygmy mouse—all associated with the presence of feminizing X chromosomes that lead to female-biased population sex ratios (Fredga et al. 1976; Gileva 1987; Bianchi 2002; Veyrunes et al. 2010).

Unbalanced sex ratios are common in angiosperms with separate sexes. However, this is probably more often because of pollen competition between lower quality Y-bearing pollen and higher-quality X-bearing pollen (but see Taylor and Ingvarsson 2003). Adaptive adjustment of offspring sex ratio occurs in some birds, and some placental and marsupial mammals, but there is no definite evidence so far of a causal

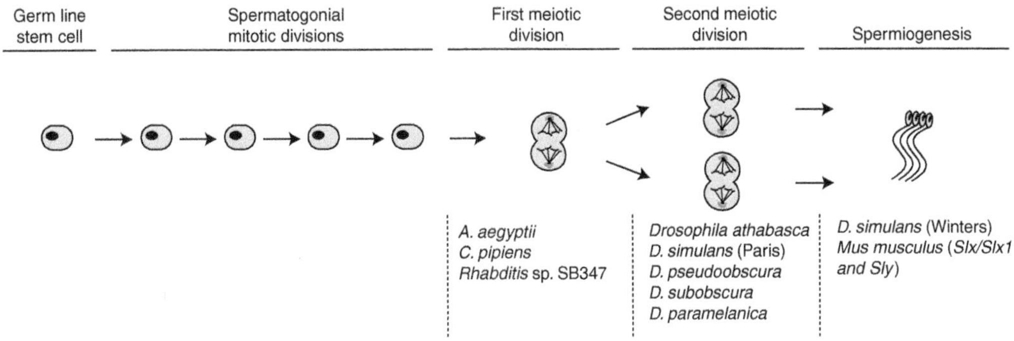

Figure 1. Spermatogenesis stages at which cellular phenotypes of sex chromosome drive are observed.

mechanism during gametogenesis (West and Sheldon 2002; Rutkowska and Badyaev 2008; Navara 2013). In contrast, strong female-biased progeny produced by males of the nematode *Rhabditis* sp. SB347 has recently been ascribed to differential transmission of X-bearing sperm. In this species with X0 males, the trait seems related to premature splitting of the X chromosome into sister chromatids during meiosis I. The cellular components that are essential for sperm motility are then partitioned almost exclusively to the X-bearing sperm (Shakes et al. 2011). Below, we summarize what is currently known about the mechanism of sex chromosome drive in rodents, *Drosophila*, and other dipterans. In *Drosophila*, we also focus on recent data and hypotheses about the molecular basis of drive.

Rodentia

Several species of rodents show X chromosome variants (usually referred to as X*) that cause XY individuals to develop into females. The most extreme case of X chromosome drive operating by feminization of XY individuals is known in the wood lemming *Myopus schisticolor*: X*Y females produce only X* eggs, owing to the nondisjunction of the sister chromatids of both sex chromosomes during some mitotic divisions in the germline. The X*X* cells enter meiosis, but the YY cells are unviable. As a result, the X* chromosome is transmitted to 100% of the progeny—all are therefore female. In the lemming *Dicrostonyx torquatus*, X*Y females produce both X*- and Y-bearing gametes. The primary drive strength is lower in this case, as the preferential transmission of the X* chromosome does not occur at the gamete stage. The sex bias occurs after fertilization; YY zygotes die, allowing the X* chromosome to be passed on to two-thirds of the progeny (X*X and X*Y females). XY females have also been found, sometimes at high frequencies, in the African pygmy mouse *Mus minutoides* (Veyrunes et al. 2010; Saunders et al. 2014) and in eight different species of *Akodon* (reviewed in Bianchi 2002). In all four of these genera, the driving X also benefits from better reproductive performance of XY females compared with XX females. For example, in *Akodon azarae*, XY females have a longer reproductive life span than XX females, and they have the same litter size even though the YY embryos are not viable (Hoekstra and Hoekstra 2001). The XY female embryos also have an accelerated rate of preimplantation development (Espinosa and Vitullo 2001).

The genetic bases of these sex-reversal mechanisms are unknown. Their recurrent evolution is likely related to the peculiar structure of the *Sry* gene in *Muroidea*. It contains an unstable microsatellite encoding a glutamine-rich domain of variable length. This domain functions as a genetic capacitor that could foster the accumulation of intragenic cryptic variation and promote the evolution of male sex determination systems that are not *Sry* dependent (Chen et al. 2013). The association with an X-linked deletion has been proven in *Myopus*, and X linkage is strongly suspected in *Dycrostonyx* and *Akodon* species where driving X chromosomes show specific rearrangements (Fredga and Bulmer 1988; Ortiz et al. 2009).

Besides these cases associated with sex reversal, experimental studies on the mouse revealed that sex chromosome drive in rodents can also occur in a very different way. Males with a deletion that removes two-thirds of the Y chromosome long arm produce female-biased progeny. This is because of an impaired fertilizing ability of Y-bearing sperm (Ward and Burgoyne 2006). The deletion causes imbalance in copy number of antagonistic X-linked (*Slx*/*Slx1*) and Y-linked (*Sly*) genes involved in the epigenetic control of postmeiotic sex chromatin (PMSC) expression (Cocquet et al. 2012). There is evidence to support the idea that a long-lasting conflict exists between these antagonistic genes in rodents (see the section Evolutionary Consequences).

Drosophila

In *Drosophila*, X drive can be very strong, sometimes leading to 100% female progeny, or close to it. X drive in *Drosophila* is often associated with abnormal behavior of the Y chromosome during meiosis II, which can lead to segregation failure of the Y chromosome's sister chromatids (direct observation in *D. simulans* Paris, *D. pseudoobscura*, and *D. athabasca*; inference in *Drosophila subobscura* and *Drosophila paramelanica* [Fig. 1]; reviewed in Cazemajor et al. 2000; Jaenike 2001). All of the cytological studies in *Drosophila* and stalk-eyed flies show evidence of spermiogenesis failure, which is thought to affect the spermatids containing the target chromosome, as shown in *D. simulans* Paris (Cazemajor et al. 2000).

Although X drive has been known for a long time, only one driver gene, *Dox* (*Distorter on the X*) has been identified molecularly. Its function is unknown. Found in *D. simulans*, *Dox* is suppressed and cryptic in its natural context. It was revealed by interspecific crosses that lifted the suppression. *Dox* is a recent gene that arose from another recent gene, *Mother of Dox* (*MDox*) (Tao et al. 2007a,b). *Dox* and *MDox* appear to be transcribed as RNAs that are noncoding or that have a very limited coding potential. *Dox* does not seem to cause meiotic defects, but rather postmeiotic failure of nuclear condensation in the Y-bearing sperm (Tao et al. 2007b; Yasuno et al. 2013). Remarkably, *Dox* gave rise through retrotransposition to an autosomal suppressor, *Not much yang* (*Nmy*). An RNA interference mechanism seems to be involved in the suppression of *Dox* by *Nmy*.

The Paris *sex-ratio* drive, also found in *D. simulans*, has the best-described cellular phenotype, but the driver elements and the suppressors are still unknown. Nevertheless, two drive loci have been mapped on the X chromosome, about 150 kb away from each other, both required to generate the distortion (Montchamp-Moreau et al. 2006). The first locus corresponds to a segmental duplication containing six genes, all expressed in the testes (Fouvry et al. 2011). The second locus is the *HP1D2* gene (*GD16106*), a member of the heterochromatin protein 1 (HP1) gene family (Q Helleu et al., unpubl.).

Mosquitoes

In *A. aegypti* and *C. pipiens*, Y drive can also result in high sex-ratio biases (Wood and Newton 1991). In both species, Y drive is associated with breakage of the X chromosome (and of the Y and autosomes to a much lesser extent) at some stage before anaphase I. This breakage may be caused by failed crossover (Newton et al. 1976). A high proportion of sperm heads in *A. aegypti* driving males contains a remarkable large amount of DNA, which likely results from incomplete anaphase during the first and second meiotic divisions (Newton et al. 1978).

In both species, the sex chromosome driver is tightly linked to the sex-determining locus on the nondegenerated Y chromosome. The sex chromosome driver in *A. aegypti* has been better studied, and several modifiers have been genetically mapped—two suppressors and an enhancer (Wood and Ouda 1987; Wood and Newton 1991). Shin et al. (2011) used comparative transcriptomics to characterize the sex chromosome drive testes transcription profile, and identified genes associated with signal transduction and cell cycle. Among these they found some Ras-related genes.

Heterochromatin, Small RNAs, and Genetic Conflict

Most of the cellular phenotypes described above—at least, those described in *Drosophila* species—suggest a failure in the chromatin state regulation. In yeast and *Drosophila*, defects in heterochromatin formation can lead to chromosome missegregation (Allshire et al. 1995; Kellum and Alberts 1995). Moreover, the interaction between the heterochromatin and the proteins involved in its regulation is one of the key steps that led to the centromere drive hypothesis, as argued by Malik and Henikoff (2009). The interesting feature about *SR* in *Drosophila* is that nearly all the Y chromosomes in this genus are ancient and highly heterochromatic (Bachtrog 2013). We suggest that for meiotic drive, the X chromosome—via DNA-binding proteins or other proteins involved in the regulation of the chromatin state—can take advantage of the singular heterochromatin state of the Y chromosome and lead to its missegregation.

The HP1 gene family is involved in heterochromatin regulation and formation (Cheutin et al. 2003; Olszak et al. 2011). The phylogenomic analysis in *Drosophila*, performed by Levine et al. (2012), has revealed a great potential for many members of this family to be involved in episodic genetic conflicts, like the germline-restricted *Oxpecker*. Some HP1s genes could suppress the selfish activity of genomic parasites, including meiotic drivers. Once suppressed, the selfish element can degenerate, then its suppressor. That could explain HP1 turnover and neofunctionalization, shown , for instance, for *Umbrea* (Levine and Malik 2013; Ross et al. 2013). The discovery that *HP1D2* in *D. simulans* is one of the distorters in the Paris system is highly significant in this respect: this is the first evidence of an HP1 protein directly involved in an ongoing genetic conflict.

Another potential actor of drive is the small RNA library that guides the DNA-binding proteins. Small RNAs are involved in the regulation of the chromatin state, in interaction with HP1s (Pal-Bhadra et al. 2004; Brower-Toland et al. 2007; Gu and Elgin 2013). Produced by satellite DNA, transposable elements, and genomic tandem repeats, small RNAs can induce heterochromatin perturbation, altering DNA condensation or segregation (Allshire et al. 1995; Grewal et al. 1998; Aravin et al. 2001; Lippman et al. 2004). Recent data support the view that small RNA can be involved in regulation of meiotic drivers. Indeed, piRNA pathway mutants enhance the drive caused by segregation distorter in *Drosophila melanogaster* (Gell and Reenan 2013). In addition, the *SR* driver Winters in *D. simulans* is thought to be suppressed through small RNA-based silencing (Tao et al. 2007b). This is also true for *Stellate* repeated genes, which lead to male sterility in the absence of a suppressor (Livak 1990). This shows that the emergence of homologous repeated sequences can resolve a conflict (Aravin et al. 2001; Stapleton et al. 2001).

POPULATION DYNAMICS OF SEX CHROMOSOME DRIVE: THE *Drosophila* MODEL

The fate of a sex-linked driver that emerges in a population can be diverse, causing population extinction, staying at a stable polymorphism, or triggering the evolution of (and possibly cycling with) suppressors of drive. Many documented examples come from the genus *Drosophila* (Fig. 2).

Population Extinction

In theory, even a low transmission advantage should allow a driver to quickly invade a population. Without any countervailing effect, the presence of a strong driver on a sex chromosome can lead to unisexual populations that will collapse (Sturtevant and Dobzhansky 1936; Hamilton 1967). If extinction can be observed (although rarely) in the laboratory (Lyttle 1977; Price et al. 2010a), it should strongly depend on life-history parameters and stochastic effects such as demographic ones (Jaenike 1996; Hatcher et al. 1999; Taylor and Jaenike 2002, 2003).

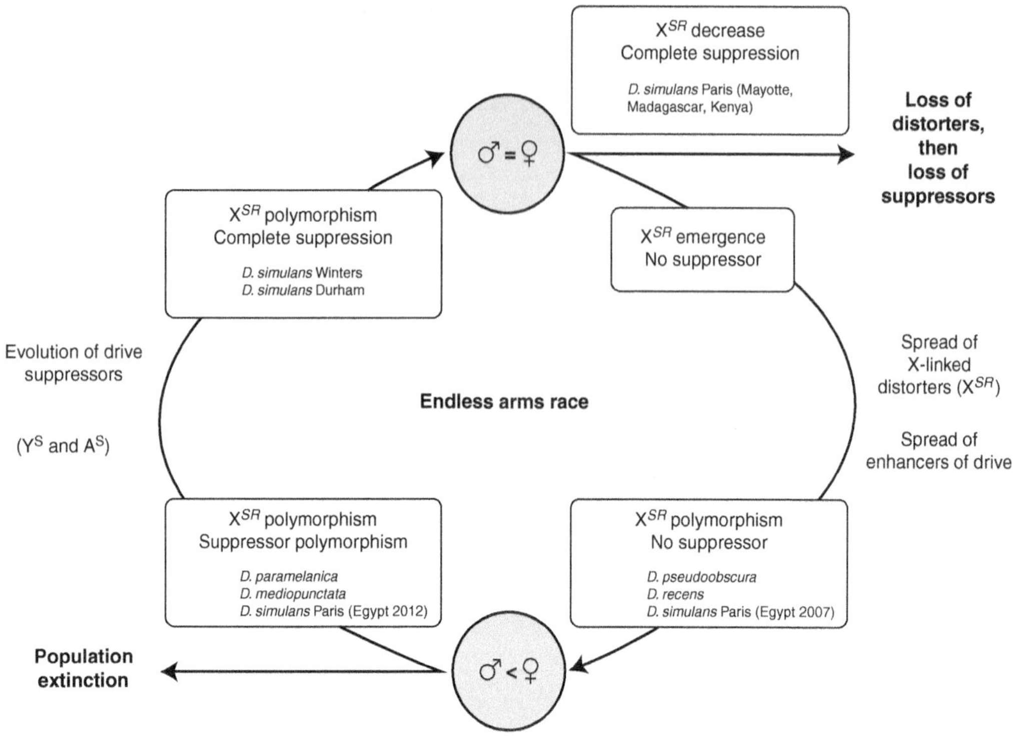

Figure 2. Model for the evolution of X-chromosome drive, with examples from *Drosophila*. The evolution can cease by population extinction or the system can stay at a stable polymorphism (without suppressors [e.g., in *D. pseudoobscura*] or with suppressors [e.g., in *Drosophila mediopunctata*]). The only driver found to cycle in natural populations is the young Paris system of *D. simulans*: the driver is currently disappearing in the Indian Ocean (Madagascar, Mayotte) where complete suppression has evolved (Bastide et al. 2011) but is increasing in frequency along with suppressors in the Middle East (e.g., in Egypt) (Bastide et al. 2013).

In fact, extinction driven by sex chromosome drive has never been observed in natural populations. Only recently, a population crash may have been related to a high *SR* frequency in *Drosophila neotestacea* (Pinzone and Dyer 2013). The absence of examples might be because, at least in part, of the short period of time during which the phenomenon can be observed. The frequency of extinctions induced by *SR* drive is probably not negligible (Carvalho and Vaz 1999; Jaenike 2001). There are few known cases of sex chromosome drive systems that are still evolving (such as the Paris system of *D. simulans*), and the vast majority of drivers that have been described so far are either cryptic (i.e., fixed and suppressed in the populations, such as the Winters system of *D. simulans*) or in a stable polymorphism (such as *D. pseudoobscura*'s *SR*). Now that we can more precisely date the origin of such elements using molecular population genetics tools, it appears that most of them are old (see below).

If the extinction effect is strong enough, it should be possible to use such selfish elements for pest management through population replacement (Hastings 1994; Sinkins and Gould 2006). For example, it might be feasible to use a killer Y for this purpose in the mosquito *A. aegypti* (Craig et al. 1960). A driving Y would be more efficient than a driving X, because it would drive in every individual carrying it, whereas the X would not drive in females (Hamilton 1967; Burt and Trivers 2006). However, naturally occurring drivers will not work if there are resistances in target populations (Wood and Newton 1991). Alternatively, it is theoretically plausible to invade a pest population with engineered sex chromosome drivers (e.g., Deredec et al. 2011) as it has been performed in experimental populations of *D. melanogaster* (e.g., Akbari et al. 2013).

Maintenance of Polymorphism

Several *SR* drivers that have been detected in natural populations show stable polymorphism. The ancient *SR* from *D. pseudoobscura*—one of the first to be discovered (Sturtevant and Dobzhansky 1936)—might be approximately one million years old (Babcock and Anderson 1996; Kovacevic and Schaeffer 2000). It has remained at the same frequencies for more than 50 years in Arizona (Beckenbach 1996). Much theoretical and empirical work has been devoted to explaining how the *SR* polymorphism is maintained in this species, given that there is no evidence of either resistance or suppression (Policansky and Ellison 1970; Beckenbach et al. 1982). Following the experimental work by Wallace (1948), a model was developed in which the selection coefficient associated with each genotype is constant, and the transmission advantage of *SR* is compensated by deleterious effects in females (Edwards 1961; Curtsinger and Feldman 1980). But considering that the driver is acting during male gametogenesis, it is unlikely that it lowers female fitness by itself. This would instead be a result of other loci that are tightly linked to *SR*, or that reside within chromosome rearrangements; several drivers, including *D. pseudoobscura SR*, are associated with inversions (Jaenike 2001). In *Drosophila recens*, the driver seems to be disappearing because of the accumulation of deleterious mutations within large X chromosome inversions associated with the driver (Dyer et al. 2007).

Alternatively, polymorphism can be maintained through frequency-dependent processes; *SR* males usually suffer from reduced sperm production to at least half of that of standard males (e.g., in *Drosophila*, *Cyrtodiopsis*, or *Aedes*) (Hickey and Craig 1966; Policansky and Ellison 1970; Hauschteck-Jungen and Maurer 1976; Wu 1983b; Montchamp-Moreau and Joly 1997; Presgraves et al. 1997). If the realized fecundity (i.e., number of offspring sired) is a sigmoidal function of the number of gametes produced, then *SR* polymorphism can be maintained by frequency-dependent selection (Hartl 1972; Jaenike 1996). When the proportion of females increases in the population because of the driver, the male mating rate increases and standard males are favored because they sire proportionally more offspring relative to their transferred sperm. Variation in population density can influence the male mating rate and contribute to the differences in *SR* abundance (Beckenbach 1978; Atlan et al. 2004). This could explain the seasonal fluctuations of *SR* as do the seasonal fluctuations of *SR* that are observed in *D. pseudoobscura* (Bryant et al. 1982). The negative effect of multiple mating on the fertility of *SR* males has been well documented in laboratory conditions (e.g., Hickey and Craig 1966; Wu 1983c; Jaenike 1996; Atlan et al. 2004; Wilkinson et al. 2006), and the effect of density on *SR* frequency has been observed in experimental populations of *D. pseudoobscura* (Beckenbach 1983). In addition, as for many other selfish genetic elements (Price and Wedell 2008), *SR* males are usually poor sperm (or pollen) competitors against standard males (e.g., Wu 1983b; Taylor et al. 1999; Wilkinson and Sanchez 2001; Wilkinson et al. 2006; Angelard et al. 2008; Price et al. 2008b). Therefore, a high female mating rate may help stabilize the *SR* polymorphism. Reduced sperm production in *SR* males could increase the female propensity to remate (Beckenbach 1981; Atlan et al. 2004), and standard males could suppress female remating in the presence of even a low number of *SR* males (Price et al. 2010b). However, when the driver increases in frequency in the population and the sex ratio becomes highly female biased, the female mating rate will tend to decrease, reducing sperm competition and therefore the *SR* disadvantage. Taylor and Jaenike (2002) modeled the dynamics of an X-linked driver under sperm competition and showed that opposing effects of male and female mating rates can lead to a stable equilibrium in certain conditions. When the *SR* male disadvantage is not too high, the driver frequency can even stabilize at ~0.5.

Population structure could influence driver distribution at the metapopulation level (Price et al. 2010a). For example, local populations that would be more sensitive to *SR* invasion would go extinct more rapidly but send out fewer migrants, increasing the probability of the driver persisting (Van Boven and Weissing 1999). Moderate gene flow between demes can greatly favor the stability of the polymorphism (Taylor and Jaenike 2003). When a stable polymorphism is observed at the metapopulation level, there are likely local dynamics behind it (Hatcher 2000). Environmental factors such as climate have long been invoked as a determinant of *SR* distribution (e.g., in *D. pseudoobscura*) because several drivers show latitudinal clines in frequency (Sturtevant and Dobzhansky 1936). Temperature-related fertility differences do not seem to impact the distribution of *D. pseudoobscura SR* (Price et al. 2012), although this could explain the high correlation between climate variables and *SR* frequency in *D. neotestacea* (Dyer 2012). Moreover, the mechanisms by which ecological variables could influence the maintenance of polymorphism is difficult to determine and may be indirect.

Evolution of Suppressors

Sex chromosome drivers are thought to trigger the evolution of resistant Y chromosomes (for X-linked drivers) and/or autosomal suppressors. When a population becomes highly female biased, any variant that produces more males will be favored through a frequency-dependent selective process to rebalance the sex ratio (Fisher 1930; Bull and Charnov 1988). Y-linked resistance should rapidly spread to fixation, especially if it is not associated with deleterious fitness effects (Thomson and Feldman 1975; Clark 1987), or it should cycle with the driving X (Hall 2004). Alternatively, it can reach a stable equilibrium that depends on frequency-dependent interactions between the driving X and the resistant Y (Carvalho et al. 1997). Autosomal suppressors of drive are also predicted to be favored and increase in frequency (Hamilton 1967), except when the driver is very strongly deleterious (Wu 1983a; Vaz and Carvalho 2004). Driver-suppressor frequencies can attain equilibrium when there is overdominance in females (Wu 1983a) or when the driver is partially deleterious (Vaz and Carvalho 2004). In the latter model, autosomal suppressors go to fixation only if the driver has no effect on fitness (Vaz and Carvalho 2004). The same qualitative prediction can be made for a resistant Y chromosome, but its fixation probability is higher, even with a deleterious *SR* (Carvalho and Vaz 1999). Indeed, Y-linked resistance should be more strongly selected than autosomal suppression (Jaenike 2001; Atlan et al. 2003). In addition to the fitness cost associated with drive and suppression in males and females (heterozygous or homozygous for *SR*), the fate of the driver and suppressor can also depend on the population structure or the time at which the suppressor arises during the spread of the driver (Carvalho and Vaz 1999; Hall 2004; Vaz and Carvalho 2004).

After the suppressor has fixed (or nearly fixed) in a population, the driver can either evolve neutrally or, if it is deleterious, decline in frequency (Burt and Trivers 2006). Neutral evolution would produce cryptic drive systems that may be revealed by chance, like the *D. simulans* Winters (Tao et al. 2007a,b; Kingan et al. 2010), whereas selection against a deleterious driver can generate evolutionary cycles of drive and suppression (Hall 2004). Resistant Y and autosomal suppression have been detected in many *Drosophila* species that show *SR* drive (e.g., Stalker 1961; Voelker 1972; De Carvalho and Klaczko 1993; Carvalho et al. 1997; Cazemajor et al. 1997; Jaenike 1999; Montchamp-Moreau et al. 2001), making the evolution of suppression more the rule than the exception in this genus. Nevertheless, the dynamics of drive and suppression has rarely been assessed in natural populations, given the paucity of known young and still-evolving systems. In *D. mediopunctata*, *SR* frequency has not changed over 10 years, suggesting that the driver has been balanced at an equilibrium frequency by partial suppression and natural selection (Carvalho and Vaz 1999). The only known example of coevolution between drive and suppression in natural populations has been observed with the *D. simulans* Paris *SR* system. The driver has been declining in East African populations (Mayotte,

Madagascar, Kenya) in which there is complete suppression (Bastide et al. 2011), but it has risen along with suppression in the Middle East within a few years, especially in Egypt (Bastide et al. 2013). In the Middle Eastern populations, suppression seems to be increasing much more quickly in one population than in the other, possibly reflecting Y-linked resistance as opposed to autosomal suppression dynamics. In both species, the increase in suppression in response to *SR* fixation had been shown in experimental populations (Carvalho et al. 1998; Capillon and Atlan 1999).

EVOLUTIONARY CONSEQUENCES

Mating System Evolution

Sex chromosome drive can have a number of consequences on the evolution of sexual reproduction processes, usually resulting from adaptations to counteract drive (Presgraves 2008) through sexual selection and sexual conflicts. In *D. pseudoobscura*, for example, *SR* promotes polyandry in experimental populations and may have evolved to prevent population extinction (Price et al. 2008a, 2010a). Accordingly, *SR* prevalence is negatively correlated with polyandry in natural populations of this species (Price et al. 2014). Selfish genetic elements in general can be common drivers of the evolution of polyandry (reviewed in Wedell 2013), and this has been suggested in other *SR* systems (e.g., Wilkinson et al. 2003; Angelard et al. 2008; but see discussion in Pinzone and Dyer 2013). In addition, if polyandry may evolve through sexual selection in females, sperm competition theory predicts that males should benefit from a reduction in female mating rate to increase their mating success (Parker 2006). Results from experimental populations of *D. pseudoobscura* tend to support this prediction for non-*SR* males (Price et al. 2010b). Driving sex chromosomes can also trigger the evolution of mate choice. In the stalk-eyed fly *Cyrtodiopsis dalmanni*, males show a dimorphism in eye-stalk span that is genetically linked with drive and suppression, and females have evolved preferences for long eye stalks in the presence of *SR* (Wilkinson et al. 1998; Cotton et al. 2014). This process appears to be rare (Price et al. 2012) and requires the suppression of recombination between meiotic drive and the male sexual character loci (Lande and Wilkinson 1999).

Sex Determination, Sex Chromosome, and Genome Evolution

Sex chromosome drivers can have consequences on the linked polymorphism, usually reducing it through (recurrent) selective sweeps (e.g., Chevin et al. 2009; Kingan et al. 2010; Dyer et al. 2013), and especially when they tend to recruit inversions (e.g., Dyer et al. 2007). However, the functional consequences of this phenomenon is not well known.

There is a great deal of speculation about the role that sex chromosome drive plays in the evolution of sex-determining mechanisms. As seen in rodents, mutations that change the sex-determining system must be costly—in particular, in species with heteromorphic chromosomes. Segregation distortion can very efficiently promote a disadvantageous mutation. For example, Charlesworth and Dempsey (2001) proposed that the X0/XY system of *Microtus oregoni* has evolved in such a way. They assumed that a single X-linked mutation causes the observed mitotic nondisjunction of X chromosomes in the male germline (which produces only 0 and Y sperm). They showed that the mutant X can spread only if it benefits from complete transmission advantage in females, as this counterbalances the transmission failure in the males. However, as far as we know, there is still no definite evidence of species with a fixed switch caused by the spread of a driving chromosome.

Alternatively, there may be changes in the sex-determining mechanism through the evolution of new feminizing or masculinizing genes to suppress the sex-ratio bias induced by a sex-linked driver. Kozielska et al. (2009) have modeled these scenarios in the case of XY sex determination. In particular, they may account for the evolution of masculinizer genes in *Diptera* (reviewed in Burt and Trivers 2006). Interestingly, Kozielska et al. (2009) point out that the driver element, which is assumed to reduce individual fitness, is often lost and no longer detectable once the new sex-determining system has become established.

The phenomenon of centromere drive that can occur in females (i.e., the preferential transmission in eggs) is thought to drive karyotype evolution (de Villena and Sapienza 2001). This specific conflict may also contribute to fixation of neo-sex chromosomes that arise from the fusion between an autosome and a sex chromosome. This was postulated by Yoshida and Kitano (2012), who found that neo-X chromosomes were more prevalent among mammalian species that have a higher percentage of metacentric chromosomes.

In a wide range of taxa (mammals, nematodes, birds), the sex chromosomes are transcriptionally silenced by epigenetic marks during meiosis (Turner 2007; Namekawa and Lee 2009). This meiotic sex chromosome inactivation (MSCI) is thought to have first evolved as a defense mechanism against selfish elements, including meiotic drivers (Meiklejohn and Tao 2010). Recent advances in the understanding of epigenetic control of postmeiotic sex chromatin (PMSC) expression in *Mus musculus* provides strong support for the idea that the protagonists of the genetic conflict over sex chromosome transmission can more broadly drive the evolution of gene expression, genome structure, and ultimately lead to speciation. In the male germline of *M. musculus*, many sex-chromosome-linked genes important for sperm differentiation and function are reactivated after meiosis (Mueller et al. 2008), but their expression remains tightly regulated by repressive epigenetic marks inherited from MSCI, via the action of *Sly*, an essential Y-linked multicopy gene. Males deficient in *Sly* show up-regulation of these genes, suffer fertility defects, and produce offspring with a sex ratio that is moderately biased toward females, because of different fertilizing ability of X- and Y-bearing sperm (Ward and Burgoyne 2006). Interestingly, the X-linked multicopy genes *Slx/Slx1*, related to *Sly*, are among the up-regulated genes. Cocquet et al. (2012) have showed that the products of *Sly* and *Slx/Slx1* genes have antagonistic effects on the epigenetic marks associated with PMSC, on the expression of sex-chromosome-linked genes, and consistently on offspring sex ratio. Imbalance of *Sly* and *Slx/Slx1* copy numbers causes sex-ratio distortion, sperm differentiation defects, and male infertility, but *Slx/slx1* deficiency rescues the defects caused by *Sly* deficiency and vice versa. The authors thus propose that the conflict between *Sly* and *Slx/Slx1* genes led to their massive amplification on the Y and X chromosomes, respectively, during rodent evolution (Fig. 3A) and also to the amplification of spermiogenic genes throughout the genome to escape the repressive effect of *Sly*. Finally, they argue that copy number imbalance may be one contributor to the evolution of hybrid sterility in mice (Fig. 3B).

Speciation

The hypothesis that meiotic drivers play a role in speciation has long been controversial. Frank (1991) and Hurst and Pomiankowski (1991) independently claimed that Haldane's rule (the preferential sterility or inviability of the heterogametic F_1 hybrid) could be explained by the divergent evolution of sex chromosome drivers and their suppressors. Coyne et al. (1991) and Charlesworth et al. (1993) contested this hypothesis, mostly because of the lack of empirical evidence (Johnson and Wu 1992; Coyne et al. 1993). At the time, the support came from observations in *D. subobscura*, where an X chromosome causing drive in the local genetic background caused male sterility when it was combined with autosomes from another population (Hauschteck-Jungen 1990). Since then, examples of a link between sex chromosome drive and hybrid male sterility have accumulated.

Tao et al. (2001) found an X-linked cryptic meiotic driver in *D. simulans* that was unleashed by the introgression of *Drosophila mauritiana* alleles at the autosomal locus *Too much yin* (*Tmy*). The introgression causes both sex chromosome drive and male sterility in *D. simulans* males. Both traits localize to a DNA fragment of 80 kb or smaller and may be pleiotropic effects from a single gene. Similarly, Yang et al. (2004) characterized a cryptic X-linked meiotic driver

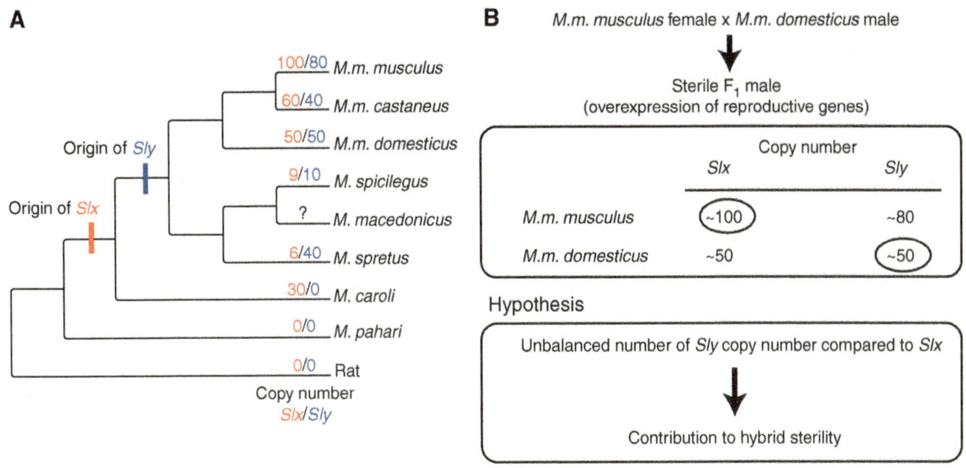

Figure 3. (*A*) Evolution of *Slx/Sly* copy number in some lineages of the genus *Mus*. (Reprinted from Good 2012 under the terms of the creative common attribution license, which permits unrestricted use, distribution, and reproduction in any medium.) (*B*) Deficiency in the number of *Sly* copies compared to *Slx* copies could contribute to F_1 male hybrid sterility in crosses between *M.m. musculus* females and *M.m. domesticus* males (Published courtesy of Julie Cocquet.)

on the neo-X chromosome of *Drosophila albomicans* expressed in F_1 male hybrids between *D. albomicans* and *Drosophila nasuta*.

Later, Phadnis and Orr (2009) identified an X-linked gene, *Overdrive* (*Ovd*), that causes both segregation distortion and sterility in hybrid males between the two subspecies *Drosophila pseudoobscura pseudoobscura* and *Drosophila pseudoobscura bogotana*. *Ovd* is predicted to encode a polypeptide with a DNA-binding domain that targets trinucleotide repeats. The hybrids become weakly fertile when they are old, and the *Ovd* allele from *D.p. bogotana* in the hybrid F_1 males leads to a sex-ratio bias, as the F_1 males sire more than 90% daughters.

Because the spread of a strong X-linked driver can lead to the evolution of Y-resistant chromosome, Hurst and Pomiankowski (1991) also proposed that the newly selected Y could be incompatible with some recently derived genomes. The Y chromosome can contribute to hybrid sterility, sometimes through an interaction with the rest of the genome (Johnson et al. 1993; Lamnissou et al. 1996; Campbell et al. 2012). Recent studies have started to shed light on the nature of these interactions (Chippindale and Rice 2001; Lemos et al. 2008; Sackton et al. 2011; Zhou et al. 2012; Branco et al. 2013). Despite the fact that there is no evidence of a resistant Y chromosome involved in a postzygotic barrier, the hypothesis remains relevant (McDermott and Noor 2010).

Sex chromosome drive can likely also lead to speciation through a change in sex-determining mechanism. This could be the case for rodents' feminizing X chromosomes and the insects *Sciara coprophila*, *Musca domestica*, and *Neococcoidea* (Haig 1993a,b; Kozielska et al. 2009).

CONCLUDING REMARKS

Like other selfish genetic elements, sex-linked meiotic drivers usually trigger genome adaptations to resolve the conflict that they induce. This can be achieved through neofunctionalization or subfunctionalization of duplicated genes (Gallach and Betrán 2011; Ross et al. 2013). For example, each cluster of the abovementioned repeated genes *Sxl/Sxl1* and *Sly* is needed when the other is present. But when both are removed, fertility is restored, suggesting that they have been amplified and conserved only because of the genetic conflict that they initiated and/or resolved. By such processes, genetic conflicts should contribute to increased genome size and complexity and be a major driver of genome evolution. Recent work on emblematic cases of genetic incompatibilities pinpoint a twofold challenge: identifying multiple genes with complex epistatic interactions and understanding the role of heterochromatic noncoding DNA repeats in sex-chromosome-associated genetic conflicts.

ACKNOWLEDGMENTS

We thank William Rice and Sergey Gavrilets for the invitation to contribute to this volume and Julie Cocquet and Frédéric Veyrunes for helpful comments. This work is supported by the Centre National de la Recherche Scientifique (UPR 9034) and a grant from the Agence Nationale de la Recherche (ANR-12-BSV7-0014-01). Q.H. is funded by a French ministerial scholarship.

REFERENCES

Akbari OS, Matzen KD, Marshall JM, Huang H, Ward CM, Hay BA. 2013. A synthetic gene drive system for local, reversible modification and suppression of insect populations. *Curr Biol* **23:** 671–677.

Allshire RC, Nimmo ER, Ekwall K, Javerzat JP, Cranston G. 1995. Mutations derepressing silent centromeric domains in fission yeast disrupt chromosome segregation. *Genes Dev* **9:** 218–233.

Angelard C, Montchamp-Moreau C, Joly D. 2008. Female-driven mechanisms, ejaculate size and quality contribute to the lower fertility of *sex-ratio* distorter males in *Drosophila simulans*. *BMC Evol Biol* **8:** 326.

Aravin AA, Naumova NM, Tulin AV, Vagin VV, Rozovsky YM, Gvozdev VA. 2001. Double-stranded RNA-mediated silencing of genomic tandem repeats and transposable elements in the *D. melanogaster* germline. *Curr Biol* **11:** 1017–1027.

Atlan A, Capillon C, Derome N, Couvet D, Montchamp-Moreau C. 2003. The evolution of autosomal suppressors of *sex-ratio* drive in *Drosophila simulans*. *Genetica* **117:** 47–58.

Atlan A, Joly D, Capillon C, Montchamp-Moreau C. 2004. *Sex-ratio* distorter of *Drosophila simulans* reduces male productivity and sperm competition ability. *J Evol Biol* **17:** 744–751.

Babcock CS, Anderson WW. 1996. Molecular evolution of the Sex-Ratio inversion complex in *Drosophila pseudoobscura*: Analysis of the *Esterase-5* gene region. *Mol Biol Evol* **13:** 297–308.

Bachtrog D. 2013. Y-chromosome evolution: Emerging insights into processes of Y-chromosome degeneration. *Nature Rev Genet* **14:** 113–124.

Bastide H, Cazemajor M, Ogereau D, Derome N, Hospital F, Montchamp-Moreau C. 2011. Rapid rise and fall of selfish *sex-ratio X* chromosomes in *Drosophila simulans*: Spatiotemporal analysis of phenotypic and molecular data. *Mol Biol Evol* **28:** 2461–2470.

Bastide H, Gérard PR, Ogereau D, Cazemajor M, Montchamp-Moreau C. 2013. Local dynamics of a fast-evolving *sex-ratio* system in *Drosophila simulans*. *Mol Ecol* **22:** 5352–5367.

Beckenbach AT. 1978. The "sex-ratio" trait in *Drosophila pseudoobscura*: Fertility relations of males and meiotic drive. *Am Nat* **112:** 97–117.

Beckenbach A. 1981. Multiple mating and the "sex-ratio" trait in *Drosophila pseudoobscura*. *Evolution* **35:** 275–281.

Beckenbach A. 1983. Fitness analysis of the "sex-ratio" polymorphism in experimental populations of *Drosophila pseudoobscura*. *Am Nat* **5:** 630–648.

Beckenbach AT. 1996. Selection and the "sex-ratio" polymorphism in natural populations of *Drosophila pseudoobscura*. *Evolution* **50:** 787–794.

Beckenbach AT, Curtsinger JW, Policansky D. 1982. Fruitless experiments with fruit flies: The "sex ratio" chromosomes of *D. pseudoobscura*. *Dros Inf Serv* **58:** 22.

Bianchi N. 2002. *Akodon* sex reversed females: The never ending story. *Cytogenet Genome Res* **96:** 60–65.

Branco AT, Tao Y, Hartl DL, Lemos B. 2013. Natural variation of the Y chromosome suppresses sex ratio distortion and modulates testis-specific gene expression in *Drosophila simulans*. *Heredity* **111:** 8–15.

Brower-Toland B, Findley SD, Jiang L, Liu L, Yin H, Dus M, Zhou P, Elgin SC, Lin H. 2007. *Drosophila* PIWI associates with chromatin and interacts directly with HP1a. *Genes Dev* **21:** 2300–2311.

Bryant SH, Beckenbach AT, Cobbs GA. 1982. "Sex-ratio" trait, sex composition, and relative abundance in *Drosophila pseudoobscura*. *Evolution* **36:** 27–34.

Bull JJ, Charnov EL. 1988. How fundamental are Fisherian sex ratios? *Oxford Surv Evol Biol* **5:** 96–135.

Burt A, Trivers R. 2006. *Genes in conflict: The biology of selfish genetic elements*. Harvard University Press, Cambridge, MA.

Campbell P, Good JM, Dean MD, Tucker PK, Nachman MW. 2012. The contribution of the Y chromosome to hybrid male sterility in house mice. *Genetics* **191:** 1271–1281.

Capillon C, Atlan A. 1999. Evolution of driving X chromosomes and resistance factors in experimental populations of *Drosophila simulans*. *Evolution* **53:** 506–517.

Carvalho AB, Vaz SC. 1999. Are *Drosophila* SR drive chromosomes always balanced? *Heredity* **83:** 221–228.

Carvalho AB, Vaz SC, Klaczko LB. 1997. Polymorphism for Y-linked suppressors of sex-ratio in two natural populations of *Drosophila mediopunctata*. *Genetics* **146:** 891–902.

Carvalho AB, Sampaio MC, Varandas FR, Klaczko LB. 1998. An experimental demonstration of Fisher's principle: Evolution of sexual proportion by natural selection. *Genetics* **148:** 719–732.

Cazemajor M, Landre C, Montchamp-Moreau C. 1997. The *sex-ratio* trait in *Drosophila simulans*: Genetic analysis of distortion and suppression. *Genetics* **147:** 635–642.

Cazemajor M, Joly D, Montchamp-Moreau C. 2000. *Sex-ratio* meiotic drive in *Drosophila simulans* is related to equational nondisjunction of the Y chromosome. *Genetics* **154:** 229–236.

Charlesworth B, Dempsey N. 2001. A model of the evolution of the unusual sex chromosome system of *Microtus oregoni*. *Heredity* **86:** 387–394.

Charlesworth B, Hartl DL. 1978. Population dynamics of the segregation distorter polymorphism of *Drosophila melanogaster*. *Genetics* **89:** 171–192.

Charlesworth B, Coyne J, Orr H. 1993. Meiotic drive and unisexual hybrid sterility: A comment. *Genetics* **133:** 421.

Chen YS, Racca JD, Sequeira PW, Phillips NB, Weiss MA. 2013. Microsatellite-encoded domain in rodent *Sry* functions as a genetic capacitor to enable the rapid evolution of biological novelty. *Proc Natl Acad Sci* **110:** E3061–E3070.

Cheutin T, McNairn AJ, Jenuwein T, Gilbert DM, Singh PB, Misteli T. 2003. Maintenance of stable heterochromatin domains by dynamic HP1 binding. *Science* **299:** 721–725.

Chevin LM, Bastide H, Montchamp-Moreau C, Hospital F. 2009. Molecular signature of epistatic selection: Interrogating genetic interactions in the *sex-ratio* meiotic drive of *Drosophila simulans*. *Genet Res* **91:** 171–182.

Chippindale AK, Rice WR. 2001. Y chromosome polymorphism is a strong determinant of male fitness in *Drosophila melanogaster*. *Proc Natl Acad Sci* **98:** 5677–5682.

Clark AG. 1987. Natural selection and Y-linked polymorphism. *Genetics* **115:** 569–577.

Cocquet J, Ellis PJ, Mahadevaiah SK, Affara NA, Vaiman D, Burgoyne PS. 2012. A genetic basis for a postmeiotic X versus Y chromosome intragenomic conflict in the mouse. *PLoS Genet* **8:** e1002900.

Cotton AJ, Földvári M, Cotton S, Pomiankowski A. 2014. Male eyespan size is associated with meiotic drive in wild stalk-eyed flies (*Teleopsis dalmanni*). *Heredity* **112:** 363–369.

Coyne JA, Rux J, David JR. 1991. Genetics of morphological differences and hybrid sterility between *Drosophila sechellia* and its relatives. *Genet Res* **57:** 113–122.

Coyne JA, Meyers W, Crittenden AP, Sniegowski P. 1993. The fertility effects of pericentric inversions in *Drosophila melanogaster*. *Genetics* **134:** 487–496.

Craig GB Jr, Hickey WA, VandeHey RC. 1960. An inherited male-producing factor in *Aedes aegypti*. *Science* **132:** 1887–1889.

Curtsinger JW, Feldman MW. 1980. Experimental and theoretical analysis of the "sex-ratio" polymorphism in *Drosophila pseudoobscura*. *Genetics* **94:** 445–466.

De Carvalho AB, Klaczko LB. 1993. Autosomal suppressors of sex-ratio in *Drosophila mediopunctata*. *Heredity* **71:** 546–546.

De Carvalho AB, Klaczko LB. 1994. Y-linked suppressors of the *sex-ratio* trait in *Drosophila mediopunctata*. *Heredity* **73:** 573–573.

Deredec A, Godfray HCJ, Burt A. 2011. Requirements for effective malaria control with homing endonuclease genes. *Proc Natl Acad Sci* **108:** E874–E880.

de Villena FPM, Sapienza C. 2001. Female meiosis drives karyotypic evolution in mammals. *Genetics* **159:** 1179–1189.

Dyer KA. 2012. Local selection underlies the geographic distribution of *sex-ratio* drive in *Drosophila neotestacea*. *Evolution* **66:** 973–984.

Dyer KA, Charlesworth B, Jaenike J. 2007. Chromosome-wide linkage disequilibrium as a consequence of meiotic drive. *Proc Natl Acad Sci* **104:** 1587–1592.

Dyer KA, Bray MJ, Lopez SJ. 2013. Genomic conflict drives patterns of X-linked population structure in *Drosophila neotestacea*. *Mol Ecol* **22:** 157–169.

Edwards AWF. 1961. The population genetics of "sex-ratio" in *Drosophila pseudoobscura*. *Heredity* **16:** 291–304.

Espinosa MB, Vitullo AD. 2001. Fast-developing preimplantation embryo progeny from heterogametic females in mammals. *Zygote* **9:** 289–292.

Fisher RA. 1930. *The genetical theory of natural selection*. Oxford University Press, Oxford.

Fouvry L, Ogereau D, Berger A, Gavory F, Montchamp-Moreau C. 2011. Sequence analysis of the segmental duplication responsible for Paris sex-ratio drive in *Drosophila simulans*. *G3 (Bethesda)* **1:** 401–410.

Frank SA. 1991. Divergence of meiotic drive-suppression systems as an explanation for sex-biased hybrid sterility and inviability. *Evolution* **45:** 262–267.

Fredga K, Bulmer M. 1988. Aberrant chromosomal sex-determining mechanisms in mammals, with special reference to species with XY females. *Phil Trans R Soc Lond B* **322:** 83–95.

Fredga K, Gropp A, Winking H, Frank F. 1976. Fertile XX- and XY-type females in the wood lemming *Myopus schisticolor*. *Nature* **261:** 225–227.

Gallach M, Betrán E. 2011. Intralocus sexual conflict resolved through gene duplication. *Trends Ecol Evol* **26:** 222–228.

Gell SL, Reenan RA. 2013. Mutations to the piRNA pathway component *Aubergine* enhance meiotic drive of segregation distorter in *Drosophila melanogaster*. *Genetics* **193:** 771–784.

Gershenson S. 1928. A new sex-ratio abnormality in *Drosophila obscura*. *Genetics* **13:** 488.

Gileva E. 1987. Meiotic drive in the sex chromosome system of the varying lemming, *Dicrostonyx torquatus Pall.* (*Rodentia, Microtinae*). *Heredity* **59:** 383–389.

Good JM. 2012. The conflict within and the escalating war between the sex chromosomes. *PLoS Genet* **8:** e1002955.

Grewal SI, Bonaduce MJ, Klar AJ. 1998. Histone deacetylase homologs regulate epigenetic inheritance of transcriptional silencing and chromosome segregation in fission yeast. *Genetics* **150:** 563–576.

Gu T, Elgin SCR. 2013. Maternal depletion of Piwi, a component of the RNAi System, impacts heterochromatin formation in *Drosophila*. *PLoS Genet* **9:** e1003780.

Haig D. 1993a. The evolution of unusual chromosomal systems in coccoids: Extraordinary sex ratios revisited. *J Evol Biol* **6:** 69–77.

Haig D. 1993b. The evolution of unusual chromosomal systems in sciarid flies: Intragenomic conflict and the sex ratio. *J Evol Biol* **6:** 249–261.

Hall DW. 2004. Meiotic drive and sex chromosome cycling. *Evolution* **58:** 925–931.

Hamilton WD. 1967. Extraordinary sex ratios. *Science* **156:** 477–488.

Hartl DL. 1972. Population dynamics of sperm and pollen killers. *Theor Appl Genet* **42:** 81–88.

Hastings I. 1994. Selfish DNA as a method of pest control. *Phil Trans R Soc Lond B* **344:** 313–324.

Hatcher MJ. 2000. Persistence of selfish genetic elements: Population structure and conflict. *Trends Ecol Evol* **15:** 271–277.

Hatcher MJ, Taneyhill DE, Dunn AM, Tofts C. 1999. Population dynamics under parasitic sex ratio distortion. *Theor Pop Biol* **56:** 11–28.

Hauschteck-Jungen E. 1990. Postmating reproductive isolation and modification of the "sex ratio" trait in *Drosophila subobscura* induced by the sex chromosome gene arrangement A2+3+5+7. *Genetica* **83:** 31–44.

Hauschteck-Jungen E, Maurer B. 1976. Sperm dysfunction in sex ratio males of *Drosophila subobscura*. *Genetica* **46:** 459–477.

Hickey WA, Craig GB Jr. 1966. Genetic distortion of sex ratio in a mosquito, *Aedes aegypti*. *Genetics* **53:** 1177.

Hoekstra HE, Hoekstra JM. 2001. An unusual sex-determination system in South American field mice (Genus *Akodon*): The role of mutation, selection, and meiotic drive in maintaining XY females. *Evolution* **55:** 190–197.

Hurst LD, Pomiankowski A. 1991. Causes of sex ratio bias may account for unisexual sterility in hybrids: A new explanation of Haldane's rule and related phenomena. *Genetics* **128:** 841–858.

Jaenike J. 1996. Sex-ratio meiotic drive in the *Drosophila quinaria* group. *Am Nat* **148:** 237–254.

Jaenike J. 1999. Suppression of sex-ratio meiotic drive and the maintenance of Y-chromosome polymorphism in *Drosophila*. *Evolution* **53:** 164–174.

Jaenike J. 2001. Sex chromosome meiotic drive. *Annu Rev Ecol Syst* **32:** 25–49.

Johnson N, Wu CI. 1992. An empirical test of the meiotic drive models of hybrid sterility: Sex-ratio data from hybrids between *Drosophila simulans* and *Drosophila sechellia*. *Genetics* **130:** 507–511.

Johnson NA, Hollocher H, Noonburg E, Wu CI. 1993. The effects of interspecific Y chromosome replacements on hybrid sterility within the *Drosophila simulans* clade. *Genetics* **135:** 443–453.

Kellum R, Alberts BM. 1995. Heterochromatin protein 1 is required for correct chromosome segregation in *Drosophila* embryos. *J Cell Sci* **108:** 1419–1431.

Kingan SB, Garrigan D, Hartl DL. 2010. Recurrent selection on the Winters sex-ratio genes in *Drosophila simulans*. *Genetics* **184:** 253–265.

Kovacevic M, Schaeffer SW. 2000. Molecular population genetics of X-linked genes in *Drosophila pseudoobscura*. *Genetics* **156:** 155–172.

Kozielska M, Weissing F, Beukeboom L, Pen I. 2009. Segregation distortion and the evolution of sex-determining mechanisms. *Heredity* **104:** 100–112.

Lamnissou K, Loukas M, Zouros E. 1996. Incompatibilities between Y chromosome and autosomes are responsible for male hybrid sterility in crosses between *Drosophila virilis* and *Drosophila texana*. *Heredity* **76:** 603–609.

Lande R, Wilkinson GS. 1999. Models of sex-ratio meiotic drive and sexual selection in stalk-eyed flies. *Genet Res* **74:** 245–253.

Larracuente AM, Presgraves DC. 2012. The selfish segregation distorter gene complex of *Drosophila melanogaster*. *Genetics* **192:** 33–53.

Lemos B, Araripe LO, Hartl DL. 2008. Polymorphic Y chromosomes harbor cryptic variation with manifold functional consequences. *Science* **319:** 91–93.

Levine MT, Malik HS. 2013. A rapidly evolving genomic toolkit of *Drosophila* heterochromatin. *Fly* **7:** 137–141.

Levine MT, McCoy C, Vermaak D, Lee YCG, Hiatt MA, Matsen FA, Malik HS. 2012. Phylogenomic analysis reveals dynamic evolutionary history of the *Drosophila* heterochromatin protein 1 (HP1) gene family. *PLoS Genet* **8:** e1002729.

Lippman Z, Gendrel AV, Black M, Vaughn MW, Dedhia N, McCombie WR, Lavine K, Mittal V, May B, Kasschau KD, et al. 2004. Role of transposable elements in heterochromatin and epigenetic control. *Nature* **430:** 471–476.

Livak KJ. 1990. Detailed structure of the *Drosophila melanogaster stellate* genes and their transcripts. *Genetics* **124:** 303.

Lyttle TW. 1977. Experimental population genetics of meiotic drive systems I. Pseudo-*Y* chromosomal drive as a means of eliminating cage populations of *Drosophila melanogaster*. *Genetics* **86:** 413–445.

Malik HS, Henikoff S. 2009. Major evolutionary transitions in centromere complexity. *Cell* **138:** 1067–1082.

McDermott SR, Noor MA. 2010. The role of meiotic drive in hybrid male sterility. *Phil Trans R Soc Lond B* **365:** 1265–1272.

Meiklejohn CD, Tao Y. 2010. Genetic conflict and sex chromosome evolution. *Trends Ecol Evol* **25:** 215–223.

Merçot H, Llorente B, Jacques M, Atlan A, Montchamp-Moreau C. 1995. Variability within the Seychelles cytoplasmic incompatibility system in *Drosophila simulans*. *Genetics* **141:** 1015.

Montchamp-Moreau C, Joly D. 1997. Abnormal spermiogenesis is associated with the X-linked *sex-ratio* trait in *Drosophila simulans*. *Heredity* **79:** 24–30.

Montchamp-Moreau C, Ginhoux V, Atlan A. 2001. The Y chromosomes of *Drosophila simulans* are highly polymorphic for their ability to suppress sex-ratio drive. *Evolution* **55:** 728–737.

Montchamp-Moreau C, Ogereau D, Chaminade N, Colard A, Aulard S. 2006. Organization of the *sex-ratio* meiotic drive region in *Drosophila simulans*. *Genetics* **174:** 1365–1371.

Morgan TH, Bridges CB, Sturtevant AH. 1925. The genetics of *Drosophila*. *Bibliogr Genet* **2:** 1–262.

Mueller JL, Mahadevaiah SK, Park PJ, Warburton PE, Page DC, Turner JM. 2008. The mouse X chromosome is enriched for multicopy testis genes showing postmeiotic expression. *Nat Genet* **40:** 794–799.

Namekawa SH, Lee JT. 2009. XY and ZW: Is meiotic sex chromosome inactivation the rule in evolution? *PLoS Genet* **5:** e1000493.

Navara KJ. 2013. Hormone-mediated adjustment of sex ratio in vertebrates. *Integr Comp Biol* **53:** 877–887.

Newton M, Wood R, Southern D. 1976. A cytogenetic analysis of meiotic drive in the mosquito, *Aedes aegypti* (L). *Genetica* **46:** 297–318.

Newton M, Southern D, Wood R. 1978. Relative DNA content of normal and sex-ratio distorting spermatozoa of the mosquito, *Aedes aegypti* (L.). *Chromosoma* **67:** 253–261.

Olszak AM, van Essen D, Pereira AJ, Diehl S, Manke T, Maiato H, Saccani S, Heun P. 2011. Heterochromatin boundaries are hotspots for *de novo* kinetochore formation. *Nat Cell Biol* **13:** 799–808.

Ortiz MI, Pinna-Senn E, Dalmasso G, Lisanti JA. 2009. Chromosomal aspects and inheritance of the XY female condition in *Akodon azarae* (Rodentia, Sigmodontinae). *Mamm Biol* **74:** 125–129.

Pal-Bhadra M, Leibovitch BA, Gandhi SG, Rao M, Bhadra U, Birchler JA, Elgin SC. 2004. Heterochromatic silencing and HP1 localization in *Drosophila* are dependent on the RNAi machinery. *Science* **303:** 669–672.

Parker G. 2006. Sexual conflict over mating and fertilization: An overview. *Phil Trans R Soc Lond B* **361:** 235–259.

Phadnis N, Orr HA. 2009. A single gene causes both male sterility and segregation distortion in *Drosophila* hybrids. *Science* **323:** 376–379.

Pinzone CA, Dyer KA. 2013. Association of polyandry and *sex-ratio* drive prevalence in natural populations of *Drosophila neotestacea*. *Proc Biol Sci* **280:** 20131397.

Policansky D, Ellison J. 1970. "sex ratio" in *Drosophila pseudoobscura*: Spermiogenic failure. *Science* **169:** 888–889.

Presgraves D. 2008. Drive and sperm: The evolution and genetics of male meiotic drive. In *Sperm biology: An evolutionary perspective* (ed. Birkhead TR, et al.), pp. 471–506. Academic, London.

Presgraves DC, Severance E, Willrinson GS. 1997. Sex chromosome meiotic drive in stalk-eyed flies. *Genetics* **147:** 1169–1180.

Price TAR, Wedell N. 2008. Selfish genetic elements and sexual selection: Their impact on male fertility. *Genetica* **134:** 99–111.

Price TAR, Hodgson D, Lewis Z, Hurst G, Wedell N. 2008a. Selfish genetic elements promote polyandry in a fly. *Science* **322:** 1241–1243.

Price TAR, Bretman AJ, Avent TD, Snook RR, Hurst GD, Wedell N. 2008b. Sex ratio distorter reduces sperm competitive ability in an insect. *Evolution* **62:** 1644–1652.

Price TAR, Hurst GD, Wedell N. 2010a. Polyandry prevents extinction. *Curr Biol* **20:** 471–475.

Price TAR, Lewis Z, Smith DT, Hurst GD, Wedell N. 2010b. Sex ratio drive promotes sexual conflict and sexual coevolution in the fly *Drosophila pseudoobscura*. *Evolution* **64:** 1504–1509.

Price TAR, Lewis Z, Smith D, Hurst G, Wedell N. 2012. No evidence of mate discrimination against males carrying a sex ratio distorter in *Drosophila pseudoobscura*. *Behav Ecol Sociobiol* **66:** 561–568.

Price TAR, Bretman A, Gradilla A, Reger J, Taylor M, Giraldo-Perez P, Campbell A, Hurst G, Wedell N. 2014. Does polyandry control population sex ratio via regulation of a selfish gene? *Proc R Soc Lond B* **281:** 20133259.

Ross BD, Rosin L, Thomae AW, Hiatt MA, Vermaak D, de la Cruz AFA, Imhof A, Mellone BG, Malik HS. 2013. Stepwise evolution of essential centromere function in a *Drosophila* Neogene. *Science* **340:** 1211–1214.

Rutkowska J, Badyaev AV. 2008. Meiotic drive and sex determination: Molecular and cytological mechanisms of sex ratio adjustment in birds. *Phil Trans R Soc Lond B* **363:** 1675–1686.

Sackton TB, Montenegro H, Hartl DL, Lemos B. 2011. Interspecific Y chromosome introgressions disrupt testis-specific gene expression and male reproductive phenotypes in *Drosophila*. *Proc Natl Acad Sci* **108:** 17046–17051.

Sandler L, Novitski E. 1957. Meiotic drive as an evolutionary force. *Am Nat* **41:** 105–110.

Saunders PA, Perez J, Rahmoun M, Ronce O, Crochet PA, Veyrunes F. 2014. XY females do better than the XX in the African pygmy mouse, *Mus minutoides*. *Evolution* doi: 10.1111/evo.12387.

Shakes DC, Neva BJ, Huynh H, Chaudhuri J, Pires-daSilva A. 2011. Asymmetric spermatocyte division as a mechanism for controlling sex ratios. *Nat Commun* **2:** 157.

Shin D, Jin L, Lobo NF, Severson DW. 2011. Transcript profiling of the meiotic drive phenotype in testis of *Aedes aegypti* using suppressive subtractive hybridization. *J Insect Physiol* **57:** 1220–1226.

Sinkins SP, Gould F. 2006. Gene drive systems for insect disease vectors. *Nature Rev Genet* **7:** 427–435.

Stalker HD. 1961. The genetic systems modifying meiotic drive in *Drosophila paramelanica*. *Genetics* **46:** 177.

Stapleton W, Das S, McKee BD. 2001. A role of the *Drosophila homeless* gene in repression of *Stellate* in male meiosis. *Chromosoma* **110:** 228–240.

Sturtevant A, Dobzhansky T. 1936. Geographical distribution and cytology of "sex ratio" in *Drosophila pseudoobscura* and related species. *Genetics* **21:** 473.

Tao Y, Hartl DL, Laurie CC. 2001. Sex-ratio segregation distortion associated with reproductive isolation in *Drosophila*. *Proc Natl Acad Sci* **98:** 13183–13188.

Tao Y, Araripe L, Kingan SB, Ke Y, Xiao H, Hartl DL. 2007a. A sex-ratio meiotic drive system in *Drosophila simulans*. II: An X-linked distorter. *PLoS Biol* **5:** e293.

Tao Y, Masly JP, Araripe L, Ke Y, Hartl DL. 2007b. A sex-ratio meiotic drive system in *Drosophila simulans*. I: An autosomal suppressor. *PLoS Biol* **5:** e292.

Taylor DR, Ingvarsson PK. 2003. Common features of segregation distortion in plants and animals. *Genetica* **117:** 27–35.

Taylor JE, Jaenike J. 2002. Sperm competition and the dynamics of X chromosome drive: Stability and extinction. *Genetics* **160:** 1721–1731.

Taylor JE, Jaenike J. 2003. Sperm competition and the dynamics of X chromosome drive in finite and structured populations. *Ann Zool Fenn* **40:** 195–206.

Taylor DR, Saur MJ, Adams E. 1999. Pollen performance and sex-ratio evolution in a dioecious plant. *Evolution* **53:** 1028–1036.

Thomson GJ, Feldman MW. 1975. Population genetics of modifiers of meiotic drive: IV. On the evolution of sex-ratio distortion. *Theor Pop Biol* **8:** 202–211.

Turner JM. 2007. Meiotic sex chromosome inactivation. *Development* **134:** 1823–1831.

Van Boven M, Weissing FJ. 1999. Segregation distortion in a deme-structured population: Opposing demands of gene, individual and group selection. *J Evol Biol* **12:** 80–93.

Vaz SC, Carvalho AB. 2004. Evolution of autosomal suppression of the sex-ratio trait in *Drosophila*. *Genetics* **166:** 265–277.

Veyrunes F, Chevret P, Catalan J, Castiglia R, Watson J, Dobigny G, Robinson TJ, Britton-Davidian J. 2010. A novel sex determination system in a close relative of the house mouse. *Proc R Soc Lond B* **277:** 1049–1056.

Voelker RA. 1972. Preliminary characterization of "sex ratio" and rediscovery and reinterpretation of "male sex ratio" in *Drosophila affinis*. *Genetics* **71:** 597–606.

Wallace B. 1948. Studies on "sex-ratio" in *Drosophila pseudoobscura*: I. Selection and "sex-ratio." *Evolution* **2:** 189–217.

Ward MA, Burgoyne PS. 2006. The effects of deletions of the mouse Y chromosome long arm on sperm function—Intracytoplasmic sperm injection (ICSI)-based analysis. *Biol Reprod* **74:** 652–658.

Wedell N. 2013. The dynamic relationship between polyandry and selfish genetic elements. *Phil Trans R Soc Lond B* **368:** 20120049.

West SA, Sheldon BC. 2002. Constraints in the evolution of sex ratio adjustment. *Science* **295:** 1685–1688.

Wilkinson GS, Sanchez MI. 2001. Sperm development, age and sex chromosome meiotic drive in the stalk-eyed fly, *Cyrtodiopsis whitei*. *Heredity* **87:** 17–24.

Wilkinson GS, Presgraves DC, Crymes L. 1998. Male eye span in stalk-eyed flies indicates genetic quality by meiotic drive suppression. *Nature* **391:** 276–279.

Wilkinson GS, Swallow JG, Christensen SJ, Madden K. 2003. Phylogeography of *sex ratio* and multiple mating in stalk-eyed flies from southeast Asia. *Genetica* **117:** 37–46.

Wilkinson G, Johns P, Kelleher E, Muscedere M, Lorsong A. 2006. Fitness effects of X chromosome drive in the stalk-eyed fly, *Cyrtodiopsis dalmanni*. *J Evol Biol* **19:** 1851–1860.

Wood RJ, Newton ME. 1991. Sex-ratio distortion caused by meiotic drive in mosquitoes. *Am Nat* **137:** 379–391.

Wood R, Ouda N. 1987. The genetic basis of resistance and sensitivity to the meiotic drive gene D in the mosquito *Aedes aegypti* L. *Genetica* **72:** 69–79.

Wu CI. 1983a. The fate of autosomal modifiers of the Sex-Ratio trait in *Drosophila* and other sex-linked meiotic drive systems. *Theor Pop Biol* **24:** 107–120.

Wu CI. 1983b. Virility deficiency and the sex-ratio trait in *Drosophila pseudoobscura*: I. Sperm displacement and sexual selection. *Genetics* **105:** 651–662.

Wu CI. 1983c. Virility deficiency and the Sex-Ratio trait in *Drosophila pseudoobscura*: II. Multiple mating and overall virility selection. *Genetics* **105:** 663–679.

Wu CI, Beckenbach AT. 1983. Evidence for extensive genetic differentiation between the sex-ratio and the standard arrangement of *Drosophila pseudoobscura* and *D. persimilis* and identification of hybrid sterility factors. *Genetics* **105:** 71–86.

Yang YY, Lin FJ, Chang HY. 2004. Sex ratio distortion in hybrids of *Drosophila albomicans* and *D. nasuta*. *Zool Stud* **43:** 622–628.

Yasuno Y, Inoue YH, Yamamoto MT. 2013. Elimination of Y chromosome-bearing spermatids during spermiogenesis in an autosomal *sex-ratio* mutant of *Drosophila simulans*. *Genes Genet Syst* **88:** 113–126.

Yoshida K, Kitano J. 2012. The contribution of female meiotic drive to the evolution of neo-sex chromosomes. *Evolution* **66:** 3198–3208.

Zhou J, Sackton TB, Martinsen L, Lemos B, Eickbush TH, Hartl DL. 2012. Y chromosome mediates ribosomal DNA silencing and modulates the chromatin state in *Drosophila*. *Proc Natl Acad Sci* **109:** 9941–9946.

Sex-Biased Gene Expression and Sexual Conflict throughout Development

Fiona C. Ingleby, Ilona Flis, and Edward H. Morrow

Evolution, Behaviour and Environment Group, School of Life Sciences, University of Sussex, John Maynard Smith Building, Falmer, Brighton BN1 9QG, United Kingdom

Correspondence: f.ingleby@sussex.ac.uk

Sex-biased gene expression is likely to account for most sexually dimorphic traits because males and females share much of their genome. When fitness optima differ between sexes for a shared trait, sexual dimorphism can allow each sex to express their optimum trait phenotype, and in this way, the evolution of sex-biased gene expression is one mechanism that could help to resolve intralocus sexual conflict. Genome-wide patterns of sex-biased gene expression have been identified in a number of studies, which we review here. However, very little is known about how sex-biased gene expression relates to sex-specific fitness and about how sex-biased gene expression and conflict vary throughout development or across different genotypes, populations, and environments. We discuss the importance of these neglected areas of research and use data from a small-scale experiment on sex-specific expression of genes throughout development to highlight potentially interesting avenues for future research.

As most of the genome is shared between males and females, differences in gene expression between the sexes are thought to underlie much of the sexual dimorphism observed in nature (Ellegren and Parsch 2007). These sex-biased genes are either expressed in one sex only (sex specific or sex limited) or expressed more strongly in one sex than the other (sex enriched). Patterns of sex-biased gene expression are widely documented and are often associated in the literature with sexual conflict (e.g., Connallon and Knowles 2006; Ellegren and Parsch 2007). The rationale behind this is based on two conceptual premises. First, sexual conflict occurs when fitness optima differ between the sexes for a shared trait, generating sexually antagonistic selection that will favor trait evolution in opposite directions between the sexes. Second, sex differences in gene expression could potentially alleviate such conflict by facilitating the evolution of sexual dimorphism.

A common assumption made, however, is that sexual dimorphism can be used as an indicator or proxy of ongoing sexual conflict, and at present, this has not been extensively studied empirically. Crucially, very few empirical studies have measured both gene expression levels and sex-specific fitness, and as a result, we have a poor understanding of how gene expression might mediate conflict. Here, we review the body of research documenting genome-wide patterns of sex-biased gene expression across a range of species, and we highlight the need for more studies that measure sex-specific fitness as

Cite this article as *Cold Spring Harb Perspect Biol* doi: 101101/cshperspect.a017632

well as gene expression. We argue that only by considering both of these aspects can patterns of sex-biased gene expression be interpreted within the context of sexual conflict.

In addition, very little is known about the progression of sex-biased gene expression throughout development and how these changes might relate to sex-specific fitness at each stage (Ellegren and Parsch 2007). Sexual dimorphism generally increases throughout development, and it has been suggested that the relationship between sexual dimorphism and sexual conflict might vary throughout ontogeny as well. Theory predicts that conflict will be strongest as the sexes differentiate and the evolutionary interests of males and females diverge (Badyaev 2002; Cox and Calsbeek 2009), yet empirical studies addressing this idea are lacking. Developmental studies on model organisms have provided in-depth examination of development (e.g., Arbeitman et al. 2002) and of the sex-specific expression of genes that are involved with particular developmental pathways of interest (Rinn and Snyder 2005), but ontogenetic studies of genome-wide sex-biased gene expression are few. We therefore consider studies of genome-wide sex-biased gene expression throughout development and present the results of a small-scale experiment on this subject.

Finally, we discuss how future research on sexual conflict could benefit from studies that bring together analysis of sex-biased gene expression, changes in the transcriptome throughout development, and sex-specific fitness to give a more comprehensive understanding of the relationship between sex-biased gene expression and sexually antagonistic selection.

QUANTIFYING SEX-BIASED GENE EXPRESSION

Genome-wide sex-biased gene expression has been studied in a wide range of species, thanks in large part to the increasing availability of high-throughput transcriptomic techniques such as microarrays and RNA sequencing (Ellegren and Parsch 2007). The research summarized in Table 1 gives some indication of the breadth of species and questions that have been studied. Whereas a lot of these studies focus on model species, it is clear that this area of research spans a diverse range of animal phyla and that high-throughput transcriptomic techniques are beginning to be applied to nonmodel species.

Perhaps unsurprisingly, there is no general consensus as to what extent the transcriptome is sex biased, and research so far has shown wide variation between species. At one end of the scale, for example, only about 2% of the transcriptome of the marine snail *Littorina saxatilis* was found to be sex biased (Martinez-Fernandez et al. 2010), whereas in *Drosophila melanogaster*, sex-biased expression for as much as 90% of transcripts has been shown (Ayroles et al. 2009; Innocenti and Morrow 2010; Wayne et al. 2012). More interestingly, the extent of sex bias can vary even between studies of the same species. There are a number of potential sources of this variation, including, but not limited to, tissue specificity of gene expression, developmental stage, intraspecific genetic and environmental variation, and the experimental design and analytical techniques specific to each study. Table 1 summarizes some of the studies that have examined each of these sources of variation.

First, tissue-specific sex-biased expression has been examined in several species. For example, in mice, patterns of sex-biased gene expression vary between different tissues (Rinn et al. 2004; Yang et al. 2006). In zebrafish, *Danio rerio*, sex bias in the transcriptome of the brain tissue was found to be quite low (~2%) (Santos et al. 2008), but a different study identified sex biases across 38% of the transcriptome that were attributed to differences in gene expression in male and female reproductive tissues (Small et al. 2009). Similarly, in the silkworm *Bombyx mori*, sex-biased gene expression varied between 2% and 30%, depending on the tissue examined (Xia et al. 2007). A number of studies have only looked at a particular tissue of interest (e.g., Malone et al. 2006; Scholz et al. 2006; Grosse-Wilde et al. 2011; Naurin et al. 2011; Pitts et al. 2011), and so to what extent the sex-biased gene expression identified in these studies is repre-

Table 1. Examples of studies with original data on sex-biased patterns of genome-wide gene expression

Species	Developmental stage(s)	Experimental design	Criteria for sex bias	Evidence for sex-biased gene expression	References
Mammals					
Mus musculus	Embryos	Microarray (5m + 5f)	$P < 0.05$-fold change > 1.1	<1% of transcripts showed sex-biased gene expression in brain tissue of embryos before gonadal formation	Dewing et al. 2003
M. musculus	Adults	Microarray (3m + 3f)	$P < 0.001$-fold change > 3	Sex-biased gene expression in kidney, liver, and reproductive tissues	Rinn et al. 2004
M. musculus	Adults	Microarray (165m + 169f)	$P < 0.01$	13%–70% of transcripts were sex biased, depending on the tissue	Yang et al. 2006
M. musculus	Adults	(114m + 93f)	$P < 0.05$	31% of eQTLs showed sex-biased expression	Bhasin et al. 2008
Primates[a]	Adults	RNA-seq (3m + 3f per species)	$P < 0.05$	Sex-specific gene expression and splicing and enrichment of sex-biased genes with RNA splicing functions	Blekhman et al. 2010
Fish					
Danio rerio	Adults	Microarray (3m + 3f)	Fold change	<1% of transcripts showed female-biased expression in both germline and somatic tissue	Wen et al. 2005
D. rerio	Adults	Microarray (6m + 5f)	$P < 0.05$-fold change > 1.2	~2% of transcripts in the adult brain tissue were sex biased	Santos et al. 2008
D. rerio	Adults	Microarray (3m + 3f)	$P < 0.05$	38% of transcripts were sex biased (all evidence of sex bias was attributable to differences between reproductive tissues)	Small et al. 2009
Gasterosteus aculeatus	Adults	Microarray (27m + 8f)	$P < 0.05$	~10% of transcripts were sex biased, concentrated on sex chromosomes	Leder et al. 2010
Poecilia reticulata	Juveniles	RNA-seq (3m + 3f)	$P < 0.05$	Sex-biased gene expression in juvenile brain and whole-body samples	Fraser et al. 2011
Birds					
Gallus gallus	Embryos	Microarray (12m + 12f)	$P < 0.0005$	Sex-biased gene expression (mostly male biased) in embryo brain tissue; sex biases present before gonad differentiation	Scholz et al. 2006

Continued

Table 1. *Continued*

Species	Developmental stage(s)	Experimental design	Criteria for sex bias	Evidence for sex-biased gene expression	References
G. gallus	Early embryo, late embryo, and adults	Microarray (3m + 3f per stage)	$P < 0.05$-fold change > 2	Sex-biased gene expression in the gonads increased throughout development from ~10% to ~50% of transcripts	Mank et al. 2010
Passerines[b]	Adults	Microarray (6–12m + 6–12f per species)	$P < 0.03$	~2% sex-biased gene expression in the brain tissue of each species, mostly male biased	Naurin et al. 2011
Amphibians					
Xenopus spp.[c]	Adults	Microarray (~4m + 4f per species)	$P < 0.05$	Extensive male-biased expression in gonad tissue of both species, with ~70% of the transcriptome showing sex bias	Malone et al. 2006
Drosophila spp.					
Drosophila spp.[d]	Adults	Microarray (3m + 3f per species)	$P < 0.03$	Many genes with divergent expression between species were also sex biased	Ranz et al. 2003
D. spp.[e]	Adults	Microarray (min. 4m + 4f per species)	$P < 0.01$	Patterns of sex-biased (especially male biased) gene expression have diverged between closely related *Drosophila* species	Zhang et al. 2007
D. spp.[f]	Adults	Microarray (3m + 3f)	$P < 0.01$	~80% of the transcriptome was sex biased in each species, and >80% of expression divergence between species was because of sex-specific expression changes	Jiang and Machado 2009
D. melanogaster	Adults (two age classes)	Microarray (12m + 12f across two genotypes)	$P < 0.0001$	Variation in gene expression between sexes, genotypes, and sex x genotype interaction (7% of genes were sex biased)	Jin et al. 2001
D. melanogaster	Larvae, pupae, and adults	Microarray (2m + 2f)	$P < 0.01$-fold change > 2-	A small number of somatic transcripts showed sex-biased expression	Arbeitman et al. 2002
D. melanogaster	Adults	Microarray (2m + 2f)	Fold change > 2	Extensive sex-biased gene expression, with underrepresentation of male-biased genes on the X chromosome	Parisi et al. 2003
D. melanogaster	Adults	Microarray (32m + 32f across eight lines)	$P < 0.05$	~50% of transcripts showed sex-biased expression, and there was evidence of sex-specific splicing patterns	McIntyre et al. 2006

D. melanogaster	Adults	Microarray (80m + 80f across 40 lines)	$P < 0.001$	~88% of the transcriptome was sex biased, and there was sex-specific variation in multiple fitness-related traits	Ayroles et al. 2009
D. melanogaster	Adults (two female age classes)	Microarray (8m + 8f per time point)	Fold change > 2	~50% of transcripts showed sex-biased expression (mostly male biased), which varied throughout female oogenesis	Baker and Russell 2009
D. melanogaster	Adults	RNA-seq (2m + 2f)	Fold change > 2	~56% of transcripts were sex biased in wild-type flies, and evidence for sex-specific splicing and chromatin structure in adult gonad tissue	Gan et al. 2010
D. melanogaster	Adults	Microarray (60m + 60f across 15 lines)	$P < 0.001$	~90% of transcripts showed sex-biased expression, whereas 8% of the transcriptome was under sexually antagonistic selection	Innocenti and Morrow 2010
D. melanogaster	Adults	Microarray (24m + 24f across two genotypes)	$P < 0.01$	Evidence for sex-biased gene expression was stronger in adults from a high-quality diet than in adults from a low-quality diet	Wyman et al. 2010
D. melanogaster	Adults	RNA-seq (6m + 6f)	$P < 0.05$	~17% of transcripts showed sex-biased expression, and there was evidence of sex-specific splicing	Chang et al. 2011
D. melanogaster	Adults (pooled-sex larval and pupal samples)	RNA-seq and microarray (3m + 3f)	$P < 0.001$	~30% of transcripts were sex-biased across the adult transcriptome, with more male- than female-biased expression	Graveley et al. 2011
D. melanogaster	Adults	RNA-seq (4m + 4f)	$P < 0.05$	Low levels of sex-biased gene expression in the brain tissue; X chromosome was enriched with male-biased genes	Catalan et al. 2012
D. melanogaster	Adults	Microarray (2m + 2f per genotype)	$P < 0.001$	~90% of the transcriptome showed sex-biased expression, and there were sex-specific patterns of inheritance of gene expression	Wayne et al. 2012
D. melanogaster	Larvae and prepupae	RNA-seq (3m + 3f)	$P < 0.05$-fold change > 2	>50% of transcripts in larvae and prepupae showed sex-biased expression in the pregonad tissue	Perry et al. 2014
Other insects					
Anopheles gambiae	Adults	Microarray (3m + 3f)	$P < 0.001$	22% of transcripts were sex biased and changes in gene expression were also observed after females had a bloodmeal	Marinotti et al. 2006

Continued

Table 1. *Continued*

Species	Developmental stage(s)	Experimental design	Criteria for sex bias	Evidence for sex-biased gene expression	References
A. gambiae	Adults	Microarray (4m + 4f)	$P < 0.05$-fold change > 2	54% of transcripts were sex biased, and adult gene expression was tissue specific	Baker et al. 2011
A. gambiae	Larvae, pupae, and adults	Microarray (3m + 3f per stage)	Fold change > 1.7	Widespread sex-biased gene expression throughout development (1%–28% sex bias from larvae to adults)	Magnusson et al. 2011
A. gambiae	Adults	RNA-seq (3m + 3f)	Fold change	Sex-biased gene expression and evidence for tissue specificity of sex bias in chemosensory tissue	Pitts et al. 2011
Bombyx mori	Larvae	Microarray (2m + 2f minimum)	$P < 0.01$	2%–30% of transcripts were sex biased, depending on the tissue, with variation between tissues in the extent of male and female bias	Xia et al. 2007
B. mori	Larvae and pupae	Microarray (2m + 2f minimum)	Fold change > 2	Sex-biased expression increased throughout development (from ~2%–13%), with more male bias than female bias, and enrichment of sex-biased genes on the sex chromosomes	Zhao et al. 2011
Manduca sexta	Adults	Microarray (4m + 4f)	$P < 0.05$	More evidence of female-biased than male-biased gene expression in antennae tissue	Grosse-Wilde et al. 2011
Solenopsis spp.[g]	Pupae and adults	Microarray (five replicates per caste per stage)	$P < 0.001$	~20% of genes were sex biased between castes (males, workers, and queens), with more male-biased genes than others, and variation in caste-biased expression between stages	Ometto et al. 2011
Tribolium castaneum	Adults	Microarray (4m + 4f)	$P < 0.01$	~20% of the transcriptome was sex biased, with enrichment of female-biased genes on the X chromosome	Prince et al. 2010
Nematodes					
Brugia malayi	Adults	RNA-seq (no replication)	$P < 0.01$	Enrichment of sex-biased expression in the germline, and ~20% of transcripts overall showed sex bias	Choi et al. 2011
Caenorhabditis elegans	Multiple larval stages and adults	Microarray (4m + 4h)	$P < 0.05$-fold change > 1.5	Differential gene expression between males and hermaphrodites throughout development in ~10% of the transcripts studied	Jiang et al. 2001

 Cite this article as *Cold Spring Harb Perspect Biol* doi: 10.1101/cshperspect.a017632

C. elegans	Multiple larval stages and adults	Microarray (4m + 4h per stage)	$P < 0.01$-fold change > 2	~1% of transcripts differed in expression between males and hermaphrodites throughout development in the somatic and germline tissue	Reinke et al. 2004
C. elegans	Multiple larval stages	Microarray (4m + 4h per stage)	$P < 0.01$-fold change > 2	More male-biased than hermaphrodite-biased gene expression, and an increase in differential gene expression throughout larval development	Thoemke et al. 2005
Platyhelminthes					
Schistosoma mansoni	Adults	Microarray (5m + 5f)	Fold change	Evidence for low levels of sex-biased gene expression in adults	Fitzpatrick et al. 2005
Crustaceans					
Daphnia pulex	Adults	Microarray (8m + 8f)	$P < 0.05$-fold change > 2	50% of assayed transcripts showed sex bias between adult males and females	Eads et al. 2007
Molluscs					
Littorina saxatilis	Adults	cDNA-AFLP (6m + 6f)	$P < 0.05$	1.8% of transcripts showed sex-biased gene expression, along with low levels of phenotypic sexual dimorphism	Martinez-Fernandez et al. 2010

This summary excludes studies which reanalyze previous data or focus only on a subset of genes. Because of the different terminology used to refer to sex-biased gene expression, this list is unlikely to be exhaustive, but it gives a good indication of the range of species studied, the widespread evidence for sex-biased gene expression, and the increase in research in this area in recent years. Studies that have examined males and females from more than one developmental stage are highlighted in dark gray, and studies that have provided data giving some indication of how gene expression relates to fitness in males and females are highlighted in light gray. "Experimental design" gives detail on the method used to quantitate gene expression and the number of male (m), female (f), and, where relevant, hermaphrodite (h) biological replicates. "Criteria for sex bias" indicates, where relevant, (i) the statistical level of significance for differential expression between the sexes and (ii) the cut-off level of fold change in expression between the sexes.

[a]*Homo sapiens, Macaca mulatta,* and *Pan troglodytes.*

[b]*Sylvia communis* and *Taeniopygia guttata.*

[c]*Xenopus laevis* and *Xenopus muelleri.*

[d]*D. melanogaster* and *Drosophila simulans.*

[e]*Drosophila. ananassae, D. melanogaster, Drosophila mojavensis, Drosophila pseudoobscura, D. simulans, Drosophila virilis,* and *Drosophila yakuba.*

[f]*Drosophila persimilis, D. pseudoobscura,* and *Drosophila pseudoobscura bogotana.*

[g]*Solenopsis invicta* and *Solenopsis richteri.*

sentative of the whole animal is yet to be determined.

Research has also found intraspecific genetic variation for patterns of sex-biased gene expression through a combination of quantitative genetics and microarray experiments in *D. melanogaster* (Jin et al. 2001; Ayroles et al. 2009; Innocenti and Morrow 2010). An environmental component of variation in sex-biased gene expression has also been shown in *D. melanogaster*, where there was more evidence of sex-biased gene expression in populations reared on a high-quality diet than in populations on a poor diet (Wyman et al. 2010). This genetic and environmental variation implies that the potential for sex-biased gene expression to evolve could vary between different populations and environments. It will be useful to examine how this variation relates to sexual conflict, and if it affects the potential for sex-biased expression to resolve conflict. Indeed, a few studies have found interactions between sexually antagonistic selection and environmental variation (e.g., Kwan et al. 2008; Long et al. 2012), and so whether or not sex-biased gene expression underlies such interactions will be an interesting direction for future research.

Finally, in terms of the experimental design and analysis, it is apparent from the summary in Table 1 that sample size and analytical criteria for identifying sex-biased expression can significantly impact on the results obtained. Clearly, sample size will affect the power of the experiment to detect significant sex-biased expression. In addition, studies have used a wide variety of analytical criteria for identifying sex-biased transcripts, involving some combination of statistical criteria (P values adjusted for false discovery rate) and relative expression levels between sexes (fold-change ratios). Usefully, Yang et al. (2006) present their data analyzed with different criteria for identifying sex bias. They examined gene expression in four mouse tissues and showed that whereas statistical criteria alone identified 72% of transcripts as sex biased in liver tissue, imposing just a 1.2 minimum fold-change threshold meant that only 13.5% of transcripts were identified as sex biased, and a fold-change threshold of 3 reduced this to only 0.5%. This trend was similar in the other tissues analyzed (Yang et al. 2006), calling into question the biological significance of fold-change criteria, and also highlighting the difficulty in comparing results between studies with different analytical methods.

To illustrate this point further, consider two *D. melanogaster* studies, Ayroles et al. (2009) and Innocenti and Morrow (2010), in which large data sets of a comparable sample size were produced, and each analysis used the same statistical criteria for identifying sex-biased transcripts. As such, the results can be compared more reliably, and both studies came to the conclusion that ~90% of the *D. melanogaster* transcriptome exhibited sex-biased expression. In fact, the correlation between levels of sex-biased gene expression identified in each study was exceptionally high ($r^2 = 0.96$) (see Innocenti 2011). The implementation of a widely accepted standard criterion for identifying sex-biased gene expression would clearly facilitate comparisons between studies (Ellegren and Parsch 2007).

SEX-BIASED GENE REGULATION AND EVOLUTION

Studies documenting patterns of sex-biased gene expression also shed some light on the potential mechanisms that might regulate sex biases. For example, sex-biased gene expression could result from alternative splicing of gene transcripts. In *D. melanogaster*, 95% of known splicing regulators were found to exhibit significant sex-biased expression (McIntyre et al. 2006; Telonis-Scott et al. 2009). Furthermore, increased use of RNA-sequencing technology in recent years has enabled the direct quantification of splicing variants, and as a result, sex-specific splicing patterns have been shown in *D. melanogaster* (Gan et al. 2010; Chang et al. 2011) and in a group of primate species (Blekhman et al. 2010). In the primate study, they also showed the enrichment of sex-biased genes with genes with known RNA splicing functions, providing further empirical support for the importance of sex-specific splicing in controlling sex-biased gene expression.

 Cite this article as *Cold Spring Harb Perspect Biol* doi: 10.1101/cshperspect.a017632

Another mechanism that might regulate sex-biased gene expression involves short (~22 nucleotides) RNA molecules known as microRNAs (miRNAs) (Morgan and Bale 2012). These miRNAs target mRNAs with complementary sequences and bind to them to regulate their expression, potentially by tagging the mRNA for destruction, or by preventing their translation (Pasquinelli 2002). Relatively little is known about sex-biased expression of miRNAs, although RNA-sequencing in the zebra finch, *Taeniopygia guttata*, has found genome-wide sex-biased expression of some miRNAs (Luo et al. 2012), and patterns of sex-biased miRNA expression have also been found in halibut (Bizuayehu et al. 2012), silkworm (Liu et al. 2010), and schistosomes (Marco et al. 2013). Taken together with the evidence for sex-biased miRNA expression in *D. melanogaster* from our unpublished data described below, it seems that further research on the sex-specific functions of miRNAs could shed some light on the mechanisms that control sex-biased gene expression. At present, very little is known about miRNA functions, and even less is known about sex-biased expression patterns of miRNAs. These small RNAs could be extremely important, however, and are thought to have a role in the regulation of gene expression in sex-biased diseases (Sharma and Eghbali 2014), but far more data are needed before this can be discussed in detail.

Sex-biased expression could also be facilitated by translocation of genes from autosomes to sex chromosomes (or alternatively, gene translocation from autosomes to sex chromosomes could promote sex-biased gene expression). In species with chromosomal (XY or ZW) sex determination, one sex is hemizygous for the X or Z chromosome, and so genes on these chromosomes will be subject to different dynamics of sex-specific selection than genes on autosomes. As a result, it is often predicted that sex chromosomes will harbor more sexually antagonistic alleles than autosomes (Rice 1984; Ellegren and Parsch 2007; Wyman et al. 2012; but see Fry 2010). Some empirical research has approached this idea by testing for enrichment of the sex chromosomes with sex-biased genes, but it may be the case that the model predictions, which concern the relative fitness of different alleles, do not directly translate to empirical work that examines sex-biased gene expression. Indeed, the data on sex-biased gene expression on the sex chromosomes are mixed. For example, the X chromosome in flour beetles is enriched with female-biased genes (Prince et al. 2010), and the X chromosome of *Caenahabditis elegans* is almost completely lacking any genes involved in sperm production (Reinke et al. 2000). In *Drosophila*, the X chromosome of both *D. melanogaster* and *D. simulans* is depleted for male-biased genes but enriched for female-biased genes (Parisi et al. 2003; Ranz et al. 2003). This pattern is apparently not limited to protein-coding genes, as the X chromosome is also enriched with male-biased miRNAs (Zhang et al. 2010). In the silkworm *B. mori*, where females are the heterogametic sex with ZW sex chromosomes, the Z chromosome is enriched with male-biased genes (Zhao et al. 2011). However, these patterns are not entirely consistent. In *Drosophila*, there is some evidence that the trend might be tissue dependent, as Catalan et al. (2012) found that the X chromosome of *D. melanogaster* was enriched for genes with male-biased expression in brain tissue. In mice, although one study found the X chromosome to be enriched with female-biased genes (Khil et al. 2004), another found a substantial number of testis-specific genes on the X (Wang et al. 2001).

If genes under sexual conflict are translocated to the sex chromosomes this might facilitate the evolution of sex-biased gene expression and sexual dimorphism and allow a release from conflict. The evidence that sex chromosomes are enriched with sex-biased genes is therefore compelling, but to directly address the theory, the chromosomal distribution of sex-biased genes needs to be associated with sexually antagonistic selection on these genes. In a quantitative genetic framework, studies have found that the X chromosome in *D. melanogaster* harbors a lot of sexually antagonistic genetic variation (Gibson et al. 2002; Pischedda and Chippindale 2006). However, to our knowledge, only two studies have explicitly addressed the relationship between sex-specific fitness and sex-specific gene expression phenotypes (Innocenti and Morrow 2010; Abbott et al. 2013). Because direct measurement of how these are linked is key to understanding the adaptive significance of sex-biased gene expression, more research in this area will be useful.

Several studies have considered the molecular evolution of sex-biased genes as a means to determine their adaptive significance. In both *Drosophila* species and the crustacean *Daphnia pulex*, male-biased genes tend to evolve more quickly than female-biased genes (Meiklejohn et al. 2003; Eads et al. 2007), suggesting that male-biased genes are under stronger selection and might contribute more to adaptive evolution. Consistent with this, there is also evidence in *Drosophila* that female-biased genes exhibit higher levels of pleiotropy than male-biased genes (Assis et al. 2012), which could limit the evolutionary potential of such genes. Indeed, it has also been shown that the majority of sex-biased expression in *Drosophila* seems to have evolved through adaptive changes in the male transcriptome (Connallon and Knowles 2006). The adaptive significance of sex-biased genes is further highlighted by studies of species divergence between *Drosophila* species. For example, 83% of the transcriptome changes between *D. simulans* and *D. melanogaster* are accounted for by sex-biased gene expression (Ranz et al. 2003), and male-biased genes have a strong and consistent signal of positive selection, indicating adaptive evolution, which is weaker in female-biased genes and lacking in non-sex-biased genes (Pröschel et al. 2006). Similarly, a large study of seven species of *Drosophila* found that male-biased genes showed the greatest divergence between species, in terms of both expression and sequence (Zhang et al. 2007), and another study between three recently diverged species in the *D. pseudoobscura* clade showed that sex-specific changes in expression explained more than 80% of transcriptome variation between species (Jiang and Machado 2009). Together, this research suggests that sex-biased gene expression, and particularly male-biased genes, play an important role in species divergence, which is interesting given the potential for sexually antagonistic selection to fuel species divergence (Parker and Partridge 1998). However, as pointed out by Rice et al. (2005), this molecular evidence is not enough by itself to conclusively show a role of sexual conflict in species divergence, and so future research will need to examine selection on sex-biased genes and how this could drive species divergence.

IDENTIFYING SEXUALLY ANTAGONISTIC GENES

It is tempting to use sex-biased gene expression (or indeed any sexual dimorphism) to make inferences about sexual conflict. However, sexual dimorphism alone does not necessarily mean a trait is under sexual conflict. To draw conclusions about sexual conflict, data on both dimorphism and sex-specific selection are needed (Cox and Calsbeek 2009). For instance, if the extent of dimorphic gene expression matches the difference between the sex-specific fitness optima, then sex-biased gene expression may signal a resolved conflict, but if not, then it is likely that there is ongoing sexual conflict despite the sex differences in gene expression. It is also important to recognize that in some cases, sexual dimorphism may have arisen without sexual conflict. For example, in butterflies, sexually dimorphic wing patterns are known to have evolved through traits that have initially arisen in one sex only (Oliver and Monteiro 2011).

Similarly, sexually monomorphic traits do not necessarily indicate an absence of sexual conflict—if fitness optima differ between the sexes for traits that do not exhibit dimorphism, then sexual conflict will exist. As such, identifying sex-biased gene expression is not enough on its own to discriminate between resolved or ongoing sexual conflict (Bonduriansky and Chenoweth 2009; Cox and Calsbeek 2009; Pennell and Morrow 2013).

Although the body of research documenting sex-biased gene expression is growing rapidly, very few of these studies have also looked at sex-specific fitness. In fact, this seems to have only been directly considered by two *D. melanogaster* studies. Ayroles et al. (2009) measured sex-biased gene expression and related this to sex-specific phenotypic variation in fitness-related traits such as lifespan, locomotor response, and mating speed. They identified modules of genes that were associated with the phenotypic variation and that were therefore likely to be under sexual conflict. Innocenti and Morrow (2010) also reported widespread sex bias in the *D. melanogaster* transcriptome, and tested how this was associated with male and female fitness using line-specific values of reproductive success. They identified about 90% of transcripts as being sex biased, whereas only 8% were subject to sexually antagonistic selection (the X chromosome being enriched for them). This indicates that almost a tenth of the transcriptome harbored sexually antagonistic genetic variation, but it also reveals a majority of genes with sex-biased expression that are not currently subject to sexually antagonistic selection during the adult stage. It is worth noting that because only adults were studied, these results could have underestimated the full extent of sexual conflict throughout an individual's life cycle. Possibly, some transcripts might only experience sexual conflict at earlier life stages, which will have been overlooked in this study. We discuss this possibility in greater depth in the next section.

In the sexual conflict literature more generally, sexually antagonistic genetic variation has been identified in a number of studies. For example, a quantitative genetic study of *D. melanogaster* showed a negative genetic correlation between sex-specific adult fitness, suggesting ongoing sexual conflict (Chippindale et al. 2001). Experimental evolution in which selection on females was suppressed for several generations resulted in masculinized populations with high male fitness and low female fitness, again indicating extensive sexually antagonistic genetic variation (Prasad et al. 2007; Abbott et al. 2010). However, similar to the studies looking at transcriptome variation, there is no clear relationship between dimorphism and conflict. Although sexually antagonistic selection on highly dimorphic traits has been found in some studies (e.g., Pischedda and Chippindale 2006; Long and Rice 2007; Bedhomme et al. 2008; Cox and Calsbeek 2009), in other studies, there is no evidence of sexually antagonistic selection on dimorphic traits (Cox and Calsbeek 2009; Bedhomme et al. 2011). These studies highlight further that sexual dimorphism cannot be equated to sexual conflict without additional information on sex-specific patterns of selection.

SEXUAL CONFLICT AND SEX-BIASED GENE EXPRESSION THROUGHOUT DEVELOPMENT

The relationship between sexual conflict and sexual dimorphism is, in fact, likely to vary throughout development. As stated briefly in the introduction, a general prediction has been made that whereas sexual dimorphism usually increases throughout development, as the sexes differentiate in morphology and behavior (Badyaev 2002; Cox and Calsbeek 2009), sexually antagonistic selection is likely to be strongest at points of development when distinct male and female traits are specified and produced. However, in terms of empirical work to support this idea, very few studies have examined the progression of sexual dimorphism and sexual conflict throughout development. Although for particular genetic pathways (e.g., sexual differentiation), sex-specific gene expression throughout development has been studied in great detail (Rinn and Snyder 2005), a genome-wide approach will be important for studies of sexual conflict, as highlighted above. Table 1 shows that many genome-wide studies of sex-biased gene expression focus on only one developmental stage. This is perhaps not surprising, because quantifying the entire transcriptome for both sexes in multiple developmental stages is, of course, no small feat, in terms of both logistics and resources. However, there are some studies that have achieved this level of detail to some extent in *D. melanogaster* (Arbeitman et al. 2002; Perry et al. 2014), the mosquito *Anopheles gambiae* (Magnusson et al. 2011), *B. mori* (Zhao et al. 2011), fire ants (Ometto et al. 2011), and *C. elegans* (Reinke et al. 2000, 2004; Jiang et al. 2001; Thoemke et al. 2005). Sex-biased gene expression appears to be dynamic throughout development in a number of species, and so focusing on the expression profile for a single developmental stage clearly risks overlooking important variation.

Furthermore, unpublished preliminary data from our laboratory clearly shows patterns of developmental changes in sex-biased expression of both mRNA and miRNA in *D. melanogaster.* We processed two biological replicates of each sex from each of three developmental stages (third-instar larvae, pupae, and adults) on (1) GeneChip *Drosophila* Genome 2.0 Affymetrix microarrays, and (2) GeneChip miRNA 3.0 Affymetrix microarrays (deposited in the GEO database, accession number GSE57322). Both mRNA and miRNA expression data were analyzed in R v.2.15.2 (www.R-project.org), using several packages on the BioConductor platform (www.bioconductor.org) (Gentleman et al. 2004), and sex-biased gene expression was identified using statistical criteria ($P < 0.01$ after correction for false discovery rate [Benjamini and Hochberg 1995]). Full details of the analysis can be provided upon request to the corresponding author (F.C.I.). We identified extensive sex-biased mRNA expression: 2638 transcripts out of the total 3754 (70%) were sex-biased during at least one developmental stage, although when comparing this to larger-scale experiments, it is likely to be an underestimate (Ayroles et al. 2009; Innocenti and Morrow 2010). Furthermore, at each developmental stage, most (in excess of 80%) of the sex-biased mRNA transcripts were male biased, consistent with previous work on *D. melanogaster* adults (Graveley et al. 2011). We also identified an increase in the overall extent of sex-biased mRNA expression throughout development, from 34% of the total transcripts being sex biased in larvae to 45% in pupae and 64% in adults. A similar trend was found in *A. gambiae* larvae, pupae, and adults (Magnusson et al. 2011).

Our preliminary data also highlight an interesting question that will have been overlooked by research that focuses on only one developmental stage: Are the same genes under sexually antagonistic selection throughout development or does selection vary? Although further research that combines measurement of gene expression and fitness throughout development will be needed to address this question fully, our results nonetheless indicate stage specificity of sex-biased expression. Approximately two-thirds of the sex-biased mRNAs we identified were either only transiently sex biased throughout development or, in some cases, changed throughout development from being male biased to female biased (or vice versa). These patterns could be explained by differences in timing of key developmental events between males and females, or they could reflect fluctuations in sexual conflict throughout development.

We also identified significant sex-biased miRNA expression: 39 miRNAs out of 213 in total (18%) were sex biased at some point during development. Each of these sex-biased miRNAs exhibited sex-biased expression in adults, but sex-biased miRNA expression was lower in pre-adults (with 5% and 4% of the total miRNAs being sex biased in larvae and pupae, respectively). Genome-wide sex biases in miRNA expression have also been identified in zebra finches (Luo et al. 2012), silkworm (Liu et al. 2010), schistosomes (Marco et al. 2013), and halibut brain and gonad tissue (Bizuayehu et al. 2012). In halibut, patterns of sex-biased expression suggest a role of miRNAs in sexual development. However, research on miRNA expression and function is a relatively new area. Although previous research has supposed that miRNAs suppress translation of target mRNAs (Pasquinelli 2002), it is also known that they can up-regulate gene expression (e.g., Carroll et al. 2012) and that networks of interacting miRNAs and mRNAs might exist. This suggests that the relationship between miRNA and mRNA could be more complex than previously assumed (Salmena et al. 2011). The small sample size of this experiment prevents us from looking at this in more detail, but clearly, sex-specific interactions between miRNAs and mRNAs could be a valuable area for future research.

Together, these studies reveal potentially interesting patterns of sex-biased gene expression fluctuating throughout development not only in protein-coding mRNA, but also in miRNAs with potential regulatory roles. More research is needed to fully understand the interactions between mRNAs and miRNAs, and how these relate to patterns of selection, thus, it will be interesting to examine sex-specific fitness throughout development, allowing interpretation of these patterns of gene expression in terms of sexual conflict.

At present, we know even less about how sexual conflict progresses throughout development than we do about patterns of sex-biased gene expression. Studies have examined sex-specific patterns of selection throughout development in some species (e.g., water striders [Preziosi and Fairbairn 2000], sticklebacks [Reimchen and Nosil 2002], and lizards [Le Galliard and Ferriere 2008]), and selection seems to vary a lot between juveniles and adults. However, a meta-analysis focusing on whether levels of sexually antagonistic selection differ between life stages produced mixed results overall (Cox and Calsbeek 2009). In addition, Chippindale et al. (2001) found a positive genetic correlation between male and female fitness in juvenile *D. melanogaster*, suggesting no conflict in pre-adults, whereas at the adult stage, a negative intersexual genetic correlation for fitness revealed sexual conflict. On the other hand, Svensson et al. (2009) showed that sexual conflict in lizards was consistent across juvenile and adult stages.

In a recent study of sex-biased gene expression in *D. melanogaster* larvae and pre-pupae, Perry et al. (2014) used the direction of selection (DoS) statistic to suggest that their results showed sex-specific selection patterns in pre-adult stages. However, although DoS has proven extremely useful in molecular evolution studies covering major evolutionary timescales (Stoletzki and Eyre-Walker 2011), it is unclear how relevant this metric is when applied to questions of current patterns of selection, and research examining sexually antagonistic selection might be better to measure relative male and female fitness within an experimental generation to directly associate patterns of sex-biased gene expression with sex-specific selection.

Theoretically, sex-biased gene expression and sexual conflict should become more prevalent as development progresses, as the sexes differentiate and the evolutionary interests of the sexes diverge after individuals have reached reproductive age (Badyaev 2002; Cox and Calsbeek 2009). This prediction might partly explain the apparent focus on adults in research on sex-biased gene expression and sexual conflict, but the little data we have across developmental stages do not provide conclusive support for the prediction. First, it is difficult to come to any general conclusion from the handful of sexual conflict studies cited above because of the conflicting results. Second, in terms of sex-biased gene expression, there are only a few studies that suggest that the prevalence of sex-biased gene expression increases throughout development (in chickens [Mank et al. 2010] and *B. mori* [Zhao et al. 2011]). Conversely, there is also a lot of evidence for widespread sex biases in the transcriptome in juvenile stages of development (e.g., Fraser et al. 2011; Magnusson et al. 2011; see below), and even before sex differentiation in embryo brain tissue of mice (Dewing et al. 2003) and chickens (Scholz et al. 2006).

It is also possible that the ontogenetic relationship between sexual dimorphism and sexual conflict will vary, depending on factors such as the timing of sex differentiation during development or the genes involved in the sex differentiation pathway. For instance, in a holometabolous animal like a fruit fly, male and female larvae are similar in terms of morphology and behavior, and their evolutionary interests more or less aligned. As such, we might expect male and female larvae to be under similar patterns of selection, and thus, there will be limited potential for sexual conflict (e.g., Chippindale et al. 2001). At the pupal stage, however, individuals undergo drastic metamorphosis as the body is completely restructured into an adult male or female. In the example of fruit flies, there is extensive sexual dimorphism in adults, and many of these differences are laid down in the pupal stage. As such, perhaps sexual conflict will be at its most intense in pupae, and the body of research largely focused on conflict in adults might provide an underestimate of the overall influence of sexual conflict. On the other hand, in species that do not develop through these distinct stages (e.g., hemimetabolous insects, birds, and mammals), sexual dimorphism generally develops more gradually as sex-specific traits are acquired. In this case, a more constant buildup of sexual conflict throughout the entirety of development may be expected. Another possibility is that in all species, sexual conflict is most prevalent in reproductive individuals (i.e., adults). A further consideration will be traits that are laid down by earlier patterns of sex-biased gene expression, but not subject to selection at the phenotypic level until later in life. In this sense, studies will need to consider how gene expression data and fitness data are coupled together. If gene expression is measured as a snapshot of a given developmental stage, should fitness data be measured in a similarly stage-specific manner or be taken as an overall lifetime fitness measurement? Given an evolutionary definition of fitness as the lifetime reproductive success of an individual relative to others in the population, stage-specific fitness measures might be difficult to define. Either way, further research will need to focus on determining how sex-biased gene expression throughout development might relate to levels of sexual conflict. Without data relating sex-specific selection and gene expression throughout development, we cannot differentiate between the predictions discussed here. Sexual conflict at any stage of development could potentially affect the expression of shared traits and overall individual fitness, and this is clearly a neglected area of research.

CONCLUDING REMARKS

Whereas it is clear that sex-biased gene expression is a widespread phenomenon, the evolutionary significance of these patterns of gene expression is less clear because of a paucity of studies that examine fitness or selection in conjunction with the molecular data. The assumption that sex-biased gene expression could be used as an indicator of sexually antagonistic selection is evidently not reliable. Where the association between sex-specific fitness and sex-biased gene expression has been directly tested, it has been shown that sex-biased genes are not necessarily subject to sexually antagonistic selection. Moreover, distinguishing between ongoing and resolved sexual conflict requires the quantification of both the extent of dimorphism in gene expression and the discrepancy in sex-specific fitness optima. Arguably, it is genes with monomorphic patterns of expression that may experience the greatest degree of sexually antagonism. As such, future research will benefit greatly from more studies that bring together measures of sex-biased gene expression and sex-specific selection. Only by doing so can sex-biased gene expression be interpreted in the context of sexual conflict, or in an evolutionary framework more generally, for example, with regards to the role of sex-biased genes in speciation.

Our review of the existing literature and our small-scale experiment also highlight potentially important avenues for future research on sex-biased gene expression. First, it will be interesting to consider how sex-biased gene expression and sexual conflict vary throughout development. Studies so far show sex-biased expression to be dynamic from one developmental stage to another, and so relating these patterns to fitness and conflict will provide useful insight. Research integrating data on gene expression, fitness, and development will be demanding, but it should be increasingly feasible as technology for examining gene expression becomes more accessible. Furthermore, sex biases seem to vary not only throughout development, but across environments and genotypes as well. This is likely to have important implications for the evolution of sex-biased gene expression and could mean that the potential for sexual conflict to be resolved through sex-biased gene expression varies between environments and populations, an idea that might also benefit from further research.

ACKNOWLEDGMENTS

Funding was provided by the European Research Council (Starting Grant Number 280632), a Royal Society University Research Fellowship and the Swedish Research Council (Number 2011-3701). The authors are grateful for useful discussions and comments on the manuscript from Jessica Abbott, Will Gilks,

Katrine Lund-Hansen, Manus Patten, and Tanya Pennell.

REFERENCES

Abbott JK, Bedhomme S, Chippindale AK. 2010. Sexual conflict in wing size and shape in *Drosophila melanogaster. J Evol Biol* **23:** 1989–1997.

Abbott JK, Innocenti P, Chippindale AK, Morrow EH. 2013. Epigenetics and sex-specific fitness: An experimental test using male-limited evolution in *Drosophila melanogaster. PLoS ONE* **8:** e70493.

Arbeitman MN, Furlong EEM, Imam F, Johnson E, Null BH, Baker BS, Krasnow MA, Scott MP, Davis RW, White KP. 2002. Gene expression during the life cycle of *Drosophila melanogaster. Science* **297:** 2270–2275.

Assis R, Zhou Q, Bachtrog D. 2012. Sex-biased transcriptome evolution in *Drosophila. Genome Biol Evol* **4:** 1189–1200.

Ayroles JF, Carbone MA, Stone EA, Jordan KW, Lyman RF, Magwire MM, Rollmann SM, Duncan LH, Lawrence F, Anholt RRH, et al. 2009. System genetics of complex traits in *Drosophila melanogaster. Nat Rev Genet* **41:** 299–307.

Badyaev AV. 2002. Growing apart: An ontogenetic perspective on the evolution of sexual size dimorphism. *Trends Ecol Evol* **17:** 369–378.

Baker DA, Russell S. 2009. Gene expression during *Drosophila melanogaster* egg development before and after reproductive diapause. *BMC Genomics* **10:** 242.

Baker DA, Nolan T, Fischer B, Pinder A, Crisanti A, Russell S. 2011. A comprehensive gene expression atlas of sex- and tissue-specificity in the malaria vector, *Anopheles gambiae. BMC Genomics* **12:** 296.

Bedhomme S, Prasad NG, Jiang PP, Chippindale AK. 2008. Reproductive behaviour evolves rapidly when intralocus sexual conflict is removed. *PLoS ONE* **3:** e2187.

Bedhomme S, Chippindale AK, Prasad NG, Delcourt M, Abbott JK, Mallet MA, Rundle HD. 2011. Male-limited evolution suggests no extant intralocus sexual conflict over the sexually dimorphic cuticular hydrocarbons of *Drosophila melanogaster. J Genetics* **90:** 443–452.

Benjamini Y, Hochberg Y. 1995. Controlling the false discovery rate: A practical and powerful approach to multiple testing. *J R Stat Soc B* **57:** 289–300.

Bhasin JM, Chakrabarti E, Peng DQ, Kulkarni A, Chen X, Smith JD. 2008. Sex-specific gene regulation and expression QTLs in mouse macrophages from a strain intercross. *PLoS ONE* **3:** e1435.

Bizuayehu TT, Babiak J, Norberg B, Fernandes JMO, Johansen SD, Babiak I. 2012. Sex-biased miRNA expression in Atlantic halibut (*Hippoglossus hippoglossus*) brain and gonads. *Sex Dev* **6:** 257–266.

Blekhman R, Marioni JC, Zumbo P, Stephens M, Gilad Y. 2010. Sex-specific and lineage-specific alternative splicing in primates. *Genome Res* **20:** 180–189.

Bonduriansky R, Chenoweth SF. 2009. Intralocus sexual conflict. *Trends Ecol Evol* **24:** 280–288.

Carroll AP, Tran N, Tooney PA, Cairns MJ. 2012. Alternative mRNA fates identified in microRNA-associated transcriptome analysis. *BMC Genomics* **13:** 561.

Catalan A, Hutter S, Parsch J. 2012. Population and sex differences in *Drosophila melanogaster* brain gene expression. *BMC Genomics* **13:** 654.

Chang PL, Dunham JP, Nuzhdin SV, Arbeitman MN. 2011. Somatic sex-specific transcriptome differences in *Drosophila* revealed by whole transcriptome sequencing. *BMC Genomics* **12:** 364.

Chintapalli VR, Wang J, Dow JAT. 2007. Using FlyAtlas to identify better *Drosophila melanogaster* models of human disease. *Nat Genet* **39:** 715–720.

Chippindale AK, Gibson JR, Rice WR. 2001. Negative genetic correlation for adult fitness between sexes reveals ontogenetic conflict in *Drosophila. PNAS* **98:** 1671–1675.

Choi YJ, Ghedin E, Berriman M, McQuillan J, Holroyd N, Mayhew GF, Christensen BM, Michalski ML. 2011. A deep sequencing approach to comparatively analyze the transcriptome of lifecycle stages of the filarial worm, *Brugia malayi. PLoS Neglect Trop D* **5:** e1409.

Connallon T, Clark AG. 2011. Association between sex-biased gene expression and mutations with sex-specific phenotypic consequences in *Drosophila. Genome Biol Evol* **3:** 151–155.

Connallon T, Knowles LL. 2006. Intergenomic conflict revealed by patterns of sex-biased gene expression. *Trends Genet* **21:** 495–499.

Cox RM, Calsbeek R. 2009. Sexually antagonistic selection, sexual dimorphism, and the resolution of intralocus sexual conflict. *Am Nat* **173:** 176–187.

Dewing P, Shi T, Horvath S, Vilain E. 2003. Sexually dimorphic gene expression in mouse brain precedes gonadal differentiation. *Mol Brain Res* **118:** 82–90.

Eads BD, Colbourne JK, Bohuski E, Andrews J. 2007. Profiling sex-biased gene expression during parthenogenetic reproduction in *Daphnia pulex. BMC Genomics* **8:** 464.

Ellegren H, Parsch J. 2007. The evolution of sex-biased genes and sex-biased gene expression. *Nat Rev Genet* **8:** 689–698.

Fitzpatrick JM, Johnston DA, Williams GW, Williams DJ, Freeman TC, Dunne DW, Hoffmann KF. 2005. An oligonucleotide microarray for transcriptome analysis of *Schistosoma mansoni* and its application/use to investigate gender-associated gene expression. *Mol Biochem Parasitol* **141:** 1–13.

Fraser BA, Weadick CJ, Janowitz, Rodd FH, Hughes KA. 2011. Sequencing and characterization of the guppy (*Poecilia reticulata*) transcriptome. *BMC Genomics* **12:** 202.

Fry JD. 2010. The genomic location of sexually antagonistic variation: Some cautionary comments. *Evolution* **64:** 1510–1516.

Gan Q, Chepelev I, Wei G, Tarayrah L, Cui K, Zhao K, Chen X. 2010. Dynamic regulation of alternative splicing and chromatin structure in *Drosophila* gonads revealed by RNA-seq. *Cell Res* **20:** 763–783.

Gautier L, Cope L, Bolstad BM, Irizarry RA. 2004. Affy-analysis of Affymetrix GeneChip data at the probe level. *Bioinformatics* **20:** 307–315.

Gentleman R, Carey V, Bates D, Bolstad B, Dettling M, et al. 2004. Bioconductor: Open software development for computational biology and bioinformatics. *Genome Biol* **5:** R80.

Gibson JR, Chippindale AK, Rice WR. 2002. The X chromosome is a hot spot for sexually antagonistic fitness variation. *Proc R Soc Lond B* **269:** 499–505.

Graveley BR, Brooks AN, Carlson JW, Duff MO, Landolin JM, Yang L, Artieri CG, van Baren MJ, Boley N, Booth BW, et al. 2011. The developmental transcriptome of *Drosophila melanogaster. Nature* **471:** 473–479.

Grosse-Wilde E, Kuebler LS, Bucks S, Vogel H, Wicher D, Hansson BS. 2011. Antennal transcriptome of *Manduca sexta. PNAS* **108:** 7449–7454.

Innocenti P. 2011. Sexual conflict and gene expression: Exploring sex-specific associations between fitness and transcriptional variation. *Fly* **5:** 10–13.

Innocenti P, Morrow EH. 2010. The sexually antagonistic genes of *Drosophila melanogaster. PLoS Biol* **8:** e1000335.

Jiang ZF, Machado CA. 2009. Evolution of sex-dependent gene expression in three recently diverged species of *Drosophila. Genetics* **183:** 1175–1185.

Jiang M, Ryu J, Kiraly M, Duke K, Reinke V, Kim SK. 2001. Genome-wide analysis of developmental and sex-regulated gene expression profiles in *Caenorhabditis elegans. PNAS* **98:** 218–233.

Jin W, Riley RM, Wolfinger RD, White KP, Passador-Gurgel G, Gibson G. 2001. The contributions of sex, genotype and age to transcriptional variance in *Drosophila melanogaster. Nat Genet* **29:** 389–395.

Khil PP, Smirnova NA, Romanienko PJ, Camerini-Otero RD. 2004. The mouse X chromosome is enriched for sex-biased genes not subject to selection by meiotic sex chromosome inactivation. *Nat Genet* **36:** 642–646.

Kwan L, Bedhomme S, Prasad NG, Chippindale AK. 2008. Sexual conflict and environmental change: Trade-offs within and between the sexes during evolution of desiccation resistance. *J Genet* **87:** 383–394.

Leder EH, Cano JM, Leinonen T, O'Hara RB, Nikinmaa M, Primmer CR, Merila J. 2010. Female-biased expression on the X chromosome as a key step in sex chromosome evolution in threespine sticklebacks. *Mol Biol Evol* **27:** 1495–1503.

Le Galliard JF, Ferriere R. 2008. Evolution of maximal endurance capacity: Natural and sexual selection across age classes in a lizard. *Evol Ecol Res* **10:** 157–176.

Liu S, Gao S, Zhang D, Yin J, Xiang Z, Xia Q. 2010. MicroRNAs show diverse and dynamic expression patterns in multiple tissues of *Bombyx mori. BMC Genomics* **11:** 85.

Long TAF, Rice WR. 2007. Adult locomotory activity mediates intralocus sexual conflict in a laboratory-adapted population of *Drosophila melanogaster. Proc R Soc Lond B* **274:** 3105–3112.

Long TA, Agrawal AF, Rowe L. 2012. The effect of sexual selection on offspring fitness depends on the nature of genetic variation. *Curr Biol* **22:** 204–208.

Luo GZ, Hafner M, Shi Z, Brown M, Feng GH, Tuschl T, Wang XJ, Li X. 2012. Genome-wide annotation and analysis of zebra finch microRNA repertoire reveal sex-biased expression. *BMC Genomics* **13:** 727.

Magnusson K, Mendes AM, Windbichler N, Papathanos PA, Nolan T, Dottorini T, Rizzi E, Christophides GK, Crisanti A. 2011. Transcription regulation of sex-biased genes during ontogeny in the malaria vector *Anopheles gambiae. PLoS ONE* **6:** e21572.

Malone JH, Hawkins DL, Michalak P. 2006. Sex-biased gene expression in a ZW sex determination system. *J Mol Evol* **63:** 427–436.

Mank JE, Nam K, Brunstrom B, Ellegren H. 2010. Ontogenetic complexity of sexual dimorphism and sex-specific selection. *Mol Biol Evol* **27:** 1570–1578.

Marco A, Kozomara A, Hui JHL, Emery AM, Rollinson D, Griffiths-Jones S, Ronshaugen M. 2013. Sex-biased expression of microRNAs in *Schistosoma mansoni. PLoS Neglect Trop D* **7:** e2402.

Marinotti O, Calvo E, Nguyen QK, Dissanayake S, Ribeiro JMC, James AA. 2006. Genome-wide analysis of gene expression in adult *Anopheles gambiae. Insect Mol Biol* **15:** 1–12.

Martinez-Fernandez M, Bernatchez L, Rolan-Alvarez E, Quesada H. 2010. Semi-quantitative differences in gene transcription profiles between sexes of a marine snail by a new variant of cDNA-AFLP analysis. *Mol Ecol* **10:** 324–330.

McIntyre LM, Bono LM, Genissel A, Westerman R, Junk D, Telonis-Scott M, Harshman L, Wayne ML, Kopp A, Nuzhdin SV. 2006. Sex-specific expression of alternative transcripts in *Drosophila. Genome Biol* **7:** R79.

Meiklejohn CD, Parsch J, Ranz JM, Hartl DL. 2003. Rapid evolution of male-biased gene expression in *Drosophila. PNAS* **100:** 9894–9899.

Morgan CP, Bale TL. 2012. Sex differences in microRNA regulation of gene expression: No smoke, just miRs. *Biol Sex Differences* **3:** 22.

Naurin S, Hansson B, Hasselquist D, Kim YH, Bensch S. 2011. The sex-biased brain: Sexual dimorphism in gene expression in two species of songbirds. *BMC Genomics* **12:** 37.

Oliver JC, Monteiro A. 2011. On the origins of sexual dimorphism in butterflies. *Proc Biol Sci* **278:** 1981–1988.

Ometto L, Shoemaker D, Ross KG, Keller L. 2011. Evolution of gene expression in fire ants: The effects of developmental stage, caste, and species. *Mol Biol Evol* **28:** 1381–1392.

Parisi M, Nuttall R, Naiman D, Bouffard G, Malley J, Andrews J, Eastman S, Oliver B. 2003. Paucity of genes on the *Drosophila* X chromosome showing male-biased expression. *Science* **299:** 697–700.

Parker GA, Partridge L. 1998. Sexual conflict and speciation. *Phil Trans R Soc Lond B* **353:** 261–274.

Pasquinelli AE. 2002. MicroRNAs: Deviants no longer. *Trends Genet* **18:** 171–173.

Pennell TM, Morrow EH. 2013. Two sexes, one genome: The evolutionary dynamics of intralocus sexual conflict. *Ecol and Evol* **3:** 1819–1834.

Perry JC, Harrison PW, Mank JE. 2014. The ontogeny and evolution of sex-biased gene expression in *Drosophila melanogaster. Mol Biol Evol* **31:** 1206–1219.

Pischedda A, Chippindale AK. 2006. Intralocus sexual conflict diminishes the benefits of sexual selection. *PLoS Biol* **4:** e356.

Pitts RJ, Rinker DC, Jones PL, Rokas A, Zwiebel LJ. 2011. Transcriptome profiling of chemosensory appendages in the malaria vector *Anopheles gambiae* revels tissue-specific signatures of odor coding. *BMC Genomics* **12:** 271.

Prasad NG, Bedhomme S, Day T, Chippindale AK. 2007. An evolutionary cost of separate genders revealed by male-limited evolution. *Am Nat* **169:** 29–37.

Preziosi RF, Fairbairn DJ. 2000. Lifetime selection on adult body size and components of body size in a waterstrider: Opposing selection and maintenance of sexual size dimorphism. *Evolution* **54:** 558–566.

Prince EG, Kirkland D, Demuth JP. 2010. Hyperexpression of the X chromosome in both sexes results in extensive female bias of X-linked genes in the flour beetle. *Genome Biol Evol* **2:** 336–346.

Proschel M, Zhang Z, Parsch J. 2006. Widespread adaptive evolution of *Drosophila* genes with sex-biased expression. *Genetics* **174:** 893–900.

Ranz JM, Castillo-Davis CI, Meiklejohn CD, Hartl DL. 2003. Sex-dependent gene expression and evolution of the *Drosophila* transcriptome. *Science* **300:** 1742–1745.

Reimchen TE, Nosil P. 2002. Temporal variation in divergent selection on spine number in threespine stickleback. *Evolution* **56:** 2472–2483.

Reinke V, Smith HE, Nance J, Wang J, Van Doren C, Begley R, Jones SJM, Davis EB, Scherer S, Ward S, et al. 2000. A global profile of germline gene expression in *C. elegans*. *Mol Cell* **6:** 605–616.

Reinke V, Gil IS, Ward S, Kazmer K. 2004. Genome-wide germline-enriched and sex-biased expression profiles in *Caenorhabditis elegans*. *Development* **131:** 311–323.

Rice WR. 1984. Sex chromosomes and the evolution of sexual dimorphism. *Evolution* **38:** 735–742.

Rice WR, Linder JE, Friberg U, Lew TA, Morrow EH, Stewart AD. 2005. Inter-locus antagonistic coevolution as an engine of speciation: Assessment with hemiclonal analysis. *PNAS* **102:** 6527–6534.

Rinn JL, Snyder M. 2005. Sexual dimorphism in mammalian gene expression. *Trends Genet* **21:** 298–305.

Rinn JL, Rozowsky JS, Laurenzi IJ, Peterson PH, Zou K, Zhong W, Gerstein M, Snyder M. 2004. Major molecular differences between mammalian sexes are involved in drug metabolism and renal function. *Dev Cell* **6:** 791–800.

Salmena L, Poliseno L, Tay Y, Kats L, Pandolfi PP. 2011. A ceRNA hypothesis: The rosetta stone of a hidden RNA language? *Cell* **146:** 353–358.

Santos EM, Kille P, Workman VL, Paull GC, Tyler CR. 2008. Sexually dimorphic gene expression in the brains of mature zebrafish. *Comp Biochem Phys* **149:** 314–324.

Scholz B, Kultima K, Mattsson A, Axelsson J, Brunstrom B, Halldin K, Stigson M, Dencker L. 2006. Sex-dependent gene expression in early brain development of chicken embryos. *BMC Neuroscience* **7:** 12.

Sharma S, Eghbali M. 2014. Influence of sex differences on microRNA gene regulation in disease. *Biol Sex Differ* **5:** 3.

Small CM, Carney GE, Mo Q, Vannucci M, Jones AG. 2009. A microarray analysis of sex- and gonad-biased gene expression in the zebrafish: Evidence for masculinization of the transcriptome. *BMC Genomics* **10:** 579.

Stoletzki N, Eyre-Walker A. 2011. Estimation of the neutrality index. *Mol Biol Evol* **28:** 63–70.

Svensson EI, McAdam AG, Sinervo B. 2009. Intralocus sexual conflict over immune defense, gender load, and sex-specific signaling in a natural lizard population. *Evolution* **63:** 3124–3135.

Telonis-Scott M, Kopp A, Wayne ML, Nuzhdin SV, McIntyre LM. 2009. Sex-specific splicing in *Drosophila*: Widespread occurrence, tissue specificity and evolutionary conservation. *Genetics* **181:** 421–434.

Thoemke K, Yi W, Ross JM, Kim S, Reinke V, Zarkower D. 2005. Genome-wide analysis of sex-enriched gene expression during *C. elegans* larval development. *Dev Biol* **284:** 500–508.

Wang PJ, McCarrey JR, Yang F, Page DC. 2001. An abundance of X-linked genes expressed in spermatonia. *Nat Genet* **27:** 422–426.

Wayne ML, Telonis-Scott M, Bono LM, Harshman L, Kopp A, Nuzhdin SV, McIntyre LM. 2012. Simpler mode of inheritance of transcriptional variation in male *Drosophila melanogaster*. *PNAS* **104:** 18577–18582.

Wen C, Zhang Z, Ma W, Xu M, Wen Z, Peng J. 2005. Genome-wide identification of female-enriched genes in zebrafish. *Dev Dynamics* **232:** 171–179.

Wyman MJ, Agrawal AF, Rowe L. 2010. Condition-dependence of the sexually dimorphic transcriptome in *Drosophila melanogaster*. *Evolution* **64:** 1836–1848.

Wyman MJ, Cutter AD, Rowe L. 2012. Gene duplication in the evolution of sexual dimorphism. *Evolution* **66:** 1556–1566.

Xia Q, Cheng D, Duan J, Wang G, Cheng T, Zha X, Liu C, Zhao P, Dai F, Zhang Z, et al. 2007. Microarray-based gene expression profiles in multiple tissues of the domesticated silkworm, *Bombyx mori*. *Genome Biol* **8:** R162.

Yang X, Schadt EE, Wang S, Wang H, Arnold AP, Ingram-Drake L, Drake TA, Lusis AJ. 2006. Tissue-specific expression and regulation of sexually dimorphic genes in mice. *Genome Res* **16:** 995–1004.

Zhang Y, Sturgill D, Parisi M, Kumar S, Oliver B. 2007. Constraint and turnover in sex-biased gene expression in the genus *Drosophila*. *Nature* **450:** 233–238.

Zhang YE, Vibranovski MD, Krinsky BH, Long M. 2010. Age-dependent chromosomal distribution of male-biased genes in *Drosophila*. *Genome Res* **20:** 1526–1533.

Zhao M, Zha XF, Liu J, Zhang WJ, He NJ, Cheng DJ, Dai Y, Xiang ZH, Xia QY. 2011. Global expression profile of silkworm genes from larval to pupal stages: Toward a comprehensive understanding of sexual differences. *Insect Science* **18:** 607–618.

Infanticide as Sexual Conflict: Coevolution of Male Strategies and Female Counterstrategies

Ryne A. Palombit

Department of Anthropology, Center for Human Evolutionary Studies, Rutgers University, New Brunswick, New Jersey 08901

Correspondence: rpalombit@anthropology.rutgers.edu

One of the earliest recognized forms of sexual conflict was infanticide by males, which imposes serious costs on female reproductive success. Here I review two bodies of evidence addressing coevolved strategies of males and females. The original sexual selection hypothesis arguing that infanticide improves male mating success by accelerating the return of females to fertilizable condition has been generally supported in some taxa—notably, some primates, carnivores, rodents, and cetaceans—but not in other taxa. One result of recent research has been to implicate other selective benefits of infanticide by males in various taxa from insects to birds to mammals, such as acquisition of breeding status or improvement of the female breeding condition. In some cases, however, the adaptive significance of male infanticide remains obscure. The second body of data I review is arguably the most important result of recent research: clarifying the possible female counterstrategies to infanticide. These potential counterstrategies span diverse biological systems, ranging from sexual behavior (e.g., polyandrous mating), to physiology (e.g., the Bruce effect), to individual behavior (e.g., maternal aggression), to social strategies (e.g., association with coalitionary defenders of either sex). Although much remains to be studied, these current data provide compelling evidence of sexually antagonistic coevolution surrounding the phenomenon of infanticide.

At its most elemental level, infanticide is the killing of a newborn individual by a conspecific. With the growing appreciation of its biological significance, however, infanticide came to be defined more broadly as any "behavior that makes a direct and significant contribution to the immediate death of an embryo or newly hatched or born member of the performer's own species" (Mock 1984, p. 4) or "any form of lethal curtailment of parental investment in offspring brought about by conspecifics" (Hrdy and Hausfater 1984, p. xv). These definitions highlight the heterogeneous and variable nature of the phenomenon, which can be perpetrated by either sex, by parents or other kin, by individuals unrelated to the victim, in a wide variety of social and mating systems, under a range of seasonal or aseasonal breeding regimes, and across diverse taxa straddling vertebrates and invertebrates.

One adaptive form of infanticide—the killing of infants by unrelated males—is arguably

Cite this article as *Cold Spring Harb Perspect Biol* doi: 101101/cshperspect.a017640

the archetype of sexual conflict. In 450 BCE, Herodotus not only documented the behavior among Egyptian cats, but explained it as a male "trick" to obtain sexual access to females otherwise preoccupied with maternal duties (Delibes et al. 2012). Among the myriad ideas inaugurating sociobiology in the 1970s, the hypothesis that infanticide is a male strategy that improves reproductive success at the expense of female fitness (Hrdy 1974) constituted one of the first demonstrations of the "battle of the sexes" theory developed by Williams (1966) and Trivers (1972). Partly because of the controversy surrounding the appearance of this hypothesis (Rees 2009), however, subsequent research focused more on male strategy than on the other party in this sexual dialectic, the female. Thus, field and laboratory research has helped to establish its many forms and conditional occurrence, describe its distribution across taxa, and clarify its adaptive significance, but it is only relatively recently that female counterstrategies have become the subjects of rigorous study, even though their potential importance was grasped early on (Hrdy 1979).

In this article, I review selected aspects of this body of data and analysis. My focus is on nonparental male infanticide targeting dependent young—in mammals, nursing individuals—as opposed to older, weaned offspring, the killing of which is variably rendered "juvenilicide," "pedicide," or "filicide" (e.g., Agoramoorthy and Mohnot 1988; Palombit 2014, in press).

THE SEXUAL CONFLICT OF INFANTICIDE: MALE STRATEGIES

Several functional hypotheses have been proposed to explain male infanticide. Foremost among them historically is the sexual selection hypothesis. I consider this hypothesis first, with special, although not exclusive, taxonomic focus on the group of animals that originally generated the hypothesis—the primates. I then move on to several alternative hypotheses for male infanticide, the discussion of which will incorporate consideration of data from other animals as well as additional data from primates.

Sexual Selection Hypothesis

Although male infanticide clearly has no unitary explanation, the biological relevance of intrasexual selection in males makes the sexual selection hypothesis a good starting point for discussion. As originally formulated for langurs (*Semnopithecus entellus*), Hrdy's (1974, 1977) argument is that the killing of an infant prematurely ends lactational amenorrhea (the cessation of ovulatory cycles during nursing) in its mother, thereby significantly advancing the time when she is available for subsequent fertilization by the perpetrator. The potential benefit of this strategy to male fitness will depend on many variables, but salient among them is the degree to which sexual access to fecund females is limited by two factors: male–male competition and the reproductive life history of females.

The latter of these factors provides a useful context for analyzing the problem of male infanticide. The L/G ratio, which compares the durations of lactation and gestation, summarizes the potential value (to males) and vulnerability (to females) of infanticide in mammals. When lactation greatly exceeds gestation, postpartum mating and early pregnancy impose potentially prohibitive costs on females in the form of simultaneous support of two sets of offspring differing in ages, requirements, and competitive capacities (van Schaik 2000b). The adaptive solution to this female problem is lactational amenorrhea, which, however, generates the reproductive problem for males that infanticide can solve. Plainly expressed: An infant is a "perfect contraceptive," and infanticide simply serves to remove this effect (Altmann et al. 1978, p. 1029).

The taxonomic distribution of sexually selected infanticide across mammals is not fully resolved, but current reports of infanticide whose patterning is consistent with this hypothesis are predictably concentrated among taxa with relatively long L/G ratios— namely, primates, fissiped carnivores, and odontocete whales (Fig. 1) (van Schaik 2000b). Several caveats warrant mention. First, this result does not mean that male infanticide is uniformly expressed within these taxa; on the contrary, there is often considerable variation over time and across populations (see below). Second, it is also not the case that male infanticide is unreported in other taxa with lower L/G ratios, but in those cases the sexually selected mechanism has been less implicated than alternative hypotheses (see below). Finally, L/G ratios are important, but other factors may explain the rarity of infanticide in some mammals, for example, ecologically determined sexual segregation, (Kunz and Ebensperger 1999), caching of infants as an antipredator strategy (Schülke 2005), need for hibernation (Thalmann 2001), or even phylogenetic inertia (Ebensperger 2001).

It is not possible here to review comprehensively the empirical tests of the sexual selection hypothesis (see Hausfater and Hrdy 1984; Parmigiani et al. 1994; Connor et al. 2000; van Schaik and Janson 2000; Ebensperger and Blumstein 2007; Palombit 2012), but some general patterns can be highlighted. The largest body of data supporting this hypothesis has come from study of the nonhuman primates. In the last 25 years, the number of species in which infanticide had been observed directly or inferred from substantive indirect evidence in the wild has risen from a dozen (Hiraiwa-Hasegawa 1988) to 54, including 84 distinct populations distributed across nine of the 14 families of strepsirrhines, monkeys, and apes (updated from Palombit 2012).

The circumstances surrounding infanticide in primates are typically (although not invariably) consistent with the sexual selection hypothesis: (1) replacement of a male occupying the position of sole or dominant breeder in the group by a male who is unlikely to have sired the infants he subsequently attacks; (2) resumption of ovulatory cycling by the mother; and (3) copulation between the mother and perpetra-

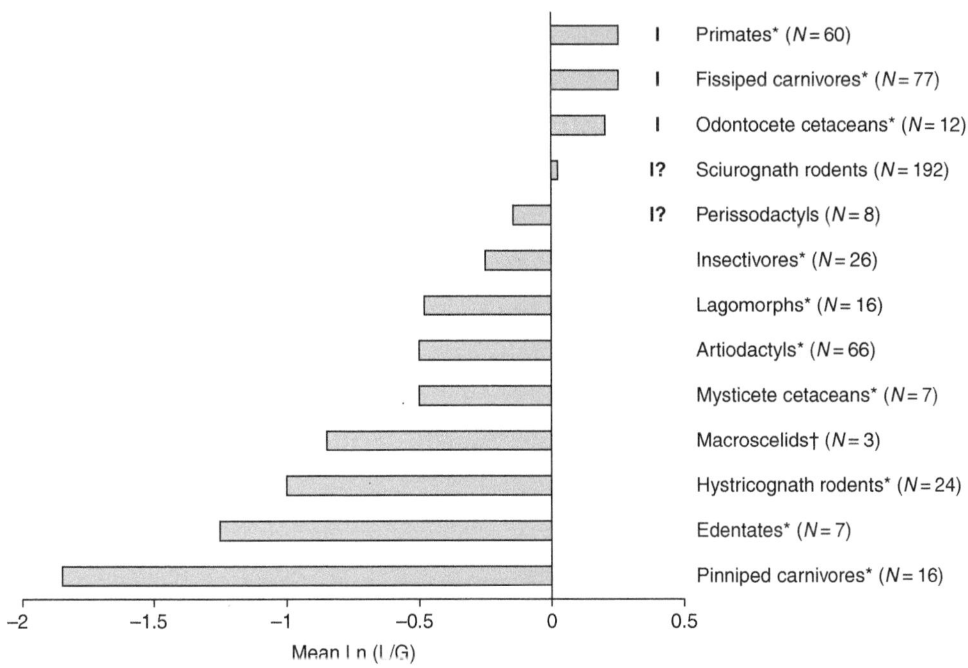

Figure 1. Mean lactation/gestation (L/G) ratios for selected mammalian groups. Log-transformed L/G ratios significantly different from 1 are indicted by *($p < 0.05$) or †($p < 0.01$). The letter "I" denotes taxa in which deliberate sexually selected infanticide by males has been recorded. Sample sizes (number of species) are indicated in parentheses. (Modified from data in van Schaik 2000b.)

tor, and thus shortening of the interbirth interval. Most but not all reports provide supportive data for all three predictions, but cases in which behavioral data are unavailable or contradictory more often concern the third prediction. Advances in field methodology have permitted collection of ancillary (nonbehavioral) data to further test the hypothesis. For example, DNA data establish that males are unrelated to the infants they attack (Pereira and Weiss 1991; Borries et al. 1999a; Soltis et al. 2000) or are fathers of the female's next infant (Morelli et al. 2009), whereas hormonal data provide direct evidence of maternal resumption of ovulatory cycling following infanticide (Harris and Monfort 2003) as well as elevated stress response in mothers following heightened infanticide risk or actual attacks (Beehner et al. 2005; see also Brockman et al. 2009). Thus, as a general explanation of infanticide, the sexual selection hypotheses has received more support than alternative hypotheses (see below) (van Schaik 2000a; Palombit 2012).

Although originally observed for primates living in unimale, polygynous groups, male infanticide has been observed in almost all of the other social systems found in this order, from dispersed societies, to multimale groups, to fission–fusion communities, and perhaps even socially monogamous pairs. Its occurrence in some multimale societies is noteworthy. The permanent presence of multiple males was expected to hinder infanticide by increasing its costs (through male defense of infants) or decreasing its potential benefits (through reduced ability to monopolize matings). Thus, across gregarious primates, infanticide rate is negatively correlated with the number of males in groups (Janson and van Schaik 2000). Nevertheless, significant male infanticide has been documented in a number of multimale societies in two general contexts: (1) by new immigrants (or, somewhat less commonly, by long-term residents) that rise to the alpha position in the male dominance hierarchy (e.g., Palombit et al. 2000; Soltis et al. 2000; Teichroeb and Sicotte 2008); or (2) by coalitionary cohorts of alien males that collectively replace the entire male membership of a group (Fedigan et al. 2008) in a manner reminiscent of lions (Pusey and Packer 1994b). One factor apparently promoting infanticide in this social context is pronounced mating skew toward dominant males (sensu Broom et al. 2004).

Compelling evidence for sexually selected infanticide in the nonprimate mammals has come from field studies of fissiped carnivores (particularly bears and well-studied lions); there are also some suggestive recent data for toothed whales (particularly dolphins) (Table 1). Among rodents, the altriciality of young and lengthy periods of female investment make sexually selected infanticide likely in many species, particularly in the murids (Wolff and MacDonald 2004). Male infanticide is known in 14–17 species of Muridae (Ebensperger and Blumstein 2007). The Norway rat (*Rattus norvegicus*), house mouse (*Mus musculus/domesticus*), and Mongolian gerbil (*Meriones unguiculatus)* have been the subjects of extensive and well-controlled laboratory research (see references in Table 1), which has shed important light on the proximate mechanisms underlying infanticide and on female counterstrategies. These data for murids are largely consistent with the sexual selection hypothesis, but almost all of them come from the laboratory; only four murid species have provided limited data on male infanticide in the wild (see Caley and Boutin 1985; Wolff and Cicirello 1991; Ylönen et al. 1997). It is partly because "virtually nothing is known" about infanticide in natural populations of murids (Berdoy and Drickamer 2007, p. 389) that the broader implications of the laboratory results are debated, particularly vis-à-vis the predicted postinfanticide reproductive benefits. Thus, Ebensperger and Blumstein (2007, p. 178) conclude that although current data reject nonadaptive explanations, the sexual selection hypothesis "remains unproven in rodents" (see also below for sciurid rodents).

Alternative Explanations for Male Infanticide

Besides the sexual selection hypothesis, there is a large number of alternative adaptive explanations for male infanticide (Table 1). Only a few of these can be addressed here. It is perhaps

Table 1. Adaptive and nonadaptive hypotheses for nonparental infanticide by males

Hypothesis	Theoretical rationale	Proposed example taxa	References
Benefits related to mating			
Sexual selection I. Accelerate female return to breeding condition	When sexual access to fecund females is limited by intrasexual competition, killing eggs or infants accelerates female return to fertilizable condition (mammals: shortens lactational amenorrhea) for subsequent insemination	Arachnids; birds (*Hirundo rustica*, *Passer domesticus*); hippos (*Hippopotamus amphibius*); horses (*Equus caballus*); dolphins (*Tursiops* spp.); lions (*Panthera leo*); bears (*Ursus arctos*); murid rodents (*M. musculus*, *R. norvegicus*, *Peromyscus* spp., *M. unguiculatus*, *Myodes glareolus*); tree squirrels (*Paraxerus cepapi*); relatively many primate species	Crook and Shields 1985; Veiga 1993; Pusey and Packer 1994b; Schneider and Lubin 1997; Lewinson 1998; Patterson et al. 1998; Swenson et al. 2001; Dunn et al. 2002; Erez et al. 2005; McLellan 2005; Bellemain et al. 2006a; Solomon and Keane 2007; Dudzinski et al. 2009; Gray 2009; Zedrosser et al. 2009; see references in Ebensperger and Blumstein 2007 and Palombit 2012
II. Improved female condition and future fertility	Premature termination of lactational amenorrhea improves female condition and thereby enhances future reproductive success following insemination by infanticidal male	Arctic ground squirrels (*Spermophilus parryii*); red deer (*Cervus elaphus*); horses (*E. caballus*, *E. ferus przewalskii*); ring-tailed lemurs (*Lemur catta*); sifakas (*Propithecus edwardsi*)	McLean 1983; Bartoš and Madlafousek 1994; Wright 1995; Jolly et al. 2000; Cameron et al. 2003; Hoesli et al. 2009; Morelli et al. 2009
Breeding social status acquisition	Infanticide facilitates acquisition of high rank or breeding status in multimale groups for resident males or new immigrants	Howler monkeys (*Alouatta arctoides*); baboons (*Papio hamadryas hamadryas*); Hanuman langurs (*S. entellus*)	Angst and Thommen 1977; Crockett 2003; Rajpurohit et al. 2008
Breeding site acquisition	Infanticide is part of a strategy to take over a breeding site by destroying existing offspring, and then mating with the resident female(s) or new immigrant female(s)	Burying beetles (*Nicrophorus* spp.); fish (*Lamprologus callipterus*); birds (*Tachycineta bicolor*, *Troglodytes aedon*)	Freed 1986, 1987; Robertson and Stutchbury 1988; Trumbo 1990a; Trumbo and Valletta 2007; Maan and Taborsky 2008
Mate acquisition			
Direct	Infanticide by an extragroup male promotes female transfer to the killer partly by advertising his superior ability to protect against infanticide	Birds (*H. rustica*); colobine monkeys (*Presbytis thomasi*, *Colobus polykomos*); mountain gorillas (*Gorilla gorilla*)	Crook and Shields 1985; Watts 1989; Steenbeek 2000; Møller 2004; Korstjens et al. 2005

Continued

Table 1. *Continued*

Hypothesis	Theoretical rationale	Proposed example taxa	References
Indirect	In species with female dispersal, infanticide by an extragroup male disrupts the mother–subadult daughter bond, facilitating the dispersal of latter to infanticidal male	Chimpanzees (*Pan troglodytes*)	Goodall 1986
Sexual monopolization of female	Infanticide induces female to restrict future copulations to infanticidal males(s) than to other male(s)	Chimpanzees (*P. troglodytes*)	Hamai et al. 1992
Removal of mating competitors	Selective killing of male infants: (a) eliminates possible future sexual competitors of infanticidal males or their sons; (b) undermines coalitionary power of rival males?	Arctic ground squirrels (*S. parryii*); Hanuman langurs (*S. entellus*); white-thighed colobus monkeys (*Colobus vellerosus*)	McLean 1983; Sommer 1994; Teichroeb and Sicotte 2008
Decrease reproductive success of rivals	Infanticide removes genes of rival male(s)	Tree squirrel (*P. cepapi*); fur seals (*Callorhinus ursinus*); patas monkey (*Erythrocebus patas*)	de Villiers 1986; Enstam et al. 2002; Kiyota and Okamura 2005
Benefits related to resources			
Removal of competitors for limiting resources	Infanticide improves access to limiting food resources to infanticidal male and/or his offspring	Gerbils (*M. unguiculatus*); purple-faced langurs (*Semnopithecus vetulus*); howler monkeys (*Alouatta arctoides, A. palliata*)	Rudran 1973b; Clarke 1983; Elwood and Ostermeyer 1984b; Agoramoorthy and Rudran 1995
Improved intergroup competition for limited resources	Killing of infants in neighboring groups enhances access to resources by: (a) causing their mothers to withdraw permanently from borders areas, thereby facilitating killers' home range to expand; (b) facilitating intergroup dominance and thereby priority of access of resources in border areas	(a) Chimpanzees (*P. troglodytes*); (b) guereza (*Colobus guereza*)	(a) Watts et al. 2002; Williams et al. 2002, 2004; Mitani et al. 2010; (b) Harris and Monfort 2003
Nutritional exploitation	Infants are cannibalized to obtain direct nutritional benefits	Fish (*Telmatherina sarasinorum*); yellow sciurid rodents (*Spermophilus beldingi,*	Paul and Kupferschmidt 1975; Svare and Bartke 1978; Sherman 1981; Elwood and

Continued

Table 1. *Continued*

Hypothesis	Theoretical rationale	Proposed example taxa	References
		S. tridecemlineatus, *Cynomys parvidens*); murid rodents (*R. norvegicus*, *M. musculus*, *M. unguiculatus*, *Microtus pennsylvanicus*); baboons (*Papio cynocephalus*)	Ostermeyer 1984a; Vestal 1991; Ebensperger et al. 2000; Hoogland 2001; Altmann et al. 2006; Gray et al. 2007
Benefits related to paternal investment			
Avoidance of misdirected paternal investment	Infanticide prevents male parental investment in offspring sired by another male	Birds (*T. aedon*, *P. domesticus*); humans (*Homo sapiens*)	Daly and Wilson 1984, 1994; Rohwer 1986; Freed 1987; Pierotti 1991; Veiga 1993
Benefits related to learning			
Acquisition of mating tactics and skills	Inexperienced subadult males lethally use pups as proxies for adult females to acquire skill in mating tactics (e.g., female stealing and herding, territory maintenance)	Otariid seals (*Mirounga* spp., *Callorhinus ursinus*)	Le Boeuf and Campagna 1994; Kiyota and Okamura 2005
Nonadaptive hypotheses			
Social pathology	Lethal attacks on infants are: (a) abnormal behavior resulting from human-induced disturbance (e.g., overcrowding) or selection; (b) results of changes in nature/rate of aggression (e.g., takeover, intergroup hostility) caused by anthropogenic disturbance and habitat degradation	Horses (*E. ferus przewalskii*); episodic in others (e.g., Hanuman langurs) (*S. entellus*); Tibetan macaques (*Macaca thibetana*); chimpanzees (*P. troglodytes*)	Jansen et al. 2002; (a) Dolhinow 1977; Curtin and Dolhinow 1978; Feh and Munkhtuya 2008; (b) Boggess 1984; Berman et al. 2007; Williams et al. 2008
Generalized aggression	Infant killing is an incidental byproduct of selection for overall intra- or intersexual aggressiveness in males or (for intergroup infanticide) of general male xenophobic territoriality	Otariid seals (*Mirounga* spp., *Otaria byronia*); horses (*E. ferus przewalskii*); episodic in others (e.g., baboons) (*Papio h. hamadryas*); macaques (*Macaca radiata*); mona monkey (*Cercopithecus campbellii*); chimpanzees (*P. troglodytes*)	Galat-Luong and Galat 1979; Rijksen 1981; Goodall 1990; Bartlett et al. 1993; Le Boeuf and Campagna 1994; Singh et al. 2006; Hoesli et al. 2009

unsurprising that the two taxa in Figure 1 that ostensibly straddle the "high L/G" and "low L/G" groupings have generated conflicting results and debate about sexually selected infanticide. These are the sciurid rodents and the perissodactyls.

Among ungulates, the perissodactyls—particularly equids—show the highest incidence of male infanticide (Pluháček and Bartoš 2005). The majority of reports have come from captivity, however, which has lead to suggestions that the behavior is an artifact of captivity or domestication (Table 1). Moreover, there is little unambiguous evidence from either the captive or few wild studies that infanticide shortens interbirth intervals in females, as predicted by the sexual selection hypothesis (e.g., Feh and Munkhtuya 2008). If there is an adaptive basis to the behavior, a variant of this hypothesis may be more relevant. This alternative hypothesis arose originally from an appreciation that the sexually selected mechanism is unlikely to operate in seasonal breeders, because infanticide cannot easily accelerate female return to fertilizable condition when mating is determined primarily by photoperiodicity or other cyclical cues. However, premature termination of lactation can potentially benefit males in seasonal breeders if it improves the female condition and thus her *future* reproductive success (Hrdy and Hausfater 1984; Hoogland 1985). Bartoš and Madlafousek's (1994) experimental study of red deer (*Cervus elaphus*) was the first to provide some empirical support for this hypothesis, which has been invoked for other mammals, including some primates, rodents, and equids (see Sexual selection II in Table 1). Another context in which sexually selected infanticide may operate among seasonal breeders is, of course, when interbirth intervals exceed 1 yr, as in langurs living at high elevations (Borries 1997). Under this scenario, infanticide can accelerate the return of females to fertilizable condition, and thereby enhance the reproductive success of a male (assuming he retains breeding status in the subsequent reproductive season).

Among the Sciuridae, male infanticide is known to occur in 11–14 species (Ebensperger and Blumstein 2007). In stark contrast to the mostly laboratory-based research on murids, almost all of the sciurid data are from the wild, particularly from studies of prairie dogs (*Cynomys* spp.), ground squirrels (*Spermophilus* spp.), and marmots (*Marmota* spp.). With few exceptions (e.g., nonseasonally breeding tree squirrels) (Solomon and Keane 2007), these naturalistic data are generally less consistent with the sexual selection hypothesis than with alternative explanations (e.g., nutritional exploitation) (see Table 1). Indeed, Hoogland (2007) has argued that his large data set on wild black-tailed prairie dogs (*Cynomys ludovicianus*) and Utah prairie dogs (*C. parvidens*) does not clearly support *any* of the existing hypotheses for male infanticide. Perhaps Blumstein's (2000, p. 195) appeal of more than a decade ago—"we simply need better data"—remains the most compelling conclusion.

Another functional hypothesis for male infanticide has arisen from avian research. Sexually selected infanticide has been suggested for at least 13 species of birds (Veiga 2000). It has been reported in polygynous species (as expected theoretically), but notably also in socially monogamous breeders as well, where more often than advancing the female's next fecund period, infanticide functions as part of a takeover strategy for achieving breeding space and/or acquiring a mate in the first place. This role of infanticide has also been suggested for several species of mammals (see Table 1).

Perhaps no single species has yielded more proposed explanations for male infanticide than the chimpanzee (*P. troglodytes*) (Table 1). Much about infanticide in this ape has long been puzzling (Arcadi and Wrangham 1999). This is particularly true of the violent attacks on the infants of *neighboring* groups, which have seemed to produce no obvious mating- or resource-related benefits for the male perpetrators. Intriguing new data, however, suggest a "range expansion" function, whereby infanticide induces victimized mothers to abandon their home ranges to the killers, which in one Ugandan community facilitated a 22% increase in their territory size (Mitani et al. 2010).

Finally, two hypotheses explain infanticide as maladaptive behavior or accidental byproduct (Table 1). Inspired partly by laboratory demonstrations of infanticide arising from debilitating overcrowding (e.g., Myers and Poole 1961; Calhoun 1962), the social pathology hypothesis links the behavior in the wild to anthropogenic disturbance of either a direct (e.g., provisioning) or indirect (e.g., habitat degradation) nature. Although this explanation may apply to some individual cases (especially in captivity), it has largely failed as a general explanation, partly because the prediction that "the infanticidal individual is behaving in a manner that indicates malfunction" (Pierotti 1991, p. 1141) has not been consistently supported. Sterck (1998) argues that rather than produce infanticide by disrupting "normal" regulatory pathways of behavior, local human disturbance often alters rates of infanticide through its effects on important causal variables and conditions, such as the occurrence of male takeovers. Some anthropogenic intervention has also provided difficult-to-obtain "experimental" support of the sexual selection hypothesis (e.g., the selective removal of males through trophy hunting in fissiped carnivores) (e.g., Wielgus and Bunnell 2000; Loveridge et al. 2007; Wielgus et al. 2013; but see Miller et al. 2003).

The second nonadaptive hypothesis, the "generalized aggression" hypothesis, suggests that infanticide arises accidentally as a byproduct of the intra- or intersexual aggression surrounding male mating competition (Bartlett et al. 1993). Again, although some individual episodes are consistent with this model (Table 1), it has not proved an adequate general explanation of larger data sets, which typically establish the highly "directed" nature of infanticidal attacks and their predictable long-term patterning. One notable exception seems to be the well-studied otariid seals, in which evidence suggests that infants (and juveniles) are trampled to death accidentally by huge males (although even in this case, adaptive hypotheses have been proffered but not tested) (see Table 1).

The Occurrence of Male Infanticide as a Potential Selective Agent for Female Counterstrategies

A defining feature of infanticide is its variability, expressed across taxa and over time. Rates of occurrence may vary significantly across populations of the same species. For example, in chacma baboons of southern Africa (*Papio hamadryas ursinus*), male infanticide accounts for an average of 38% of infant mortality, but among east African yellow (*P. h. cynocephalus*) and olive baboons (*P. h. anubis*), it is not only much less demographically significant but less clearly a male reproductive strategy (Palombit 2003; for other examples, see also Butynski 1990; Arcadi and Wrangham 1999). Even in populations with relatively high rates of infanticide, it is not invariably expressed by *all* males experiencing the putative causal conditions. This has led some researchers to propose that infanticide is a facultative strategy (a behavioral polymorphism?) (e.g., McLean 1983; Palombit et al. 2000; Cords and Fuller 2010), as also suggested by theoretical models (e.g., Chapman and Hausfater 1979; Glass et al. 1985; Yamamura et al. 1990; Broom et al. 2004). Consequently, rates of occurrence in an "infanticidal" population may vary considerably over time. For example, although male infanticide accounted for an average of 38% of infant mortality in the previously mentioned chacma baboon population, this figure ranged from nearly 0% to as high as 75% in certain years (Cheney et al. 2004). The reasons for variation in infanticide are not well understood, but one relevant variable for interpopulation differences in primates as well as carnivores is the rate of breeding male turnover (Packer et al. 1988; Sterck 1998; Crockett and Janson 2000).

Long-term studies of populations in which male infanticide occurs have generally suggested it is a major—if not *the* major—source of infant mortality. Among Serengeti lions (*P. leo*), it accounts for 27% of cub mortality overall, but nearly 100% of the mortality of cubs under the age of 9 mo (Pusey and Packer 1994b). In a sample of 29 primate populations in which infanticide is known, 19 had relative rates $\geq$30% (Palombit 2012). In some of these taxa, male replacement produces a 3-, 11-, or 32-fold increase in infant mortality (Beehner and Bergman 2008; Fedigan et al. 2008; Pavé et al. 2012). In sciurid rodents, male (as well as female) infanticide is a major cause of infant and juvenile mortality (Ebensperger and Blumstein 2007).

In total, the empirical evidence for adaptive strategies of male infanticide provides a compelling rationale for evaluating its possible importance as a selective agent on the evolution of female biology.

THE SEXUAL CONFLICT OF INFANTICIDE: FEMALE COUNTERSTRATEGIES

A very large number of possible female counterstrategies to male infanticide have been proposed and studied to varying degrees. These can be grouped into three broad categories addressing: (1) sexual behavior and reproduction; (2) individual prevention and deterrence; and (3) social counterstrategies. Space limitations preclude discussion of all of these phenomena, and so I focus on some of the major empirically supported patterns.

Sexual and Reproductive Counterstrategies of Females

Among the numerous potential counterstrategies related to sex and reproduction (Table 2), arguably the best known is "confusing paternity" through copulation with males (Hrdy 1979). Direct tests of this hypothesis are limited primarily to laboratory murids (see below), but field data have generally suggested that infanticide is perpetrated by males who have no sexual history with the victimized infants' mothers. In one of the few direct tests in primates, DNA paternity data showed that male attacks on infants were directed at eight times the rate toward unrelated individuals than toward offspring (Soltis et al. 2000).

A logical extension of this hypothesis is that females vulnerable to infanticide will pursue a strategy of "promiscuous" or "multimale" mating, which Hrdy (1979, 1981) invoked to explain the pronounced sexual assertiveness of haplorrhine primates. Broad comparative analyses provide indirect support for this hypothesis. For example, in eutherian mammals, multimale mating is more pronounced in species characterized by male infanticide; among primates alone, 62% of species with female susceptibility to infanticide show promiscuity, compared to 11% of nonvulnerable species (van Noordwijk and van Schaik 2000). Recent data have even suggested a heretofore unappreciated potential for this benefit to accrue to females living in *unimale* groups, through copulations with extragroup males (Palombit 2012). Similarly, Wolff and MacDonald (2004) argue on the basis of their correlative analysis that multimale mating in rodents is best understood as anti-infanticidal confusion of paternity, but it is important to remember that multimale mating has many other possible functions which are implicated in numerous species (e.g., Bryja et al. 2008; Lane et al. 2008). Notably, Hoogland (2007) rejects this hypothesis for Utah prairie dogs based on observations of males attacking *some* of the pups of females with whom they previously mated. However, students of species in which females produce multiply sired litters have speculated that males may be able to recognize kin phenotypically and then selectively target unrelated infants in a litter (e.g., Vestal 1991; Berdoy and Drickamer 2007). If true, this outcome would imply a possible male counterstrategy to counteract the female anti-infanticide counterstrategy of promiscuity (Table 2).

A large number of laboratory tests of this hypothesis in murid rodents suggest that copulation inhibits subsequent infanticidal behavior of males in at least seven species (reviewed in Ebensperger and Blumstein 2007). These results have not been replicated in several studies, however, which sometimes indicate the importance of not just copulation but a period of postcopulatory cohabitation in facilitating inhibition of infanticidal tendencies (e.g., Elwood and Kennedy 1994).

More intriguingly, a series of experiments on the house mouse suggest that this female counterstrategy of promiscuity may have itself selected for additional male counterstrategies. Copulation is known to inhibit infanticide in males (vom Saal and Howard 1982). That inhibition, however, does not commence until about 12 d after the sexual encounter, and it terminates spontaneously 60 d later, a schedule that corresponds roughly to the periods of gestation and lactation, respectively (vom Saal

Table 2. Sexual and reproductive counterstrategies of females with selected example taxa

Proposed counterstrategy	Theoretical rationale	Proposed example taxa	References
Strategies obscuring paternity among males			
Promiscuity/multimale mating	Elevated sexual receptivity and proceptivity and/or polyestrous cycling promotes insemination by multiple males, thereby confusing paternity	Birds (*Tachycineta bicolor*); murid rodents (*Mus musculus*); prairie dogs (*Cynomys parvidens*); bears (*Ursus arctos*); lions (*Panthera leo*); badgers (*Meles meles*); dolphins (*Tursiops* spp.); primates	Bertram 1975; Robertson 1990; Connor et al. 1996; 2000; Dahle and Swenson 2003; Ebensperger and Blumstein 2007; Hoogland 2007; Dugdale et al. 2011; see references in Palombit 2012
Multiply sired litters	In polytocous breeders, females achieve fertilizations by multiple males in single litters	Mice (*M. musculus*); voles (*Myodes rufocanus*); bears (*U. arctos*)	Bellemain et al. 2006a; Ishibashi and Saitoh 2008; Thonhauser et al. 2013
Extended estrus period	Lengthened periods of estrus or polyestrous cycling increase opportunities for copulation with multiple males	Dolphins (*Tursiops*); catarrhine primates with sexual swellings, langurs (*S. entellus*)	Connor et al. 1996; van Schaik et al. 2000; Heistermann et al. 2001
Concealed ovulation	Paternity is obscured by absence of reliable signals of ovulation and/or by variability in the timing of ovulation	Hanuman langurs (*S. entellus*); vervet monkeys (*Chlorocebus pygerythrus*); capuchin monkeys (*Cebus capucinus*)	Andelman 1987; Heistermann et al. 2001; Carnegie et al. 2006
Postconception sexual behavior	Anovulatory copulations (and when relevant, signaling of "ovulation") occur regularly in early gestation	Many catarrhine primates, some platyrrhine primates	See references in Palombit 2012
Situation-dependent sexual receptivity ("pseudoestrus")	Exposure to new (or extragroup) males facultatively causes (a) "deceptive" anovulatory sexual behavior in pregnant females; or (b) longer receptive periods among cycling females	(a) Lemurs (*Propithecus verreauxi*); capuchin monkeys (*Cebus apella*); blue monkeys (*Cercopithecus mitis*); baboons (*Papio hamadryas hamadryas*); red colobus (*Procolobus badius*); (b) cercopithecine monkeys (*C. mitis, Macaca fuscata*)	(a) Struhsaker and Leland 1985; Fairgrieve 1995; Brockman and Whitten 1996; Zinner and Deschner 2000; Pazol 2003; Ramírez-Llorens et al. 2008; (b) Cords 1984; Takahata et al. 1994; Pazol 2003
Strategies focusing copulations and/or conceptions on particular male(s)			
Female mate choice targeting infanticidal male(s)	Females mate preferentially with males most likely to commit infanticide because of factors such as dominance status or spatial proximity	Murid rodents (*Lemmus, Microtus, Mus, Myodes*); bears (*U. arctos*); capuchin monkeys (*C. nigritus, C. olivaceous*)	O'Brien 1991; Agrell et al. 1998; Bellemain et al. 2006b; Izar et al. 2009

Continued

Table 2. *Continued*

Proposed counterstrategy	Theoretical rationale	Proposed example taxa	References
Female mate choice targeting protector male(s) (direct)	Females limit periovulatory matings to a male likely to be a future protector against infanticide	Hamadryas baboons (*P. h. hamadryas*)	Swedell and Saunders 2006
Female mate choice targeting protector male(s) (indirect)	Females incite male mate guarding by advertising ovulation or copulations (e.g., through sexual swellings, vocalizations), which reduces polyandry at ovulation and thereby concentrates paternity in future, high-quality male defender of infant	Catarrhine primates living in multimale societies	See references in Palombit 2012
Strategies manipulating reproduction			
Pregnancy block and pregnancy termination (Bruce effect)	Direct or indirect exposure to a new replacement male results in failure of implantation or absorption of fetus	Equids (*E. burchelli*, *E. caballus*); baboons (*P. h. hamadryas*, *P. h. cynocephalus*); langurs (*S. entellus*); geladas (*Theropithecus gelada*)	Berger 1983; Pereira 1983; Agoramoorthy et al. 1988; Colmenares and Gomendio 1988; Pluháček and Bartoš 2000; Bartoš et al. 2011; Roberts et al. 2012
Superfetation	Conception occurs during gestation accompanied by fertilized eggs entering embryonic diapause	European badger (*Meles meles*)	Yamaguchi et al. 2006
Abandonment of infant	Mothers abandon a healthy infant or an infant injured in infanticidal attack	Colobine monkeys (*Colobus vellerosus*, *S. entellus*)	Hrdy 1977; Sicotte et al. 2007
Acceleration of weaning	Following male takeover, mothers wean older infants sooner than usual	Vervet monkeys (*Chlorocebus pygerythrus*); baboons (*P. h. hamadryas*)	Fairbanks and McGuire 1987; Colmenares and Gomendio 1988
Temporarily reduced fecundity	Following male takeover, females temporarily suspend ovulatory cycles until new male(s) are established and unlikely to be replaced again	Lions (*Panther leo*); leaf monkeys (*P. thomasi*)	Pusey and Packer 1994b; Steenbeek 1999b
Breeding synchrony	Among group-living females, greater ovulatory synchronization prevents single males from monopolizing matings, generating multimale groups that are generally less vulnerable to takeover infanticide	Lions (*P. leo*); lemurs (*Eulemur fulvus rufus*); baboons (*Papio* spp.); blue monkeys (*Cercopithecus mitis*)	Bertram 1975; Cords 1986; Altmann 1990; Ostner and Kappeler 2004

1985; Perrigo et al. 1990, 1992). This timing suggests that the female (sexual) counterstrategy has, in turn, selected for counteradaptations in males that "fine-tune" the expression of infanticidal inhibition to periods when pups are most likely to be their genetic offspring, and that thereby avoid the "evolutionary mistake" of failing to attack unrelated pups. A further step in this coevolutionary scenario is suggested by the fact that aggressive mothers protecting their pups attack males who have copulated with them very recently (within the last 24 h) with the same severity they direct toward males who have not copulated with them at all (Parmigiani et al. 1988b). Taken together, these patterns imply the spiraling, antagonistic dynamic characterizing sexual conflict.

Recognition of the potential importance of male protectors in deterring infanticide in some species (see below) has led to modification of this hypothesized counterstrategy of paternity confusion. Under these conditions, females may benefit most from pursuing a dual sexual strategy that concentrates paternity in the best available male defender and also dilutes (confuses) it in other males (e.g., Clarke et al. 2009). Cues of ovulation, such as the sexual swellings characterizing many species of Old World monkeys and apes, may provide a mechanism for achieving what seem at first glance to be incompatible reproductive goals. The "graded signal hypothesis" holds that: (1) the probability of ovulation is highest at the time of maximal tumescence, and thus sexual swellings can promote fertilization by high-quality (protector) males; but (2) the probability of ovulation at other times is inherently *variable* and *nonzero*, and, hence, paternity can be confused among other males (Nunn 1999). Hormonal data from langurs confirmed that the timing of ovulation was sufficiently variable that low-ranking males occasionally achieved fertilizations despite the dominant male's sexual monopolization of females during periods when ovulation was statistically more likely (Heistermann et al. 2001). Further support for this hypothesis is provided by observations that female primates of several species mate selectively at periovulatory periods (coinciding with increased probability of ovulation) and promiscuously at other times in their cycle (when ovulation is less likely to occur) (reviewed Palombit 2012). Moreover, in catarrhine primates exhibiting sexual swellings, the follicular phase of the menstrual cycle—during which follicles mature in the ovary and ending with ovulation—is significantly longer in species in which females are subject to higher risk of male infanticide (van Schaik et al. 2000). One interpretation of this pattern is that longer phases provide greater opportunities for matings to confuse paternity.

A recently suggested angle on this coevolutionary dialectic is that *males* may themselves exploit this female counterstrategy to protect their offspring. Evidence from chacma baboons suggests that alpha males may "cede" a proportion of mating opportunities to other males and thereby reduce the risk of infanticidal attack (Boyko and Marshall 2009; Henzi et al. 2010).

Individual Counterstrategies of Females

A number of possible deterrent counterstrategies are based on individual action (Table 3). Maternal (or postpartum) aggression, the "fierce and persistent" hostility of pregnant or lactating females to intruders is a conspicuous feature of rodent reproductive biology (Lonstein and Gammie 2002, p. 869) and is argued to underlie female territoriality in some species (Wolff and Peterson 1998; but see Ebensperger 1998a). Although maternal aggression is a heterogeneous phenomenon subserving multiple functions, a link to infanticide prevention is currently suggested in at least eight murid species and possibly several sciurid species (reviewed Ebensperger and Blumstein 2007; McGuire and Bemis 2007; Weber and Olsson 2008). Once again, numerous detailed data have emerged from laboratory investigations of murids. In species such as the house mouse, the aggressiveness of mothers is striking in light of the virtual absence of aggression when females are in other reproductive states (Heiming et al. 2013). Aggression is sensitive to the presence of pups (Elwood et al. 1990), increases with litter size (Maestripieri and Alleva 1990; see also

Table 3. Individual counterstrategies of females with selected example taxa

Proposed counterstrategy	Theoretical rationale	Proposed example taxa	References
Maternal aggression	Females increase in heterosexual aggressiveness during lactation and/or directly defend infants from infanticidal attacks by males; females become territorial during periods of offspring rearing	Birds (*H. rustica*, *Tachycineta bicolor*); many murid rodents, gerbils (*Gerbilliscus leucogaster*); shrews (*Sorex araneus*)	Møller 1988; Robertson 1990; Wolff and Peterson 1998; Ebensperger and Blumstein 2007; Lötter and Pillay 2012; Oleinchenko 2012
Female dominance	Female maintains a relationship of social dominance over males either during lactation or more permanently	Gerbils (*M. unguiculatus*); ring-tailed lemurs (*Lemur catta*)	Pereira and Weiss 1991; Elwood and Kennedy 1994; Ichino 2005
Maternal protectiveness	Under conditions of heightened infanticide risk, mothers increase visual monitoring of potentially infanticidal male and/or offspring, and restrain and/or maintain proximity to infants more	Horses (*E. caballus*); many primates	Cameron et al. 2003, 2009; see references in Palombit 2012
Chemical suppression	Pregnant females produce pheromone that disrupts neurochemical pathways mediating infanticidal behavior in males	Rats (*R. norvegicus*)	Mennella and Moltz 1988
Sexual segregation	Lactating females (a) alter individual ranging to avoid areas with potentially infanticidal males; or (b) for gregarious females, adjust intragroup spatial relations to decrease proximity to infanticidal males or reduce participation in intergroup encounters	(a) Lions (*Panthera pardus*); bears (*U. arctos*); spider monkeys (*Ateles* spp.); chimpanzees (*P. troglodytes*); (b) baboons (*P. h. ursinus*); capuchin monkeys (*C. capucinus*)	(a) Wrangham and Smuts 1980; Packer and Pusey 1983, 1994b; McComb et al. 1993; Sakura 1994; Matsumoto-Oda 1999; Wielgus and Bunnell 2000; Dahle and Swenson 2003; Ben-David et al. 2004; Aureli et al. 2008; (b) Palombit et al. 2001; Crofoot 2007

Koskela et al. 2000), and wanes as pups mature (Parmigiani et al. 1994). Moreover, in their defensive responses to the infanticidal threat posed by sexually naïve males, lactating females direct "severe" bites to vulnerable body parts (such as the head and ventrum) at 3–5 times the rate they do so in aggressive interactions with (less infanticidal) virgin females (Parmigiani et al. 1988a, 1990). Mothers can also differentiate among males on the basis of infanticidal potential and respond accordingly (Elwood et al. 1990). A possible indication of sexually antago-

 Cite this article as *Cold Spring Harb Perspect Biol* doi: 10.1101/cshperspect.a017640

nistic coevolution in this domain is that the magnitudes of male infanticidal behavior and maternal aggression are positively correlated with one another in genetic strains of mice (Parmigiani et al. 1999).

The actual effectiveness of maternal aggression in deterring male infanticide in rodents is debated, however. In the laboratory, male mice retaliate with extreme aggression, particularly against females with whom they have not mated (or have mated only recently), often overcoming maternal defense and successfully committing infanticide, leading Parmigiani et al. (1989) to conclude that maternal aggression may simply "delay," rather than prevent, infanticide (see also Ebensperger 1998b). Ebensperger and Blumstein (2007), however, suggest that under more naturalistic conditions, such a delay would probably often translate into maternal success. On the other hand, these researchers point out that ineffectiveness does seem to characterize maternal aggression in several species of squirrel, and they explicitly raise the possibility that this may reflect different outcomes or phases in sexually antagonistic coevolution.

Similarly, primate mothers mount concerted and protracted defense of their infants against infanticidal males, but once again evidence suggests that such efforts succeed in only a handful of species (Palombit 2012), some of which are characterized by the relatively rare social arrangement of stable female dominance over males (which also applies to some rodent examples, Elwood and Kennedy 1994). The reasons for this general ineffectiveness remain unclear, but they may partly reflect a greater reliance of female primates on socially based counterstrategies (see below).

A final individual counterstrategy has been suggested in several mammalian species characterized by solitariness or fission–fusion/dispersed gregariousness. In these species, lactating females may pursue a strategy of evasion based on sexual segregation (Table 2). There is evidence that even females in highly gregarious species may adopt this counterstrategy on a smaller spatial scale by adjusting behavior to decrease proximity to potentially infanticidal males residing in their groups (see Table 2).

Social Counterstrategies of Females

A large set of potential counterstrategies involve deterrence through cooperation with conspecifics. These can be grouped into two sets of strategies involving social interactions with other females and with males, respectively (Table 4).

Female gregariousness has traditionally been viewed as an adaptive solution to the problems of predation and feeding ecology (e.g., Alexander 1974; Wrangham 1980), and it is only recently that the selective role of infanticide risk has begun to be formally incorporated into these models (e.g., Sterck et al. 1997). There are several possible mechanisms through which group living potentially reduces infanticide risk: dilution effects (sensu Hamilton 1971), early detection through communal vigilance and alarm-based avoidance of infanticidal males; female–female coalitionary aggression to extragroup males to prevent takeover or to resident males to thwart attacks on infants. As with ecological models, related females may be more likely to join together in this manner (e.g., lions, see below), but aggregations of unrelated females to reduce infanticidal threat have also been suggested (e.g., Maestripieri and Rossi-Arnaud 1991; Cameron et al. 2009).

Although female gregariousness has been invoked as a counterstrategy for a number of species (Table 4), direct evidence for the predicted inhibitory effect on infanticide is still relatively rare. In a study of a seminatural population of house mice, Manning et al. (1995) reported that female communal nests experienced less than half the rate of infanticide (by both sexes) than did single-mother nests. A possible mechanism underlying this outcome is suggested by laboratory experiments demonstrating that paired females effectively repel male (and female) intruders (Parmigiani 1986; Maestripieri and Rossi-Arnaud 1991). Similarly, when food was experimentally manipulated to promote increased aggregation of female bank voles (*Myodes glareolus*), pup survival increased, which Rémy et al. (2013) attributed to improved anti-infanticide protection.

There are several lines of evidence that group living among lionesses (*Panthera leo*) is

Table 4. Social counterstrategies of females with selected example taxa

Proposed counterstrategy	Theoretical rationale	Proposed example taxa	References
Social strategies involving other females			
Female–female association	(a) Females living in permanent groups with other females reduce infanticide risk through derive dilution, vigilance, and/or defensive benefits; (b) lactating members of aggregate temporarily with other females	(a) Voles (*Microtus oeconomus*, *Myodes glareolus*); house mice (*M. musculus*); Arctic ground squirrels (*S. parryii*); horses (*E. caballus*); lions (*P. leo*); dolphins (*Tursiops* spp.); langurs (*S. entellus*); primates generally; (b) baboons (*P. hamadryas ursinus*)	(a) McLean 1982; Pusey and Packer 1994a,b; Brereton 1995; Manning et al. 1995; Treves and Chapman 1996; Connor 2000; Le Galliard et al. 2006; Cameron et al. 2009; Rémy et al. 2013; (b) Cowlishaw 1999; Palombit et al. 2001
Female coalitionary defense	Females collectively attack infanticidal males and/or deter male immigration with relative success	Lions (*Panthera leo*); sifakas (*Propithecus edwardsi*, *P. verreauxi*); chimpanzees (*P. troglodytes*); bonobo (*Pan paniscus*); blue monkeys (*Cercopithecus mitis*); red colobus (*Procolobus badius*); bonobos (*P. paniscus*)	Butynski 1982; Packer et al. 1990; Richard et al. 1993; Starin 1994; Grinnell and McComb 1996; de Waal 1997; Morelli et al. 2009; Cords and Fuller 2010
Residency in small (female) groups	If risk of male takeover increases with female group size, females maintain small groups via secondary transfer, group fission, and/or aggression to female immigrants	Lions (*P. leo*); many species of colobine monkeys, howler monkeys, gorillas	Pusey and Packer 1987; see references in Palombit 2012
Social strategies involving males			
Transfer with evicted male	In species in which infanticide is preceded by male takeover, lactating females temporarily accompany ousted male until infant is weaned	Capuchin monkeys (*Cebus capucinus*); colobine monkeys (*Semnopithecus vetulus*, *S. entellus*, *Presbytis rubicunda*)	Rudran 1973a; Hrdy 1977; Davies 1987; Jack and Fedigan 2009
Male–female association	Females associate with individual male defender(s) as: (a) relatively loose association in dispersed societies; (b) temporary relationship with particular males in multimale groups during lactation; or (c) permanent male–female pairs or unimale groups	Burying beetles (*Nicrophorus* spp.); birds (*Troglodytes aedon*); Arctic ground squirrels (*S. parryii*); muroid rodents; bears (*U. arctos*); lions (*P. leo*); dolphins (*Tursiops* spp.); many primates including mouse lemurs (*Cheirogaleus major*); howler monkeys (*Alouatta* spp.); mangabeys	McLean 1983; Freed 1986; Watts 1989; Trumbo 1990b, 2007; Gust 1994; Pusey and Packer 1994b; Kappeler 1997; Sommer 1997; Agrell et al. 1998; Borries et al. 1999b, 2011; Steenbeek 1999a; Connor et al. 2000; Crockett and Janson 2000; Delgado and van Schaik 2000; Weingrill

Continued

Table 4. *Continued*

Proposed counterstrategy	Theoretical rationale	Proposed example taxa	References
		(*Cercocebus atys*, *C. galeritus*); baboons (*P. hamadryas hamadryas*, *P. h. ursinus*); colobine monkeys (*Colobus vellerosus*, *Procolobus badius*, *P. thomasi*, *S. entellus*); gibbons (*Hylobates* spp.); great apes (*Gorilla gorilla* subspp., *Pongo pygmaeus*)	2000; Stokes 2004; Bellemain et al. 2006b; Swedell and Saunders 2006; Teichroeb and Sicotte 2008; Palombit 2009; Yamagiwa et al. 2009; Fruteau et al. 2010; Opie et al. 2013
Residency in multimale groups	In female-dispersal systems, females immigrate to groups with more males; in female-philopatric systems, females promote multimale group structure	Lemurs (*Eulemur fulvus*, *P. verreauxi*); howler monkeys (*Alouatta arctoidea*, *A. caraya*, *A. palliata*); cercopithecine monkeys (*Chlorocebus pygerythrus*, *Macaca fuscata*); gorillas (*Gorilla gorilla beringei*)	Crockett and Janson 2000; Treves 2001; Isbell et al. 2002; Ostner and Kappeler 2004; Yamada and Nakamichi 2006; Lewis 2008; Robbins et al. 2009a; Pavé et al. 2012
Residency in groups with better male defender	Primarily among females living in unimale societies, cycling females immigrate to a new group with a better protector male	Colobine monkeys (*Colobus vellerosus*, *Procolobus badius*, *P. rufomitratus*, *P. thomasi*); gorillas (*Gorilla gorilla* subspp.)	Marsh 1979; Starin 1994; Sterck 1997; Stokes et al. 2003; Robbins et al. 2009b; Teichroeb et al. 2009; Yamagiwa et al. 2009

an effective deterrent to infanticide (Packer and Pusey 1983). Cub mortality is much lower for females living in groups of two or more than for solitary mothers (Pusey and Packer 1994b). This is partly achieved through coalitionary attack on alien males, which successfully protects cubs from attack by single individuals or small groups of males, who "may have to fight for their lives" against lionesses capable of inflicting fatal injuries (Pusey and Packer 1994b, p. 297). The counterstrategic response of males is to use vocal cues to avoid groups of females (Grinnell and McComb 1996) or to travel in large male cohorts, against which female defense is less effective. Like house mouse dams, however, lionesses appear capable of discriminating among males on the basis of infanticide threat and may use evasion to reduce such interactions (McComb et al. 1993). Finally, socioecological data suggest that pride size does not maximize individual food intake through communal hunting, but rather infanticidal protection (Packer et al. 1990).

Somewhat surprisingly, female coalitionary defense against infanticidal males in most primates appears as generally unsuccessful as individual maternal defense (Palombit 2012). It is possible that the anti-infanticide advantage conferred by gregariousness involves the harder-to-test mechanism of improved vigilance and nonaggressive intervention (e.g., retrieving infants from dangerous situations before attacks can be launched, as has been noted in a number of field studies). In some primate species, however, an alternative version of this female–female counterstrategy may operate.

This alternative version is based on the idea that larger groups of females are predicted to be more attractive as takeover targets to potentially infanticidal males. Thus, females may theoretically garner increased protection for their infants by breeding in relatively small groups (Janson and Goldsmith 1995; Treves and Chapman 1996). This goal can be achieved via several possible mechanisms, such as the fissioning of large groups, individual female transfer to smaller groups, or the aggressive deterrence of alien female immigration by resident females. Supportive evidence for this counterstrategy has been found in a number of primate species in which: (1) the predicted negative correlation between infant mortality or infanticide rates and female group size has been documented; and (2) one or more of the above three mechanisms has been identified (reviewed Palombit 2012; see also Pusey and Packer 1987 for lions). This counterstrategy carries a potentially important implication that may help to explain the surprising ineffectiveness of female coalitionary aggression against infanticide in many primates. The individual female transfer that underlies this counterstrategy in primates tends to produce small groups of females who share relatively low genetic relatedness with one another (cf. lions). Such conditions may impede the formation of the kinds of female–female coalitions that are effective deterrents to infanticidal males (Palombit 2012). This hypothesis has not been tested directly, but it underscores the possibility that various female counterstrategies may be incompatible with one another, such that adoption of one may preclude another.

A final set of counterstrategies has been inspired by a growing body of data demonstrating the important role males may play in preventing infanticide. The majority of relevant data on these counterstrategies has come from studies of nonhuman primates, possibly because other animals, such as rodents, are less likely to rely on socially based counterstrategies to infanticide (Blumstein 2000). There are several variants on this strategic theme, which can be distinguished by general features of social system, such as the feasibility of transfer as a strategic option for females at risk of infanticide (reviewed in Palombit 2012).

The ecological and social costs of intergroup transfer vary considerably for females, but they are generally expected to be higher for frugivorous species than for folivores. This is because the greater rarity of fruits (relative to leaves) in most environments engenders elevated feeding competition and greater difficulties in finding food in the unfamiliar habitats that a transferring female would necessarily encounter. Moreover, the social costs of female transfer are potentially higher for frugivores because the "female-bonded" social organization of many fruit-eating primates (particularly Old World monkeys) is centered on networks of social relationships among related females. Thus, female transfer may necessarily require cessation of cooperative interactions with female kin, which can impose costs on both individual and inclusive fitness. It is for these reasons that transfer is less of an option for females in the female-philopatric, multimale societies. In this case, an anti-infanticide counterstrategy based on cooperation with resident males would be more likely. One such possible counterstrategy is to maintain a temporary relationship with a male protector (or a small number of adult males) during the period of infant vulnerability (Table 4), as exemplified by the chacma baboon of southern Africa. When a female gives birth, she establishes such a relationship, or "friendship," with a particular male, which lasts as long as she has her nursing infant; if the infant weans or dies, the relationship ends. Observational and experimental playback data suggest that male friends are significantly more predisposed than other males to come to the aid of females under attack by the potentially infanticidal alpha male (Palombit et al. 1997), possibly because of paternity of the infants involved (Moscovice et al. 2010). Immigration of a new male who attains alpha status results in elevated cortisol among the group's mothers (but not among cycling females); cortisol levels climb even higher if an infanticide is successfully perpetrated or, notably, among the (very few) mothers who lack a male friend (Beehner et al. 2005; Engh et al. 2006). Lactating females compete for male friends, such that a mother's rank is positively correlated with the rank of her male friend (Palombit et al. 2001).

It is partly because the vast majority of mothers have male friends that it is difficult to assess directly the fitness benefits of friendship for females, and data are still too few to evaluate how variation in the characteristics of friendships or male friends affect infant survival. Nevertheless, the evidence is suggestive that this system reflects an anti-infanticidal counterstrategy.

For folivorous primates, the costs of female transfer are predicted to be lower, opening the way for dispersal-based counterstrategies, such as the previously described transfer to smaller female groups. Many of these species are characterized by unimale breeding units, which are particularly prone to takeover infanticide. Females in these societies may reduce the risk of infanticide by choosing a group with a high-quality male protector. This hypothesis has not been subject to rigorous testing, partly because of the difficulties in measuring variation in male quality as protector. There are suggestive supportive data for a few species, however, notably Thomas's leaf monkeys (*P. thomasi*), in which infanticide is not perpetrated as part of a male "takeover" by a new immigrant, but rather by extragroup males. Victimized females respond to infanticide by abandoning their group and joining the killer. And, as predicted, cycling females who transfer from older males to "prime" (younger) males subsequently experience fewer infanticidal attacks.

For relatively folivorous primates in which female transfer remains an ecologically viable option, but in which the number of males in groups varies, an alternative counterstrategy is to preferentially target groups with more males. This hypothesis is supported in several species, such as howler monkeys (*Alouatta* spp.) and mountain gorillas (*G. gorilla beringei*) in which there is empirical evidence that infanticide occurs at predictably lower rates in groups with more males (independently of female group size). Even so, puzzles remain, such as the apparent lack of preference among female gorillas for multimale groups as destinations for transfer (Robbins et al. 2009a).

The above examples have focused on gregarious animals, but even in dispersed societies in which the sexes are relatively solitary and interact sporadically, female association with a male may nevertheless afford some protection against infanticide (Table 4). For example, ousted male lions will re-establish or maintain residency near a pride that offers few mating opportunities if it contains vulnerable cubs they likely sired (Pusey and Packer 1994b). Male–female pairs of house mice repulse infanticidal intruders (Palanza et al. 1996) so successfully that Parmigiani et al. (1994) suggest that the apparent failure of maternal aggression to thwart infanticide in previous laboratory experiments may derive from the artificial absence of the stud male.

Finally, infanticide has been suggested to be the primary selective force behind the evolution of social monogamy (Table 4) (reviewed in Palombit 2000). Perhaps the most compelling evidence for this hypothesis comes from research on burying beetles (*Nicrophorus* spp.). In a manner reminiscent of avian monogamy, a male–female pair of beetles maintains a long period of postcopulatory association with one another on a vertebrate carcass that provides nourishment for developing larvae (Eggert and Müller 1997). Both parents provision the larvae by regurgitating predigested carrion obtained from this carcass, although the paternal contribution is not crucial for offspring success (Trumbo 1991; Müller et al. 1998). On the other hand, male presence significantly reduces the risk of takeover infanticide by another male, suggesting an anti-infanticide benefit of social monogamy in some populations (Trumbo 1990a,b, 1991; Robertson 1993; Trumbo and Valletta 2007). Similarly, in some birds such as tropical house wrens (*Troglodytes aedon*), protection against infanticide appears to be a more important advantage of social monogamy than biparental provisioning of offspring (Freed 1986; see also Veiga 2000). This hypothesis as it applies to primates, notably the monogamous gibbons (*Hylobates* spp.) (van Schaik and Dunbar 1990), has been questioned on the basis of comparative data (Palombit 1999; Fuentes 2000; Brockelman 2009), but recent indirect evidence of infanticide in one population of white-handed gibbons suggests an intriguing possibility (Borries et al. 2011). Contradictory conclusions about the selective role of infanticide in the evolution of mammalian social monogamy have emerged from recent phylogenetic and life history analyses (Lukas and Clutton-Brock 2013; Opie et al. 2013).

CONCLUSIONS

Male infanticide appears to be an arresting example of interlocus conflict (sensu Arnqvist and Rowe 2005). There is still a great deal we do not understand about the array of behaviors constituting this system. The adaptive significance, occurrence, and variability of male infanticide remains unclear in many species, and we still lack enough data to compare directly the lifetime reproductive success of the "infanticidal" versus alternative "noninfanticidal" strategies. Nevertheless, evidence accumulated over the last several decades yields several general conclusions. First, in the well-studied populations, it seems clear that infanticide is not simply an aberration or byproduct of other biological characters or conditions, but represents an adaptive strategy for males in some circumstances. Second, the adaptive significance of the strategy is multifarious; male infanticide is not unitary. The sexual selection hypothesis may be the best known explanation, but there is sufficient evidence to consider a multiplicity of other functions (Table 1). Third, whatever the benefit of infanticide to males, there is little doubt that it usually imposes a substantive fitness cost on females (as well as infants, of course), particularly in the slowly reproducing species in which one form of infanticide—sexually selected—is expected to operate. This provides a strong rationale for expecting that male infanticide has acted as a selective agent in the evolution of female biology in these species, perhaps operating as strongly as the conventionally recognized contingencies related to predation and feeding ecology.

Female counterstrategies to infanticide have been the subject of rigorous study for a relatively short time, however. What is perhaps most striking about current data is the great variety of possibilities that have been proposed (Tables 2–4). These mechanisms range from counterstrategies that increase the potential costs of infanticide to would-be perpetrators (e.g., promiscuity, maternal aggression, coalitionary defense with other females or males), to those that decrease its benefits (promoting small female group sizes or multimale social structure), to some that simply "cut the losses" of infanticide (e.g., pregnancy termination). It is also striking that females in the well-studied populations appear multiple times throughout Tables 2–4, suggesting that females at risk of infanticide rely not on one counterstrategy but on an array of counterstrategies. Lions provide an excellent example of this principle. When cycling, lionesses mate promiscuously to confuse paternity (Table 2). During lactation, females with older cubs ($\geq$10 mo) capable of independent locomotion opt for a strategy of evasion, living largely solitarily on the periphery of their home ranges (sexual segregation) (Table 3). Should alien males be encountered visually or aurally, females attempt to escape or resort to individual maternal aggression (Table 3) (although the latter counterstrategy carries potentially high costs). For females with younger, less mobile cubs, however, sexual segregation is less feasible, and they rely on membership in a pride of female relatives (especially if it is a larger pride) (Table 4). They thereby benefit from the coalitionary defense that is "always able to overcome or drive off one or a few males" (Pusey and Packer 1994b, p. 297). But if prides become too large and thus more attractive for male takeover, then females may disperse (Table 4). The more general conclusion suggested here is that sexually antagonistic coevolution may be a mechanism accounting for much of the great diversity of animal behavior.

There is an unquestionable need for further research on possible female counterstrategies. There are alternative explanations for most of the behaviors listed in Tables 2–4, which must be carefully tested (and rejected) to substantiate the anti-infanticide hypothesis. Analyses are additionally complicated by the fact that predictions of the anti-infanticide hypotheses are often similar to those of alternatives based on ecology. Finally, although it seems likely that female counterstrategies will help to explain variation in the occurrence of male infanticide,

 Cite this article as *Cold Spring Harb Perspect Biol* doi: 10.1101/cshperspect.a017640

data from the wild are insufficient to address this issue quantitatively.

It should also be noted that the implications for sexually antagonistic coevolution may extend well beyond infanticide and its immediate counterstrategies. A case in point concerns the suite of behaviors collectively known as male "sexual coercion" of females. Those aspects of sexual coercion that involve restricting multimale mating by females may be best understood as the male evolutionary response to promiscuity as a female counterstrategy to the male strategy of infanticide (Wolff and MacDonald 2004; Palombit 2014). Similarly, female promiscuity may have exerted selection on systems of sperm competition among males in some taxa, which in turn may have then "counter selected" for mechanisms of cryptic female choice (Kappeler 2012). Thus, the potential cascading effects of the sexual conflict originating with infanticide may be extensive, broad, and diverse.

In many ways, the study of male infanticide and female counterstrategies does not lend itself well to the kinds of methodologies and reproductive profiles that allow direct assessment of sexually antagonistic coevolution (Palombit 2010). The relevant organisms often reproduce slowly, are not easily amenable to experimental manipulation, and are difficult to study in the wild. Despite these difficulties, however, these systems offer opportunities to study the action of sexual conflict in the context of highly gregarious life styles and, more generally, to broaden our appreciation of the diversity of pathways over which sexually antagonistic coevolution has operated.

ACKNOWLEDGMENTS

I thank William Rice and Sergey Gavrilets for inviting me to contribute to this volume. I am also grateful to S. Alavi, M. Janiak, S. Kivai, A. Moldawer, D. Prasetyo, and E. Vogel for discussion of the material for this manuscript.

REFERENCES

Agoramoorthy G, Mohnot SM. 1988. Infanticide and juvenilicide in hanuman langurs (*Presbytis entellus*) around Jodhpur, India. *Hum Evol* **3:** 279–296.

Agoramoorthy G, Mohnot SM, Sommer V, Srivastava A. 1988. Abortions in free-ranging Hanuman langurs (*Presbytis entellus*)—A male induced strategy? *Hum Evol* **3:** 297–308.

Agoramoorthy G, Rudran R. 1995. Infanticide by adult and subadult males in free-ranging red howler monkeys, *Alouatta seniculus*, in Venezuela. *Ethology* **99:** 75–88.

Agrell J, Wolff JO, Ylönen H. 1998. Counter-strategies to infanticide in mammals: Costs and consequences. *Oikos* **83:** 507–517.

Alexander RD. 1974. The evolution of social behavior. *Annu Rev Ecol Syst* **5:** 325–383.

Altmann J. 1990. Primate males go where the females are. *Anim Behav* **39:** 192–195.

Altmann J, Altmann SA, Hausfater G. 1978. Primate infant's effects on mother's future reproduction. *Science* **201:** 1028–1029.

Altmann J, Sayialel S, Bayes M, Bruford MW, Alberts SC. 2006. Cannibalism in baboons: Sexual selection versus hungry hunters. *Int J Primatol* **27** (Suppl 1): 315.

Andelman SJ. 1987. Evolution of concealed ovulation in vervet monkeys (*Cercopithecus aethiops*). *Am Nat* **129:** 785–799.

Angst W, Thommen D. 1977. New data and discussion of infant killing in Old World monkeys and apes. *Folia Primatol* **27:** 198–229.

Arcadi CA, Wrangham RW. 1999. Infanticide in chimpanzees: Review of cases and a new within-group observation from Kanyawara study group in Kibale National Park. *Primates* **40:** 337–351.

Arnqvist G, Rowe L. 2005. *Sexual conflict*. Princeton University Press, Princeton, NJ.

Aureli F, Schaffner CM, Boesch C, Bearder SK, Call J, Chapman CA, Connor RC, di Fiore A, Dunbar RIM, Henzi SP, et al. 2008. Fission-fusion dynamics. *Curr Anthropol* **49:** 627–654.

Bartlett TQ, Sussman RW, Cheverud JM. 1993. Infant killing in primates: A review of observed cases with specific reference to the sexual selection hypothesis. *Am Anthropol* **95:** 958–990.

Bartoš L, Madlafousek J. 1994. Infanticide in a seasonal breeder: The case of the red deer. *Anim Behav* **47:** 217–219.

Bartoš L, Bartošová J, Pluháček J, Šindelářová J. 2011. Promiscuous behaviour disrupts pregnancy block in domestic horse mares. *Behav Ecol Sociobiol* **65:** 1567–1572.

Beehner JC, Bergman TJ. 2008. Infant mortality following male takeovers in wild geladas. *Am J Primatol* **70:** 1152–1159.

Beehner JS, Bergman J, Cheney DL, Seyfarth RM, Whitten PL. 2005. The effect of new alpha males on female stress in free-ranging baboons. *Anim Behav* **69:** 111–1221.

Bellemain E, Swenson JE, Taberlet P. 2006a. Mating strategies in relation to sexually selected infanticide in a non-social carnivore: The brown bear. *Ethology* **112:** 238–246.

Bellemain E, Zedrosser A, Manel S, Waits LP, Taberlet P, Swenson JE. 2006b. The dilemma of female mate selection in the brown bear, a species with sexually selected infanticide. *Proc Roy Soc Lond B* **273:** 283–291.

Ben-David M, Titus K, Beier L. 2004. Consumption of salmon by Alaskan brown bears: A trade-off between nutritional requirements and the risk of infanticide? *Oecologia* **138:** 465–474.

Berdoy M, Drickamer LC. 2007. Comparative social organization and life history of *Rattus* and *Mus*. In *Rodent societies: An ecological and evolutionary perspective* (ed. Wolff JO, Sherman PA), pp. 380–392. University of Chicago Press, Chicago.

Berger J. 1983. Induced abortion and social factors in wild horses. *Nature* **303:** 59–61.

Berman C, Li J, Ogawa H, Ionica C, Yin H. 2007. Primate tourism, range restriction, and infant risk among *Macaca thibetana* at Mt. Huangshan, China. *Int J Primatol* **28:** 1123–1141.

Bertram BC. 1975. Social factors influencing reproduction in wild lions. *J Zool* **177:** 463–482.

Blumstein DT. 2000. The evolution of infanticide in rodents: A comparative analysis. In *Infanticide by males and its implications* (ed. van Schaik CP, Janson CH), pp. 178–197. Cambridge University Press, Cambridge.

Boggess JE. 1984. Infant killing and male reproductive strategies in langurs (*Presbytis entellus*). In *Infanticide: Comparative and evolutionary perspectives* (ed. Hausfater G, Hrdy SB), pp. 283–310. Aldine, New York.

Borries C. 1997. Infanticide in seasonally breeding multimale groups of Hanuman langurs (*Presbytis entellus*) in Ramnagar (South Nepal). *Behav Ecol Sociobiol* **41:** 139–150.

Borries C, Laundhard K, Epplen C, Epplen JT, Winkler P. 1999a. DNA analyses support the hypothesis that infanticide is adaptive in langur monkeys. *Proc Roy Soc Lond B* **266:** 901–904.

Borries C, Launhardt K, Epplen C, Epplen JT, Winkler P. 1999b. Males as infant protectors in Hanuman langurs (*Presbytis entellus*) living in mulitmale groups: Defense pattern, paternity, and sexual behavior. *Behav Ecol Sociobiol* **46:** 350–356.

Borries C, Savini T, Koenig A. 2011. Social monogamy and the threat of infanticide in larger mammals. *Behav Ecol Sociobiol* **65:** 685–693.

Boyko RH, Marshall AJ. 2009. The willing cuckold: Optimal paternity allocation, infanticide and male reproductive strategies in mammals. *Anim Behav* **77:** 1397.

Brereton AR. 1995. Coercion-defense hypothesis: The evolution of primate sociality. *Folia Primatol* **64:** 207–214.

Brockelman WY. 2009. Ecology and the social system of gibbons. In *The gibbons* (ed. Lappan S, Whittaker D), pp. 211–239. Springer, New York.

Brockman DK, Whitten PL. 1996. Reproduction in free-ranging *Propithecus verreauxi*: Estrus and the relationship between multiple partner matings and fertilization. *Am J Phys Anthropol* **100:** 57–69.

Brockman DK, Cobden AK, Whitten PL. 2009. Birth season glucocorticoids are related to the presence of infants in sifaka (*Propithecus verreauxi*). *Proc Roy Soc Lond B* **276:** 1855–1863.

Broom M, Borries C, Koenig A. 2004. Infanticide and infant defence by males—Modelling the conditions in primate multi-male groups. *J Theor Biol* **231:** 261–270.

Bryja J, Patzenhauerová H, Albrecht T, Mošanský L, Stanko M, Stopka P. 2008. Varying levels of female promiscuity in four *Apodemus* mice species. *Behav Ecol Sociobiol* **63:** 251–260.

Butynski TM. 1982. Harem-male replacement and infanticide in the blue monkey (*Cercopithecus mitis stuhlmanni*) in the Kibale Forest, Uganda. *Am J Primatol* **3:** 1–22.

Butynski TM. 1990. Comparative ecology of blue monkeys (*Cercopithecus mitis*) in high and low density subpopulations. *Ecol Monogr* **60:** 1–26.

Caley J, Boutin S. 1985. Infanticide in wild populations of *Ondatra zibethicus* and *Microtus pennsylvanicus*. *Anim Behav* **33:** 1036–1037.

Calhoun JB. 1962. Population density and social pathology. *Sci Am* **206:** 139–148.

Cameron EZ, Linklater WL, Stafford KJ, Minot EO. 2003. Social grouping and maternal behaviour in feral horses (*Equus caballus*): The influence of males on maternal protectiveness. *Behav Ecol Sociobiol* **53:** 92–101.

Cameron EZ, Setsaas TH, Linklater WL. 2009. Social bonds between unrelated females increase reproductive success in feral horses. *Proc Natl Acad Sci* **106:** 13850–13853.

Carnegie SD, Fedigan LM, Ziegler TE. 2006. Post-conceptive mating in white-faced capuchins, *Cebus capucinus*: Hormonal and sociosexual patterns of cycling, noncycling, and pregnant females. In *New perspectives in the study of Mesoamerican primates* (ed. Estrada A, et al.), pp. 387–409. Springer, New York.

Chapman M, Hausfater G. 1979. The reproductive consequences of infanticide in langurs: A mathematical model. *Behav Ecol Sociobiol* **5:** 227–240.

Cheney DL, Seyfarth RM, Fischer J, Beehner J, Bergman T, Johnson SE, Kitchen DM, Palombit RA, Rendall D, Silk JB. 2004. Factors affecting reproduction and mortality among baboons in the Okavango Delta, Botswana. *Int J Primatol* **25:** 401–428.

Clarke MR. 1983. Infant-killing and infant disappearance following male takeovers in a group of free-ranging howling monkeys (*Alouatta palliata*) in Costa Rica. *Am J Primatol* **5:** 241–247.

Clarke PMR, Pradhan GR, van Schaik CP. 2009. Intersexual conflict in primates: Infanticide, paternity allocation, and the role of coercion. In *Sexual coercion in primates and humans* (ed. Muller MN, Wrangham RW), pp. 42–77. Harvard University Press, Cambridge, MA.

Colmenares F, Gomendio M. 1988. Changes in female reproductive condition following male take-overs in a colony of hamadryas and hybrid baboons. *Folia Primatol* **50:** 157–174.

Connor RC. 2000. Group living in whales and dolphins. In *Cetacean societies: Field studies of dolphins and whales* (ed. Mann J, et al.), pp. 199–218. University of Chicago Press, Chicago.

Connor RC, Richard AF, Smolker RA, Mann J. 1996. Patterns of female attractiveness in Indian Ocean bottlenosed dolphins. *Behaviour* **133:** 37–69.

Connor RC, Read AJ, Wrangham RW. 2000. Male reproductive strategies and social bonds. In *Cetacean societies: Field studies of dolphins and whales* (ed. Mann J, et al.), pp. 247–269. University of Chicago Press, Chicago.

Cords M. 1984. Mating patterns and social structure in redtail monkeys (*Cercopithecus ascanius*). *Z Tierpsychol* **64:** 313–329.

Cords M. 1986. Promiscuous mating among blue monkeys in the Kakamega Forest, Kenya. *Ethology* **72:** 214–226.

Cords M, Fuller JL. 2010. Infanticide in *Cercopithecus mitis stuhlmanni* in the Kakamega Forest, Kenya: Variation in the occurrence of an adaptive behavior. *Int J Primatol* **31:** 409–431.

Cowlishaw G. 1999. Ecological and social determinants of spacing behaviour in desert baboon groups. *Behav Ecol Sociobiol* **45:** 67–77.

Crockett CM. 2003. Re-evaluating the sexual selection hypothesis for infanticide by *Alouatta* males. In *Sexual selection and reproductive competition in primates: New perspectives and directions* (ed. Jones CB), pp. 327–365. American Society of Primatologists, Norman, OK.

Crockett CM, Janson CH. 2000. Infanticide in red howlers: Female group size, male membership, and a possible link to folivory. In *Infanticide by males and its implications* (ed. van Schaik CP, Janson CH), pp. 75–98. Cambridge University Press, Cambridge.

Crofoot MC. 2007. Mating and feeding competition in white-faced capuchins (*Cebus capucinus*): The importance of short- and long-term strategies. *Behaviour* **144:** 1473–1495.

Crook JR, Shields WM. 1985. Sexually selected infanticide by adult male barn swallows. *Anim Behav* **33:** 754–761.

Curtin R, Dolhinow P. 1978. Primate social behavior in a changing world. *Am Sci* **66:** 468–475.

Dahle B, Swenson JE. 2003. Seasonal range size in relation to reproductive strategies in brown bears *Ursus arctos*. *J Anim Ecol* **72:** 660–667.

Daly M, Wilson M. 1984. A sociobiological analysis of human infanticide. In *Infanticide: Comparative and evolutionary perspectives* (ed. Hausfater G, Hrdy SB), pp. 487–502. Aldine, New York.

Daly M, Wilson M. 1994. Stepparenthood and the evolved psychology of discriminative parental solicitude. In *Infanticide and parental care* (ed. Parmigiani S, vom Saal F), pp. 121–134. Harwood, Chur, Switzerland.

Davies G. 1987. Adult male replacement and group formation in *Presbytis rubicunda*. *Folia Primatol* **49:** 111–114.

Delgado RA, van Schaik CP. 2000. The behavioral ecology and conservation of the orangutan (*Pongo pygmaeus*): A tale of two islands. *Evol Anthropol* **9:** 201–218.

Delibes M, Blázquez MC, Delibes-Mateos M. 2012. Historical books in ethology: Sexual purpose used to explain two ancient documentations of infanticide by males. *Ethol Ecol Evol* **24:** 294–300.

de Villiers DJ. 1986. Infanticide in the tree squirrel, *Paraxerus cepapi*. *S Afr J Zool* **21:** 183–184.

de Waal FBM. 1997. *Bonobo: The forgotten ape*. University of California Press, Berkeley, CA.

Dolhinow P. 1977. Normal monkeys? *Am Sci* **65:** 266.

Dudzinski KM, Gregg JD, Ribic CA, Kuczaj SA. 2009. A comparison of pectoral fin contact between two different wild dolphin populations. *Behav Proc* **80:** 182–190.

Dugdale HL, Griffiths A, MacDonald DW. 2011. Polygynandrous and repeated mounting behaviour in European badgers, *Meles meles*. *Anim Behav* **82:** 1287–1297.

Dunn DG, Barco SG, Pabst A, McLellan WA. 2002. Evidence for infanticide in bottlenose dolphins of the western north Atlantic. *J Wildlife Dis* **38:** 505–510.

Ebensperger LA. 1998a. Strategies and counterstrategies to infanticide in mammals. *Biol Rev Camb Philos Soc* **73:** 321–346.

Ebensperger LA. 1998b. The potential effects of protected nests and cage complexity on maternal aggression in house mice. *Agg Behav* **24:** 385–396.

Ebensperger LA. 2001. No infanticide in the hystricognath rodent, *Octodon degus*: Does ecology play a role? *Acta Ethologica* **3:** 89–93.

Ebensperger LA, Blumstein DT. 2007. Nonparental infanticide. In *Rodent societies: An ecological and evolutionary perspective* (ed. Wolff JO, Sherman PA), pp. 267–279. University of Chicago Press, Chicago.

Ebensperger LA, Botto-Mahan C, Tamarin RH. 2000. Nonparental infanticide in meadow voles, *Microtus pennsylvanicus*: The influence of nutritional benefits. *Ethol Ecol Evol* **12:** 149–160.

Eggert A-K, Müller JK. 1997. Biparental care and social evolution in burying beetles: Lessons from the larder. In *The evolution of social behavior in insects and arachnids* (ed. Choe JC, Crespi BJ), pp. 216–236. Cambridge University Press, Cambridge.

Elwood RW, Kennedy HF. 1994. Selective allocation of parental and infanticidal responses in rodents: A review of mechanisms. In *Infanticide and parental care* (ed. Parmigiani S, vom Saal F), pp. 397–425. Harwood, Chur, Switzerland.

Elwood RW, Ostermeyer MC. 1984a. The effects of food deprivation, aggression, and isolation on infanticide in the male Mongolian gerbil. *Agg Behav* **10:** 293–301.

Elwood RW, Ostermeyer MC. 1984b. Infanticide by male and female Mongolian gerbils: Ontogeny, causation, and function. In *Infanticide: Comparative and evolutionary perspectives* (ed. Hausfater G, Hrdy SB), pp. 367–386. Aldine, New York.

Elwood RW, Nesbitt AA, Kennedy HF. 1990. Maternal aggression in response to the risk of infanticide by male mice, *Mus musculus*. *Anim Behav* **40:** 1080–1086.

Engh AL, Beehner JC, Bergman TJ, Whitten PL, Hoffmeier RR, Seyfarth RM, Cheney DL. 2006. Female hierarchy instability, male immigration and infanticide increase glucocorticoid levels in female chacma baboons. *Anim Behav* **71:** 1227–1237.

Enstam KL, Isbell LA, de Maar TW. 2002. Male demography, female mating behavior, and infanticide in wild patas monkeys (*Erythrocebus patas*). *Int J Primatol* **23:** 85–104.

Erez T, Schneider JM, Lubin Y. 2005. Is male cohabitation costly for females of the spider *Stegodyphus lineatus* (Eresidae)? *Ethology* **111:** 693–704.

Fairbanks LA, McGuire MT. 1987. Mother–infant relationships in vervet monkeys: Response to new adult males. *Int J Primatol* **8:** 351–366.

Fairgrieve C. 1995. Infanticide and infant eating in the blue monkey (*Cercopithecus mitis stuhlmanni*) in the Budongo Forest Reserve, Uganda. *Folia Primatol* **64:** 69–72.

Fedigan LM, Carnegie SD, Jack KM. 2008. Predictors of reproductive success in female white-faced capuchins (*Cebus capucinus*). *Am J Phys Anthropol* **137:** 82–90.

Feh C, Munkhtuya B. 2008. Male infanticide and paternity analyses in a socially natural herd of Przewalski's horses: Sexual selection? *Behav Proc* **78:** 335–339.

Freed LA. 1986. Territory takeover and sexually selected infanticide in tropical house wrens. *Behav Ecol Sociobiol* **19:** 197–206.

Freed LA. 1987. Prospective infanticide and protection of genetic paternity in tropical house wrens. *Am Nat* **130:** 948–954.

Fruteau C, Range F, Noë R. 2010. Infanticide risk and infant defence in multi-male free-ranging sooty mangabeys, *Cercocebus atys*. *Behav Proc* **83:** 113–118.

Fuentes A. 2000. Hylobatid communities: Changing views on pair bonding and social organization in Hominoids. *Am J Phys Anthropol* **43:** 33–60.

Galat-Luong A, Galat G. 1979. Conséquences comportementales des perturbations sociales repetées sure une troupe de Mones de Lowe, *Cercopithecus campbelli lowei*, de Côte d'Ivoire [Behavioral consequences of repeated social disturbance in a group of *Cercopithecus campbelli lowei* monkeys in the Ivory Coast]. *Terre et Vie* **33:** 4–57.

Glass GE, Holt R, Slade NA. 1985. Infanticide as an evolutionarily stable strategy. *Anim Behav* **33:** 384–391.

Goodall J. 1986. *The chimpanzees of Gombe: Patterns of behavior.* Belknap, Cambridge, MA.

Goodall J. 1990. *Through a window.* Houghton Mifflin, Boston.

Gray ME. 2009. An infanticide attempt by a free-roaming feral stallion (*Equus caballus*). *Biol Lett* **5:** 23–25.

Gray SM, Dill LM, McKinnon JS. 2007. Cuckoldry incites cannibalism: Male fish turn to cannibalism when perceived certainty of paternity decreases. *Am Nat* **169:** 258–263.

Grinnell J, McComb K. 1996. Maternal grouping as a defense against infanticide by males: Evidence from field playback experiments on African lions. *Behav Ecol* **7:** 55–59.

Gust D. 1994. Alpha-male sooty mangabeys differentiate between females' fertile and their postconception maximal swellings. *Int J Primatol* **15:** 289–301.

Hamai M, Nishida T, Takasaki H, Turner LA. 1992. New records of within-group infanticide in wild chimpanzees. *Primates* **33:** 151–162.

Hamilton WD. 1971. Geometry for the selfish herd. *J Theor Biol* **31:** 295–311.

Harris TR, Monfort SL. 2003. Behavioral and endocrine dynamics associated with infanticide in a black and white colobus monkey (*Colobus guereza*). *Am J Primatol* **61:** 135–142.

Hausfater G, Hrdy SB. 1984. *Infanticide: Comparative and evolutionary perspectives.* Aldine, New York.

Heiming RS, Mönning A, Jansen F, Kloke V, Lesch K-P, Sachser N. 2013. To attack, or not to attack? The role of serotonin transporter genotype in the display of maternal aggression. *Behav Brain Res* **242:** 135–141.

Heistermann M, Ziegler TE, van Schaik CP, Launhardt K, Winkler P, Hodges JK. 2001. Loss of oestrus, concealed ovulation and paternity confusion in free-ranging Hanuman langurs. *Proc Roy Soc Lond B* **268:** 2445–2451.

Henzi SP, Clarke PMR, van Schaik CP, Pradhan GR, Barrett L. 2010. Infanticide and reproductive restraint in a polygynous social mammal. *Proc Natl Acad Sci* **107:** 2130–2135.

Hiraiwa-Hasegawa M. 1988. Adaptive significance of infanticide in primates. *Trends Ecol Evol* **3:** 102–105.

Hoesli T, Nikowitz T, Walzer C, Kaczensky P. 2009. Monitoring of agonistic behaviour and foal mortality in free-ranging Przewalski's horse harems in the Mongolian Gobi. *Equus*, September, 113–138.

Hoogland JL. 1985. Infanticide in prarie dogs: Lactating females kill offspring of close kin. *Science* **230:** 1037–1040.

Hoogland JL. 2001. Black-tailed, Gunninson's, and Utah prairie dogs reproduce slowly. *J Mammal* **82:** 917–927.

Hoogland JL. 2007. Alarm calling, multiple mating, and infanticide among black-tailed, Gunnison's, and Utah prairie dogs. In *Rodent societies: An ecological and evolutionary perspective* (ed. Wolff JO, Sherman PA), pp. 438–450. University of Chicago Press, Chicago.

Hrdy SB. 1974. Male-male competition and infanticide among the langurs (*Presbytis entellus*) of Abu, Rajasthan. *Folia Primatol* **22:** 19–58.

Hrdy SB. 1977. *The langurs of Abu: Female and male strategies of reproduction.* Harvard University Press, Cambridge, MA.

Hrdy SB. 1979. Infanticide among animals: A review, classification, and examination of the implications for reproductive strategies of females. *Ethol Sociobiol* **1:** 13–40.

Hrdy SB. 1981. *The woman that never evolved.* Harvard University Press, Cambridge, MA.

Hrdy SB, Hausfater G. 1984. Comparative and evolutionary perspectives on infanticide: Introduction and overview. In *Infanticide: Comparative and evolutionary perspectives* (ed. Hausfater G, Hrdy SB), pp. xiii–xxxv. Aldine, New York.

Ichino S. 2005. Attacks on a wild infant ring-tailed lemur (*Lemur catta*) by immigrant males at Berenty, Madagascar: Interpreting infanticide by males. *Am J Primatol* **67:** 267–272.

Isbell LA, Cheney DL, Seyfarth RD. 2002. Why vervet monkeys (*Cercopithecus aethiops*) live in multimale groups. In *The guenons: Diversity and adaptation in African monkeys* (ed. Glenn ME, Cords M), pp. 173–187. Kluwer, New York.

Ishibashi Y, Saitoh T. 2008. Effect of local density of males on the occurrence of multimale mating in gray-sided voles (*Myodes rufocanus*). *J Mammal* **89:** 388–397.

Izar P, Stone A, Carnegie S, Nakai ES. 2009. Sexual selection, female choice and mating systems. In *South American primates: Comparative perspectives in the study of behavior, ecology, and conservation* (ed. Garber PA, et al.), pp. 157–189. Springer, New York.

Jack KM, Fedigan LM. 2009. Female dispersal in a female-philopatric species, *Cebus capucinus*. *Behaviour* **146:** 471–497.

Jansen T, Forster P, Levine MA, Oelke H, Hurles M, Renfrew C, Weber J, Olek K. 2002. Mitochondrial DNA and the origins of the domestic horse. *Proc Natl Acad Sci* **99:** 10905–10910.

Janson CH, Goldsmith ML. 1995. Predicting group size in primates: Foraging costs and predation risks. *Behav Ecol* **6:** 326–336.

Janson CH, van Schaik CP. 2000. The behavioral ecology of infanticide by males. In *Infanticide by males and its implications* (ed. van Schaik CP, Janson CH), pp. 469–494. Cambridge University Press, Cambridge.

Kappeler PM. 1997. Determinants of primate social organization: Comparative evidence and new insights from Malagasy lemurs. *Biol Rev Camb Philos Soc* **72:** 111–151.

Kappeler PM. 2012. Mate choice. In *The evolution of primate societies* (ed. Mitani JC, et al.), pp. 367–386. University of Chicago Press, Chicago.

Kiyota M, Okamura H. 2005. Harassment, abduction, and mortality of pups by nonterritorial male northern fur seals. *J Mammal* **86:** 1227–1236.

Korstjens A, Nijssen E, Noë R. 2005. Intergroup relationships in western black-and-white colobus, *Colobus polykomos polykomos*. *Int J Primatol* **26:** 1267–1289.

Koskela E, Juutistenaho P, Mappes T, Oksanen T. 2000. Offspring defence in relation to litter size and age: Experiment in the bank vole *Clethrionomys glareolus*. *Evol Ecol* **14:** 99–109.

Kunz TH, Ebensperger LA. 1999. Why does non-parental infanticide seem so rare in bats? *Acta Chiropterol* **1:** 17–29.

Lane JE, Boutin S, Gunn MR, Slate J, Coltman DW. 2008. Female multiple mating and paternity in free-ranging North American red squirrels. *Anim Behav* **75:** 1927–1937.

Le Boeuf BJ, Campagna C. 1994. Protection and abuse of young in pinnipeds. In *Infanticide and parental care* (ed. Parmigiani S, vom Saal F), pp. 257–276. Harwood, Chur, Switzerland.

Le Galliard J-F, Gundersen G, Andreassen HP, Stenseth NC. 2006. Natal dispersal, interactions among siblings and intrasexual competition. *Behav Ecol* **17:** 733–740.

Lewinson R. 1998. Infanticide in the hippopotamus: Evidence for polygynous ungulates. *Ethol Ecol Evol* **10:** 277–286.

Lewis RJ. 2008. Social influences on group membership in *Propithecus verreauxi verreauxi*. *Int J Primatol* **29:** 1249–1270.

Lonstein JS, Gammie SC. 2002. Sensory, hormonal, and neural control of maternal aggression in laboratory rodents. *Neurosci Biobehav Rev* **26:** 869–838.

Lötter T, Pillay N. 2012. Social interactions associated with reproduction in the bushveld gerbil *Gerbilliscus leucogaster*. *Acta Theriol* **57:** 29–39.

Loveridge AJ, Searle AW, Murindagomo F, MacDonald DW. 2007. The impact of sport-hunting on the population dynamics of an African lion population in a protected area. *Biol Conserv* **134:** 548–558.

Lukas D, Clutton-Brock TH. 2013. The evolution of social monogamy in mammals. *Science* **341:** 526–530.

Maan ME, Taborsky M. 2008. Sexual conflict over breeding substrate causes female expulsion and offspring loss in a cichlid fish. *Behav Ecol* **19:** 302–308.

Maestripieri D, Alleva E. 1990. Maternal aggression and litter size in the female house mouse. *Ethology* **84:** 27–34.

Maestripieri D, Rossi-Arnaud C. 1991. Kinship does not affect litter defence in pairs of communally nesting female house mice. *Agg Behav* **17:** 223–228.

Manning CJ, Dewsbury DA, Wakeland EK, Potts WK. 1995. Communal nesting and communal nesting in house mice, *Mus musculus domesticus*. *Anim Behav* **50:** 741–751.

Marsh CW. 1979. Female transference and mate choice among Tana River red colobus. *Nature* **281:** 568–569.

Matsumoto-Oda A. 1999. Mahale chimpanzees: Grouping patterns and cycling females. *Am J Primatol* **47:** 197–207.

McComb K, Pusey A, Packer C, Grinnell J. 1993. Female lions can identify potentially infanticidal males from their roars. *Proc Roy Soc Lond B* **252:** 59–64.

McGuire B, Bemis WE. 2007. Parental care. In *Rodent societies: An ecological and evolutionary perspective* (ed. Wolff JO, Sherman PA), pp. 231–242. University of Chicago Press, Chicago.

McLean IG. 1982. The association of female kin in the Arctic ground squirrel *Spermophilus parryii*. *Behav Ecol Sociobiol* **10:** 91–99.

McLean IG. 1983. Paternal behaviour and killing of young in arctic ground squirrels. *Anim Behav* **31:** 32–44.

McLellan BN. 2005. Sexually selected infanticide in grizzly bears: The effects of hunting on cub survival. *Ursus* **16:** 141–156.

Mennella JA, Moltz H. 1988. Infanticide in rats: Male strategy and female counter-strategy. *Physiol Behav* **42:** 19–28.

Miller SD, Sellers RA, Keay JA. 2003. Effects of hunting on brown bear cub survival and litter size in Alaska. *Ursus* **14:** 130–152.

Mitani JC, Watts DP, Amsler SJ. 2010. Lethal intergroup aggression leads to territorial expansion in wild chimpanzees. *Cur Biol* **20:** R507–R508.

Mock DW. 1984. Infanticide, siblicide, and avian nestling mortality. In *Infanticide: Comparative and evolutionary perspectives* (ed. Hausfater G, Hrdy SB), pp. 3–30. Aldine, New York.

Møller AP. 1988. Infanticidal and anti-infanticidal strategies in the swallow *Hirundo rustica*. *Behav Ecol Sociobiol* **22:** 365–371.

Møller AP. 2004. Rapid temporal change in frequency of infanticide in a passerine bird associated with change in population density and body condition. *Behav Ecol* **15:** 462–468.

Morelli TL, King SJ, Pochron ST, Wright PC. 2009. The rules of disengagement: Takeovers, infanticide, and dispersal in a rainforest lemur, *Propithecus edwardsi*. *Behaviour* **146:** 499–523.

Moscovice LR, Di Fiore A, Crockford C, Kitchen DM, Wittig R, Seyfarth RM, Cheney DL. 2010. Hedging their bets? Male and female chacma baboons form friendships based on likelihood of paternity. *Anim Behav* **79:** 1007–1015.

Müller JK, Eggert AK, Sakaluk SK. 1998. Carcass maintenance and biparental brood care in burying beetles: Are males redundant? *Ecol Entomol* **23:** 195–200.

Myers K, Poole WE. 1961. A study of the biology of the wild rabbit, *Orytolagus cuniculus* (L.), in confined populations: II. The effects of season and population increase on behaviour. *CSIRO Wildlife Res* **6:** 1–41.

Nunn CL. 1999. The evolution of exaggerated sexual swellings in primates and the graded-signal hypothesis. *Anim Behav* **58:** 229–246.

O'Brien TG. 1991. Female-male social interactions in wedge-capped capuchin monkeys: Benefits and costs of group living. *Anim Behav* **41:** 555–567.

Oleinchenko VY. 2012. Behavioral interactions of adult females of the common shrew (*Sorex araneus*) with conspecifics on familiar territory. *Biol Bull* **39:** 351–359.

Opie C, Atkinson QD, Dunbar RIM, Shultz S. 2013. Male infanticide leads to social monogamy in primates. *Proc Natl Acad Sci* **110:** 13328–13332.

Ostner J, Kappeler PM. 2004. Male life history and the unusual adult sex ratios of redfronted lemur, *Eulemur fulvus rufus*, groups. *Anim Behav* **67:** 249–259.

Packer C, Pusey AE. 1983. Adaptations of female lions to infanticide by incoming males. *Am Nat* **121:** 716–728.

Packer C, Herbst L, Pusey AE, Bygott JD, Hanby JP, Cairns SJ, Mulder MB. 1988. Reproductive success in lions. In *Reproductive success: Studies of individual variation in contrasting breeding systems* (ed. Clutton-Brock TH), pp. 363–383. University of Chicago Press, Chicago.

Packer C, Scheel D, Pusey AE. 1990. Why lions form groups: Food is not enough. *Am Nat* **136:** 9–19.

Palanza P, Re L, Mainardi D, Brain PF, Parmigiani S. 1996. Male and female competitive strategies of wild house mice pairs (*Mus musculus domesticus*) confronted with intruders of different sex and age in artificial territories. *Behaviour* **133:** 863–882.

Palombit RA. 1999. Infanticide and the evolution of pair bonds in nonhuman primates. *Evol Anthropol* **7:** 117–129.

Palombit RA. 2000. Infanticide and the evolution of male-female bonds in animals. In *Male infanticide and its implications* (ed. van Schaik CP, Janson CH), pp. 239–268. Cambridge University Press, Cambridge.

Palombit RA. 2003. Male infanticide in wild savanna baboons: Adaptive significance and intraspecific variation. In *Sexual selection and reproductive competition in primates: New perspectives and directions* (ed. Jones CB), pp. 364–411. American Society of Primatologists, Norman, OK.

Palombit RA. 2009. Friendships with males: A female counterstrategy to infanticide in the Okavango chacma baboons. In *Sexual coercion in primates and humans* (ed. Muller MN, Wrangham RW), pp. 377–409. Harvard University Press, Cambridge, MA.

Palombit RA. 2010. Conflict and bonding between the sexes. In *Mind the gap: Tracing the origin of human universals* (ed. Kappeler PM, Silk JB), pp. 53–84. Springer, Berlin.

Palombit RA. 2012. Infanticide: Male strategies and female counterstrategies. In *The evolution of primate societies* (ed. Mitani JC, et al.), pp. 432–468. University of Chicago Press, Chicago.

Palombit RA. Infanticide. In *Encyclopedia of human sexuality* (ed. Whelehan P, Bolin A). Wiley, Hoboken, NJ (in press).

Palombit RA. 2014. Sexual conflict in nonhuman primates. *Adv Stud Behav* **46:** 191–280.

Palombit RA, Seyfarth RM, Cheney DL. 1997. The adaptive value of "friendships" to female baboons: Experimental and observational evidence. *Anim Behav* **54:** 599–614.

Palombit RA, Cheney DL, Fischer J, Johnson S, Rendall D, Seyfarth RM, Silk JB. 2000. Male infanticide and defense of infants in wild chacma baboons. In *Infanticide by males and its implications* (ed. van Schaik CP, Janson CH), pp. 123–152. Cambridge University Press, Cambridge.

Palombit RA, Cheney DL, Seyfarth RM. 2001. Female–female competition for male "friends" in wild chacma baboons (*Papio cynocephalus ursinus*). *Anim Behav* **61:** 1159–1171.

Parmigiani S. 1986. Rank order in pairs of communally nursing female mice (*Mus musculus domesticus*) and maternal aggression towards conspecific intruders of differing sex. *Agg Behav* **12:** 377–386.

Parmigiani S, Brain PF, Mainardi D, Brunoni V. 1988a. Different patterns of biting attack generated when lactating female mice (*Mus domesticus*) encounter male and female conspecific intruders. *J Comp Psychol* **102:** 287–293.

Parmigiani S, Sgoifo A, Mainardi D. 1988b. Parental aggression displayed by female mice in relation to the sex, reproductive status, and infanticidal potential of conspecific intruders. *Monit Zool Ital* **22:** 193–201.

Parmigiani S, Palanza P, Brain PF. 1989. Intraspecific maternal aggression in the house mouse (*Mus domesticus*): A counterstrategy to infanticide by male? *Ethol Ecol Evol* **1:** 341–352.

Parmigiani S, Palanza P, Mainardi M, Mainardi D. 1990. Fear and defensive components of maternal aggression in mice. In *Fear and defense* (ed. Brain PF, et al.), pp. 109–126. Harwood, London.

Parmigiani S, Palanza P, Mainardi D, Brain PF. 1994. Infanticide and protection of young in house mice (*Mus domesticus*): Female and male strategies. In *Infanticide and parental care* (ed. Parmigiani S, vom Saal F), pp. 341–363. Harwood, Chur, Switzerland.

Parmigiani S, Palanza P, Rodgers J, Ferrari PF. 1999. Selection, evolution of behavior and animal models in behavioral neuroscience. *Neurosci Biobehav Rev* **23:** 957–970.

Patterson IAP, Reid RJ, Wilson B, Grellier K, Ross HM, Thompson PM. 1998. Evidence for infanticide in bottlenose dolphins: An explanation for violent interactions with harbour porpoises. *Proc Roy Soc Lond B* **265:** 1167–1170.

Paul L, Kupferschmidt J. 1975. Killing of conspecific and mouse young by male rats. *J Comp Physiol Psychol* **88:** 755–763.

Pavé R, Kowalewski M, Garber P, Zunino G, Fernandez V, Peker S. 2012. Infant mortality in black-and-gold howlers (*Alouatta caraya*) living in a flooded forest in northeastern Argentina. *Int J Primatol* **33:** 937–957.

Pazol K. 2003. Mating in the Kakamega forest blue monkeys (*Cercopithecus mitis*): Does female sexual behavior function to manipulate paternity assessment? *Behaviour* **140:** 473–499.

Pereira M. 1983. Abortion following the immigration of an adult male baboon (*Papio cynocephalus*). *Am J Primatol* **4:** 93–98.

Pereira ME, Weiss ML. 1991. Female mate choice, male migration, and the threat of infanticide in ringtailed lemurs. *Behav Ecol Sociobiol* **28:** 141–152.

Perrigo G, Bryant WC, vom Saal FS. 1990. A unique neural timing system prevents male mice from harming their own offspring. *Anim Behav* **39:** 535–539.

Perrigo G, Belvin L, vom Saal F. 1992. Time and sex in the male mouse: Temporal regulation of infanticide and parental behavior. *Chronobiol Int* **9:** 421–433.

Pierotti R. 1991. Infanticide versus adoption: An intergenerational conflict. *Am Nat* **138:** 1140–1158.

Pluháček J, Bartoš L. 2000. Male infanticide in captive plains zebra, *Equus caballus*. *Anim Behav* **59:** 689–694.

Pluháček J, Bartoš L. 2005. Further evidence for male infanticide and feticide in captive zebra, *Equus burchelli*. *Folia Zool* **54:** 258–262.

Pusey AE, Packer C. 1987. The evolution of sex-biased dispersal in lions. *Behaviour* **101:** 275–310.

Pusey AE, Packer C. 1994a. Non-offspring nursing in social carnivores: Minimizing the costs. *Behav Ecol* **5:** 362–374.

Pusey AE, Packer C. 1994b. Infanticide in lions: Consequences and counterstrategies. In *Infanticide and parental care* (ed. Parmigiani S, vom Saal F), pp. 277–299. Harwood, Chur, Switzerland.

Rajpurohit IS, Chhangani AK, Rajpurohit RS, Bhaker NR, Rajpurohit DS, Sharma G. 2008. Recent observations on resident male change followed by infanticide in Hanuman langurs (*Semnopithecus entellus*) around Jodhpur. *Primate Rep* **75:** 33–40.

Ramírez-Llorens P, Di Bitetti MS, Baldovino MC, Janson CH. 2008. Infanticide in black capuchin monkeys (*Cebus apella nigritus*) in Iguazú National Park, Argentina. *Am J Primatol* **70:** 473–484.

Rees A. 2009. *The infanticide controversy*. University of Chicago Press, Chicago.

Rémy A, Odden M, Richard M, Stene MT, Le Galliard J-F, Andreassen HP. 2013. Food distribution influences social organization and population growth in a small rodent. *Behav Ecol* **24:** 832–841.

Richard AF, Rakotomanga P, Schwartz M. 1993. Dispersal by *Propithecus verreauxi* at Beza Mahafaly, Madagascar: 1984–1991. *Am J Primatol* **30:** 1–20.

Rijksen HD. 1981. Infant killing: A possible consequence of a disputed leader role. *Behaviour* **78:** 138–167.

Robbins AM, Stoinski T, Fawcett K, Robbins M. 2009a. Socioecological influences on the dispersal of female mountain gorillas—Evidence of a second folivore paradox. *Behav Ecol Sociobiol* **63:** 477–489.

Robbins AM, Stoinski T, Fawcett K, Robbins MM. 2009b. Leave or conceive: Natal dispersal and philopatry of female mountain gorillas in the Virunga volcano region. *Anim Behav* **77:** 831–838.

Roberts EK, Lu A, Bergman TJ, Beehner JC. 2012. A Bruce effect in wild geladas. *Science* **335:** 1222–1225.

Robertson RJ. 1990. Tactics and counter-tactics of sexually selected infanticide in tree swallows. In *Population biology of passerine birds: An integrated approach* (ed. Blondel J, et al.), pp. 381–390. Springer, Berlin.

Robertson IC. 1993. Nest intrusions, infanticide, and parental care in the burying beetle, *Nicrophorus orbicollis* (Coleoptera, Silphidae). *J Zool* **231:** 583–593.

Robertson RJ, Stutchbury BJ. 1988. Experimental evidence for sexually selected infanticide in tree swallows. *Anim Behav* **36:** 749–753.

Rohwer S. 1986. Selection for adoption versus infanticide by replacement mates in birds. *Cur Ornithol* **3:** 353–395.

Rudran R. 1973a. The reproductive cycles of two subspecies of purple-faced langurs (*Presbytis senex*) with relation to enviornmental factors. *Folia Primatol* **19:** 41–60.

Rudran R. 1973b. Adult male replacement in one-male troops of purple faced langurs (*Presbytis senex senex*) and its effect on population structure. *Folia Primatol* **19:** 166–192.

Sakura O. 1994. Factors affecting party size and composition of chimpanzees (*Pan troglodytes verus*) Bossou, Guinea. *Int J Primatol* **15:** 167–183.

Schneider JM, Lubin Y. 1997. Infanticide by males in a spider with suicidal maternal care, *Stegodyphus lineatus* (Eresidae). *Anim Behav* **54:** 305–312.

Schülke O. 2005. Evolution of pair-living in *Phaner furcifer*. *Int J Primatol* **26:** 903–919.

Sherman PW. 1981. Reproductive competition and infanticide in Belding's ground squirrels and other animals. In *Natural selection and social behavior* (ed. Alexander RD, Tinkle DW), pp. 311–331. Chiron, New York.

Sicotte P, Teichroeb J, Saj T. 2007. Aspects of male competition in *Colobus vellerosus*: Preliminary data on male and female loud calling, and infant deaths after a takeover. *Int J Primatol* **28:** 627–636.

Singh M, Kumara HN, Ananda KM, Singh M, Cooper M. 2006. Male influx, infanticide, and female transfer in *Macaca radiata*. *Int J Primatol* **27:** 515–528.

Solomon NG, Keane B. 2007. Reproductive strategies in female rodents. In *Rodent societies: An ecological and evolutionary perspective* (ed. Wolff JO, Sherman PA), pp. 42–56. University of Chicago Press, Chicago.

Soltis J, Thomsen R, Matusabayshi K, Takenaka O. 2000. Infanticide by resident males and female counter-strategies in wild Japanese macaques (*Macaca fuscata*). *Behav Ecol Sociobiol* **48:** 195–202.

Sommer V. 1994. Infanticide among the langurs of Jodhpur: Testing the sexual selection hypothesis with a long-term record. In *Infanticide and parental care* (ed. Parmigiani S, vom Saal F), pp. 155–198. Harwood, Chur, Switzerland.

Sommer S. 1997. Monogamy in *Hypogeomys antimena*, an endemic rodent of the deciduous dry forest in western Madagascar. *J Zool* **241:** 301–314.

Starin ED. 1994. Philopatry and affiliation among red colobus. *Behaviour* **130:** 253–270.

Steenbeek R. 1999a. Tenure related changes in wild Thomas's langurs: I. Between-group interactions. *Behaviour* **136:** 595–625.

Steenbeek R. 1999b. "Female choice and male coercion in wild Thomas's langurs." PhD dissertation, Utrecht University, Netherlands.

Steenbeek R. 2000. Infanticide by males and female choice in wild Thomas's langurs. In *Infanticide by males and its implications* (ed. van Schaik CP, Janson CH), pp. 153–177. Cambridge University Press, Cambridge.

Sterck EHM. 1997. Determinants of female dispersal in Thomas langurs. *Am J Primatol* **42:** 179–198.

Sterck EHM. 1998. Variation in langur social organization in relation to the socioecological model, human habitat alteration, and phylogenetic constraints. *Primates* **40:** 199–213.

Sterck EHM, Watts DP, van Schaik CP. 1997. The evolution of female social relationships in nonhuman primates. *Behav Ecol Sociobiol* **41:** 291–309.

Stokes EJ. 2004. Within-group social relationships among females and adult males in wild western lowland gorillas (*Gorilla gorilla gorilla*). *Am J Primatol* **64:** 233–246.

Stokes E, Parnell RJ, Olejniczak C. 2003. Female dispersal and reproductive success in wild western lowland gorillas. *Behav Ecol Sociobiol* **54:** 329–339.

Struhsaker TT, Leland L. 1985. Infanticide in a patrilineal society of red colobus monkeys. *Z Tierpsychol* **69:** 89–132.

Svare B, Bartke A. 1978. Food deprivation induces conspecific pup-killing in mice. *Agg Behav* **4:** 253–261.

Swedell L, Saunders J. 2006. Infant mortality, paternity certainty, and female reproductive strategies in hamadryas baboons. In *Reproduction and fitness in baboons: Behavioral, ecological, and life history perspectives* (ed. Swedell L, Leigh SR), pp. 19–51. Springer, New York.

Swenson JE, Brunberg S, Segerström P. 2001. Factors associated with loss of brown bear cubs in Sweden. *Ursus* **12:** 69–80.

Takahata Y, Sprague DS, Suzuki S, Okayasu N. 1994. Female competition, co-existence, and the mating structure of wild Japanese macaques on Yakushima Island, Japan. *Physiol Ecol* **29:** 163–179.

Teichroeb JA, Sicotte P. 2008. Infanticide in ursine colobus monkeys (*Colobus vellerosus*) in Ghana: New cases and a test of the existing hypotheses. *Behaviour* **145:** 727–755.

Teichroeb JA, Wikberg EC, Sicotte P. 2009. Female dispersal patterns in six groups of ursine colobus (*Colobus vellerosus*): Infanticide avoidance is important. *Behaviour* **146:** 551–582.

Thalmann U. 2001. Food resource characteristics in two nocturnal lemurs with different social behavior: *Avahi occidentalis* and *Lepilemur edwardsi*. *Int J Primatol* **22:** 287–324.

Thonhauser KE, Raveh S, Hettyey A, Beissmann H, Penn DJ. 2013. Why do female mice mate with multiple males? *Behav Ecol Sociobiol* **67:** 1961–1970.

Treves A. 2001. Reproductive consequences of variation in the composition of howler monkey (*Alouatta* spp.) groups. *Behav Ecol Sociobiol* **50:** 61–71.

Treves A, Chapman CA. 1996. Conspecific threat, predation avoidance, and resource defense: Implications for grouping in langurs. *Behav Ecol Sociobiol* **39:** 43–53.

Trivers RL. 1972. Parental investment and sexual selection. In *Sexual selection and the descent of man, 1871–1971* (ed. Campbell B), pp. 136–179. Aldine, Chicago.

Trumbo ST. 1990a. Reproductive benefits of infanticide in a biparental burying beetle *Nicrophorus orbicollis*. *Behav Ecol Sociobiol* **27:** 269–273.

Trumbo ST. 1990b. Interference competition among burying beetles (*Silphidae, Nicrophorus*). *Ecol Entomol* **15:** 347–355.

Trumbo ST. 1991. Reproductive benefits and the duration of paternal care in a biparental burying beetle, *Necrophorus orbicollis*. *Behaviour* **117:** 82–115.

Trumbo ST. 2007. Defending young biparentally: Female risk-taking with and without a male in the burying beetle, *Nicrophorus pustulatus*. *Behav Ecol Sociobiol* **61:** 1717–1723.

Trumbo ST, Valletta RC. 2007. The costs of confronting infanticidal intruders in a burying beetle. *Ethology* **113:** 386–393.

van Noordwijk MA, van Schaik CP. 2000. Reproductive patterns in eutherian mammals: Adaptations against infanticide. In *Infanticide by males and its implications* (ed. van Schaik CP, Janson CH), pp. 322–360. Cambridge University Press, Cambridge.

van Schaik CP. 2000a. Infanticide by male primates: The sexual selection hypothesis revisited. In *Infanticide by males and its implications* (ed. van Schaik CP, Janson CH), pp. 27–60. Cambridge University Press, Cambridge.

van Schaik CP. 2000b. Vulnerability to infanticide by males: Patterns among mammals. In *Infanticide by males and its implications* (ed. van Schaik CP, Janson CH), pp. 61–71. Cambridge University Press, Cambridge.

van Schaik CP, Dunbar RIM. 1990. The evolution of monogamy in large primates: A new hypothesis and some crucial tests. *Behaviour* **115:** 30–62.

van Schaik CP, Janson CH (eds.). 2000. *Infanticide by males and its implications*. Cambridge University Press, Cambridge.

van Schaik CP, Hodges JK, Nunn CL. 2000. Paternity confusion and the ovarian cycles of female primates. In *Infanticide by males and its implications* (ed. van Schaik CP, Janson CH), pp. 361–387. Cambridge University Press, Cambridge.

Veiga JP. 1993. Prospective infanticide and ovulation retardation in free-living sparrows. *Anim Behav* **45:** 43–46.

Veiga JP. 2000. Infanticide by male birds. In *Infanticide by males and its implications* (ed. van Schaik CP, Janson CH), pp. 198–220. Cambridge University Press, Cambridge.

Vestal BM. 1991. Infanticide and cannibalism by male thirteen-lined ground squirrels. *Anim Behav* **41:** 1103–1104.

vom Saal FS. 1985. Time-contingent change in infanticide and parental behavior induced by ejaculation in male mice. *Physiol Behav* **34:** 7–15.

vom Saal FS, Howard LS. 1982. The regulation of infanticide and parental behavior: Implications for reproductive success in male mice. *Science* **215:** 1270–1272.

Watts DP. 1989. Infanticide in mountain gorillas: New cases and a reconsideration of the evidence. *Ethology* **81:** 1–18.

Watts DP, Mitani JC, Sherrow HM. 2002. New cases of intercommunity infanticide by male chimpanzees at Ngogo, Kibale National Park, Uganda. *Primates* **43:** 263–270.

Weber EM, Olsson IAS. 2008. Maternal behaviour in *Mus musculus* sp.: An ethological review. *Appl Anim Behav Sci* **114:** 1–22.

Weingrill T. 2000. Infanticide and the value of male-female relationships in mountain chacma baboons (*Papio cynocephalus ursinus*). *Behaviour* **137:** 337–359.

Wielgus RB, Bunnell FL. 2000. Possible negative effects of adult male mortality on female grizzly bear reproduction. *Biol Conserv* **93:** 145–154.

Wielgus RB, Morrison DE, Cooley HS, Maletzke B. 2013. Effects of male trophy hunting on female carnivore population growth and persistence. *Biol Conserv* **167:** 69–75.

Williams GC. 1966. *Adaptation and natural selection.* Princeton University Press, Princeton, NJ.

Williams JM, Pusey AE, Carlis JV, Farm BP, Goodall J. 2002. Female competition and male territorial behaviour influence female chimpanzees' ranging patterns. *Anim Behav* **63:** 347–360.

Williams JM, Oehlert GW, Carlis JV, Pusey AE. 2004. Why do male chimpanzees defend a group range? *Anim Behav* **68:** 523–532.

Williams JM, Lonsdorf EV, Wilson ML, Schumacher-Stankey J, Goodall J, Pusey AE. 2008. Causes of death in the Kasekela chimpanzees of Gombe National Park, Tanzania. *Am J Primatol* **70:** 766–777.

Wolff JO, Cicirello DM. 1991. Comparative paternal and infanticidal behavior of white-footed mice (*Peromyscus leucopus noveboracensis*) and deermice (*P. maniculatus nubiterrae*). *Behav Ecol* **2:** 38–45.

Wolff JO, MacDonald DW. 2004. Promiscuous females protect their offspring. *Trends Ecol Evol* **19:** 127–134.

Wolff JO, Peterson JA. 1998. An offspring-defense hypothesis for territoriality in female mammals. *Ethol Ecol Evol* **10.** 227–239.

Wrangham RW. 1980. An ecological model of female-bonded primate groups. *Behaviour* **75:** 262–299.

Wrangham RW, Smuts BB. 1980. Sex differences in the behavioural ecology of chimpanzees in the Gombe National Park. *J Reprod Fertil Suppl* **28:** 13–31.

Wright PC. 1995. Demography and life history of free-ranging *Propithecus diadema edwardsi* in Ranomafana National Park, Madagascar. *Int J Primatol* **16:** 835–854.

Yamada K, Nakamichi M. 2006. A fatal attack on an unweaned infant by a non-resident male in a free-ranging group of Japanese macaques (*Macaca fuscata*) at Katsuyama. *Primates* **47:** 165–169.

Yamagiwa J, Kahekwa J, Basabose AK. 2009. Infanticide and social flexibility in the genus *Gorilla*. *Primates* **50:** 293–303.

Yamaguchi N, Dugdale H, MacDonald DW. 2006. Female receptivity, embryonic diapause, and superfetation in the European badger (*Meles meles*): Implications for the reproductive tactics of males and females. *Q Rev Biol* **81:** 33–48.

Yamamura N, Hasegawa T, Ito Y. 1990. Why mothers do not resist infanticide: A cost-benefit genetic model. *Evolution* **44:** 1346–1357.

Ylönen H, Koskela E, Mappes T. 1997. Infanticide in the bank vole (*Clethrionomys glareolus*): Occurrence and the effect of familiarity on female infanticide. *Ann Zoo Fenn* **34:** 259–266.

Zedrosser A, Dahle B, Støen O-G, Swenson J. 2009. The effects of primiparity on reproductive performance in the brown bear. *Oecologia* **160:** 847–854.

Zinner D, Deschner T. 2000. Sexual swellings in female hamadryas baboons after male take-overs: "Deceptive" swellings as a possible female counter-strategy against infanticide. *Am J Primatol* **52:** 157–168.

Human Homosexuality: A Paradigmatic Arena for Sexually Antagonistic Selection?

Andrea Camperio Ciani[1], Umberto Battaglia[1], and Giovanni Zanzotto[2]

[1]Dipartimento di Filosofia, Sociologia, Pedagogia e Psicologia Applicata, Università di Padova, 35100 Padova, Italy

[2]Dipartimento di Psicologia Generale, Università di Padova, 35100 Padova, Italy

Correspondence: andrea.camperio@unipd.it

Sexual conflict likely plays a crucial role in the origin and maintenance of homosexuality in our species. Although environmental factors are known to affect human homosexual (HS) preference, sibling concordances and population patterns related to HS indicate that genetic components are also influencing this trait in humans. We argue that multilocus, partially X-linked genetic factors undergoing sexually antagonistic selection that promote maternal female fecundity at the cost of occasional male offspring homosexuality are the best candidates capable of explaining the frequency, familial clustering, and pedigree asymmetries observed in HS male proband families. This establishes male HS as a paradigmatic example of sexual conflict in human biology. HS in females, on the other hand, is currently a more elusive phenomenon from both the empirical and theoretical standpoints because of its fluidity and marked environmental influence. Genetic and epigenetic mechanisms, the latter involving sexually antagonistic components, have been hypothesized for the propagation and maintenance of female HS in the population. However, further data are needed to truly clarify the evolutionary dynamics of this trait.

Sexual conflict, the general theme of this collection, has emerged in the last decades as a main research topic in evolutionary biology. Its possible ubiquity and key role have become clear in many phenomena, from speciation to genetic diversity maintenance in populations, sexual phenotypic dimorphism, asymmetric mating and parenting strategies, antagonistic coevolution, and others (Bonduriansky and Chenoweth 2009; Rice et al. 2013; see also Gavrilets 2014). A most interesting example of sexual conflict is given by homosexual (HS) behavior and mate preference in humans, a topic whose societal impact is widely felt at present.

Homosexuality is not confined just to our species, but is also occasionally present in numerous animals (Bagemihl 1999; Sommer and Vasey 2006), with permanent, long-term same-sex pair bonding having been reported in some birds and ungulates (Bagemihl 1999; Ngun et al. 2011). The presence of not solely sporadic and opportunistic HS behavior has also been observed in nonhuman primates, which suggests our hominid ancestry might also have had HS individuals (Vasey 1995).

The main evolutionary questions about HS in our species do not concern occasional HS behavior, which might have, in various con-

Cite this article as *Cold Spring Harb Perspect Biol* doi: 101101/cshperspect.a017657

texts, a fitness advantage (Bagemihl 1999), but rather refer to the origin, prevalence, frequency, population distribution, maintenance, and selective values of long-term HS preference. First and foremost, the stable permanence of HS in human populations constitutes a puzzling "Darwinian paradox" for any putative genetic loading in this apparently detrimental trait. Even a partial genetic predisposition would seemingly contradict the assumption that natural selection will progressively eliminate any genetic factors contributing to reduce individual fitness. Understanding HS is not only a challenging scientific conundrum related to the evolutionary conflicts between sexes; because of the estimated prevalence of this trait involving a nontrivial fraction of the human population (see below), these issues also potentially concern several hundred million people.

Since the 1990s, the role of biological factors, both genetic and nongenetic, has come to light in relation to HS (see Camperio Ciani et al. 2004; Gavrilets and Rice 2006; Jannini et al. 2010). The possible genetic components influencing human HS have since been explored empirically and theoretically. However, we emphasize at the outset that a number of environmental factors (zygotic, prenatal, and postnatal) are known to affect HS preference in either or both genders, such as maternal stress, mother estrogen-progestinics assumption during fetal development, fraternal birth order, as well as environmental and social influences (see Ward and Weisz 1980; Dörner et al. 1983; Whisman 1996; Rosenbluth 1997; Stein 1997; Blanchard 2001; Garcia-Falgueras and Swaab 2010; Långström et al. 2010; Balthazart 2011; Mychasiuck et al. 2012). Thus, genetic components can never be considered as exclusive determinants of HS, and they should always be interpreted as acting within the frame of an environmental background that may also significantly contribute to same-sex preference in individuals (Balthazart 2011).

Besides environmental effects, incomplete penetrance and variable expressivity are also expected for any complex behavioral traits with less-than-Mendelian ratios in pedigrees such as HS. Although these factors contribute to blur the phenotypic expression of any genetic basis on behavior (Plomin et al. 2001), for HS they do not entirely obfuscate the recognizable effect of genetic influences and polymorphism in human populations.

Population genetics, demographic distributions, and biometric properties are increasingly recognized as main tools needed for the understanding of complex traits for which (epi)-genetic roots are hypothesized (Sham 2001; Boomsma et al. 2002; Lynch 2007). It is important to stress that population patterns should provide a set of constraints against which any propagation or maintenance hypotheses for HS-related genetic factors must be explicitly tested, as lacking such information, any theoretical results and predictions remain largely impossible to falsify. Thus, the efforts to explain HS should consider a series of specific empirical aspects that this trait presents in human populations, namely: (1) low prevalence and stable, universal distribution; (2) familial aggregation; (3) specific distribution within families; and (4) possible balancing fecundity effects associated with the traits. We summarize, below, the main findings concerning such population patterns. We observe that some general trends emerge in the replications by independent research groups, with data collected through diverse methods and protocols, but much work is still necessary, especially in relation to female HS.

A number of hypotheses have been posited in the evolutionary investigation of male HS, whose selection routes have been studied in more detail than for female HS. Below, we report on the results regarding the likely evolutionary mechanisms for male HS (Camperio Ciani et al. 2004, 2008; Gavrilets and Rice 2006). As we will discuss, research in this direction has indicated that among the studied mechanisms, sexually antagonistic selection, an important aspect of sexual conflict, appears the best hypothesis, which can correctly account for the available population data related to male HS. Population surveys and mathematical modeling have thus concurred to establish male HS as a paradigmatic example of sexual conflict in human biology.

 Cite this article as *Cold Spring Harb Perspect Biol* doi: 10.1101/cshperspect.a017657

In contrast, female HS remains to this day a much more elusive phenomenon from both the empirical and theoretical standpoints. Several marked differences exist, which make female HS, in many ways, unlike its counterpart in males. Some of the dissimilarities appear fully elucidated, others await further work for clarification. For instance, environmental effects are known to play a stronger role in female HS, especially the shared-environmental influences, which are virtually absent in male HS (Bailey et al. 2000; Långström et al. 2010). A further obstacle lies in the fact that it is not clear whether HS in females has entailed a nontrivial fecundity reduction in the evolutionary past, as opposed to present times. The coexistence of HS with other powerful behavioral patterns may have masked the effects of any genetic propensity to HS behavior in females with its possible implied reduced fitness. This could pose difficulties for thoroughly ascertaining the evolutionary dynamics of this trait in human populations.

Nonetheless, genetic and epigenetic mechanisms (Pattatucci and Hamer 1995; Rice et al. 2008) have been hypothesized for the evolutionary dynamics of female HS. In particular, to account for some peculiar patterns observed within the families of HS females, it has been hypothesized that sexual conflict be at play through (epi)genetic factors known as sexually antagonistic zygotic drive, which might contribute to the transmission (Rice et al. 2008; see also Friberg and Rice 2014). This would posit sexual conflict as a basic selection mechanism also for HS in females. Such models, however, have yet to be fully tested against well-established data on the population patterns connected to female HS, which also remain, to this day, relatively scarce.

For both sexes, and especially for females, the current evidence on population data limits, but does not rule out, the influence of other (epi)genetic factors on HS, perhaps unrelated to sexual conflict. Their possible action, however, falls within a more opaque and stochastic background, to which also environmental effects largely contribute, and in which clear patterns have not yet been properly identified, neither for male nor for female HS.

EMPIRICAL EVIDENCE ON POPULATION PATTERNS RELATED TO HS IN HUMAN POPULATIONS

Stability of HS in Human Populations

Homosexuality is a stable trait that has been documented in human societies over several millennia (Herdt 1997; Nash 2001; Crompton 2006). It has been confirmed through archaeological evidence that both female and male HS occurred also prehistorically and cannot be considered a product of modern times (Ruse 1988; Herdt 1997; Nash 2001).

Trait Prevalence and Features

HS is present in almost all human populations, with relatively low frequency, or prevalence (Diamond 1993; but see Hewlett and Hewlett (2010) and Barthes et al. (2013) for cases of presumed absence of HS in some cultures). The prevalence of HS behavior may largely vary over time and geographical region. Population-based surveys have suggested prevalences of ~2%–9% for lesbians, compared with 0.5%–15% for gays, with a more general agreement on 2%–7% in most Western countries for both genders (Bell and Weinberg 1978; Bailey and Benishay 1993; Diamond 1993; Johnson et al. 1994; Sell et al. 1995; Herek et al. 1997, 1999; Bullough 1998; Bailey et al. 2000; Bogaert 2004; Camperio Ciani et al. 2004, 2009; Butler 2005; Caceres et al. 2006; Wright 2009; Hewlett and Hewlett 2010). However, prejudice, persecution, homophobia, and religious orthodoxy might induce individuals not to reveal, even in anonymous questionnaires, their own sexual preference, creating an "obscure number" of individuals who will never explicitly disclose such information (Savin-Williams 2006; Camperio Ciani and Battaglia 2014). Also, sampling problems affect surveys on sexual preference, as well as the location, whether urban or rural, and the method used, such as Internet surveys, etc. The above figures should thus be considered as fairly uncertain estimates producing no final consensus.

Better ascertained is the fact that the distribution of HS preference on the Kinsey scale is polarized and U-shaped for males, with a relative higher frequency of complete or almost complete heterosexuals or homosexuals and relatively few bisexuals. In contrast, females show, on the same scale, a progressively declining frequency from complete heterosexuality toward complete homosexuality (Pillard and Weinrich 1986; Bailey and Pillard 1991; Pattatucci 1998; Bailey et al. 2000; Camperio Ciani et al. 2009). Bisexuality is, thus, more common in women than in men (Bailey and Pillard 1991; Hamer et al. 1993; Whitman et al. 1993; Pattatucci and Hamer 1995; Diamond 2008; Camperio Ciani et al. 2009; Alanko et al. 2010). Also, HS females seem more flexible in response to shared-environmental factors compared with HS males (Bailey et al. 2000; Rahman 2005; Lippa 2006; Alanko et al. 2010; Långström et al. 2010). Females might express sexual preference for females also because of feminist, political, egalitarian reasons, in addition to being driven to HS by erotic or emotional attraction (Whisman 1996; Rosenbluth 1997; Stein 1997). More females than males appear to have switched from heterosexual to HS preference, even after a satisfactory period of heterosexuality (Blumstein and Schwartz, 1976), and vice versa. None of these patterns are common in HS men. The sexuality of males becomes relatively fixed by young adulthood, contrary to females who may maintain their sexual fluidity well into adulthood (Blumstein and Schwartz 1977; Rust 1992; Bailey and Benishay 1993; Bailey et al. 1993; Pattatucci and Hamer 1995; Esterberg 1997; Baumeister 2000).

Family Clustering and Asymmetries in Trait Frequency

Almost all studies have shown that both male and female HS are not evenly distributed within the population, but rather run in families (Bailey and Zucker 1995; Pattatucci and Hamer 1995; Pattatucci 1998; Bailey et al. 2000; Kendler et al. 2000; Långström et al. 2010; Camperio Ciani and Pellizzari 2012).

In females, HS concordance probability for sisters was found to be higher than chance (Bailey and Benishay 1993; Bailey et al. 1993; Pattatucci and Hamer 1995; Pattatucci 1998). Also, Pattatucci and Hamer (1995) recorded elevated rates of nonheterosexuality in four classes of female relatives of lesbian probands: sisters, daughters, nieces, and female cousins through a paternal uncle. However, new preliminary data suggest elevated rates of HS in all female classes of lesbian probands' relatives, in both parental lines (Camperio Ciani et al. 2014a).

Male HS also runs in families, and gay probands show elevated frequencies of HS brothers, cousins, and uncles, mostly in the maternal line (see also the section Asymmetries in Pedigrees), and do not show more female HS in their pedigrees if compared with heterosexual males (Hamer et al. 1993; Camperio Ciani et al. 2004; Blanchard and Lippa 2007; Rahman et al. 2008; Iemmola and Camperio Ciani 2009). Also, Pillard and Weinrich (1986) found that sisters of male HS are not more likely to be lesbian than the sisters of heterosexual males.

These data indicate that male and female HS orientation clusters in families, but in a different manner in the two genders.

Fecundity of Male HS

HS males reproduce significantly less than heterosexual males, an evolutionary cost that has been confirmed by several studies (Bell and Weinberg 1978; Iemmola and Camperio Ciani 2009; Rieger et al. 2012). The average present-day fecundity of HS males as a population's subgroup, normalized to the fecundity of heterosexuals, has been quantitatively estimated in Western European countries or North America approximately between 0.2 and 0.7 (Moran 1972; Bell and Weinberg 1978; Camperio Ciani et al. 2008). This interval is wide enough that it may be reasonably assumed to encompass also the reduced fitness of male HS under the varying circumstances of our evolutionary past.

Fecundity of Female HS

Modern Western female HS show low normalized fecundity, estimated below 0.5 (Pattatucci and Hamer 1995; Gates et al. 2007; Patterson and Riskind 2010; Camperio Ciani et al.

2014a). Such strongly reduced value may be a result of, not only to same-sex attraction or choice, but also to female HS being victims of homophobia, prejudice, and laws against same-sex households. Indeed, recent investigations appear to indicate that more liberal laws for same-sex couples, maternity through alternative means, such as in vitro fertilization, and adoption by same-sex couples, might be responding to a maternal drive to motherhood in HS females (Butler 2005; Patterson and Riskind 2010; Farr and Patterson 2013), possibly raising, in turn, also their fecundity. Because maternal drive arises later in adulthood than erotic attraction, it has been predicted that fecundity in lesbians should grow later in age, a fact that has partly been confirmed by Farr and Patterson (2013). Other recent data, however, question the latter conclusion, showing low fecundity values below 0.2 also in aged Italian lesbians living in contemporary relatively liberal environments (Camperio Ciani et al. 2014a).

In contrast, a low or very low fecundity of female HS in the past might be questionable for multiple reasons: (1) patriarchal societies, (2) sexual flexibility, (3) strong maternal drive, and (4) as a counterstrategy against male family desertion and infanticide. The convergent effects of such causes might have eliminated any fecundity reduction in female HS (caused by same-sex attraction) in our evolutionary past. The vast majority of past societies were patriarchal and females have been ruled, traded and exchanged, destined to opportunistic marriages, and forced to reproduce, with substantial indifference to their sexual preferences (Kraemer 1991; Bar On 1994; Vrissimtzis 1997; Apostolou 2013). Also, an inner drive to nurture offspring may often surpass one's own other needs at a physiological, hormonal, and behavioral level, including HS preference (Mezey 2008). Furthermore, maternal drive is independent of sexual orientation, and before in vitro fertilization, attaining fecundity in response to a growing maternal instinct might have been possible also through natural means by accepting to reproduce with the other sex (Mezey 2008; Patterson and Riskind 2010). Contrary to males, in which a lack of sexual attraction makes fertilization physiologically demanding, females could use sexuality more widely to opportunistically achieve fertilization, independently of explicit erotic attraction to males (Baumeister 2000). Finally, lesbian behavior might have enhanced same-sex pair bonds to adequately nurture offspring, and to protect it in case of male desertion of the family, which may have been universally common through human history (Eubank 1916; Colcord 1919; May 1988; Lamb 2004). A lesbian strategy to exclude the intrusion in the household of a new potentially infanticidal male might have produced strong fitness benefits on the female side. Recent studies show that female counterstrategies to avoid male infanticide in humans have been powerful in determining the adaptive value of a trait (de Waal and Gavrilets 2013; Lukas and Clutton Brock 2013; Opie et al. 2013). All of the above causes might have significantly contributed to balance past HS females' lack of erotic attraction to males, increasing their fecundity compared with the present day, to possibly match the average heterosexual fecundity. This may have contributed to making female HS an evolutionarily neutral trait (see the section Candidate Selection Mechanisms for Female HS).

Asymmetries in Pedigrees

In male HS, empirical observations indicate the existence of characteristic "pedigree asymmetries" concerning female fecundity within the gay probands' families. Camperio Ciani et al. (2004) found that, in general, female maternal relatives of HS males have significantly higher fecundity than female maternal relatives of heterosexual males, and this difference was not found among female paternal relatives. Specifically, increased female fecundity regards mothers (with normalized fecundity, $\alpha > 1.2$), aunts, and grandmothers in the maternal line of male HS, as well as male bisexuals (Camperio Ciani et al. 2009; Jannini et al. 2010; Camperio Ciani and Pellizzari 2012). These findings were replicated in a number of studies (see Blanchard and Lippa 2007; Rahman et al. 2008; Iemmola and Camperio Ciani 2009; VanderLaan and Vasey 2011; Blanchard 2012; Rieger et al. 2012). It was also found that mothers and aunts of male HS recorded fewer abortions, fewer gestational and parturition complications, and higher extraversion (Camperio Ciani et al. 2012). It should be noted, however, that other investigators have reported a generalized fecundity increase in other members in the pedigrees of male HS probands, independent of the maternal or paternal line (King et al. 2005; Schwartz et al. 2010), and Rieger et al. (2012), based on the dataset by Schwartz et al. (2010), found excess fecundity in brothers of male HS. Subsequently, the data by Schwartz et al. (2010) were reanalyzed by Camperio Ciani and Pellizzari (2012), highlighting, also in this dataset, a previously unnoticed larger maternal-line female fecundity. Finally, a population dynamics analysis of male HS (see Camperio Ciani et al. 2008 and the section Genetic Models for Male HS below), showed a rather singular, previously unconsidered, fecundity increase in paternal aunts of heterosexuals, as compared with homosexuals. This effect was later detected in a reanalysis of empirical data by Camperio Ciani and Pellizzari (2012).

For female HS, Pattatucci and Hamer (1995) did not report, in the pedigrees, any significant, possibly asymmetric, fecundity increases as mentioned above regarding male HS. However, preliminary data on families of HS females show they have a larger pedigree size (more relatives in the three antecedent generations) compared with control heterosexual females' families, indicating, if confirmed, a general fecundity increase (Camperio Ciani et al. 2014a). This effect is particularly evident in the subsample of families with at least one lesbian member other than the proband, which show an excess of males in the paternal line versus the maternal line.

Fraternal Birth Order Effect

Several studies have found that having older biological brothers (not sisters) increases the probability of HS in later-born males (Blanchard and Bogaert 1996; Blanchard 2001, 2004; Cantor et al. 2002). This is a maternal influence, referred to as the fraternal birth order (FBO) effect, which has been hypothesized to relate to a progressive immunization of the mother to successive male fetus exposure (Blanchard 2004; Bogaert and Skorska 2011). The effect holds also if the siblings are raised in separate households, whereas step- or adoptive brothers have no effect on sexual orientation (Bogaert 2006). Notably, in case of later-born monozygotic (MZ) twins, the FBO could contribute to increasing the HS concordance rates in males (Camperio Ciani et al. 2004). However, not all researchers confirmed a significant FBO effect in male HS (Frisch and Hviid 2006; Blanchard and Lippa 2007; see also LeVay 2011, for questions on the FBO). No birth order effect has been reported in female HS (Ellis and Blanchard 2001; Blanchard 2004). McConaghy et al. (2006) have suggested that both females and males with HS feelings are most likely born later, not for biological reasons, but rather because of social processes; however, this hypothesis ignores the evidence on adoptive and separated siblings.

Twin Studies and Estimates of Genetic Heritability

Empirical evidence from family studies, comparing pairs of adoptive brothers, biological brothers, dizygotic (DZ), and MZ twins, have shown that the probability of HS sibling concordance in all such classes to progressively increase if the proband is HS, which strongly suggests the presence of a genetic influence (Kendler et al. 2000; Plomin et al. 2001; Posthuma et al. 2003). Also, the much lower rate of HS concordance between adopted sisters as compared with biological sisters is in line with the genetic heritability paradigm (Bailey et al. 1993; Plomin et al. 2001; Posthuma et al. 2003).

Earlier heritability estimates for HS used the values of either pairwise or probandwise HS concordance rates in MZ and DZ twins. Studies with small sample sizes, with inherent ascertainment bias, found either high pairwise concordance of HS between 40% and 75%, both in male and female MZ twins (Pillard and Weinrich 1986; Kendler and Eaves 1989; Bailey and Pillard 1991; Bailey et al. 1993; Withman et al.

1993; Hershberger 1997), or very low rates (Eckert et al. 1986; Hershberger 1997). Subsequently, it was clarified (McGue 1992) that the probandwise concordance is the most adequate metric to interpret individual probabilities when comparing different studies because it better accounts for familiar risk-probability factors, different ascertainment biases, and different population prevalence rates. Moreover, the more recent population-wide twin studies on sexual orientation have also provided estimates on the heritability of homosexuality through the identification of sample variance components, that is, additive genetics, common (shared) environment, and unique (nonshared) environment (ACE model) (Martin et al. 1997), rather than the sole concordance, hence, providing further results on heritability in homosexuality.

Thus, in a population-wide twin study in Australia, Bailey et al. (2000) reported probandwise HS concordance values 20%–37% for males and 24%–30% for females, depending on more or less restrictive definitions of homosexuality. They estimated that, for females, the genetic influence accounted for less than 0.10, the nonshared environment for 0.50, and the shared environment for 0.40 of the observed variance. For males, they estimated a much higher genetic influence of 0.45, a nonshared environment effect of 0.55, and an almost absent effect of shared environment. Also, in a population-wide twin study in Sweden, Långström et al. (2010) reported 18% probandwise concordance for males and 22% for females for any lifetime same-sex partners. They estimated, for females, 0.18–0.19 of the same-sex sexual behavior variance was explained by genetic factors, 0.64–0.66 by nonshared environmental factors, with shared-environmental effects accounting for 0.16–0.17 of the variance. For males, they estimated that genetic effects explained 0.34–0.39 of the variance, individual environment accounted for 0.61–0.66, and, again, found an absence of shared-environment effects in males. Finally, in a population-wide twin study in Finland, Alanko et al. (2010) reported probandwise concordance of 50% for males and 47% for females. They also gave a higher heritability estimate for sexual orientation of 0.45 for females and 0.50 for males, the rest being influenced by a nonshared environment, with an absence of shared-environment effect in both sexes.

Notably, heritability is population specific because it depends on the prevalence among all members of the sampled population; thus, it is not surprising that different studies, based on different populations, have yielded diverse heritability estimates. However, as more researchers replicate studies and findings, results should become more generalizable (Alanko et al. 2010).

Overall, twin studies, despite their large heterogeneity, do suggest human HS to be influenced by a significant genetic component, with a likely larger effect in males than females (Bailey et al. 2000; Alanko et al. 2010; Långström et al. 2010), and with most investigations converging on a high influence of shared (social, educational) environment only in females (Bailey et al. 2000; Långström et al. 2010). However, see Hershberger (1997) for a higher genetic influence in female HS than in males.

Interestingly, a multiple regression pedigree analysis of possible genetic influences on male HS was found to account for a significant 21% of sexual orientation variance in an Italian population sample of male HS (Camperio Ciani et al. 2004), fairly in agreement with the heritability values obtained from the twin studies just described.

CRITICISMS TO GENETIC MODEL INFLUENCING HS

Recently, the hypothesis of a genetic influence on human HS has been challenged, advancing a possible sole role for epigenetics in HS (Rice et al. 2012, 2013). Two main reasons given for this are that (1) the estimated MZ twin concordance for HS appears to be too low, and (2) specific genes influencing HS have not yet been determined.

Regarding the latter statement (2), although direct connections to specific polymorphic genes have yet to be uncontroversially determined for HS, we notice that a wide array of phenotypic expressions are hypothesized to have some genetic influences still await the direct polymorphism identifications. However, the literature has given indications on the possible location of marker loci for a genetic substrate of HS, at least in males. Two linkage studies reported DNA marker loci for male HS on the X chromosome (Hamer et al. 1993; Hu et al. 1995), at the level of the region q28. A later investigation found no such linkage (Rice et al. 1999). However, a meta-analysis of the results across such studies yielded an estimated level of Xq28 allele sharing between HS brothers of 64% instead of the 50% expected by chance (Hamer 1999, 2002). In later work, a genome-wide scan reported several new autosomal markers for male sexual orientation (Mustanski et al. 2005). These investigators looked at the entire human genome of males who had two or more HS brothers, and found linkage with polymorphisms on three chromosomes—7, 8, and 10—that were shared by approximately 60% of the HS brothers in the study. Later studies also showed extremely skewed X inactivation in mothers of male HS as compared with mothers of heterosexuals, further suggesting an involvement of the X chromosome (Bocklandt et al. 2006), as well as other maternally imprinted polymorphisms on the autosomes (Bocklandt and Hamer 2003). More recently, a new method supplemented linkage analysis by investigating higher resolution genome-wide association of single nucleotide polymorphism with phenotypic traits. The first studies using this method reported nonsignificant results (Ramagopalan et al. 2010; Drabant et al. 2012) with just a relevant, albeit not significant, effect, possibly a result of a low power of the tests, in region q12.3 of chromosome 8, associated with male HS subjects (Drabant et al. 2012). However, Sanders et al. (2012), in a more recent large and systematic multicenter study (see also Bailey 2014), found a significant association with the region Xq28 and a region in chromosome 8, substantially confirming the earliest findings of Hamer et al. (1993) and the findings, with the earlier methods, by Mustanski et al. (2005) and Bocklandt et al. (2006).

Overall, it seems that most of the present empirical evidence confirm genomic polymorphisms associated with male HS in at least two loci, including both the X chromosome and selected autosomes. Interestingly, as discussed below in the section Candidate Mechanisms for Balancing Selection in Male HS, this evidence concurs well with the predictions in Camperio Ciani et al. (2008) obtained by comparing theoretical population-genetics models with population data for male HS.

Concerning the possible doubts on the genetic influence of HS caused by too-low MZ twin concordances, for example, see the above point (1), we have recalled in the section Twin Studies and Estimates of Genetic Heritability that the estimates derived from the presently known population-wide empirical concordances in MZ twins all indicate a measurable genetic influence on HS, although these twin data may not be incompatible also with the epigenetic influences theoretically proposed by Rice et al. (2012, 2013). Notably, a direct estimate of the probandwise concordance in MZ twins does not appear to indicate the observed concordances to be too low to be consistent with a genetic loading for HS. Because of partial expressivity and penetrance of the underlying genotype, the prevalence of the HS phenotype in the population results from less-than-Mendelian ratios. In these cases, the probandwise concordance in MZ twins (computed as in McGue 1992) depends on the value of: (1) the prevalence of the phenotypic trait (either from genetic or other origin) in the population; (2) the partial expressivity of the genetic factor as a result of a multicausal origin of HS preference in our species, given by the percentage of individuals showing the trait and who carry the genetic factors (Jannini et al. 2010); and (3) the partial penetrance, that is, the proportion of individuals carrying the genetic factors and expressing the phenotypic trait, possibly attributable to epigenetic, developmental, or environmental modifiers (Balthazart 2011). Then, by considering the HS trait as dichotomous for simplicity (i.e., present or absent), and for reasonable values of these three parameters, for instance, in the range 0.06–0.08 for the prevalence, between 0.3–05 for the expressivity, and 0.3–0.5 for the penetrance, the probandwise concordance in

MZ twins is computed to be in the range between 12% and 27% (Camperio Ciani et al. 2014a). These values are well compatible with, or even lower than, the probandwise concordances reported in the MZ twins studies recalled in the section Twin Studies and Estimates of Genetic Heritability (giving ranges 17%–50% for HS concordance in males and 21%–47% in females) (Bailey et al. 2000; Alanko et al. 2010; Långström et al. 2010).

POPULATION GENETICS MODELS OF HS

Candidate Mechanisms for Balancing Selection in Male HS

The empirical findings recalled in the section Empirical Evidence on Population Patterns Related to HS in Human Populations indicate that male HS is influenced, as mentioned earlier, by both environmental and genetic factors. The empirical evidence suggests that the frequency of HS is too large to be only because of mutation selection, even if a multiplicity of genetic factors might be involved (Gavrilets and Rice 2006; see Yoder 2011). Furthermore, the known population patterns for male HS described in the section Empirical Evidence on Population Patterns Related to HS in Human Populations also suggest that candidate genetic factors are not undergoing directional, stabilizing, or divergent selection in human populations, but rather are under some form of balancing selection, with mixed evolutionary strategies occurring within stable polymorphic populations.

Kin selection has been advanced (Trivers 1974; Wilson 1975; Pillard and Bailey 1998) as a mechanism by which HS males compensate for their reduced fecundity, enhancing their inclusive fitness through investment toward their most immediate relatives, such as nephews and nieces. Because of the rather low normalized fecundity, γ, of HS males (in the range $0.2 < \gamma < 0.7$; see the section Fecundity of Male HS), any such avuncularity in HS males should be very pronounced to achieve a true balancing effect. However, most studies, conducted in a number of different populations, have refuted this hypothesis (Bobrow and Bailey 2001; Rahman and Hull 2005; Forrester et al. 2011, Camperio Ciani et al. 2014b; but see Vasey et al. 2007; Vasey and VanderLaan 2012).

Other candidate balancing mechanisms for male HS are a priori possible. Hereafter, we follow the systematic analysis of Camperio Ciani et al. (2008), denoted CCZ for brevity. The results derived in each case were framed by CCZ against the constraints given by the relevant empirical data highlighted in the section Empirical Evidence on Population Patterns Related to HS in Human Populations to test the compatibility of the models' outputs with the known population patterns for HS.

Genetic Models for Male HS

After earlier work on the population genetics of male HS (Getz 1993; MacIntyre and Estep 1993), the more recent debate has considered diverse selection mechanisms, including direct selection modes in which the fitness is solely influenced by an individual's genotype, based on overdominance (OD), or sexually antagonistic (SA) selection, and other selection modes for male HS including maternal effects and genomic imprinting, possibly involving epigenetic activity.

Inspired by some of these competing hypotheses, Gavrilets and Rice (2006) and CCZ have studied a number of models for a single diallelic locus at either an autosomal or X-linked location, and various selection modes in each case, under the assumption of random mating. Their analysis characterized the range of selection parameters guaranteeing a stable polymorphism and the persistence of the trait in the population. The results showed that single-locus models are largely unstable as their parameter ranges are too narrow to allow for stable polymorphism under the normal variability of population conditions. This leads to either fixation or extinction of the genetic factors, contrary to the main stability requirement of the section Empirical Evidence on Population Patterns Related to HS in Human Populations.

Multilocus genetic factors for male HS should therefore be considered, with the empirical constraints suggesting that at least one of the components should be X-linked or maternally imprinted. CCZ thus systematically investigated (again, under the hypothesis of random mating) all the relevant (1- and) 2-locus models, that is, involving (AX) one autosomal locus and one X-linked locus, (XX) two X-linked loci for different values of the linkage, or (AA) two autosomal loci. All models in CCZ included tunable effects of possible antagonistic selection, favoring females, but decreasing the fitness of male carriers, by considering normalized average fecundities γ and α, respectively, for male and female genotypes, within windows of variability such as $0.2 < \gamma < 0.7$ and $1.0 < \alpha < 1.8$. This analysis yielded the following results.

1. OD, that is, heterozygous advantage in males. In this hypothesis (Hutchinson 1959; Weinrich 1987; MacIntyre and Estep 1993; King et al. 2005; Gavrilets and Rice 2006), a homozygous form would be maintained attributable to a selective advantage in the corresponding heterozygous form in males. As considered in CCZ, such OD mechanism does not coincide with the classical assumption in which heterozygous advantage is manifested in all carriers regardless of sex, because a tunable SA component is always present, the examined genetic factors being favorable to females. Thus, to a value $\gamma < 1$ in homozygous males, there correspond values $\gamma' > 1$ of the average fecundity for heterozygous males, as well as $\alpha > 1$ for female carriers. All of these OD models were shown to generate polymorphic equilibria compatible with the stability requirements highlighted in the section Empirical Evidence on Population Patterns Related to HS in Human Populations for male HS. However, despite the SA ingredient they contain, they always account incorrectly for the pedigree asymmetries highlighted in the section Empirical Evidence on Population Patterns Related to HS in Human Populations, as they predict increased fecundities in all pedigree lines and regardless of sex. As fecundity enrichments in the families of male HS seem not to occur in such a generalized way, but only asymmetrically in maternal-line females (see the section Asymmetries in Pedigrees), the OD selection mode should be excluded from the possible maintenance mechanisms for male HS in the population.

2. SA selection. In this hypothesis (Camperio Ciani et al. 2004, 2009; Gavrilets and Rice 2006; Rahman et al. 2008; Iemmola and Camperio Ciani 2009; VanderLaan and Vasey 2011), mothers of male HS have a higher-than-baseline fecundity, so that values $\gamma < 1$ and $\alpha > 1$ are assigned, respectively, to the normalized fecundities of male and female carriers. From the analysis in CCZ, the autosomal SA-AA model was expectedly found unsuitable. However, the mixed 2-locus model AX-SA, with an autosomal and an X-linked allele, and sexually antagonistic selection, emerged as the simplest prototypical population dynamics capable of meeting all the requirements of the section Empirical Evidence on Population Patterns Related to HS in Human Populations. Also, the fully X-linked XX-SA model was found to be in accordance with such requirements, although less accurately than the AX-SA hypothesis. In both cases, polymorphism stability at low frequency was predicted, with no extinction or fixation of the genetic factors in wide ranges of increased female fecundity and reduced reproductive success of HS males. Both models also describe all the correct pedigree asymmetries in the maternal line of HS males, with a higher-than-patriline frequency of HS in the males and a higher-than-patriline fecundity in the females. Interestingly, a fecundity increase in paternal aunts of heterosexuals, as compared with homosexuals, was also predicted (later observed in empirical data, see the section Asymmetries in Pedigrees, strengthening the heuristic of these models). An interesting variation on the theme of SA selection was recently explored by Barthes et al. (2013), who showed how male HS can evolve through the dynamics of a SA hypergyny factor within a stratified society; although, in such treatment, no asymmetries were investigated.

 Cite this article as *Cold Spring Harb Perspect Biol* doi: 10.1101/cshperspect.a017657

3. Other selection modes on male offspring. CCZ have also analyzed a number of other selection mechanisms based on 2-locus AX and AA models. These included (1) maternal effects, accounted for by assuming that the fitness does not depend on the individual's genotype, but rather on the mother's genotype (Blanchard 1997, 2001, 2004). An interesting instance of this effect would be a possible genetic substrate for the induced immunization hypothesized to be at the basis of the FBO effect in the section Fraternal Birth Order Effect (Blanchard and Bogaert, 1996; Blanchard 2001, 2004; Cantor et al. 2002; Bogaert and Skorska, 2011). The analysis in CCZ also included (2) a selection model describing a form of maternal genomic imprinting (Bocklandt and Hamer 2003; Mustanski et al. 2005; Bocklandt et al. 2006), according to which the fitness depends on alleles that are active in a son only if inherited from the mother. This is described by distinguishing two male genotypes depending on the provenience of the gametes. For the same reasons as with the OD models in the above point (2), these further models contain a tunable SA component. The systematic analysis of their propagation and equilibria showed that all such transmission mechanisms behave in an unstable way, leading too easily to extinction or fixation of the genetic factors, and not allowing for the creation of reported pedigree asymmetries. These conclusions suggest, for instance, that a possible genetic influence on the FBO effect cannot, alone, explain the persistence and patterns of male HS in humans.

The models (1) and (2) may involve epigenetic activity as the origin or concurrent cause of the considered maternal effects or genomic-imprinting mechanisms. The recent hypothesis of epigenetic factors influencing HS caused by possible erasure errors for the epi-marks aiding the canalization of sexual differentiation (Rice et al. 2012, 2013), can likewise be accounted for by this framework. The analysis of CCZ adapted to such epigenetic mechanisms shows its instability, a conclusion confirmed also by the modeling in Rice et al. (2012). Any such fixated (epi)genetic factors, with no stable polymorphism, would mostly have stochastic expression once diffused into the population, creating a stochastic noise that accompanies the familial patterns and asymmetries established for male HS (see the section Empirical Evidence on Population Patterns Related to HS in Human Populations), contributing to the apparent nonshared environment components for this trait (Bailey et al. 2000; Plomin et al. 2001; Alanko et al. 2010; Långström et al. 2010). Some fixated factors may include SA epigenetic activity with occasional transmission across generations (Rice et al. 2008, 2012, 2013), creating few-generation familial patterns possibly analogous to those produced by the polymorphic SA genetic components in point (2). Further analysis and data focusing on the familial clustering of the HS trait across numerous generations may contribute to clarify and discern between these different and possibly coexisting transmission mechanisms.

Candidate Selection Mechanisms for Female HS

Also for female HS, although scarcer than for males, the results from twin and population-pattern studies (see the section Empirical Evidence on Population Patterns Related to HS in Human Populations) indicate that, besides environmental factors, there is a nontrivial genetic influence for this trait (Bailey et al. 2000; Alanko et al. 2010; Långström et al. 2010). Because of the differences in pedigree asymmetries and the frequency of transmission, researchers suggested that different genetic factors affect the expression of sexual orientation in females as compared with males (Pattatucci and Hamer 1995; Alanko et al. 2010).

As discussed in the section Fecundity of Female HS, lesbian fecundity might not always have been, in our evolutionary past, as low as currently observed, and might have (almost) matched the heterosexual average levels. This would have made female HS or, indeed, its coexistence with other female behavioral patterns, an evolutionarily neutral trait, for which any genetic substratum could have persevered and drifted sheltered from natural selection. At present, were female HS fecundity to remain at the low levels suggested by current data (see the section Fecundity of Female HS), the lack of an efficient balancing fecundity effect would imply a disequilibrium producing a progressive reduction of any genetic component for this trait in future generations.

Pattatucci and Hamer (1995), based on their own wide pedigree analysis of female HS, suggested a possible model for female HS entailing a sex-limited, autosomal, dominant genetic factor, with reduced penetrance and a net negative effect on female reproduction ($\alpha < 1$). They suggested that such a model could explain most of their findings about population patterns of female HS, including its less-than-Mendelian observed ratios and higher rates of HS in female compared with male relatives of female HS, the mother-to-daughter transmission with a lack of generation skipping, and the higher rates in daughters than in mothers, as well as the elevated rates in female cousins through paternal uncles (Pattatucci and Hamer 1995; Pattatucci 1998). No modeling, however, has yet been performed to confirm their dynamics in the population, and clarify how it could be maintained/balanced in the population with $\alpha < 1$.

In a recent proposal, Rice et al. (2008) reinterpreted female HS and part of its familial distribution reported in Pattatucci and Hamer (1995) in light of a SA zygotic drive, which is a competition of the paternal Y chromosome, against selected alleles on the X chromosome, to enhance male vigor ($\gamma > 1$) at the expense of female offspring ($\alpha < 1$). The SA zygotic drive is a form of sexual conflict that would reduce viability of female fetuses because the Y chromosome, through epigenetic inactivation, would disrupt daughter-specific developmental pathways. In some cases, this could produce female HS ($\alpha < 1$), whereas in others, it may favor female fetus abortion ($\alpha = 0$). An increase of abortion frequency in mothers of lesbians has indeed been observed (Ellis and Blanchard 2001). Such a SA zygotic drive would, in any case, be expected to be temporary, because, although lacking compensatory effects, it would be rapidly repressed by Fisherian sex-ratio selection (see Friberg and Rice 2014).

Rice et al. (2008) indicate such a mechanism could account for some of the asymmetries reported for HS females by Pattatucci and Hamer (1995), showing elevated frequencies of HS in female relatives whose fathers shared the same Y chromosome, as nieces from paternal uncles. The average HS rates reported for paternal aunts would then be compatible with the fact that elevated rates of female HS should be associated to the X-Y combination of the father, rather than the Y alone. Also, the observed elevated rates of HS in female HS's daughters would be consistent with the paternal epigenetic effects on the X chromosome being sometimes transmitted across generations. The data in Camperio Ciani et al. (2014a), if confirmed, might be however problematic for the hypothesis of a SA zygotic drive for this trait, as elevated rates of HS in all female classes were recorded, unlike with the earlier data in Pattatucci and Hamer (1995).

CONCLUSIONS

Multiple independent causes, genetic and environmental, concur to establish same-sex preference in humans, with marked differences in the expression of this phenotype in males and females. Empirically observed population patterns, such as frequency distributions, family clustering, pedigree asymmetries, and sibling concordances, point to genetic heritability of HS, indicating a low-prevalence underlying genetic influence in both sexes with a low-penetrance and partial-expressivity genetic influence on both males and females. Further differences also point to a much higher shared (social, educational) environmental influence in female HS.

For male HS, about which there exist adequate population data, the systematic mathematical analysis of the evolutionary propagation mechanisms eliminates the possible Darwinian paradox associated with HS, resolving it within the framework of sexual conflict, suggesting the following conclusions.

1. OD seems excluded from the putative balancing mechanisms of male HS because the

patterns of generalized increased fecundity that it indifferently generates in all the relatives of HS males are not empirically confirmed.

2. Sexual antagonism for a multilocus factor with at least an X-linked locus is the selection mode providing closest adherence of the models to the empirically known patterns for both HS sexual orientation of males and higher-than-average fecundity for females in their maternal line. These findings point, with a particularly relevant example, to the occurrence of a first well-identified sexually antagonistic character in humans. This perspective may help shift the focus away from male HS preference per se. Rather than concentrating on the sole aspect of the reduced male fitness that it entails, this places male HS within the more general sexual-conflict framework of a genetic trait with gender-specific benefits, which may have evolved by increasing the fecundity of females, and neither disappears nor completely invades the gene pool.

3. The presence of other genetic components that may have evolved to fixation (such as maternal or genomic-imprinting effects, possibly involving epigenetic activity and sexual conflict) cannot be excluded to coexist independently of and in conjunction with the genetic sexually antagonistic factors in point (2). Being fixated, any such factors would be expressed in the population in a mostly stochastic manner and might non-negligibly contribute both to what we presently consider non-shared environment effects. When transmitted through a few generations (Rice et al. 2008, 2012, 2013), their expression may also create some familial patterns possibly resembling the effect of polymorphic genes.

Female HS has not been studied to the same degree as male HS. Although the latter is associated with matrilines, female HS does not yet appear to be associated with certainty with either parental line, although empirical population and pedigree data are still incomplete. Possible sexually antagonistic (epi)genetic effects and a fecundity compensation, if any, are still to be confirmed, and systematic mathematical modeling has not been performed on this trait.

In conclusion, human HS has likely different inheritance systems in the two genders, although sexually antagonistic effects are likely driving forces for any (epi)genetic HS heritable components affecting their transmission. The imprint of sexual conflict appears well detectable in males, whereas in females it seems clouded by a number of other unresolved questions.

REFERENCES

*Reference is also in this collection.

Alanko K, Santtila P, Harlaar N, Witting K, Varjonen M, Jern P, Johansson A, von der Pahlen B, Sandnabba NK. 2010. Common genetic effects of gender atypical behavior in childhood and sexual orientation in adulthood: A study of Finnish twins. *Arch Sex Behav* **39:** 81–92.

Apostolou M. 2013. The parental choice branch of sexual selection: Re-examining the evolution of mating behavior. *J Integr Soc Sci* **3:** 37–54.

Bagemihl B. 1999. *Biological exuberance. Animal homosexuality and natural diversity.* Profile, London.

Bailey JM. 2014. *Discussion event held in conjunction with the Annual Meeting of the American Association for the Advancement of Science.* Chicago, www.aaas.org/meetings.

Bailey JM, Benishay DS. 1993. Familial aggregation of female sexual orientation. *Am J Psychiatry* **150:** 272–277.

Bailey JM, Pillard RC. 1991. A genetic study of male sexual orientation. *Arch Gen Psychiatry* **48:** 1089–1096.

Bailey JM, Zucker KJ. 1995. Childhood sex typed behavior and sexual orientation: A conceptual analysis and quantitative review. *Dev Psychol* **31:** 43–56.

Bailey JM, Pillard RC, Neale MC, Agyei Y. 1993. Heritable factors influence sexual orientation in women. *Arch Gen Psychiatry* **50:** 217–223.

Bailey JM, Dunne MP, Martin NG. 2000. Genetic and environmental influences on sexual orientation and its correlates in an Australian twin sample. *J Pers Soc Psychol* **78:** 524–536.

Balthazart J. 2011. *The biology of homosexuality.* Oxford University Press, Oxford.

Bar On B-A. 1994. *Engendering origins: Critical feminist readings in Plato and Aristotle.* SUNY Press, New York.

Barthes J, Godelle G, Raymond M. 2013. Human social stratification and hypergyny: Toward an understanding of male homosexual preference. *Evol Hum Behav* **34:** 155–163.

Baumeister RF. 2000. Gender differences in erotic plasticity: The female sex drive as socially flexible and responsive. *Psychol Bull* **126:** 347–374.

Bell AP, Weinberg M. 1978. *Homosexualities: A study of diversity among men and women.* Simon and Schuster, New York.

Blanchard R. 1997. Birth order and sibling sex ratio in homosexual versus heterosexual males and females. *Annu Rev Sex Res* **8:** 27–67.

Blanchard R. 2001. Fraternal birth order and the maternal immune hypothesis of male homosexuality. *Horm Behav* **40:** 105–114.

Blanchard R. 2004. Quantitative and theoretical analyses of the relation between older brothers and homosexuality in men. *J Theor Biol* **230:** 173–178.

Blanchard R. 2012. Fertility in the mothers of first born homosexual and heterosexual men. *Arch Sex Behav* **41:** 551–556.

Blanchard R, Bogaert AF. 1996. Homosexuality in men and number of older brothers. *Am J Psychiatry* **153:** 27–31.

Blanchard R, Lippa RA. 2007. Birth order, sibling sex ratio, handedness, and sexual orientation of male and female participants in a BBC Internet research project. *Arch Sex Behav* **36:** 163–176.

Blumstein PW, Schwartz P. 1976. Bisexuality in women. *Arch Sex Behav* **5:** 171–181.

Blumstein PW, Schwartz P. 1977. Bisexuality: Some social psychological issues. *J Soc Issues* **33:** 30–45.

Bobrow D, Bailey JM. 2001. Is male homosexuality maintained via kin selection? *Evol Hum Behav* **22:** 361–368.

Bocklandt S, Hamer DH. 2003. Beyond hormones: A novel hypothesis for the biological basis of male sexual orientation. *J Endocrinol Invest* **26:** 8–12.

Bocklandt S, Horvath S, Vilain E, Hamer DH. 2006. Extreme skewing of X chromosome inactivation in mothers of homosexual men. *Hum Genet* **118:** 691–694.

Bogaert A. 2004. The prevalence of male homosexuality: The effect of fraternal birth order and variations in family size. *J Theor Biol* **230:** 733–770.

Bogaert AF. 2006. Biological versus non-biological older brothers and men's sexual orientation. *Proc Natl Acad Sci* **103:** 10771–10774.

Bogaert AF, Skorska M. 2011. Sexual orientation, fraternal birth order, and the maternal immune hypothesis: A review. *Front Neuroendocrinol* **32:** 247–254.

Bonduriansky R, Chenoweth F. 2009. Intralocus sexual conflict. *Trends Ecol Evol* **24:** 280–288.

Boomsma D, Busjahn A, Peltonen L. 2002. Classical twin studies and beyond. *Nat Rev Genet* **3:** 872–882.

Bullough V. 1998. Alfred Kinsey and the Kinsey report: Historical overview and lasting contribution. *J Sex Res* **35:** 127–131.

Butler AC. 2005. Gender differences in the prevalence of same-sex sexual partnering: 1988–2002. *Soc Forces* **84:** 421–449.

Cáceres C, Konda K, Pecheny M, Chatterjee A, Lyerla R. 2006. Estimating the number of men who have sex with men in low and middle income countries. *Sex Transm Infect* **82:** 3–9.

Camperio Ciani A, Battaglia U. 2014. Implicit measurements of sexual preference in self-declared heterosexual men: A pilot study on the rate of androphilia in Italy. *J Sex Med* doi: 10.1111/jsm.12565.

Camperio Ciani A, Pellizzari E. 2012. Fecundity of paternal and maternal non-parental female relatives of homosexual and heterosexual men. *PLoS ONE* **7:** e51088.

Camperio Ciani A, Corna F, Capiluppi C. 2004. Evidence for maternally inherited factors favouring male homosexuality and promoting female fecundity. *Proc Biol Sci* **271:** 2217–2221.

Camperio Ciani A, Cermelli P, Zanzotto G. 2008. Sexually antagonistic selection in human male homosexuality. *PLoS ONE* **3:** 1–8.

Camperio Ciani A, Iemmola F, Blecher S. 2009. Bisexuals and not exclusive homosexuals show evidence of the same genetic factors that promote a female fecundity increase on the maternal line. *J Sex Med* **6:** 449–455.

Camperio Ciani A, Fontanesi L, Iemmola F, Giannella E, Ferron C. 2012. Factors associated with higher fecundity in female maternal relatives of homosexual men. *J Sex Med* **9:** 2878–2887.

Camperio Ciani A, Pellizzari E. 2014a. Sexually antagonistic selection explains homosexuality evolutionary conundrum (abstract). In *Cooperation and Conflict in the Family Conference.* Sydney, February 2–5.

Camperio Ciani A, Battaglia U, Liotta M. 2014b. Societal norms rather than sexual orientation influences kin altruism and avuncularity in tribal Urak-lawoi, Italian and Spanish adult males. *J Sex Res* (in press).

Cantor JM, Blanchard R, Paterson AD, Bogaert AF. 2002. How many gay men owe their sexual orientation to fraternal birth order? *Arch Sex Behav* **31:** 63–71.

Colcord JC. 1919 *Broken homes: A study of family desertion and its social treatment project*, pp. 1–212. Gutenberg, Salt Lake City, UT (reprinted in 2008).

Crompton L. 2006. *Homosexuality and civilization.* Harvard University Press, Boston.

de Waal FBM, Gavrilets S. 2013. Monogamy with a purpose. *Proc Natl Acad Sci* **110:** 15167–15168.

Diamond M. 1993. Homosexuality and bisexuality in different populations. *Arch Sex Behav* **22:** 291–310.

Diamond LM. 2008. Female bisexuality from adolescence to adulthood: Results from a 10-year longitudinal study. *Dev Psychol* **44:** 5–14.

Dörner G, Rohde W, Schott G, Schnabl C. 1983. On the LH response to oestrogen and LH-RH in transsexual men. *Exp Clin Endocrinol* **82:** 257–267.

Drabant EM, et al. 2012. Genome-wide linkage association study of sexual orientation in a large web-based cohort. *American Society of Human Genetics Conference* (abstract 1957T). San Francisco.

Eckert ED, Bouchard TJ, Bohlen J, Heston LL. 1986. Homosexuality in monozygotic twins reared apart. *Br J Psychiatry* **148:** 421–425.

Ellis L, Blanchard R. 2001. Birth order, sibling sex ratio, and maternal miscarriages in homosexual and heterosexual men and women. *Pers Individ Differ* **30:** 543–552.

Esterberg KG. 1997. Lesbian and bisexual identities: Constructing communities. In *Constructing selves.* Temple University Press, Philadelphia.

Eubank EE. 1916. *A study of family desertion*, pp. 1–74. The Department of Public Welfare, University of Chicago Libraries, Chicago.

Farr RH, Patterson CJ. 2013. Coparenting among lesbian, gay, and heterosexual couples: Associations with adopted children's outcomes. *Child Dev* 84: 1226–1240.

Forrester D, VanderLaan DP, Parker JL, Vasey PL. 2011. Male sexual orientation and avuncularity in Canada: Implications for the kin selection hypothesis. *J Cognit Cult* **11:** 3–4.

* Friberg U, Rice WR. 2014. Sexually antagonistic zygotic drive: A new form of genetic conflict between the sex chromosomes. *Cold Spring Harb Perspect Biol* doi: 10.1101/cshperspect.a017608.

Frisch M, Hviid A. 2006. Childhood family correlates of heterosexual and homosexual marriages: A national cohort study of two million Danes. *Arch Sex Behav* **35:** 533–547.

Garcia-Falgueras A, Swaab DF. 2010. Sexual hormones and the brain: An essential alliance for sexual identity and sexual orientation in pediatric neuroendocrinology. *Endocr Dev* **17:**, 22–35.

Gates GJ, Badget M, Macomber JE, Chambers K. 2007. *Adoption and foster care by gay and lesbian parents in the United States*. The Williams Institute, Los Angeles.

* Gavrilets S. 2014. Is sexual conflict an "engine of speciation?" *Cold Spring Harb Perspect Biol* doi: 10.1101/cshperspect.a017723.

Gavrilets S, Rice WR. 2006. Genetic models of homosexuality: Generating testable predictions. *Proc Biol Sci* **273:** 3031–3038.

Getz WM. 1993. Invasion and maintenance of alleles that influence mating and parental success. *J Theor Biol* **162:** 515–537.

Hamer D. 1999. Genetics and male sexual orientation. *Science* 285–803.

Hamer D. 2002. Genetics of sexual behavior. In *Molecular genetics and the human personality* (ed. Benjamin J, et al.), pp. 257–272. American Psychiatric Publishing, Washington, DC.

Hamer DH, Hu S, Magnuson VL, Hu N, Pattatucci AM. 1993. A linkage between DNA markers on the X-chromosome and male sexual orientation. *Science* **261:** 321–327.

Herdt GH. 1997. *Same sex, different cultures: Gays and lesbians across cultures*, pp. 1–204. Westview, Boulder, CO.

Herek GM, Gillis JR, Cogan JC, Glunt EK. 1997. Hate crime victimization among lesbian, gay, and bisexual adults: Prevalence, psychological correlates, and methodological issues. *J Interpers Violence* **12:** 195–215.

Herek GM, Gillis JR, Cogan JC. 1999. Psychological sequelae of hate-crime victimization among lesbian, gay, and bisexual adults. *J Consult Clin Psychol* **67:** 945–951.

Hershberger SL. 1997. A twin registry study of male and female sexual orientation. *J Sex Res* **34:** 212–222.

Hewlett BS, Hewlett BL. 2010. Sex and searching for children among Aka foragers and Ngandu farmers of central Africa. *Afr Study Monogr* **31:** 107–125.

Hu S, Pattatucci AML, Patterson C, Li L, Fulkner DW, Cherney SS, Kruglyak L, Hamer DH. 1995. Linkage between sexual orientation and chromosome Xq28 in males but not in females. *Nat Genet* **11:** 248–256.

Hutchinson GE. 1959. A speculative consideration of certain forms of sexual selection in men. *Am Nat* **93:** 81–91.

Iemmola F, Camperio Ciani A. 2009. New evidence of genetic factors influencing sexual orientation in men: Female fecundity increase in the maternal line. *Arch Sex Behav* **38:** 393–399.

Jannini EA, Blanchard R, Camperio Ciani A, Bancroft J. 2010. Male homosexuality: Nature or culture? *J Sex Med* **7:** 3245–3253.

Johnson AM, Wadsworth J, Wellings K, Field J. 1994. *Sexual attitudes and lifestyles*. Blackwell Scientific, Cambridge, MA.

Kendler KS, Eaves LJ. 1989. The estimation of probandwise concordance in twins: The effect of unequal ascertainment. *Acta Genet Med Gemellol (Roma)* **38:** 253–270.

Kendler KS, Thornton LM, Gilman SE, Kessler RC. 2000. Sexual orientation in a US national sample of twin and non-twin sibling pairs. *Am J Psychiatry* **157:** 1843–1846.

King M, Green J, Osborn DPJ, Arkell J, Hetherton J, Pereira E. 2005. Family size in white gay and heterosexual men. *Arch Sex Behav* **34:** 117–122.

Kraemer S. 1991. The origins of fatherhood: An ancient family process. *Fam Proc* **30:** 377–392.

Lamb ME (ed.). 2004. *The role of the father in child development*. Wiley, Hoboken, NJ.

Långström N, Rahman Q, Carlström E, Lichtenstein P. 2010. Genetic and environmental effects on same-sex sexual behavior: A population study of twins in Sweden. *Arch Sex Behav* **39:** 75–80.

LeVay S. 2011. *Gay, straight, and the reason why: The science of sexual orientation*. Oxford University Press, Oxford.

Lippa RA. 2006. Finger lengths, 2D:4D ratios, and their relation to gender-related personality traits and the Big Five. *Biol Psychol* **71:** 116–121.

Lukas D, Clutton-Brock TH. 2013. The evolution of social monogamy in mammals. *Science* **341:** 526–530.

Lynch M. 2007. The frailty of adaptive hypotheses for the origins of organismal complexity. *Proc Natl Acad Sci* **104:** 8597–8604.

MacIntyre F, Estep KW. 1993. Sperm competition and the persistence of genes for male homosexuality. *Biosystems* **31:** 223–233.

Martin N, Boomsma D, Machin G. 1997. A twin-pronged attack on complex trait. *Nat Genet* **17:** 387–392.

May M. 1988. The "problem of duty": Family desertion in the progressive era. *Soc Serv Rev* **62:** 40–60.

McConaghy N, Hadzi-Pavlovic C, Stevens C, Manicavasagar V, Buhrich N, Vollmer-Conna U. 2006. Fraternal birth order and ratio of heterosexual/homosexual feelings in women and men. *J Homosex* **51:** 161–74.

McGue N. 1992. When assessing twin concordance, use the probandwise not the pairwise rate. *Schizophr Bull* **18:** 171–176.

Mezey NJ. 2008. *New choices, new families: How lesbians decide about motherhood*, pp. 1–208. Johns Hopkins University Press, Baltimore.

Moran PAP. 1972. Familial effects in schizophrenia and homosexuality. *Aust NZ Psychiatry* **6:** 116–119.

Mustanski BS, DuPree MG, Nievergelt CM, Bocklandt S, Schork NJ, Hamer DH. 2005. A genomewide scan of male sexual orientation. *Hum Genet* **116:** 272–278.

Mychasiuk R, Gibb R, Kolb B. 2012. Prenatal stress produces sexually dimorphic and regionally specific changes in gene expression in hippocampus and frontal cortex of developing rat offspring. *Dev Neurosci* **33:** 531–538.

Nash G. 2001. The subversive male: Homosexual and bestial images on European mesolithic rock art. In *Indecent exposure: Sexuality, society and the archaeological record*, pp. 43–55. Cruithne, Oxford.

Ngun TC, Ghahramani N, Sánchez FJ, Bocklandt S, Vilain E. 2011. The genetics of sex differences in brain and behavior. *Front Neuroendocrinol* **32:** 227–246.

Opie C, Atkinson QD, Dunbar RI, Shultz S. 2013. Male infanticide leads to social monogamy in primates. *Proc Natl Acad Sci* **110:** 13328–13332.

Pattatucci AML. 1998. Molecular investigations into complex behavior: Lessons from sexual orientation studies. *Hum Biol* **70:** 367–386.

Pattatucci AML, Hamer DH. 1995. Development and familiality of sexual orientation in females. *Behav Genet* **25:** 407–419.

Patterson CJ, Riskind RG. 2010. To be a parent: Issues in family formation among gay and lesbian adults. *J GLBT Fam Stud* **6:** 326–340.

Pillard RC, Bailey JM. 1998. Human sexual orientation has a heritable component. *Hum Biol* **70:** 347–365.

Pillard RC, Weinrich JD. 1986. Evidence of familial nature of male homosexuality. *Arch Gen Psychiatry* **43:** 808–812.

Plomin R, DeFries JC, McClearn GE, McGuffin P. 2001. *Behavioral genetics*, 4th ed. Worth, New York.

Posthuma D, Beem AL, de Geus EJC, van Baal G M, von Hjelmborg JB, Iachine I, Boomsma DI. 2003. Theory and practice in quantitative genetics. *Twin Res* **6:** 361–376.

Rahman Q. 2005. The neurodevelopment of human sexual orientation. *Neurosci Biobehav Rev* **29:** 1057–1066.

Rahman Q, Hull MS. 2005. An empirical test of the kin selection hypothesis for male homosexuality. *Arch Sex Behav* **34:** 461–467.

Rahman Q, Collins A, Morrison M, Orrells JC, Cadinouche K, Greenfield S, Begum S. 2008. Maternal inheritance and familial fecundity factors in male homosexuality. *Arch Sex Behav* **37:** 962–969.

Ramagopalan SV, Dyment DA, Handunnetthi L, Rice GP, Ebers GC. 2010. A genomewide scan of male sexual orientation. *J Hum Genet* **55:** 131–132.

Rice G, Anderson C, Risch N, Ebers G. 1999. Male homosexuality: Absence of linkage to microsatellite markers at Xq28. *Science* **284:** 665–667.

Rice WR, Gavrilets S, Friberg U. 2008. Sexually antagonistic "zygotic drive" of the sex chromosomes. *PLoS Genet* **4:** e1000313.

Rice WR, Friberg U, Gavrilets S. 2012. Homosexuality as a consequence of epigenetically canalized sexual development. *Q Rev Biol* **87:** 343–368.

Rice WR, Friberg U, Gavrilets S. 2013. Homosexuality via canalized development: A testing protocol for a new epigenetic model. *Bioessays* **35:** 764–770.

Rieger G, Blanchard R, Schwartz G, Bailey JM, Sanders AR. 2012. Further data concerning Blanchard's (2011) fertility in the mothers of firstborn homosexual and heterosexual men. *Arch Sex Behav* **41:** 529–531.

Rosenbluth S. 1997. Is sexual orientation a matter of choice? *Psychol Women Q* **21:** 595–610.

Ruse M. 1988. *Philosophy of biology today.* SUNY Press, New York.

Rust PC. 1992. The politics of sexual identity: Sexual attraction and behavior among lesbian and bisexual women. *Soc Probl* **39:** 366–386.

Sanders AR, et al. 2012. Genome-wide linkage scan of male sexual orientation. In *American Society of Human Genetics Conference* (abstract 1957T). San Francisco.

Savin-Williams RC. 2006 Who's gay? Does it matter? *Curr Dir Psychol Sci* **15:** 40–44.

Schwartz G, Kim RM, Kolundzija AB, Rieger G, Sanders AR. 2010. Biodemographic and physical correlates of sexual orientation in men. *Arch Sex Behav* **39:** 93–109.

Sell RL, Wells JA, Wypij D. 1995. The prevalence of homosexual behavior and attraction in the United States, the United Kingdom and France: Results of national population-based samples. *Arch Sex Behav* **24:** 235–248.

Sham P. 2001. Shifting paradigms in gene-mapping methodology for complex traits. *Pharmacogenomics* **2:** 195–202.

Sommer V, Vasey PL. 2006. *Homosexual behaviour in animals: An evolutionary perspective.* Cambridge University Press, Cambridge.

Stein A. 1997. *Sex and sensibility: Stories of a lesbian generation.* University of California Press, Oakland, CA.

Trivers RL. 1974. Parent–offspring conflict. *Am Zool* **14:** 249–264.

VanderLaan DP, Vasey PL. 2011. Male sexual orientation in independent Samoa: Evidence for fraternal birth order and maternal fecundity effects. *Arch Sex Behav* **40:** 495–503.

Vasey PL. 1995. Homosexual behavior in primates: A review of evidence and theory. *Int J Primatol* **16:** 173–204.

Vasey PL, VanderLaan DP. 2012. Sexual orientation in men and avuncularity in Japan: Implication for the kin selection hypothesis. *Arch Sex Behav* **41:** 209–215.

Vasey PL, Pocock DS, VanderLaan DP. 2007. Kin selection and male androphilia in Samoan Fa'afafine. *Evol Human Behav* **28:** 159–167.

Vrissimtzis NA. 1997. *Love, sex and marriage in ancient Greece*, 3rd ed. Polygrama, Athens, Greece.

Ward IL, Weisz J. 1980. Maternal stress alters plasma testosterone in fetal males. *Science* **207:** 328–329.

Weinrich JD. 1987. *Sexual landscapes.* Scribner's, New York.

Whisman V. 1996. *Queer by choice: Lesbians, gay men, and the politics of identity.* Routledge, New York.

Whitman FL, Diamond M, Martin J. 1993. Homosexual orientation in twins: A report on 61 pairs and three triplet sets. *Arch Sex Behav* **22:** 187–206.

Wilson EO. 1975. *Sociobiology: The new synthesis.* Harvard University Press, Cambridge, MA.

Wright J, ed. 2009. Homosexuality in the U.S., 1998–2000. In *The New York Times Almanac*, pp. 1–314. Penguin, New York.

Yoder J. 2011. The intelligent homosexual's guide to natural selection and evolution, with a key to many complicating factors. *Scientific American*, blogs.scientific american.com/guest-blog/2011/06/21.

Sexual Conflict between Parents: Offspring Desertion and Asymmetrical Parental Care

Tamás Székely

Department of Biology & Biochemistry, University of Bath, Bath BA2 7AY, United Kingdom

Correspondence: T.Szekely@bath.ac.uk

Parental care is an immensely variable social behavior, and sexual conflict offers a powerful paradigm to understand this diversity. Conflict over care (usually considered as a type of postzygotic sexual conflict) is common, because the evolutionary interests of male and female parents are rarely identical. I investigate how sexual conflict over care may facilitate the emergence and maintenance of diverse parenting strategies and argue that researchers should combine two fundamental concepts in social behavior to understand care patterns: cooperation and conflict. Behavioral evidence of conflict over care is well established, studies have estimated specific fitness implications of conflict for males or females, and experiments have investigated specific components of conflict. However, studies are long overdue to reveal the full implications of conflict for both males and females. Manipulating (or harming) the opposite sex seems less common in postzygotic conflicts than in prezygotic conflicts because by manipulating, coercing, or harming the opposite sex, the reproductive interest of the actor is also reduced. Parental care is a complex trait, although few studies have yet considered the implications of multidimensionality for parental conflict. Future research in parental conflict will benefit from understanding the behavioral interactions between male and female parents (e.g., negotiation, learning, and coercion), the genetic and neurogenomic bases of parental behavior, and the influence of social environment on parental strategies. Empirical studies are needed to put sexual conflict in a population context and reveal feedback between mate choice, pair bonds and parenting strategies, and their demographic consequences for the population such as mortalities and sex ratios. Taken together, sexual conflict offers a fascinating avenue for understanding the causes and consequences of parenting behavior, sex roles, and breeding system evolution.

Sexual conflict over care is a type of evolutionary conflict that emerges from the different interests of males and females in regard to parental care (Trivers 1972; Clutton-Brock 1991; Chapman et al. 2003; Arnqvist and Rowe 2005). The conflict arises when the young benefit from the effort of either parent, but each parent pays only the cost of its own effort, so that each parent would have higher fitness if the other parent provides more care (Houston et al. 2005; Lessells 2006; Klug et al. 2012). Conflict refers to the way selection acts on the two sexes that have different optimum values in parental provisioning; between the two optima, sexually antagonistic selection operates (Lessells 2012). Sexual conflict over care can be seen as tug-of-war, be-

Cite this article as *Cold Spring Harb Perspect Biol* doi: 101101/cshperspect.a017665

cause each parent is tempted to pull out of care leaving the other parent to provide more care for the young (Székely et al. 1996; Arnqvist and Rowe 2005; Lessells 2012).

Sexual conflict over care seems to be the rule rather than the exception. The conflict may be resolved by one or both parents failing to adopt the optimal parenting for their mate and nonetheless remaining in conflict, or by both parents adopting the optima that suit their mate (i.e., exhibit the maximum provisioning possible). Examples of the latter conflict resolution (whereby the conflict is completely wiped out) are exceedingly rare and seem to be limited to three scenarios. First, conflict over care is not expected in obligate monogamy by both males and females so that the lifetime reproductive successes of both parents are identical. This may occur in semelparous organisms (i.e., both the male and the female put their resources into a single breeding event) or in iteroparous organisms with lifelong exclusive monogamy. Second, males and females might be genetically identical, so even though one or both sexes are polygamous, polygamy would benefit the same genome whether it is in the male or the female phenotype. Third, parental care is cost-free and thus parents provide maximum level of care (P Smiseth, pers. comm.). However, few, if any, organisms fit these restrictive assumptions, and thus conflict-free parenting seems exceedingly rare in nature: (1) some level of polygamy (by males, females, or both sexes) appears to be widespread; (2) the reproduction by genetically identical individuals (clones) as separate sexes (males and females) seems unlikely although not impossible if sex is determined environmentally; and (3) care provisioning, as far as we are aware, does have costs that discourage parents from providing their absolute maxima for a given batch of offspring.

Parents may have conflicting interest over caring or deserting the young, the amount of care provided for each young, the number of simultaneous mates, the size and sex ratio of their brood, and the synchronization of birth for a clutch or litter of young (Westneat and Sargent 1996; Houston et al. 2005; Klug et al. 2012; Lessells 2012). Conflict between parents over care is usually labeled as a postzygotic conflict although resources had been already allocated into the gametes before fertilization as part of parental provisioning (Clutton-Brock 1991); other examples of postzygotic conflicts include infanticide and genomic imprinting (Chapman et al. 2003; Tregenza et al. 2006; Lessells 2012; see Palombit 2014).

Studies of conflict over care are fascinating for at least four major reasons. First, parental care is diverse. There is great variation both between and within species in the types of care provided, duration of care, and the sex of the care-providing parent (Wilson 1975; Clutton-Brock 1991; McGraw et al. 2010; Royle et al. 2012), and sexual conflict is thought to be one of the main drivers of this diversity. Second, parental care is one of the core themes in breeding systems and sex role evolution, and it is increasingly evident that parental care can only be understood by dissecting the entangled relationships between ecological and life-history settings, and the variety of mating and parenting behavior (Székely et al. 2000; Webb et al. 2002; Wedell et al. 2006; Jennions and Kokko 2010; Klug et al. 2012). Third, parental care was (and is) one of the test beds of evolutionary game theory. Numerous models have been developed to understand how parents interact with each other and with their offspring (Trivers 1972; Maynard Smith 1977; Houston and Davies 1985; Balshine-Earn and Earn 1998; McNamara et al. 1999, 2000; Webb et al. 1999; Johnstone and Hinde 2006; Johnstone et al. 2014). Parental care research is one field in which empiricists are extensively testing the predictions of evolutionary game theoretic models in both the laboratory and wild populations (Székely et al. 1996; Balshine-Earn and Earn 1998; Harrison et al. 2009; Klug et al. 2012; Lessells 2012; van Dijk et al. 2012), although the congruence between theoretical and empirical work is not as tight as often assumed (Houston et al. 2013). Finally, parental care—wherever it occurs—is often a major component of fitness, because whether the offspring are cared for or abandoned has a large impact on their survival, maturation, and reproduction (Smiseth et al. 2012). Therefore, parental care (or the lack of it) may have an impact on population productivity and population growth and influences the resilience of populations to various threats (Bessa-Gomes et al. 2004; Veran and Beissinger 2009; Blumstein 2010). Thus, understanding the behavioral interactions between parents and the fitness implications of these interactions is highly relevant for population dynamics and biodiversity conservation (Alonzo and Sheldon 2010; Blumstein 2010).

Sexual conflict over care has been reviewed recently (van Dijk and Székely 2008; Lessells 2012; Houston et al. 2013). Here, I focus on three issues that have not been extensively covered by previous reviews: (1) why sexual conflict over care occurs in some environments, whereas in others parental cooperation appears to dominate; (2) how can one detect sexual conflict over care; and (3) what are the implications of sexual conflict over care for macroevolution. I view causes and implications of parental care primarily from empirical perspectives; there are excellent reviews on the rich theoretical literature (Lessells 2006, 2012; Klug et al. 2012; Houston et al. 2013). My intention is not to be comprehensive; instead, I use selected examples to illustrate salient features of conflict over care. I focus on ecological and evolutionary aspects; for a discussion of the genetic and neuroendocrine bases of parental care, see Adkins-Regan (2005), McGraw et al. (2010), and Champagne and Curley (2012). I prefer to use the term "parental care" instead of "parental investment," because the latter, as admitted by Trivers (1985), is extremely difficult to estimate empirically and thus may have a limited use in empirical studies (Mock and Parker 1997; McGraw et al. 2010). The term "parental investment" can be deceptive, if used without directly demonstrating the full costs of care. The term "parental care" is less restrictive, because it refers to any form of parental behavior that appears to increase the fitness of an offspring and is likely to have evolved for this function (Clutton-Brock 1991; Smiseth et al. 2012). In this review, I focus on families in the narrow sense (i.e., two parents and their offspring), although in numerous organisms the families are more extensive and may include several generations of offspring living together and/or unrelated individuals that assist the parents rearing the young.

SEXUAL CONFLICT AND PARENTAL BEHAVIOR

Conflict between parents may occur in species with identical sex roles or with different sex roles driven by the different cost and benefits of care for males and females arising from the sex differences in physiology, ecology, and life history (Fairbairn et al. 2007; King et al. 2013). Conflict may occur in organisms that have no parental care, assuming that at least some parental care (by the male, the female, or both parents) would improve offspring survival and thus parental fitness. Conflict may also occur in organisms in which only the male, the female, or both parents provide care. I start this section by emphasizing the diversity parental care strategies and then explore how conflict over care could influence the emergence and maintenance of this diversity.

Diversity of Care Strategies

Parental care is one of the most diverse social behaviors (Wilson 1975; Reynolds et al. 2002; McGraw et al. 2010; Smiseth et al. 2012). The type of care, the duration of care, and the involvement of one or both parents in various care activities are all highly variable both within and between animal taxa. Conflicts between parents and the resolution of these conflicts offer powerful approaches to understand this diversity (Trivers 1972; Maynard Smith 1977; Arnqvist and Rowe 2005; Lessells 2012). Whereas the majority of invertebrates and many vertebrates do not provide any care for the fertilized embryo beyond supplying the eggs with nutrients, the species that do exhibit care have amazing adaptations. There are excellent recent reviews on parental care in both vertebrates and invertebrates (Balshine 2012; Trumbo 2012), and, thus, here I focus on selected examples that illustrate some of this variation and comment on their relevance to the study of sexual conflict.

Studies have discovered an immense variety of care strategies, and discoveries of novel forms and types of parental care are still being made (Fig. 1). For instance, tropical frogs have some of the most diverse reproductive and parental care systems among animals (Wells 2007), and parental care is thought to have evolved independently at least 41 times (Balshine 2012). Some species of frogs prepare a foam nest for their eggs, whereas others attend the eggs laid on leaves that overhang streams or are viviparous and give birth to small froglets. Parents in other frog species brood the eggs on their back, in their vocal sac, or in their stomach, transport the tadpoles and froglets, or urinate on the eggs to prevent them from desiccating (Summers et al. 2006). Parental care can be a major occupation for male and female frogs for substantial periods of time when they regularly check the development of tadpoles, and the female may lay trophic eggs to nourish the tadpoles (Brown et al. 2010). Parents may also seek out pools that are free from competitor larvae and cannibalistic tadpoles and move their own tadpoles to predator-free pools if necessary (Summers et al. 2006; Brown et al. 2010).

Because the reproduction of a vast number of species has not been studied in detail, especially those that live in difficult-to-access habitats such as tropical forest canopies, the deep sea, or caves or soil, new forms of parental care are waiting to be discovered. For example, in caecilians, a little-studied subterrestrial amphibian group, it was not known until recently that mothers incubate their eggs in underground burrows and that the altricial hatchlings feed for an extended period of time on the modified and lipid-rich outer layer of the skin of their mother using specialized dentition (Kupfer et al. 2006; Wilkinson et al. 2013).

Figure 1. Diverse forms of parental care. (*A*) Begging behavior in *Nicrophorus vespillioides* burying beetle. (*B*) Midas cichlid (*Amphilophus citrinellus*) couple looking after the young. (*C*) Male Peruvian mimic frog (*Ranitomeya imitator*) carrying a tadpole to a pool on his back. (*D*) Female rock sparrow (*Petronia petronia*) feeding her nestlings. (*E*) Saddle-backed tamarin (*Saguinus fuscicollis*) carrying twins. (From Per Smiseth, *A*; Ad Konings, *B*; Rainer Schulte, *C*; Adriano De Faveri, *D*; and Mojca Sojan-Dolar, *E*.)

Care can be provided for a long time not only in whales, elephants, and primates, including humans, in which it may last up to several decades (de Waal and Tyack 2003; Mitani et al. 2012), but also in invertebrates. Mother whip spiders *Phyrnus marginemaculatus* protect their young for at least 11 months (Trumbo 2012). During such an extended period, the parent–infant relationship that is initially driven by offspring demand and the parents' readiness to provide care may shift toward an alliance between the parent(s) and the siblings. For instance, in scorpions, groups of young can help the mother to subdue large prey (Trumbo 2012).

It is usually assumed that evolutionary transitions in duration of care or type of care are driven by sex-specific costs and benefits, and, thus, the transitions are the result of changes in ecology and life history of males, females, or both sexes. However, it is plausible that some of these transitions are largely (or entirely) the result of changes in male and/or female behavior as they are trying to resolve the conflict. First, for a given set of costs and benefits, multiple patterns of care may occur in a population; this may be the result of mixed evolutionarily stable strategies (ESSs) (Webb et al. 1999), to different behavioral interactions between parents (McNamara et al. 1999; Johnstone et al. 2014), or to social interactions between parents and other members of a population that can stabilize different ESSs in a coevolutionary process that involves mate choice, mating behavior, and parental care (McNamara et al. 2000; van Dijk et al. 2012). Second, an important insight from evolutionary game theory is that as males (or females) attempt to attain their respective fitness optima, they may change the cost and benefit functions for their mates, and thus influence the fitness landscape for the opposite sex (McNamara and Weissing 2010). Although it is not known whether the first and/or the second processes are more likely generating multiple care patterns in wild populations, it does seem that conflict resolution at both ecological and evolutionary timescales are influencing the transitions between different parenting behaviors.

Why (Not) Care?

Sexually antagonistic selection is one of the theories that behavioral ecologists use to understand the emergence and maintenance of diverse care patterns (Trivers 1972; Maynard Smith 1977; Houston et al. 2005; Lessells 2012). As each sex is moving toward its parental care optima given the behavior of its mate and the rest of population, it elicits a response from the other parent and vice versa. Selection may operate on these changes. Actions and responses that lead to higher reproductive success are likely retained in the population. However, similarly to other types of sexual conflict, it is difficult to infer conflict from behavioral patterns alone, because the parental behavior exhibited by the sexes does not tell anything about the difference between the optima of males and females (Lessells 2012). Therefore, observations that one parent deserts and all care is provided by the other parent or that both parents share care equally do not tell too much about the extent of conflict. Furthermore, because sexual conflict refers to selection process, conflict may not have behavioral signs; for instance, aggression between parents may (or may not) indicate the intensity of conflict over care.

The direct evidence for sexual conflict over care is scarce, because few studies have estimated the fitness outcomes of parenting behavior from the perspectives of both the male and the female (see the section Detecting Conflict over Care). The hedonistic breeding systems of Eurasian penduline tits (*Remiz pendulinus*) may illustrate fitness consequences of various parenting options (Fig. 2). In this small passerine bird (body mass of ~9 g), either the male or the female abandons the clutch and renests with a new mate shortly after desertion. Remating is common; both males and females may have up to five different mates in a single breeding season (Persson and Öhrström 1989). A puzzling aspect of penduline tit breeding biology is the large number of deserted clutches. About 30%–40% of clutches are abandoned by both parents and these clutches produce no young. High frequencies of biparental desertion have been observed in all studied populations to date, so most

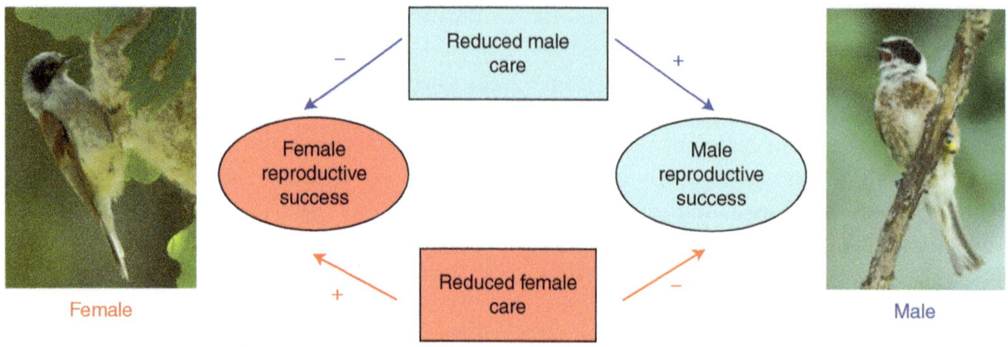

Figure 2. Fitness consequences of various parenting options. The impact of reduced care (i.e., offspring desertion) on reproductive success of Eurasian penduline tits (based on Szentirmai et al. 2007). Positive and negative signs indicate improvement and reduction, respectively, on reproductive success of males (blue) and females (pink). (From René van Dijk.)

likely it is part of natural breeding behavior (van Dijk et al. 2010a). Whereas in many animal populations predation of eggs or young is the major sources of breeding failure, in penduline tits, biparental desertion is a substantially more common cause of breeding failure than predation of eggs or young, or any other cause of nest failure.

Szentirmai et al. (2007) estimated the reproductive success separately for caring and deserting penduline tits using data from an intensely studied population in Hungary. Deserting the clutch increased the reproductive success of the deserting males, because many of these males found a new mate and renested. Desertion, however, was costly to deserted females, because they either deserted the clutch themselves and thus doomed the offspring to death, or stayed with the offspring for about one month and cared for them until they became independent (Szentirmai et al. 2007; van Dijk et al. 2012). The fitness consequences of desertion and caring in males are mirrored by fitness consequences in females. Deserting the clutch increases a female's own reproductive success but reduces her mate's reproductive success. Therefore, there is a temptation for both male and female penduline tits to abandon the mate and seek a new partner, although en route to increase their own reproductive success by deserting and subsequently remating, they risk the survival of their existing brood. The high frequency of biparentally deserted clutches suggests that the parents' gamble often did not work out. Extra-pair paternity does occur in penduline tits (van Dijk et al. 2010b), although the frequency of extra-pair young (EPY) is comparable between male-cared and male-deserted young, suggesting that EPY does not bias the estimated reproductive success of deserted versus cared nests.

Species with variable care patterns, like penduline tits, offer great opportunities to quantify fitness implications of care and desertion and to assess the extent of sexual conflict. Similarly, the highly variable care patterns (both within and between species) in assassin bugs, cichlid fishes, poison dart frogs, and shorebirds may emerge via conflict over care whereby a shift in costs and benefits of care for one sex (or for both sexes) can flip from one pattern of care to another. Although different costs and benefits of care for male and female, and thus difference in sex roles, are not essential for sexual conflict over care, these taxa offer biological systems in which the fitness implications can be evaluated. Because selection is expected to produce male behavior that is the best response to female parental behavior, the changes in cost and/or benefits of care in one sex are likely to induce change in parental behavior of the other sex, somehow analogous to the sexually antagonistic prezygotic selection (see Gavrilets 2014; Sirot et al. 2014).

Theoretical models suggest that social behavior itself can generate shifts. There are situations in which both uniparental care and biparental care are ESSs, and they can coexist in a population (McNamara et al. 2000; Kokko and Jennions 2008; Klug et al. 2012; van Dijk et al. 2012). The presence of several care patterns in a single population is consistent with theoretical results (see the section Diversity of Care Strategies), although alternative explanations of coexisting caring strategies are also possible—for instance, age-dependent care strategies and/or temporal or spatial variation in costs and benefits of care for different members of the population.

Parental care, however, is a complex trait even though theoretical models and comparative studies often reduce care to a single (or a few) variable(s). Representing care as a single trait is problematic, because parents may provide different types of care and these different components may evolve independently from each other (Smiseth et al. 2012; Székely et al. 2013). Parents may also divide the tasks so that each parent specializes on particular tasks; male dung beetles, for instance, excavate the ground under the dung ball, whereas the female covers the ball with soil (Trumbo 2012). Such specialization can reduce conflict between males and females and maintain biparental care (Lessells 2012; Barta et al. 2014).

Manipulation and Parental Tactics

Males and females may use a variety of tactics to entice (or coerce) their mate to increase their care. In biparental species, a female may attempt to monopolize the parental care of its mate (Chapman et al. 2003; Arnqvist and Rowe 2005). Females may solicit superfluous copulations from their mates (Eens and Pinxten 1996) or interfere with their mate to prevent them from attracting new females (Slagsvold and Lifjeld 1994). To impose monogamy on her mate, female burying beetles *Nicrophorus defodiens* bite and attempt to push the male off his signaling perch and interfere with his attempt to release pheromones attracting additional mates (Arnqvist and Rowe 2005). Similarly, females may be hostile toward other females so as to keep their mate's care provisioning for their own offspring (Sandell and Smith 1996; Liker and Székely 1997). Males, however, can counteract female strategies by intervening directly and keeping peace between females (Walter and Trillmich 1994), or by attracting a new female away from their current family so that their existing female cannot interfere with the new female.

Parents may manipulate their mates' behavior to extract more care using two strategies. First, paternally imprinted genes in placental mammals may facilitate embryonic growth so that the developing embryos extract more resources from the mother than would be optimal for her (genomic imprinting; see Haig et al. 2014). An analogous manipulation has been proposed for birds. Females are hypothesized to deposit elevated levels of androgens in the eggs to increase chick begging behavior, so that the chicks extract more care from the male (Schwabl 1996; Groothuis et al. 2005). The latter hypothesis has been tested by several studies, and, currently, little evidence supports it (Lessells 2006; Laaksonen et al. 2011). Instead of improving offspring viability, elevated androgen level appears to reduce offspring viability in the long term (Ruuskanen et al. 2012).

Second, parents may strategically handicap themselves to extract more care from their mate (Barta et al. 2002). By reducing their own body condition, females can put their mate in a difficult situation. If the male abandons, then the female alone cannot rear the young so the brood would die ("credible threat"; Barta et al. 2002). Although body condition has been shown to relate to parental care (e.g., males in low body condition reduce their care [Steinegger and Taborsky 2007]), the existence of strategic handicapping by lowering body condition has not been shown.

Although the larger (or physically stronger) individuals in a pair can "force" the smaller parent to care, reports of physical coercion are rare (Awata and Kohda 2004). There may be three reasons for this. First, harmful behavior, coercion, and manipulation are expected to be weaker in conflict over care than in conflict over mating (Lessells 2006), because, in the former, a harmful behavior to a mate would reduce not only the mate's reproductive success, but also the actor's reproductive success. Second, the manipulated parent could retaliate and harm (or eat) the offspring and thus defeat the objective of the manipulation in the first place. Third, enforcing a complex behavior such as care that may be tuned to a specific offspring age and demand seems exceedingly difficult. In contrast to other forms of coercion that seem straightforward (e.g., keeping another animal away from a resource [e.g., food or water] or forcing another individual to copulate) and specific behaviors that appear to have evolved to achieve these objectives (e.g., aggression), forcing an animal to carry out a complex and fine-tuned behavior such as parental care seems less likely.

DETECTING SEXUAL CONFLICT OVER CARE

Because sexual conflict may involve adaptation and counteradaptation, it is thought that these processes and their results will be difficult to observe (Chapman et al. 2003; Arnqvist and Rowe 2005). Theoretically, the extent of conflict can be estimated in two ways: (1) by quantifying the parental optima for males and females and then estimating the difference between the two optima (the conflict "battleground"; Godfray 1995); or (2) by estimating the fitness reduction in males, females, or both sexes caused by conflict ("conflict load"; Lessells 2006). Ideally, both battleground and conflict load should be estimated simultaneously to reveal both the behavioral differences caused by conflict and their fitness implications; however, no study appears to have performed both. Much of our current knowledge is based on either of these estimates or on indirect inferences of the conflict.

Observations

Fitness implications of different duration, type, or intensity of care can be established by studying wild or laboratory populations. Studies have compared the reproductive success of different care patterns (e.g., male-cared vs. female-cared, uniparental vs. biparental families, no care vs. care; Clutton-Brock 1991; Eldegard and Sonerud 2009; Pogány et al. 2012), assuming that a difference between the two estimates indicates the lost reproductive success because of the unwillingness of one or both parents to provide care.

Offspring desertion by the male, female, or both parents is a common behavioral strategy that occurs in wide range of taxa (insects, fish, frogs, birds, and mammals) (Clutton-Brock 1991; Székely et al. 1996; Korpimäki et al. 2011), and studies suggest that conflict over care is involved (Houston et al. 2005; Griggio et al. 2008; King et al. 2013). The social environment may modulate the benefit of desertion. High density of potential mates is expected to favor desertion, whereas low density may temper desertion (Owens 2002). Social environment, however, may offer biased mating opportunities given that adult sex ratio (ASR) is skewed in numerous organisms (Donald 2007; Székely et al. 2014), and the biased ASR favors desertion by one sex over the other. For instance, male-biased ASR was thought to explain female-biased desertion in birds (Box 1) (Pilastro et al. 2001; Kosztolányi et al. 2011; Liker et al. 2013; Parra et al. 2014). Furthermore, the benefit of desertion may differ between the sexes, if one sex needs more time to recover from breeding than the other (Gubernick et al. 1993; Balshine-Earn and Earn 1998).

In principle, comparing the two strands of benefits (care vs. desertion) that lie at the core of parental care evolution (Clutton-Brock 1991; Székely et al. 1996) should indicate the fitness consequences for males and females, and thus tell to what extent these fitness peaks differ between males and females (the "battleground"). The timing of desertion and the sex of the deserting parent are highly variable between species and populations (see the section Diversity of Care Strategies) (McGraw et al. 2010; Balshine 2012; Trumbo 2012), and these variations provide excellent opportunities to assess the benefits of care versus desertion. However, there are caveats. Although the examples in the next paragraph are framed in the context of discrete parental decisions (i.e., care vs. desertion; Székely et al. 1996), similar arguments can be made for other aspects of care such as duration of care and type of care.

BOX 1. CONFLICT AND COOPERATION BETWEEN KENTISH PLOVER PARENTS

The Kentish plover, *Charadrius alexandrinus*, is a small shorebird (body mass of 40–44 g) that exhibits a variable mating system and parental care (Lessells 1984; Székely and Lessells 1993; Amat et al. 1999). Two to three eggs are laid in a small scrape on the ground. Both parents incubate the eggs, although, after hatching, one of the parents usually abandons the brood and renests with a new mate. Thus, in a single population, three types of brood care (biparental, male-only, and female-only) and three types of mating system (social monogamy, sequential polygamy by the male, and sequential polygamy by the female) are observed. DNA fingerprinting shows that extra-pair paternity is rare (<4% of chicks) (Küpper et al. 2004); and, thus, the social mating system reflects genetic mating system. A series of experiments and observational studies have been performed in the natural habitat to clarify the roles of environmental and life-history variables that may influence the extent of conflict and parental cooperation (Table 1).

Kentish plovers live in temperate and subtropical habitats, and their breeding habitat includes salt flats around Mediterranean lakes and deserts in which ground surface temperature may exceed 50°C (Amat and Masero 2004). Preventing the eggs from overheating is a major challenge, and Al-Rashidi et al. (2011) hypothesized that exposure to heat may select for biparental incubation. To test this hypothesis, they experimentally manipulated the exposure of eggs to sun by shading the nest with a small bush (Fig. 3). The results were consistent with the hypothesis, because the extent of biparental incubation was reduced in the shaded nests.

To compare the benefits of deserting for male and female plovers, unmated plovers were experimentally created, and the time they took to find a new mate was observed (Székely et al. 1999). Unmated females remated after a substantially shorter time than unmated males, suggesting that mating opportunities are significantly better for female than male plovers. This experimental result is consistent with a demographic study that estimated a heavily male-biased adult sex ratio (Kosztolányi et al. 2011).

A recent experimental study in closely related plover species found different mating opportunities. In white fronted plover *Charadrius marginatus* males and females mating opportunities were not different, whereas in Kittlitz's plover *Charadrius pecuarius*, the remating time was longer in females than in males (Parra et al. 2014). The latter results suggest that mating opportunities—and thus the benefits from deserting—can be substantially different in closely related species.

Table 1. Different components of parental care in the Kentish plover

	Uniparental care	Biparental care	Type of evidence	Reference
Life history				
Family size	Small	Large	Experiment	Székely and Cuthill 2000
Timing of breeding	Early	Late	Observation	Székely and Cuthill 2000
Ambient environment				
Environmental harshness	Mild	Harsh	Experiment	Al-Rashidi et al. 2011
Food abundance	Poor	Rich	Observation	Kosztolányi et al. 2006
Predation rate	Low	High	Observation	Amat et al. 1999
Social environment				
Adult sex ratio	Strongly biased	Less biased	Observation	Kosztolányi et al. 2011

Both parents cooperate in incubating the eggs, although after hatching, most broods are cared for by a single parent (either the male or the female). Desertion is beneficial only for the deserting parent and costly to its mate.

Figure 3. Female Kentish plover incubating the eggs (*left*) and experimental manipulation of nest cover (*right*). (*Left*, from Hugo Amador; *right*, from Al-Rashidi et al. 2011; reprinted, with permission, from Elsevier © 2011.)

First, comparing the fitness consequences of caring and deserting for a selected group of animals may not represent the population as a whole. Thus, the best fathers may decide to care, whereas the most attractive fathers may decide to desert and find another mate. Similarly, a single parent may be able to provision the young on a territory with abundant food, whereas both parents may be needed to feed the young on a poor territory (Eldegard and Sonerud 2009). Second, the benefit for a given parent, let us say the male, from deserting depends on his mate's response. Will she continue rearing the offspring or desert herself? Therefore, estimating the fitness consequences of caring and deserting should be performed at various response levels of the other parent. This is rarely feasible, because wild populations rarely exhibit all behavioral strategies. Third, the benefits of caring and deserting may manifest over a long time period, whereas studies usually estimate short-term fitness consequences (van Dijk et al. 2012). One may need to investigate several generations to reveal the full scale of costs and benefits. This can be challenging especially in long-lived animals or in polygamous species, in which the number of mates may proliferate into an extensive network of breeders for which reproductive success estimates are required.

How males and females play out these conflicts is rarely studied in detail. Unlike divorce in humans that can be an extended and convoluted process, desertion in nonhuman animals can be rapid (van Dijk et al. 2012). Studies are needed to work out on a behavioral scale how parents interact, whether they may escalate or converge in response to each other's behavior (Johnstone et al. 2014).

Experiments

To overcome the limitations of observational studies, two kinds of manipulations were used to perturb parental behavior and seek the consequences of perturbation on parental behavior and reproductive success. First, experimenters manipulated the benefits of matings—for instance, making males (or females) more attractive to the opposite sex (Smith 1995; Griggio et al. 2010). For example, by setting up an additional nest box close to a pair of common starlings (*Sturnus vulgaris*), male starlings reduced their involvement in care and sang to attract a new mate (Smith 1995).

Second, researchers manipulated parental attendance (e.g., by removing or handicapping one parent) to investigate the consequent changes in partner's behavior and fitness (Harrison et al. 2009). By experimentally removing one parent, Royle et al. (2002) created uniparental and biparental broods in zebra finches *Taeniopygia guttata*. They showed that male chicks reared by a single parent were more attractive as adults to females than males reared by two parents, suggesting that the conflict between male and female parents results in lower-quality offspring.

Males (or females) were handicapped (or removed) in various biparental organisms (insects: Rauter and Moore 2004; Smiseth et al. 2005; Suzuki and Nagano 2009; fish: Mrowka 1982; Itzkowitz et al. 2001; birds: Sanz et al. 2000; Harrison et al. 2009; mammals: Wynne-Edwards and Lisk 1989; Gubernick and Teferi 2000). These experiments produced two main insights. First, although there is large variation between species in response to manipulation of parents, mates of handicapped parents tend to compensate; however, the compensation is usually not complete (Harrison et al. 2009; but see Mrowka 1982). This is consistent with theoretical arguments; partial compensation is necessary to maintain biparental care (Houston and Davies 1985; McNamara et al. 2002; Lessells 2012). Second, although parental care tends to be asymmetric in that females usually take a larger share than males (Queller 1997), across the species compensatory responses of males and females are not different (Harrison et al. 2009). This is in contrast with three species of *Nicrophorus* beetles, in which the males but not females compensated for the lost care of their mate (Lessells 2012). Presumably, in the latter species, the females are already working close to their maximum capacity while they are still attended by their mate, and once their mate is removed they cannot improve their workload (Lessells 2012).

In burying beetles, male desertion may be actually beneficial for the female, because staying males eat some of the carcass that would be available for the larvae (Boncoraglio and Kilner 2012). Therefore, females may have coevolved to anticipate desertion by their partner so that they now benefit from the male's absence. However, in wild populations in which males and females roam freely, the male presence at the carcass may be beneficial by helping to protect both the carcass and the developing larvae from intruding males that may kill the offspring (Trumbo 2007).

Although experimental studies have intrinsic advantages over observational studies, these experimental studies also suffer from drawbacks. First, manipulations of a focal sex (let us say, males) should be designed to reveal the fitness implications at various levels of response (here, by the female). Existing experiments, however, usually manipulate one sex and estimate fitness implications at the self-selected level care of its mate. Because parents cannot be forced to care, manipulating systematically the care of both parents simultaneously seems to be exceedingly difficult. Second, complex mating patterns (such as the one exhibited by Eurasian penduline tits) exacerbate this difficulty, because full exploration of parental behavior and their fitness implications would mean experimental manipulation not only in one family, but in all subsequent families of the focal individuals. Therefore, the best empirical system for experimental evaluation of battleground and conflict load would be short-lived, predominantly monogamous animals that only breed a few times throughout their life.

Comparative Analyses

Phylogenetic comparative analyses have been used to detect the tug-of-war between males and females (Arnqvist and Rowe 2005). In shorebirds, which exhibit unusually diverse parental care strategies ranging from male-only care to biparental and female-only care, the duration of male care is traded off against the duration of female care (Reynolds and Székely 1997), so that evolutionary decreases in male care are associated with increases in female care and vice versa. Although the duration of care is not necessarily a good indicator of parental effort, this pattern is consistent with theoretical predictions that each parent should reduce its involvement in care, whereas its mate is selected to compensate the lost care. An implication of the phylogenetic trade-off between male and female care is that males and females do adjust their care to the care of their mate, consistent with experimental manipulations in contemporary populations (Harrison et al. 2009; Kosztolányi et al. 2009; Trumbo 2012).

Whether the tug-of-war over care occurs may depend on phylogenetic plasticity. Webb et al. (2010) showed that care strategies are more variable in species with short rather than long offspring development. Furthermore, taxa with high variation in care duration exhibit variable care patterns, because male involvement in care is associated with an extensive period of parental care (Webb et al. 2010).

Artificial selection and experimental evolution are powerful approaches to investigate both the causes and the implications of prezygotic sexual conflict (Holland and Rice 1999; Chapman et al. 2003; Arnqvist and Rowe 2005; Rowe and Day 2005), although there might be confounding effects because of different sources of conflict (e.g., interlocus or intralocus conflict). Despite their widespread use in studies of prezygotic conflict, they have not been used to investigate conflict over care. There may be two reasons for this gap. First, fast-breeding laboratory species often do not exhibit care (e.g., many flies, beetles, and nematode worms), or if they do have some care (e.g., mice *Mus musculus*), there is little flexibility in males' and females' involvement that would capture the parental care variation seen in numerous populations in nature. Second, although care-related behaviors have been artificially selected in poultry (e.g., Champagne and Curley 2012), such work usually targets only females, so it is not straightforward to extrapolate from these single-sex selection experiments to fitness implications for both sexes.

THE IMPLICATIONS OF SEXUAL CONFLICT OVER CARE FOR MACROEVOLUTION

Sexual conflict over care, similarly to other types of sexual conflict, is expected to facilitate diversification, speciation, and rapid evolution (see Gavrilets 2014). Although a specific theory has not been developed to investigate macroevolutionary consequences of conflict over care, the sexually antagonistic selection—the adaptations and counteradaptations to increase male or female fitness—is thought to influence care strategies, somehow analogous to prezygotic sexual conflict (Chapman et al. 2003). For instance, the diverse reproductive modes in frogs and fish are related to parental care (Goodwin et al. 1998; Gomez-Mestre et al. 2012), and conflict between parents is a candidate driver of this diversity (see the section Diversity of Care Strategies). However, a caveat is that it is extremely difficult to nail down whether speciation (or diversification) is largely (or exclusively) a result of conflict over care, because phylogenetic studies are correlational and they do not identify causation. The most convincing approach would seem to be experimental evolution, although as I noted above, this approach has not been used in studies of conflict over care.

Conversely, offspring desertion and uniparental care may reduce productivity of the population (Holman and Kokko 2013). As an extreme example, the 30%–40% clutch loss in Eurasian penduline tits because of conflict (Szentirmai et al. 2007) is likely to impact on population productivity and increase the risk of population decline and/or extinction. Furthermore, reduced productivity because of the reluctance of males to provide care appears to be costly across wide range of bird species, because female-only care leads to a 20% reduction of productivity compared with other types of care (Sibly et al. 2012).

ECOLOGY OF CONFLICT AND COOPERATION

Although theory tells us that family life is rife with conflicts, the intricate pair-bonding ceremonies and the tender (often affectionate) relationships between males and females observed in some insects, fish, birds, and mammals leave no doubt that males and females coordinate their behavior; they do cooperate not only to copulate and make temporary liaisons but also to rear their young for extended time periods and to prepare them for the adult life. Studies of cooperation have a venerable history (Hamilton 1964; Wilson 1975; West et al. 2007; Székely et al. 2010), and one may view parental behavior as a balance between factors that facilitate conflict and those that facilitate cooperation (see below). The various costs and benefits of care can be translated into that of the former or the latter. There are excellent reviews on the costs and benefits of care (Clutton-Brock 1991; Wedell et al. 2006; Reichard and Boesch 2007; Royle et al. 2012); thus, here I focus on one issue: how ambient environment and social environment may interactively shape cooperation and conflict between parents.

On the one hand, parents tend to cooperate in rearing the young in a harsh physical environment (i.e., extreme cold or heat, very humid or dry) (Wilson 1975; Clutton-Brock 1991). In such environments, the fitness consequences of not cooperating fully are severe, so there will be little conflict. However, convincing evidence for this hypothesis is scarce, because manipulating the physical environment can be challenging. One of the few experimental studies manipulated the exposure of parents and their eggs in a desert environment by shading some nests or removing the shade from other nests (Al-Rashidi et al. 2011). The results were consistent with the harsh environment hypothesis. At the experimentally exposed nests, parental cooperation increased, whereas at shaded nests, parental cooperation was reduced. The need of the young may also influence parental cooperation. When substantial parental resources are needed (e.g., food or protection), biparental care tends to evolve (Thomas et al. 2005; Brown et al. 2010).

Different skills of the parents may also facilitate parental cooperation (parental role specialization; Lessells 2012; Trumbo 2012). If one parent specializes in providing one type of care (e.g., feeding the young), whereas the other parent specializes in a different task (e.g., defending the offspring), this stabilizes biparental care (Barta et al. 2014). Finally, long lifespan and durable pair bonds correlate with biparental care of the young (Ens et al. 1996; Port and Cant 2013), suggesting that in long-lived animals retaining the mate for future breeding is beneficial.

On the other hand, high mating opportunities and promiscuity tend to destabilize parental cooperation and reduce the incentive to provide care in one or both parents. Adult sex ratios are often unbalanced, and in female-biased populations like in many mammals, males are at an advantage in finding a new mate (Donald 2007; Kokko and Jennions 2008; Székely et al. 2014). When one sex has higher mating opportunity than the other, the favored sex tends to capitalize from the opportunity to leave the other parent to look after the young (Balshine-Earn and Earn 1998; Pilastro et al. 2001; Kosztolányi et al. 2011; Liker et al. 2013). In addition, promiscuity tends to temper the males' willingness to provide care (Queller 1997; Kokko and Jennions 2008), although caring for somebody else's offspring may still be the best option if finding a new mate is difficult or if the male can somehow selectively assist his own offspring (Møller and Cuervo 2000; Kamel and Grosberg 2012).

The breeding system of a ground-nesting small shorebird, the Kentish plover, illustrates how conflicts and cooperation may interactively influence parental care (Box 1). In the Kentish plover (as in most shorebirds), both parents incubate the eggs, and biparental care of the eggs for a substantial part of incubation seems essential for any young to hatch. Because hatchlings are precocial and capable of feeding themselves, the demand for care is reduced so that one parent (the male or the female) may desert the family after hatching. Brood desertion benefits only the deserting parent, because the deserting parent typically remates and renests with a new mate, whereas the deserted parent stays with the chicks until the chicks die or fledge at ~4 wk of age. Desertion is costly for both parents because brood survival is lower after desertion than before desertion (Székely and Williams 1995). On top of this cost of desertion, there are additional costs for the deserted parent because it can be killed by predators while it defends its young, and he/she loses mating opportunities.

The fitness implications of Kentish plover parental care may illustrate two important points (Box 1). First, the costs and benefits of care are often complex and they may change over the course of breeding season. In addition, costs and benefits may differ between populations because of different ambient environments and social environments. To reveal these costs and benefits, one may need to combine observations with experiments to expose the full magnitude of fitness implications for both males and females. As I argued above, fitness implications alone do not determine the outcome of parental conflict, so that the behavioral interactions between male and female parents need to be investigated as well. Second, the conflict and cooperation between parents are played out in a breeding population in which breeding and nonbreeding individuals interact, and thus it is important to understand how parenting decisions relate to mate choice, competition for mates, and pair bonding. Thus, conflict over care needs to be approached as influencing (and being influenced by) other aspects of breeding behavior (McNamara et al. 2000; Houston et al. 2005; Alonzo and Sheldon 2010; Jennions and Kokko 2010; Székely et al. 2014).

FUTURE DIRECTIONS

Although immense progress has been made recently in studies of conflict over care, certain areas are ripe for major advances. First, care is a complex and multidimensional social behavior, and analyzing the implications of the different components is likely to give novel insights into conflicts over care and conflict resolution in regard to both parental strategies (battleground) and fitness (conflict load). Theoretical, experimental, and phylogenetic analyses of this multidimensionality are long overdue.

Second, parental conflict is played out in a population, and it is essential to reveal the interactions between a focal pair (or family) and the rest of the population (Székely et al. 2000; Alonzo 2010; Jennions and Kokko 2010). Because behavioral actions have demographic consequences (e.g., maturation, reproduction, and mortalities), rolling up the complex relationships between sexual selection, cost of care, sex-specific mortalities, and breeding system dynamics seems especially important (Liker et al. 2013). The causes of parenting behavior and their consequences at the population level are likely to be complex, although detailed studies of populations that are monitored for a long time now offer the opportunities to tease apart these feedback relationships (Klug et al. 2012; Liker et al. 2013).

Third, although studies established some of the genetic, hormonal, and neuroendocrine causes of parental care, much remains to be discovered about the mechanisms of care and parental interactions.

1. New advances in genomics and neuroscience are offering unparalleled insights into the genetic, genomic, and neural causes of parental behavior (McGraw et al. 2010; O'Connell and Hoffman 2012), and using these approaches, one may work out how each sex is attempting to maximize its own reproductive success and/or how they influence (or manipulate) their mate's behavior.
2. Learning of parenting and sex roles should receive more attention—for instance, when, how, and why young animals learn to be a parent. Learning and genetic mechanisms of care are probably interwoven in a complex way (Székely et al. 2010), and revealing these interdependencies will establish the pathways of how males and females may respond to each other's behavior.
3. Male and female parents may monitor each other's behavior, and alter their own strategy in response to the other parent. Such repeated interactions ("negotiation") (McNamara et al. 1999; Lessells 2012) have received much theoretical attention, although how these interactions play out in nature is not known in detail (but see Johnstone et al. 2014).
4. Parents may manipulate their mate's behavior or take on self-imposed handicaps. Experimental studies are needed to test these theoretical scenarios.

 Cite this article as *Cold Spring Harb Perspect Biol* doi: 10.1101/cshperspect.a017665

Fourth, whereas various studies provide behavioral evidence of conflict over care, few studies have quantified the magnitude of conflict in terms of reproductive outcomes for both males and females.

1. Studies are needed that contrast the theoretically expected versus observed behavior and reproductive success of males and females. Also, detailed studies are needed to work out the assumptions and fitness implications such as the costs and benefits of care. For instance, a common assumption of theoretical models is that the benefit of care follows a saturating curve, whereas the cost function is accelerating (Houston and Davies 1985; Lessells 2012). Although these assumptions seem perfectly plausible, I am not aware of any empirical study that has estimated the shape of these functions over a range of realistic parameter values. This may be important because theoretical predictions may depend on the shape of the cost and benefit functions. Using behavioral norms seems to be one of the promising approaches to quantify these functions (Westneat et al. 2011).
2. Artificial selection and experimental evolution are powerful approaches, although no study seems to use these approaches for sexual conflict over care. Using appropriate model organisms, selection experiments can reveal immense information about the genetic and behavioral structures and fitness implications that underlie conflict over care and conflict resolution.
3. Furthermore, theoretical, comparative, and experimental studies are needed to work out the potential implications of sexual conflict over care for speciation and extinction, somehow analogous to comparable works in prezygotic conflict.

Finally, parental care has only been investigated in detail in $<1\%$ of animal species that populate the planet, and our knowledge is biased toward taxa that are readily studied (McGraw et al. 2010; Balshine 2012; Trumbo 2012). The parental care of organisms living in extreme environments (e.g., deep sea, extreme heat, or freezing temperatures) deserve further attention. Information from a wider range of organisms coupled with recent advances in phylogenies and phylogenetic comparative analyses are likely to reveal novel aspects of conflict over care.

CONCLUDING REMARKS

Conflict over care is central to the understanding of sex roles and breeding systems. Research in this field is rapidly advancing because of the tractability of many biological systems to experimental manipulations, long-term field studies, and phylogenetic comparative analyses. Parental care is also one of the well-known behaviors used as real-life examples in evolutionary game theory, and empiricists are making laudable efforts to match theoretical predictions with experimental results.

Research on parental care and on its implications is making a fundamental contribution to evolutionary biology as exemplified by the study of Trivers (1972) on parental investment that ignited a paradigm shift and became one of the most cited studies produced in this broad field. Parental care research will continue to benefit from new theoretical advances, methods and techniques, and is likely to remain a prosperous field in evolutionary biology.

ACKNOWLEDGMENTS

I appreciate the comments of Alasdair Houston, Kate Lessells, Per Smiseth, and an anonymous reviewer on previous versions of the manuscript. During the writing of this paper, I held a Humboldt Award at the Georg-August University at Göttingen and was supported by a Mercator Visiting Professorship at Bielefeld University.

REFERENCES

Reference is also in this collection.

Adkins-Regan E. 2005. *Hormones and animal social behavior.* Princeton University Press, Princeton, NJ.

Alonzo SH. 2010. Social and coevolutionary feedbacks between mating and parental investment. *Trends Ecol Evol* **25:** 99–108.

Alonzo SH, Sheldon BC. 2010. Population density, social behaviour and sex allocation. In *Social behaviour: Genes, ecology and evolution* (ed. Székely T, et al.), pp. 474–488. Cambridge University Press, Cambridge.

Al-Rashidi M, Kosztolányi A, Shobrak M, Küpper C, Székely T. 2011. Parental cooperation in an extreme hot environment: Natural behaviour and experimental evidence. *Anim Behav* **82:** 235–243.

Amat JA, Masero JA. 2004. How Kentish plovers *Charadrius alexandrinus* cope with heat stress during incubation. *Behav Ecol Sociobiol* **56:** 26–33.

Amat JA, Fraga RM, Arroyo GM. 1999. Brood desertion and polygamous breeding in the Kentish Plover *Charadrius alexandrinus. Ibis* **141:** 596–607.

Arnqvist G, Rowe L. 2005. *Sexual conflict.* Princeton University Press, Princeton, NJ.

Awata S, Kohda M. 2004. Parental roles and the amount of care in a bi-parental substrate brooding cichlid: The effect of size differences within pairs. *Behaviour* **141:** 1135–1149.

Balshine S. 2012. Patterns of parental care in vertebrates. In *The evolution of parental care* (ed. by Royle NP, et al.), pp. 62–80. Oxford University Press, Oxford.

Balshine-Earn S, Earn DJD. 1998. On the evolutionary pathway of parental care in cichlid fishes. *Proc R Soc B* **265:** 2217–2222.

Barta Z, Houston AI, McNamara JM, Székely T. 2002. Sexual conflict about parental care: The role of reserves. *Am Nat* **159:** 687–705.

Barta Z, Székely T, Liker A, Harrison F. 2014. Social role specialization promotes cooperation between parents. *Am Nat* **183:** 747–761.

Bessa-Gomes C, Legendre S, Clobert J. 2004. Allee effects, mating systems and the extinction risk in populations with two sexes. *Ecol Lett* **7:** 802–812.

Blumstein DT. 2010. Social behavior in conservation. In *Social behaviour: Genes, ecology and evolution* (ed. Székely T, et al.), pp. 520–534. Cambridge University Press, Cambridge.

Boncoraglio G, Kilner RM. 2012. Female burying beetles benefit from male desertion: Sexual conflict and counter-adaptation over parental investment. *PLoS ONE* **7:** e31713.

Brown JL, Morales V, Summers K. 2010. A key ecological trait drove the evolution of biparental care and monogamy in an amphibian. *Am Nat* **175:** 436–446.

Champagne FS, Curley JP. 2012. Genetics and epigenetics of parental care. In *The evolution of parental care* (ed. by Royle NP, et al.), pp. 304–324. Oxford University Press, Oxford.

Chapman T, Arnqvist G, Bangham J, Rowe L. 2003. Sexual conflict. *Trends Ecol Evol* **18:** 41–47.

Clutton-Brock TH. 1991. *The evolution of parental care.* Princeton University Press, Princeton, NJ.

de Waal FBM, Tyack PL, eds. 2003. *Animal social complexity.* Harvard University Press, Cambridge, MA.

Donald P. 2007. Adult sex ratios in wild bird populations. *Ibis* **149:** 671–692.

Eens M, Pinxten R. 1996. Female European starlings increase their copulation solicitation rate when faced with the risk of polygyny. *Anim Behav* **51:** 1141–1147.

Eldegard K, Sonerud GA. 2009. Female offspring desertion and male-only care increase with natural and experimental increase in food abundance. *Proc R Soc B* **276:** 1713–1721.

Ens BJ, Choudhury S, Black JM. 1996. Mate fidelity in monogamous birds. In *Partnerships in birds* (ed. Black JM), pp. 344–401. Oxford University Press, Oxford.

Fairbairn DJ, Blanckenhorn WU, Szekely T editors. 2007. *Sex, size and gender roles: Evolutionary studies of sexual size dimorphism.* Oxford University Press, Oxford.

* Gavrilets S. 2014. Is sexual conflict an "engine of speciation?" *Cold Spring Harb Perspect Biol* doi: 10.1101/cshperspect.a017723.

Godfray HCJ. 1995. Evolutionary theory of parent–offspring conflict. *Nature* **376:** 133–138.

Gomez-Mestre I, Pyron RA, Wiens JJ. 2012. Phylogenetic analyses reveal unexpected patterns in the evolution of reproductive modes in frogs. *Evolution* **66:** 3687–3700.

Goodwin NB, Balshine-Earn S, Reynolds JD. 1998. Evolutionary transitions in parental care in cichlid fish. *Proc R Soc B* **265:** 2265–2272.

Griggio M, Pilastro A. 2008. Sexual conflict over parental care in a species with female and male brood desertion. *Anim Behav* **74:** 779–785.

Griggio M, Zanollo V, Hoi H. 2010. Female ornamentation, parental quality, and competitive ability in the rock sparrow. *J Ethology* **28:** 455–462.

Groothuis TGG, Müller W, von Engelhardt N, Carere C, Eising C. 2005. Maternal hormones as a tool to adjust offspring phenotype in avian species. *Neurosci Biobehav Rev* **29:** 329–352.

Gubernick DJ, Teferi T. 2000. Adaptive significance of male parental care in a monogamous mammal. *Proc R Soc B* **267:** 147–150.

Gubernick DJ, Wright SL, Brown RE. 1993. The significance of father's presence for offspring survival in the monogamous California mouse, *Peromyscus californicus. Anim Behav* **46:** 539–546.

* Haig D, Úbeda F, Patten MM. 2014. Specialists and generalists: The sexual ecology of the genome. *Cold Spring Harb Perspect Biol* doi: 10.1101/cshperspect.a017525.

Hamilton WD. 1964. The genetical evolution of social behaviour I-II. *J Theor Biol* **7:** 1–52.

Harrison F, Barta Z, Cuthill I, Székely T. 2009. Conflict and cooperation between parents over care: A meta-analysis. *J Evol Biol* **22:** 1800–1812.

Holland B, Rice WR. 1999. Experimental removal of sexual selection reverses intersexual antagonistic coevolution and removes reproductive load. *Proc Natl Acad Sci* **96:** 5083–5088.

Holman L, Kokko H. 2013.The consequences of polyandry for population viability, extinction risk and conservation. *Phil Trans R Soc B* **368:** SI 20120053.

Houston AI, Davies NB. 1985. The evolution of cooperation and life history in the dunnock, *Prunella modularis.* In *Behavioural ecology* (ed. Sibly RM and Smith RM), pp. 471–487. Blackwell Scientific, Oxford.

Houston AI, Székely T, McNamara JM. 2005. Conflict over parental care. *Trends Ecol Evol* **20:** 33–38.

Houston AI, Székely T, McNamara JM. 2013. The parental investment models of Maynard Smith: A retrospective and prospective view. *Anim Behav* **86:** 667–674.

Itzkowitz M, Santangelo N, Richter M. 2001. Parental division of labour and the shift from minimal to maximal role specializations: An examination using a biparental fish. *Anim Behav* **61:** 1237–1245.

Jennions MD, Kokko H. 2010. Sexual selection. In *Evolutionary behavioral ecology* (ed. Westneat D, Fox C), pp. 343–364. Oxford University Press, New York.

Johnstone RA, Hinde CA. 2006. Negotiation over offspring care—How should parents respond to each other's efforts? *Behav Ecol* **17:** 818–827.

Johnstone RA, Manica A, Fayet AL, Stoddard MC, Rodriguez-Girones MA, Hinde CA. 2014. Reciprocity and conditional cooperation between great tit parents. *Behav Ecol* **25:** 216–222.

Kamel SJ, Grosberg RK. 2012. Exclusive male care despite extreme female promiscuity and low paternity in a marine snail. *Ecol Lett* **15:** 1167–1173.

King EDA, Banks PB, Brooks RC. 2013. Sexual conflict in mammals: Consequences for mating systems and life history. *Mamm Rev* **43:** 47–58.

Klug H, Alonzo SH, Bonsall MB. 2012. Theoretical foundations of parental care. In *The evolution of parental care* (ed. Royle NP, et al.), pp 21–39. Oxford University Press, Oxford.

Kokko H, Jennions M. 2008. Parental investment, sexual selection and sex ratios. *J Evol Biol* **21:** 919–948.

Korpimäki E, Salo P, Valkama J. 2011. Sequential polyandry by brood desertion increases female fitness in a bird with obligatory bi-parental care. *Behav Ecol Sociobiol* **65:** 1093–1102.

Kosztolányi A, Székely T, Cuthill IC, Tulluham Yilmaz K, Berberoğlu S. 2006. The influence of habitat on brood-rearing behavior in the Kentish plover. *J Anim Ecol* **75:** 257–265.

Kosztolányi A, Cuthill IC, Székely T. 2009. Negotiation between parents over care: Reversible compensation during incubation. *Behav Ecol* **20:** 446–452.

Kosztolányi A, Barta Z, Küpper C, Székely T. 2011. Persistence of an extreme male-biased adult sex ratio in a natural population of polyandrous bird. *J Evol Biol* **24:** 1842–1846.

Kupfer A, Müller H, Antoniazzi MM, Jared C, Greven H, Nussbaum RA, Wilkinson M. 2006. Parental investment by skin feeding in a caecilian amphibian. *Nature* **440:** 926–929.

Küpper C, Kis J, Kosztolanyi A, Szekely T, Cuthill IC, Blomqvist D. 2004. Genetic mating system and timing of extra-pair fertilizations in the Kentish plover. *Behav Ecol Sociobiol* **57:** 32–39.

Laaksonen T, Adamczyk F, Ahola M, Möstl E, Lessells CM. 2011. Yolk hormones and sexual conflict over parental investment in the pied flycatcher. *Behav Ecol Sociobiol* **65:** 257–264.

Lessells CM. 1984. The mating system of Kentish plovers *Charadrius alexandrinus*. *Ibis* **126:** 474–483.

Lessells CM. 2006. The evolutionary outcome of sexual conflict. *Phil Trans R Soc B* **361:** 301–317.

Lessells CM. 2012. Sexual conflict. In *The evolution of parental care* (ed. Royle NP, et al.), pp. 150–170. Oxford University Press, Oxford.

Liker A, Székely T. 1997. Aggression among female lapwings *Vanellus vanellus*. *Anim Behav* **54:** 797–802.

Liker A, Freckleton RP, Székely T. 2013. The evolution of sex roles in birds is related to adult sex ratio. *Nat Commun* **4:** 1587.

Maynard Smith J. 1977. Parental investment: A prospective analysis. *Anim Behav* **25:** 1–9.

McGraw L, Székely T, Young LJ. 2010. Pair bonds and parental behaviour. In *Social behaviour: Genes, ecology and evolution* (ed. Székely T, et al.), pp. 271–301. Cambridge University Press, Cambridge.

McNamara JM, Weissing FJ. 2010. Evolutionary game theory. In *Social behaviour: Genes, ecology and evolution* (ed. Székely T, et al.), pp. 88–106. Cambridge University Press, Cambridge.

McNamara JM, Gasson CE, Houston AI. 1999. Incorporating rules for responding into evolutionary games. *Nature* **401:** 368–371.

McNamara JM, Székely T, Webb JN, Houston AI. 2000. A dynamic game-theoretic model of parental care. *J Theor Biol* **205:** 605–623.

McNamara JM, Houston AI, Székely T, Webb JN. 2002. Do parents make independent decisions about desertion? *Anim Behav* **64:** 147–149.

Mitani JC, Call J, Kappeler PM, Paolmbit RA, Silk JB eds. 2012. *The evolution of primate societies*. University Chicago Press, Chicago.

Mock DW, Parker GA. 1997. *The evolution of sibling rivalry*. Oxford University Press, New York.

Møller AP, Cuervo JJ. 2000. The evolution of paternity and parental care in birds. *Behav Ecol* **11:** 472–485.

Mrowka W. 1982. Effect of removal of the mate on the parental care behavior of the biparental cichlid *Aequidens paraguayensis*. *Anim Behav* **30:** 295–297.

O'Connell LA, Hofmann HA. 2012. Evolution of a vertebrate social decision-making network. *Science* **336:** 1154–1157.

Owens IPF. 2002. Male-only care and classical polyandry in birds: Phylogeny, ecology and sex differences in remating opportunities. *Phil Trans R Soc B* **357:** 283–293.

* Palombit RA. 2014. Infanticide as sexual conflict: Coevolution of male strategies and female counterstrategies. *Cold Spring Harb Perspect Biol* doi: 10.1101/cshperspect.a017640.

Parra JE, Betrán M, Zefania S, Dos Remedios N, Székely T. 2014. Experimental assessment of mating opportunities in three shorebird species. *Anim Behav* **90:** 83–90.

Persson O, Öhrström P. 1989. A new avian mating system: Ambisexual polygamy in the Penduline Tit *Remiz pendulinus*. *Ornis Scand* **20:** 105–111.

Pilastro A, Biddau L, Marin G, Mingozzi T. 2001. Female brood desertion increases with number of available mates in the rock sparrow. *J Avian Biol* **32:** 68–72.

Pogány Á, Van Dijk R, Horváth P, Székely T. 2012. Parental behavior and reproductive output in male-only and female-only cared clutches in the Eurasian penduline tit *Remiz pendulinus*. *Auk* **29:** 773–781.

Port M, Cant MA. 2013. Longevity suppresses conflict in animal societies. *Biol Lett* **9:** 20130680.

Queller DC. 1997. Why do females care more than males? *Proc R Soc B* **264:** 1555–1557.

Rauter CM, Moore AJ. 2004. Time constraints and trade-offs among parental care behaviours: Effects of brood size, sex and loss of mate. *Anim Behav* **68:** 695–702.

Reichard UH, Boesch C editors. 2007. *Monogamy: Mating strategies and partnerships in birds, humans and other mammals*. Cambridge University Press, Cambridge.

Reynolds JD, Székely T. 1997. The evolution of parental care in shorebirds: Life histories, ecology and sexual selection. *Behav Ecol* **8:** 126–134.

Reynolds JD, Goodwin NB, Freckleton RP. 2002. Evolutionary transitions in parental care and live-bearing in vertebrates. *Phil Trans R Soc B* **357:** 269–281.

Rowe L, Day T. 2005. Detecting sexual conflict and sexually antagonistic coevolution. *Phil Trans R Soc B* **361:** 277–285.

Royle NJ, Hartley IR, Parker GA. 2002. Sexual conflict reduces offspring fitness in zebra finches. *Nature* **416:** 733–736.

Royle N, Smiseth P, Kölliker M. 2012. *The evolution of parental care*. Oxford University Press, Oxford.

Ruuskanen S, Doligez B, Pitala N, Gustafsson L, Laaksonen T. 2012. Long-term fitness consequences of high yolk androgen levels: Sons pay the costs. *Funct Ecol* **26:** 884–894.

Sandell MI, Smith HG. 1996. Already mated females constrain male mating success in the European starling. *Proc R Soc B* **263:** 743–747.

Sanz JJ, Kranenbarg S, Tinbergen JM. 2000. Differential response by males and females to manipulation of partner contribution in the great tit (*Parus major*). *J Anim Ecol* **69:** 74–84.

Schwabl H. 1996. Maternal testosterone in the egg enhances postnatal growth. *Comp Biochem Physiol* **114:** 271–276.

Sibly RM, Witt CC, Wright NA, Venditti C, Jetz W, Brown JH. 2012. Energetics, lifestyle, and reproduction in birds. *Proc Natl Acad Sci* **109:** 10937–10941.

* Sirot LK, Wong A, Chapman T, Wolfner MF. 2014. Sexual conflict and seminal fluid proteins: A dynamic landscape of sexual interactions. *Cold Spring Harb Perspect Biol* doi: 10.1101/cshperspect.a017533.

Slagsvold T, Lifjeld JT. 1994. Polygyny in birds: The role of competition between females for male parental care. *Am Nat* **143:** 59–94.

Smiseth PT, Dawson C, Varley E, Moore AJ. 2005. How do caring parents respond to mate loss? Differential response by males and females. *Anim Behav* **69:** 551–559.

Smiseth P, Kölliker M, Royle N. 2012. What is parental care? In *The evolution of parental care* (ed. Royle NP, et al.), pp. 1–17. Oxford University Press, Oxford.

Smith HG. 1995. Experimental demonstration of a trade-off between mate attraction and parental care. *Proc R Soc B* **260:** 45–51.

Steinegger M, Taborsky B. 2007. Asymmetric sexual conflict over parental care in a biparental cichlid. *Behav Ecol Sociobiol* **61:** 933–941.

Summers K, McKeon CS, Heying H. 2006. The evolution of parental care and egg size: A comparative analysis in frogs. *Proc R Soc B* **273:** 687–692.

Suzuki S, Nagano M. 2009. To compensate or not? Caring parents respond differentially to mate removal and mate handicapping in the burying beetle, *Nicophorus quadripunctatus*. *Ethology* **115:** 1–6.

Székely T, Cuthill IC. 2000. Trade-off between mating opportunities and parental care: Brood desertion by female Kentish plovers. *Proc R Soc B* **267:** 2087–2092.

Székely T, Lessells CM. 1993. Mate change by Kentish Plovers *Charadrius alexandrinus*. *Ornis Scand* **24:** 317–322.

Székely T, Williams TD. 1995. Costs and benefits of brood desertion in female Kentish plovers, *Charadrius alexandrinus*. *Behav Ecol Sociobiol* **37:** 155–161.

Székely T, Webb JN, Houston AI, McNamara JM. 1996. An evolutionary approach to offspring desertion in birds. In *Current ornithology*, Vol. 13 (ed. Nolan V, Ketterson DE), pp. 271–330. Plenum, New York.

Székely T, Cuthill IC, Kis J. 1999. Brood desertion in Kentish plover: Sex differences in remating opportunities. *Behav Ecol* **10:** 185–190.

Székely T, Webb JN, Cuthill IC. 2000. Mating patterns, sexual selection and parental care: An integrative approach. In *Vertebrate mating systems* (ed. Apollonio M, et al.), pp. 194–223. World Science, London.

Székely T, Moore AJ, Komdeur J editors. 2010. *Social behaviour: Genes, ecology and evolution*. Cambridge University Press, Cambridge.

Székely T, Remeš V, Freckleton RP, Liker A. 2013. Why care? Inferring the evolution of complex social behaviour. *J Evol Biol* **26:** 1381–1391.

Székely T, Weissing FJ, Komdeur J. 2014. Adult sex ratio variation: Implications for breeding system evolution. *J Evol Biol* doi: 10.1111/jeb.12415.

Szentirmai I, Székely T, Komdeur J. 2007. Sexual conflict over care: Antagonistic effects of clutch desertion on reproductive success of male and female penduline tits. *J Evol Biol* **20:** 1739–1744.

Thomas GH, Székely T. 2005. Evolutionary pathways in shorebird breeding systems: Sexual conflict, parental care and chick development. *Evolution* **59:** 2222–2230.

Tregenza T, Wedell N, Chapman T. 2006. Introduction. Sexual conflict: A new paradigm? *Phil Trans R Soc B* **361:** 229–234.

Trivers R. 1972. Parental investment and sexual selection. In *Sexual selection and the descent of man* (ed. Campbell B), pp. 136–179. Aldine, Chicago.

Trivers R. 1985. *Social evolution*. Benjamin/Cummings, Menlo Park, CA.

Trumbo ST. 2007. Defending young biparentally: Female risk-taking with and without a male in the burying beetle, *Nicophorus pustulatus*. *Behav Ecol Sociobiol* **61:** 1717–1723.

Trumbo ST. 2012. Patterns of parental care in invertebrates. In *The evolution of parental care* (ed. Royle NP, et al.), pp. 81–100. Oxford University Press, Oxford.

van Dijk R, Székely T. 2008. Post-fertilization reproductive strategies. In *Encyclopedia of life sciences.* Wiley, Hoboken, NJ.

van Dijk RE, Brinkhuizen D, Székely T, Komdeur J. 2010a. Parental care strategies in Eurasian penduline tit are not related to breeding densities and mating opportunities. *Behaviour* **147:** SI1551–SI1565.

van Dijk RE, Mészáros LA, van der Velde M, Székely T, Pogány A, Szabad J, Komdeur J. 2010b. Nest desertion is not predicted by cuckoldry in the Eurasian penduline tit. *Behav Ecol Sociobiol* **64:** 1425–1435.

van Dijk RE, Székely T, Komdeur J, Pogány A, Fawcett TW, Weissing FJ. 2012. Individual variation and the resolution of conflict over parental care in penduline tits. *Proc Biol Sci* **279:** 1927–1936.

Veran S, Beissinger SR. 2009. Demographic origins of skewed operational and adult sex ratios: Perturbation analyses of two-sex models. *Ecol Lett* **12:** 129–143.

Walter B, Trillmich F. 1994. Female aggression and male peace-keeping in a cichlid fish harem: Conflict between and within the sexes in *Lamprologus ocellatus. Behav Ecol Sociobiol* **34:** 105–112.

Webb JN, Houston AI, McNamara JM, Székely T. 1999. Multiple patterns of parental care. *Anim Behav* **58:** 983–993.

Webb JN, Székely T, Houston AI, McNamara JM. 2002. A theoretical analysis of the energetic costs and consequences of parental care decisions. *Phil Trans Roy Soc* **357:** 331–340.

Webb TJ, Olson V, Székely T, Freckleton RP. 2010. Who cares? Quantifying the evolution of division of parental effort. *Meth Ecol Evol* **1:** 221–230.

Wedell N, Kvarnemo C, Lessells CM, Tregenza T. 2006. Sexual conflict and life histories. *Anim Behav* **71:** 999–1011.

Wells KD. 2007. *The ecology and behavior of amphibians.* University of Chicago Press, Chicago.

West SA, Griffin AS, Gardner A. 2007. Social semantics: Altruism, cooperation, mutualism, strong reciprocity and group selection. *J Evol Biol* **20:** 415–432.

Westneat DF, Sargent RC. 1996. Sex and parenting: The effects of sexual conflict and parentage on parental strategies. *Trends Ecol Ecol* **11:** 87–91.

Westneat DF, Hatch MI, Wetzel DP, Ensminger AL. 2011. Individual variation in parental care reaction norms: Integration of personality and plasticity. *Am Nat* **178:** 652–667.

Wilkinson M, Sherratt E, Starace F, Gower DJ. 2013. A new species of skin-feeding caecilian and the first report of reproductive mode in *Microcaecilia* (Amphibia: Gymnophiona: Siphonopidae). *PLoS ONE* **8:** e57756.

Wilson EO. 1975. *Sociobiology: The new synthesis.* Harvard University Press, Cambridge, MA.

Wynne-Edwards KE, Lisk RD. 1989. Differential effects of paternal presence on pup survival in two species of Dwarf hamster (*Phodopus sungorus* and *Phodopus campbelli*). *Physiol Behav* **45:** 465–469.

Sexual Conflict in Hermaphrodites

Lukas Schärer[1], Tim Janicke[2], and Steven A. Ramm[3]

[1]Evolutionary Biology, Zoological Institute, University of Basel, 4051 Basel, Switzerland

[2]Centre d'Écologie Fonctionnelle et Évolutive, CNRS UMR 5175, 34293 Montpellier Cedex 05, France

[3]Evolutionary Biology, Bielefeld University, 33615 Bielefeld, Germany

Correspondence: lukas.scharer@unibas.ch

Hermaphrodites combine the male and female sex functions into a single individual, either sequentially or simultaneously. This simple fact means that they exhibit both similarities and differences in the way in which they experience, and respond to, sexual conflict compared to separate-sexed organisms. Here, we focus on clarifying how sexual conflict concepts can be adapted to apply to all anisogamous sexual systems and review unique (or especially important) aspects of sexual conflict in hermaphroditic animals. These include conflicts over the timing of sex change in sequential hermaphrodites, and in simultaneous hermaphrodites, over both sex roles and the postmating manipulation of the sperm recipient by the sperm donor. Extending and applying sexual conflict thinking to hermaphrodites can identify general evolutionary principles and help explain some of the unique reproductive diversity found among animals exhibiting this widespread but to date understudied sexual system.

Conceptual and empirical work on sexual conflict is dominated by studies on gonochorists (species with separate sexes) (e.g., Parker 1979, 2006; Rice and Holland 1997; Holland and Rice 1998; Rice and Chippindale 2001; Chapman et al. 2003; Arnqvist and Rowe 2005; Rice et al. 2006), and the same is true for other evolutionary aspects of sexual reproduction, such as sexual selection (e.g., Andersson 1994; Shuster and Wade 2003) and, to a lesser extent, sex allocation (e.g., Charnov 1982; Hardy 2002; West 2009). The fundamental asymmetry from which sexual conflict emerges is ultimately rooted in the evolution of anisogamy, which may itself have resulted from a primordial sexual conflict over allocation to gamete provisioning. Specifically, the (proto)male sexual strategy of making more but smaller gametes—driven by (proto)sperm competition—likely forced the (proto)female sexual strategy into investing more resources per gamete (Parker et al. 1972; Parker 1978, 1979, 2011, 2014). In different species, the male and female sexual strategies can be distributed over individuals in various ways (Hamilton 1967; Charnov 1982; Munday et al. 2006a; Schärer 2009; Weeks 2012). In the most familiar—gonochorism—they are always confined to different individuals, namely, males and females, as exemplified by some of the best-studied animal groups (e.g., insects, birds, and mammals). But another way—hermaphroditism, in which every individual can exhibit both sexual strategies—is in fact very widespread, occurring in ~70% of

Cite this article as *Cold Spring Harb Perspect Biol* doi: 10.1101/cshperspect.a017673

animal phyla (Jarne and Auld 2006) and >90% of plant genera (Renner and Ricklefs 1995). Although models for the evolution of anisogamy have tended to assume gonochorism (e.g., Parker et al. 1972; Lessells et al. 2009; Lehtonen and Kokko 2011; Togashi and Cox 2011), which of these "sexual systems" is actually ancestral remains an open question (Ghiselin 1969; Eppley and Jesson 2008; Iyer and Roughgarden 2008; Ryan et al. 2013; Riesgo et al. 2014).

Here, we show that sexual conflict thinking has equal validity when applied to hermaphroditic taxa (Table 1). In fact, some of the earliest insights about sexual conflict have come from the hermaphrodite literature. In a far-reaching paper published in the same year as Parker's (1979) influential book chapter, Charnov (1979) pointed out that hermaphrodites experience sexual ("male–female") conflicts over mating and fertilization, favoring, for example, the evolution of (1) manipulative seminal fluids that stimulate egg laying and prevent remating (including the disruption of the partner's male function), (2) female control of incoming ejaculate components, (3) postcopulatory female choice (later called cryptic female choice; Thornhill 1983; Eberhard 1985), and (4) male strategies—including traumatic insemination—to bypass female control. Clearly, these insights have very general relevance to all sexual systems and likely arose from the realization around this time that conflicts are an integral part of the evolution of sexual reproduction (e.g., Parker 1970, 1978, 1979; Trivers 1972, 1974; Dawkins 1976; see also Parker 2006). Unfortunately, many people working on sexual conflict in gonochorists still appear surprisingly unaware of Charnov's landmark paper (see also Schärer and Janicke 2009).

We wish to advocate a more inclusive approach to studying evolutionary reproductive biology, which aims at understanding the sexes—and how they are distributed over different individuals—from first principles (see also Schärer et al. 2012; Parker and Birkhead 2013). Thus, rather than seeing sexual conflict in hermaphrodites as a special case, we believe that studying sexual conflict in a range of different sexual systems is essential for deriving general principles of the evolution of anisogamy and its consequences.

The current bias in sexual conflict thinking probably in part reflects an uncertainty among researchers over how to think about the male and female sexual strategies in reproduction when they appear in one and the same individual. We therefore begin by outlining how some of the established (gonochorist-specific) sexual conflict concepts can be broadened to encompass not only hermaphrodites, but arguably all anisogamous sexual systems (Table 1), before briefly introducing how hermaphrodites parti-

Table 1. Important concepts of sexual conflict as commonly defined in the literature

Concept	Common definition	Broadened definition	Note for hermaphrodites
Sexual conflict[a] (Parker 1979, p. 124)	"a conflict between the evolutionary interests of **individuals of the two sexes**"[b]	A conflict between the evolutionary interests of **a sperm donor and a sperm recipient**[c]	Most individuals act in both sexual roles, either at different life stages, in alternating matings, or simultaneously in reciprocal matings
Primordial sexual conflict[d] (Parker 1978, p. 25f)	"[Because of the evolution of anisogamy] **females** are forced into providing all of the parental investment for the zygote. This represents the primordial sexual conflict"	Because of the evolution of anisogamy **the sperm recipient is** forced into providing all of the parental investment for the zygote. This represents the primordial sexual conflict	It is currently unclear whether the evolution of anisogamy occurred in gonochorists, hermaphrodites, or both Given current sex allocation theory, the "cost of anisogamy" might often be lower in hermaphrodites

Continued

Table 1. *Continued*

Concept	Common definition	Broadened definition	Note for hermaphrodites
Mating conflict (Parker 2006, p. 237)	"Mating conflicts relate to asymmetries between **the sexes** in the benefits of mating versus its costs"[e]	Mating conflicts relate to asymmetries between **acting as a sperm donor or sperm recipient** in the benefits of mating versus its costs	Asymmetries may lead to identical sex role preferences and thus incompatible mating interests In unilateral matings, one mate has to assume the nonpreferred role, resulting in a behavioral conflict[f]
Intralocus sexual conflict (Rice et al. 2006, p. 287)	"occurs when there is a negative correlation between the selection coefficients of the same allele when expressed in **males and females**"	Occurs when there is a negative correlation between the selection coefficients of the same allele when expressed in **a sperm donor and a sperm recipient**	Individuals often act as both donors and recipients, so much of this represents intraindividual trade-offs Variance in reproductive success and/or linkage to sex allocation might maintain antagonistic variation
Interlocus sexual conflict (Arnqvist and Rowe 2005, p. 217)	"[occurs] when an allelic substitution at one locus expressed in **one sex** compromises the genetic interests of **individuals of the other sex** and thus selects for a new allele at an interacting locus expressed in **the latter sex**"[g]	Occurs when an allelic substitution at one locus expressed in **one individual** compromises the genetic interests of **its mates** and thus selects for a new allele at an interacting locus expressed in **those mates**	Given that most individuals act as both sperm donors and recipients, avoiding exposure to such sexual conflicts is difficult

The concepts are broadened (bold indicates changed text) to include all anisogamous sexual systems. Note that the terms sperm donor and sperm recipient can be used to talk about mating interactions in all anisogamous sexual systems (including those with external fertilization), as each mating necessarily involves a donor and a recipient (for plants, "sperm" can be replaced by "pollen").

[a]Parker (1970), although not using the term "sexual conflict," clearly adopts sexual conflict thinking, stating: "Provided that the [mating] plug confers a sexual selective advantage on the male which outweighs its natural selective disadvantage [on both male and female], it should evolve or be maintained. Resultant modifications within the female to prevent or reduce the disadvantageous effects of the plug might be expected; these adaptations may conflict with the line of adaptation in the male sex" (p. 559).

[b]Parker (2006) suggests retaining this "broad" definition, stating: ". . . sexual conflict is present in all forms of female choice involving the rejection of some males, whether rejection occurs because they are not attractive enough or because of the costs they impose" and ". . . it is best to retain the simple broad definition . . ., and to recognize that it applies to many different biological circumstances" (p. 236).

[c]Charnov (1979) describes "male–female conflict" in hermaphrodites, stating: "There must often exist a conflict of interest between mating partners–as a recipient each should be inclined to accept sperm (not necessarily for fertilization of its own eggs) to give its sperm away. As a donor, one should be selected to induce one's partner to use the new sperm in fertilization" (p. 2482).

[d]Parker et al. (1972) posit that the evolution of anisogamy involves a (primordial) sexual conflict, stating: "Ovum producers are forced into an evolutionary impasse" and "The relationship between sperm and ova is not dissimilar from the relationship between parasite and host—the parasitic sperm (producers) dependent upon and propagating at the expense of the host ovum (producers)" (p. 549).

[e]Parker (2006) goes on to say that "[Mating conflicts] may be precopulatory, for instance, concerning mating decisions (Trivers 1972; Dawkins 1976; Parker 1979), or postcopulatory, for instance, concerning sperm use (Thornhill 1983; Eberhard 1985)" (p. 237), both of which were also pointed out by Charnov (1979).

[f]Wethington and Dillon (1996) called this "gender conflict" (p. 1108); we see no advantage in this term, because similar conflicts may also apply to reciprocally mating species.

[g]Arnqvist and Rowe (2005) attribute this definition to Rice and Holland (1997), but this precise wording does not appear in that paper.

tion resources between the male and female sex functions (i.e., sex allocation), which is important to understand sexual conflict in these organisms. We then take stock of what we do know and do not know about sexual conflict—first in sequential hermaphrodites, and then in simultaneous hermaphrodites—with the aim of illuminating general principles and attempting to understand what is similar and what is different about sexual conflict when it occurs in different sexual systems. Given limitations on space and our own expertise, we confine our discussion to hermaphroditic animals, but emphasize here that plants undoubtedly also offer tremendous promise to expand our understanding of sexual conflict (e.g., Charnov 1979; Bernasconi et al. 2004; Arnqvist and Rowe 2005; Lankinen and Larsson 2009; Madjidian et al. 2012).

SEXUAL CONFLICT THINKING IN HERMAPHRODITES

Conventionally, evidence for both intra- and interlocus sexual conflict relies on measuring fitness effects in two types of individuals that must, by definition, have opposite sexes (Table 1). So, a difficulty for thinking about sexual conflicts in hermaphrodites is that individuals are not of two types, but either change from one type to the other (sequential hermaphrodites) or are both types at the same time (simultaneous hermaphrodites), and thus each individual has two potential routes to fitness (Fig. 1). The definitions can readily be adapted by simply replacing "male" with "sperm donor" and "female" with "sperm recipient" (Table 1), but this alone does not suffice to understand these phenomena in hermaphrodites, because (1) an individual often acts as both sperm donor and recipient, and (2) its actions often affect its partners as both sperm donors and recipients. This means there are some fundamental differences in the nature of, and likely responses to, sexual conflict compared to gonochorists.

Thinking about interlocus sexual conflict, at least initially, seems less problematic, as this simply represents conflict between two interacting mating partners, the sperm donor and the sperm recipient (Charnov 1979), and much of the literature we discuss below follows this rationale. A necessary qualification for hermaphrodites, however, is that—because of the dual routes to fitness (Fig. 1)—interactions with a mate could affect an individual's male fitness, its female fitness, or both (e.g., via reduced survival). Moreover, sex allocation trade-offs and other evolutionary links between male and female fitness, which occur in both sequential and simultaneous hermaphrodites (see below), mean that one cannot just assume that an individual is affected by sexual conflict in only one

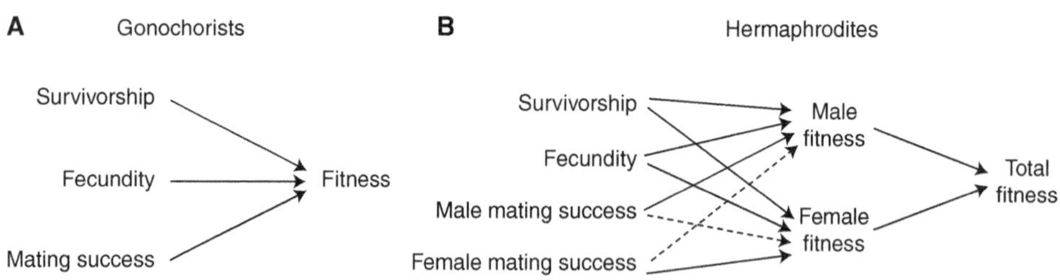

Figure 1. Two routes to fitness in hermaphrodites. (*A*) In gonochorists (species with separate sexes), fitness can be divided into three main components, namely, survivorship, fecundity, and mating (including fertilization) success (modified from Arnqvist and Rowe 2005). (*B*) In hermaphrodites, these components contribute to total fitness through both the male and female sex function. Note that in hermaphrodites, fecundity could further be split into separate male and female components that may trade off with each other. And, although in sequential hermaphrodites male and female mating success happens during nonoverlapping time periods, they may be constrained to be identical in simultaneous hermaphrodites with reciprocal mating. Finally, male mating success could affect female fitness (and vice versa, dashed arrows) and such "cross-sex effects" can probably be negative (Anthes et al. 2010).

of its sex functions. This clearly complicates interpretations compared to gonochorists, but it also presents some interesting targets for sexual conflict, including conflicts over which sex role the partner assumes and its sex allocation, as we discuss in the main sections below.

In contrast to the long history of research on interlocus sexual conflict, the possibility of intralocus sexual conflict in hermaphrodites has only been explored very recently (Abbott 2011; see also Arnqvist and Rowe 2005; Bedhomme et al. 2009). Consider a sexually antagonistic allele arising in a gonochorist. When expressed in (say) males, the allele benefits its bearer, whereas when expressed in females, the same allele is selected against. The ultimate fate of the allele in the population depends on how these benefits and costs balance out (and whether potential modifiers arise at other loci in the meantime) (Rice and Chippindale 2001). The situation in hermaphrodites is similar, in that one is still interested in the fate of a segregating allele that has opposing fitness effects on the carrier when expressed in sperm donors and sperm recipients, but in this case individuals (sequentially or simultaneously) express both sex functions, and so will often act as both sperm donor and sperm recipient within their own lifetime. Although we retain the familiar terminology for continuity, in a strict sense, this means that intralocus sexual conflict cannot exist, because the allele is not simply confined to individuals of one type (male/sperm donor) or another (female/sperm recipient). A useful alternative approach may be to think of such an allele simply in terms of antagonistic pleiotropy, with the opposing fitness effects in the male and female function representing a life-history trade-off, and with the accounting of those fitness effects playing out on a more immediate scale (i.e., within the lifetime of each and every individual in the population). However, this approach may be too simplistic for a number of reasons, which we consider in more detail in the main sections.

Any type of sex-specific selection must obey the frequency-dependent processes of the evolution of sex allocation, which differ between gonochorists and hermaphrodites in ways relevant to sexual conflict, as we outline next. Each zygote results from the fusion of a sperm and an egg, which, given anisogamy, contribute different amounts of resources to the embryo. Yet despite this initial asymmetry, the sexes make an equal genetic contribution (i.e., each offspring has exactly one father and one mother), and so exactly half of the fitness in the population results from male and female reproduction, respectively (the Fisher condition) (Houston and McNamara 2005). Under common assumptions of random mating (i.e., any sperm in the population has the same probability of fusing with any egg) and large population size (i.e., a gamete never encounters a related gamete), this predicts equal resource investment to male and female reproduction, as an unequal investment increases the fitness returns per unit of resource invested into the underrepresented sex function and frequency-dependent selection redresses the balance (Düsing 1884; Fisher 1930).

As a result of interactions between related mates and/or related gametes, however, deviations from these strong assumptions may arguably be the rule rather than the exception. Specifically, brothers sometimes compete for access to mates because of limited dispersal ("local mate competition") (Hamilton 1967; Charnov 1982), and related sperm (e.g., from within the same ejaculate) will often compete for access to eggs because of limited sperm mixing and/or internal fertilization ("local sperm competition") (Schärer 2009; Schärer and Pen 2013). Such local competition will tend to limit returns on investment through the male function, which given the two routes to fitness in hermaphrodites will often favor a reallocation of resources toward the own female function and lead to an overall female-biased sex allocation, as reflected by both theoretical models and empirical data (Charnov 1982; Munday et al. 2006a; Schärer 2009). In contrast, males in gonochorists are more constrained, as they of course cannot directly reallocate to a female function, and so they instead appear to invest in other male routes to fitness, such as gaining more matings (see also Parker 2014). So although these common assumptions may also be broken in gonochorists, these seem less prone to exhibit biased sex allocation. Once sex allocation is biased toward one sex function, a higher fitness return per unit resource invested is obtained from the other sex function (i.e., half of the fitness is obtained with less than half of the resource investment), but the still-broken assumptions mean that this situation can be stable despite the fact the Fisher condition of course still holds. We believe that such biased sex allocation could contribute to preferred sex roles and other sources of sexual conflict between the mating partners, as we discuss specifically for sequential and simultaneous hermaphrodites in the respective main sections below (see also Schärer 2009).

In summary, we suggest that sexual conflict thinking can clearly be extended to hermaphrodites. However, a crucial difference is that the two sex functions simply do not map onto individuals in the way they do in gonochorists, likely complicating the picture for predicting the evolutionary dynamics of sexual conflict. The field would greatly benefit from theoretical work to fully understand the implications of whether and how the evolution of sexually antagonistic alleles is expected to differ between sexual systems, and we outline many relevant factors in the following sections. But in the meantime, we take the concepts we have developed here and apply them first to sequential and then to simultaneous hermaphrodites.

SEXUAL CONFLICTS IN SEQUENTIAL HERMAPHRODITES

Sequential Hermaphrodites: The Basics

Sequential hermaphroditism occurs when individuals can increase their reproductive value by changing sex. This requires that (1) the expected reproductive success is higher for one sex at one life stage and later becomes higher for the other sex, and (2) the fitness costs of changing sex (e.g., in terms of time out and/or physiological reorganization) do not exceed the expected fitness gains (potentially favoring simple reproductive systems in sex changers). Such changes in reproductive value are often linked to changes in body size and/or age (the size-advantage model) (Ghiselin 1969; Warner 1975; Munday et al. 2006a).

One scenario, as represented by anemonefishes, involves a local group inhabiting a spatially restricted habitat that forces group size to be very small. Here, a male cannot monopolize many females and thus male reproductive value depends only weakly on body size. In contrast, as is true for many fishes, female reproductive value increases strongly with body size. This scenario predicts protandrous sex change, in which individuals mature first as males and later change sex to become females. Another scenario, as represented by labroid fishes, involves a local group inhabiting a somewhat larger habitat that permits group size to reach half a dozen or more individuals. Here, a large dominant territorial male can monopolize several females and male reproductive value depends strongly on body size, and more so than for females. This scenario predicts protogynous sex change, in which individuals mature as females and later change sex to become males. In both scenarios, the sex exhibited by other individuals in the local group—and their sex change decisions—are major determinants of an individual's reproductive value (Munday et al. 2006a), so sex-specific selection will be highly frequency-dependent.

Intralocus Sexual Conflict

A sequentially hermaphroditic individual could carry an allele that shows antagonistic pleiotropy for its male and female fitness (Schultz and Warner 1991; Abbott 2011). Imagine an allele in a protogynous hermaphrodite that increases an individual's aggression level at all life stages. Such an allele may make its carrier more successful when male by facilitating territorial defense, while having negative effects when female, for example, because aggression among females decreases fecundity. This scenario was recently suggested for the protogynous reef fish *Parapercis cylindrica* (Sprenger et al. 2012). Females that showed high aggression levels before sex change turned into more aggressive males. Independent evidence further suggested that high aggression levels may indeed be beneficial for males and detrimental for females in this sexual system (Sprenger et al. 2012, and references therein), which, assuming a genetic basis for the aggression level, might therefore be indicative of a sexually antagonistic allele (more specifically, a male-benefit/female-detriment allele). Insofar as the effect of such an allele is measured as lifetime fitness of its carrier (e.g., as the total number of offspring produced), we believe it can simply be seen as mediating a life history trade-off between the male and female function (before and after sex change).

As far as we know, there have been no systematic empirical or theoretical investigations of the evolutionary dynamics of sexually antagonistic alleles in sequential hermaphrodites, which might well be more complex than the simple trade-off scenario outlined above implies (see also Abbott 2011). One important consideration concerns the distribution of fitness across individuals in the population. In the protogynous bluehead wrasse, *Thalassoma bifasciatum*, mating success of territorial males varies considerably and can reach up to 100 females per day (mean $\pm$ S.D. = 36 $\pm$ 21), whereas females only mate about every 2 out of 3 days (Warner et al. 1995). So a male-benefit/female-detriment allele can potentially achieve a high fitness if it resides in a successful male (who may be successful as a male precisely because it carries this allele). Moreover, few individuals live long enough to eventually change sex (with mortality presumably involving a significant stochastic component) and the population sex ratio can be extremely biased (Warner and Hoffman 1980), which also means that very few individuals obtain most of the male fitness (old and large individuals can be extremely successful). Another consideration is how such a sexually antagonistic allele will affect the sex-specific and total fitness of the carrier's offspring bearing that allele (affecting the total number of grandchildren produced). Although such an allele may allow a father to produce many offspring, the above-mentioned skews and stochasticity may mean that none of its offspring will live to become a successful male that goes on to reap the same male-benefit effects, whereas all offspring may still suffer the female-detriment effects while being females. Also, the sex-specific effects of a spreading sexually antagonistic allele are likely to be strongly affected by frequency-dependent selection (Abbott 2011).

One way that a sexually antagonistic allele can increase the probability of reaping the benefits and avoiding the costs of being expressed in a specific sex would be to either (1) influence, or (2) become associated with the sex change decision (Arnqvist and Rowe 2005; Abbott 2011). In the reef fish example above, the former might occur if a high aggression level were achieved by a high titer of a circulating hormone that also causes an earlier sex change (i.e., an earlier sex change could be seen as a pleiotropic effect of a trait for high aggression). The latter could result from linkage disequilibrium, in which alleles for early sex change become statistically associated with male-benefit alleles (i.e., alleles for early sex change may tend to be inherited together with male-benefit alleles) or, more permanently, from actual physical linkage of such alleles in the genome, both resulting in a genetic correlation. Relatively little is currently known about the genetics of sex change decisions in sequential hermaphrodites (but see Hodgkin and Barnes 1991), but it is clear that these decisions show a high degree of plasticity in many species (Munday et al. 2006a; see also next section), so it is not evident how easily sexually antagonistic alleles could become linked to the change decisions. Interestingly, any allele that advances the timing of sex change in a protogynous hermaphrodite will likely make its carrier a more successful sperm donor and a less successful sperm recipient, and so could itself be viewed as a sexually antagonistic allele. Thus, we can rephrase the idea of this paragraph to say that male-benefit/female-detriment alleles may benefit from becoming evolutionarily associated with other male-benefit/female-detriment alleles, leading to sex-role specialization.

Interlocus Sexual Conflict

As mentioned above, sex change decisions can be highly phenotypically plastic, and it is therefore interesting to consider what kinds of constraints might restrict an individual's freedom to decide whether and when to change sex. For example, in a protogynous hermaphrodite, the fitness of a large territorial male depends strongly on the sex change decisions of the other individuals in its local group, for two separate reasons. When a female decides to change sex, the territorial male potentially (1) loses a female mate and any future fitness that might have derived from mating with her, and (2) gains a male competitor that may start to compete for matings (or fertilizations) in the local group. It therefore appears plausible that there could be sexual conflicts over the occurrence and timing of sex change between the territorial male and the resident females, and that a male may actively try to prevent the females from changing sex. This could involve some form of social control of sex change, which is a widespread (although by no means universal) phenomenon in sequential hermaphrodites (e.g., Robertson 1972; Warner and Swearer 1991; Collin et al. 2005; reviewed in Munday et al. 2006a), and often involves stunning phenotypic plasticity. In several species, experimental removal of the territorial male leads to the largest female taking over the dominant role within the group very rapidly, sometimes even within a few minutes (e.g., Warner and Swearer 1991), although it takes several days or even weeks for them to become functional males (Warner and Swearer 1991; Sprenger et al. 2012).

Social control of sex change could result from at least two different scenarios. In the first, a female essentially evaluates the probability of increasing her reproductive value by changing sex, and concludes that she is better off remaining an active female, and waiting for a territory to become available. In this case, we can view the female's poor prospects for changing sex to be the result of male–male competition in that, by investing in territoriality, the territorial male leaves no room for a prospective male to gain more fitness than it could currently do as a female. This seems not to involve sexual conflict, at least in the classical sense, as the conflict involves male–male competition and the individual's interests are harmed only indirectly (by denying it the potential benefits of changing sex). In the second scenario, the female concludes that changing sex would indeed be the preferred option to maximize her reproductive value (e.g., because reduced spending on making eggs could lead to a growth spurt and increase her chances of eventually becoming a successful male; cf. Warner 1984), but she is actively prevented from executing that choice by the territorial male, possibly by aggression-induced stress hormones that physiologically prevent sex change (cf. Marsh-Hunkin et al. 2013). Such aggression might cause direct fitness costs to the female, reducing her egg production below the level she would have achieved without it, which one would clearly view as sexual conflict. A third, possibly intermediate, scenario could result if the territorial male, by directing aggression toward male intruders, somehow causes collateral harm to females. An interesting study that could be seen in this context is that of Warner et al. (1995), who showed that the most successful territorial males release so few sperm that they provide the lowest fertilization rates to their females, presumably because territorial defense is so costly that they need to economize on how much sperm they can produce and allocate to each female.

As a result of hermaphrodites having two routes to fitness (Fig. 1), it may be difficult to determine empirically whether sexual conflict affects the male or female fitness of the partner (who is ultimately trying to optimize total fitness), thus making it difficult to decide which of the above scenarios represents "sexual conflict." One could argue that in all three scenarios it is the same other individual whose (male, female, or both) genetic interests are harmed, thus making it somewhat artificial to consider certain scenarios as involving sexual conflict and others not. Ultimately, this reflects the difficulty of mapping the notions of "males" and "females" onto sequential hermaphrodites (because all individuals can be both). Nevertheless, we believe that the last two scenarios could readily be thought of as interlocus sexual conflict, predicting sexually antagonistic coevolution between a (male-benefit inducing) male persistence trait (e.g., aggression) and a (female-detriment avoiding) female resistance or tolerance trait (e.g., insensitivity to the stress hormone). We are not aware that this possibility has been explored, either empirically or theoretically. The first scenario might also lead to similar coevolutionary dynamics, but here driven by male–male competition between male-benefit inducing alleles and male-detriment avoiding alleles (e.g., small males trying to mimic females, as observed in some diandric hermaphrodites) (Munday et al. 2006b).

SEXUAL CONFLICTS IN SIMULTANEOUS HERMAPHRODITES

Simultaneous Hermaphrodites: The Basics

Simultaneous hermaphroditism could be favored (1) by facilitating reproductive assurance when individuals (or their gametes) risk failed reproduction for lack of encountering partners (Darwin 1876; Tomlinson 1966; Ghiselin 1969), (2) if the male and female reproductive functions do not (strongly) trade off, because they have nonoverlapping resource requirements or can share costs (Charnov et al. 1976; Heath 1979), or (3) if the investment to a sex function shows diminishing fitness returns (e.g., because of low mobility) (Ghiselin 1969), restricted "mating group size" (Charnov 1980), and/or local sperm competition (Schärer 2009; Schärer and Pen 2013), thus favoring reallocation of remaining reproductive resources to the other sex function. The balance of evidence suggests that all these factors may play some role, but the last is likely the most important, at least in animals (reviewed in Schärer 2009).

The best-studied determinant of sex allocation in simultaneous hermaphrodites is the mating group size, which affects the strength of male–male competition in a local group. If mating group size is very small, then little is gained from male investment beyond some minimum needed to ensure full fertility because sperm from a single ejaculate largely compete among themselves for fertilization (termed "local sperm competition" in Schärer 2009) and more is gained from reallocating resources to the female function, resulting in a stable female-biased sex allocation. As mating group size increases, however, success in sperm competition (between sperm from different donors) (Parker 1970) becomes increasingly important to determining fitness through the male sex function, favoring a greater male investment and thus a more even sex allocation. Of course, mating group size can vary in space and time, which in many species has favored the ability to plastically adjust sex allocation to the prevailing (social) conditions (reviewed in Schärer 2009). Such sex allocation adjustment mechanisms could potentially be manipulated by mating partners and are therefore a putative target of sexual conflict in simultaneous hermaphrodites (Charnov 1979; Michiels 1998; Schärer 2009; Schärer and Janicke 2009; see also below).

Intralocus Sexual Conflict

Most work on simultaneous hermaphrodites concerns interlocus sexual conflict (see below), and with respect to intralocus sexual conflict, there are many parallels with sequential hermaphrodites, so this section will be brief. In simultaneous hermaphrodites, male and female fitness is probably also often distributed differently over individuals in a population, and although individuals generally exhibit both sex functions simultaneously, the emphasis often shifts with size (or age), leading to size-dependent sex allocation (Klinkhamer and de Jong 1997; Schärer et al. 2001; Schärer 2009). So here again we can expect linkage to arise between sexually antagonistic alleles and alleles that influence sex allocation (and note here that the latter can again be viewed as largely equivalent to the former). Although there is now ample evidence that sex allocation can be extremely plastic in many simultaneous hermaphrodites (Schärer 2009), there is also some evidence for standing genetic variation in sex allocation (e.g., Hughes 1989; Yund et al. 1997; S Ramm and L Schärer, unpubl.), thus such linkage is clearly possible. Moreover, many traits in simultaneous hermaphrodites may evolve as a result of interlocus sexual conflict and antagonistic coevolution (see below), so there could be many sexually antagonistic alleles with asymmetric fitness effects. For example, a novel seminal fluid (male persistence) allele could have a weak negative (pleiotropic) effect on the female function of its carrier (e.g., because of general toxicity or negative effects during selfing), which is to say that an allele that arises in the context of interlocus sexual conflict (aimed at manipulating the mating partner) can simultaneously generate intralocus sexual conflict (by imposing harm on the carrier's own female function). Similar effects could result from any allele involved in sexually antagonistic coevolution, and most if not all such alleles likely involve some degree of intralocus sexual conflict, if only because investment in the relevant trait draws resources from a common resource pool (and thus likely comes at some cost to the individual, including its female sex function). In the remaining subsections, we focus on different aspects of interlocus sexual conflict.

Premating Conflicts: The Battleground

Sexual conflicts over mating ("mating conflicts") in simultaneous hermaphrodites relate to asymmetries between acting as a sperm donor or sperm recipient in the benefits of mating versus its costs (Table 1) and manifest themselves in unique conflicts over mating interactions (Charnov 1979; Michiels 1998). The evolutionary battleground for such mating conflicts can be visualized as a matrix of compatible and incompatible behavioral interactions, which are defined by the interest of two individuals to be a sperm donor, a sperm recipient, or both (Fig. 2A) (Michiels 1998). Mating conflicts occur whenever the interacting individuals want to adopt incompatible sex roles. We

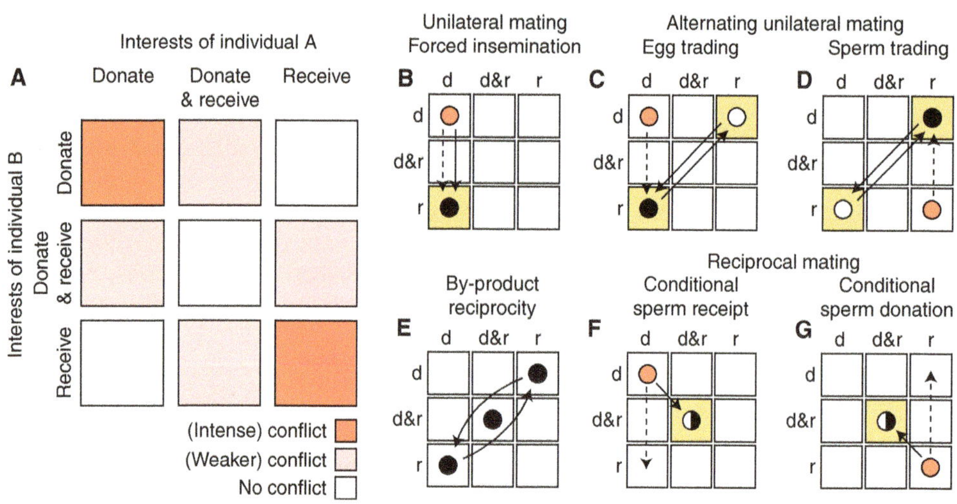

Figure 2. Evolutionary battleground of mating conflicts in simultaneous hermaphrodites (modified from Michiels 1998) and some potential routes to resolution. (*A*) Mating conflicts between individuals A and B can arise from incompatible interests to donate and/or receive sperm. (*B–G*) The routes to resolution of mating conflicts over incompatible interests (i.e., donate sperm, d; receive sperm, r; or both donate and receive sperm, d&r) encompass the following possibilities (viewed from individual A's perspective): forced unilateral insemination (*B*); gamete trading with alternating unilateral matings, namely, egg trading (*C*) or sperm trading (*D*); by-product reciprocity (*E*); and gamete trading during reciprocal matings, namely, conditional sperm receipt (*F*) or conditional sperm donation (*G*). Red squares in *A* and the circles in *B–G* characterize the conflict over mating. Dashed arrows indicate the optimal behavioral outcome for individual A and solid arrows denote the realized behavioral outcome for a given resolution of mating conflict. Yellow squares in *B–G* point to the type of interaction by which the conflict is resolved reflecting benefits (black circles) and costs (open circles) for individual A. Note that by-product reciprocity (*E*) does not actually involve conflict (no yellow squares), but it looks superficially similar to cases that do and so is included here for comparison. Specifically, the curved arrows connecting the *upper-right* and *bottom-left* squares in *E* are similar to the cases of alternating unilateral matings in *C* and *D*, and the central squares in *E* are similar to the cases of reciprocal matings in *F* and *G*.

Cite this article as *Cold Spring Harb Perspect Biol* doi: 10.1101/cshperspect.a017673

first review the theoretical framework and empirical evidence for sex-role preferences, as they represent a key assumption for mating conflicts to arise, and then discuss hypotheses developed for, and evidence bearing on, the resolution of such conflicts.

Premating Conflicts: Sex-Role Preferences

The theoretical foundations of sex-role preferences date back to Bateman (1948), who compared the fitness benefits of mating between male and female fruit flies. The study revealed a "greater dependence of males for their fertility on frequency of insemination" (Fig. 3) (later coined "Bateman's principle" by Charnov 1979) and suggested this pattern is "an almost universal attribute of sexual reproduction" in both animals and plants (Bateman 1948), thus presumably including hermaphrodites. Three decades later, Charnov (1979) explored the hypothesis that Bateman's principle also applies to simultaneous hermaphrodites, arguing that hermaphroditic individuals may "copulate not so much to gain sperm to fertilize eggs as to give sperm away," which he expected to result in an overall preference for the male role and frequent conflicts of interest between mating partners (Fig. 2A, upper-left panel of the matrix).

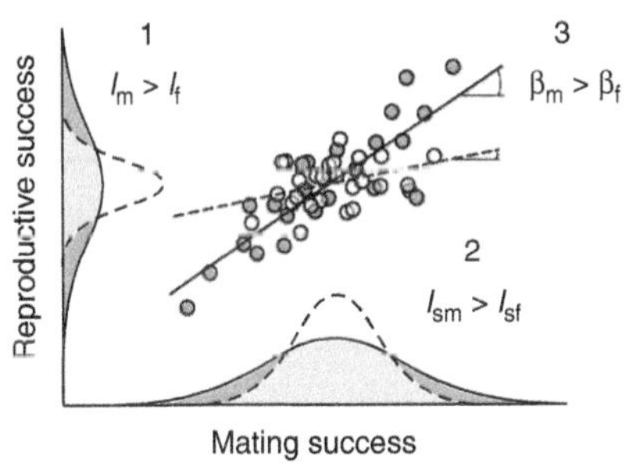

Figure 3. Schematic illustration of Bateman's three principles as proposed by Arnold (1994). The sex that experiences the stronger sexual selection (usually the male or sperm donor) is predicted to show (1) a higher relativized variance in reproductive success (also called opportunity for selection, I), (2) a higher relativized variance in mating success (also called opportunity for sexual selection, I_s), and (3) a steeper Bateman gradient (i.e., the slope of an ordinary least squares regression of reproductive success on mating success; also called sexual selection gradient, β; Arnold and Duvall 1994). The graph shows arbitrary data for the male (*m*; gray circles and density curves; solid line) and female (*f*; open circles and density curves; dashed line) sex. Note that in simultaneous hermaphrodites, the male and female estimates can be obtained from the same individual, so that reproductive success of one sex function can be regressed on mating success of the other sex function. The slopes of such regressions denote "cross-sex effects" that are indicative of trade-offs in sex allocation or sexual antagonism. (Adapted from Anthes et al. 2010.)

Empirical data in terms of so-called Bateman gradients (Fig. 3) are still limited, but they tend to support Charnov's contention. In the freshwater gastropods, *Biomphalaria glabrata* and *Physa acuta*, the male gradients are steeper than the female gradients (Anthes et al. 2010; Pélissié et al. 2012). It is important to spell out why this may often be the case, particularly in species with internal fertilization and sperm storage. The shallow female gradient mainly results from what happens after the first mating in the female role, at which point the female function may have sufficient stored sperm to maintain fertility (i.e., a single mating can potentially switch the female function from infertile to fully fertile). So if one counts the number of "profitable" matings that can be achieved through the male and female function, there are clearly more to be had through the male function (i.e., the male function may obtain some fitness each time the individual mates as a male). The overall male sex-role preference could then simply be seen as a sampling effect. A randomly sampled individual in the population will normally gain more from mating next as a male, because it likely has mated as a female already. The empirically observed Bateman gradients thus probably mean that these snails show a male sex-role preference (Anthes et al. 2010; Pélissié et al. 2012), and two studies on sea slugs, in which only the female function was examined, even suggested costs for high female mating rates (Sprenger et al. 2008; Lange et al. 2012).

This perspective suggests that sex-role preferences need not be fixed, as emphasized by the "gender-ratio hypothesis" (Anthes et al. 2006b). Even if one sex function can usually benefit more from an additional mating, the actual fitness gains from mating as a male or female during any particular encounter will depend on a range of factors, notably the mating history (and thus sperm reserves as both donor and recipient) (e.g., Wethington and Dillon 1996; Anthes et al. 2006a; Ludwig and Walsh 2008), as well as other life-history traits, such as fecundity, survival, and sex allocation of the sperm donor and/or the recipient (e.g., DeWitt 1996; Gianguzza et al. 2004; Janicke and Schärer 2009b; Dillen et al. 2010). This provides a more flexible framework for thinking about variation in sex-role preferences (Anthes et al. 2006b), and all that is required for conflict is that at least some encounters lead to incompatible mating interests.

Measuring male and female Bateman gradients is clearly a promising approach to predict sex-role preferences and the intensity of the resulting mating conflicts (expressed as the slope difference) (see also Anthes et al. 2010). However, Bateman gradients only quantify the average fitness returns an individual can expect from an additional mating in a given sex role, whereas they ignore the costs that have accrued to obtain that fitness (Jennions and Kokko 2010; Kokko et al. 2012). These costs might also be affected as a result of sex allocation biases (Schärer 2009). Specifically, a female-biased sex allocation will, given the Fisher condition, on average lead to a higher fitness return per unit resource investment to the male sex function. Investing one unit of resource into mating in the male role could therefore potentially be more profitable than investing the same amount in the female role. Given that a female-biased sex allocation is predicted under many conditions and actually observed in many species (reviewed in Schärer 2009; Schärer and Pen 2013), this could potentially further contribute to a preference for the male sex role among simultaneous hermaphrodites.

Alternative ideas about sex-role preferences instead predict a female preference. For example, many simultaneous hermaphrodites possess specific organs to digest sperm and/or other ejaculate components (reviewed in Baur 1998; Michiels 1998; Nakadera and Koene 2013) and so direct (nongenetic) benefits from mating in the female role could potentially result from ejaculate digestion, which might generate a female sex-role preference. However, to our knowledge, evidence that the digestion of ejaculates is actually energetically beneficial to the recipient is still lacking (see Postmating Conflicts sections below). Other direct benefits of multiple mating, such as nuptial gifts or paternal care, are presumably also of minor importance as the former, to our knowledge, do not exist (other than possibly via ejaculates; see below) and the latter is rare in simultaneous hermaphrodites (but see Sella and Ramella 1999).

An additional hypothesis posits that a critical factor for sex-role preferences is the risk of one's gametes remaining unfertilized. Specifically, the "gamete trading hypothesis" (Leonard and Lukowiak 1984, 1985; Leonard 1990) assumes that (1) in internally fertilizing hermaphrodites, it is the female role that has more control over the fate of the invested gametes and that this role may therefore be preferred (although there is accumulating evidence that sperm donors often retain at least some control over the fate of their gametes; see also the Postmating Conflicts sections below). Moreover, the hypothesis assumes that (2) in externally fertilizing species, the male role should be preferred as eggs are often released before sperm, which carries the risk they will not be fertilized. However, this implies that a sperm donor would decline to fertilize, at presumably small cost, the (costly) eggs spawned by the partner, even after having invested in precopulatory interactions, which seems implausible (and would probably have to be seen as spiteful behavior). And finally, the gamete trading hypothesis is based on the assumption that the low-risk strategy has a higher mean fitness, which cannot hold given the Fisher condition, as previously pointed out by Greeff and Michiels (1999).

Overall, the currently available empirical evidence suggests a preference to copulate in the male sex role to be generally more likely, but the conditions thought to favor a female sex-role preference remain largely unexplored. Continued efforts to measure the strength of sexual selection in a range of different hermaphroditic systems (Lorenzi and Sella 2008; Anthes et al.

2010; Pélissié et al. 2012) will likely permit a better understanding of Bateman's three principles (Fig. 3) and enable these and some additional proposed hypotheses for sex-role preferences to be properly evaluated (e.g., Crowley et al. 1998; Leonard 1999; reviewed in Leonard 2005, 2006).

Premating Conflicts: Routes to Resolution

Mating conflicts in simultaneous hermaphrodites appear to be resolved in various ways, ranging from the selfish imposition of the nonpreferred role on the mating partner to putatively cooperative mating interactions (Fig. 2B–G). A common approach to predict behavioral strategies arising from conflicts over preferred sex roles is game theory. For example, Axelrod and Hamilton (1981) used an iterated Prisoner's Dilemma to model the intriguing mating behavior of a reef fish, the black hamlet *Hypoplectrus nigricans*—first observed and termed "egg trading" by Fischer (1980)—and they suggested that it involves conditional reciprocity via a tit-for-tat strategy, thus potentially resolving mating conflicts arising from a shared interest to donate sperm in a cooperative way (see also Fischer 1988; Leonard 1990; Crowley et al. 1998). However, we discuss egg trading in some detail below and conclude that the payoff matrices of these models probably do not properly capture the biology of the black hamlet. This example clearly illustrates the need for empirical data that enable the possible strategies to be properly defined and thereby parameterize the payoff matrix of a given mating interaction to predict how mating conflicts could be resolved. This requires not just extensive knowledge about the reproductive biology of a given system, but also about how environmental conditions (such as density and predation risk) may alter the payoffs for reciprocation and defection (Crowley and Hart 2007; Hart et al. 2010). Ultimately, one probably needs manipulative experiments to assess the costs and benefits of the different strategies, to corroborate the importance of particular traits and mating states, and to substantiate claims of strategies involving conditionality (see, e.g., Anthes et al. 2005).

To illustrate the great diversity of behavioral adaptations putatively arising from mating conflicts we start with an apparently simple form of conflict resolution—forced unilateral insemination (Fig. 2B)—which seems to be associated with high costs of sperm receipt, making cooperative behaviors presumably unlikely to evolve. Such overt conflict characterizes the mating behavior of the marine polyclad flatworm *Pseudoceros bifurcus*, in which both interacting individuals attempt to inject sperm hypodermically into their opponent, while trying to avoid being stabbed themselves, as the latter incurs costs of injury (Michiels and Newman 1998; Arnqvist and Rowe 2005). This so-called "penis-fencing" supposedly originates from a higher benefit of sperm donation, although experimental evidence for a male preference to copulate is also still lacking for this species (for other examples of forced unilateral matings, see Michiels 1998, and references therein).

Unilateral matings may also occur consensually when the mating interests of both partners are compatible. For example, many freshwater gastropods mate unilaterally without it resembling the hit-and-run strategy outlined before for flatworms. In such systems, the mating status and thus the amount of available allosperm (i.e., received sperm to fertilize own eggs) and autosperm (i.e., produced sperm and/or seminal fluid for donation) might predict a preference for playing the male or female role at any one time (Wethington and Dillon 1996; Koene and ter Maat 2005), even in the presence of an overall sex-role preference (see "gender-ratio hypothesis" above) (Anthes et al. 2006b). In these systems, occasional conflicts arising from a shared male preference might simply be resolved by an evasive behavior of the would-be female-acting snail (DeWitt 1991; Wethington and Dillon 1996).

Several unilaterally copulating simultaneous hermaphrodites exhibit pronounced sex-role alternation when mating, which is considered one of the best examples for cooperative behavior via reciprocity (Dugatkin 1997). However, for many of the studied systems it remains unclear whether the observed alternation of sex roles actually involves conditionality (i.e., actual trading of eggs or sperm; see Fig. 2C and 2D, respectively), or whether it is just a side effect of a mutual willingness (1) to copulate in both sex roles, or (2) to copulate in one sex role while not objecting to mating in the other (herein called "by-product reciprocity," curved arrows connecting the upper-right and bottom-left panels in Fig. 2E) (Michiels 1998; Anthes and Michiels 2005; Facon et al. 2008).

As briefly outlined above, alternating unilateral exchange of gametes was first described by Fischer (1980), who, based on behavioral observations, suggested that in the black hamlet (and some other serranid reef fishes), a conflict over mating in the male role is resolved by "egg trading." Egg trading involves the parceling of the clutch into small subunits, which are then conditionally exchanged by alternating between adopting the preferred sex role, sperm donation, and the nonpreferred sex role, egg donation (or sperm receipt) (Fig. 2C) (Fischer 1980, 1984, 1987; Pressley 1981; Petersen 1995, 2006). Early approaches modeled this as a two-player game with no option to desert and mate with another individual (Axelrod and Hamilton 1981; Fischer 1988; Leonard 1990). However, Connor (1992) showed that incorporating costs of desertion (e.g., for mate searching or higher predation risk) and allowing the size of the egg parcels to be adjusted may in fact make it more profitable to offer the next batch of eggs to the current partner than to leave in search of other mating opportunities. Moreover, Petersen (2006) stressed that the model assumption of equivalent players and symmetric payoffs are often violated and that we therefore urgently need field estimates of parcel size in egg-trading fish. Similarly, egg trading has been proposed to occur during mating interactions of externally fertilizing polychaetes of the genus *Ophryotrocha*, which also show serial alternation of sex roles (Sella 1985, 1988; Sella et al. 1997; Sella and Lorenzi 2000). However, for both of these putative egg-trading systems, firm experimental corroboration of conditionality (e.g., by showing punishment of experimental cheaters) is still lacking and an overall preference for mating in the male sex function has, so far, only been assumed rather than tested.

Another form of alternating unilateral exchange of gametes, sperm trading (Fig. 2D), was suggested to occur in sea slugs (Leonard and Lukowiak 1984, 1991; Karlsson and Haase 2002; Michiels et al. 2003; Anthes et al. 2005; but see Anthes and Michiels 2005). However, in most of these empirical studies, sperm trading was only inferred from the alternating unilateral exchange of sperm, which might in theory also arise as a by-product of a shared preference to donate with no objection to receive sperm (see Fig. 2E) (Michiels 1998; Puurtinen and Kaitala 2002; Schärer and Pen 2013). Convincing support for conditional sperm trading comes from just a single study. Anthes et al. (2005) experimentally manipulated the ability to donate sperm in the sea slug *Chelidonura hirundinina*, showing that individuals are reluctant to continue mating with such a nonreciprocating partner. Nevertheless, this observed reluctance could still be interpreted as a rejection of nonpreferred mating partners, rather than a punishment of cheaters, if sperm donation serves as a sexual signal indicating the viability of the donor, as hypothesized by Landolfa (2002).

Initially, sperm trading was proposed to occur in internally fertilizing hermaphrodites with an overall preference to copulate in the female sex function based on the "gamete-trading hypothesis" (see previous subsection; Leonard and Lukowiak 1984). In contrast, it has also been argued that sperm trading can be associated with a male sex-role preference even if ejaculates are costly for the donor and provide an energetic benefit to the receiver (Greeff and Michiels 1999). According to this so-called "opportunistic male hypothesis" sperm donors trade their nutrient-rich ejaculates to get access to the eggs of their opponents (Michiels et al. 2003). Here, the ejaculate may function as a nuptial gift and potentially as paternal investment in addition to its usual function in mediating sperm donation to fertilize eggs (Yamaguchi et al. 2012), as has been proposed for many gonochorists (e.g., Vahed 1998). We, however, see problems with this nuptial gift argument, namely, that there must be a considerable energy conversion loss when sperm is first produced and then digested (see also Leonard 2005) and that (just as with nuptial gifts in general) there is no guarantee that the partner will actually use the received energy to make eggs (or that the already existing eggs would preferentially be fertilized by that donor). Both these caveats mean that fitness could usually be obtained much more reliably by instead investing the resources into the own female function. Indeed, when considering a commonly observed energy transfer efficiency between trophic levels of 10% (Pauly and Christensen 1995), rather than the probably unrealistically high value of 50% used by Yamaguchi et al. (2012), then the increase in male allocation compared to their model without a nuptial gift effect is indeed minute (<1%–2% in most cases). And finally, the opportunistic male hypothesis relies on the thus far unproven assumption that sperm digestion actually yields a net resource benefit to the sperm recipient (see the Postmating Conflicts sections below).

Finally, many simultaneous hermaphrodites such as flatworms, annelids, and gastropods show reciprocal mating by donating and receiving sperm at the same time (e.g., Vreys and Michiels 1997; Baur et al. 1998; Michiels and Streng 1998; Michiels and Bakovski 2000; Schärer et al. 2004; Jordaens et al. 2005; Domínguez and Velando 2013), which may or may not result from a conflict over the preferred sex role (see also Puurtinen and Kaitala 2002). In practice, it will usually be difficult to distinguish conditional sperm receipt (Fig. 2F) and conditional sperm donation (Fig. 2G) from by-product reciprocity (Fig. 2E) in reciprocally mating hermaphrodites (Anthes and Michiels 2005), unless sperm transfer can be measured and experimentally manipulated simultaneously in a given copulation, clearly a very challenging task.

To conclude, mating conflicts in simultaneous hermaphrodites are apparently resolved in ways which seem to reflect, at least partially, the proposed asymmetries in the net benefit of acting as a sperm donor and recipient. However, even for well-studied systems we usually still lack detailed information on the payoffs of the observed behavioral strategies, which are needed to fully understand how the mating conflict is resolved. Consequently, it is too early to speculate on the relative incidence of the different types of conflict resolution.

Postmating Conflicts: The Battleground

Even after mating, there remains scope for conflicts of interest between the sperm donor and sperm recipient. Many of these are similar to the phenomena occurring between males and females in gonochorists. Often, however, we expect postmating conflicts to be especially relevant in simultaneous hermaphrodites, for two main reasons: (1) because of the apparent scarcity of overt premating sexually selected traits and the frequent willingness of both partners to mate (or their inability to avoid doing so), meaning we can expect a general shift in sexual selection from the pre- to the postmating arena, and (2) because simultaneous hermaphroditism creates unique targets for sexual conflict, most notably via manipulating also the male sex function and overall sex allocation of the mating partner (Charnov 1979; Michiels 1998; Arnqvist and Rowe 2005; Schärer 2009; Schärer and Janicke 2009). Below, we examine postmating conflicts first from the perspective of the sperm donor attempting to manipulate the sperm recipient and then from the perspective of the sperm recipient attempting to retain or regain control of events occurring after mating.

Postmating Conflicts: Interests of the Sperm Donor

Just as for males in gonochorists (see Edward et al. 2014), several adaptations in simultaneous hermaphrodites likely function to maximize the fertilization success of sperm donors under sperm competition, in a manner that may not be in the recipient's interest. Relevant traits include the number (e.g., Velando et al. 2008) and morphology (e.g., Schärer et al. 2011) of sperm transferred; mechanisms for the efficient transfer of sperm (e.g., Janicke and Schärer 2009a), or for physical displacement of previously deposited sperm (a proposed function of the "disposable" penis of the sea slug *Chromodoris reticulata*, Sekizawa et al. 2013); suppression of remating by sperm recipients (see below); and maximization of the amount of transferred sperm that are retained in storage and used for fertilization (e.g., Landolfa et al. 2001; Rogers and Chase 2002; Chase and Blanchard 2006; Dillen et al. 2009; Garefalaki et al. 2010; Kimura and Chiba 2013). In the following, we discuss first one especially relevant target of manipulation in simultaneous hermaphrodites, namely, aspects of the recipient's sex allocation (Charnov 1979; Michiels 1998), and then consider the various ways in which manipulative substances can be transferred from the donor to the recipient.

Two main types of sex allocation manipulation can be envisaged (Schärer 2009, 2014; Schärer and Janicke 2009). First, the sperm donor could directly affect the female function of the sperm recipient in whatever manner maximizes its reproductive success, and this is essentially analogous to a male manipulating a female in gonochorists. Alternatively, the sperm donor could indirectly boost the mating partner's female function by attacking that individual's male function, assuming that this actually leads to a reallocation of resources to the recipient's female sex function (i.e., assuming plasticity and a sex allocation trade-off), and further assuming that the sperm donor itself gains some advantage from this relative to its competitors, which cannot just be assumed (Schärer 2014). Manipulating the partner's sex allocation should therefore be particularly advantageous in species with low levels of multiple mating, in which manipulation would likely yield benefits exclusively or mainly to the manipulating individual. In this context, a second potential advantage of attacking the recipient's male function may also be to boost the donor's paternity share, as an impeded male function may prevent the recipient from remating in either sex role, which might lower the level of sperm competition experienced by the manipulating focal (see Schärer 2014).

A major route through which manipulative substances (collectively termed "allohormones") (Koene and Chase 1998; Koene and ter Maat 2001) can be transferred from donor to the recipient is in the ejaculate (Zizzari et al. 2014). Seminal-fluid-mediated effects are increasingly recognized as a common source of sexual conflict (see Sirot et al. 2014), also in simultaneous hermaphrodites, as originally predicted by Charnov (1979). For example, seminal fluid investment (as indicated by variation in prostate gland size) covaries with female reproductive tract morphology among opisthobranch sea slugs in a manner consistent with sexually antagonistic coevolution (Anthes et al. 2008). Moreover, the receipt of possibly manipulative seminal fluid substances could explain why *Macrostomum lignano* flatworms mated to virgin partners (that have more seminal fluid in storage to transfer) are less likely to perform the postcopulatory "suck" behavior, hypothesized to function in influencing the fate of received ejaculate components (Marie-Orleach et al. 2013).

By far the best-studied hermaphroditic system with respect to seminal fluid function is the pond snail *Lymnaea stagnalis*, in which several seminal fluid components have recently been identified that influence the partner's reproductive behavior and/or resource allocation. One protein called LyAcp10 (also called ovipostatin) suppresses short-term egg production (Koene et al. 2010), by reducing the number of eggs produced (see also Van Duivenboden et al. 1985; Koene et al. 2006, 2009), while apparently increasing the biomass associated with each egg (Hoffer et al. 2012), although precisely what advantage the sperm donor may gain from this remains unclear. Moreover, an intriguing study by Nakadera et al. (2014) recently found an even more surprising effect, namely, that two proteins (LyAcp5 and LyAcp8b) significantly decrease the amount of sperm a recipient transfers in its next mating (note that LyAcp10 may actually have a similar effect). These studies are clearly highly relevant in the context of the sex allocation manipulations discussed above, but given that they focused on measuring short-term effects on female and male reproduction, respectively, it is currently difficult to interpret these findings. Future studies should aim at simultaneously measuring effects on both sexes, preferably over longer time periods (Schärer 2014).

A second route through which manipulative substances can be transferred from sperm donor to recipient, and one that appears especially common in simultaneous hermaphrodites (Lange et al. 2013a), is via accessory structures that appear to have evolved specifically for this purpose (Zizzari et al. 2014). Striking adaptations in this context are the so-called "love darts" of various land snails (reviewed in Koene 2006; Chase 2007). Experimental evidence in *Cornu aspersum* (formerly *Helix aspersa*) indicates that these sharp, calcareous structures—injected into the mating partner's body before copulation—transfer bioactive substances that cause muscular contractions of the female reproductive tract (Koene and Chase 1998), promoting sperm storage (Rogers and Chase 2001), and enhancing paternity success (Landolfa et al. 2001; Rogers and Chase 2002; Chase and Blanchard 2006). In contrast, the mucus transferred via the love dart of the snail *Euhadra quaesita* appears to suppress remating in recipient individuals (Kimura et al. 2013), suggesting that dart shooting can serve multiple functions (see also Adamo and Chase 1990). Love darts may be counter to sperm recipient's interests for two reasons: by (1) wounding them (Baur 1998, but see Chase and Vaga 2006), and (2) manipulating their reproductive physiology or behavior. As might then be expected from the resulting interlocus sexual conflicts, love darts are highly variable across species (Koene and Schulenburg 2005; Koene et al. 2013) and display signatures of antagonistic coevolution and counteradaptation by sperm recipients in the form of modified reproductive organ morphologies (Koene and Schulenburg 2005; see also Davison et al. 2005; Beese et al. 2009; Sauer and Hausdorf 2009). Similar functions may be performed by the copulatory setae of earthworms (Koene et al. 2002, 2005; König et al. 2006) and devices such as the stylet-like penis appendage found in *Siphopteron* sea slugs (Anthes and Michiels 2007; Lange et al. 2014) and the penial gland that is everted following external sperm exchange in *Deroceras* land slugs (Reise 2007; Reise et al. 2007; Benke et al. 2010; for a recent review of "traumatic secretion transfer," see also Beese et al. 2009 and Lange et al. 2013a). In all of these cases, however, the exact costs and benefits to donor and recipient because of either wounding or receipt of a specific (manipulative) substance under naturalistic conditions remain to be determined.

Irrespective of their method of delivery, allohormones in simultaneous hermaphrodites have been predicted to differ from those in gonochorists. In the former, both male and female substances are expressed in all individuals, which might permit donors to more easily "borrow" their own female substances to manipulate their mating partners, whereas in the latter, such substances will usually have sex-specific expression, which might require more drastic changes in expression (Koene 2005). Too few allohormones have so far been identified to permit a meaningful comparison between sexual systems, but even in hermaphrodites such female substances are normally expressed at different levels in different tissues and would still presumably need to be expressed more highly in novel tissues before they could be coopted by the male function. So the differences between sexual systems may not be so profound in this respect.

Postmating Conflicts: Interests of the Sperm Recipient

As we have seen, individual simultaneous hermaphrodites may often either agree to mate reciprocally or be unable to avoid mating in the role of sperm recipient. This means that they will typically receive many ejaculates. Receiving large amounts of sperm and/or other ejaculate components could be problematic for several reasons, some of which can be seen as either (1) naturally selected (e.g., owing to an increased risk of polyspermy, an increased exposure to sexually transmitted pathogens, and the need to remove a large amount of foreign material from the site of ejaculate receipt), (2) driven by antagonistic coevolution (e.g., avoiding ejaculate-mediated manipulation by the sperm donor), or (3) by sexual selection (e.g., choosing sperm of specific donors). There is thus likely to be a strong incentive for the sperm recipients to control the fate of received substances, and one possible solution is to evolve means of sperm (ejaculate) digestion. There can be no doubt that significant sperm digestion occurs, with well-developed sperm-digesting organs described for example in gastropods (Beeman 1970; Brandriff and Beeman 1973; Beeman 1977; Beese et al. 2006), flatworms (Sluys 1989), and oligochaetes (Westheide 1999). Similarly, even in species with traumatic insemination (see below) that lack specific sperm receiving organs, received sperm appear to be rapidly cleared from recipients (e.g., *Macrostomum hystrix*; S Ramm, A Schlatter, M Poirier, and L Schärer, unpubl.). Although we expect a net fitness benefit to sperm digestion (otherwise it could not have evolved), what is less clear to date is whether sperm recipients can actually recoup any net nutritional benefit from sperm digestion (it could, at least initially, also occur at a net energetic cost). This is an important empirical question, because some recent arguments for mating conflict resolution based on sperm trading assume such a nutritional benefit ("opportunistic male hypothesis") (Greeff and Michiels 1999; Michiels et al. 2003; see also Yamaguchi et al. 2012).

Any hurdle to fertilization that the sperm recipient erects (i.e., any resistance trait, sensu Rowe et al. 2005) is expected to lead to counterselection on sperm donors to overcome these defenses and is likely to lead to paternity being biased toward particular (persistent) sperm donors (leading to cryptic female choice, sensu Pitnick and Brown 2000). A number of potential mechanisms for cryptic female choice have been proposed. These include the differential storage and usage of sperm from multiple sperm storage tubules (Baur 2007; but see Koemtzopoulos and Staikou 2007), the postmating "suck" behavior of *M. lignano* already mentioned above (Schärer et al. 2004; Vizoso et al. 2010; Marie-Orleach et al. 2013), and the blocking and phagocytosis of received self-sperm and certain allosperm in the oviduct, and thus preferential fertilization by allosperm from "preferred" clones in the spermcasting ascidian *Diplosoma listerianum* (Bishop 1996; Bishop et al. 1996; Pemberton et al. 2004). Sperm recipients may also retain more control of sperm uptake, and hence paternity outcomes, when transfer occurs via the external exchange of spermatophores (e.g., in the nudibranch *Aeolidiella glauca*) (Haase and Karlsson 2000; Karlsson and Haase 2002; see also Bishop and Pemberton 2006). Again, though, it is not clear in all of these cases whether or precisely how the sperm recipient gains direct or indirect fitness benefits from exerting control.

Finally, should sperm recipients gain considerable control over received ejaculates, this could, as originally predicted by Charnov (1979), set the stage for sperm donors to attempt to bypass the ordinary route to fertilization and instead transfer sperm via traumatic insemination into some other part of the recipient's body (e.g., Angeloni 2003; Anthes and Michiels 2007; Schmitt et al. 2007; Smolensky et al. 2009; Schärer et al. 2011; Lange et al. 2012, 2013b; see also Premating Conflicts sections above). Wounding during insemination, and the subsequent need to process/remove ejaculate components transferred in this manner, likely results in significant costs of traumatic insemination to the sperm recipient (e.g., Smolensky et al. 2009; see also Reinhardt et al. 2014). The taxonomic distribution of traumatic insemination suggests that the evolution of this fertilization route is common in simultaneous hermaphrodites (Lange et al. 2013a), but it is currently unclear whether simultaneous hermaphroditism is itself a cause of this bias, as might be expected from theoretical predictions of a greater propensity toward escalated mate harm in hermaphrodites (Michiels and Koene 2006; see also Preece et al. 2009), or whether the pattern arises as a consequence of other common features of simultaneous hermaphrodites such as their tendency to be soft-bodied (Lange et al. 2013a) and/or to possess a high capacity for regeneration.

CONCLUDING REMARKS

Applying sexual conflict thinking to hermaphrodites is a potentially highly rewarding exercise, both for establishing general principles about sexual conflict and for explaining reproductive diversity among the wide range of taxa exhibiting hermaphroditic sexual systems.

One of our main aims was to show how concepts and ideas normally applied to gonochorists can be adapted to all sexual systems that have resulted from the evolution of anisogamy, and to explore the evolutionary consequences of sexual conflict in sequentially and simultaneously hermaphroditic animals. We hope to have made some progress on at least the first point, but one clear conclusion is that there is considerable scope for further theoretical work exploring just how intralocus and interlocus sexual conflict dynamics differ when they occur in hermaphrodites; we have pointed out some of the most pressing issues.

A recurrent theme throughout this review has been that our ideas about sexual conflict in hermaphrodites and the sort of patterns we would predict as a result currently run some way ahead of the empirical evidence to support or refute them. As we have discussed, even some of the best-established notions—such as gamete trading or sperm digestion—actually have surprisingly limited experimental support. A second major conclusion is thus an appeal for more rigorous testing of the assumptions and predictions of these and other ideas we have discussed across as many different model systems as possible. Specifically, we need to focus future efforts on measuring all relevant costs and benefits to both individuals involved in a particular mating interaction. Moreover, we should focus on measuring these costs and benefits under naturalistic conditions that result in meaningful measures of male, female, and total fitness, ideally using experimental manipulations that permit the collection of causal and not just correlational evidence. This amounts to a significant empirical challenge.

We believe that the combined theoretical and empirical challenge is well worth the effort, given the importance of understanding sexual conflict—not just in the special case of gonochorists—but in the full range of sexual systems in which it actually occurs in nature.

ACKNOWLEDGMENTS

We thank Bill Rice and Sergey Gavrilets for the opportunity to contribute to this interesting project. We would further like to thank all the participants of a series of hermaphrodite meetings over the last several years, with whom we have discussed many of the ideas we have reviewed here, especially Jessica Abbott, Nils Anthes, Patrice David, Joris Koene, Nico Michiels and Dita Vizoso. Moreover, we thank Nils Anthes, Patrice David, Jussi Lehtonen, Lucas Marie-Orleach, Jutta Schneider, Nikolas Vellnow, and Dita Vizoso for providing comments on earlier versions of this article. While writing this article, our work was supported by the Swiss National Science Foundation (grant 31003A-143732 to L.S. and grant PA00P3-145375 to T.J.), the Seventh Framework Programme of the European Union (Marie Curie Career Integration Grant to S.A.R.) and the Deutsche Forschungsgemeinschaft (grant RA 2468/1-1 to S.A.R.).

REFERENCES

*Reference is also in this collection.

Abbott JK. 2011. Intra-locus sexual conflict and sexually antagonistic genetic variation in hermaphroditic animals. *Proc R Soc B* **278:** 161–169.

Adamo SA, Chase R. 1990. The "love dart" of the snail *Helix aspersa* injects a pheromone that decreases courtship duration. *J Exp Zool* **255:** 80–87.

Andersson M. 1994. *Sexual selection*. Princeton University Press, Princeton.

Angeloni L. 2003. Sexual selection in a simultaneous hermaphrodite with hypodermic insemination: Body size, allocation to sexual roles and paternity. *Anim Behav* **66:** 417–426.

Anthes N, Michiels NK. 2005. Do "sperm trading" simultaneous hermaphrodites always trade sperm? *Behav Ecol* **16:** 188–195.

Anthes N, Michiels NK. 2007. Precopulatory stabbing, hypodermic injections and unilateral copulations in a hermaphroditic sea slug. *Biol Lett* **3:** 121–124.

Anthes N, Putz A, Michiels NK. 2005. Gender trading in a hermaphrodite. *Curr Biol* **15:** R792–R793.

Anthes N, Putz A, Michiels NK. 2006a. Hermaphrodite sex role preferences: The role of partner body size, mating history and female fitness in the sea slug *Chelidonura sandrana*. *Behav Ecol Sociobiol* **60:** 359–367.

Anthes N, Putz A, Michiels NK. 2006b. Sex role preferences, gender conflict, and sperm trading in simultaneous hermaphrodites: A new framework. *Anim Behav* **72:** 1–12.

Anthes N, Schulenburg H, Michiels NK. 2008. Evolutionary links between reproductive morphology, ecology and mating behavior in opisthobranch gastropods. *Evolution* **62:** 900–916.

Anthes N, David P, Auld JR, Hoffer JNA, Jarne P, Koene JM, Kokko H, Lorenzi MC, Pélissié B, Sprenger D, et al. 2010. Bateman gradients in hermaphrodites: An extended approach to quantify sexual selection. *Am Nat* **173:** 249–263.

Arnold SJ. 1994. Bateman's principles and the measurement of sexual selection in plants and animals. *Am Nat* **144:** S126–S149.

Arnold SJ, Duvall D. 1994. Animal mating systems: A synthesis based on selection theory. *Am Nat* **143:** 317–348.

Arnqvist G, Rowe L. 2005. *Sexual conflict.* Princeton University Press, Princeton, NJ.

Axelrod R, Hamilton WD. 1981. The evolution of cooperation. *Science* **211:** 1390–1396.

Bateman AJ. 1948. Intra-sexual selection in *Drosophila. Heredity* **2:** 349–368.

Baur B. 1998. Sperm competition in molluscs. In *Sperm competition and sexual selection* (ed. Birkhead TR, Møller AP), pp. 255–305. Academic, New York.

Baur B. 2007. Reproductive biology and mating conflict in the simultaneously hermaphroditic land snail *Arianta arbustorum. Am Malacol Bull* **23:** 157–172.

Baur B, Locher R, Baur A. 1998. Sperm allocation in the simultaneously hermaphroditic land snail *Arianta arbustorum. Anim Behav* **56:** 839–845.

Bedhomme S, Bernasconi G, Koene JM, Lankinen A, Arathi HS, Michiels NK, Anthes N. 2009. How does breeding system variation modulate sexual antagonism? *Biol Lett* **5:** 717–720.

Beeman RD. 1970. An autoradiographic study of sperm exchange and storage in a sea hare, *Phyllaplysia taylori*, a hermaphroditic gastropod (Opisthobranchia: Anaspidea). *J Exp Zool* **175:** 125–132.

Beeman RD. 1977. Gastropoda: Opisthobranchia. In *Reproduction in marine invertebrates* (ed. Giese AC, Pearse JA), pp. 115–179. Academic, New York.

Beese K, Beier K, Baur B. 2006. Bursa tract diverticulum in the hermaphroditic land snail *Arianta arbustorum* (Stylommatophora: Helicidae): Morphology, function, and evolutionary implications. *J Morphol* **267:** 940–953.

Beese K, Armbruster GFJ, Beier K, Baur B. 2009. Evolution of female sperm-storage organs in the carrefour of stylommatophoran gastropods. *J Zool Syst Evol Res* **47:** 49–60.

Benke M, Reise H, Montagne-Wajer K, Koene JM. 2010. Cutaneous application of an accessory-gland secretion after sperm exchange in a terrestrial slug (Mollusca: Pulmonata). *Zoology* **113:** 118–124.

Bernasconi G, Ashman TL, Birkhead TR, Bishop JDD, Grossniklaus U, Kubli E, Marshall DL, Schmid B, Skogsmyr I, Snook RR, et al. 2004. Evolutionary ecology of the prezygotic stage. *Science* **303:** 971–975.

Bishop JDD. 1996. Female control of paternity in the internally fertilizing compound ascidian *Diplosoma listerianum*: I. Autoradiographic investigation of sperm movements in the female reproductive tract. *Proc R Soc B* **263:** 369–376.

Bishop JDD, Pemberton AJ. 2006. The third way: Spermcast mating in sessile marine invertebrates. *Integr Comp Biol* **46:** 398–406.

Bishop JDD, Jones CS, Noble LR. 1996. Female control of paternity in the internally fertilizing compound ascidian *Diplosoma listerianum*: II. Investigation of male mating success using RAPD markers. *Proc R Soc B* **263:** 401–407.

Brandriff B, Beeman R. 1973. Observations on the gametolytic gland in the anaspidean opisthobranchs, *Phyllaplysia taylori* and *Aplysia californica. J Morphol* **141:** 395–409.

Chapman T, Arnqvist G, Bangham J, Rowe L. 2003. Sexual conflict. *Trends Ecol Evol* **18:** 41–47.

Charnov EL. 1979. Simultaneous hermaphroditism and sexual selection. *Proc Natl Acad Sci* **76:** 2480–2484.

Charnov EL. 1980. Sex allocation and local mate competition in barnacles. *Mar Bio Lett* **1:** 269–272.

Charnov EL. 1982. *The theory of sex allocation.* Princeton University Press, Princeton, NJ.

Charnov EL, Maynard Smith J, Bull JJ. 1976. Why be an hermaphrodite? *Nature* **263:** 125–126.

Chase R. 2007. The function of dart shooting in helicid snails. *Am Malacol Bull* **23:** 183–189.

Chase R, Blanchard KC. 2006. The snail's love-dart delivers mucus to increase paternity. *Proc R Soc B* **273:** 1471–1475.

Chase R, Vaga K. 2006. Independence, not conflict, characterizes dart-shooting and sperm exchange in a hermaphroditic snail. *Behav Ecol Sociobiol* **59:** 732–739.

Collin R, McLellan M, Gruber K, Bailey-Jourdain C. 2005. Effects of conspecific associations on size at sex change in three species of calyptraeid gastropods. *Mar Ecol Prog Ser* **293:** 89–97.

Connor RC. 1992. Egg-trading in simultaneous hermaphrodites: An alternative to tit-for-tat. *J Evol Biol* **5:** 523–528.

Crowley PH, Hart MK. 2007. Evolutionary stability of egg trading and parceling in simultaneous hermaphrodites: The chalk bass revisited. *J Theor Biol* **246:** 420–429.

Crowley PH, Cottrell T, Garcia T, Hatch M, Sargent RC, Stokes BJ, White JM. 1998. Solving the complementarity dilemma: Evolving strategies for simultaneous hermaphroditism. *J Theor Biol* **195:** 13–26.

Darwin C. 1876. *The effects of cross and self-fertilisation in the vegetable kingdom.* John Murray, London.

Davison A, Wade CM, Mordan PB, Chiba S. 2005. Sex and darts in slugs and snails (Mollusca: Gastropoda: Stylommatophora). *J Zool* **267:** 329–338.

Dawkins R. 1976. *The selfish gene.* Oxford University Press, Oxford.

DeWitt TJ. 1991. Mating behavior of the freshwater pulmonate snail *Physa gyrina. Am Malacol Bull* **9:** 81–84.

DeWitt TJ. 1996. Gender contests in a simultaneous hermaphrodite snail: A size-advantage model for behaviour. *Anim Behav* **51:** 345–351.

Dillen L, Jordaens K, Backeljau T. 2009. Sperm transfer, sperm storage, and sperm digestion in the hermaphroditic land snail *Succinea putris* (Gastropoda, Pulmonata). *Invertebr Biol* **128:** 97–106.

Dillen L, Jordaens K, van Dongen S, Backeljau T. 2010. Effects of body size on courtship role, mating frequency and sperm transfer in the land snail *Succinea putris. Anim Behav* **79:** 1125–1133.

Domínguez J, Velando A. 2013. Sexual selection in earthworms: Mate choice, sperm competition, differential allocation and partner manipulation. *Appl Soil Ecol* **69:** 21–27.

Dugatkin LA. 1997. *Cooperation among animals: An evolutionary perspective.* Oxford University Press, Oxford.

Düsing C. 1884. The regulation of the sex ratio. *Jenaische Zeitschrift für Naturwissenschaft* **17:** 593–940.

Eberhard WG. 1985. *Sexual selection and animal genitalia.* Harvard University Press, Cambridge, MA.

* Edward DA, Stockley P, Hosken DJ. 2014. Sexual conflict and sperm competition. *Cold Spring Harb Perspect Biol* doi: 10.1101/cshperspect.017707.

Eppley SM, Jesson LK. 2008. Moving to mate: The evolution of separate and combined sexes in multicellular organisms. *J Evol Biol* **21:** 727–736.

Facon B, Ravigné V, Goudet J. 2008. Gender-role alternation in the simultaneously hermaphroditic freshwater snail *Physa acuta*: Not with the same partner. *Behav Ecol Sociobiol* **62:** 713–720.

Fischer EA. 1980. The relationship between mating system and simultaneous hermaphroditism in the coral reef fish, *Hypoplectrus nigricans* (Serranidae). *Anim Behav* **28:** 620–633.

Fischer EA. 1984. Egg trading in the chalk bass, *Serranus tortugarum*, a simultaneous hermaphrodite. *Z Tierpsychol* **66:** 143–151.

Fischer EA. 1987. Mating behavior in the black hamlet gamete trading or egg trading. *Env Biol Fish* **18:** 143–148.

Fischer EA. 1988. Simultaneous hermaphroditism, tit-for-tat, and the evolutionary stability of social systems. *Ethol Sociobiol* **9:** 119–136.

Fisher RA. 1930. *The genetical theory of natural selection.* Oxford University Press, Oxford.

Garefalaki M-E, Triantafyllidis A, Abatzopoulos TJ, Staikou A. 2010. The outcome of sperm competition is affected by behavioural and anatomical reproductive traits in a simultaneously hermaphroditic land snail. *J Evol Biol* **23:** 966–976.

Ghiselin MT. 1969. The evolution of hermaphroditism among animals. *Q Rev Biol* **44:** 189–208.

Gianguzza P, Badalamenti F, Jensen KR, Chemello R, Cannicci S, Riggio S. 2004. Body size and mating strategies in the simultaneous hermaphrodite *Oxynoe olivacea* (Mollusca, Opisthobranchia, Sacoglossa). *Funct Ecol* **18:** 899–906.

Greeff JM, Michiels NK. 1999. Sperm digestion and reciprocal sperm transfer can drive hermaphrodite sex allocation to equality. *Am Nat* **153:** 421–430.

Haase M, Karlsson A. 2000. Mating and the inferred function of the genital system of the nudibranch, *Aeolidiella glauca* (Gastropoda: Opisthobranchia: Aeolidioidea). *Invertebr Biol* **119:** 287–298.

Hamilton WD. 1967. Extraordinary sex ratios. *Science* **156:** 477–488.

Hardy ICW, editor. 2002. *Sex ratios: Concepts and research methods.* Cambridge University Press, Cambridge.

Hart MK, Kratter AW, Svoboda A-M, Lawrence CL, Sargent RC, Crowley PH. 2010. Sex allocation in a group-living simultaneous hermaphrodite: Effects of density at two different spatial scales. *Evol Ecol Res* **12:** 189–202.

Heath DJ. 1979. Brooding and the evolution of hermaphroditism. *J Theor Biol* **81:** 151–155.

Hodgkin J, Barnes TM. 1991. More is not better: Brood size and population growth in a self-fertilizing nematode. *Proc R Soc B* **246:** 19–24.

Hoffer JNA, Schwegler D, Ellers J, Koene JM. 2012. Mating rate influences female reproductive investment in a simultaneous hermaphrodite, *Lymnaea stagnalis. Anim Behav* **84:** 523–529.

Holland B, Rice WR. 1998. Perspective: Chase-away sexual selection: Antagonistic seduction versus resistance. *Evolution* **52:** 1–7.

Houston AI, McNamara JM. 2005. John Maynard Smith and the importance of consistency in evolutionary game theory. *Biol Phil* **20:** 933–950.

Hughes DJ. 1989. Variation in reproductive strategy among clones of the bryozoan *Celleporella hyalina. Ecol Monogr* **59:** 387–403.

Iyer P, Roughgarden J. 2008. Dioecy as a specialization promoting sperm delivery. *Evol Ecol Res* **10:** 867–892.

Janicke T, Schärer L. 2009a. Determinants of mating and sperm-transfer success in a simultaneous hermaphrodite. *J Evol Biol* **22:** 405–415.

Janicke T, Schärer L. 2009b. Sex allocation predicts mating rate in a simultaneous hermaphrodite. *Proc R Soc B* **276:** 4247–4253.

Jarne P, Auld JR. 2006. Animals mix it up too: The distribution of self-fertilization among hermaphroditic animals. *Evolution* **60:** 1816–1824.

Jennions MD, Kokko H. 2010. Sexual selection. In *Evolutionary behavioral ecology* (eds. Westneat DF, Fox CW), pp. 343–364. Oxford University Press, Oxford.

Jordaens K, Pinceel J, Backeljau T. 2005. Mate choice in the hermaphroditic land snail *Succinea putris* (Stylommatophora: Succineidae). *Anim Behav* **70:** 329–337.

Karlsson A, Haase M. 2002. The enigmatic mating behaviour and reproduction of a simultaneous hermaphrodite, the nudibranch *Aeolidiella glauca* (Gastropoda, Opisthobranchia). *Can J Zool* **80:** 260–270.

Kimura K, Chiba S. 2013. Delayed spermatophore removal in the land snail *Euhadra peliomphala. Biol J Linn Soc* **108:** 806–811.

Kimura K, Shibuya K, Chiba S. 2013. The mucus of a land snail love-dart suppresses subsequent matings in darted individuals. *Anim Behav* **85:** 631–635.

Klinkhamer PGL, de Jong TJ. 1997. Size-dependent allocation to male and female reproduction. In *Plant resource allocation* (ed. Bazzaz FA, Grace J), pp. 211–229. Academic, New York.

Koemtzopoulos E, Staikou A. 2007. Variation in spermathecal morphology is independent of sperm competition intensity in populations of the simultaneously hermaphroditic land snail *Cornu aspersum. Zoology* **110:** 139–146.

Koene JM. 2005. Allohormones and sensory traps: A fundamental difference between hermaphrodites and gonochorists? *Invertebr Reprod Dev* **48:** 101–107.

Koene JM. 2006. Tales of two snails: Sexual selection and sexual conflict in *Lymnaea stagnalis* and *Helix aspersa. Integr Comp Biol* **46:** 419–429.

Koene JM, Chase R. 1998. Changes in the reproductive system of the snail *Helix aspersa* caused by mucus from the love dart. *J Exp Biol* **201:** 2313–2319.

Koene JM, Schulenburg H. 2005. Shooting darts: Co-evolution and counter-adaptation in hermaphroditic snails. *BMC Evol Biol* **5:** 25.

Koene JM, ter Maat A. 2001. "Allohormones": A class of bioactive substances favoured by sexual selection. *J Comp Physiol A* **187:** 323–326.

Koene JM, ter Maat A. 2005. Sex role alternation in the simultaneously hermaphroditic pond snail *Lymnaea stagnalis* is determined by the availability of seminal fluid. *Anim Behav* **69:** 845–850.

Koene JM, Sundermann G, Michiels NK. 2002. On the function of body piercing during copulation in earthworms. *Invertebr Reprod Dev* **41:** 35–40.

Koene JM, Pförtner T, Michiels NK. 2005. Piercing the partner's skin influences sperm uptake in the earthworm *Lumbricus terrestris*. *Behav Ecol Sociobiol* **59:** 243–249.

Koene JM, Montagne-Wajer K, ter Maat A. 2006. Effects of frequent mating on sex allocation in the simultaneously hermaphroditic great pond snail (*Lymnaea stagnalis*). *Behav Ecol Sociobiol* **60:** 332–338.

Koene JM, Brouwer A, Hoffer JNA. 2009. Reduced egg laying caused by a male accessory gland product opens possibility for sexual conflict in a simultaneous hermaphrodite. *Anim Biol* **59:** 435–448.

Koene JM, Sloot W, Montagne-Wajer K, Cummins SF, Degnan BM, Smith JS, Nagle GT, ter Maat A. 2010. Male accessory gland protein reduces egg laying in a simultaneous hermaphrodite. *PLoS ONE* **5:** e10117.

Koene JM, Liew T-S, Montagne-Wajer K, Schilthuizen M. 2013. A syringe-like love dart injects male accessory gland products in a tropical hermaphrodite. *PLoS ONE* **8:** e69968.

Kokko H, Klug H, Jennions MD. 2012. Unifying cornerstones of sexual selection: Operational sex ratio, Bateman gradient and the scope for competitive investment. *Ecol Lett* **15:** 1340–1351.

König S, Mehlich A-M, Bullesbach J, Michiels NK. 2006. Allohormones in *Lumbricus terrestris*? Mass spectrometry of the setal gland product indicates possible role of ubiquitin. *Invertebr Reprod Dev* **49:** 103–112.

Landolfa MA. 2002. On the adaptive function of gamete trading in the black hamlet *Hypoplectrus nigricans*. *Evol Ecol Res* **4:** 1191–1199.

Landolfa MA, Green DM, Chase R. 2001. Dart shooting influences paternal reproductive success in the snail *Helix aspersa* (Pulmonata, Stylommatophora). *Behav Ecol* **12:** 773–777.

Lange R, Gerlach T, Beninde J, Werminghausen J, Reichel V, Anthes N. 2012. Female fitness optimum at intermediate mating rates under traumatic mating. *PLoS ONE* **8:** 8.

Lange R, Reinhardt K, Michiels NK, Anthes N. 2013a. Functions, diversity, and evolution of traumatic mating. *Biol Rev* **88:** 585–601.

Lange R, Werminghausen J, Anthes N. 2013b. Does traumatic secretion transfer manipulate mating roles or reproductive output in a hermaphroditic sea slug? *Behav Ecol Sociobiol* **67:** 1239–1247.

Lange R, Werminghausen J, Anthes N. 2014. Cephalo-traumatic secretion transfer in a hermaphrodite sea slug. *Proc R Soc B* **281:** 20132424.

Lankinen Å, Larsson MC. 2009. Conflicting selection pressures on reproductive functions and speciation in plants. *Evol Ecol* **23:** 147–157.

Lehtonen J, Kokko H. 2011. Two roads to two sexes: Unifying gamete competition and gamete limitation in a single model of anisogamy evolution. *Behav Ecol Sociobiol* **65:** 445–459.

Leonard JL. 1990. The hermaphrodite's dilemma. *J Theor Biol* **147:** 361–372.

Leonard J. 1999. Modern portfolio theory and the prudent hermaphrodite. *Invertebr Reprod Dev* **36:** 129–135.

Leonard JL. 2005. Bateman's principle and simultaneous hermaphrodites: A paradox. *Integr Comp Biol* **45:** 856–873.

Leonard JL. 2006. Sexual selection: Lessons from hermaphrodite mating systems. *Integr Comp Biol* **46:** 349–367.

Leonard JL, Lukowiak K. 1984. Male-female conflict in a simultaneous hermaphrodite resolved by sperm trading. *Am Nat* **124:** 282–286.

Leonard JL, Lukowiak K. 1985. Courtship, copulation, and sperm trading in the sea slug, *Navanax inermis* (Opisthobranchia: Cephalaspidea). *Can J Zool* **63:** 2719–2729.

Leonard JL, Lukowiak K. 1991. Sex and the simultaneous hermaphrodite: Testing models of male–female conflict in a sea slug, *Navanax intermis* (Opisthobranchia). *Anim Behav* **41:** 255–266.

Lessells CM, Snook RR, Hosken DJ. 2009. The evolutionary origin and maintenance of sperm: Selection for a small, motile gamete mating type. In *Sperm biology: An evolutionary perspective* (ed. Birkhead TR, Hosken DJ, Pitnick S), pp. 69–149. Academic, San Diego.

Lorenzi MC, Sella G. 2008. A measure of sexual selection in hermaphroditic animals: Parentage skew and the opportunity for selection. *J Evol Biol* **21:** 827–833.

Ludwig AN, Walsh PJ. 2008. Multiple mating, sperm storage, and mating preference in *Aplysia californica*. *Biol Bull* **215:** 365–271.

Madjidian JA, Hydbom S, Lankinen A. 2012. Influence of number of pollinations and pollen load size on maternal fitness costs in *Collinsia heterophylla*: Implications for existence of a sexual conflict over timing of stigma receptivity. *J Evol Biol* **25:** 1623–1635.

Marie-Orleach L, Janicke T, Schärer L. 2013. Effects of mating status on copulatory and postcopulatory behaviour in a simultaneous hermaphrodite. *Anim Behav* **85:** 453–461.

Marsh-Hunkin KE, Heinz HM, Hawkins MB, Godwin J. 2013. Estrogenic control of behavioral sex change in the bluehead wrasse, *Thalassoma bifasciatum*. *Integr Comp Biol* **53:** 951–959.

Michiels NK. 1998. Mating conflicts and sperm competition in simultaneous hermaphrodites. In *Sperm competition and sexual selection* (ed. Birkhead TR, Møller AP), pp. 219–254. Academic, San Diego.

Michiels NK, Bakovski B. 2000. Sperm trading in a hermaphroditic flatworm: Reluctant fathers and sexy mothers. *Anim Behav* **59:** 319–325.

Michiels NK, Koene JM. 2006. Sexual selection favors harmful mating in hermaphrodites more than in gonochorists. *Integr Comp Biol* **46:** 473–480.

Michiels NK, Newman LJ. 1998. Sex and violence in hermaphrodites. *Nature* **391:** 647.

Michiels NK, Streng A. 1998. Sperm exchange in a simultaneous hermaphrodite. *Behav Ecol Sociobiol* **42:** 171–178.

Michiels NK, Raven-Yoo-Heufes A, Brockmann KK. 2003. Sperm trading and sex roles in the hermaphroditic opisthobranch sea slug *Navanax inermis*: Eager females or opportunistic males? *Biol J Linn Soc* **78:** 105–116.

Munday PL, Buston PM, Warner RR. 2006a. Diversity and flexibility of sex-change strategies in animals. *Trends Ecol Evol* **21:** 89–95.

Munday PL, White JW, Warner RR. 2006b. A social basis for the development of primary males in a sex-changing fish. *Proc R Soc B* **273:** 2845–2851.

Nakadera Y, Koene JM. 2013. Reproductive strategies in hermaphroditic gastropods: Conceptual and empirical approaches. *Can J Zool* **91:** 367–381.

Nakadera Y, Swart EM, Hoffer JNA, den Boon O, Ellers J, Koene JM. 2014. Receipt of seminal fluid proteins causes reduction of male investment in a simultaneous hermaphrodite. *Curr Biol* **24:** 859–862.

Parker GA. 1970. Sperm competition and its evolutionary consequences in the insects. *Biol Rev* **45:** 525–567.

Parker GA. 1978. Selection on nonrandom fusion of gametes during evolution of anisogamy. *J Theor Biol* **73:** 1–28.

Parker GA. 1979. Sexual selection and sexual conflict. In *Sexual selection and reproductive competition in insects* (ed. Blum MS, Blum NA), pp. 123–166. Academic, San Diego.

Parker GA. 2006. Sexual conflict over mating and fertilisation: An overview. *Phil Trans R Soc B* **361:** 235–260.

Parker GA. 2011. The origin and maintenance of two sexes (anisogamy), and their gamete sizes by gamete competition. In *The evolution of anisogamy: A fundamental phenomenon underlying sexual selection* (ed. Togashi T, Cox PA), pp. 17–74. Cambridge University Press, Cambridge.

* Parker GA. 2014. The sexual cascade and the rise of pre-ejaculatory (Darwinian) sexual selection, sex roles, and sexual conflict. *Cold Spring Harb Perspect Biol* doi: 10.1101/cshperspect.017509.

Parker GA, Birkhead TR. 2013. Polyandry: The history of a revolution. *Phil Trans R Soc B* **368:** 20120335.

Parker GA, Smith VGF, Baker RR. 1972. Origin and evolution of gamete dimorphism and male-female phenomenon. *J Theor Biol* **36:** 529–553.

Pauly D, Christensen V. 1995. Primary production required to sustain global fisheries. *Nature* **374:** 255–257.

Pélissié B, Jarne P, David P. 2012. Sexual selection without sexual dimorphism: Bateman gradients in a simultaneous hermaphrodite. *Evolution* **66:** 66–81.

Pemberton AJ, Sommerfeldt AD, Wood CA, Flint HC, Noble LR, Clarke KR, Bishop JDD. 2004. Plant-like mating in an animal: Sexual compatibility and allocation trade-offs in a simultaneous hermaphrodite with remote transfer of sperm. *J Evol Biol* **17:** 506–518.

Petersen CW. 1995. Reproductive behavior, egg trading, and correlates of male mating success in the simultaneous hermaphrodite, *Serranus tabacarius*. *Env Biol Fish* **43:** 351–361.

Petersen CW. 2006. Sexual selection and reproductive success in hermaphroditic seabasses. *Integr Comp Biol* **46:** 439–448.

Pitnick S, Brown WD. 2000. Criteria for demonstrating female sperm choice. *Evolution* **54:** 1052–1056.

Preece T, Mao Y, Garrahan JP, Davison A. 2009. Harmful mating tactics in hermaphrodites. *Am Nat* **173:** 632–639.

Pressley PH. 1981. Pair formation and joint territoriality in a simultaneous hermaphrodite: The coral reef fish *Serranus tigrinus*. *Z Tierpsychol* **56:** 33–46.

Puurtinen M, Kaitala V. 2002. Mate-search efficiency can determine the evolution of separate sexes and the stability of hermaphroditism in animals. *Am Nat* **160:** 645–660.

* Reinhardt K, Anthes N, Lange R. 2014. Copulatory wounding and traumatic insemination. *Cold Spring Harb Perspect Biol* doi: 10.1101/cshperspect.017582.

Reise H. 2007. A review of mating behavior in slugs of the genus *Deroceras* (Pulmonata: Agriolimacidae). *Am Malacol Bull* **23:** 137–156.

Reise H, Visser S, Hutchinson JMC. 2007. Mating behaviour in the terrestrial slug Deroceras gorgonium: Is extreme morphology associated with extreme behaviour? *Anim Biol* **57:** 197–215.

Renner SS, Ricklefs RE. 1995. Dioecy and its correlates in the flowering plants. *Am J Bot* **82:** 596–606.

Rice WR, Chippindale AK. 2001. Intersexual ontogenetic conflict. *J Evol Biol* **14:** 685–693.

Rice WR, Holland B. 1997. The enemies within: Intergenomic conflict, interlocus contest evolution (ICE), and the intraspecific Red Queen. *Behav Ecol Sociobiol* **41:** 1–10.

Rice WR, Stewart AD, Morrow EH, Linder JE, Orteiza N, Byrne PG. 2006. Assessing sexual conflict in the *Drosophila melanogaster* laboratory model system. *Phil Trans R Soc B* **361:** 287–299.

Riesgo A, Novo M, Sharma PP, Peterson M, Maldonado M, Giribet G. 2014. Inferring the ancestral sexuality and reproductive condition in sponges (Porifera). *Zool Scr* **43:** 1–17.

Robertson DR. 1972. Social control of sex reversal in a coral reef fish. *Science* **177:** 1007–1009.

Rogers DW, Chase R. 2001. Dart receipt promotes sperm storage in the garden snail *Helix aspersa*. *Behav Ecol Sociobiol* **50:** 122–127.

Rogers DW, Chase R. 2002. Determinants of paternity in the garden snail *Helix aspersa*. *Behav Ecol Sociobiol* **52:** 289–295.

Rowe L, Cameron E, Day T. 2005. Escalation, retreat, and female indifference as alternative outcomes of sexually antagonistic coevolution. *Am Nat* **165:** S5–S18.

Ryan JF, Pang K, Schnitzler CE, Nguyen A-D, Moreland RT, Simmons DK, Koch BJ, Francis WR, Havlak P, Program NCS, et al. 2013. The genome of the ctenophore *Mnemiopsis leidyi* and its implications for cell type evolution. *Science* **342:** 1242592.

Sauer J, Hausdorf B. 2009. Sexual selection is involved in speciation in a land snail radiation on Crete. *Evolution* **63:** 2535–2546.

Schärer L. 2009. Tests of sex allocation theory in simultaneously hermaphroditic animals. *Evolution* **63:** 1377–1405.

Schärer L. 2014. Evolution: Don't be so butch, dear! *Curr Biol* **24:** R311–R313.

Schärer L, Janicke T. 2009. Sex allocation and sexual conflict in simultaneously hermaphroditic animals. *Biol Lett* **5:** 705–708.

Schärer L, Pen I. 2013. Sex allocation and investment into pre- and postcopulatory traits in simultaneous hermaphrodites: The role of polyandry and local sperm competition. *Phil Trans R Soc Lond B* **368:** 20120052.

Schärer L, Karlsson LM, Christen M, Wedekind C. 2001. Size-dependent sex allocation in a simultaneous hermaphrodite parasite. *J Evol Biol* **14:** 55–67.

Schärer L, Joss G, Sandner P. 2004. Mating behaviour of the marine turbellarian *Macrostomum* sp.: These worms *suck*. *Mar Biol* **145:** 373–380.

Schärer L, Littlewood DTJ, Waeschenbach A, Yoshida W, Vizoso DB. 2011. Mating behaviour and the evolution of sperm design. *Proc Natl Acad Sci* **108:** 1490–1495.

Schärer L, Rowe L, Arnqvist G. 2012. Anisogamy, chance and the evolution of sex roles. *Trends Ecol Evol* **27:** 260–264.

Schmitt V, Anthes N, Michiels NK. 2007. Mating behaviour in the sea slug *Elysia timida* (Opisthobranchia, Sacoglossa): Hypodermic injection, sperm transfer and balanced reciprocity. *Front Zool* **4:** 17.

Schultz ET, Warner RR. 1991. Phenotypic plasticity in life-history traits of female *Thalassoma bifasciatum* (Pisces: Labridae): 2. Correlation of fecundity and growth rate in comparative studies. *Env Biol Fish* **30:** 333–344.

Sekizawa A, Seki S, Tokuzato M, Shiga S, Nakashima Y. 2013. Disposable penis and its replenishment in a simultaneous hermaphrodite. *Biol Lett* **9:** 20121150.

Sella G. 1985. Reciprocal egg trading and brood care in a simultaneous polychaete worm. *Anim Behav* **33:** 938–944.

Sella G. 1988. Reciprocation, reproductive success, and safeguards against cheating in a hermaphroditic polycheate worm, *Ophryotrocha diadema* Åkesson, 1976. *Biol Bull* **175:** 212–217.

Sella G, Lorenzi MC. 2000. Partner fidelity and egg reciprocation in the simultaneously hermaphroditic polychaete worm *Ophryotrocha diadema*. *Behav Ecol* **11:** 260–264.

Sella G, Ramella L. 1999. Sexual conflict and mating systems in the dorvilleid genus *Ophryotrocha* and the dinophilid genus *Dinophilus*. *Hydrobiologia* **402:** 203–213.

Sella G, Premoli MC, Turri F. 1997. Egg trading in the simultaneously hermaphroditic polychaete worm *Ophryotrocha gracilis* (Huth). *Behav Ecol* **8:** 83–86.

Shuster SM, Wade MJ. 2003. *Mating systems and strategies*. Princeton University Press, Princeton, NJ.

* Sirot LK, Wong A, Chapman T, Wolfner MF. 2014. Sexual conflict and seminal fluid proteins: A dynamic landscape of sexual interactions. *Cold Spring Harb Perspect Biol* doi: 10.1101/cshperspect.017533.

Sluys R. 1989. Sperm resorption in triclads (Platyhelminthes, Tricladida). *Invertebr Reprod Dev* **15:** 89–95.

Smolensky N, Romero MR, Krug PJ. 2009. Evidence for costs of mating and self-fertilization in a simultaneous hermaphrodite with hypodermic insemination, the opisthobranch *Alderia willowi*. *Biol Bull* **216:** 188–199.

Sprenger D, Faber J, Michiels NK, Anthes N. 2008. Natural female mating rate maximizes hatchling size in a marine invertebrate. *J Anim Ecol* **77:** 696–701.

Sprenger D, Dingemanse NJ, Dochtermann NA, Theobald J, Walker SPW. 2012. Aggressive females become aggressive males in a sex-changing reef fish. *Ecol Lett* **15:** 986–992.

Thornhill R. 1983. Cryptic female choice and its implications in the scorpionfly *Harpobittacus nigriceps*. *Am Nat* **122:** 765–788.

Togashi T, Cox PA. 2011. *The evolution of anisogamy: A fundamental phenomenon underlying sexual selection*. Cambridge University Press, Cambridge.

Tomlinson J. 1966. The advantage of hermaphroditism and parthenogenesis. *J Theor Biol* **11:** 54–58.

Trivers RL. 1972. Parental investment and sexual selection. In *Sexual selection and the descent of man 1871–1971* (ed. Campbell B), pp. 136–179. Aldine, Chicago.

Trivers RL. 1974. Parent-offspring conflict. *Amer Zool* **14:** 249–264.

Vahed K. 1998. The function of nuptial feeding in insects: A review of empirical studies. *Biol Rev* **73:** 43–78.

Van Duivenboden YA, Pieneman AW, ter Maat A. 1985. Multiple mating suppresses fecundity in the hermaphrodite freshwater snail *Lymnaea stagnalis*: A laboratory study. *Anim Behav* **33:** 1184–1191.

Velando A, Eiroa J, Domínguez J. 2008. Brainless but not clueless: Earthworms boost their ejaculates when they detect fecund non-virgin partners. *Proc R Soc B* **275:** 1067–1072.

Vizoso DB, Rieger G, Schärer L. 2010. Goings-on inside a worm: Functional hypotheses derived from sexual conflict thinking. *Biol J Linn Soc* **99:** 370–383.

Vreys C, Michiels NK. 1997. Flatworms flatten to size up each other. *Proc R Soc B* **264:** 1559–1564.

Warner RR. 1975. The adaptive significance of sequential hermaphroditism in animals. *Am Nat* **109:** 61–82.

Warner RR. 1984. Deferred reproduction as a response to sexual selection in a coral reef fish: A test of the life historical consequences. *Evolution* **38:** 148–162.

Warner RR, Hoffman SG. 1980. Local population size as a determinant of mating system and sexual composition in two tropical marine fishes (*Thalassoma* spp.). *Evolution* **34:** 508–518.

Warner RR, Swearer SE. 1991. Social control of sex change in the bluehead wrasse, *Thalassoma bifasciatum* (Pisces: Labridae). *Biol Bull* **181:** 199–204.

Warner RR, Shapiro DY, Marconato A, Petersen CW. 1995. Sexual conflict: Males with highest mating success convey the lowest fertilization benefits to females. *Proc R Soc Lond B* **262:** 135–139.

Weeks SC. 2012. The role of androdioecy and gynodioecy in mediating evolutionary transitions between dioecy and hermaphroditism in the Animalia. *Evolution* **66:** 3670–3686.

West SA. 2009. *Sex allocation.* Princeton University Press, Princeton, NJ.

Westheide W. 1999. Ultrastructure and functional significance of intestinojunctional spermathecae in enchytraeids (Oligochaeta, Annelida). *Hydrobiologia* **406:** 199–211.

Wethington AR, Dillon RT Jr. 1996. Gender choice and gender conflict in a non-reciprocally mating simultaneous hermaphrodite, the freshwater snail, *Physa*. *Anim Behav* **51:** 1107–1118.

Yamaguchi S, Sawada K, Nakashima Y, Takahashi S. 2012. Sperm as a paternal investment: A model of sex allocation in sperm-digesting hermaphrodites. *Theor Ecol* **5:** 99–103.

Yund PO, Marcum Y, Stewart-Savage J. 1997. Life-history variation in a colonial ascidian: Broad-sense heritabilities and tradeoffs in allocation to asexual growth and male and female reproduction. *Biol Bull* **192:** 290–299.

Zizzari ZV, Smolders I, Koene JM. 2014. Alternative delivery of male accessory gland products. *Front Zool* **11:** 32.

Patterns and Mechanisms of Evolutionary Transitions between Genetic Sex-Determining Systems

G. Sander van Doorn

Centre for Ecological and Evolutionary Studies, University of Groningen, 9700 CC Groningen, The Netherlands

Correspondence: g.s.van.doorn@rug.nl

The diversity and patchy phylogenetic distribution of genetic sex-determining mechanisms observed in some taxa is thought to have arisen by the addition, modification, or replacement of regulators at the upstream end of the sex-determining pathway. Here, I review the various evolutionary forces acting on upstream regulators of sexual development that can cause transitions between sex-determining systems. These include sex-ratio selection and pleiotropic benefits, as well as indirect selection mechanisms involving sex-linked sexually antagonistic loci or recessive deleterious mutations. Most of the current theory concentrates on the population–genetic aspects of sex-determination transitions, using models that do not reflect the developmental mechanisms involved in sex determination. However, the increasing availability of molecular data creates opportunities for the development of mechanistic models that can clarify how selection and developmental architecture interact to direct the evolution of sex-determination genes.

Biparental sexual reproduction is a common mode of reproduction in higher organisms. It is found in gonochorous animals (Bull 1983; Barnes et al. 2001), heterothallic fungi (Heitman et al. 2013), and dioecious flowering plants and algae (Ainsworth 2000; Umen 2011). Species belonging to this diverse group of organisms have distinct sexes, and their development typically passes through a critical stage at which the zygote commits irreversibly to either the male or the female sexual fate (Valenzuela 2008) (except in sequential hermaphrodites, which change sex during their life). This ontogenetic process, known as sex determination, triggers the differentiation of specialized male or female reproductive organs and organizes many sex-specific differences in gene expression, physiology, morphology, and behavior (sex differentiation) (Badyaev 2002; Ellegren and Parsch 2007).

Despite the universal and simple dichotomous outcome of sex determination, sex-determining mechanisms are highly diverse across taxa. Some species use a specific environmental cue (e.g., temperature, photoperiod, or population density) as the primary sex-determining signal (environmental sex determination; ESD), whereas others rely on various types of genetic sex determination (GSD), including male or female heterogamety, haplodiploidy or multilocus sex-determining mechanisms (Bull 1983;

Cite this article as *Cold Spring Harb Perspect Biol* doi: 10.1101/cshperspect.a017681

Marshall Graves 2008; Janousek and Mrackova 2010). In addition, sex determination can depend on epigenetic factors such as imprinting or maternal gene products deposited in the egg (Verhulst et al. 2010a). The apparent variability of sex determination is even more puzzling given that other processes acting in mid-development are evolutionarily conserved, presumably as a result of strong ontogenetic constraints (Marín and Baker 1998; Kalinka and Tomancak 2012). Considerable effort has therefore been directed at explaining the function and evolutionary origin of diversity in the mechanisms determining sex.

Here I review this literature, concentrating on transitions between genetic sex determination systems (for other recent reviews, see Beukeboom and Perrin 2014; van Doorn 2014). An important ultimate cause of such transitions is sexually antagonistic selection, which interacts with various other evolutionary forces shaping the sex-determining system. To disentangle these factors, I will first discuss the current paradigm for how sex-determination pathways have been modified, before reviewing the various mechanisms thought to be responsible for transitions in sex determination. The final part of this review will straddle the proximate/ultimate divide by exploring how adaptive mechanisms interact with the developmental architecture of sex determination.

PATTERNS OF EVOLUTION IN SEX-DETERMINATION CASCADES

Genetic and developmental analyses have revealed that the gene-regulatory networks responsible for sex determination in various model organisms have a clear hierarchical structure. For example, in the round worm *Caenorhabditis elegans*, the primary sex-determination signal is relayed by a cascade of inhibitory regulatory interactions involving the genes *xol-1*, *sdc-1,2,3*, *her-1*, *tra-2,3*, *fem-1,2,3*, and *tra-1* (Stothard and Pilgrim 2003). The final gene in the cascade inhibits *egl-1*, *mab-3* (*male abnormal 3*), and other genes responsible for male development. In the fruit fly *Drosophila melanogaster*, sex determination involves a cascade of regulatory genes whose expression is controlled by sex-specific alternative splicing. The topmost gene, *Sxl* (*Sex-lethal*), responds to the primary sex-determination system early in development and then maintains the female state by means of an autoregulatory feedback loop (Schütt and Nöthiger 2000). The gene *dsx* (*doublesex*), at the bottom of the cascade, shares a characteristic cysteine rich DNA-binding domain (Zhu et al. 2000) with the *mab-3* gene of *C. elegans*. This finding provided a first clue that sex-determination pathways in distantly related organisms derive from a common ancestral pathway. Consistent with their sequences being similar, *dsx* and *mab-3* share biological functions important for male development and the protein product of the male-specific splice form of *dsx* can partly restore the wild-type phenotype in male worms mutant for *mab-3* (Raymond et al. 1998). Other *dsx-mab-3* (DM)-family genes with a role in sex determination or differentiation have later been discovered in crustaceans, vertebrates, and cnidarians, supporting a single origin of the pathways controlling sexual development in animals (Hodgkin 2002; Haag and Doty 2005; Shukla and Nagaraju 2010).

A striking pattern that is shared by the well-studied sex-determination pathways of insects (Nöthiger and Steinmann-Zwicky 1985; Gempe and Beye 2011), nematodes (Zarkower 2001; Stothard and Pilgrim 2003) and vertebrates (Marshall Graves 2008; Mawaribuchi et al. 2012), is that the downstream genes are more conserved than upstream ones (Zarkower 2001). For example, *Sxl*, the first gene in the *D. melanogaster* cascade, has no known function in sex determination outside the Drosophilidae (Sánchez 2008). However, its downstream target, *tra* (*transformer*), is found in several orders of the holometabolous insects (Verhulst et al. 2010b; Geuverink and Beukeboom 2014). Functional orthologs of *tra* have been identified, for instance, in other Dipterans (*Ceratitis capitata*, *Musca domestica*, *Anastrepha sp.*, *Bactrocera oleae*, *Lucilia cuprina*) (Lagos et al. 2007; Ruiz et al. 2007; Concha and Scott 2009; Hediger et al. 2010), Coleoptera (*Tribolium castaneum*) (Shukla and Palli 2012) and in the Hymenoptera (*Apis mellifera* and *Nasonia vitripennis*)

(Hasselmann et al. 2008; Verhulst et al. 2010a), the most basal order of the Holometabola. Thus far, no transformer homologs have been found in the Lepidoptera or outside the insects (Geuverink and Beukeboom 2014). The regulation of *tra* by upstream factors is highly diverse (Verhulst et al. 2010b). In the diploid Mediterranean fruitfly *C. capitata*, for example, the production of functional Tra protein is repressed by a dominant masculinizing factor; in the haplodiploid species *A. mellifera*, it is under control of the complementary sex-determiner gene *csd*, and in *N. vitripennis* (another haplodiploid species), its expression is regulated by imprinting of the maternal copy of *tra* (Verhulst et al. 2010a,b; Gempe and Beye 2011). However, in all cases, *tra* acts as a splice regulator of *dsx*, a putative ancient sex-determination gene that has been conserved across phyla.

The observation that downstream effectors (such as the DM-family genes in animals) are conserved, whereas the top most genes of sex-determination cascades are highly variable, supports the hypothesis that sex-determination pathways have evolved from the bottom up by the successive addition of novel upstream regulators (Nöthiger and Steinmann-Zwicky 1985; Wilkins 1995; Marín and Baker 1998). Theoretical studies on sex determination therefore tend to concentrate on the evolutionary forces responsible for the recruitment of upstream modifiers (implicitly), assuming that the downstream parts of the pathway remain fixed. A clear example of this approach is provided by Pomiankowski et al. (2004), who give a detailed step-by-step reconstruction of the recruitment of first *tra* and then *Sxl* to the sex-determining pathway of *D. melanogaster*.

The original motivation of the "bottom-up" hypothesis has been the idea that an ongoing evolutionary conflict between feminizing and masculinizing genetic factors may have driven the sequential invasion of novel master-sex determining switches, each one reversing the action of the previous master sex-determining gene (Wilkins 1995). The result would be a cascade of inhibitory interactions, as observed in the *C. elegans* pathway. Although the existence of a tug-of-war between the sexes is plausible, it is worth noting that a cascade of inhibitory interactions could also be explained simply by the fact that the invasion of a novel sex-determining gene is more likely to be successful if it is epistatically dominant over the ancestral one (van Doorn and Kirkpatrick 2010). Epistatically dominant modifiers can either replace the ancestral upstream regulator or lead to its fixation (Bull and Charnov 1977), causing the length of the pathway to remain equal or to grow by one step, respectively. Fixation of the ancestral gene occurs only when the novel and the ancestral gene have opposite effects, such as when a feminizing allele invades in a population with a masculinizing Y chromosome (XX♀/XY♂ heterogamety). So, any mechanism favoring the repeated invasion of epistatically dominant upstream regulators would result in a cascade of inhibitory interactions, irrespective of which selective forces are operating or the order in which feminizing and masculinizing alleles appear. Data on evolutionary patterns must therefore be complemented with mechanistic models to clarify how sex-determination pathways evolve.

NEUTRAL EVOLUTION

The first theoretical studies attempting to explain evolutionary transitions between different modes of sex determination were published by Bull and Charnov (1977) and Bull (1981), well before the molecular structure of sex-determination pathways was elucidated. As a consequence, these classical studies, as well as many later theoretical models building on them, are not explicit about the developmental mechanisms involved in sex determination, nor are they framed in terms of a specific hypothesis for the mode of pathway evolution (Pomiankowski et al. 2004; Uller and Helanterä 2011). Nevertheless, interpreted in the light of Wilkins' bottom-up model, the early population-genetic models help clarify what different evolutionary forces act on upstream sex-determining genes when they first appear in a population.

A crucial insight offered by the early theoretical studies is that evolutionary transitions between sex-determining mechanisms can occur via selectively neutral intermediate states in which multiple polymorphic sex-determination loci determine sex (Scudo 1967). For example, one of the scenarios analyzed by Bull and Charnov (1977) concerns a population with XX♀/XY♂ GSD (male heterogamety) as the ancestral state, in which some individuals carry an epistatically dominant feminizing mutation *W* on one of the autosome pairs. When this mutant allele spreads in the population, it can establish a new ZZ♂/ZW♀ GSD system. Bull and Charnov (1977) showed that the new state of female heterogamety could be reached gradually through a connected set of intermediate states with multifactorial sex determination. They also showed that there is no tendency for selection to favor the ancestral sex-determining system, the novel one, or any one of the intermediate states once the sex ratio has equilibrated, unless some sex-determination genotypes are assumed to be intrinsically more fit than others. In the absence of other selective forces, it is thus conceivable that demographic stochasticity or other sources of random variation induce a transition from male to female heterogamety by causing the population to drift through genotype space along a line of equilibria connecting the alternative modes of sex determination (Fig. 1A). A similar con-

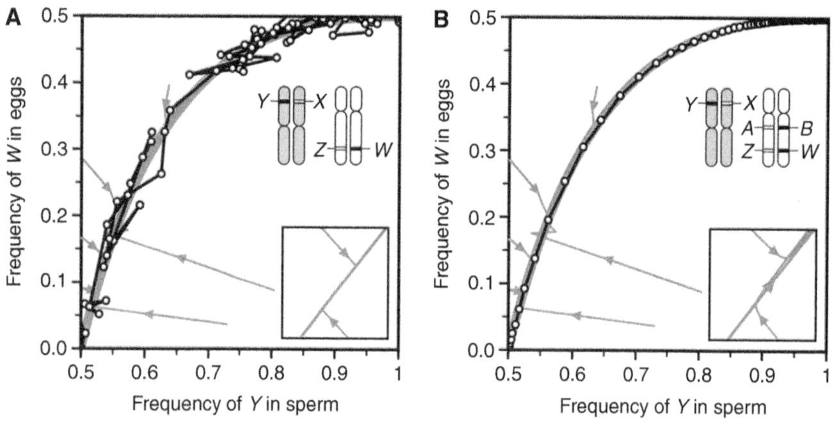

Figure 1. Neutral and adaptive models of a transition from male to female heterogamety. The two panels illustrate the change of the gamete frequencies of the ancestral (horizontal axes) and novel (vertical axes) sex-determining alleles during a transition from XX♀/XY♂ GSD (male heterogamety) to ZZ♂/ZW♀ GSD (female heterogamety). For each panel, genetic assumptions are illustrated by means of schematic representations of the ancestral sex chromosome pair (gray) with the sex-determining locus, and a pair of autosomes (white) carrying a sex-determination mutation (*A*,*B*) and a sexually antagonistic locus (*B* only). (*A*) Invasion of an autosomal, epistatically dominant feminizing mutation *W*. In the absence of intrinsic fitness differences between sex-determination genotypes, the allele *W* can drift to fixation as the population moves stochastically along a line of equilibria (thick gray line) (Bull and Charnov 1977), away from the ancestral state in the lower left corner of the diagram. By the time variation at the ancestral sex-determining locus is lost and the allele *W* reaches fixation (in the upper right corner of genotype space), the former X chromosome has disappeared from the population. Populations initialized with arbitrary combinations of allele frequencies quickly evolve toward the line of equilibria (thin gray trajectories), under the influence of selection for a balanced sex ratio. The process of drift along the line of equilibria is illustrated by the results from a stochastic, individual-based implementation of the population-genetic model (black trajectory with open circles; gamete frequencies are plotted every 50 generations for a population of 1000 individuals). (*B*) A sexually antagonistic gene located on the same autosome pair as the feminizing mutation causes the *W* allele to spread as a result of indirect selection, supported by the development of linkage disequilibrium between *W* and the female beneficial allele *B* (van Doorn and Kirkpatrick 2010). A simulation of a large population of $1 \cdot 10^6$ individuals illustrates the slow deterministic movement of the allele frequencies along a nearly neutral path close to the former line of equilibria (thick grey line). Small insets in *A* and *B* present a close-up view of evolutionary trajectories in the vicinity of the line of equilibria, confirming that movement along the line is no longer neutral in *B*.

Cite this article as *Cold Spring Harb Perspect Biol* doi: 10.1101/cshperspect.a017681

tinuous path of neutral equilibria between different modes of sex determination has been observed in population genetic analyzes of multifactorial GSD (Scudo 1967), one-locus multiallele sex determination (Karlin and Lessard 1984), and in models of transitions between genetic and environmental sex determination (Bull 1981).

The existence of a line of equilibria in models without intrinsic fitness differences implies that generic sex-determining mechanisms, like male and female heterogamety or ESD, do not necessarily represent different peaks in the adaptive landscape. So, populations in transition from one mechanism to another do not necessarily have to pass through a fitness valley (cf. Gavrilets 1997; Valenzuela 2008). Moreover, only small fitness differences between genotypes suffice to induce a consistent movement of the population along the line of equilibria (Fig. 1B), toward one of the alternative single-factor sex-determining mechanisms (Bull and Charnov 1977; Bull 1981). Sex-determination transitions may therefore be driven by subtle sources of selection, such as those acting indirectly via linkage disequilibria with other loci (van Doorn and Kirkpatrick 2007), or by a bias generated by the interaction of sex-ratio selection and demographic stochasticity around the line of equilibria (Vuilleumier et al. 2007).

Another implication is that neutral transitions serve mainly as a null model for the evolutionary turnover of sex-determination systems. In fact, any source of additional selection, however weak, acting on the sex-determining genes transforms the line of equilibria into a nearly neutral path along which populations will evolve deterministically in one direction or the other (Fig. 1B). This may be one reason why multifactorial sex determination is not often found in nature (Bull 1983) (even though its actual frequency is hard to estimate, given that the presence of a multifactorial system is only revealed by elaborate crossing experiments). The best-studied current examples among species with genetic sex determination are the housefly *Musca domestica* (Kozielska et al. 2006) and the platyfish *Xiphophorus maculatus* (Orzack et al. 1980). In both cases, rather than representing a temporary transitional state, multifactorial sex determination appears to be maintained as a consequence of specific relative fitness differences between genotypes (Orzack et al. 1980; Feldmeyer et al. 2008). Therefore, both the rarity of multifactorial sex determination and its apparent maintenance by selection in species with unusual sex determination provide an argument for exploring the role of adaptive mechanisms in sex determination transitions.

SOURCES OF DIRECT SELECTION ON NOVEL SEX-DETERMINING GENES

Both sex-ratio selection and pleiotropic fitness benefits can drive the spread of a novel sex-determining mutation by means of direct selection. This type of selection results from fitness differences between alternative alleles at the sex-determining locus that exist independently of alleles carried at other loci. In contrast, indirect selection mechanisms (which are reviewed after this section) rely on a nonrandom statistical association (linkage disequilibrium) between sex-determination alleles and loci experiencing direct selection.

Sex-Ratio Selection

Because of the immediate connection between sex determination and the sex ratio, any deviation of the sex ratio from its optimum generates selection on sex-determination genes. A sex-determination transition is a likely response to sex-ratio selection if there is genetic variation for an alternative sex-determining mechanism capable of producing a sex ratio closer to the optimum. Bulmer and Bull (1982) were the first to point out that this mechanism could lead to transitions from ESD to GSD. In particular, the accumulation of genetic variation at loci affecting the reaction norm for ESD would help to dampen maladaptive fluctuations in the population sex ratio induced by spatial or temporal variation in environmental conditions. If the cost of producing a suboptimal sex ratio outweighs the fitness benefit of using ESD (i.e., being able to adaptively match offspring sex to the state of the environment) (Charnov and Bull 1977), pure ESD, using a switch-like threshold reaction norm, is no longer evolutionarily stable. Selection then favors the establishment of genetic polymorphism at loci that control the location of the threshold of the ESD reaction norm. Eventually, a gene with major effect on sex determination may invade, marking the transition to GSD (Bulmer and Bull 1982; van Dooren and Leimar 2003).

Although species with ESD are more prone to exhibit suboptimal sex ratio fluctuations, they might enjoy a fitness advantage in situations in which selection favors a biased sex ratio, for example, because of local mate competition or different energetic costs to the parents for the production of sons versus daughters (Charnov 1982). According to this argument, ESD is more flexible and permits faster evolution toward an optimal biased sex ratio than common GSD mechanisms, which are constrained by the segregation ratio in the heterogametic sex (Bull 1983; Charnov and Bull 1989; Freedberg and Wade 2001). Empirical studies in species with ESD offer no support for the hypothesis that selection for a biased sex ratio has been a key driving factor in the origin or maintenance of ESD (Shine 1999; Janzen and Phillips 2006). Several lines of evidence do suggest, however, that male or female heterogamety in species with GSD cannot easily be modified into a flexible mechanism capable of producing a biased sex ratio (Uller et al. 2007). For example, Kozielska et al. (2006) calculated the response of the multifactorial sex-determining mechanism of the housefly *M. domestica* to sex-ratio selection and found that the sex-ratio bias produced by the system was much smaller than predicted by sex-allocation theory owing to population-genetic constraints. Similarly, Werren et al. (2002) showed that even a flexible GSD mechanism could evolve to lose its capacity to produce a biased sex ratio, as a result of an arms race between maternally and zygotically expressed genes induced by selection for a skewed sex ratio. Consistent with these findings, there is overall little evidence of modifications to the sex-determining mechanism in response to sex-ratio selection among diplo-diploid species with sex chromosomes (Krackow 2002). The same species also show almost no heritable genetic variation for sex ratio (Bulmer and Bull 1982; Bull and Charnov 1988). Nevertheless, it is unlikely that the lack of adaptive control over sex determination poses more than a mild evolutionary constraint for species with GSD, given that several other mechanisms to accommodate a biased sex ratio (e.g., sperm selection or selective resorption of embryos) have evolved (Krackow 2002).

When selection favors a balanced sex ratio and the sex-determining mechanism is capable of producing equal number of males and females, population sex ratios may still deviate markedly from 1:1, owing to the manipulation of sex determination by selfish genetic elements. The resulting distortion of the sex ratio generates sustained sex-ratio selection that may evoke counteradaptations of the sex-determining system (Werren and Beukeboom 1998). For example, in the Spanish mole, *Talpa occidentalis* (McVean and Hurst 1996), the creeping vole, *Microtus oregoni* (Charlesworth and Dempsey 2001), and several other rodent species, genetic conflict between sex-linked segregation distorters and their modifiers seems to have driven the evolution of unusual sex chromosome systems. Genetic conflicts can also drive transitions between male and female heterogamety, as shown by various models. For example, under sex-chromosome meiotic drive (Jaenike 2001), sex-ratio selection favors the invasion of a novel sex-determining allele that produces individuals of the underrepresented sex (Kozielska et al. 2010). As the new sex-determining locus spreads, polymorphism at the ancestral locus is lost, eventually leading to the demise of the sex-ratio distorter. Similarly, the repression of cytoplasmic sex-ratio distorters by host masculinizing alleles can induce a switch from female heterogamety to a multifactorial system dominated by male heterogamety (Caubet et al. 2000).

Pleiotropic Benefits

If a novel sex-determining mutation confers a higher viability or other intrinsic fitness advantage to its carrier, the mutation can spread as the result of simple natural selection. Bull and Charnov (1977) observed that models incorporating such direct selection behaved similar to their equal fitness counterparts in that the genotype frequencies initially approached a region near the line of equilibria of the corresponding equal-fitness model. In that region of genotype space, the invading sex-determining mutant is unaffected by sex-ratio selection, but still experiences positive natural selection. The frequency of the mutant therefore increases, whereas the genotype frequencies continue to follow approximately the neutral line of equilibria, until genetic polymorphism at the ancestral sex-determining locus is lost.

A serious problem for the pleiotropic-benefit hypothesis of sex-determination transitions, already noted by Bull and Charnov (1977), is that there are few obvious biological reasons why a novel sex-determining allele would have an intrinsic fitness advantage over the ancestral one. One realistic possibility, that is nearly indistinguishable from pleiotropy in practice, is that the invading sex determiner is in strong positive linkage disequilibrium with an allele that is favored by selection (Bull 1983). For example, some sex-determination variants in the housefly appear to be linked to pesticide-resistance alleles (Werren and Beukeboom 1998). As the beneficial allele sweeps to fixation, it drags along the genetically associated sex-determining allele, giving it the appearance of being favored by selection itself (Barton 2000). However, such "genetic hitchhiking" proceeds for as long as there is genetic variation at the locus under selection and recombination has not broken down the linkage equilibrium. Accordingly, hitchhiking can only facilitate a sex-determination transition if the sex determiner is tightly physically linked to a polymorphic locus under strong positive selection. This is quite unlikely, except perhaps when both the sex-determining mutation and the novel beneficial allele arise on a chromosomal inversion (Kirkpatrick 2010).

The alternative possibility that the novel sex determiner itself is favored by direct selection has sometimes been considered in theoretical studies, but is then usually discarded on the basis of parsimony, because it requires a fortuitous type of pleiotropy (e.g., Rice 1986). A notable exception is provided by Kraak and De Looze (1993), who proposed a verbal model for the evolution of sex determination in vertebrates, in which a gene under selection gains control over sex determination. Their hypothesis builds on the observation that the growth rate of the undifferentiated gonads in the developing embryo acts as a proximate cue for sex determination in several species with ESD. Therefore, a gene causing the undifferentiated gonads to develop more rapidly would be able to substitute for the role of the environment and act as a trigger for sex determination. Individual-based computer simulations show that the proposed transition from ESD to GSD is feasible if genetic variation for growth rate is concentrated at a single locus (Kraak et al. 2000). However, when genetic variation can accumulate at multiple loci, growth-rate accelerating alleles become genetically associated with each other only when size benefits differ maximally between males and females and several additional restrictive conditions are met (Kraak and Pen 2002).

Without empirical evidence for major-effect genes that modify sex determination and that also differentially affect the fitness of males and females by their impact on embryonic growth rate (or any other fitness-relevant process involved in sex determination), the role for pleiotropic benefits in triggering sex-determination transitions appears rather limited. Still, this conclusion does not rule out the possibility that pleiotropic fitness effects are important to stabilize an established sex-determination system. For example, the *Sxl* gene in *Drosophila* regulates both *tra*, its downstream gene in the sex-determination cascade, and *msl-2*, a gene involved in dosage compensation. Because of the latter interaction, which became critical after the establishment of *Sxl* and the differentiation of its sex chromosome (Pomiankowski et al. 2004), replacement of *Sxl* by another sex-determining switch is likely to have severe detrimental effects.

SOURCES OF INDIRECT SELECTION ON NOVEL SEX-DETERMINING GENES

Novel sex-determination factors can enjoy an indirect fitness advantage because of their association with sexually antagonistic genes or sex-linked deleterious mutations. In general, such indirect selection is much weaker than direct selection, because the linkage disequilibria between loci are often only modest (Kirkpatrick et al. 2002). However, many sex-determination genes occur in genomic regions of reduced recombination, which facilitate the maintenance of linkage disequilibrium. As a result, indirect selection is an important factor in the evolution of GSD systems.

Intralocus Sexual Conflict

Many traits that are expressed in the context of sexual selection and reproduction have sex-specific optima, owing to the functional divergence of male and female gender roles (Parker 1979). Because males and females share a common gene pool, the genetic conflict between the sexes over the expression of these traits (intralocus sexual conflict) cannot always be resolved by sex-specific gene regulation or other means of evolving sexual dimorphism (van Doorn 2009; Stewart et al. 2010; Pennell and Morrow 2013). As a result, sexually antagonistic variation in fitness is maintained in populations (Brommer et al. 2007; Foerster et al. 2007), with some individuals carrying alleles that are beneficial in males but deleterious in females, and others carrying alleles with the opposite fitness effects.

Intralocus sexual conflict has major consequences for the evolution of the sex chromosomes. In particular, selection favors the accumulation of sexually antagonistic genes on the sex chromosomes (Charlesworth and Charlesworth 1980; Rice 1984), as well as the evolution of reduced recombination between the sex-determining gene and sexually antagonistic loci in its vicinity (Bull 1983; Rice 1987; Otto et al. 2011). Both processes inhibit the breakdown of adaptive linkage disequilibria, which partially resolve the sexual conflict. These linkage disequilibria arise because alleles at a sex-determination locus automatically become statistically associated with sexually antagonistic alleles segregating at loci nearby on the same chromosome, in such a way that the transmission of each sexually antagonistic allele is biased toward the sex in which it is beneficial. For example, a young (recombining) sex chromosome carrying a dominant feminizing allele (e.g., a proto-W chromosome) is predicted to become enriched for female-beneficial alleles, particularly in the region surrounding the sex-determination locus, which is transmitted almost exclusively from mother to daughter (Clark 1988; Jordan and Charlesworth 2011). The same region on the homologous proto-Z chromosome spends up to two-thirds of its time in males, and is therefore expected to become enriched for male-beneficial alleles.

The genetic association between sex-determination genes and sex-linked sexually antagonistic loci enables evolution to implement a conditional sex-determination strategy that adaptively matches the sex of a zygote to the sexually antagonistic alleles it carries. Hence, in the presence of sexually antagonistic variation, GSD enjoys a clear selective advantage over alternative sex-determination mechanisms that assign sex randomly with respect to genetic background. To see this, consider once more the example of a dominant feminizing allele that is linked to a sexually antagonistic locus. If the recombination between the loci is low, selection can maintain strong linkage disequilibrium between them. Zygotes carrying a female-beneficial allele are then highly likely to carry the feminizing allele as well, in which case they will develop as females. Conversely, most of the zygotes carrying two male-beneficial alleles will develop as males, because they are unlikely to have inherited the feminizing allele. GSD thus results in offspring with, on average, a higher fitness than when the sex antagonistic alleles would be transmitted randomly to male and female offspring.

In line with this argument, Rice (1986) showed that a sexually antagonistic gene, tightly linked to a dominant masculinizing mutation, facilitated the spread of the mutant sex-determining allele in a population with polygenic sex determination as ancestral state. As expected, the transition from polygenic to single-locus GSD was critically supported by the genetic association between the masculinizing mutation and the sexually antagonistic gene. Individuals that inherited the mutation were more likely to have inherited the male-beneficial allele as well, which gave them a selective advantage over other types of males. This indirect benefit allowed the major sex-determining gene to spread, eventually resulting in the loss of polygenic sex determination.

The same mechanism can cause transitions between different single-locus genetic sex-determining systems, resulting in the establishment of a new sex-determining locus, the evolution of a new sex chromosome pair, or a transition between male and female heterogamety. The evolutionary dynamic of such GSD transitions is complicated by the fact that also the ancestral sex-determining locus might be linked to sexually antagonistic genes. van Doorn and Kirkpatrick (2007) therefore analyzed the relatively simple scenario of a novel masculinizing mutation arising on an autosome, in a population with male heterogamety. In their model, both sex-determining alleles were assumed to be linked to a sexually antagonistic gene. The autosomal sexually antagonistic gene favored the invasion of the masculinizing mutation, in the same way as discussed by Rice (1986). However, sexually antagonistic variation segregating on the ancestral sex chromosomes inhibited the spread of the mutant, because males carrying the novel masculinizing allele also carried two X chromosomes enriched for female-beneficial alleles. Mutant males were therefore more likely to inherit the autosomal male-beneficial allele, but also more likely to inherit the female-beneficial allele on the ancestral sex chromosomes than males carrying an ancestral Y chromosome.

Depending on the net fitness effect of these two indirect selection forces, van Doorn and Kirkpatrick (2007) identified two generic evolutionary outcomes. Either the masculinizing mutation was lost or it spread and replaced the ancestral sex-determining gene, leading to the loss of the ancestral Y chromosome and the establishment of a new sex chromosome pair. In some cases, selection could maintain a stable sex-chromosome polymorphism, causing a transition from single-locus GSD to a multifactorial sex determining system. However, this outcome required rather specific combinations of sexually antagonistic selection coefficients, as well as tight linkage. The same conclusions were later shown to apply to more complicated transitions between genetic sex-determining mechanisms (van Doorn and Kirkpatrick 2010), including the invasion of a novel sex-determining allele at the ancestral sex-determining locus, the establishment of a novel sex-determining gene located on the ancestral sex chromosomes, and heterogamety transitions involving an arbitrary number of sexually antagonistic loci or recessive deleterious alleles on the ancestral Y chromosome (Fig. 2). In all cases, fixation of the mutant gene occurred when the sexually antagonistic loci associated with the sex-determination modifier harbored more genetic variation, when their alleles had stronger sexually antagonistic fitness effects, or when they were more tightly linked to their nearby sex-determining gene than the sexually antagonistic genes on the ancestral sex chromosome.

One way to interpret these results is that evolution favors whichever GSD mechanism responds most effectively to predictors of sex-specific fitness at the time of sex determination (Leimar 2005). In the case of sex-chromosome turnovers, these predictors are represented by (groups of) sexually antagonistic loci segregating at different linkage groups. However, in a broader context, they may also include environmental factors (Charnov and Bull 1977) or variation in parental condition with sex-differential fitness effects (Trivers and Willard 1973; Langkilde and Shine 2005). In all of these cases, similar principles underlie the evolution of the sex-determining mechanism (van Doorn 2014). When individuals differ in their genetic background or experience different environmental conditions during development, and not all components of such variation affect males and females equally (e.g., Chippindale et al. 2001; Warner and Shine 2008), selection favors a conditional sex-determination strategy. More-

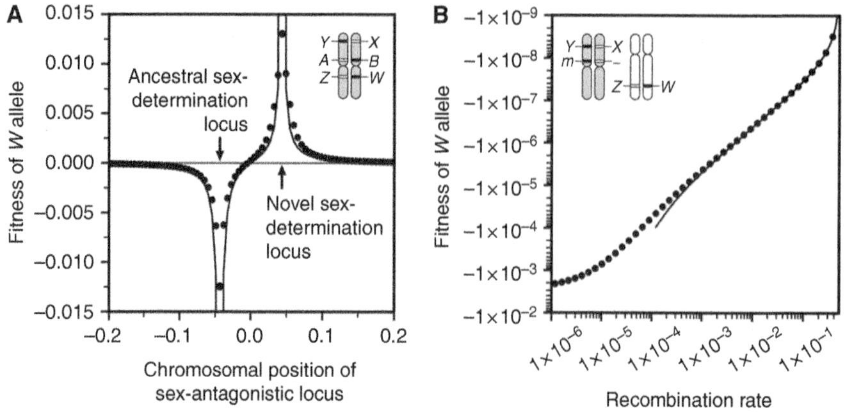

Figure 2. Contributions of linkage and sex-differential selection to sex-determination transitions. If selection is weak relative to the force of recombination, the effect of a sexually antagonistic locus on the fitness of a sex-determining mutation (measured as its exponential rate of spread, λ) is quantified by the expression

$$\lambda = \frac{1}{2}(G_{\text{intra}} - G_{\text{inter}})\left(\frac{1-2r'}{2r'} - \frac{1-2r}{2r}\right), \tag{1}$$

irrespective of many details of the ancestral sex-determining mechanism (van Doorn and Kirkpatrick 2007, 2010). Here, G_{intra} is the intrasexual additive genetic variance for fitness at the sexually antagonistic locus and G_{inter} is the corresponding genetic covariance between fitness in males and females. The parameters r and r' denote the rates of recombination between the sexually antagonistic gene and the ancestral/novel sex-determining locus, respectively. (*A*) Equation 1 diverges when one of the recombination rates approaches zero, indicating that the direction of sex-determination transitions is strongly affected by patterns of linkage. This is illustrated for the fitness effect of a sexually antagonistic gene on an epistatically dominant feminizing mutation *W* that has originated on an ancestral XY sex chromosome pair. The fitness of the *W*-allele varies depending on the chromosomal position of the sexually antagonistic locus (horizontal axis). The strength of indirect selection is highest when the sexually antagonistic gene is tightly linked to either the novel or the ancestral sex-determining locus (the rate of recombination between the sex-determination genes was kept fixed at 0.08). The prediction based on Equation 1 (solid line) agrees with the results of simulations based on the exact population-genetic recursions (closed circles), except when linkage is tight. Various refinements of Equation 1 can be developed to resolve this problem (van Doorn and Kirkpatrick 2010), but these depend on details of the GSD system. (*B*) Equation 1 also indicates that transitions in the sex-determination system can occur under conditions that are more general than those typically considered in discussions of sexually antagonistic selection. It suffices for the relative fitness effects of the selected locus to be different in males and females; it is not necessary that the allele favored in males is detrimental to females (or vice versa). The latter condition is stricter, as it implies that the intersexual covariance in fitness G_{inter} has to be negative, whereas indirect selection on the sex-determination locus is already generated if the intersexual covariance is positive but smaller than G_{intra}. Therefore, also transiently polymorphic loci and deleterious alleles maintained by migration or mutation-selection balance contribute to the indirect selection force on sex-determination loci if they have different (but not necessarily antagonistic) fitness effects in males and females. This effect is illustrated for a locus on the ancestral sex chromosomes that carries recessive deleterious alleles (m). The deleterious alleles are assumed to be expressed only in females, such that the mutation initially can accumulate freely on the Y chromosome. The invasion of an autosomal, epistatically dominant feminizing allele *W* gives rise to a subpopulation of homozygous YY females in which the deleterious alleles are expressed, reducing the fitness of the *W* allele. The figure shows the magnitude of this fitness reduction as a function of the rate of recombination between the locus with deleterious alleles and the ancestral sex-determination locus. The prediction based on Equation 1 (solid line) matches again with the result of simulations (closed circles) except for very low recombination rates.

over, the optimal conditional strategy adaptively switches development into the male or female trajectory contingent on whichever genetic or environmental cue proves to be the most reliable predictor of sex-specific fitness (Charnov and Bull 1977; Rice 1986; van Dooren and Leimar 2003; Pen et al. 2010).

Simulations of sex-chromosome transitions with a large number of sexually antagonistic loci suggest that the few sexually antagonistic genes that are most tightly linked to a sex-determining gene have a decisive influence on the direction of sex-determination transitions. This would imply that the sexually antagonistic variation effectively responsible for sex-determination transitions segregates at a small subset of the sexually antagonistic loci. Therefore, even a small inversion that captures both a sex-determining mutation and a single fitness-relevant sexually antagonistic locus on an autosome, could conceivably hijack sex-determination from the ancestral sex chromosomes. Consistent with this suggestion, the sex-determining region on the recently evolved Y chromosome in the three-spined stickleback has experienced multiple inversions (Ross and Peichel 2008; Kirkpatrick 2010), and these chromosomal rearrangements appear to have occurred differently between the Japan Sea lineage (which has a novel sex-chromosome system) (Kitano et al. 2009) and its Pacific Ocean ancestor (Natri et al. 2013). However, the molecular data allow for multiple interpretations (Natri et al. 2013), and it remains unclear whether inversions have triggered the sex-chromosome shifts that have occurred in sticklebacks, or occurred after the sex-determining locus was established (as suggested by the traditional theory for sex-chromosome evolution) (Rice 1987; Charlesworth et al. 2005). This can, in principle, be tested by analyzing the patterns of neutral variation on the inversion and the neo-sex chromosome (Kirkpatrick et al. 2010; Guerrero et al. 2012a). Further empirical support for a role of sexual antagonism in sex-determination transitions comes from a growing number of cases of recently derived sex chromosomes that carry sexually selected loci (Kallman 1970; Wada et al. 1998; Lindholm and Breden 2002; Fernandez and Morris 2008; Kitano et al. 2009). A particularly interesting example is provided by a mutation in the *Pax7* gene in lake Malawi cichlid fish, which causes an orange-blotched phenotype subject to sexually antagonistic selection. This mutation originated once in lake Malawi cichlids, and appears to have spread in close association with a feminizing allele that is epistatically dominant over the ancestral XX♀/XY♂ sex-determining system (Roberts et al. 2009).

Deleterious Mutations Accumulating on the Heterogametic Sex Chromosome

The theory reviewed in this section so far considers evolutionarily young sex chromosomes that have not yet stopped recombining. Older, heteromorphic sex chromosomes typically contain large regions in which recombination is suppressed (Rice 1987; Otto et al. 2011). The nonrecombining portion of the heterogametic sex chromosome tends to accumulate deleterious mutations, which can eventually lead to a complete degeneration of the chromosome (Charlesworth et al. 2005). Similar to sexually antagonistic selection, sex-chromosome degeneration does not affect both sexes equally, and it therefore generates qualitatively similar indirect selection pressures on the sex-determination system (Fig. 2B). A potential evolutionary response to sex-chromosome degeneration is that the old sex chromosome is replaced by a new one after a transposition of the sex-determining gene to an autosome pair, or the origin of a new autosomal sex-determining gene. Blaser et al. (2013) recently showed that such transitions are favored when the deleterious mutation load outweighs the disadvantage of losing beneficial sexually antagonistic alleles fixed on the decaying chromosome. This cost/benefit argument ignores the disruptive effect the transition might have on the expression of genes on the ancestral sex chromosomes, which would occur if specific sex-linked gene regulation (e.g., dosage compensation) had evolved. As these effects might often be substantial, sex-chromosome transitions involving heteromorphic sex chromosomes are expected to be rare.

Besides, if such transitions occur, they are predicted to conserve the ancestral pattern of heterogamety (consistent with the pattern observed in some clades of vertebrates) (Woram et al. 2003; Miura 2007), because turnovers leading to the fixation of a degenerated sex chromosome are particularly likely to be associated with severe fitness costs.

The accumulation of sexually antagonistic variation on the sex chromosomes, the subsequent arrest of recombination and the resulting accumulation of deleterious mutations on the heterogametic sex chromosome can combine, over extended periods of time, to generate repeated sex-chromosome turnovers (Blaser et al. 2014). Several studies indicate that some autosomes are more likely than others to be co-opted as sex chromosomes in this process (Miura 2007; Graves and Peichel 2010; O'Meally et al. 2012; Brelsford et al. 2013), suggesting that the involvement of a chromosome pair in sex determination leaves a long-lasting signature. In particular, former sex chromosomes could be enriched for modifiers of recombination and potential targets of mutations with sexually antagonistic effects or effects on sex determination (Blaser et al. 2014; Beukeboom and Perrin 2014), all of which would increase the probability of recapturing the role of sex chromosome in a turnover event. A possible example of a chromosome that carries signatures of a previous role in sex determination is the dot chromosome of *Drosophila*, which has lost its ancestral function as an X chromosome after a sex-chromosome turnover in early drosophilid evolution, but which still has a minor feminizing role in sex determination and is subject to a specific mechanism of gene regulation similar to that of sex-linked genes (Vicoso and Bachtrog 2013).

GENETIC AND DEVELOPMENTAL ARCHITECTURE

As the structure of sex-determining pathways is being elucidated in a growing number of species, it becomes increasingly clear that the genetic and developmental architecture of a species shapes the selection pressures acting on its sex-determining genes. Understanding the interplay between adaptation and developmental mechanisms is a key challenge for future theoretical and empirical work on sex determination.

One obvious question to ask in this context is why sex determination appears to be labile in some taxa, but stable in others. For example, the evolutionary stability of sex determination in birds and mammals contrasts sharply with the high rate of transitions observed in some clades of cold-blooded vertebrates (Marín and Baker 1998; Kraak and Pen 2002; Ezaz et al. 2006). Part of this difference arises because the evolution of heteromorphic sex chromosomes constrains future sex-chromosome transitions, owing to the accumulation of sexually antagonistic genes on the sex chromosomes, the reduction of recombination rates in the vicinity of the master sex-determining gene, the loss of functional genes from the Y chromosome, the evolution of dosage compensation, and the translocation of genes essential for male fitness to the Y chromosome (Bull 1983; Rice 1987; Charlesworth et al. 2005). Once the sex chromosomes have become heteromorphic, XX males, XY females, or YY individuals of either sex, which arise when sex determination is hijacked by another chromosome pair, are likely to suffer from reduced fertility and/or viability, inhibiting the spread of new sex-determining genes (Bull and Charnov 1985; van Doorn and Kirkpatrick 2007; Mawaribuchi et al. 2012).

Sex-chromosome heteromorphism can thus explain why sex determination appears to be locked into its current state in birds and mammals. Yet, so far, the argument leaves unaddressed why sex determination should continue to be labile in taxa lacking heteromorphic sex chromosomes. The lower vertebrates, which generally have undifferentiated sex chromosomes, show evidence of frequent sex-chromosome transitions. For example, a phylogenetic analysis of genetic sex determination in teleost fishes found that eight of 26 families include both species with XY and species with ZW sex determination (Mank et al. 2006), and within the subfamily Salmoninae (including char, trout, and salmon), at least four different chromosomes determine sex in different species

(Woram et al. 2003). A further example of frequent turnovers between male and female heterogamety is provided by a survey in amphibians, which identified eight transitions in a phylogenetically diverse sample of 63 species (Hillis and Green 1990). However, the same data also show that many species of fish and amphibians have not undergone a change in sex determination since more than several mya, and yet, almost none of these species show evidence of sex-chromosome differentiation (Stöck et al. 2011).

A plausible hypothesis for the maintenance of homomorphic sex chromosomes in lower vertebrates is that the sex chromosomes of cold-blooded vertebrates are prevented from degenerating because of rare recombination events prompted by a low rate of sex reversal (Perrin 2002). Because of the general effect of temperature on the development of ectotherms, such sex reversal events occur once in a while under natural conditions. Consistent with this idea, a recent analysis of neutral genetic variation on the sex chromosomes provides statistical support for a low rate of X-Y recombination in European tree frogs (*Hyla* spp.) (Guerrero et al. 2012b), a group of species with cryptic sex chromosomes and stable sex determination (Stöck et al. 2011). Furthermore, a theoretical study of mutation accumulation on the sex chromosomes shows that occasional sex reversal events provide sufficient opportunity for recombination to purge the Y chromosome from its deleterious mutation load (Grossen et al. 2012). The evolution of sex chromosomes (or the lack thereof in species with ESD) has several other consequences for the genetic architecture that can interact with the evolution of sex determination. Some of these are discussed by Ewert and Nelson (1991) (inbreeding), Reeve and Pfennig (2003) (sexual selection), and Kitano and Peichel (2012) (speciation).

A central challenge for future work is to integrate our understanding of the population genetics of sex-determination transitions with molecular data (Pomiankowski et al. 2004). Models that incorporate the available knowledge of sex-determination pathways are essential to complement the black-box approach of population-genetic analysis, to validate its implicit assumptions and to generate testable predictions. For example, combining evolutionary and systems biology modeling approaches can help clarify why it is so rare to observe "leaky" sex determination. Is it because selection generally disfavors (partially) random sex determination, or is a canalized all-or-nothing switch easier to implement, considering that the developmental decision triggered by the initial sex-determination signal must be stabilized and retained during development (Valenzuela 2008)? This issue is relevant for some sex-ratio selection models of sex-determination transitions, as these assume that developmental constraints prevent optimization of the sex ratio in species with GSD. A second possibility is to look into how sex determination interacts with downstream developmental processes, and how these interactions generate potential pleiotropic fitness effects. The results of such an analysis may validate the common assumption that sex determination can be studied in isolation from other developmental processes (which means that sex-determination modifiers can safely be assumed to have no other fitness effects). Alternatively, they could support the pleiotropic benefits hypothesis for the evolution of sex-determining genes.

Finally, realistic models of evolving sex-determining pathways are needed to clarify how selection and molecular mechanisms interact to generate the variation of evolutionary rates between genes at different positions in the sex-determining cascade (MacCarthy et al. 2010). Despite that population genetic models tend to ignore the complexity of development, they do offer suggestions for why the flexibility of sex determination may predominantly rely on the recruitment of novel upstream regulators to the pathway. In particular, it seems reasonable to suppose that a novel upstream regulator that interacts directly with the original master switch (for example, by mimicking or blocking the primary sex-determination signal) can reverse the outcome of the original switch for a certain set of environmental or genetic conditions and, in this way, act as an epistatically dominant modifier of sex determination. In contrast, regulators interacting with downstream sex-determination genes would perhaps interfere with how the sex-determination signal is processed, which would be likely to have negative pleiotropic effects or result in a breakdown of canalization (Uller and Helanterä 2011). Selection against leaky sex determination may therefore be one of the factors preventing the spread of downstream sex-determination modifiers. A thorough test of this hypothesis requires investigating how genes in the sex-determination pathway coevolve. There is growing evidence that such coevolution is occurring (Haag and Ackerman 2005; Stothard and Pilgrim 2006), and that it is associated with rapid evolution of sex-determining genes, often without sex-determination transitions or structural pathway rearrangements (Ferris et al. 1997; O'Neill and O'Neill 1999; Chandler et al. 2012). The evolutionary flexibility of sex determination may therefore be even higher than is currently appreciated.

ACKNOWLEDGMENTS

The author thanks Michael Wade, Sergey Gavrilets, and an anonymous referee for their comments on this manuscript, and gratefully acknowledges financial support from the European Research Council (ERC Starting grant 30955) and the Netherlands Organization for Scientific Research (NWO VIDI grant 864.11.012).

REFERENCES

Ainsworth C. 2000. Boys and girls come out to play: The molecular biology of dioecious plants. *Ann Bot* **86:** 211–221.

Badyaev AV. 2002. Growing apart: An ontogenetic perspective on the evolution of sexual size dimorphism. *Trends Ecol Evol* **17:** 369–378.

Barnes RSK, Calow P, Olive PJW, Golding DW, Spicer JI. 2001. *The Invertebrates: A Synthesis.* Blackwell Science, Oxford.

Barton NH. 2000. Genetic hitchhiking. *Phil Trans R Soc B* **355:** 1553–1562.

Beukeboom LW, Perrin N. 2014. Transitions among sex-determination systems. In *Evolution of sex determination* (ed. Beukeboom LW, Perrin N). Oxford University Press, Oxford.

Blaser O, Grossen C, Neuenschwander S, Perrin N. 2013. Sex-chromosome turnovers induced by deleterious mutation load. *Evolution* **67:** 635–645.

Blaser O, Neuenschwander S, Perrin N. 2014. Sex-chromosome turnovers: The hot-potato model. *Am Nat* **183:** 140–146.

Brelsford A, Stöck M, Betoo-Colliard C, Dubey S, Dufresnes C, Jourdan-Pineau H, Rodrigues N, Savary R, Sermier R, Perrin N. 2013. Homologous sex chromosomes in three deeply divergent anuran species. *Evolution* **67:** 2434–2440.

Brommer JE, Kirkpatrick M, Qvarnström A, Gustafsson L. 2007. The intersexual genetic correlation for lifetime fitness in the wild and its implications for sexual selection. *PLoS ONE* **2:** e744.

Bull JJ. 1981. Evolution of environmental sex determination from genotypic sex determination. *Heredity* **47:** 173–184.

Bull JJ. 1983. *Evolution of sex determining mechanisms.* Benjamin Cummings, Menlo Park, CA.

Bull JJ, Charnov EL. 1977. Changes in the heterogametic mechanism of sex determination. *Heredity* **39:** 1–14.

Bull JJ, Charnov EL. 1985. On irreversible evolution. *Evolution* **39:** 1149–1155.

Bull JJ, Charnov EL. 1988. How fundamental are Fisherian sex ratios? *Oxford Surv Evol Biol* **5:** 98–135.

Bulmer MG, Bull JJ. 1982. Models of polygenic sex determination and sex ratio control. *Evolution* **36:** 13–26.

Caubet Y, Hatcher MJ, Mocquard J-P, Rigaud T. 2000. Genetic conflict and changes in heterogametic mechanisms of sex determination. *J Evol Biol* **13:** 766–777.

Chandler CH, Chadderdon GE, Phillips PC, Dworkin I, Janzen FJ. 2012. Experimental evolution of the *Caenorhabditis elegans* sex determination pathway. *Evolution* **66:** 82–93.

Charlesworth D, Charlesworth B. 1980. Sex differences in fitness and selection for centric fusions between sex-chromosomes and autosomes. *Genet Res Camb* **35:** 205–214.

Charlesworth B, Dempsey ND. 2001. A model of the evolution of the unusual sex chromosome system of *Microtus oregoni. Heredity* **86:** 387–394.

Charlesworth D, Charlesworth B, Marais G. 2005. Steps in the evolution of heteromorphic sex chromosomes. *Heredity* **95:** 118–128.

Charnov EL. 1982. The theory of sex allocation. *Monogr Popul Biol* **18:** 1–355.

Charnov EL, Bull J. 1977. When is sex environmentally determined? *Nature* **266:** 828–830.

Charnov EL, Bull JJ. 1989. The primary sex ratio under environmental sex determination. *J Theor Biol* **139:** 431–436.

Chippindale AK, Gibson JR, Rice WR. 2001. Negative genetic correlation for adult fitness between sexes reveals ontogenetic conflict in *Drosophila. Proc Natl Acad Sci* **98:** 1671–1675.

Clark AG. 1988. The evolution of the Y chromosome with X-Y recombination. *Genetics* **119:** 711–720.

Concha C, Scott MJ. 2009. Sexual development in *Lucilia cuprina* (Diptera, Calliphoridae) is controlled by the transformer gene. *Genetics* **182:** 785–798.

Ellegren H, Parsch J. 2007. The evolution of sex-biased genes and sex-biased gene expression. *Nat Rev Genet* **8:** 689–698.

Ewert MA, Nelson CE. 1991. Sex determination in turtles: Diverse patterns and some possible adaptive values. *Copeia* **1991:** 50–69.

Ezaz T, Stiglec R, Veyrunes F, Marshall Graves JA. 2006. Relationships between vertebrate ZW and XY sex chromosome systems. *Curr Biol* **16:** R736–R743.

Feldmeyer B, Kozielska M, Kuijper B, Weissing FJ, Beukeboom LW, Pen I. 2008. Climatic variation and the geographical distribution of sex-determining mechanisms in the housefly. *Evol Ecol Res* **10:** 797–809.

Fernandez AA, Morris MR. 2008. Mate choice for more melanin as a mechanism to maintain a functional oncogene. *Proc Natl Acad Sci* **105:** 13503–13507.

Ferris PJ, Pavlovic C, Fabry S, Goodenough UW. 1997. Rapid evolution of sex-related genes in *Clamydomonas*. *Proc Natl Acad Sci* **94:** 8634–8639.

Foerster K, Coulson T, Sheldon BC, Pemberton JM, Clutton-Brock TH, Kruuk LEB. 2007. Sexually antagonistic genetic variation for fitness in red deer. *Nature* **447:** 1107–1110.

Freedberg S, Wade MJ. 2001. Cultural inheritance as a mechanism for population sex-ratio bias in reptiles. *Evolution* **55:** 1049–1055.

Gavrilets S. 1997. Evolution and speciation on holey adaptive landscapes. *Trends Ecol Evol* **12:** 307–312.

Gempe T, Beye M. 2011. Function and evolution of sex-determination mechanisms, genes and pathways in insects. *Bioessays* **33:** 52–60.

Geuverink E, Beukeboom LW. 2014. Phylogenetic distribution and evolutionary dynamics of the sex determination genes *doublesex* and *transformer* in insects. *Sex Dev* **8:** 38–49.

Graves JAM, Peichel CL. 2010. Are homologies in vertebrate sex determination due to shared ancestry or to limited options? *Genome Biol* **11:** 205.

Grossen C, Neuenschwander S, Perrin N. 2012. The evolution of XY recombination: Sexually antagonistic selection versus deleterious mutation load. *Evolution* **66:** 3155–3166.

Guerrero RF, Rousset F, Kirkpatrick M. 2012a. Coalescent patterns for chromosomal inversions in divergent populations. *Phil Trans R Soc B* **367:** 430–438.

Guerrero RF, Kirkpatrick M, Perrin N. 2012b. Cryptic recombination in the ever-young sex chromosomes of Hylid frogs. *J Evol Biol* **25:** 1947–1954.

Haag ES, Ackerman AD. 2005. Intraspecific variation in *fem-3* and *tra-2*, two rapidly coevolving nematode sex-determining genes. *Gene* **349:** 35–42.

Haag ES, Doty AV. 2005. Sex determination across evolution: Connecting the dots. *PLoS Biol* **3:** 21–24.

Hasselmann M, Gempe T, Schiott M, Nunes-Silva CG, Otte M, Beye M. 2008. Evidence for the evolutionary nascence of a novel sex determination pathway in honeybees. *Nature* **454:** 519–522.

Hediger M, Henggeler C, Meier N, Perez R, Saccone G, Bopp D. 2010. Molecular characterization of the key switch F provides a basis for understanding the rapid divergence of the sex-determining pathway in the housefly. *Genetics* **184:** 155–170.

Heitman J, Sun S, James TY. 2013. Evolution of fungal sexual reproduction. *Mycologia* **105:** 1–27.

Hillis DM, Green DM. 1990. Evolutionary changes of heterogametic sex in the phylogenetic history of amphibians. *J Evol Biol* **3:** 49–64.

Hodgkin J. 2002. The remarkable ubiquity of DM domain factors as regulators of sexual phenotype: Ancestry or aptitude? *Genes Dev* **16:** 2322–2326.

Jaenike J. 2001. Sex chromosome meiotic drive. *Annu Rev Ecol Syst* **32:** 25–49.

Janousek B, Mrackova M. 2010. Sex chromosomes and sex-determination pathway dynamics in plant and animal models. *Biol J Linn Soc* **100:** 737–752.

Janzen FJ, Phillips PC. 2006. Exploring the evolution of environmental sex determination, especially in reptiles. *J Evol Biol* **19:** 1775–1784.

Jordan CY, Charlesworth D. 2011. The potential for sexually antagonistic polymorphism in different genome regions. *Evolution* **66:** 505–516.

Kalinka AT, Tomancak P. 2012. The evolution of early animal embryos: Conservation or divergence? *Trends Ecol Evol* **27:** 385–393.

Kallman KD. 1970. Different genetic basis of identical pigment patterns in two populations of platyfish, *Xiphophorus maculatus*. *Copeia* **1970:** 472–487.

Karlin S, Lessard S. 1984. On the optimal sex-ratio: A stability analysis based on a characterization for one-locus multiallele viability models. *J Math Biol* **20:** 15–38.

Kirkpatrick M. 2010. How and why chromosome inversions evolve. *PLoS Biol* **8:** e1000501.

Kirkpatrick M, Johnson T, Barton N. 2002. General models of multilocus evolution. *Genetics* **161:** 1727–1750.

Kirkpatrick M, Guerrero RF, Scarpino SV. 2010. Patterns of neutral genetic variation on recombining sex chromosomes. *Genetics* **184:** 1141–1152.

Kitano J, Peichel CL. 2012. Turnover of sex chromosomes and speciation in fishes. *Environ Biol Fish* **94:** 549–558.

Kitano J, Ross JA, Mori S, Kume M, Jones FC, Chan YF, Absher DM, Grimwood J, Schmutz J, Myers RM, et al. 2009. A role for a neo-sex chromosome in stickleback speciation. *Nature* **461:** 1079–1083.

Kozielska M, Pen I, Beukeboom LW, Weissing FJ. 2006. Sex ratio selection and multi-factorial sex determination in the housefly: A dynamic model. *J Evol Biol* **19:** 879–888.

Kozielska M, Weissing FJ, Beukeboom LW, Pen I. 2010. Segregation distortion and the evolution of sex-determining mechanisms. *Heredity* **104:** 100–112.

Kraak SBM, de Looze EMA. 1993. A new hypothesis on the evolution of sex determination in vertebrates: Big females ZW, big males XY. *Neth J Zool* **43:** 260–273.

Kraak SBM, Pen I. 2002. Sex-determining mechanisms in vertebrates. In *Sex ratios: Concepts and research methods* (ed. Hardy ICW), pp. 158–177. Cambridge University Press, Cambridge.

Kraak SBM, Pen I, Weissing FJ. 2000. "Joint evolution of environmental and genetic sex determination, in Pen I: Sex allocation in a life history context." PhD thesis, University of Groningen, Groningen, Netherlands, pp. 149–160.

Krackow S. 2002. Why parental sex ratio manipulation is rare in higher vertebrates. *Ethology* **108:** 1041–1056.

Lagos D, Koukidou M, Savakis C, Komitopoulou K. 2007. The transformer gene in *Bactrocera oleae*: The genetic switch that determines its sex fate. *Insect Mol Biol* **16:** 221–230.

Langkilde T, Shine R. 2005. Different optimal offspring sizes for sons versus daughters may favor the evolution of temperature-dependent sex determination in viviparous lizards. *Evolution* **59:** 2275–2280.

Leimar O. 2005. The evolution of phenotypic polymorphism: Randomized strategies versus evolutionary branching. *Am Nat* **165:** 669–681.

Lindholm A, Breden F. 2002. Sex chromosomes and sexual selection in poeciliid fishes. *Am Nat* **160:** S214–S224.

MacCarthy T, Seymour RM, Pomiankowski A. 2010. Differential regulation drives plasticity in sex determination gene networks. *BMC Evol Biol* **10:** 388.

Mank JE, Promislow DEL, Avise JC. 2006. Evolution of alternative sex-determining mechanisms in teleost fishes. *Biol J Linn Soc* **87:** 83–93.

Marín I, Baker BS. 1998. The evolutionary dynamics of sex determination. *Science* **281:** 1990–1994.

Marshall Graves JA. 2008. Weird animal genomes and the evolution of vertebrate sex and sex chromosomes. *Annu Rev Genet* **42:** 565–586.

Mawaribuchi S, Yoshimoto S, Ohashi S, Takamatsu N, Ito M. 2012. Molecular evolution of vertebrate sex-determining genes. *Chromosome Res* **20:** 139–151.

McVean G, Hurst LD. 1996. Genetic conflicts and the paradox of sex determination: Three paths to the evolution of female intersexuality in a mammal. *J Theor Biol* **179:** 199–211.

Miura I. 2007. An evolutionary witness: The frog *Rana rugosa* underwent change of heterogametic sex from XY male to ZW female. *Sex Dev* **1:** 323–331.

Natri HM, Shikano T, Merilä J. 2013. Progressive recombination suppression and differentiation in recently evolved neo-sex chromosomes. *Mol Biol Evol* **30:** 1131–1144.

Nöthiger R, Steinmann-Zwicky M. 1985. A single principle for sex determination in insects. *Cold Spring Harb Symp Quant Biol* **50:** 615–621.

O'Meally D, Ezaz T, Georges A, Sarre SD, Graves JAM. 2012. Are some chromosomes particularly good at sex? Insights from amniotes. *Chromosome Res* **20:** 7–19.

O'Neill MJ, O'Neill RJW. 1999. Whatever happened to *SRY*? *Cell Mol Life Sci* **56:** 883–893.

Orzack SH, Sohn JJ, Kallman KD, Levin SA, Johnston R. 1980. Maintenance of the three sex chromosome polymorphism in the platyfish *Xiphophorus maculatus*. *Evolution* **34:** 663–672.

Otto SP, Pannell JR, Peichel CL, Ashman TL, Charlesworth D, Chippindale AK, Delph LF, Guerrero RF, Scarpino SV, McAllister BF. 2011. About PAR: The distinct evolutionary dynamics of the pseudoautosomal region. *Trends Genet* **27:** 358–367.

Parker GA. 1979. Sexual selection and sexual conflict. In *Sexual selection and reproductive competition in insects* (ed. Blum MS, Blum NA), pp. 123–166. Academic, London.

Pen I, Uller T, Feldmeyer B, Harts A, While GM, Wapstra E. 2010. Climate-driven population divergence in sex-determining systems. *Nature* **468:** 436–438.

Pennell TM, Morrow EH. 2013. Two sexes, one genome: The evolutionary dynamics of intralocus sexual conflict. *Ecol Evol* **3:** 1819–1834.

Perrin N. 2009. Sex reversal: A fountain of youth for sex chromosomes? *Evolution* **63:** 3043–3049.

Pomiankowski A, Nöthiger R, Wilkins A. 2004. The evolution of the *Drosophila* sex-determination pathway. *Genetics* **166:** 1761–1773.

Raymond CS, Shamu CE, Shen MM, Seifert KJ, Hirsch B, Hodgkin J, Zarkower D. 1998. Evidence for evolutionary conservation of sex-determining genes. *Nature* **391:** 691–695.

Reeve HK, Pfennig DW. 2003. Genetic basis for showy males: Are some genetic systems especially conducive to sexual selection? *Proc Natl Acad Sci* **100:** 1089–1094.

Rice WR. 1984. Sex chromosomes and the evolution of sexual dimorphism. *Evolution* **38:** 735–742.

Rice WR. 1986. On the instability of polygenic sex determination: The effect of sex-specific selection. *Evolution* **40:** 633–639.

Rice WR. 1987. The accumulation of sexually antagonistic genes as a selective agent promoting the evolution of reduced recombination between primitive sex chromosomes. *Evolution* **41:** 911–914.

Roberts RB, Ser JR, Kocher TD. 2009. Sexual conflict resolved by invasion of a novel sex determiner in lake Malawi cichlid fishes. *Science* **326:** 998–1001.

Ross JA, Peichel CL. 2008. Molecular cytogenetic evidence of rearrangements on the Y chromosome of the threespine stickleback fish. *Genetics* **179:** 2173–2182.

Ruiz MF, Milano A, Salvemini M, Eirín-López JM, Perondini ALP, Selivon D, Polito C, Saccone G, Sánchez L. 2007. The gene *transformer* of *Anastrepha* fruit flies (Diptera, Tephritidae) and its evolution in insects. *PLoS ONE* **2:** e1239.

Sánchez L. 2008. Sex-determining mechanisms in insects. *Int J Dev Biol* **52:** 837–856.

Schütt C, Nöthiger R. 2000. Structure, function and evolution of sex-determining systems in Dipteran insects. *Development* **127:** 667–677.

Scudo FM. 1967. Criteria for the analysis of multifactorial sex determination. *Monitore Zool Ital* **1:** 1–21.

Shine R. 1999. Why is sex determined by nest temperature in many reptiles? *Trends Ecol Evol* **14:** 186–189.

Shukla JN, Nagaraju J. 2010. Doublesex: A conserved downstream gene controlled by diverse upstream regulators. *J Genet* **89:** 341–356.

Shukla JN, Palli SR. 2012. Sex determination in beetles: Production of all male progeny by parental RNAi knockdown of *transformer*. *Sci Rep* **2:** 602.

Stewart AD, Pischedda A, Rice WR. 2010. Resolving intralocus sexual conflict: Genetic mechanisms and time frame. *J Hered* **101:** S94–S99.

Stöck M, Horn A, Grossen C, Lindtke D, Sermier R, Betto-Colliard C, Dufresnes C, Bonjour E, Dumas Z, Luquet E, et al. 2011. Ever-young sex chromosomes in European tree frogs. *PLoS Biol* **9:** e1001062.

Stothard P, Pilgrim D. 2003. Sex-determination gene and pathway evolution in nematodes. *Bioessays* **25:** 221–231.

Stothard P, Pilgrim D. 2006. Conspecific and interspecific interactions between the FEM-2 and the FEM-3 sex-determining proteins despite rapid sequence divergence. *J Mol Evol* **62:** 281–291.

Trivers RL, Willard DE. 1973. Natural selection of parental ability to vary the sex ratio of offspring. *Science* **179:** 90–92.

Uller T, Helanterä H. 2011. From the origin of sex-determining factors to the evolution of sex-determining systems. *Q Rev Biol* **86:** 163–180.

Uller T, Pen I, Wapstra E, Beukeboom LW, Komdeur J. 2007. The evolution of sex ratios and sex-determining systems. *Trends Ecol Evol* **22:** 292–297.

Umen JG. 2011. Evolution of sex and mating loci: An expanded view from Volvocine algae. *Curr Opin Microbiol* **14:** 634–641.

Valenzuela N. 2008. Sexual development and the evolution of sex determination. *Sex Dev* **2:** 64–72.

van Dooren TJM, Leimar O. 2003. The evolution of environmental and genetic sex determination in fluctuating environments. *Evolution* **57:** 2667–2677.

van Doorn GS. 2009. Intralocus sexual conflict. *Ann NY Acad Sci* **1168:** 52–71.

van Doorn GS. 2014. Evolutionary transitions between sex-determining mechanisms: A review of theory. *Sex Dev* **8:** 7–19.

van Doorn GS, Kirkpatrick M. 2007. Turnover of sex chromosomes induced by sexual conflict. *Nature* **449:** 909–912.

van Doorn GS, Kirkpatrick M. 2010. Transitions between male and female heterogamety caused by sex-antagonistic selection. *Genetics* **186:** 629–645.

Verhulst EC, Beukeboom LW, van de Zande L. 2010a. Maternal control of haplodiploid sex determination in the wasp *Nasonia*. *Science* **328:** 620–623.

Verhulst EC, van de Zande L, Beukeboom LW. 2010b. Insect sex determination: It all evolves around *transformer*. *Curr Opin Genet Dev* **20:** 376–383.

Vicoso B, Bachtrog D. 2013. Reversal of an ancient sex chromosome to an autosome in *Drosophila*. *Nature* **499:** 332–335.

Vuilleumier S, Lande R, van Alphen JJM, Seehausen O. 2007. Invasion and fixation of sex-reversal genes. *J Evol Biol* **20:** 913–920.

Wada H, Shimada A, Fukamachi S, Naruse K, Shima A. 1998. Sex-linked inheritance of the *lf* locus in the medaka fish (*Oryzias latipes*). *Zool Sci* **15:** 123–126.

Warner DA, Shine R. 2008. The adaptive significance of temperature-dependent sex determination in a reptile. *Nature* **451:** 566–568.

Werren JH, Beukeboom LW. 1998. Sex determination, sex ratios, and genetic conflict. *Annu Rev Ecol Syst* **29:** 233–261.

Werren JH, Hatcher MJ, Godfray HCJ. 2002. Maternal-offspring conflict leads to the evolution of dominant zygotic sex determination. *Heredity* **88:** 102–111.

Wilkins AS. 1995. Moving up the hierarchy: A hypothesis on the evolution of a genetic sex determination pathway. *Bioessays* **17:** 71–77.

Woram RA, Gharbi K, Sakamoto T, Hoyheim B, Holm LE, Naish K, McGowan C, Ferguson MM, Phillips RB, Stein J, et al. 2003. Comparative genome analysis of the primary sex-determining locus in salmonid fishes. *Genome Res* **13:** 272–280.

Zarkower D. 2001. Establishing sexual dimorphism: Conservation amidst diversity? *Nat Rev Genet* **2:** 175–185.

Zhu L, Wilken J, Phillips NB, Narendra U, Chan G, Stratton SM, Kent SB, Weiss MA. 2000. Sexual dimorphism in diverse metazoans is regulated by a novel class of intertwined zinc fingers. *Genes Dev* **14:** 1750–1764.

Reproductive Parasitism: Maternally Inherited Symbionts in a Biparental World

Gregory D.D. Hurst and Crystal L. Frost

Institute of Integrative Biology, University of Liverpool, Liverpool L69 7ZB, United Kingdom

Correspondence: g.hurst@liv.ac.uk

Most species of insect, and many other plants and animals, carry maternally heritable microorganisms—viruses, bacteria, unicellular eukaryotes, and fungi that pass from a female host to her progeny. Maternal inheritance establishes a correlation between the fitness of symbiont and host female, which can select for the symbiont to contribute to host fitness. Nevertheless, its lack of transmission through male hosts places the symbiont in conflict with biparentally inherited nuclear genes. In this review, we first examine how this conflict is manifest in selection to promote the production and survival of infected female hosts and gametes. We then examine how the distorted population sex ratios that they produce may affect host reproductive ecology, and thus the intensity of other conflicts associated with sexual reproduction. Finally, we examine evolved host responses to symbiont manipulation. We argue that the unusual intensity of symbiont–host conflict generates extreme selection pressures that can drive changes in sex-determination systems, the basic pathway through which males and females are constructed.

Between 1909 and 1932, Hubert Simmonds bred individuals of the great eggfly, *Hypolimnas bolina*, seemingly as a pet project to distract him from his daily work as an agricultural entomologist in Fiji. What Simmonds discovered, by accident, was the presence of a trait in which female butterflies produced only daughters. He found that many females in his population had this trait, such that few males were present in the wild. He could maintain his all-female lines indefinitely, irrespective of the male used in the cross, and concluded that it was a trait acquired from the mother alone. Finally, he noted that half of the eggs laid by the female died during embryogenesis, and concluded that the all-female trait was associated with the death of sons (Simmonds 1923a,b, 1926, 1927, 1928, 1930). He had, by accident and very careful observation, discovered the first maternally inherited male-killing agent.

Simmonds's observation was the first of a number of cases of unusual maternally inherited traits. In 1941, Lewis built the first model of evolution under cytoplasmic inheritance, which recognized that male sterility in plants spread into populations much more easily when the trait was maternally inherited (Lewis 1941). In the same year, maternally inherited feminization of woodlice hosts was observed by Vandel (1941). The next 20 years saw numerous

Cite this article as *Cold Spring Harb Perspect Biol* doi: 10.1101/cshperspect.a017699

records of maternally inherited male lethality in ladybird beetles and flies (Lus 1947; Shull 1948; Magni 1953; Cavalcanti et al. 1957; Malogolowkin 1958). Further, unusual cross-incompatibilities between members of different populations of *Culex* mosquitoes were also noted as having a maternally inherited basis (Laven 1951). Collectively, these traits have been termed "reproductive parasitism"—because the impacts on reproduction are deleterious to host fitness.

Later work in the "molecular genetics age" examined the cause of these traits. "Reproductive parasitic" behaviors in plants are commonly associated with mutations in the mitochondrial genome, but have recently also been ascribed to heritable fungi (Gorischek et al. 2013). Reproductive parasitism in animals is associated with a wide variety of microbial symbionts: viruses, bacteria, unicellular eukaryotes, and fungi that pass from a female to her progeny either inside eggs or through smearing on the outside of eggs followed by trophic acquisition (Terry et al. 2004; Nakanishi et al. 2008; Duron and Hurst 2013; Gorischek et al. 2013). Although many heritable microbes are simply "beneficial symbionts," reproductive parasitism of the type described above is widespread (and may be combined with beneficial impacts on the female host).

We first review the rationale underlying the selection for reproductive parasitic phenotypes, the mechanisms by which the manipulations are achieved, and the impact of the phenotypes on the symbiont's population biology. We then examine the impacts of reproductive parasitism on host reproductive ecology and other conflicts associated with reproduction. We then examine nuclear responses to the presence of heritable parasites, first for sex-ratio/sex-allocation distorters and then for incompatibility-inducing symbionts. We argue that reproductive parasites can create the strongest selective pressures observed in nature, and these will likely be associated with changes in reproductive systems. We speculate that these strong forces may be important in driving "unusual" evolutionary events, such as those underlying the evolution of sex-determination systems.

MANIFESTATIONS OF CYTOPLASM–NUCLEAR CONFLICT; POPULATION BIOLOGY OF REPRODUCTIVE PARASITES

The selection for reproductive parasitic traits derives from the symbiont's uniparental (maternal) inheritance (Cosmides and Tooby 1981). Maternally inherited symbiont fitness depends solely on the symbiont's impact on the survival or production of female hosts. A symbiont that is purely maternally inherited will only spread if infected individuals on average produce more infected female offspring than uninfected hosts produce uninfected daughters. The "drive" that creates this increase in daughter production can simply be symbiont traits that increase the fitness of infected females over that of their uninfected counterparts. Indeed, many of these symbionts are beneficial in both sexes, providing "services" such as protection against natural enemies (Hamilton and Perlman 2013) or provision of anabolic capability (Moran et al. 2008). Notwithstanding this, drive can additionally (and not mutually exclusively [Himler et al. 2011]) be achieved through skewing sex allocation/sex ratio such that infected females produce more daughters. Alternately, the symbiont may leave males unharmed in terms of survival but modify male host gametogenesis, producing sperm that are modified to selectively eliminate uninfected zygotes. This phenomenon of cytoplasmic incompatibility promotes symbiont spread through selectively reducing the fitness of uninfected females. Altogether, these phenotypes represent cases of the wider phenomenon of intragenomic conflict, with the pattern of selection on uniparentally inherited elements differing from selection on nuclear genes with respect to sex allocation, sex ratio, and male fertility (Hurst et al. 1996).

The drive created by a symbiont must be sufficient to overcome any segregrational loss—progeny that fail to inherit them—and any direct physiological/pathological costs of carrying a symbiont. Theory predicts that symbionts with perfect transmission (no segregational loss) that can invade populations will, in a panmictic population, become fixed (Hurst 1991). Indeed, reproductive parasitic symbionts are known to infect all female individuals in some species (Stouthamer et al. 1990; Rigaud and Juchault 1992; Dyson and Hurst 2004). However, polymorphism of reproductive-parasite-carrying uninfected individuals is more common. Polymorphism is stabilized by three factors: (1) imperfect maternal inheritance, in which segregational loss prevents fixation of symbionts; (2) additional cost of symbiont carriage at high prevalence, for example, unmatedness of symbiont-carrying individuals (Groenenboom and Hogeweg 2002); and (3) evolution of host factors that affect the phenotype of the symbiont, for example, evolution of host suppressor genes.

The manipulation phenotypes above evolve because male hosts are a dead end for a maternally inherited symbiont. However, some symbionts can combine maternal transmission with infectious transmission, such that male hosts represent a viable source of onward infection. In a number of cases, female hosts are used exclusively for maternal transmission and male hosts for onward infectious transmission. Infectious transmission through sons often occurs relatively late in host development and is associated with the death of male hosts. This discrepancy in timing has led to these examples being termed "late male killing," differentiating them from the more common embryonic male-killing interactions in which male death is not associated with infectious transmission (Hurst 1991).

Below we outline the three main manifestations of symbiont–host conflict (distortion of sex ratio/sex allocation, cytoplasmic incompatibility, and late male killing) in terms of mechanism, diversity, and population biology. For simplicity, we assume in our treatment of population biology that a symbiont has a single phenotypic impact on its host; the ramifications of contravening this assumption are discussed later.

Distortions of Sex Ratio/Sex Allocation

Maternally inherited symbionts in host species with separate sexes can spread either through biasing sex ratio to the production of daughters or, where there are antagonistic interactions between male and female siblings, through killing males and reducing the intensity of competition suffered by infected females. Within hermaphroditic host species, biasing sex allocation to female reproductive function is favored where there is a trade-off between male and female reproductive activity. Altogether, four different distortions of sex ratio/sex allocations have evolved to aid symbiont/organelle spread.

Feminization

Symbiont-induced feminization is commonly observed in amphipod and isopod Crustacea (Bouchon et al. 1998) and also in mites, leafhoppers, and butterflies. The phenotype is the conversion of genetic males into phenotypic females (Table 1). In the best-studied case of symbiont-induced feminization, *Wolbachia* infection of *Armidillidium vulgare*, *Wolbachia* infection inhibits the formation of the male-determining androgenic gland during early development, and in so doing results in female development of individuals that are karyotypically male (ZZ, the species being female heterogametic). Converted females (neofemales) are reproductively functional although with occasional intersex properties. In mate-choice experiments, males prefer true females to feminized neofemales (Moreau et al. 2001).

The dynamics of feminizers depends on the transmission efficiency of the symbiont, the efficiency of feminization, and the host sex–determination system. In *A. vulgare*, feminization is efficient, but 10% of the progeny of a female fail to inherit the infection and develop as male (Rigaud and Juchault 1992). Because the species is female heterogametic, spread of the *Wolbachia* may be associated with fixation of the Z chromosome (Taylor 1990; but see Kobayashi and Telschow 2010). Indeed, *A. vulgare* populations exist in which all females are ZZ and carry feminizing *Wolbachia*, and population sex ratio is determined by the proportion of zygotes that do not inherit the microbe and develop as males. In other feminization systems, some males escape feminization, which may also stabilize population sex ratio and persistence (Moret et al. 2001).

Table 1. Incidence and mechanism of symbiont-induced feminization

Symbiont	Host	Suggested mechanism	References
Wolbachia	*Armadillidium vulgare* (Isopoda)	Inhibition of androgenic gland differentiation and/or inhibition of androgenic hormone reception	Bouchon et al. 1998
	Zyginidia pullula (Hemiptera)	Interference with male DNA methylation, resulting in female imprinting patterns across the genome	Negri et al. 2009
	Eurema hecabe (Lepidoptera)	Unknown; continuous during larval development	Hiroki et al. 2002; Narita et al. 2007
Cardinium	*Brevipalpus phoenicis* (Acari)	Feminization of haploid males	Weeks et al. 2001
Microspora sp.	*Gammarus duebeni* (Amphipoda)	Inhibition of androgenic gland differentiation	Jahnke et al. 2013
Nosema granulosis		Inhibition of androgenic gland differentiation and potential partial refractoriness to androgenic hormone	Jahnke et al. 2013

Parthenogenesis Induction

In haplodiploid arthropods, males develop from unfertilized embryos. In *Trichogramma* wasps, manipulation of ploidy by microbes (doubling of chromosome constitution) leads to development of a female host from these unfertilized eggs (Table 2). Various mechanisms of parthenogenesis induction are now recognized, and three microbes (*Wolbachia*, *Cardinium*, and *Rickettsia*) are recognized as able to induce asexuality. Negative consequences of these symbionts on the overall reproductive success of infected females have been shown in some laboratory comparisons (Stouthamer and Luck 1993; Huigens et al. 2004), but field studies have found either little impact or a beneficial impact of symbiont presence on host fecundity (Segoli et al. 2013). Whichever is the case, the number of females produced by symbiont-infected females is always greater than for uninfected hosts, and thus the drive for an increase in the number of "transmitters" over "nontransmitters" is achieved.

Outside of haplodiploid taxa, the involvement of microbes in parthenogenesis is uncertain. There are certainly many parthenogenetic species that harbor symbionts (Koivisto and Braig 2003), and in some of these there is also evidence that antibiotic treatment alters the capacity of the host to reproduce (e.g., Timmermans and Ellers 2009). However, in nonhaplodiploid taxa, symbiont-free individuals do not produce males and antibiotic treatment may interfere with reproduction in other ways (e.g., because the symbionts are required for oogenesis as seen in *Asobara tabida* [Dedeine et al. 2001]) rather than being required for parthenogenesis, thus making definitively testing the role of symbionts in inducing parthenogenesis very difficult.

Embryonic Male Killing

Killing of zygotes that would develop as males is commonly observed, with multiple symbionts exhibiting this trait across a wide range of arthropods (Table 3). The mechanistic basis of male killing is also diverse, with male killing occurring via interference in male sex determination in some cases but as a result of host sex–specific symbiont responses in others.

The evolutionary advantage for male-killing behavior derives from the impact that death of males may have on the survival/reproduction of their sisters, who will share the symbiont infection through descent. In ladybird beetles, female larvae consume the eggs containing their dead male brothers, which increases their chance of survival (Elnagdy et al. 2011). In other cases, male and female siblings compete for re-

Table 2. The diversity and mechanism of parthenogenesis induction by heritable symbionts

Symbiont	Host	Suggested mechanism	References
Wolbachia	*Trichogramma* spp. and *Leptopilina clavipes* (Hymenoptera)	Doubling of chromosome number (haploid males) by failure of chromosomes to segregate at the first mitotic division	Stouthamer and Kazmer 1994; Pannebakker et al. 2004
	Muscidifurax uniraptor (Hymenoptera)	Doubling of haploid male chromosome number via the fusion of two cell nuclei following the initial mitosis	Gottlieb et al. 2002
	Franklinothrips vespiformis (thrips)	Not known	Arakaki et al. 2001
	Bryobia praetiosa and *Bryobia* sp. (Acari)	Apomictic parthenogenesis suggested, whereby a lack of meiosis in eggs means that doubling of chromosome number is not required or else automictic parthenogenesis with central fusion	Weeks and Breeuwer 2001
Cardinium	*Encarsia hispida* (Hymenoptera)	Both automictic parthenogenesis (chromosome doubling) and feminization is required for a female phenotype, but the mechanism for these is unknown	Giorgini et al. 2009
Rickettsia—transitional group	*Neochrysocharis formosa*	Occurs via achiasmatic apomictic parthenogenesis, whereby a lack of the creation of chiasma during egg meiosis means that doubling of chromosome number is not required	Adachi-Hagimori et al. 2008
Rickettsia—bellii group	*Pnigalio soemius* (Hymenoptera)	Not known	Giorgini et al. 2010

This list is exemplars, and many other examples exist.

sources, and the death of males reduces the intensity of competition suffered by female hosts (Koop et al. 2009). Finally, where females in a species suffer from inbreeding, male killing may be advantageous to the symbiont as a mechanism of promoting outbreeding (Werren 1987; but see Dannowski et al. 2009).

The magnitude of "drive" provided by male killing to the symbiont is inevitably less than that achieved through conversion of individuals from one sex to another. Under the most compliant biology, the viviparous harlequin pseudoscorpion *Cordylochernes scorpioides*, male killing reduces the intensity of resource competition in utero and infected females produce 26% more daughters, with these daughters also being larger in size than those from uninfected mothers (Koop et al. 2009). However, the drive achieved by male killers is likely to be commonly much lower than this. This low level of drive makes the frequency of male killers quite variable and sensitive to vertical transmission efficiency (Fig. 1). Although male killers with perfect maternal inheritance do obtain very high prevalence (>80% of females infected) (Jiggins et al. 2002; Dyson and Hurst 2004), imperfect maternal inheritance combined with modest benefits of male killing (and a lack of other benefits to female hosts) is associated with low prevalence. In many cases, male killers exist in <20% of female hosts (Williamson and Poulson 1979; Hurst et al. 1993, 2000). At this prevalence, they may still create strong selection against their action, but ecological impacts are more limited.

In some host species, male killing has no obvious adaptive benefits. Examples of symbionts that appear to "accidentally" impact male

Table 3. The diversity and mechanism of male-killing symbionts in arthropods

Symbiont	Host taxa	Suggested mechanism	References
Wolbachia	*Ostrinia scapulalis* (Lepidoptera)	Interference with the sex-specific splicing of *doublesex*, with mismatch between the phenotypic and genetic sex resulting in the death of males	Sugimoto and Ishikawa 2012; Sugimoto et al. 2012
	Drosophila bifasciata (Diptera)	Modulation of male chromatin remodeling followed by disruption of normal mitotic spindle formation	Riparbelli et al. 2012
	Various Lepidoptera, Diptera, Coleoptera, and harlequin pseudoscorpions	Unknown	Hurst et al. 1999a; Zeh et al. 2005
Rickettsia	Ladybird and buprestid beetles	Unknown	Werren et al. 1994; Lawson et al. 2001
Arsenophonus nasoniae	*Nasonia vitripennis* (Hymenoptera)	Inhibition of formation of maternal centrosomes required for early male development in haplodiploids	Ferree et al. 2008
Hamiltonella related	*Cheilomenes sexmaculata* (ladybird beetle)	Unknown	Majerus and Majerus 2010
Spiroplasma poulsoni	*Drosophila melanogaster*/ *Drosophila nebulosa*	Widespread apoptosis in male-killed embryos; neural tissue is inhibited from properly packing into the neural chord and from forming axons; male killing requires the male form of the dosage compensation complex	Veneti et al. 2005; Bentley et al. 2007; Martin et al. 2013
Spiroplasma ixodetis group	Various ladybird beetles, Lepidoptera, and Hemiptera	Unknown	Hurst et al. 1999b; Jiggins et al. 2000a; Sanada-Morimura et al. 2013

This list is exemplars, and many other examples exist.

fitness include protective *Spiroplasma* in aphids, which causes lethality in sexual males only (Simon et al. 2011), and *Spiroplasma* embryonic male killing in *Danaus chrysippus*, in which females lay solitary eggs, such that sibling–sibling interference is limited (Jiggins et al. 2000a). There are two possible logics underlying male lethality in these cases. First, male killing may be a relict following switch of a symbiont from a host in which male killing is adaptive into a novel host species in which it is not, but is nevertheless not disadvantageous. Second, it may derive from lack of selection against deleterious impacts observed in male hosts only (the "mother's curse" as discussed in Wade [2014]).

Deviations in Sex Allocation in Hermaphrodite Plants

Cytoplasmic male sterility (CMS), in which male reproductive function is selectively destroyed, is widespread in flowering plants. It is commonly associated with particular mitochondrial types, often associated with unusual open reading frames. Plant mitochondrial genomes commonly exist as minicircles that recombine. This intragenomic recombination creates new open reading frames, which are chimeric fusions, a fraction of which induces male sterility. The mechanism of CMS is known in a number of cases (Kubo et al. 2011). Typically,

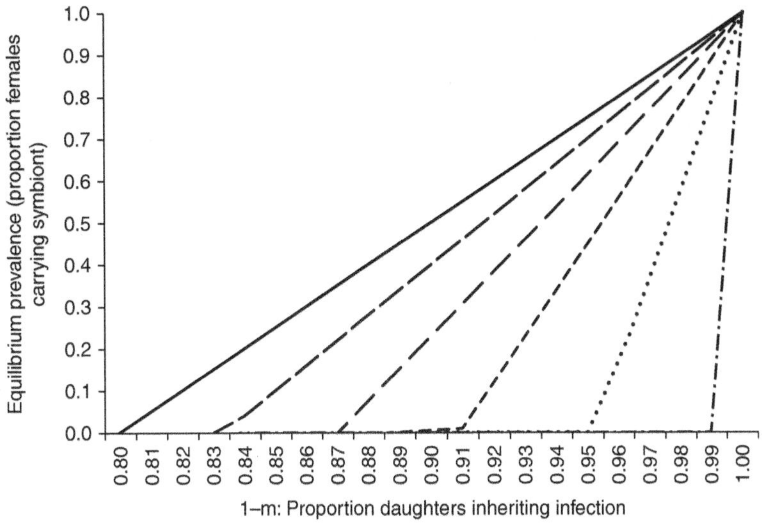

Figure 1. Equilibrium prevalence of a cost-free maternally inherited male-killing bacterium that is transmitted to a fraction $1 - \mu$ of daughters, and whose daughters have increased chance of survival $1 + b$ associated with the death of their brothers. Calculated from $p^* = [(1 - \mu)(1 + b) - 1]/b$, following, Hurst (1991). Lines represent, from *left* to *right*, $b = 0.25$, $b = 0.2$, $b = 0.15$, $b = 0.1$, $b = 0.05$, and $b = 0.01$.

CMS involves mitochondrial genes that encode hydrophobic proteins, and commonly derives from genes encoding ATPase subunit. Interestingly, CMS mitochondria do not appear to have strongly deleterious effects outside of anther tissue. Notwithstanding major patterns like these, mechanisms are variable. In rice, CMS is associated with a novel mitochondrial gene whose product accumulates in anther tapetum, where it interferes with the function of a nuclear-encoded factor causing anther death (Luo et al. 2013).

The spread of CMS mitochondria relies on a compensation process, in which energy that would be dedicated to production of pollen instead is allocated to ovules (Poot 1997). Unlike heritable microbes, mitochondria have perfect vertical transmission, which can allow very high frequency of CMS types if CMS plants do not suffer from pollen limitation (McCauley and Brock 1998). The dynamics of CMS thus depend in part on pollen availability, which varies with the spatial setting of the plant (Olson et al. 2005) and whether hermaphrodite plants can self-fertilize. Stabilization of polymorphism for CMS mitochondria will be seen in a metapopulation, as female individuals cannot found new populations (owing to lack of pollen) and can only invade existing ones (McCauley and Taylor 1997). As CMS spreads through a local population, pollen availability becomes more and more limiting, such that local patches may go extinct through CMS, again creating the conditions for CMS and wild-type mitotypes to coexist. Finally, selfing may provide a means for non-CMS individuals to avoid pollen limitation when CMS is common, and can produce a polymorphic equilibrium for CMS/wild-type mitochondria.

More recently, a shift in sex allocation from male to female function has been observed to be associated with maternally inherited fungal endophytes in *Elymus virginicus* grass. Although fungal endophytes are commonly considered defensive mutualists, Gorischek et al. (2013) observed that hermaphrodites infected with fungal symbionts had female-biased sex allocation; this is in contrast to the male-biased allocation pattern observed in non-endophyte-infected conspecific individuals obtained through segregation (imperfect maternal inheritance). They also showed greater vegetative propagation, again permitting symbiont maintenance.

Incompatibility Associated with Symbiosis

In the reproductive phenotypes discussed thus far, the symbiont/organelle spreads by compromising the rate at which male hosts are produced or survive, or in hermaphrodites, investment into male gametes. However, in the most common reproductive phenotype, cytoplasmic incompatibility (CI), males are produced in normal numbers and develop to reproductive maturity in normal numbers. The symbiont has its impact in male gametogenesis, whereby it modifies sperm so that when they fuse with uninfected or differentially infected eggs, the offspring are not viable (Fig. 2). This trait is known for two symbionts, *Wolbachia* and *Cardinium* (Yen and Barr 1971; Hunter et al. 2003).

In diploids the CI phenotype is typified by mortality of both sexes. The mechanism of CI involves condensation of paternally derived chromatin in uninfected zygotes, through unknown microbial effectors (Callaini et al. 1997; Tram and Sullivan 2002; Tram et al. 2006; Landmann et al. 2009). In haplodiploid species, such as the Hymenoptera, CI results in male-biased sex ratios. Here male individuals are haploid and not the product of fertilization. Two types of CI are observed—ones in which diploid individuals are returned to full haploid status and develop as functional males (all male broods of normal size), and ones in which diploid individuals that are the product of incompatible crosses are characterized by widespread aneuploidy and death (all male broods corresponding in size to the number of males produced normally) (Breeuwer and Werren 1990; Vavre et al. 2000).

In causing cytoplasmic incompatibility, the symbiont selectively decreases the fitness of uninfected females. The degree of decrease in fitness of uninfected lineages is positively frequency-dependent—where infected males are common, uninfected females suffer more commonly. Where the symbiont is not perfectly transmitted or imposes a cost on its female bearer, there is a threshold frequency of infection that must be attained before deterministic spread of the symbiont (Fine 1978; Turelli et al. 1992). Below this frequency the CI drive phenotype is too rare to compensate for segregational loss/cost of carriage. Above this frequency it has a positive drive and will invade. The threshold frequency may be achieved through stochastic processes or through alternative drive mechanisms such as direct benefit (Fenton et al. 2011). Indeed, strains of *Wolbachia* that induce CI are also commonly known to make the host tolerant to viral attack (Hedges et al. 2008; Teixeira et al. 2008; Osborne et al. 2009), suggesting that both direct benefit and reproductive parasitism may contribute to invasion/dynamics.

Following invasion, CI strains can spread to very high prevalence. Spatial waves of advance of CI symbionts have been observed on multiple occasions (Turelli and Hoffmann 1991; Hoshizaki and Shimada 1995; Kriesner et al. 2013), indicating that there are ongoing and repeated waves of symbiont spread in insects. The ability to drive rapidly to high prevalence is useful in that it potentiates *Wolbachia*-mediated control-

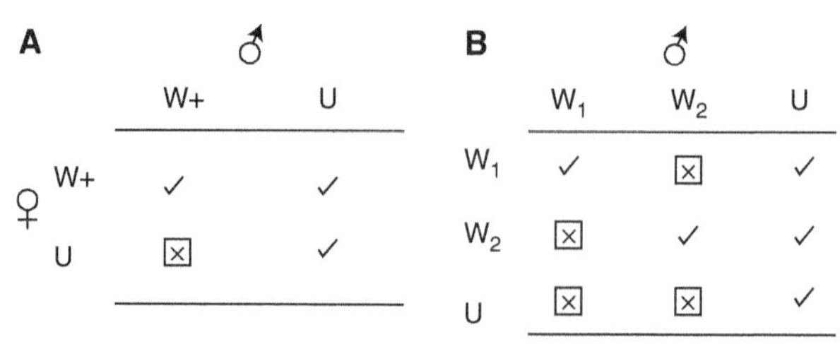

Figure 2. Cross-compatibility relationships in the presence of *Wolbachia*/*Cardinium* inducing CI. (*A*) Compatibility (viable zygotes) for the case of unidirectional incompatibility (one symbiont)—crosses are compatible save between infected male and uninfected female. (*B*) Compatibility for bidirectional incompatibility, with two symbionts, W_1 and W_2, alongside uninfected individuals. *Wolbachia*-infected males are incompatible with both uninfected females and females carrying the alternate strain.

of-vector competence following introduction (Walker et al. 2011; Yeap et al. 2011). This is currently being trialed in *Aedes aegypti*, into which the *Wolbachia* strain *w*Mel has been transinfected (Hoffmann et al. 2011). Mosquitoes carrying *w*Mel are refractory to onward transmission of dengue, and the *Wolbachia* also induces incompatibility, allowing its spread into populations following release. High-prevalence infections have been established in natural populations through release and CI-mediated spread, raising the possibility of interrupting dengue transmission in nature.

Male Lethality/Virulence Associated with Infectious Transmission

Selective killing of male hosts during larval phases is commonly associated with symbionts that can be both maternally and infectiously transmitted. It was first described in mosquitoes, associated with pathogenic activity of microsporidian symbionts. These symbionts are commonly vertically transmitted through females, and a subset of these are also pathological to male hosts during the fourth larval instar (Kellen et al. 1965; Hurst 1991). Spore transmission occurs through the aquatic phase and commonly requires a secondary host (e.g., copepod) (Andreadis 1991). "Late" male killing of this type has also recently been observed in a terrestrial insect, the Oriental tea tortrix, *Homona magna*. Here an infectious RNA virus exhibits vertical transmission in female hosts and male-specific lethality in late instar caterpillars (Morimoto et al. 2001; Nakanishi et al. 2008). Parallel limitation of virulence associated with female hosts is observed in helminth parasites of mammals. Shoop (1991) described how helminths show sex-specific tropism and pathogenesis, achieving vertical transmission through milk/accumulation in mammary tissue in females and infectious transmission through the gut in males.

Male death (in contrast to cases of embryonic male-killing) is associated with liberation of infectious spores into the environment and onward infectious transmission. Intuitively, efficient infectious transmission would select for killing of both sexes, and late male killing will be thus favored where infectious transmission rates are low. It is notable in this context that symbionts can be plastic, exhibiting vertical transmission in a growing population in which opportunities for heritable transmission are good, and infectious transmission in one at carrying capacity in which opportunities are rare (Agnew and Koella 1999; Kaltz and Koella 2003). They may also evolve to be benign where vertical transmission opportunities are common (Magalon et al. 2010).

IMPACT OF CYTOPLASMIC SEX-RATIO DISTORTERS ON HOST REPRODUCTIVE ECOLOGY AND PATTERNS OF INTRAGENOMIC AND SEXUAL CONFLICT

The spread of a symbiont that affects host sex ratio, sex allocation, or sexuality has obvious impacts on the host mating system. These include obligate loss of sex, alteration of host mating system from one of dioecy to gynodioecy, and alteration in operational sex ratio. These alterations may then impact on other conflicts associated with sexual reproduction, such as those exhibited between selfish genetic elements and their hosts, and between male and female.

Symbionts and the Evolution of Obligate Asexuality

Species and populations of haplodiploid arthropods that are obligate asexuals as a result of symbiont activity are common. Indeed, the only male specimens ever obtained in many of these species are those produced following treatment of the host with antibiotics (Stouthamer et al. 1990). However, for a symbiont alone to produce loss of sexual reproduction, it would be required to be transmitted with near perfect fidelity and to be completely accurate in inducing female development in unfertilized males. Given that symbionts commonly show some segregational loss/imperfect manipulation abilities, the transition into obligate asexuality is more commonly a complex process that involves host evolution in addition to symbiont spread.

Stouthamer et al. (2010) propose a scenario in which symbiont spread promotes the loss of sexual function in females. In haplodiploid species, symbiont induction of asexuality commonly involves diploidization of unfertilized male embryos. Where the symbiont becomes common and males are rare, selection can promote the production of sons through "virginity" mutations, which inhibit females from mating (Stouthamer et al. 2010). Uninfected females then produce male offspring exclusively, and infected females produce more sons if there is inefficient transmission of the symbiont or if some haploid individuals escape diploidization. The spread of virginity mutations will drive up the frequency of the asexuality-inducing symbiont, as uninfected females produce fewer daughters. The process can thus lead to the fixation of the symbiont and the loss of all sexual reproduction.

Empirical studies bear out the observation that sexual function degrades rapidly and, as predicted by the virginity hypothesis, female reproductive function is lost first. In three species (*Leptopilina clavipes*, *Asobara japonica*, and *Apoanagyrus diversicornis*), interpopulation variation in the presence of a symbiont that induces asexuality has allowed the functionality of males and females derived from asexual lineages to be tested. In all three cases this revealed female function to have deteriorated before male, with infected females unable to use sperm from males, while derived males were partially fertile (Pijls et al. 1996; Pannebakker et al. 2005; Kremer et al. 2009). In a variety of cases the loss of sexual traits has been shown to have a simple genetic basis (Jeong and Stouthamer 2005; Russell and Stouthamer 2011).

Alteration in Mating System/Reproductive Ecology

Symbiont-induced sex-ratio biases have been observed to alter the pattern of reproductive competition in a number of arthropod species. In *Acraea encedon* butterflies, 95% of females in some populations carry a *Wolbachia* male killer. These populations are sex role reversed, with female reproduction limited by access to males and female–female competition for mates in evidence (Jiggins et al. 2000b). In *H. bolina* butterflies, male killer prevalence across islands varies from absence (1:1 population sex ratio) to 100 females per male. Interestingly, comparison of reproductive systems between islands suggests that the impacts on host reproduction are nonlinear (Charlat et al. 2007b). Female mating rate is highest in populations in which males are moderately (but not very) scarce, with polyandry most profound at 50%–75% male-killer prevalence. It is believed that increases in polyandry where males are quite scarce are associated with an increased desire to remate for females that receive small spermatophores. The size of spermatophore transferred by a male declines with increasing male mating rate. Strongly female-biased sex ratios are thus associated with the transfer of small spermatophores and elevated female mating rate.

The mating system changes observed to date in the butterfly sex-ratio distorter systems are plastic—in *Hypolimnas* the reproductive investment patterns of males do not vary between island populations when the males are maintained in a common garden environment (Charlat et al. 2007b). This may be partly explained by the weak efficacy of selection where male killers are common (see above). However, there is evidence in feminizer–host systems for evolution of male reproductive behavior. In a comparative analysis, the male ability to fertilize multiple females was higher in two species carrying a feminizer than in five other species in which there was no feminizer present (Moreau and Rigaud 2003). This is consistent with the action of selection to partition ejaculates to enable a greater mating rate in males. Further, in *Gammarus duebeni*, males are sperm-limited and allocate larger ejaculates to females that do not carry feminizing symbionts (Dunn et al. 2006). Although this may be the expression of a historic preference for high-quality females, it is also consistent with adaptive preferences evolved to counter the problem of feminization.

Consequences for Other Conflicts Associated with Reproduction

Changes in mating system are very likely to alter the intensity of other conflicts. Selfish genetic elements, for instance, possess the highest drive in outbreeding populations. In plants the spread of cytoplasmic male sterility alters the host mating system from one of dioecy (all individuals hermaphrodite) to gynodioecy (a mix of female plants and hermaphrodites). CMS mitochondria enforce outbreeding and will reduce the rate of selfing within the population. Selfing both creates homozygosity for particular elements (preventing drive) and exposes selfish genetic elements to the damage they cause (Burt and Trivers 1998). Thus, the spread of CMS, and consequent reduction in frequency of dioecious individuals and inbreeding, may facilitate the spread of other selfish genetic elements. An expectation is that in species where selfing is possible, the presence of CMS mitochondria at high frequency would be associated with a higher load of B chromosomes and transposable elements.

The intensity of sexual conflict is likely to be affected by sex-ratio distortion. In biasing population sex ratio toward females, the conflict between male and female is likely to change in both intensity and direction. Further, the intensity of male–male precopulatory conflict will diminish. Although not examined for any particular natural case, the impact of sex-ratio manipulation on the intensity of these conflicts is well-known from experimental evolution studies (Wigby and Chapman 2004; Linklater et al. 2007). Where insects are maintained in female-biased sex ratios, selection results in a reduction in the intensity with which males manipulate female physiology, and females are less selected to resist harm and are more susceptible to damage. Further, males are selected to partition ejaculates in the "expectation" of future mating opportunities.

HOST RESPONSES TO SEX-RATIO DISTORTER ACTIVITY, AND THE EVOLUTION OF CORE FEATURES OF MALE/FEMALE BIOLOGY

Both sex ratio–distorting microbes and mitotypes that alter sex allocation via CMS can generate unusually strong selective forces. Where they become common, males/male gametes become very rare. This places them in conflict with selection on biparentally inherited nuclear genes. The nature of Fisherian selection on the sex ratio is such that modifiers of the host nuclear genome that restore male production are very strongly selected as the male host/function they rescue has very high fitness. Thus, the impact of the parasite on sex ratio makes the conflict between the parties potentially the most intense in nature.

The evolution of restoration of male function is established widely for cases of cytoplasmic male sterility induced by mitochondrial variants, in which nuclear genes are commonly termed *Rf* genes, or "restorers of fertility." Restorer genes in these systems often encode proteins with pentatricopeptide motifs, which bind RNA and function through posttranscriptional suppression of the mitochondrial transcript that would create CMS. However, alternate mechanisms of restoration exist (Luo et al. 2013). Restoration of male host survival is also known in five cases for male killing induced by heritable microbes, in which the genes are commonly termed "suppressors" (Table 4). However, the mechanisms of restoration are not known in any case. Nuclear genes that reduce the vertical transmission of feminizing microbes are also known (Rigaud and Juchault 1992).

The evolutionary dynamics of suppressor/restorer genes are quite variable. The strength of selection will depend on the initial frequency of the distorting element, the effectiveness with which the symbionts eliminate male hosts/function (e.g., all infected males die or are feminized vs. some), and the degree to which the nuclear gene restores phenotype (e.g., partial vs. complete restoration of male fertility). Following invasion, the equilibrium prevalence of the microbe/mitochondria and frequency of the suppressor gene depend on whether the microbe has an alternate source of "drive," as well as the "cost" of carrying a suppressor.

Where the symbionts have no alternate form of drive, a cost-free suppressor will eliminate the symbiont, whereas suppressor and male killer are expected to remain polymorphic when suppressors are costly (Randerson et al. 2000), as

Table 4. Nuclear-encoded factors that prevent the action or transmission of sex ratio–distorting symbionts

Natural host	Symbiont	Nature of suppression	Notes	References
Hypolimnas bolina	*Wolbachia* *w*Bol1	Dominant zygotic suppression of male killing	Spatially variable; evidence of recent rapid spread in some populations	Hornett et al. 2006, 2009; Charlat et al. 2007a
Cheilomenes sexmaculata	*Hamiltonella* relative	Dominant zygotic suppression of male killing	Suppressor polymorphic within populations	Majerus and Majerus 2010
Drosophila recens	*Wolbachia*	Suppression suggested by emergence of male killing in some lines of *Drosophila subquinaria* following transfer from *D. recens*		Jaenike 2007
Ephestia cautella	*Wolbachia*	Suppression suggested by emergence of male killing following transfer into *Ephestia kuehniella*		Sasaki et al. 2002
Drosophila prosaltans	Unknown	Polymorphic recessive factor affecting male killing		Cavalcanti et al. 1957
Armadillidium vulgare	*Wolbachia* *w*Vul1	Nuclear-mediated resistance to transmission of feminizing symbiont	Sex jointly determined by *Wolbachia* presence and nuclear factor influencing its transmission	Rigaud and Juchault 1992

observed in the ladybird *Cheilomenes sexmaculata* (Majerus and Majerus 2010). However, many symbionts have alternate drive mechanisms such that they may be retained even when suppression is complete. For instance, in *H. bolina* the suppressed male killer induces cytoplasmic incompatibility in rescued males (Hornett et al. 2008), which drives the symbiont to fixation following spread of the suppressor (Hornett et al. 2010). At this point, the suppressor becomes required for male host survival in the population and can become a fixed aspect of the host biology—albeit one whose evolutionary origins are in maintaining survival against heritable parasites alone.

The spread of suppression of male killing has been observed in natural populations of the butterfly *H. bolina*. As discussed previously, the male-killing *Wolbachia* in this species can attain very high prevalence. This is most obvious on Samoa, where four records over a 100-year period up to 2001 reported population sex ratios of 100 females per male. This extreme sex ratio had disappeared completely by 2006 (Charlat et al. 2007a). The heritable microbe was still present in the population, but the spread of a nuclear gene that rescued male embryos (a suppressor of male killing) had restored individual and population sex ratios to parity. The spread was observed in real time over Savai'i, one of the two large islands in Independent Samoa, with population sex ratios of 100 females per male in 2005 and parity of sex ratio in 2006, 8–10 generations later.

In this system, the strength of selection at the point of origin of the suppressor is extreme. Where the sex ratio is 100 females per male, essentially all surviving male hosts will carry the suppressor in the generation after its introduction (male survival without the suppressor is negligible). Where the suppressor activity is dominant (the case in *H. bolina*), all males in the population are heterozygous for the suppressor element after one generation. Thus, the frequency of the suppressor increases to 25% in a single generation.

The strength of selection that suppression can engender is important for two reasons.

First, it determines the pace of change in reproductive systems of its host—in the case of *H. bolina*, a switch from a reproductive system in which males are the "in-demand" sex to a female-limited one in potentially a single generation. Second, it determines what type of mutation can spread. Where selection is intense, mutations that would otherwise be very deleterious can actually spread. For the case of a 100:1 female/male sex ratio, a suppressor that rescued males could spread even if it reduced female host fitness by 99%.

The ability to drive the spread of mutations that would otherwise be highly deleterious is perhaps one of the most interesting features of sex ratio–distorting symbionts. The intensity of conflict is such that it may drive unusual, and otherwise forbidden, alterations in the way in which male and female individuals are constructed (their sex-determination system). Although we do not know the molecular nature of suppressor mutations in any system, sex-ratio distorters interact with host sex–determination systems in multiple ways (Table 3), and thus an obvious avenue for selection is change in these systems. Our thesis is that the evolution of suppression of sex ratio–distorting activity will frequently involve loci associated with sex determination. Theory has proposed that the evolution of haplodiploidy and paternal genome elimination may be driven by sex ratio–distorting activity (Engelstadter and Hurst 2006; Ubeda and Normark 2006). More widely, more subtle changes in the construction of sex-determination systems are likely. We would argue that although extreme prevalence of sex-ratio distorters may be rare, they create the circumstances that can drive change in genetic systems that otherwise are rarely selected because mutations within them are otherwise deleterious. Thus, they may be a rare force that drives revolutionary changes of the type classic theory finds hard to explain.

There is already some evidence for the involvement of sex-ratio distorters in the evolution of sex-determination systems. Trivially, microbes that feminize are novel sex-determining loci. In the female heterogametic pill woodlouse, *A. vulgare*, spread of the feminizing agent is also associated with loss of the female-determining chromosome, such that sex becomes primarily determined by the presence/absence of *Wolbachia*-induced feminization. Nuclear factors that affect *Wolbachia* transmission are then, de facto, novel sex-determination factors, as all progeny that do not inherit *Wolbachia* develop as males.

Evolution of sex-determination systems has also been implied in recent study of the moth *Ostrinia scapulalis*. In this species *w*Sca *Wolbachia* induces male killing. Male killing is associated with alteration of the splicing of the gene *doublesex*, such that all individuals have the female splice form of *dsx* and those that are ZZ (destined to be males) die, presumably from an imbalance between sex chromosome karyotype and developmental sex (Sugimoto and Ishikawa 2012). Interestingly, when ZW individuals with *Wolbachia* (destined to be female) are cured using antibiotics, all male sex ratios are observed associated with male-form splicing of *dsx* in all progeny, and the death of ZW progeny (Kageyama and Traut 2004). The investigators conclude that the *Wolbachia* kills through feminization and that female sex–determining factors on the W chromosome have degraded in the presence of the male killer, as their expression was made redundant by the *Wolbachia* feminization.

CONFLICTS ASSOCIATED WITH CYTOPLASMIC INCOMPATIBILITY–INDUCING SYMBIONTS MAY DRIVE THE EVOLUTION OF SPERMATOGENESIS AND MATE-CHOICE PATTERNS

The spread of cytoplasmic incompatibility–inducing symbionts produces different selection on male and female hosts. Because possession of a CI-inducing symbiont confers protection against CI, female hosts are selected to maintain and transmit the symbiont. In contrast, the incompatibility produced by the symbiont imposes a cost on the male host. There may also be additional costs associated with the presence of a symbiont in developing sperm bundles (e.g., Snook et al. 2000). Thus, selection favors modifiers that restrict the proliferation of the symbiont in males and prevent entry to, and modification of, the germline (Koehncke et al. 2009). Importantly, selection on the symbiont to maintain the CI phenotype is very weak when the symbiont is common (Orr 1996), such that an antagonistic arms race is unlikely and symbiont loss thus probable. This logic creates an evolutionary trajectory where the spread of a CI symbiont is followed by host selection to prevent incompatibility, leading to the loss of the symbiont over time.

Two observations support this evolutionary trajectory. First, there is considerable variation between host species in the expression of CI by given symbiont strains (Veneti et al. 2003). Second, cytoplasmic incompatibility–inducing symbionts do not persist within a host species over long periods of evolutionary time—sibling species pairs commonly either have different infection status (infected vs. uninfected) or carry different symbiont strains that do not share a common ancestor within the host matriline in which they are currently found (Raychoudhury et al. 2009). Although there are alternate explanations for this pattern (e.g., Turelli 1994), we believe they are most likely a manifestation of selection on male hosts to prevent entry into testes/germline modification.

In the above models, males are selected not to express CI, to avoid the cost of incompatible matings. Where cross-incompatibility exists, there may also be selection on females to avoid incompatible matings (Telschow et al. 2005). An example of this process can be seen in *Drosophila subquinaria*. *Drosophila recens* and *D. subquinaria* are sympatric in part of their range. *D. recens* carries a *Wolbachia* that causes incompatibility in crosses to *D. subquinaria* males. *D. subquinaria* females show behavioral avoidance of mating with *D. recens* males, but only where the two species are sympatric (Jaenike et al. 2006). The thesis is that *Wolbachia*-induced cross-incompatibility selects for "reinforcement" through females' refusing incompatible matings in the zone where hybridization is possible. Indeed, the evolution of premating barriers in *D. subquinaria* in sympatry with *D. recens* also creates behavioral isolation against allopatric males of the same species.

AREAS FOR FUTURE RESEARCH

We currently understand the rationale underlying reproductive parasitism and have a good working knowledge of the population biology of these symbionts when they exhibit a single phenotype (e.g., male killing). However, the impact of these agents in natural populations is not well established. Further, there are good reasons, based on the strength of selection they produce, to believe they will drive rapid evolution in the host, but the systems under selection remain unclear. Finally, our comprehension of the biology of these symbionts has been impeded by the historical tendency to separate symbionts into "beneficial" and "reproductive parasite," whereas it is now clear that a particular symbiont can have multiple impacts, directly benefiting their host in terms of providing "function" and manipulating host reproduction to enhance their spread. Below we outline how these areas of research may be approached in future studies.

The Mechanisms of Reproductive Parasite Action

Outside of cytoplasmic male sterility, our knowledge of the mode of action of reproductive parasites is weak. There is currently little functional knowledge of the microbial effectors important in inducing phenotype, with research limited to candidate genes expressed at appropriate phases of host development (Sinkins et al. 2005). There is more (but still limited) knowledge of the host pathways with which these effectors interact (see Tables 1–3), but in no case have the host genes affecting reproductive parasite action been defined.

Study of microbial effectors has been limited by the fastidious nature of most heritable symbionts, which have generally proven recalcitrant to transgenic manipulation (despite having been shown to have recombining genomes). Study of host systems that alter symbiont action has been limited by the nonmodel nature of the species. However, it is now straightforward and cheap to obtain genome sequences for arthropods, and we expect this to allow at least the genes involved in evolved host responses to be identified. Further, transgenic manipulation of nonmodel species is becoming more routine and may allow more-defined functional testing.

Mechanistic work of this kind has a high priority because it will both potentiate the use of symbionts in control and also uncover the foci for selection in host species that are subject to a reproductive parasite. Currently, the nature of host responses to reproductive parasites is at the level of high-level phenotypic description: "modifies CI strength," "rescues males from death," and "prevents symbiont transmission." We should be able to establish the extent to which suppression of phenotype is associated with changes in sex determination, which genes involved in testes biology are altered to reduce symbiont impact in testes, and whether processes of oogenesis/egg formation alter to prevent symbiont transmission. This will require a mix of functional and genetic analysis.

The Importance of Multiple Phenotypes

The population and evolutionary ecology of heritable symbionts has traditionally been viewed through a single-phenotype prism—a symbiont has one impact on a host. It is clear that this is not generally true. Symbionts may combine nonprotective benefit with sex-ratio distortion (Himler et al. 2011), combine protection and reproductive manipulation (Osborne et al. 2009), and may also combine reproductive manipulation with increased susceptibility to natural enemies (Graham et al. 2012).

Symbionts that have more than one phenotypic effect will be expected to have different dynamics to our current understanding under the single-phenotype models, and the intensity of the conflict with the nuclear genome will also differ. For symbionts that produce both natural enemy resistance and reproductive parasitism, two consequences seem likely:

1. A reproductive manipulation phenotype would allow a symbiont to persist even in the absence of an ecologically contingent benefit. The advantage gained from many symbiont phenotypes is ecologically contingent—they provide benefit only in the presence of a particular parasite or in a particular abiotic environment. This renders them susceptible to loss when the pressure that produces the benefit is relaxed (Oliver et al. 2008). A symbiont with additional reproductive parasitism may be maintained indefinitely, as it additionally possesses a drive mechanism that does not rely on ecological circumstance.
2. Symbiont frequency, and thus population sex ratio, may be altered by natural enemy pressure. The link between natural enemy pressure and population sex ratio is likely to create unusual dynamics, as population sex ratio can affect population growth/density, which will have an impact on the success of the natural enemy.

The multiphenotype model also changes the intensity, and potentially the direction, of cytoplasm–nuclear conflict. First, it alters whether the host is selected to remove or retain the symbiont. Beneficial symbionts do not engender selection to remove them and may produce selection for active maintenance. Fixation and maintenance is best recognized for symbionts that make an anabolic contribution to the host (Moran et al. 1993), but is also observed for some cases of natural enemy protection (Nakabachi et al. 2013). Combining beneficial impacts with sex-ratio distortion will alter the direction of selection when a symbiont is common; the nature of Fisherian selection on the host's sex ratio is such that a symbiont that alters host sex ratio will promote the spread of modifiers that prevent sex-ratio distortion/symbiont presence when the symbiont is at high frequency, notwithstanding any benefits of symbiont presence.

At longer evolutionary timescales, protective symbionts will clearly be vulnerable to mutation that leads to the acquisition of reproductive manipulation phenotypes. It is notable that even where "dual phenotypes" are not known, related symbiont microbes are parasitic or protective. Given the ease of movement of traits between bacteria, it is likely that it will be "easy" for symbionts to acquire additional phenotypes.

The acquisition of parasitic traits may change the direction of selection on the host from maintenance to removal, and may in part explain why protective symbionts, despite being beneficial, commonly do not have long evolutionary histories with their host. One prediction from this is that protection is likely to be maintained for longer periods of evolutionary time in host taxa in which common reproductive manipulations (e.g., male killing) are not beneficial. Conversely, protective symbiosis should be less common in groups in which reproductive parasitic phenotypes are common, as here protection does not contribute to long-term maintenance.

REFERENCES

**Reference is also in this collection.*

Adachi-Hagimori T, Miura K, Stouthamer R. 2008. A new cytogenetic mechanism for bacterial endosymbiont-induced parthenogenesis in Hymenoptera. *Proc Biol Sci* **275:** 2667–2673.

Agnew P, Koella JC. 1999. Constraints on the reproductive value of vertical transmission for a microsporidian parasite and its female-killing behaviour. *J Anim Ecol* **68:** 1010–1019.

Andreadis TG. 1991. Epizootiology of *Amblyospora connecticus* (Microsporida) in field populations of the saltmarsh mosquito, *Aedes cantator*, and the cyclopoid copepod, *Acnthocyclops vernalis*. *J Protozool* **37:** 174–182.

Arakaki N, Miyoshi T, Noda H. 2001. *Wolbachia*-mediated parthenogenesis in the predatory thrips *Franklinothrips vespiformis* (Thysanoptera: Insecta). *Proc Biol Sci* **268:** 1011–1016.

Bentley JK, Veneti Z, Heraty J, Hurst GDD. 2007. The pathology of embryo death caused by the male-killing *Spiroplasma* bacterium in *Drosophila nebulosa*. *BMC Biol* **5:** 9.

Bouchon D, Rigaud T, Juchault P. 1998. Evidence for widespread *Wolbachia* infection in isopod crustaceans: Molecular identification and host feminization. *Proc Biol Sci* **265:** 1081–1090.

Breeuwer JAJ, Werren JH. 1990. Microorganisms associated with chromosome destruction and reproductive isolation between two insect species. *Nature* **346:** 558–560.

Burt A, Trivers R. 1998. Selfish DNA and breeding systems in flowering plants. *Proc Biol Sci* **265:** 141–146.

Callaini G, Dallai R, Riparbelli MG. 1997. *Wolbachia*-induced delay of paternal chromatin condensation doe not prevent maternal chromosomes from entering anaphase in incompatible crosses of *Drosophila simulans*. *J Cell Sci* **110:** 271–280.

Cavalcanti AGL, Falcao DN, Castro LE. 1957. "Sex-ratio" in *Drosophila prosaltans*—A character due to interaction between nuclear genes and cytoplasmic factors. *Am Nat* **91:** 327–329.

Charlat S, Hornett EA, Fullard JH, Davies N, Roderick GK, Wedell N, Hurst GDD. 2007a. Extraordinary flux in sex ratio. *Science* **317:** 214–214.

Charlat S, Reuter M, Dyson EA, Hornett EA, Duplouy A, Davies N, Roderick GK, Wedell N, Hurst GDD. 2007b. Male-killing bacteria trigger a cycle of increasing male fatigue and female promiscuity. *Curr Biol* **17:** 273–277.

Cosmides L, Tooby J. 1981. Cytoplasmic inheritance and intragenomic conflict. *J Theor Biol* **89:** 83–129.

Dannowski J, Flor M, Telschow A, Hammerstein P. 2009. The effect of sibmating on the infection dynamics of male-killing bacteria. *Evolution* **63:** 2525–2534.

Dedeine F, Vavre F, Fleury F, Loppin B, Hochberg ME, Boulétreau M. 2001. Removing symiobitic *Wolbachia* bacteria specifically inhibits oogenesis in a parasitic wasp. *Proc Natl Acad Sci* **98:** 6247–6252.

Dunn AM, Andrews T, Ingrey H, Riley J, Wedell N. 2006. Strategic sperm allocation under parasitic sex-ratio distortion. *Biol Lett* **2:** 78–80.

Duron O, Hurst G. 2013. Arthropods and inherited bacteria: From counting the symbionts to understanding how symbionts count. *BMC Biol* **11:** 45.

Dyson EA, Hurst GDD. 2004. Persistence of an extreme sex-ratio bias in a natural population. *Proc Natl Acad Sci* **101:** 6520–6523.

Elnagdy S, Majerus MEN, Handley LJL. 2011. The value of an egg: Resource reallocation in ladybirds (Coleoptera: Coccinellidae) infected with male-killing bacteria. *J Evol Biol* **24:** 2164–2172.

Engelstadter J, Hurst GDD. 2006. Can maternally transmitted endosymbionts facilitate the evolution of haplodiploidy? *J Evol Biol* **19:** 194–202.

Fenton A, Johnson KN, Brownlie JC, Hurst GDD. 2011. Solving the *Wolbachia* paradox: Modeling the tripartite interaction between host, *Wolbachia*, and a natural enemy. *Am Nat* **178:** 333–342.

Ferree PM, Avery A, Azpurua J, Wilkes T, Werren JH. 2008. A bacterium targets maternally inherited centrosomes to kill males in *Nasonia*. *Curr Biol* **18:** 1409–1414.

Fine PEM. 1978. On the dynamics of symbiote-dependent cytoplasmic incompatibility in culicine mosquitoes. *J Invertebr Pathol* **31:** 10–18.

Giorgini M, Monti MM, Caprio E, Stouthamer R, Hunter MS. 2009. Feminization and the collapse of haplodiploidy in an asexual parasitoid wasp harboring the bacterial symbiont *Cardinium*. *Heredity (Edinb)* **102:** 365–371.

Giorgini M, Bernardo U, Monti MM, Nappo AG, Gebiola M. 2010. *Rickettsia* symbionts cause parthenogenetic reproduction in the parasitoid wasp *Pnigalio soemius* (Hymenoptera: Eulophidae). *Appl Environ Microbiol* **76:** 2589–2599.

Gorischek AM, Afkhami ME, Seifert EK, Rudgers JA. 2013. Fungal symbionts as manipulators of plant reproductive biology. *Am Nat* **181:** 562–570.

Gottlieb Y, Zchori-Fein E, Werren JH, Karr TL. 2002. Diploidy restoration in *Wolbachia*-infected *Muscidifurax uniraptor* (Hymenoptera: Pteromalidae). *J Invertebr Pathol* **81:** 166–174.

Graham RI, Grzywacz D, Mushobozi WL, Wilson K. 2012. *Wolbachia* in a major African crop pest increases susceptibility to viral disease rather than protects. *Ecol Lett* **15:** 993–1000.

Groenenboom MAC, Hogeweg P. 2002. Space and the persistence of male-killing endosymbionts in insect populations. *Proc Biol Sci* **269:** 2509–2518.

Hamilton PT, Perlman SJ. 2013. Host defense via symbiosis in *Drosophila*. *PLoS Pathog* **9:** e1003808.

Hedges LM, Brownlie JC, O'Neill SL, Johnson KN. 2008. *Wolbachia* and virus protection in insects. *Science* **322:** 702–702.

Himler AG, Adachi-Hagimori T, Bergen JE, Kozuch A, Kelly SE, Tabashnik BE, Chiel E, Duckworth VE, Dennehy TJ, Zchori-Fein E, et al. 2011. Rapid spread of a bacterial symbiont in an invasive whitefly is driven by fitness benefits and female bias. *Science* **332:** 254–256.

Hiroki M, Kato Y, Kamito T, Miura K. 2002. Feminization of genetic males by a symbiotic bacterium in a butterfly, *Eurema hecabe* (Lepidoptera: Pieridae). *Naturwissenschaften* **89:** 167–170.

Hoffmann AA, Montgomery BL, Popovici J, Iturbe-Ormaetxe I, Johnson PH, Muzzi F, Greenfield M, Durkan M, Leong YS, Dong Y, et al. 2011. Successful establishment of *Wolbachia* in *Aedes* populations to suppress dengue transmission. *Nature* **476:** 454–457.

Hornett EA, Charlat S, Duplouy AMR, Davies N, Roderick GK, Wedell N, Hurst GDD. 2006. Evolution of male killer suppression in a natural population. *PLoS Biol* **4:** 1643–1648.

Hornett EA, Duplouy AMR, Davies N, Roderick GK, Wedell N, Hurst GDD, Charlat S. 2008. You can't keep a good parasite down: Evolution of a male-killer suppressor uncovers cytoplasmic incompatibility. *Evolution* **62:** 1258–1263.

Hornett EA, Charlat S, Wedell N, Jiggins CD, Hurst GDD. 2009. Rapidly shifting sex ratio across a species range. *Curr Biol* **19:** 1628–1631.

Hornett EA, Engelstadter J, Hurst GDD. 2010. Hidden cytoplasmic incompatibility alters the dynamics of male-killer/host interactions. *J Evol Biol* **23:** 479–487.

Hoshizaki S, Shimada T. 1995. PCR-based detection of *Wolbachia*, cytoplasmic microorganisms, infected in natural populations of *Laodelphax striatellus* (Homoptera: Delphacidae) in central Japan: Has the distribution of *Wolbachia* spread recently? *Insect Mol Biol* **4:** 237–243.

Huigens ME, Hohmann CL, Luck RF, Gort G, Stouthamer R. 2004. Reduced competitive ability due to *Wolbachia* infection in the parasitoid wasp *Trichogramma kaykai*. *Entomol Exp Appl* **110:** 115–123.

Hunter MS, Perlman SJ, Kelly SE. 2003. A bacterial symbiont in the *Bacteroidetes* induces cytoplasmic incompatibility in the parasitoid wasp *Encarsia pergandiella*. *Proc Biol Sci* **270:** 2185–2190.

Hurst LD. 1991. The incidences and evolution of cytoplasmic male killers. *Proc Biol Sci* **244:** 91–99.

Hurst GDD, Majerus MEN, Walker LE. 1993. The importance of cytoplasmic male killing elements in natural populations of the two spot ladybird, *Adalia bipunctata* (Linnaeus) (Coleoptera: Coccinellidae). *Biol J Linn Soc* **49:** 195–202.

Hurst LD, Atlan A, Bengtsson B. 1996. Genetic conflicts. *Q Rev Biol* **71:** 317–364.

Hurst GDD, Jiggins FM, von der Schulenburg JHG, Bertrand D, West SA, GoriachevaII, Zakharov IA, Werren JH, Stouthamer R, Majerus MEN. 1999a. Male-killing *Wolbachia* in two species of insect. *Proc Biol Sci* **266:** 735–740.

Hurst GDD, von der Schulenburg JHG, Majerus TMO, Bertrand D, Zakharov IA, Baungaard J, Volkl W, Stouthamer R, Majerus MEN. 1999b. Invasion of one insect species, *Adalia bipunctata*, by two different male-killing bacteria. *Insect Mol Biol* **8:** 133–139.

Hurst GDD, Johnson AP, von der Schulenburg JHG, Fuyama Y. 2000. Male-killing *Wolbachia* in *Drosophila*: A temperature-sensitive trait with a threshold bacterial density. *Genetics* **156:** 699–709.

Jaenike J. 2007. Spontaneous emergence of a new *Wolbachia* phenotype. *Evolution* **61:** 2244–2252.

Jaenike J, Dyer KA, Cornish C, Minhas MS. 2006. Asymmetrical reinforcement and *Wolbachia* infection in *Drosophila*. *PLoS Biol* **4:** 1852–1862.

Jahnke M, Smith JE, Dubuffet A, Dunn AM. 2013. Effects of feminizing microsporidia on the masculinizing function of the androgenic gland in *Gammarus duebeni*. *J Invertebr Pathol* **112:** 146–151.

Jeong G, Stouthamer R. 2005. Genetics of female functional virginity in the parthenogenesis-*Wolbachia* infected parasitoid wasp *Telenomus nawai* (Hymenoptera: Scelionidae). *Heredity (Edinb)* **94:** 402–407.

Jiggins FM, Hurst GDD, Jiggins CD, von der Schulenburg JHG, Majerus MEN. 2000a. The butterfly *Danaus chrysippus* is infected by a male-killing *Spiroplasma* bacterium. *Parasitology* **120:** 439–446.

Jiggins FM, Hurst GDD, Majerus MEN. 2000b. Sex-ratio-distorting *Wolbachia* causes sex-role reversal in its butterfly host. *Proc Biol Sci* **267:** 69–73.

Jiggins FM, Randerson JP, Hurst GDD, Majerus MEN. 2002. How can sex ratio distorters reach extreme prevalences? Male-killing *Wolbachia* are not suppressed and have near-perfect vertical transmission efficiency in *Acraea encedon*. *Evolution* **56:** 2290–2295.

Kageyama D, Traut W. 2004. Opposite sex-specific effects of *Wolbachia* and interference with the sex determination of its host *Ostrinia scapulalis*. *Proc Biol Sci* **271:** 251–258.

Kaltz O, Koella JC. 2003. Host growth conditions regulate the plasticity of horizontal and vertical transmission in *Holospora undulata*, a bacterial parasite of the protozoan *Paramecium caudatum*. *Evolution* **57:** 1535–1542.

Kellen WR, Chapman HC, Clark TB, Lindegren JE. 1965. Host-parasite relationships of some *Thelohania* from mosquitoes (Nosematidae: Microsporidia). *J Invertebr Pathol* **7:** 161–166.

Kobayashi Y, Telschow A. 2010. Cytoplasmic feminizing elements in a two-population model: Infection dynamics, gene flow modification, and the spread of autosomal suppressors. *J Evol Biol* **23:** 2558–2568.

Koehncke A, Telschow A, Werren JH, Hammerstein P. 2009. Life and death of an influential passenger: *Wolbachia* and the evolution of CI-modifiers by their hosts. *PLoS ONE* **4:** e4425.

Koivisto RKK, Braig HK. 2003. Microorganisms and parthenogenesis. *Biol J Linn Soc* **79:** 43–58.

Koop JL, Zeh DW, Bonilla MM, Zeh JA. 2009. Reproductive compensation favours male-killing *Wolbachia* in a live-bearing host. *Proc Biol Sci* **276:** 4021–4028.

Kremer N, Charif D, Henri H, Bataille M, Prevost G, Kraaijeveld K, Vavre F. 2009. A new case of *Wolbachia* dependence in the genus *Asobara*: Evidence for parthenogenesis induction in *Asobara japonica*. *Heredity (Edinb)* **103:** 248–256.

Kriesner P, Hoffmann AA, Lee SF, Turelli M, Weeks AR. 2013. Rapid sequential spread of two *Wolbachia* variants in *Drosophila simulans*. *PLoS Pathog* **9:** e1003607.

Kubo T, Kitazaki K, Matsunaga M, Kagami H, Mikami T. 2011. Male sterility-inducing mitochondrial genomes: How do they differ? *Crit Rev Plant Sci* **30:** 378–400.

Landmann F, Orsi GA, Loppin B, Sullivan W. 2009. *Wolbachia*-mediated cytoplasmic incompatibility is associated with impaired histone deposition in the male pronucleus. *PLoS Pathog* **5:** e1000343.

Laven H. 1951. Crossing experiments with *Culex* strains. *Evolution* **5:** 370–375.

Lawson ET, Mousseau TA, Klaper R, Hunder MD, Werren JH. 2001. *Rickettsia* associated with male-killing in a buprestid beetle. *Heredity (Edinb)* **86:** 497–505.

Lewis D. 1941. Male sterility in natural populations of hermaphrodite plants: The equilibrium between females and hermaphrodites to be expected with different types of inheritance. *New Phytol* **40:** 56–63.

Linklater JR, Wertheim B, Wigby S, Chapman T. 2007. Ejaculate depletion patterns evolve in response to experimental manipulation of sex ratio in *Drosophila melanogaster*. *Evolution* **61:** 2027–2034.

Luo DP, Xu H, Liu ZL, Guo JX, Li HY, Chen LT, Fang C, Zhang QY, Bai M, Yao N, et al. 2013. A detrimental mitochondrial-nuclear interaction causes cytoplasmic male sterility in rice. *Nat Genet* **45:** 573–577.

Lus YY. 1947. Some rules of reproduction in populations of *Adalia bipunctata*: II. Non-male strains in populations. *Dokl Akad Nauk SSSR* **57:** 951–954.

Magalon H, Nidelet T, Martin G, Kaltz O. 2010. Host growth conditions influence experimental evolution of life history and virulence of a parasite with vertical and horizontal transmission. *Evolution* **64:** 2126–2138.

Magni GE. 1953. "Sex-ratio": A non-Mendelian character in *Drosophila bifasciata*. *Nature* **172:** 81.

Majerus TMO, Majerus MEN. 2010. Intergenomic arms races: Detection of a nuclear rescue gene of male-killing in a ladybird. *PLoS Pathog* **6:** e1000987.

Malogolowkin C. 1958. Maternally inherited "sex-ratio" condition in *Drosophila willistoni* and *Drosophila paulistorum*. *Genetics* **43:** 274–286.

Martin J, Chong T, Ferree PM. 2013. Male killing *Spiroplasma* preferentially disrupts neural development in the *Drosophila melanogaster* embryo. *PLoS ONE* **8:** e79368.

McCauley DE, Brock MT. 1998. Frequency-dependent fitness in *Silene vulgaris*, a gynodioecious plant. *Evolution* **52:** 30–36.

McCauley DE, Taylor DR. 1997. Local population structure and sex ratio: Evolution in gynodioecious plants. *Am Nat* **150:** 406–419.

Moran NA, Munson MA, Baumann P, Ishikawa H. 1993. A molecular clock in endosymbiotic bacteria is calibrated using the insect hosts. *Proc Biol Sci* **253:** 167–171.

Moran NA, McCutcheon JP, Nakabachi A. 2008. Genomics and evolution of heritable bacterial symbionts. *Annu Rev Genet* **42:** 165–190.

Moreau J, Rigaud T. 2003. Variable male potential for reproduction: High male mating capacity as an adaptation to prevent parasite-induced excess of females. *Proc Biol Sci* **270:** 1535–1540.

Moreau J, Bertin A, Caubet Y, Rigaud T. 2001. Sexual selection in an isopod with *Wolbachia*-induced sex reversal: Males prefer real females. *J Evol Biol* **14:** 388–394.

Moret Y, Juchault P, Rigaud T. 2001. *Wolbachia* endosymbiont responsible for cytoplasmic incompatibility in a terrestrial crustacean. *Heredity (Edinb)* **86:** 325–332.

Morimoto S, Nakai M, Ono A, Kunimi Y. 2001. Late male-killing phenomenon found in a Japanese population of the oriental tea tortrix, *Homona magnanima* (Lepidoptera: Tortricidae). *Heredity (Edinb)* **87:** 435–440.

Nakabachi A, Ueoka R, Oshima K, Teta R, Mangoni A, Gurgui M, Oldham NJ, van Echten-Deckert G, Okamura K, Yamamoto K, et al. 2013. Defensive bacteriome symbiont with a drastically reduced genome. *Curr Biol* **23:** 1478–1484.

Nakanishi K, Hoshino M, Nakai M, Kunimi Y. 2008. Novel RNA sequences associated with late male killing in *Homona magnanima*. *Proc Biol Sci* **275:** 1249–1254.

Narita S, Kageyama D, Nomura M, Fukatsu T. 2007. Unexpected mechanism of symbiont-induced reversal of insect sex: Feminizing *Wolbachia* continuously acts on the butterfly *Eurema hecabe* during larval development. *Appl Environ Microbiol* **73:** 4332–4341.

Negri I, Franchini A, Gonella E, Daffonchio D, Mazzoglio PJ, Mandrioli M, Alma A. 2009. Unravelling the *Wolbachia* evolutionary role: The reprogramming of the host genomic imprinting. *Proc Biol Sci* **276:** 2485–2491.

Oliver KM, Campos J, Moran NA, Hunter MS. 2008. Population dynamics of defensive symbionts in aphids. *Proc Biol Sci* **275:** 293–299.

Olson MS, McCauley DE, Taylor D. 2005. Genetics and adaptation in structured populations: Sex ratio evolution in *Silene vulgaris*. *Genetica* **123:** 49–62.

Orr MR. 1996. Host manipulation by *Wolbachia* is a neutral trait within single populations. *Anim Behav* **51:** 1183–1185.

Osborne SE, Leong YS, O'Neill SL, Johnson KN. 2009. Variation in antiviral protection mediated by different *Wolbachia* strains in *Drosophila simulans*. *PLoS Pathog* **5:** e1000656.

Pannebakker BA, Pijnacker LP, Zwaan BJ, Beukeboom LW. 2004. Cytology of *Wolbachia*-induced parthenogenesis in *Leptopilina clavipes* (Hymenoptera: Figitidae). *Genome* **47:** 299–303.

Pannebakker BA, Schidlo NS, Boskamp GJF, Dekker L, Van Dooren TJM, Beukeboom LW, Zwaan BJ, Brakefield PM, Van Alphen JJM. 2005. Sexual functionality of *Leptopilina clavipes* (Hymenoptera: Figitidae) after reversing *Wolbachia*-induced parthenogenesis. *J Evol Biol* **18:** 1019–1028.

Pijls JWAM, van Steenbergen HJ, van Alphen JJM. 1996. Asexuality cured: The relations and differences between sexual and asexual *Apoanagyrus diversicornis*. *Heredity (Edinb)* **76:** 506–513.

Poot P. 1997. Reproductive allocation and resource compensation in male-sterile and hermaphroditic plants of *Plantago lanceolata* (Plantaginaceae). *Am J Bot* **84:** 1256.

Randerson JP, Smith NGC, Hurst LD. 2000. The evolutionary dynamics of male-killers and their hosts. *Heredity (Edinb)* **84:** 152–160.

Raychoudhury R, Baldo L, Oliveira D, Werren JH. 2009. Modes of acquisition of *Wolbachia*: Horizontal transfer, hybrid introgression, and codivergence in the *Nasonia* species complex. *Evolution* **63:** 165–183.

Rigaud T, Juchault P. 1992. Genetic control of the vertical transmission of a cytoplasmic sex factor in *Armadillidium vulgare*. *Heredity (Edinb)* **68:** 47–52.

Riparbelli MG, Giordano R, Ueyama M, Callaini G. 2012. Wolbachia-mediated male killing is associated with defective chromatin remodeling. *PLoS ONE* **7:** e30045.

Russell JE, Stouthamer R. 2011. The genetics and evolution of obligate reproductive parasitism in *Trichogramma pretiosum* infected with parthenogenesis-inducing *Wolbachia*. *Heredity (Edinb)* **106:** 58–67.

Sanada-Morimura S, Matsumura M, Noda H. 2013. Male killing caused by a *Spiroplasma* symbiont in the small brown planthopper, *Laodelphax striatellus*. *Heredity* **104:** 821–829.

Sasaki T, Kubo T, Ishikawa H. 2002. Interspecific transfer of *Wolbachia* between two lepidopteran insects expressing cytoplasmic incompatibility: A *Wolbachia* variant naturally infecting *Cadra cautella* causes male killing in *Ephestia kuehniella*. *Genetics* **162:** 1313–1319.

Segoli M, Stouthamer R, Stouthamer CM, Rugman-Jones P, Rosenheim JA. 2013. The effect of *Wolbachia* on the lifetime reproductive success of its insect host in the field. *J Evol Biol* **26:** 2716–2720.

Shoop WL. 1991. Vertical transmission of helminths: Hypobiosis and amphiparatenesis. *Parasitol Today* **7:** 51–54.

Shull HF. 1948. An all-female strain of lady beetles with reversion to normal sex ratios. *Am Nat* **82:** 241–251.

Simmonds HW. 1923a. All female families of *Hypolimnas bolina* L., bred in Fiji by H. W. Simmonds. *Proc R Entomol Soc Lond* **1923:** x–xii.

Simmonds HW. 1923b. Families of *Hypolimnas bolina* L. bred from known female parents from different Fijian islands. *Trans R Entomol Soc Lond* **1923:** 651–662.

Simmonds HW. 1926. Sex ratio of *Hypolimnas bolina* L., in Viti Levu, Fiji. *Proc R Entomol Soc Lond* **1:** 29–32.

Simmonds HW. 1927. Sex-ratio of *Hypolimnas bolina* L. in Viti Levu, Fiji. *Proc R Entomol Soc Lond* **2:** 5–10.

Simmonds HW. 1928. Mr. H. W. Simmonds' conclusion that all-female-producing females form a persistent strain in Suva. *Proc R Entomol Soc Lond* **3:** 43–44.

Simmonds HW. 1930. Further notes on *Hypolimnas bolina* L. in Fiji. *Proc R Entomol Soc Lond* **5:** 75–77.

Simon J-C, Boutin S, Tsuchida T, Koga R, Le Gallic J-F, Frantz A, Outreman Y, Fukatsu T. 2011. Facultative symbiont infections affect aphid reproduction. *PLoS ONE* **6:** e21831.

Sinkins SP, Walker T, Lynd AR, Steven AR, Makepeace BL, Godfray HCJ, Parkhill J. 2005. *Wolbachia* variability and host effects on crossing type in *Culex* mosquitoes. *Nature* **436:** 257–260.

Snook RR, Cleland SY, Wolfner MF, Karr TL. 2000. Offsetting effects of *Wolbachia* infection and heat shock on sperm production in *Drosophila simulans*: Analysis of fecundity, fertility and accessory gland proteins. *Genetics* **155:** 167–178.

Stouthamer R, Kazmer D. 1994. Cytogenetics of microbe-associated parthenogenesis and its consequence for gene flow in *Trichogramma* wasps. *Heredity (Edinb)* **73:** 317–327.

Stouthamer R, Luck RF. 1993. Influence of microbe-associated parthenogenesis on the fecundity of *Trichogramma deion* and *T. pretiosum*. *Entomol Exp Appl* **67:** 183–192.

Stouthamer R, Luck RF, Hamilton WD. 1990. Antibiotics cause parthenogenetic *Trichogramma* (Hymenoptera, Trichogrammatidae) to revert to sex. *Proc Natl Acad Sci* **87:** 2424–2427.

Stouthamer R, Russell JE, Vavre F, Nunney L. 2010. Intragenomic conflict in populations infected by parthenogenesis inducing *Wolbachia* ends with irreversible loss of sexual reproduction. *BMC Evol Biol* **10:** 229.

Sugimoto TN, Ishikawa Y. 2012. A male-killing *Wolbachia* carries a feminizing factor and is associated with degradation of the sex-determining system of its host. *Biol Lett* **8:** 412–415.

Sugimoto TN, Fujii T, Kayukawa T, Sakamoto H, Ishikawa Y. 2012. Expression of a *doublesex* homologue is altered in sexual mosaics of *Ostrinia scapulalis* moths infected with *Wolbachia*. *Insect Biochem Mol Biol* **40:** 847–854.

Taylor DR. 1990. Evolutionary consequences of cytoplasmic sex-ratio distorters. *Evol Ecol* **4:** 235–248.

Teixeira L, Ferreira A, Ashburner M. 2008. The bacterial symbiont *Wolbachia* induces resistance to RNA viral infections in *Drosophila melanogaster*. *Plos Biol* **12:** 2753–2763.

Telschow A, Hammerstein P, Werren JH. 2005. The effect of *Wolbachia* versus genetic incompatibilities on reinforcement and speciation. *Evolution* **59:** 1607–1619.

Terry RS, Smith JE, Sharpe RG, Rigaud T, Littlewood DTJ, Ironside JE, Rollinson D, Bouchon D, MacNeil C, Dick JTA. 2004. Widespread vertical transmission and associated host sex-ratio distortion within the eukaryotic phylum Microspora. *Proc Biol Sci* **271:** 1783–1789.

Timmermans MTN, Ellers J. 2009. *Wolbachia* endosymbiont is essential for egg hatching in a parthenogenetic arthropod. *Evol Ecol* **23:** 931–942.

Tram U, Sullivan W. 2002. Role of delayed nuclear envelope breakdown and mitosis in *Wolbachia*-induced cytoplasmic incompatibility. *Science* **296:** 1124–1126.

Tram U, Fredrick K, Werren JH, Sullivan W. 2006. Paternal chromosome segregation during the first mitotic division determines *Wolbachia*-induced cytoplasmic incompatibility phenotype. *J Cell Sci* **119:** 3655–3663.

Turelli M. 1994. Evolution of incompatibility-inducing microbes and their hosts. *Evolution* **48:** 1500–1513.

Turelli M, Hoffmann AA. 1991. Rapid spread of an inherited incompatibility in California *Drosophila*. *Nature* **353:** 440–442.

Turelli M, Hoffmann AA, McKechnie SW. 1992. Dynamics of cytoplasmic incompatibility and mtDNA variation in natural *Drosophila simulans* populations. *Genetics* **132:** 713–723.

Ubeda F, Normark BB. 2006. Male killers and the origins of paternal genome elimination. *Theor Popul Biol* **70:** 511–526.

Vandel A. 1941. Recherches sur la génétique et la sexualité des Isopodes terrestre: VI. Les phénomènes de monologénie chez Onisocides [Research on genetics and sexuality of the terrestrial isopods: VI. The phenomena of unisexual broods in Oniscidea]. *Bull Biol de la France et Belgique* **75:** 316–363.

Vavre F, Fleury F, Varaldi J, Fouillet P, Boulétreau M. 2000. Evidence for female mortality in *Wolbachia*-mediated cytoplasmic incompatibility in haplodiploid insects: Epidemiologic and evolutionary consequences. *Evolution* **54:** 191–200.

Veneti Z, Clark ME, Zabalou S, Karr TL, Savakis B, Bourtzis K. 2003. Cytoplasmic incompatibility and sperm cyst infection in different *Drosophila*/host associations. *Genetics* **164:** 545–552.

Veneti Z, Bentley JK, Koana T, Braig HR, Hurst GDD. 2005. A functional dosage compensation complex required for male killing in *Drosophila*. *Science* **307:** 1461–1463.

* Wade MJ. 2014. Paradox of mother's curse and the maternally provisioned offspring microbiome. *Cold Spring Harb Perspect Biol* doi: 10.1101/cshperspect.a017541.

Walker T, Johnson PH, Moreira LA, Iturbe-Ormaetxe I, Frentiu FD, McMeniman CJ, Leong YS, Dong Y, Axford J, Kriesner P, et al. 2011. The *w*Mel *Wolbachia* strain blocks dengue and invades caged *Aedes aegypti* populations. *Nature* **476:** 450–453.

Weeks AR, Breeuwer JAJ. 2001. *Wolbachia* induced parthenogenesis in a genus of phytophagous mite. *Proc Biol Sci* **268:** 2245–2251.

Weeks AR, Marec F, Breeuwer JAJ. 2001. A mite species that consists entirely of haploid females. *Science* **292:** 2479–2482.

Werren JH. 1987. The coevolution of autosomal and cytoplasmic sex ratio factors. *J Theor Biol* **124:** 317–334.

Werren JH, Hurst GDD, Zhang W, Breeuwer JAJ, Stouthamer R, Majerus MEN. 1994. Rickettsial relative associated with male killing in the ladybird beetle (*Adalia bipunctata*). *J Bacteriol* **176:** 388–394.

Wigby S, Chapman T. 2004. Female resistance to male harm evolves in response to manipulation of sexual conflict. *Evolution* **58:** 1028–1037.

Williamson DL, Poulson DF. 1979. Sex ratio organisms (Spiroplasmas) of *Drosophila*. In *The mycoplasmas* (ed. Whitcomb RF, Tully JG), pp. 175–208. Academic, New York.

Yeap HL, Mee P, Walker T, Weeks AR, O'Neill SL, Johnson P, Ritchie SA, Richardson KM, Doig C, Endersby NM, et al. 2011. Dynamics of the "popcorn" *Wolbachia* infection in outbred *Aedes aegypti* informs prospects for mosquito vector control. *Genetics* **187:** 583–595.

Yen JH, Barr AR. 1971. New hypothesis of the cause of cytoplasmic incompatibility in *Culex pipiens*. *Nature* **232:** 657–658.

Zeh DW, Zeh JA, Bonilla MM. 2005. *Wolbachia*, sex ratio bias and apparent male killing in the harlequin beetle riding pseudoscorpion. *Heredity (Edinb)* **95:** 41–49.

Sexual Conflict and Sperm Competition

Dominic A. Edward[1], Paula Stockley[1], and David J. Hosken[2]

[1]Mammalian Behaviour and Evolution Group, Institute of Integrative Biology, University of Liverpool, Leahurst Campus, Neston CH64 7TE, United Kingdom

[2]Centre for Ecology and Conservation, University of Exeter, Cornwall Campus, Penryn, Cornwall TR10 9FE, United Kingdom

Correspondence: D.J.Hosken@exeter.ac.uk

Traits that increase a male's fertilization success during sperm competition can be harmful to females and therefore represent a source of sexual conflict. In this review, we consider the variety of male adaptations to sperm competition (MASC) that may give rise to sexual conflict—including mate guarding, prolonged copulations, the transfer of large numbers of sperm, and the manipulation of females through nonsperm components of the ejaculate. We then reflect on the fitness economics influencing the escalation of these sexual conflicts, considering the likelihood of females evolving traits to offset the negative effects of MASC when compared with the strong selection on males that lead to MASC. We conclude by discussing the potential evolutionary outcomes of sexual conflict arising from MASC, including the opportunities for females to mitigate conflict costs and the prospects for conflict resolution.

Male mate competition can lead to sexual conflict because male traits that increase the chances of securing a mate can reduce female fitness. For example, intense male–male competition for females in yellow dungflies can lead to females drowning in dung (Parker 1970a), or the large mandibles used to win male–male competition in *Gnathocerus* flour beetles reduce female fitness. This occurs via a chain of intra- and intersexual genetic correlations that results in female body morphology deviating from a fitness-optimizing feminine shape (Harano et al. 2010). Male–male competition can also continue after mating in the form of sperm competition, which arises whenever the ejaculates of two or more males are in direct competition to fertilize the same set of ova (Parker 1970b, 1998). Sperm competition will select for male traits that increase a male's fertilization success, yet these traits can also be harmful to females and therefore represent a source of sexual conflict, even in situations in which females initiate polyandry (Stockley 1997).

In this article, we begin by briefly reviewing some of the many male adaptations to sperm competition (MASC) and how these can give rise to sexual conflict. We then go on to discuss factors that may govern the extent of sexual conflict arising from MASC. In particular, we reflect on the importance of the relative costs and benefits of these traits to males and females, as understanding the fitness economics of sexual conflict is central to determining the potential evolutionary trajectories of MASC. We con-

Cite this article as *Cold Spring Harb Perspect Biol* doi: 10.1101/cshperspect.a017707

clude by discussing avenues through which females could mitigate costs and the prospects for conflict resolution.

CAUSES OF SEXUAL CONFLICT

If the expression of a trait increases the fitness of one sex while reducing fitness of the other, sexual conflict occurs. In principle, it is straightforward to decide whether there is sexual conflict over a trait by simply asking, would trait values change if expression levels were completely controlled by only one sex. For example, if female influence on copulation duration or sperm storage is removed, does copulation duration increase/decrease or do sperm storage patterns differ (e.g., Hellriegel and Bernasconi 2000)?

Within this general framework there exists a wide variety of MASC that can lead to sexual conflict. Male adaptations include mating behaviors, such as mate guarding and extended or repeated copulation; alterations to ejaculate size or composition, including sperm number; and the use of copulatory plugs and strategic allocation of mating effort, including male mate choice (Parker 1970b, 1998; Edward and Chapman 2011). Some MASC may reduce either the risk or intensity of sperm competition (Parker 1998). For example, mate guarding or mating plugs made from ejaculate components (e.g., Fenton 1984) or detached male genitals (e.g., Nessler et al. 2007; Kuntner et al. 2009) can prevent or delay female remating, and a similar function has been suggested for copulatory stimulation (Stockley 2002). Indeed, any male trait that reduces female remating rates may have evolved, in part at least, through selection arising from the risk of sperm competition. Other MASC, rather than reducing competition, can increase male fitness when sperm directly compete. For example, adaptations that function in physically removing or debilitating rival sperm (e.g., Cordoba-Aguilar et al. 2003). MASC can also be classified as benefiting either the earlier mating male over subsequent competitors (i.e., defensive traits), or benefiting later mating males when a female has previously mated (i.e., offensive traits). Such adaptations can each increase male fertilization success when sperm competition does arise and can result in sexual conflict.

The majority of examples of sexual conflict arising from MASC reviewed here relate to interlocus conflict (i.e., conflicting alleles are present at different loci in males and females). For example, a sexual conflict over mating duration may be determined by male persistence traits that are expressed at one locus and female resistance traits expressed at a different locus. However, MASC can potentially also give rise to intralocus sexual conflict (i.e., when an allele at a single locus has contrasting fitness effects when expressed in either males or females). Although there is currently limited direct evidence of this (Lewis et al. 2008), in *Callosobruchus maculatus*, males that are more harmful to their mates, which may increase male fitness, sire daughters that are more susceptible to harm (Gay et al. 2011).

Male Mating Behavior

Mate Guarding

Males can prevent or reduce the risk of sperm competition by guarding their mates. Mate guarding is found in many taxa and is manifest as either extended physical contact (including extended copulation—see below) or remaining close to recent partners so that copulations with other males can be delayed or prevented.

The prolonged association between males and females that occurs with mate guarding must generate opportunity costs for both sexes, as the time involved might otherwise be spent on other fitness-related activities such as foraging (e.g., Westneat 1994; Komdeur 2001). Mate guarding may be particularly costly for females if guarding males have low fertility, as reported for cockroaches, *Nauphoeta cinerea*, in which males that have become sperm depleted will continue to enforce female monogamy (Montrose et al. 2004). In some cases, mate guarding can also limit a female's ability to exercise cryptic mate choice, for example by delaying sperm ejection (Helfenstein et al. 2003), or may limit other aspects of behavior such as choice of oviposition sites (Smith et al. 2002). However, mate guarding can also be beneficial for females if it reduces sexual harassment. Indeed, female blackbirds, *Turdus merula*, appear to actively prolong sexual activity to induce mate guarding, as they benefit from reduced harassment by other males and also receive greater paternal care of offspring from their guarders as a result (Wysocki and Halupka 2004).

Nonetheless, if mate guarding is costly for females, selection will favor female traits that reduce guarding duration. For example, females may struggle to escape from males, as in sepsid flies (Parker 1972; Hosken et al. 2003). However, this can generate additional costs for females (e.g., physical injury) and the extent of female resistance will depend on the relative costs of guarding and resistance. Where mate guarding is important to male reproductive success, as it is in sepsids (guarding males only mate after egg laying) (Parker 1972; Hosken et al. 2003), then males should respond to female resistance, potentially leading to coevolutionary arms races between male guarding and female resistance strategies. An excellent example of this is the correlated evolution of male grasping and female antigrasping structures found across 15 species of gerrid water striders (Arnqvist and Rowe 2002).

Extended or Repeated Copulations

Another MASC is to prolong mating, either by extending copula duration, pre- or postejaculation (e.g., Cordoba-Aguilar et al. 2009), or copulating repeatedly with the same female (e.g., Laird et al. 2004; Stockley and Preston 2004; Preston and Stockley 2006). There are many examples of males strategically altering copulation to benefit either from the guarding effect of this behavior (see above) or from transferring larger ejaculates to potentially achieve greater fertilization success (e.g., Wedell 1998; Martin and Hosken 2002).

Extended or repeated copulations typically allow males to pass more sperm or other nonsperm ejaculate components to their mates (see the section Ejaculate Variation below) and can also allow males to displace the ejaculates of previous rivals (Parker and Simmons 2000). This is clearly beneficial in situations in which sperm number is the primary determinant of fertilization success during sperm competition (Parker 1984, 1998). Extended or repeated copulations may also give males greater opportunity to assess female mating status and to strategically adjust mating effort accordingly (see the section Strategic Allocation of Male Mating Effort below). Additionally, it has been suggested that males may repeatedly attempt to mate with the same female to assess receptivity to further copulations (Sato and Kohama 2007).

Prolonged copulations are likely to generate costs of mate guarding for females because the time might otherwise be spent on other activities. In addition, prolonged and repeated copulations may increase the risk of physical injury or disease transmission associated with mating (Hosken et al. 2003). If increased copulation duration or frequency is costly for females, then selection will favor traits that reduce these costs. For example, females may struggle to terminate copulation sooner, although males extend copulation beyond the female optima (Mazzi et al. 2009) and may even control kicking duration (Wilson and Tomkins 2014). Where males have evolved to physically prevent females escaping, struggles such as these may lead to physical injury of females (e.g., Hotzy and Arnqvist 2009; Johns et al. 2009). Nevertheless, an assessment of the level of sexual conflict arising from prolonged copulations should also consider potential benefits to females. Costs may be mitigated if females benefit from, for example, an increased ability to assess male quality or reduced costs of harassment from other males (e.g., Eberhard 1996). Females may also directly benefit from the receipt of more sperm or nuptial gifts (Arnqvist and Nilsson 2000).

Ejaculate Variation

Sperm Number, Quality, and Morphology

When sperm compete numerically, a male may increase his fertilization success simply by ejaculating more sperm (Parker 1998). This response to sperm competition risk has been shown repeatedly (e.g., Gage 1991; Simmons et al. 1993; Wedell 1998; Martin and Hosken 2002). Males can increase sperm number by increasing copulation duration, ejaculating repeatedly with the same female (see section above) or by increasing sperm transfer rates. This male strategy has many potential costs to females, including reducing female control of paternity and increasing the risk of polyspermy. This could explain why the female reproductive tract is typically hostile to sperm (Birkhead et al. 1993) and females may eject unwanted sperm (e.g., Manier et al. 2010; Dean et al. 2011b). Alternatively, when the intensity of sperm competition is high, or anticipated payoffs of mating are relatively low, males may benefit from transferring fewer sperm (Parker et al. 1996), or searching for alternative mating opportunities (Schwagmeyer and Parker 1990). In rainbow darters, *Etheostoma caeruleum*, larger males often forego mating opportunities in which there is a high intensity of sperm competition, instead conserving resources in anticipation of less competitive future matings (Fuller 1998). This may be costly for females if they benefit from polyandry or from mating with larger males.

In addition to varying sperm numbers, adjusting sperm quality may also be a MASC (Ball and Parker 1996). For example, sperm competition risk can influence sperm morphology and performance (Gage and Morrow 2003). However, selection for increased sperm swimming speed/fertilization efficiency is thought to be a key factor in the incidence and impact of polyspermy (Snook et al. 2011). So although highly competitive sperm may be advantageous for males, they may result in ova loss for females. If so, increased sperm competition, and associated MASC, may select for tolerance to polyspermy, which could explain why in some taxa, such as birds, polyspermy does not result in fertilization failure (Snook et al. 2011).

Greater investment in sperm number or sperm quality will ultimately trade off against other traits (Parker 1998). Sperm number and quality may also trade off against each other, in which case selection to increase either trait will depend on the mechanism of sperm competition (Immler et al. 2011). Alternatively, both sperm number and sperm quality may increase in response to sperm competition (Locatello et al. 2007) potentially forcing a trade-off with nonejaculate traits. For example, ejaculate quality may trade off with attractiveness (Evans 2010; Rowe et al. 2010) or immune function (Simmons 2012), and testes size has been shown to trade off with traits such as flight capability (Saglam et al. 2008) and weaponry (Simmons and Emlen 2006). It is therefore possible that increased investment in ejaculates could trade off with traits that directly enhance female fitness, such as nuptial gift size or provision of parental care (e.g., Simmons et al. 1993). For example, male secondary sexual traits can trade off with paternal care provision (Smith 1995) and such trade-offs could be mediated by testosterone production (Wingfield et al. 1990; Mascaro et al. 2013).

Greater investment in each ejaculate, without a concomitant increase in sperm production, could contribute to female sperm limitation (Wedell et al. 2002). For example, in two heteropterans with different mating systems, male investment in spermatogenesis differs and this has profound effects on female fitness (Franco et al. 2011). In the monoandrous *Macrolophus pygmaeus*, females receive enough sperm to fertilize the majority of their ova. In contrast, in the polyandrous *Nesidiocoris tenuis*, females run out of sperm soon after mating, and so must readily accept copulations and spend longer in copula to maintain fertility. Ultimately, if the cost of producing a competitive ejaculate is sufficiently high, male mating rates may be reduced. This could influence the direction of sexual selection as female–female competition for mates increases (Lorch 2002; Wedell et al. 2002).

In many taxa males produce more than one sperm type, but usually only one sperm class are fertilizing sperm (eusperm) (Pitnick et al. 2009). Nonetheless, the nonfertilizing (apyrene or para-) sperm can affect the outcome of sperm competition and their receipt can be costly for females. In *Drosophila pseudoobscura*, for example, the survival of eusperm within the female reproductive tract is highest when greater numbers of apyrene sperm are also inseminated (Holman and Snook 2008). Additionally, in the green-veined white butterfly, *Pieris napi*, nonfertile sperm fill the female sperm storage organ, reducing female receptivity to further copulations, which is clearly in male interests (Wedell 2001). The receipt of nonfertile sperm is costly for females as it reduces fertility, which will be compounded if remating is delayed. The precise function of most apyrene sperm is unclear, but larger apyrene sperm may fill female sperm storage organs or block female genital tracts, excluding sperm from other males (see the section Copulatory Plugs below). Again, this is likely to be beneficial to males but costly to females if polyandry is in female interests.

Nonsperm Ejaculate Substances

In addition to sperm, ejaculates can contain a wide variety of substances that influence sperm competitiveness. These have been particularly well studied in *Drosophila melanogaster* (e.g., Wolfner 2002; Chapman 2008), although they are taxonomically widespread (Perry et al. 2013). Seminal peptides found in the ejaculate of male *D. melanogaster* influence many aspects of female physiology and behavior, including female remating rate, egg production rate, and sperm storage. By manipulating these female traits, males increase their fertilization success. Furthermore, it has been shown that male *D. melanogaster* strategically adjust allocation of seminal peptides in accordance with female mating status and the perceived risk of sperm competition (Wigby et al. 2009; Fedorka et al. 2011; Sirot et al. 2011) as predicted by theory (Hodgson and Hosken 2006). However, male manipulation comes at a cost to females, as seminal fluid proteins reduce female longevity (Chapman et al. 1995) and alter egg production schedules to reduce female lifetime reproductive success (Wigby and Chapman 2005). Furthermore, if female evolutionary responses to male manipulation are halted, males evolve to be even more manipulative and females pay even greater costs (Rice 1996).

There is evidence that females evolve to mitigate the costs of nonsperm ejaculate components. For example, resistance to seminal fluid proteins evolves when female *D. melanogaster* are subjected to higher mating rates (Wigby and Chapman 2004). Studies have also begun to identify candidate genes expressed in the female that interact with male seminal proteins (Swanson et al. 2004; Chow et al. 2010). Male adzuki bean beetles (*Callosobruchus chinensis*), for example, transfer substances to females during mating that reduce female receptivity. Female beetles from strains in which the potency of these substances is greater were found to have evolved greater resilience to their effects, suggestive of coevolution between male manipulation and female resistance (Yamane and Miyatake 2012).

Nonsperm ejaculate substances can also interfere with the female immune response (Morrow and Innocenti 2012). These traits may have evolved because sperm are foreign bodies and females often show an immune response after insemination that may indirectly, or directly, target sperm. In the cricket, *Allonemobius socius*, seminal components interrupt phenoloxidase activity, a key pathway in insect immunity, thus compromising female immune function (Fedorka and Zuk 2005). Although this may benefit males by ensuring that a greater number of their sperm survive to compete, it may also leave females more vulnerable to disease.

Copulatory Plugs

Another MASC is to block the female genital tract with a copulatory plug. Copulatory plugs are found in many taxa (e.g., Lung and Wolfner 2001; Dixson and Anderson 2002; Moreira and Birkhead 2004; Ramm et al. 2005) and can prevent or delay female remating (e.g., Takami et al. 2008) or facilitate sperm transport within the female reproductive tract (e.g., Dean 2013). Copulatory plugs are often derived from the nonsperm portion of the ejaculate, although the male genitalia themselves can also act as plugs (e.g., Herberstein et al. 2012). Plugs can be costly to females if they prevent remating but they can also prevent females from using or storing sperm efficiently. There is indirect evidence that females find copulatory plugs costly to some extent as they are often removed shortly after copulation (e.g., Takami et al. 2008) and females can produce proteases that degrade the structure of the plug (Dean et al. 2011a). This has the potential to generate antagonistic coevolution between males and females, with males selected to improve plug efficacy and females to remove or degrade them. There is evidence for this in rodents, in which a number of different ejaculatory proteins form the plug along with a variety of protease inhibitors that disrupt the action of female degradation enzymes (Dean et al. 2011a). A further female-borne cost of plugs can arise if males attempt to mate with "plugged" females (e.g., Moreira and Birkhead 2004). Females could incur physical harm, particularly if males have evolved traits such as penile spines to assist in plug removal.

Again, however, an assessment of the level of sexual conflict arising from copulatory plugs should also consider potential benefits to females. Plugs can be large enough to ensure that some of the plugging cost can be offset if the female is able to consume the plug on removal or absorb nutrients from the plug as it is degraded. A plug may also be beneficial if this increases fertility (Timmermeyer et al. 2010; Dean 2013) or protects the female from male harassment (e.g., Kuntner et al. 2012).

Strategic Allocation of Male Mating Effort

Another way for males to maximize fitness returns when faced with the risk of sperm competition is to either not mate with, or allocate fewer resources to, already mated females (e.g., Simmons et al. 2003; Friberg 2006). Saved resources might then be allocated elsewhere. It is now increasingly accepted that males can be choosy, particularly if, in the face of intense sperm competition, males are under strong selection to produce larger or more complex ejaculates that may limit mating rates (Lorch 2002; Wedell et al. 2002; Edward and Chapman 2011). If the energy required to mate is high, the prospects of securing paternity are low and if other females are likely to be available, then males might be better off searching for alternative mates (Schwagmeyer and Parker 1990). However, it should be remembered that strategic allocation of mating effort will only benefit males if alternative, more economical mating opportunities are available to which saved resources can be reallocated (e.g., Fuller 1998; Barry and Kokko 2010).

If males allocate fewer resources (i.e., fewer sperm or smaller ejaculates), or are less willing to mate with nonvirgin females, this could reduce female fertility. Females could also incur costs of reduced mate choice, for example by not being able to "trade up" to a better mate, or be unable to secure additional fitness-enhancing nuptial gifts. Female adaptations to avoid costs of strategic allocation of male mating effort might include disguising mating status or perhaps even signaling competitively to attract males (e.g., Clutton-Brock 2009; Stockley et al. 2013). Female signals that indicate greater fecundity might convince a male that the potential rewards of mating outweigh potential costs of sperm competition (e.g., LeBas et al. 2003).

THE EXTENT OF SEXUAL CONFLICT

Costs and Benefits to Males

MASC are often costly for males and thus may limit other reproductive investment. For example, increasing ejaculate production or producing larger copulatory plugs may require significant energetic investment and trade-off against other fitness components (Hosken 2001). Similarly, prolonged guarding or copulatory behaviors can be costly in terms of both time and energy. This can potentially result in trade-offs such as that between mate guarding and territory size in male stitchbirds, *Notiomystis cincta* (Low 2005). The costs of MASC are further evidenced by examples of strategic ejaculate allocation and variation in male mating behavior and mate guarding behavior according to perceived levels of sperm competition (Komdeur 2001; Wedell et al. 2002; Bretman et al. 2009). If MASC were not costly there would be little or no selection for plasticity in their expression. Costs may also be more obvious than this. As an extreme example, the ability of males to remate is limited when genitals are broken off during mating to form a copulatory plug (e.g., Nessler et al. 2009).

The costs of expressing MASC will ultimately impose a limit on the evolution of these traits and hence the potential to inflict harm on females could be self-limiting. However, the harm inflicted on females may still be significant as male costs are often outweighed by very strong selection for these characters. If the risk or intensity of sperm competition is high, selection can thus still favor the expression of MASC that are very costly to males. For example, in the orb-web spider, *Argiope lobata*, males break off part of their genitalia to form a copulatory plug that increases their share of paternity if the female should remate. This occurs despite the very significant cost of a reduction in the male's ability to remate, and an increased risk of sexual cannibalism (Nessler et al. 2009). Nevertheless, the fitness benefits of depositing the plug are sufficient to outweigh even these extreme costs, which is in part owing to the paucity of future mating opportunities (Nessler et al. 2009). This shows that the strength of selection on males through sperm competition can be sufficient to select for costly characters, which in turn can generate sexual conflict.

Costs and Benefits to Females

Although MASC can be under strong selection, corresponding selection in females to resist them may be weaker as MASC are unlikely to cause a complete loss of female fitness. Basically, the variance in male reproductive success is typically higher than that in females. Thus in general we might predict, at least initially, greater selection for males to express MASC than for females to evolve resistance to them. The balance of selection may therefore tend to favor the evolution of MASC, with weaker selection for female resistance to them. Consequently, females probably suffer a greater conflict load (i.e., females may be further from their fitness optima as a result of sexual conflict). However, this prediction is complicated by the costs of female resistance (see the section Mitigation and Resolution of Sexual Conflict below).

Finally, there are potential female benefits from MASC. An assessment of these benefits is important not only for accurately determining the conflict load, but also the precise trait, or aspect of a trait, over which there may be a sexual conflict. For example, a copulatory plug can improve fertilization success and thus benefit both sexes (e.g., Timmermeyer et al. 2010; Dean 2013). Consideration should then be given to whether there is conflict over plug formation per se, or only over the optimal size or consistency of the plug.

It may also be important to consider how fitness costs of MASC are measured. Male manipulation of female egg production may function as a MASC as it will benefit males if more fertilized eggs are produced before females remate. This manipulation can have a detrimental effect on female lifespan and/or lifetime reproductive success (Chapman et al. 1995; Wigby and Chapman 2005). However, the production of more offspring at a young age will also contribute to an individual's intrinsic rate of offspring production. This can enhance female fitness, particularly if other females in a population are also being stimulated to produce more offspring so that population size is increasing (Edward et al. 2011). Thus, female fitness may, under certain conditions, be increased despite a reduction in overall lifetime reproductive success. This highlights the importance of not only considering the potential fitness benefits of MASC to females but also greater consideration of the way that fitness is measured.

In summary, to understand the precise nature and extent of sexual conflict arising from MASC it is important to understand the costs and benefits of conflict in both sexes. The effects of MASC can be complex, and although there is often an obvious cost to females, potential benefits of these traits can easily be overlooked. A key challenge for future research is a more complete consideration of the fitness economics of sexual conflict and improved measurement of the costs involved.

MITIGATION AND RESOLUTION OF SEXUAL CONFLICT

Although it is clear that MASC can give rise to sexual conflict, the evolutionary trajectories of such conflicts are less clear. Although there is potential for selection arising from MASC to favor female traits that directly counter them, conflict need not lead to selection. Outcomes depend in part on how costly it is to resist MASC and the strength of selection imposed on females. An important factor here is that despite strong selection acting on males, female resistance traits could be relatively cheap to produce and express (Parker 1979, 1984). For example, although a copulatory plug may be very costly to produce, it could be relatively easy for a female to remove. This kind of asymmetry could tilt evolutionary outcomes of these conflicts in favor of females, despite stronger selection acting on males. However, although female cost-avoiding responses to one MASC could potentially resolve a specific conflict, this could have the potential to select for an alternative MASC, thus kick-starting new rounds of sexually antagonistic coevolution.

Sperm competition is intimately related to polyandry, and because of this, the causes and patterns of polyandry may also be particularly important in determining the nature of sexual conflict that results. It is clear that females can benefit from polyandry (Arnqvist and Nilsson 2000; Fedorka and Mousseau 2002; Slatyer et al. 2012), yet mating too often can be costly (Arnqvist and Nilsson 2000). Hence, the net benefits or costs of polyandry should influence specific sexual conflicts arising from MASC.

If actual rates of polyandry are greater than the female optima, selection will favor female traits that reduce mating rates. Selection for reduced rates of polyandry should in turn be expected to reduce levels of sperm competition and thus potentially also reduce the costs arising from MASC—although it could spark other sexual conflicts if males are moved below their optimal mating rates. Alternatively, if rates of polyandry are below the female optima, females will be selected to increase multiple mating. In this case females can generate a conflict load from MASC via selection for polyandry. As long as fitness benefits of polyandry exceed subsequent fitness costs of MASC we would expect selection for rates of polyandry to be maintained. However, it is not easy to disentangle cause and effect. Females may be seen to have higher fitness with increased mating because they have evolved to reduce costs by storing fewer sperm. That is, because sperm are so easy to acquire females do not need to store them. Thus, it is not always clear whether female multiple mating is because males transfer few sperm or females simply do not want to store them.

Additionally, as rates of polyandry are often only indirectly associated with sexual conflict arising from MASC it is not inevitable that an individual reduction in mating rates will reduce costs of this conflict. If a majority of females in a population are polyandrous and/or if MASC are relatively cheap to produce, it is likely that they will be expressed even at low risk of sperm competition. Consequently, a female trait that reduces mating rates is unlikely to confer an immediate benefit to its bearer through reduced costs of MASC. However, if males assess female mating status to determine the risk of sperm competition and adjust the expression of MASC accordingly, then an individual female with a lower rate of polyandry could benefit from a reduction in costs associated with MASC.

In summary, a resolution of sexual conflict over MASC may be reached directly or indirectly via reduced rates of polyandry. However, this will be dependent on the benefits of polyandry to females, including whether rates of polyandry are above or below optima, and whether MASC are tailored according to the immediate risk of sperm competition. Unfortunately at this point in time it is not even clear how sexual conflict generates selection generally, and not just how often MASC actually reduce female fitness. These are areas that need empirical work on a broad range of taxa to allow us to have some feel for general patterns.

ACKNOWLEDGMENTS

We are grateful to Bill Rice and Sergey Gavrilets for the opportunity to write this article. We appreciate support from National Environmental Research Council (NERC) grant NE/I013008/1 (P.S. and D.A.E.).

REFERENCES

Arnqvist G, Nilsson T. 2000. The evolution of polyandry: Multiple mating and female fitness in insects. *Anim Behav* **60:** 145–164.

Arnqvist G, Rowe L. 2002. Correlated evolution of male and female morphologies in water striders. *Evolution* **56:** 936–947.

Ball MA, Parker GA. 1996. Sperm competition games: External fertilization and "adaptive" infertility. *J Theor Biol* **180:** 141–150.

Barry KL, Kokko H. 2010. Male mate choice: Why sequential choice can make its evolution difficult. *Anim Behav* **80:** 163–169.

Birkhead TR, Moller AP, Sutherland WJ. 1993. Why do females make it so difficult for males to fertilize their eggs. *J Theor Biol* **161:** 51–60.

Bretman A, Fricke C, Chapman T. 2009. Plastic responses of male *Drosophila melanogaster* to the level of sperm competition increase male reproductive fitness. *Proc R Soc Lond B* **276:** 1705–1711.

Chapman T. 2008. The soup in my fly: Evolution, form and function of seminal fluid proteins. *PLoS Biol* **6:** 1379–1382.

Chapman T, Liddle LF, Kalb JM, Wolfner MF, Partridge L. 1995. Cost of mating in *Drosophila melanogaster* females is mediated by male accessory-gland products. *Nature* **373:** 241–244.

Chow CY, Wolfner MF, Clark AG. 2010. The genetic basis for male × female interactions underlying variation in reproductive phenotypes of *Drosophila*. *Genetics* **186:** 1355–1365.

Clutton-Brock T. 2009. Sexual selection in females. *Anim Behav* **77:** 3–11.

Cordoba-Aguilar A, Uhia E, Rivera AC. 2003. Sperm competition in Odonata (Insecta): The evolution of female sperm storage and rivals' sperm displacement. *J Zool* **261:** 381–398.

Cordoba-Aguilar A, Serrano-Meneses MA, Cordero-Rivera A. 2009. Copulation duration in nonterritorial odonate species lasts longer than in territorial species. *Ann Entomol Soc Am* **102:** 694–701.

Dean MD. 2013. Genetic disruption of the copulatory plug in mice leads to severely reduced fertility. *PLoS Genet* **9:** e1003185.

Dean MD, Findlay GD, Hoopmann MR, Wu CC, MacCoss MJ, Swanson WJ, Nachman MW. 2011a. Identification of ejaculated proteins in the house mouse (*Mus domesticus*) via isotopic labeling. *BMC Genomics* **12:** 306.

Dean R, Nakagawa S, Pizzari T. 2011b. The risk and intensity of sperm ejection in female birds. *Am Nat* **178:** 343–354.

Dixson AF, Anderson MJ. 2002. Sexual selection, seminal coagulation and copulatory plug formation in primates. *Folia Primatol* **73:** 63–69.

Eberhard WG. 1996. *Female control: Sexual selection by cryptic female choice*. Princeton University Press, Princeton.

Edward DA, Chapman T. 2011. The evolution and significance of male mate choice. *Trends Ecol Evol* **26:** 647–654.

Edward DA, Fricke C, Gerrard DT, Chapman T. 2011. Quantifying the life-history response to increased male exposure in female *Drosophila melanogaster*. *Evolution* **65:** 564–573.

Evans JP. 2010. Quantitative genetic evidence that males trade attractiveness for ejaculate quality in guppies. *Proc Biol Sci* **277:** 3195–3201.

Fedorka KM, Mousseau TA. 2002. Material and genetic benefits of female multiple mating and polyandry. *Anim Behav* **64:** 361–367.

Fedorka KM, Zuk M. 2005. Sexual conflict and female immune suppression in the cricket, *Allonemobious socius*. *J Evol Biol* **18:** 1515–1522.

Fedorka KM, Winterhalter WE, Ware B. 2011. Perceived sperm competition intensity influences seminal fluid protein production prior to courtship and mating. *Evolution* **65:** 584–590.

Fenton MB. 1984. Sperm competition: The case of vespertilionid and rhinolophid bats. In *Sperm competition and the evolution of animal mating systems* (ed. Smith RL), pp. 573–588. Academic, Orlando, FL.

Franco K, Jauset A, Castane C. 2011. Monogamy and polygamy in two species of mirid bugs: A functional-based approach. *J Insect Physiol* **57:** 307–315.

Friberg U. 2006. Male perception of female mating status: Its effect on copulation duration, sperm defence and female fitness. *Anim Behav* **72:** 1259–1268.

Fuller RC. 1998. Sperm competition affects male behaviour and sperm output in the rainbow darter. *Proc Biol Sci* **265:** 2365–2371.

Gage MJG. 1991. Risk of sperm competition directly affects ejaculate size in the Mediterranean fruit-fly. *Anim Behav* **42:** 1036–1037.

Gage MJG, Morrow EH. 2003. Experimental evidence for the evolution of numerous, tiny sperm via sperm competition. *Curr Biol* **13:** 754–757.

Gay L, Hosken DJ, Eady P, Vasudev R, Tregenza T. 2011. The evolution of harm-effect of sexual conflicts and population size. *Evolution* **65:** 725–737.

Harano T, Okada K, Nakayama S, Miyatake T, Hosken DJ. 2010. Intralocus sexual conflict unresolved by sex-limited trait expression. *Curr Biol* **20:** 2036–2039.

Helfenstein F, Wagner RH, Danchin E. 2003. Sexual conflict over sperm ejection in monogamous pairs of kittiwakes *Rissa tridactyla*. *Behav Ecol Sociobiol* **54:** 370–376.

Hellriegel B, Bernasconi G. 2000. Female-mediated differential sperm storage in a fly with complex spermathecae, *Scatophaga stercoraria*. *Anim Behav* **59:** 311–317.

Herberstein ME, Wignall AE, Nessler SH, Harmer AMT, Schneider JM. 2012. How effective and persistent are fragments of male genitalia as mating plugs? *Behav Ecol* **23:** 1140–1145.

Hodgson DJ, Hosken DJ. 2006. Sperm competition promotes the exploitation of rival ejaculates. *J Theor Biol* **243:** 230–234.

Holman L, Snook RR. 2008. A sterile sperm caste protects brother fertile sperm from female-mediated death in *Drosophila pseudoobscura*. *Curr Biol* **18:** 292–296.

Hosken DJ. 2001. Sex and death: Microevolutionary tradeoffs between reproductive and immune investment in dung flies. *Curr Biol* **11:** R379–R380.

Hosken DJ, Martin OY, Born J, Huber F. 2003. Sexual conflict in *Sepsis cynipsea*: Female reluctance, fertility and mate choice. *J Evol Biol* **16:** 485–490.

Hotzy C, Arnqvist G. 2009. Sperm competition favors harmful males in seed beetles. *Curr Biol* **19:** 404–407.

Immler S, Pitnick S, Parker GA, Durrant KL, Lupold S, Calhim S, Birkhead TR. 2011. Resolving variation in the reproductive tradeoff between sperm size and number. *Proc Natl Acad Sci* **108:** 5325–5330.

Johns JL, Roberts JA, Clark DL, Uetz GW. 2009. Love bites: Male fang use during coercive mating in wolf spiders. *Behav Ecol Sociobiol* **64:** 13–18.

Komdeur J. 2001. Mate guarding in the Seychelles warbler is energetically costly and adjusted to paternity risk. *Proc Biol Sci* **268:** 2103–2111.

Kuntner M, Kralj-Fiser S, Schneider JM, Li D. 2009. Mate plugging via genital mutilation in nephilid spiders: An evolutionary hypothesis. *J Zool* **277:** 257–266.

Kuntner M, Gregoric M, Zhang SC, Kralj-Fiser S, Li DQ. 2012. Mating plugs in polyandrous giants: Which sex produces them, when, how and why? *PLoS ONE* **7:** e40939.

Laird G, Gwynne DT, Andrade MCB. 2004. Extreme repeated mating as a counter-adaptation to sexual conflict? *Proc Biol Sci* **271:** S402–S404.

LeBas NR, Hockham LR, Ritchie MG. 2003. Nonlinear and correlational sexual selection on "honest" female ornamentation. *Proc R Soc Lond B* **270:** 2159–2165.

Lewis Z, Price TAR, Wedell N. 2008. Sperm competition, immunity, selfish genes and cancer. *Cell Mol Life Sci* **65:** 3241–3254.

Locatello L, Pilastro A, Deana R, Zarpellon A, Rasotto MB. 2007. Variation pattern of sperm quality traits in two gobies with alternative mating tactics. *Funct Ecol* **21:** 975–981.

Lorch PD. 2002. Understanding reversals in the relative strength of sexual selection on males and females: A role for sperm competition? *Am Nat* **159:** 645–657.

Low M. 2005. Factors influencing mate guarding and territory defence in the stitchbird (hihi) *Notiomystis cincta*. *NZ J Ecol* **29:** 231–242.

Lung O, Wolfner MF. 2001. Identification and characterization of the major *Drosophila melanogaster* mating plug protein. *Insect Biochem Mol Biol* **31:** 543–551.

Manier MK, Belote JM, Berben KS, Novikov D, Stuart WT, Pitnick S. 2010. Resolving mechanisms of competitive fertilization success in *Drosophila melanogaster*. *Science* **328:** 354–357.

Martin OY, Hosken DJ. 2002. Strategic ejaculation in the common dung fly *Sepsis cynipsea*. *Anim Behav* **63:** 541–546.

Mascaro JS, Hackett PD, Rilling JK. 2013. Testicular volume is inversely correlated with nurturing-related brain activity in human fathers. *Proc Natl Acad Sci* **110:** 15746–15751.

Mazzi D, Kesaniemi J, Hoikkala A, Klappert K. 2009. Sexual conflict over the duration of copulation in *Drosophila montana*: Why is longer better? *BMC Evol Biol* **9:** 132.

Montrose VT, Harris WE, Moore PJ. 2004. Sexual conflict and cooperation under naturally occurring male enforced monogamy. *J Evol Biol* **17:** 443–452.

Moreira PL, Birkhead TR. 2004. Copulatory plug displacement and prolonged copulation in the Iberian rock lizard (*Lacerta monticola*). *Behav Ecol Sociobiol* **56:** 290–297.

Morrow EH, Innocenti P. 2012. Female postmating immune responses, immune system evolution and immunogenic males. *Biol Rev Camb Philos Soc* **87:** 631–638.

Nessler SH, Uhl G, Schneider JM. 2007. Genital damage in the orb-web spider *Argiope bruennichi* (Araneae: Araneidae) increases paternity success. *Behav Ecol* **18:** 174–181.

Nessler S, Uhl G, Schneider J. 2009. Sexual cannibalism facilitates genital damage in *Argiope lobata* (Araneae: Araneidae). *Behav Ecol Sociobiol* **63:** 355–362.

Parker GA. 1970a. The reproductive behaviour and nature of sexual selection in *Scatophaga stercoraria* L (Diptera: Scatophagidae): 4. Epigamic recognition and competition between males for possession of females. *Behaviour* **37:** 113.

Parker GA. 1970b. Sperm competition and its evolutionary consequences in insects. *Biol Rev Camb Philos Soc* **45:** 525.

Parker GA. 1972. Reproductive behavior of *Sepsis cynipsea* (L) (Diptera: Sepsidae): 1. Preliminary analysis of reproductive strategy and its associated behavior patterns. *Behaviour* **41:** 172.

Parker GA. 1979. Sexual selection and sexual conflict. In *Sexual selection and reproductive competition in insects* (ed. Blum MS, Blum NA), pp. 123–166. Academic, New York.

Parker GA. 1984. Sperm competition and the evolution of animal mating strategies. In *Sperm competition and the evolution of animal mating systems* (ed. Smith RL), pp. 2–61. Academic, Orlando, FL.

Parker GA. 1998. Sperm competition and the evolution of ejaculates: Towards a theory base. In *Sperm competition and sexual selection* (ed. Birkhead TR, Moller AP), pp. 3–54. Academic, Orlando, FL.

Parker GA, Simmons LW. 2000. Optimal copula duration in yellow dung flies: Ejaculatory duct dimensions and size-dependent sperm displacement. *Evolution* **54:** 924–935.

Parker GA, Ball MA, Stockley P, Gage MJG. 1996. Sperm competition games: Individual assessment of sperm competition intensity by group spawners. *Proc R Soc Lond B* **263:** 1291–1297.

Perry JC, Sirot L, Wigby S. 2013. The seminal symphony: How to compose an ejaculate. *Trends Ecol Evol* **28:** 414–422.

Pitnick S, Hosken DJ, Birkhead TR. 2009. Sperm morphological diversity. In *Sperm biology: An evolutionary perspective* (ed. Pitnick S, Hosken DJ, Birkhead TR), pp. 69–183. Academic, Orlando, FL.

Preston BT, Stockley P. 2006. The prospect of sexual competition stimulates premature and repeated ejaculation in a mammal. *Curr Biol* **16:** R239–R241.

Ramm SA, Parker GA, Stockley P. 2005. Sperm competition and the evolution of male reproductive anatomy in rodents. *Proc Biol Sci* **272:** 949–955.

Rice WR. 1996. Sexually antagonistic male adaptation triggered by experimental arrest of female evolution. *Nature* **381:** 232–234.

Rowe M, Swaddle JP, Pruett-Jones S, Webster MS. 2010. Plumage coloration, ejaculate quality and reproductive phenotype in the red-backed fairy-wren. *Anim Behav* **79:** 1239–1246.

Saglam IK, Roff DA, Fairbairn DJ. 2008. Male sand crickets trade-off flight capability for reproductive potential. *J Evol Biol* **21:** 997–1004.

Sato Y, Kohama T. 2007. Post-copulatory mounting behavior of the West Indian sweetpotato weevil, *Euscepes postfasciatus* (Fairmaire) (Coleoptera: Curculionidae). *Ethology* **113:** 183–189.

Schwagmeyer PL, Parker GA. 1990. Male mate choice as predicted by sperm competition in 13-lined ground-squirrels. *Nature* **348:** 62–64.

Simmons LW. 2012. Resource allocation trade-off between sperm quality and immunity in the field cricket, *Teleogryllus oceanicus*. *Behav Ecol* **23:** 168–173.

Simmons LW, Emlen DJ. 2006. Evolutionary trade-off between weapons and testes. *Proc Natl Acad Sci* **103:** 16346–16351.

Simmons LW, Craig M, Llorens T, Schinzig M, Hosken D. 1993. Bush-cricket spermatophores vary in accord with sperm competition and parental investment theory. *Proc Biol Sci* **251:** 183–186.

Simmons LW, Alcock J, Reeder A. 2003. The role of cuticular hydrocarbons in male attraction and repulsion by female Dawson's burrowing bee, *Amegilla dawsoni*. *Anim Behav* **66:** 677–685.

Sirot LK, Wolfner MF, Wigby S. 2011. Protein-specific manipulation of ejaculate composition in response to female mating status in *Drosophila melanogaster*. *Proc Natl Acad Sci* **108:** 9922–9926.

Slatyer RA, Mautz BS, Backwell PRY, Jennions MD. 2012. Estimating genetic benefits of polyandry from experimental studies: A meta-analysis. *Biol Rev Camb Philos Soc* **87:** 1–33.

Smith HG. 1995. Experimental demonstration of a trade-off between mate attraction and paternal care. *Proc Biol Sci* **260:** 45–51.

Smith C, Douglas A, Jurajda P. 2002. Sexual conflict, sexual selection and sperm competition in the spawning decisions of bitterling, *Rhodeus sericeus*. *Behav Ecol Sociobiol* **51:** 433–439.

Snook RR, Hosken DJ, Karr TL. 2011. The biology and evolution of polyspermy: Insights from cellular and functional studies of sperm and centrosomal behavior in the fertilized egg. *Reproduction* **142:** 779–792.

Stockley P. 1997. Sexual conflict resulting from adaptations to sperm competition. *Trends Ecol Evol* **12:** 154–159.

Stockley P. 2002. Sperm competition risk and male genital anatomy: Comparative evidence for reduced duration of female sexual receptivity in primates with penile spines. *Evol Ecol* **16:** 123–137.

Stockley P, Preston BT. 2004. Sperm competition and diversity in rodent copulatory behaviour. *J Evol Biol* **17:** 1048–1057.

Stockley P, Bottell L, Hurst JL. 2013. Wake up and smell the conflict: Odour signals in female competition. *Philos Trans R Soc Lond B* **368:** 20130082.

Swanson WJ, Wong A, Wolfner MF, Aquadro CF. 2004. Evolutionary expressed sequence tag analysis of *Drosophila* female reproductive tracts identifies genes subjected to positive selection. *Genetics* **168:** 1457–1465.

Takami Y, Sasabe M, Nagata N, Sota T. 2008. Dual function of seminal substances for mate guarding in a ground beetle. *Behav Ecol* **19:** 1173–1178.

Timmermeyer N, Gerlach T, Guempel C, Knoche J, Pfann JF, Schliessmann D, Michiels NK. 2010. The function of copulatory plugs in *Caenorhabditis remanei*: Hints for female benefits. *Front Zool* **7:** 28.

Wedell N. 1998. Sperm protection and mate assessment in the bushcricket *Coptaspis* sp. 2. *Anim Behav* **56:** 357–363.

Wedell N. 2001. Female remating in butterflies: Interaction between female genotype and nonfertile sperm. *J Evol Biol* **14:** 746–754.

Wedell N, Gage MJG, Parker GA. 2002. Sperm competition, male prudence and sperm-limited females. *Trends Ecol Evol* **17:** 313–320.

Westneat DF. 1994. To guard mates or go forage: Conflicting demands affect the paternity of male red-winged blackbirds. *Am Nat* **144:** 343–354.

Wigby S, Chapman T. 2004. Female resistance to male harm evolves in response to manipulation of sexual conflict. *Evolution* **58:** 1028–1037.

Wigby S, Chapman T. 2005. Sex peptide causes mating costs in female *Drosophila melanogaster*. *Curr Biol* **15:** 316–321.

Wigby S, Sirot LK, Linklater JR, Buehner N, Calboli FCF, Bretman A, Wolfner MF, Chapman T. 2009. Seminal fluid protein allocation and male reproductive success. *Curr Biol* **19:** 751–757.

Wilson CJ, Tomkins JL. 2014. Countering counteradaptations: Males hijack control of female kicking behavior. *Behav Ecol* doi: 10.1093/beheco/aru022.

Wingfield JC, Hegner RE, Dufty AM, Ball GF. 1990. The challenge hypothesis: Theoretical implications for patterns of testosterone secretion, mating systems, and breeding strategies. *Am Nat* **136:** 829–846.

Wolfner MF. 2002. The gifts that keep on giving: Physiological functions and evolutionary dynamics of male seminal proteins in *Drosophila*. *Heredity* **88:** 85–93.

Wysocki D, Halupka K. 2004. The frequency and timing of courtship and copulation in blackbirds, turdus merula, reflect sperm competition and sexual conflict. *Behaviour* **141:** 501–512.

Yamane T, Miyatake T. 2012. Evolutionary correlation between male substances and female remating frequency in a seed beetle. *Behav Ecol* **23:** 715–722.

Conflict on the Sex Chromosomes: Cause, Effect, and Complexity

Judith E. Mank[1], David J. Hosken[2], and Nina Wedell[2]

[1]Department of Genetics, Evolution and Environment, University College London, London WC1E 6BT, United Kingdom

[2]Centre for Ecology & Conservation, University of Exeter, Cornwall, Tremough, Penryn TR10 9EZ, United Kingdom

Correspondence: judith.mank@ucl.ac.uk

Intralocus sexual conflict and intragenomic conflict both affect sex chromosome evolution and can in extreme cases even cause the complete turnover of sex chromosomes. Additionally, established sex chromosomes often become the focus of heightened conflict. This creates a tangled relationship between sex chromosomes and conflict with respect to cause and effect. To further complicate matters, sexual and intragenomic conflict may exacerbate one another and thereby further fuel sex chromosome change. Different magnitudes and foci of conflict offer potential explanations for lineage-specific variation in sex chromosome evolution and answer long-standing questions as to why some sex chromosomes are remarkably stable, whereas others show rapid rates of evolutionary change.

Compared to the autosomes, the unique inheritance pattern of the sex chromosomes is often thought to intensify evolutionary conflict (Rice 1984; Frank 1991; Jaenike 2001). Sex chromosomes are therefore hot spots for two specific types of conflict: intralocus conflict, in which an allele confers different fitness effects depending on the sex in which it is found, and intragenomic conflict, in which selfish genetic elements (SGEs) promote their own transmission at the expense of unlinked regions of the genome. These conflicts act in distinct but complementary ways. Not only do they shape sex chromosome and genome evolution, but in some cases, they also have the power to cause complete turnover of sex chromosomes.

Unlike the autosomes, because the sex chromosomes are unevenly transmitted between males and females and are also unevenly distributed between the sexes (Fig. 1), the relative effect of male- and female-specific selection acting on them is unbalanced. The inherent differences in sex-specific selection on the sex chromosomes themselves, and between the sex chromosomes and the autosomes, form the basis of a large and often compelling body of evolutionary theory that predicts the ways that intralocus sexual conflict will arise, play out, and in some cases potentially be resolved. This theory predicts that, under some conditions, the sex chromosomes are hot spots of intralocus sexual conflict (Rice 1984; Albert and Otto 2005; Con-

Cite this article as *Cold Spring Harb Perspect Biol* doi: 10.1101/cshperspect.a017715

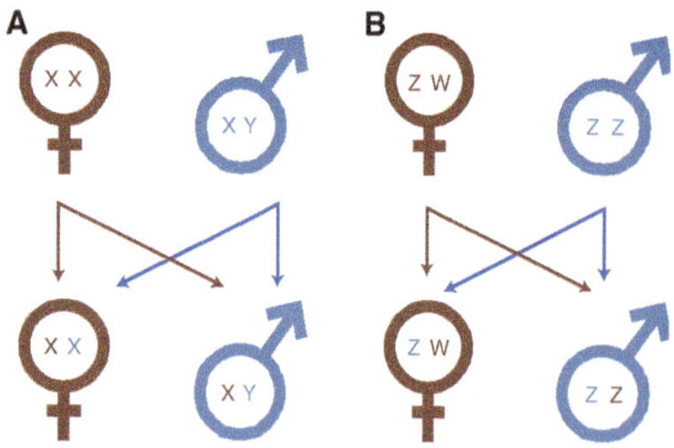

Figure 1. Transmission of the sex chromosomes. Females are shown in red, males in blue. In male heterogamety (*A*), the Y chromosome is passed through the patriline and limited to males. The maternal X chromosome is passed from mother to both sons and daughters, but the paternal X can only be transmitted to daughters. Additionally, the X is present two-thirds of the time in females. In female heterogamety (*B*), the W chromosome is limited to females and passed solely from mother to daughter. The paternal Z chromosome can be passed from father to both daughters and sons; however, the maternal Z chromosome is only passed to sons. Converse to the X chromosome, the Z chromosome is resident in males two-thirds of the time.

nallon and Clark 2010), and in some cases alleles that harm one sex more than they benefit the other can still reach high frequencies if they are sex-linked (Rice 1984; Dean et al. 2012). All this theory predicts that although the sex chromosomes generally represent a small proportion of the genome, they should play a disproportionately large role in sexual conflict, sexual dimorphism, and sexual selection. There is substantial empirical evidence supporting at least some of this theory (Table 1, and references therein), although there remains something of a disconnect between molecular genetic and phenotypic findings (Dean and Mank 2014).

In addition to their role in sexual conflict, the sex chromosomes also experience high levels of intragenomic conflict resulting from SGEs that promote their own transmission at the expense of the rest of the genome (Burt and Trivers 2006). Of particular importance are sex-linked segregation distorters, as these generate strong selection favoring genes that suppress their action. Theory predicts that intragenomic conflict should be particularly intense when involving the sex chromosomes (Hurst and Pomiankowski 1991). This is because sex chromosomes can generate antagonistic coevolution between sex-linked segregation distorters and their suppressors (Partridge and Hurst 1998).

In addition to shaping the evolutionary properties of existing sex chromosomes, in some situations sexual and intragenomic conflict may also actually catalyze the formation of sex chromosomes (Fisher 1931; Charlesworth and Charlesworth 1980; Rice 1987; Werren and Beukeboom 1998). Conflict may also explain the rate of degradation of the sex-limited Y and W chromosomes (Bachtrog et al. 2011) and turnover of sex chromosomes (Hall 2004; van Doorn and Kirkpatrick 2007, 2010), suggesting that conflict plays a causal role in sex chromosome evolution.

There is also growing evidence of a direct relationship between SGEs and sexual conflict involving the sex chromosomes. Many SGEs increase their transmission advantage by targeting sperm, which can reduce male fertility owing to reduced overall sperm production. Lower sperm production can in turn result in reduced siring success during sperm competition (Price and Wedell 2008) and potentially favor polyandry as a female strategy to bias paternity against SGE-carrying males (Wedell 2013). As a consequence, SGEs may also influence the potential for sexual conflict as they can favor increased female mating rates, which increases the potential for conflict between the sexes.

Sexual and intragenomic conflict are therefore critical for sex chromosome evolution and, once sex chromosomes are established, can further shape their evolutionary and genomic properties. Disentangling cause and effect is difficult but crucial to understanding the role of sex chromosomes in sexual and intragenomic conflict and vice versa.

CONFLICT AS AN AGENT OF SEX CHROMOSOME TURNOVER AND CHANGE

One of the major unanswered questions in sex chromosome evolution is why some systems

Table 1. Studies showing a disproportionate role of the sex chromosomes in sexual dimorphism, fitness, or fertility

	Male heterogamety		Female heterogamety	
	X chromosome	Y chromosome	Z chromosome	W chromosome
Associations with sexually dimorphic phenotypic traits	Size dimorphism in red deer (Foerster et al. 2007) Sexually antagonistic fitness variation in *Drosophila* (Gibson et al. 2002; Innocenti and Morrow 2010) Variation in stalk-eyed fly eye span (Wolfenbarger and Wilkinson 2001)	Male coloration in guppies (Winge 1927; Postma et al. 2011) Male mating behavior in sticklebacks (Kitano et al. 2009; although see Natri et al. 2013)	Female mating preference in moths (Iyengar et al. 2002) Female mate choice in flycatchers (Saether et al. 2007) Male plumage traits in flycatchers (Saetre et al. 2003)	Female benefit coloration in cichlids (Roberts et al. 2009)
Associations with reproduction and fertility	Male reproductive genes in mice (Mueller et al. 2008)	Male fertility genes in mammals (Lange et al. 2009) Male fitness in *Drosophila* (Lemos et al. 2010; Sackton et al. 2011)	Genes related to male reproductive function in silk moth (Arunkumar et al. 2009)	Genes related to female fecundity and fertility in chickens (Moghadam et al. 2012)

show long periods of relative evolutionary stasis, as exemplified by the conservation of orthologous sex chromosomes within the therian mammal (Potrzebowski et al. 2008), avian (Mank and Ellegren 2007), and *Drosophila* (Vicoso and Bachtrog 2013) lineages, whereas many others, described below, show rapid and repeated turnover. We also do not understand why some sex chromosome pairs diverge markedly from each other over time, whereas others exist in a state of arrested decay.

Although homologous sex chromosomes in different systems share many properties (Bachtrog et al. 2011), they also show many lineage-specific patterns, many of which have proved difficult to explain. Different conflict strengths and foci offer potential explanations for lineage-specific variation in sex chromosome evolution, as conflict can affect both rates of sex chromosome turnover and sex chromosome divergence.

Intralocus Conflict and Sex Chromosome Change

The genetic architecture and extent of intralocus sexual conflict may affect the origin of sex chromosomes in two ways. First, sexual conflict may affect the rate at which the W and Y chromosomes degenerate. The primary model for sex chromosome divergence is based on selection for suppression of recombination between the X and Y, or Z and W, attributable to linkage between the primary sex-determining gene and any nearby gene(s) with sexually antagonistic properties (Fisher 1931; Charlesworth and Charlesworth 1980; Rice 1987). Selection will act to prevent the breakup of a male-determining gene on the nascent Y chromosome from any nearby male-benefit alleles by stopping recombination between the X and Y in this region. Similarly, reduced recombination will be favored between a female-determining gene on an emergent W chromosome and nearby female-benefit alleles.

The suppression of recombination causes the Y and W chromosomes to degenerate at many loci in both gene sequence and expression, as well as allowing them to accumulate repetitive elements (Bachtrog 2013), and this is what causes the cytological and functional differences between X and Y and between Z and W chromosomes. The likelihood that a sex-determining gene will be proximate to a locus with sexually antagonistic alleles is in many ways defined by the proportion of loci in the genome with sexually antagonistic effects. The degree of conflict, and therefore the proportion of the genome with sexually antagonistic effects, may be in turn largely defined by the mating system (Charlesworth and Mank 2010). If the mating system has a major effect on sex chromosome degeneration, then we would expect monogamous species with minimal levels of sexual conflict to show slower rates of sex chromosome divergence than polyandrous or polygamous systems. However, this is complicated and may depend in part on whether the monogamy is male- or female-driven (Hosken et al. 2009).

Additionally, although sex chromosomes are relatively stable and conserved in some lineages, such as *Drosophila*, the therian mammals, and birds (Mank and Ellegren 2007; Potrzebowski et al. 2008; Vicoso and Bachtrog 2013), sex chromosomes in other lineages exhibit rapid turnover. For example, the sex chromosomes are not orthologous among closely related species of *Oryzias* rice fish (Matsuda et al. 2002; Kondo et al. 2003; Miura 2007), populations of *Rana rugosa* frogs (Nanda et al. 2002; Ogata et al. 2008), or in *Musca domestica*. In the latter case, the male-determining factor (M) has relocated in some populations from the Y to any of the five autosomes or even the X (Dübendorfer et al. 2002).

There are still many unanswered questions as to why some sex chromosome systems are highly conserved across long periods of evolutionary time, whereas others change quickly and even vary across populations of the same species. However, there are some predictions suggesting that sex chromosome turnover itself may in fact be driven by sexual conflict (van Doorn and Kirkpatrick 2007, 2010), and SGEs (see below) as the emergence of a new sex chromosome system can resolve conflict over sexually antagonistic alleles by creating a male-limited (Y) or a female-limited (W) chromosome (Roberts et al. 2009; Postma et al. 2011). Sexual conflict may also cause the formation of neosex chromosomes, via autosomal fusion to a sex chromosome. This is because sexual conflict on the autosome can select for sex-linkage, which may explain the formation of a neosex chromosome in the Japan Sea population of three-spined sticklebacks, *Gasterosteus aculeatus* (Kitano et al. 2009; but see Natri et al. 2013).

Population Size, Linkage, and Turnover

Population size is a major determinant in types and rates of evolution. Low effective population size (N_E) reduces the efficacy of selection, and therefore a larger proportion of mutations are effectively neutral in smaller populations. Recombination affects N_E, and therefore molecular evolution, as higher rates of recombination within a specific region result in a local increase in N_E.

Rates of evolution of sex chromosomes also vary because of differences in relative N_E. In general, and assuming an XY system and an equal population sex ratio, the census population size of the chromosomes is such that there are four copies of each autosome for every three X chromosomes and every single Y. This means that in truly monogamous populations (populations with limited sexual conflict), in which male and female reproductive skew are equal, the relative effective population size of the autosomes (N_{E^A}) is greater than that of the X chromosome (N_{E^X}), which is in turn greater than that of the Y chromosome (N_{E^Y}) (Fig. 2).

In populations experiencing greater sexual selection and conflict—typically those with increased reproductive variance in males—relative N_{E^Y} decreases even further, and N_{E^X} increases (Bachtrog et al. 2011). This is because with strong sexual selection the effective male population size is very small and autosomes are transmitted more through males than the

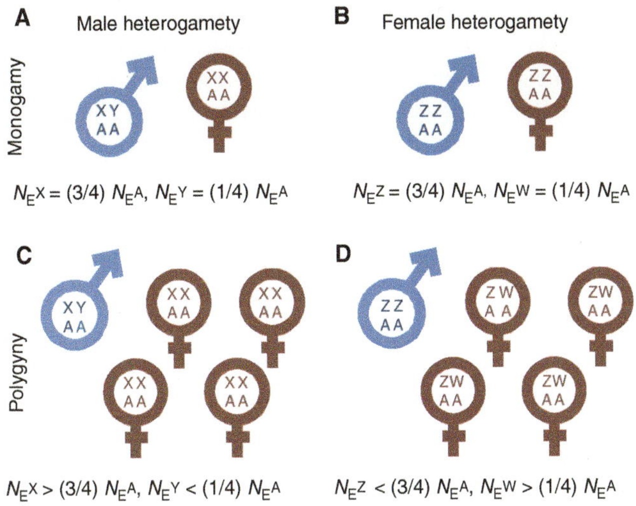

Figure 2. The effect of sexual selection on the relative effective population size. Population size of X (N_{EX}), Y (N_{EY}), Z (N_{EZ}), and W (N_{EW}) chromosomes are compared to the autosomes. Under strict monogamy (*A*,*B*), N_{EX}/N_{EA} is broadly equal to N_{EX}/N_{EA}. Sexual selection (*C*,*D*) increases relative N_{EX} and N_{EW}at the expense of N_{EZ} and N_{EY}. These differences in effective population size may have profound effects on sex chromosome evolution and turnover.

X chromosome is. In some cases in which recombination rates are higher on the X than the autosomes because recombination is restricted to females as with *Drosophila* (Andolfatto 2001), N_{EX} can even exceed N_{EA}.

Sexual selection and its influence on the relative effective population size of the sex chromosomes is known to affect rates of evolution for X-linked genes (Charlesworth et al. 1987; Vicoso and Charlesworth 2010). However, increased sexual selection and conflict can also affect Y chromosome evolution. Sexual selection decreases N_{EY} (Fig. 2), and this may lower the mutational input on Y chromosomes and reduce the efficacy of selection on Y-linked genes. However, the population size effects on Y-mutational input could be compensated for by male-biased mutation rates in some taxa (Sayres and Makova 2011), and the Y chromosomes' ability to respond to selection may be offset by intrachromosome recombination and gene conversion (Lange et al. 2009; Marais et al. 2010)

Additionally, the rate of evolution of alleles on the sex chromosomes also depends on the dominance characteristics of any beneficial mutations as this affects the strength of drift the mutation is subject to (on the X), and whether or not newly beneficial alleles are already part of the standing genetic variation, which is in turn dependent on population size. For example, Y-linkage can boost the rate of increase of rare beneficial alleles and increase the purging of deleterious alleles, but it can also result in lower standing variation on which selection can act (Charlesworth et al. 1987), notwithstanding possible higher male mutation rates. Thus, it is the balance of population size, mutation rate, and dominance that determine allelic fates and sex chromosome evolution.

Similarly, the outcome of new male-determining factors depends on linkage associations between male-benefit alleles and old and neo-Y chromosomes. Depending on the detail of these associations, new sex-determining factors can be lost, retained with the loss of the old Y chromosome, or new and old can both be retained or lost (van Doorn and Kirkpatrick 2007). For example, if genes essential for male sexual fitness are closely linked to the old Y chromosome, then the spread of a new sex-determining factor can be retarded as sex chromosomes differentiate and turnover can be even more unlikely. However, with the advent of sex-limited gene expression, invasion by new sex-determining factors is easier because male/female-benefit alleles will not be inappropriately expressed. That is, female-benefit alleles (for example) will not be expressed in males with a neo-Y if the allele is already sex-limited in expression, and hence this potential avenue of selection against the neo-Y will not exist.

Sex chromosome turnover may also be facilitated by the fact that nonrecombining regions of the sex chromosomes accumulate deleterious mutations (Blaser et al. 2013). The ease of turnover depends on the magnitude of male benefits derived from alleles on the old Y chromosome, as this chromosome is typically lost in turnover events. If the Y chromosome carries numerous benefit alleles to males or, in the case of the eutherian Y, carries several genes essential to male fertility (Lange et al. 2009), loss of these genes prevents turnover as the resulting males would be sterile. With small cumulative benefits on the Y chromosome, turnover events are more likely, and this may suggest that younger Y chromosomes, which have had less time to accumulate male-benefit alleles, are more likely to undergo turnover than older Y chromosomes.

It could therefore be that small populations in general, or populations with intense sexual selection and conflict, are most likely to experience sex chromosome turnovers. However, the strong linkage effects and Hill–Robertson interference that act on Y chromosomes (Bachtrog 2013) may obscure the role of effective population size in shaping Y chromosome evolution. To date, there is little empirical data on the role of relative effective population size in sex chromosome turnover, but comparison of island versus mainland populations could be revealing.

Conflict and Sex Chromosome Dosage Compensation

Sexual conflict may also explain the curious pattern that has emerged recently regarding the difference between X and Z chromosomes in dosage compensation (Mank 2009, 2013). Once recombination is halted between the sex chromosome pair, the sex-limited Y or W degenerates in gene content and activity (Bachtrog 2013). This creates a gene imbalance between males and females, with the heterogametic sex having just one copy of X- or Z-linked genes. There is selection in the heterogametic sex to up-regulate this single copy to equilibrate dose with the two copies in the homogametic sex. Because gene expression is highly correlated between males and females (Griffin et al. 2013), dosage compensation in the heterogametic sex can cause overexpression in the homogametic sex leading to conflict between the sexes over optimal gene expression (Mank et al. 2011).

It is increasingly clear that this conflict plays out differently for X and Z chromosomes, with many male heterogametic species showing effective dosage compensation in males (Hahn and Lanzaro 2005; Ercan et al. 2007; Straub and Becker 2007; Gelbart and Kuroda 2009; Julien et al. 2012; Muyle et al. 2012; Wilkinson et al. 2013). In stark contrast, complete Z chromosome dosage compensation has yet to be seen in any female heterogametic species (Itoh et al. 2007; Vicoso and Bachtrog 2011; Harrison et al. 2012; Vicoso et al. 2013). This pattern may be attributed to the fact that sexual selection is usually stronger on males. This would mean that selection in XY males to up-regulate dosage-sensitive X-linked genes would override the cost of overexpression in XX females, but selection in ZW females to hyperexpress similar genes on the Z chromosome would be retarded by selection against overexpression in ZZ males (Mank 2013). If this is true, then sexual conflict may explain fundamental genomic properties such as sex chromosome dosage compensation.

SGEs and Sex Chromosome Turnover

SGEs can also influence sex chromosome evolution and turnover, and in many ways the effects of SGEs on sex chromosome evolution are similar to those for sexual conflict. However, the causes can differ quite substantially. For example, SGEs can also cause sex chromosome turnover. Segregation distorters such as meiotic

 Cite this article as *Cold Spring Harb Perspect Biol* doi: 10.1101/cshperspect.a017715

drivers, involve two loci—a killer (drive) locus and a sensitivity (target) locus—being in strong linkage repulsion to ensure that the drive and sensitive alleles do not form a suicidal combination resulting in the driver killing itself. The reduced recombination of sex chromosomes makes them particularly prone to invasion by segregation distorters as any gene on the X/Z can drive against the Y/W (and vice versa) (Hamilton 1967; Frank 1991). Unlike distorters on autosomes, segregation distorters on sex chromosomes result in biased sex ratios, as a preponderance of X chromosome–bearing sperm results in more female offspring, or Y-bearing sperm more male offspring. This creates strong selection to rebalance sex ratios, which in turn may promote the evolution of neosex chromosomes (cf. female meiotic drive in mammals) (Yoshida and Kitano 2012) and even in the formation of entirely new sex-determination systems that supplant existing ones (Cordaux et al 2011).

Just as selection can favor reduced recombination between sex-determining genes and sexually antagonistic alleles, segregation distorters can also catalyze the silencing of sex-linked genes. However, the mechanisms by which they do this are somewhat different. One potent silencing mechanism is meiotic sex chromosome inactivation (MSCI), in which most of the genes are silenced on the sex chromosomes during meiosis in the heterogametic sex. MSCI occurs in mammals (Fernandez-Capetill et al. 2003), but remains contentious in *Drosophila* (Hense et al. 2007; Meiklejohn et al. 2011) and birds (Schoenmakers et al. 2009; Guioli et al. 2012; Livernois et al. 2013). One hypothesis put forward is that a key function of MSCI is to prevent SGEs from invading sex chromosomes, which may be particularly vulnerable in the heterogametic sex where they are unpaired. MSCI may therefore act to protect sex chromosomes from the invasion of sex ratio distorters (Meiklejohn and Tao 2010), although this remains to be shown. However, silencing of meiotic drivers on the sex chromosomes may also contribute to the degeneration of the Y or the W. In the fly *Drosophila albomicans*, for example, it is suggested that gene silencing on the Y chromosome may initiate the process of degeneration (Zhou and Bachtrog 2012). Segregation distorters may therefore have the potential to fuel the rate of sex chromosome turnover by invading and subsequently contributing to their degeneration—and hence being a potent promoter of new sex chromosomes.

SEX CHROMOSOMES AS AGENTS OF CONFLICT

Once formed, sex chromosomes may become a flashpoint within the genome for conflict. This means that conflict not only influences sex chromosome turnover and divergence but also acts more powerfully on established sex chromosomes to shape their gene content and evolution.

Sexualization of the X and Z Chromosomes

Once the X and Y (or Z and W) chromosomes have ceased recombining with each other, sexually antagonistic forces can shape both the genomic properties and even the gene content of the sex chromosomes. In theory, the genes on each of the sex chromosome types (X, Y, Z, and W) should be sexualized in unique and specific ways. Although the sex-limited Y and W chromosome often contain relatively few genes, those that remain are thought to be crucial for male (for the Y chromosome) and female (for the W chromosome) fitness, as strong sex-specific selection is needed to maintain genes on these chromosomes once recombination is suppressed (Bachtrog 2013). Evidence consistent with this is found for the *D. melanogaster* Y (Chippindale and Rice 2001) and the chicken W (Moghadam et al. 2012) chromosomes. Additionally, gene duplicates with sex-specific effects may be maintained on Yand W chromosomes, as relocation to the sex-limited chromosome instantly resolves sexual conflict for genes that are dispensable in the homogametic sex. This may explain the fact that the rate of gene gain on the *Drosophila* Y chromosome is 10 times higher than the rate of gene loss (Koerich et al. 2008), although the primate Y chromosome is more conserved for gene content (Hughes et al. 2012).

X and Z chromosomes are thought to be masculinized and feminized in different ways depending on whether alleles with sex-specific effects are dominant or recessive. In male heterogamety, the hemizygosity experienced by males may lead to the accumulation of recessive alleles with male benefits (Rice 1984; Connallon and Clark 2010; Dean et al. 2012), even if there is a substantial cost to females, as X-linked genes are always exposed to selection in males. Alternatively, the fact that the X chromosome is more often present in females may give it a larger role in dominant female-beneficial traits because the X chromosome is selected for female-benefit effects twice as often as male effects (Rice 1984; Connallon and Clark 2010). The converse predictions are expected for Z chromosomes, which should accumulate recessive female-benefit mutations or dominant male-benefit mutations.

Chromosome partitioning approaches to studies of conflict bear up these theoretical predictions, with the X chromosome in *Drosophila* carrying the vast majority of sexually antagonistic variation (e.g., Gibson et al. 2002). Similarly, quantitative genetic approaches show similar results, with the X chromosome contributing more to the variance than expected for sexual dimorphism in Portuguese water dogs (Chase et al. 2005) and sexually antagonistic fitness variation in red deer (Foerster et al. 2007). However, the sex chromosomes do not always seem to carry a disproportional abundance of sexually antagonistic variation (Poissant et al. 2010; Husby et al. 2013).

Predictions about the sexualization of sex chromosomes have also been used to interpret gene expression data that show nonrandom distributions of sex-biased genes. Sex-biased expression is often thought to result from selection for different transcription optima for females and males (Innocenti and Morrow 2010; Connallon and Clark 2011a; Pointer et al. 2013) and represent at least partially resolved sexual conflict (Connallon and Knowles 2005; Connallon and Clark 2011b). In this framework, genes expressed at higher levels in females (female-biased genes) are thought to confer female benefits, and genes expressed at higher levels in males (male-biased genes) are thought to confer benefits to males (Connallon and Clark 2011b; Wright et al. 2012; Pointer et al. 2013).

Using this framework, the X chromosome in both mice and *Drosophila* has been feminized for expression (Khil et al. 2004; Meisel et al. 2012) and the Z chromosome has been masculinized in birds (Wright et al. 2012). The *Drosophila* X chromosome shows weaker and less consistent evidence of demasculinization (Meisel et al. 2012), although this may be owing to the mechanism of dosage compensation in this clade (Vicoso and Charlesworth 2009). The lack of dosage compensation in birds makes it difficult to assess the degree of defeminization. Even more interesting, species with alternative reproductive cycles show deviations from the general pattern of sexualization that are entirely consistent with expectations, as observed with the masculinization of the pea aphid X chromosome, which is caused by its unique inheritance between males, sexual females, and parthenogenic females (Jaquiery et al. 2013).

Despite these broadly consistent patterns, we do not yet know the rate of sexualization of the sex chromosomes, or how the sexualization rates might be affected by feminizing or male killing endosymbionts like *Wolbachia*, which could be expected to act as accelerants. However, many sex chromosomes have multiple strata, or regions, that have been effective sex chromosomes for different lengths of time (Lahn and Page 1999; Bergero et al. 2007; Wright et al. 2012; Vicoso et al. 2013). Strata studies in birds suggest that masculinization of Z chromosome gene expression is much slower than the rate of mating system evolution (Wright et al. 2012).

Similarly, the sex chromosomes can become sexualized via gene duplications, as gene duplicates may find the sex chromosomes either a more favorable or a more hostile environment depending on their sex-specific effects (Gallach et al. 2010; Connallon and Clark 2011b; Gallach and Betran 2011). This may explain the nonrandom patterns of gene movement observed from and to the X chromosome in mammals

(Potrzebowski et al. 2008) and *Drosophila* (Meisel et al. 2009; Vibranovski et al. 2009). However, no such pattern has been observed for the Z chromosome in birds or Lepidoptera (Toups et al. 2011). Furthermore, the out of X migration seen in *Drosophila* and mammals is a slow process (Potrzebowski et al. 2008; Vibranovski et al. 2009), suggesting that significant feminization of the X chromosome only occurs over a long period of time.

Sex Chromosomes Are Magnets for SGEs

Sex chromosomes can act as attractors of segregation distorters, and this can result in sex-ratio biases promoting rapid changes in sex-determination systems, with a concomitant selective sweep of the genomic region surrounding the new sex-determining locus (Charlesworth and Hartl 1978; Hall 2004). Evidence consistent with this scenario is that sex ratio distorters may be responsible for very rapid selective sweeps (e.g., Nolte et al 2013). Spread of segregation distorters may also result in the hitchhiking of tightly linked genes that can potentially have dramatic fitness consequences. Loci tightly linked to segregation distorters (e.g., contained within an inversion) will spread as a consequence of drive, even if they are associated with fitness costs, and can remain at high frequencies even after suppression of the distorter. Conversely, it is possible that such genetic hitchhiking also has the potential to directly affect the rate of spread of SGEs in general, providing they are tightly linked to beneficial high fitness alleles under positive selection (i.e., not opposed by the parliament of genes).

THE INTERPLAY BETWEEN SGEs AND SEXUALLY ANTAGONISTIC ALLELES

The selfish nature of SGEs generates conflict with the rest of the genome that will select for suppression and silencing. SGEs and their corresponding suppressors are therefore locked in an ongoing antagonistic arms race. For SGEs on sex chromosomes that cause sex ratio distortion, this arms race is particularly acute owing to selection to rebalance sex ratios. Counter-selection to restore the sex ratio to unity in response to sex-ratio distorters can promote evolution of novel sex-determination pathways (Cordaux et al. 2011), resulting in new and different ways to determine sex with direct consequences for sexual conflict. Changes to sex determination, such as going from male heterogamety to female heterogamety (XY to ZW, or vice versa) in response to selfish distorters, will simultaneously alter the opportunity for sexual conflict (see Rice 1984). Any SGEs that are already present on a sex chromosome (or on a former autosome now involved in sex determination [e.g., Kaiser and Bachtrog 2010]) will experience a shift in the strength of sex-specific selection. In addition, they may also have different fitness effects when expressed in males and females, as many SGEs (e.g., segregation distorters, male killers, feminizers, and parthenogenesis-inducing endosymbionts) are associated with sex-ratio distortion that clearly favors one sex at the expense of the other. For example, a genome that has experienced extensive periods of feminizing selection (e.g., by feminizing, male killing, or parthenogenesis-inducing bacteria) may have accumulated female-benefit alleles that lower male fitness when expressed in "rescued" males after the evolution of suppressors of sex-ratio distortion. Sex-specific selection under periods of sex-ratio distortion may result in the accumulation of female/male-benefit alleles that are now being exposed in the opposite sex in which they were rarely expressed before suppression of sex-ratio distortion. Over time, the cost of expressing such newly exposed sexually antagonistic alleles in the "rescued" sex may be ameliorated (Bonduriansky and Chenoweth 2009). One might therefore predict the resurgence of sexually antagonistic alleles to be more prominent in populations experiencing a recent spread of a segregation-distorting suppressor allele or experiencing a shift in sex determination.

Ongoing antagonistic arms races between segregation distorters and their suppressors involving the sex chromosomes can contribute to population reproductive failure by causing male sterility and can also potentially contribute to reproductive divergence and even speciation

(Meiklejohn and Tao 2010; Pregraves 2010). Reduced gene flow between populations can occur via hybrid sterility, which itself may arise for different reasons. First, it could be the consequence of a heteropopulation lacking the appropriate suppressor for segregation distorters. Second, hybrid dysfunction could arise because of coevolution between the sex ratio distorters and the MSCI silencing mechanisms. The molecular basis of MSCI may vary between populations in response to suppressing newly emerged distorters or those that have evolved to escape suppression. As a result of the conflict over transmission, incompatibilities can evolve between different components of the MSCI machinery that can result in male hybrid sterility. For example, in populations of the house mouse (*Mus musculus*), there is conflict between the sex chromosomes involving the Y-linked multicopy *Sly* gene, and the X-linked *Slx* and *Slx1* gene complex. These genes act antagonistically during spermatogenesis with *Sly* promoting the transmission of the Y chromosome during MSCI, whereas *Slx*/*Slx1* promote the transmission of the X (Cocquet et al. 2012). Interestingly, this antagonism depends on the relative expression of these genes and not on their total abundance. Nonetheless, there is evidence of a rapid expansion in copy number of these sex-linked genes during the past 3 million years, and it is suggested this is caused by the antagonistic conflict that the X and Y chromosomes are locked into (Good 2012). This is further corroborated by recurrent hybrid male sterility that is associated with overexpression of *Slx*/*Slxl1* and *Sly* because of copy number variation and a concomitant biased sex ratio. There are of course other reasons for sterility such as being caused by compensatory evolution of other genes involved in gametogenesis, but it is clear that conflict involving SGEs is a major sterility source.

The rapid turnover of sex chromosomes generated by segregation-distorting SGEs and their suppressor(s) will simultaneously affect any existing sex-linked allele by altering their exposure to selection, and this will also potentially contribute to sexual conflict. There is therefore likely to be a direct link between the recurrent intragenomic conflicts generated by SGEs and their corresponding suppressors and the resurgence and exposure of sexually antagonistic alleles on the sex chromosomes. Seeing the prevalence of SGEs in nature, this source of sexual conflict is likely to be greatly overlooked.

CONCLUSIONS

It seems clear that conflict at one level can cause conflict at other levels, from genes to organelles and upward, that there can also be feedback, and that all of this interaction has implications for sex chromosome evolution. For example, SGEs such as segregation distorters, male killers, feminizers, and parthenogenesis-inducing endosymbionts can directly target sex chromosomes, thereby generating additional intragenomic and sexual conflict, whereas sexual selection can generate sexual conflict and also select against SGEs and impose strong selection on sex chromosomes. Indeed, the recognition that SGEs can be crucial for sex chromosome evolution is a recent insight that has yet to be fully explored. This is also true of population size and sex chromosome evolution. However, although there is still much to understand, the basic logic of sexual conflict and sex chromosomes has proved to be remarkably prescient and still provides a powerful lens through which sex chromosome evolution can be studied. Equally, these chromosomes provide a unique window into the genetic basis of sexual conflict.

ACKNOWLEDGMENTS

We thank Bill Rice and Sergey Gavrilets for the invitation to participate in this collection and two anonymous referees for their valuable input. J.E.M. is supported by the European Research Council (grant agreement 260233) and a short-term fellowship from the Wissenschaftskolleg zu Berlin. D.J.H. is supported by the University of Exeter, and N.W. by the University of Exeter and The Royal Society (Wolfson Award).

REFERENCES

Albert AYK, Otto SP. 2005. Sexual selection can resolve sex-linked sexual antagonism. *Science* **310:** 119–121.

Andolfatto P. 2001. Contrasting patterns of X-linked and autosomal nucleotide variation in *Drosophila melanogaster* and *Drosophila simulans*. *Mol Biol Evol* **18:** 279–290.

Arunkumar KP, Mita K, Nagaraju J. 2009. The silkworm Z chromosome is enriched in testis-specific genes. *Genetics* **182:** 493–501.

Bachtrog D. 2013. Y-chromosome evolution: Emerging insights into processes of Y-chromosome degeneration. *Nat Rev Genet* **14:** 113–124.

Bachtrog D, Kirkpatrick M, Mank JE, McDaniel SF, Pires JC, Rice WR, Valenzuela N. 2011. Are all sex chromosomes created equal? *Trends Genet* **27:** 350–357.

Bergero R, Forrest A, Kamau E, Charlesworth D. 2007. Evolutionary strata on the X chromosomes of the dioecious plant *Silene latifolia*: Evidence from new sex-linked genes. *Genetics* **175:** 1945–1954.

Blaser O, Grossen C, Neuenschwander S, Perrin N. 2013. Sex-chromosome turnovers induced by deleterious mutation load. *Evolution* **67:** 635–645.

Bonduriansky R, Chenoweth SF. 2009. Intralocus sexual conflict. *Trends Ecol Evol* **24:** 280–288.

Burt A, Trivers R. 2006. *Genes in conflict: The biology of selfish genetic elements*. Harvard University Press, Cambridge MA.

Charlesworth D, Charlesworth B. 1980. Sex differences in fitness and selection for centric fusions between sex chromosomes and autosomes. *Genet Res* **35:** 205–214.

Charlesworth B, Hartl DL. 1978. Population dynamics of segregation distorter polymorphism of *Drosophila melanogaster*. *Genetics* **89:** 171–192.

Charlesworth D, Mank JE. 2010. The birds and the bees and the flowers and the trees: Lessons from genetic mapping of sex determination in plants and animals. *Genetics* **186:** 9–31.

Charlesworth B, Coyne JA, Barton NA. 1987. The relative rates of evolution of sex chromosomes and autosomes. *Am Nat* **130:** 113–146.

Chase K, Carrier DR, Adler FR, Ostrander EA, Lark KG. 2005. Interaction between the X chromosome and an autosome regulations size sexual dimorphism in Portuguese water dogs. *Genome Res* **15:** 1820–1824.

Chippindale AK, Rice WR. 2001. Y chromosome polymorphism is a strong determinant of male fitness in *Drosophila melanogaster*. *Proc Natl Acad Sci* **98:** 5677–5682.

Cocquet J, Ellis PJI, Mahadevaiah SK, Affara NA, Vaiman D, Burgoyne PS. 2012. A genetic basis for a postmeiotic X versus Y chromosome intragenomic conflict in the mouse. *PLoS Genet* **8:** e1002900.

Connallon T, Clark AG. 2010. Sex linkage, sex-specific selection, and the role of recombination in the evolution of sexually dimorphic gene expression. *Evolution* **64:** 3417–3442.

Connallon T, Clark AG. 2011a. Association between sex-biased gene expression and mutations with sex-specific phenotypic consequences in *Drosophila*. *Genome Biol Evol* **3:** 151–155.

Connallon T, Clark AG. 2011b. The resolution of sexual antagonism by gene duplication. *Genetics* **187:** 919–937.

Connallon T, Knowles LL. 2005. Intergenomic conflict revealed by patterns of sex-biased gene expression. *Trends Genet* **21:** 495–499.

Cordaux R, Bouchon D, Grève P. 2011. The impact of endosymbionts on the evolution of host sex-determination mechanisms. *Trends Genet* **27:** 332–341.

Dean R, Mank JE. 2014. The role of sex chromosomes in sexual dimorphism: Discordance between molecular and phenotypic data. *J Evol Biol* doi: 10.1111/jeb.12345.

Dean R, Perry JC, Pizzari T, Mank JE, Wigby S. 2012. Experimental evolution of a novel sexually antagonistic allele. *PLoS Genet* **8:** e1002917.

Dübendorfer A, Hediger M, Burghardt G, Bopp D. 2002. *Musca domestica*: A window on the evolution of sex-determining mechanisms in insects. *Int J Dev Biol* **46:** 75–79.

Ercan S, Giresi PG, Whittle CM, Zhang XM, Green RD, Lieb JD. 2007. X chromosome repression by localization of the *C. elegans* dosage compensation machinery to sites of transcription initiation. *Nat Genet* **39:** 403–408.

Fernandez-Capetillo O, Mahadevaiah SK, Celeste A, Romanienko PJ, Camerini-Otero RD, Bonner WM, Manova K, Burgoyne P, Nussenzweig A. 2003. H2AX is required for chromatin remodeling and inactivation of sex chromosomes in male mouse meiosis. *Dev Cell* **4:** 497–508.

Fisher RA. 1931. The evolution of dominance. *Biol Rev* **6:** 345–368.

Foerster K, Coulson T, Sheldon BC, Pemberton JM, Clutton-Brock TH, Kruuk LEB. 2007. Sexually antagonistic genetic variation for fitness in red deer. *Nature* **447:** 1107–U9.

Frank SA. 1991. Divergence of meiotic drive suppression systems as an explanation for sex-biased hybrid sterility and inviability. *Evolution* **45:** 262–267.

Gallach M, Betran E. 2011. Intralocus sexual conflict resolved through gene duplication. *Trends Ecol Evol* **26:** 222–228.

Gallach M, Chandrasekaran C, Betran E. 2010. Analyses of nuclearly encoded mitochondrial genes suggest gene duplication as a mechanism for resolving intralocus sexually antagonistic conflict in *Drosophila*. *Genome Biol Evol* **2:** 835–850.

Gelbart ME, Kuroda MI. 2009. *Drosophila* dosage compensation: A complex voyage to the X chromosome. *Development* **136:** 1399–1410.

Gibson JR, Chippindale AK, Rice WR. 2002. The X chromosome is a hot spot for sexually antagonistic fitness variation. *Proc Biol Sci* **269:** 499–505.

Good JM. 2012. The conflict within and the escalating war between the sex chromosomes. *PLoS Genet* **8:** e1002955.

Griffin RM, Dean R, Grace JL, Ryden P, Friberg U. 2013. The shared genome is a pervasive constraint on the evolution of sex-biased gene expression. *Mol Biol Evol* **30:** 2168–2176.

Guioli S, Lovell-Badge R, Turner JMA. 2012. Error-prone ZW pairing and no evidence for meiotic sex chromosome inactivation in the chicken germ line. *PLoS Genet* **8:** e1002560.

Hahn MW, Lanzaro GC. 2005. Female-biased gene expression in the malaria mosquito *Anopheles gambiae*. *Curr Biol* **15:** R192–R193.

Hall DW. 2004. Meiotic drive and sex chromosome cycling. *Evolution* **58:** 925–931.

Hamilton WD. 1967. Extraordinary sex ratios. *Science* **156:** 477–488.

Harrison P, Mank JE, Wedell. 2012. Incomplete sex chromosome dosage compensation in the Indian meal moth, *Plodia interpunctella*, based on de novo transcriptome assembly. *Genome Biol Evol* **4:** 1118–1126.

Hense W, Baines JF, Parsch J. 2007. X chromosome inactivation during *Drosophila* spermatogenesis. *PLoS Biol* **5:** 2288–2295.

Hosken DJ, Stockley P, Tregenza T, Wedell N. 2009. Monogamy and the battle of the sexes. *Annu Rev Entomol* **54:** 361–378.

Hughes JF, Skaletsky H, Brown LG, Pyntikova T, Graves T, Fulton RS, Dugan S, Ding Y, Buhay CJ, Kremitzki C, et al. 2012. Strict evolutionary conservation followed rapid gene loss on human and rhesus Y chromosomes. *Nature* **483:** 82–124.

Hurst LD, Pomiankowski A. 1991. Causes of sex-ratio bias may account for unisexual sterility in hybrids. A new explanation of Haldane's Rule and related phenomena. *Genetics* **128:** 841–845.

Husby A, Schielzeth H, Forstmeier W, Gustafsson L, Qvarnstrom A. 2013. Sex chromosome linked genetic variance and the evolution of sexual dimorphism of quantitative traits. *Evolution* **67:** 609–619.

Innocenti P, Morrow EH. 2010. The sexually antagonistic genes of *Drosophila melanogaster*. *PLoS Biol* **8:** e1000335.

Itoh Y, Melamed E, Yang X, Ka PK, Wang S, Yehya N, van Nas A, Replogle K, Band MR, Clayton DF, et al. 2007. Dosage compensation is less effective in birds than in mammals. *J Biol* **6:** 2.

Iyengar VK, Reeve HK, Eisner T. 2002. Paternal inheritance of a female moth's mating preference. *Nature* **419:** 830–832.

Jaenike J. 2001. Sex chromosome meiotic drive. *Annu Rev Ecol Syst* **32:** 25–49.

Jaquiery J, Rispe C, Roze D, Legeai F, Le Trionnaire G, Stoeckel S, Mieuzet L, Da Silva C, Poulain J, Prunier-Leterme N, et al. 2013. Masculinization of the X chromosome in the pea aphid. *PLoS Genet* **9:** e1003690.

Julien P, Brawand D, Soumillon M, Necsulea A, Liechti A, Schutz F, Daish T, Grutzner F, Kaessmann H. 2012. Mechanisms and evolutionary patterns of mammalian and avian dosage compensation. *PLoS Biol* **10:** e100128.

Kaiser B, Bachtrog D. 2010. Evolution of sex chromosomes in insects. *Annu Rev Genet* **44:** 91–112.

Khil PP, Smirnova NA, Romanienko PJ, Camerini-Otero RD. 2004. The mouse X chromosome is enriched for sex-biased genes not subject to selection by meiotic sex chromosome inactivation. *Nat Genet* **36:** 642–646.

Kitano J, Ross JA, Mori S, Kume M, Jones FC, Chan YF, Absher DM, Grimwood J, Schmutz J, Myers RM, et al. 2009. A role for a neo-sex chromosome in stickleback speciation. *Nature* **461:** 1079–1083.

Koerich LB, Wang XY, Clark AG, Carvalho AB. 2008. Low conservation of gene content in the *Drosophila* Y chromosome. *Nature* **456:** 949–951.

Kondo M, Nanda I, Hornung U, Asakawa S, Shimizu N, Mitani H, Schmid M, Shima A, Schartl M. 2003. Absence of the candidate male sex-determining gene *dmrt1b(Y)* of medaka from other fish species. *Curr Biol* **13:** 416–420.

Lahn BT, Page DC. 1999. Four evolutionary strata on the human X chromosome. *Science* **286:** 964–967.

Lange J, Skaletsky H, van Daalen SKM, Embry SL, Korver CM, Brown LG, Oates RD, Silber S, Repping S, Page DC. 2009. Isodicentric Y chromosomes and sex disorders as byproducts of homologous recombination that maintains palindromes. *Cell* **138:** 855–869.

Lemos B, Branco AT, Hartl DL. 2010. Epigenetic effects of polymorphic Y chromosomes modulate chromatin components, immune response, and sexual conflict. *Proc Natl Acad Sci* **107:** 15826–15831.

Livernois AM, Waters SA, Deakin JE, Graves JAM, Waters PD. 2013. Independent evolution of transcriptional inactivation on sex chromosomes in birds and mammals. *PLoS Genet* **9:** e1003635.

Mank JE. 2009. The W, X, Y and Z of sex chromosome dosage compensation. *Trends Genet* **25:** 226–233.

Mank JE. 2013. Sex chromosome dosage compensation, definitely not for everyone. *Trends Genet* **29:** 677–683.

Mank JE, Ellegren H. 2007. Parallel divergence and degradation of the avian W sex chromosome. *Trends Ecol Evol* **22:** 389–391.

Mank JE, Hosken DJ, Wedell N. 2011. Some inconvenient trusts about sex chromosome dosage compensation and the potential role of sexual conflict. *Evolution* **65:** 2133–2144.

Marais GAB, Campos PRA, Gordo I. 2010. Can intra-Y conversion oppose the degeneration of the human Y chromosome: A simulation study. *Genome Biol Evol* **2:** 347–357

Matsuda M, Nagahama Y, Shinomiya A, Sato T, Matsuda C, Kobayashi T, Morrey CE, Shibata N, Asakawa S, Shimizu N, et al. 2002. *DMY* is a Y-specific DM-domain gene required for male development in the medaka fish. *Nature* **417:** 559–563.

Meiklejohn CD, Tao Y. 2010. Genetic conflict and sex chromosome evolution. *Trends Ecol Evol* **25:** 215–223.

Meiklejohn CD, Landeen EL, Cook JM, Kingan SB, Presgraves DC. 2011. Sex chromosome-specific regulation in the *Drosophila* male germline but little evidence for chromosomal dosage compensation or meiotic inactivation. *PLoS Biol* **9:** e1001126.

Meisel RP, Han MV, Hahn MW. 2009. A complex suite of forces drives gene traffic from *Drosophila* X chromosomes. *Genome Biol Evol* **1:** 176–188.

Meisel RP, Malone JH, Clark AG. 2012. Disentangling the relationship between sex-biased gene expression and X-linkage. *Genome Res* **22:** 1255–1265.

Miura I. 2007. An evolutionary witness: The frog *Rana rugosa* underwent change of heterogametic sex from XY male to ZW female. *Sex Dev* **1:** 323–331.

Moghadam HK, Pointer MA, Wright AE, Berlin S, Mank JE. 2012. W chromosome expression responds to female-specific selection. *Proc Natl Acad Sci* **109:** 8207–8211.

Mueller JL, Mahadevaiah SK, Park PJ, Warburton PE, Page DC, Turner JMA. 2008. The mouse X chromosome is enriched for multicopy testis genes showing postmeiotic expression. *Nat Genet* **40:** 794–799.

Muyle A, Zemp N, Deschamps C, Mousset S, Widmer A, Marais GAB. 2012. Rapid de novo evolution of X chromosome dosage compensation in *Silene latifolia*, a plant with young sex chromosomes. *PLoS Biol* **10:** e1001308.

Nanda I, Kondo M, Hornung U, Asakawa S, Winkler C, Shimizu A, Shan ZH, Haaf T, Shimizu N, Shima A, et al. 2002. A duplicated copy of *DMRT1* in the sex-determining region of the Y chromosome of the medaka, *Oryzias latipes*. *Proc Natl Acad Sci* **99:** 11778–11783.

Natri HM, Shikano T, Merila J. 2013. Progressive recombination suppression and differentiation in recently evolved neo-sex chromosomes. *Mol Biol Evol* **30:** 1131–1144.

Nolte V, Pandey RM, Kofler R, Schlotterer C. 2013. Genome-wide patterns of natural variation reveal string selective sweeps and ongoing genomic conflict in *Drosophila mauritiana*. *Genome Res* **23:** 99–110.

Ogata M, Hasegawa Y, Ohtani H, Mineyama M, Miura I. 2008. The ZZ/ZW sex-determining mechanism originated twice and independently during evolution of the frog, *Rana rugosa*. *Heredity* **100:** 92–99.

Partridge L, Hurst LD. 1998. Sex and conflict. *Science* **281:** 2003–2008.

Pointer MA, Harrison PW, Wright AE, Mank JE. 2013. Masculinization of gene expression is associated with exaggeration of male sexual dimorphism. *PLoS Genet* **9:** e1003697.

Poissant J, Wilson AJ, Coltman DW. 2010. Sex-specific genetic variance and the evolution of sexual dimorphism: A systematic review of cross-sex genetic correlations. *Evolution* **64:** 97–107.

Postma E, Spyrou N, Rollins LA, Brooks RC. 2011. Sex-dependent selection differentially shapes genetic variation on and off the guppy Y chromosome. *Evolution* **65:** 2145–2156.

Potrzebowski L, Vinckenbosch N, Marques AC, Chalmel F, Jegou B, Kaessmann H. 2008. Chromosomal gene movements reflect the recent origin and biology of therian sex chromosomes. *PLoS Biol* **6:** 709–716.

Presgraves, DC. 2010. The molecular evolutionary basis of species formation. *Nat Rev Genet* **11:** 175–180.

Price TAR, Wedell N. 2004. Selfish genetic elements and sexual selection: Their impact on male fertility. *Genetica* **132:** 295–307.

Rice WR. 1984. Sex chromosomes and the evolution of sexual dimorphism. *Evolution* **38:** 735–742.

Rice WR. 1987. The accumulation of sexually antagonistic genes as a selection agent promoting the evolution of reduced recombination between primitive sex chromosomes. *Evolution* **41:** 911–914.

Roberts RB, Ser JR, Kocher TD. 2009. Sexual conflict resolved by invasion of a novel sex determiner in Lake Malawi cichlid fishes. *Science* **326:** 998–1001.

Sackton TB, Montenegro H, Hartl DL, Lemos B. 2011. Interspecific Y chromosome introgressions disrupt testis-specific gene expression and male reproductive phenotypes in *Drosophila*. *Proc Natl Acad Sci* **108:** 17046–17051.

Saether SA, Saetre GP, Borge T, Wiley C, Svedin N, Andersson G, Veen T, Haavie J, Servedio MR, Bures S, et al. 2007. Sex chromosome-linked species recognition and evolution of reproductive isolation in flycatchers. *Science* **318:** 95–97.

Saetre GP, Borge T, Lindroos K, Haavie J, Sheldon BC, Primmer C, Syvanen AC. 2003. Sex chromosome evolution and speciation in *Ficedula* flycatchers. *Proc Biol Sci* **270:** 53–59.

Sayres MAW, Makova KD. 2011. Genome analyses substantiate male mutation bias in many species. *Bioessays* **33:** 938–945.

Schoenmakers S, Wassenaar E, Hoogerbrugge JW, Laven JSE, Grootegoed JA, Baarends WM. 2009. Female meiotic sex chromosome inactivation in chicken. *PLoS Genet* **5:** e100466.

Straub T, Becker PB. 2007. Dosage compensation: The beginning and end of generalization. *Nat Rev Genet* **8:** 47–57.

Toups MA, Pease JB, Hahn MW. 2011. No excess gene movement is detected off the avian or lepidopteran Z chromosome. *Genome Biol Evol* **3:** 1463–1472.

van Doorn GS, Kirkpatrick M. 2007. Turnover of sex chromosomes induced by sexual conflict. *Nature* **449:** 909–912.

van Doorn GS, Kirkpatrick M. 2010. Transitions between male and female heterogamety caused by sex-antagonistic selection. *Genetics* **186:** 629–U294.

Vibranovski MD, Zhang Y, Long MY. 2009. General gene movement off the X chromosome in the *Drosophila* genus. *Genome Res* **19:** 897–903.

Vicoso B, Bachtrog D. 2011. Lack of global dosage compensation in *Schistosoma mansoni*, a female-heterogametic parasite. *Genome Biol Evol* **3:** 230–235.

Vicoso B, Bachtrog D. 2013. Reversal of an ancient sex chromosome to an autosome in *Drosophila*. *Nature* **499:** 332.

Vicoso B, Charlesworth B. 2009. The deficit of male-biased genes on the *D. melanogaster* X chromosome is expression-dependent: A consequence of dosage compensation? *J Mol Evol* **68:** 576–583.

Vicoso B, Charlesworth B. 2010. Effective population size and the faster-X effect: An extended model. *Evolution* **63:** 2413–2426.

Vicoso B, Emerson JJ, Zektser Y, Mahajan S, Bachtrog D. 2013. Comparative sex chromosome genomics in snakes: Differentiation, evolutionary strata and lack of global dosage compensation. *PLoS Biol* **11:** e1001643.

Wedell N. 2013. The dynamic relationship between polyandry and selfish genetic elements. *Phil Trans R Soc* **368:** 1–10.

Werren JH, Beukeboom LW. 1998. Sex determination, sex ratios, and genetic conflict. *Annu Rev Ecol Syst* **29:** 233–261.

Wilkinson GS, Johns PM, Metheny JD, Baker RH. 2013. Sex-biased gene expression during head development in a sexually dimorphic stalk-eyed fly. *PLoS ONE* **8:** e59826.

Winge Ö. 1927. The location of eighteen genes in *Lebistes reticulata. J Genet* **18:** 1–43.

Wolfenbarger LL, Wilkinson GS. 2001. Sex-linked expression of a sexually selected trait in the stalk-eyed fly, *Cyrtodiopsis dalmanni. Evolution* **55:** 103–110.

Wright AE, Moghadam HK, Mank JE. 2012. Trade-off between selection for dosage compensation and masculinization on the avian Z chromosome. *Genetics* **192:** 1433–1445.

Yoshida K, Kitano J. 2012. The contribution of female meiotic drive to the evolution of neo-sex chromosomes. *Evolution* **66:** 3198–3208.

Zhou Q, Bachtrog D. 2012. Chromosome-wide gene silencing initiate Y-degeneration in *Drososphila. Curr Biol* **22:** 522–525.

Is Sexual Conflict an "Engine of Speciation"?

Sergey Gavrilets

Department of Ecology and Evolutionary Biology, Department of Mathematics, National Institute for Mathematical and Biological Synthesis, University of Tennessee, Knoxville, Tennessee 37996

Correspondence: gavrila@tiem.utk.edu

At the end of the last century, sexual conflict was identified as a powerful engine of speciation, potentially even more important than ecological selection. Earlier work that followed—experimental, comparative, and mathematical—provided strong initial support for this assertion. However, as the field matures, both the power of sexual conflict and constraints on the evolution of reproductive isolation as driven by sexual conflict are becoming better understood. From theoretical studies, we now know that speciation is only one of several possible evolutionary outcomes of sexual conflict. In line with these predictions, both experimental evolution studies and comparative analyses of fertilization proteins and of species richness show that sexual conflict leads to, or is associated with, reproductive isolation and speciation in some cases but not in others. Increased genetic variation (especially in females) without reproductive isolation is an underappreciated consequence of sexually antagonistic selection.

By the end of 1990s, studies of sexual conflict and sexually antagonistic coevolution moved to the forefront of experimental and theoretical research in evolutionary biology (Rice and Holland 1997; Holland and Rice 1998; Rice 1998). Although the potential evolutionary importance of sexual conflict was anticipated and articulated from a theoretical point of view by Geoff Parker 20 years earlier (Parker 1979), the explosive interest in this topic was a result of groundbreaking experimental work with *Drosophila melanogaster* by Bill Rice (1993, 1996), which directly showed high potential for sexually antagonistic coevolution.

Sexual conflict is a special case of intragenomic conflict (Rice and Holland 1997; Rice 1998; Crespi and Nosil 2013). Sexual conflict occurs if the interests of the sexes with regard to certain aspects of reproduction differ (Parker 1979; Arnqvist and Rowe 2005). Ultimately, sexual conflict arises because of the differences in the roles played by the sexes in the process of reproduction, which in turn lead to the differences between the sexes in the costs and benefits of mating and reproduction (Bateman 1948; Trivers 1972; Parker 1979). Sexual conflict can occur over mating rate (Rice and Holland 1997; Holland and Rice 1998; Rice 1998), offspring size (Haig 2000), parental care (Smith and Härdling 2000; Barta et al. 2002), the use of sperm (Ball and Parker 2003), epigenetic control of development (Rice et al. 2012), etc.

Sexual conflict can occur through two genetic routes (Chapman and Partridge 1996; Par-

Cite this article as *Cold Spring Harb Perspect Biol* doi: 10.1101/cshperspect.a017723

ker and Partridge 1998). Within-locus conflict occurs when the locus controls a trait expressed in both sexes and the optimum trait values differ between the sexes. As a result, optimizing the trait value in one sex will lead to a fitness reduction in the other sex. Within-locus conflict can be resolved via a number of mechanisms, including the evolution of sex linkage, sex-specific expression of genes, gene duplication, and condition dependence (Bonduriansky and Chenoweth 2009; van Doorn 2009). Between-locus conflict occurs when there are two different (sets of) traits each expressed in one sex only but affecting the fitness of both sexes in opposite directions. In this case, adaptive changes in a trait of one sex cause deleterious fitness consequences for the other sex, which can be negated by the evolution in a trait of the other sex, which in turn will cause deleterious fitness consequences for the first sex. For example, males can evolve adaptations increasing their mating rate, which would be detrimental for females who would then evolve some counteradaptations to decrease the mating rate (Rice 1996).

One particularly exciting idea that has emerged from studies of sexual conflict and sexually antagonistic coevolution is that sexual conflict can be an important "engine of speciation" (Rice 1996, 1998; Howard et al. 1998; Parker and Partridge 1998). In standard modern perspective, speciation is a result of genetic divergence between populations accompanied by the evolution of reproductive isolation (Howard and Berlocher 1998; Schluter 2000; Coyne and Orr 2004; Dieckmann et al. 2004; Gavrilets 2004). Genetic divergence can be driven by a variety of evolutionary factors, including mutation, random genetic drift, and natural, sexual, and social selection. Reproductive isolation can follow from a variety of mechanisms, resulting in incompatibilities (including genetic, developmental, morphological, ecological, and behavioral) of males and females from diverging populations or in a reduced fitness of their offspring. As was argued by Rice (1998), Parker and Partridge (1998), and others (e.g., Howard et al. 1998), sexual conflict can contribute to these processes in a number of ways.

Below, I briefly summarize several, mostly verbal, theories of biological diversification caused by sexual conflict and then move to discussing some of the more concrete mathematical models and empirical data and patterns.

WITHIN-LOCUS CONFLICT AND SPECIATION

In general, if within-locus conflict is not resolved (e.g., via sex-specific expression of genes), one would expect that observed trait values in both sexes will deviate from their respective optima. Also, there will be genetic correlations between the traits expressed in males and females so that selection on one sex can cause a correlated response in the other. Unresolved within-locus conflict can impede any form of genetic divergence, reduce the mean fitness of populations, and increase their extinction rate (Bonduriansky and Chenoweth 2009). These effects will also hinder speciation. At the same time, genetic correlations between the traits expressed in males and females can simplify evolution across a valley of maladaptation toward a new optimum for one sex if the other sex is under strong directional selection. This effect can promote speciation (Bonduriansky and Chenoweth 2009).

If within-locus conflict is resolved and male and female traits have evolved to their own new optima, additional effects can occur (Parker and Partridge 1998). First, independent evolution of sexual dimorphism or sex-limited traits by sexual conflict in two isolated populations can result in hybrid breakdown and postzygotic reproductive isolation (Michalak and Noor 2003). Second, such evolution can, as a by-product, affect mate choice traits, which would lead to premating isolation when secondary contact is made. Note that these two effects are not specific to sexual conflict and can happen as a result of genetic divergence driven by any evolutionary factor (Coyne and Orr 2004). At the same time, one can argue that sexually antagonistic genes are more likely than other genes to contribute to incompatibilities between species, because they are responsible for a large part of the standing additive genetic variation and are therefore likely to diverge between populations in response to divergent selection.

Recently, SP De Lisle and L Rowe (unpubl.) argued that sexual dimorphism evolving to resolve the sexual conflict may represent a key innovation driving, rather than constraining, diversification. This can happen if the newly acquired ability of the sexes to evolve independently allows them to explore novel parts of the morphospace, which would be maladaptive for a sexually monomorphic species.

BETWEEN-LOCUS CONFLICT AND SPECIATION

The main emphasis of Rice and Holland (1997) and Rice (1998) was between-locus conflict over mating rate and its consequences for the evolution of reproductive isolation between allopatric populations. They argued that direct deleterious effects of high mating rates on female fitness (e.g., Rice 1996; Rice et al. 2005) will continuously select for increased "resistance" of females to mating attempts from males. Increased resistance in females will in turn result in selection for males able to overcome this resistance. As a consequence, sexual conflict over mating rate will drive perpetual and rapid coevolution of male and female traits controlling reproduction. As a by-product of this coevolution, different allopatric populations will diverge in these traits quickly, losing mutual reproductive compatibility and, thus, becoming different species.

REINFORCEMENT AND SEXUAL CONFLICT

Parker and Partridge (1998) have also offered a new perspective on a classical scenario of speciation with gene flow: reinforcement. The standard line of reasoning (Dobzhansky 1940; Butlin 1987; Howard 1993) is that if two partially diverged populations start hybridizing and hybrids happen to have reduced fitness, then selection will favor evolution of reinforced premating isolating mechanisms, which would result in decreasing the frequency of unfit hybrids. An implicit assumption here is that both males and females will benefit from reinforcement. In contrast, Parker and Partridge (1998) argued that although, in general, females are indeed selected to resist such hybridization, males can benefit from it even if offspring has somewhat reduced fitness. As a result, differences in selection acting on females and on males for acceptance of matings in interpopulation encounters can produce a role for sexual conflict in determining whether reinforcement of mating barriers does or does not occur. Parker and Partridge (1998) made three interesting predictions. First, speciation will be more extensive in groups in which females generally win mating conflicts than in those groups in which males usually win. Second, where there is evidence for reinforcement, it will be the females that are responsible for the resulting premating isolation. Third, in those groups in which females tend to win mating conflicts, the genetic variability within species will be less than that in comparable groups in which males tend to win. Some empirical evidence supports these predictions (Parker and Partridge 1998). At the same time, more recent modeling work has shown that in many cases sexual conflict does not have a winner but rather results in an evolutionary compromise between the sexes (see below). This might limit the generality of the above predictions.

ECOLOGICAL CO-OPTATION

Bonduriansky (2011) has recently suggested that sexual conflict (and more generally sexual selection) can play an important role in the evolution and diversification of ecological traits and strategies through the co-optation of sexual traits for viability-related functions. He identifies three mechanisms through which this can happen. First, sexual conflict can displace populations from their local ecological optima. Second, sexual traits can serve as a preadaptation for novel ecological functions. For example, sexual selection in an insect species can favor elongated legs in males for grasping females. Longer legs in males can then open a possibility to invade a new ecological niche, such as foraging on wet surfaces. Third, traits that were sex-limited ancestrally (e.g., horns in beetles) can transfer between the sexes. A fourth mechanism modeled by van Doorn et al. (2009) is when sexual traits serve as a signal of ecological performance, allowing females to resist mating with a nonlocal male, reducing gene flow between locally adapted populations. Each of these four mechanisms allows populations to explore broader areas of phenotypic space leading to an increase in overall phenotypic diversification and potentially to the emergence of new species.

MODELING CONSEQUENCES OF SEXUAL CONFLICT FOR DIVERSIFICATION

There are three general approaches that have been used for modeling evolutionary consequences of sexual conflict: classical population genetic models of within-locus conflict, game-theoretic models, and dynamic models of between-locus conflict.

Population Genetic Models of Within-Locus Conflict

The very first models of within-locus conflict were built by Owen (1953), Bodmer (1965), Karlin (1972), and Kidwell et al. (1977), who studied one-locus two-allele models of a diploid population subject to sex-differential selection. In their models, fitnesses assigned to three different genotypes—say, AA, Aa, and aa—were different for males and females. These models can be viewed as describing the conflict over traits controlling viability or fertility. The results of mathematical analyses show that depending on the strength of selection, the population can evolve to an equilibrium that is optimal for one sex only or an equilibrium describing an evolutionary compromise between the sexes. The models also show that multiple equilibria can be stable simultaneously so that the evolutionary outcome will strongly depend on initial conditions (Owen 1953; Bodmer 1965). A general conclusion from the population genetic models of within-locus conflict is that the conditions for the maintenance of polymorphism at autosomal sexually antagonistic loci are quite strict, much more so than the conditions for the maintenance of sex-linked polymorphism. In some situations, the population will not settle to an equilibrium but rather will instead cycle (Gavrilets 1998). More recently, these models were adopted for describing some enigmatic human traits, such as homosexuality (Gavrilets and Rice 2006; Ciani et al. 2008; Rice et al. 2012), and extended to include more loci (Connallon and Clark 2010) and additional evolutionary factors, such as random genetic drift (Connallon and Clark 2012).

Game-Theoretic Models

Parker (1979) coined the term sexual conflict in his theoretical study of the possibility of a conflict over mating decision and its possible resolutions. Parker was building on an earlier idea of the "battle-of-the-sexes" game about the costs of raising offspring, which was introduced and studied by Dawkins (1976). Following Parker's groundbreaking work, most earlier modeling studies of sexual conflict used game-theoretical methods to show its existence, provide measures of its intensity, and identify conditions under which alleles or behaviors causing sexual conflict could invade the population (e.g., Jormalainen et al. 1994; Parker and Partridge 1998; Härdling et al. 1999, 2001; Mylius 1999; Alonzo and Warner 2000; Smith and Härdling 2000; Reuter and Keller 2001; Barta et al. 2002; Andres and Morrow 2003; Ball and Parker 2003). Parker and Partridge (1998) applied this approach to the problem of reinforcement as discussed above.

Evolutionary Dynamics of Between-Locus Conflict

A major component of modeling the evolutionary consequences of between-locus conflict has been done for a conflict over mating rate, an issue playing a prominent role in Rice's arguments. To illustrate the logic of the theoretical approaches used, assume that mating and/or fertilization are controlled by the interaction of trait x in females and trait y in males. Let us specify a mating compatibility function $\psi(x, y)$ giving a probability of mating (or fertilization) between a female with trait x and a male with trait y. Function ψ is analogous to the preference function in models of sexual selection (Lande 1981; Gavrilets 2004). Let x and y be the average female and male traits in the population. Then function $P = \psi(x, y)$ can be viewed as a measure of how well a female with trait x is compatible with the typical males in her population. Analogously, function $Q = \psi(x, y)$ can be viewed as a measure of how well a male with trait y is compatible with the typical females in his population. With everything else held the same, P and Q can also be viewed as measures of mating rates for the two sexes. For males, reproductive success is expected to increase monotonically with Q (e.g., linearly as in Fig. 1A). In contrast, for females, mating rates that are too high can be detrimental, meaning female fitness is maximized at an intermediate value of P_{opt} as in Figure 1B.

Turning to the function of mating compatibility $\psi(x, y)$, there are two major types capturing different biological mechanisms. First, if mating requires complementarity or matching of the male and female traits, then ψ can be modeled as a unimodal function of the difference $y - x$ (Fig. 1C). In this case, females can achieve an optimum mating rate by evolving their trait x to be larger or smaller than the male trait y. Second, if mating is more of a contest between male "offense" and female "defense" traits in which extreme "offense" trait values would always be beneficial for males, then ψ can be modeled as an increasing function of the difference $y - x$ (Fig. 1D). In this case, females can achieve an optimum mating rate by evolving stronger and stronger "defense" trait x.

Rice's original discussion of the evolutionary consequences of sexual conflict over mating has mostly concentrated on the regime of perpetual coevolutionary chase between the sexes (Rice 1996, 1998; Holland and Rice 1998). Mathematical models have confirmed that such a chase between the average traits x and y in the population is indeed possible. However, mathematical models also show that coevolutionary chase is but one of several possibilities. Overall, there are at least six different dynamic regimes observed in models of between-locus

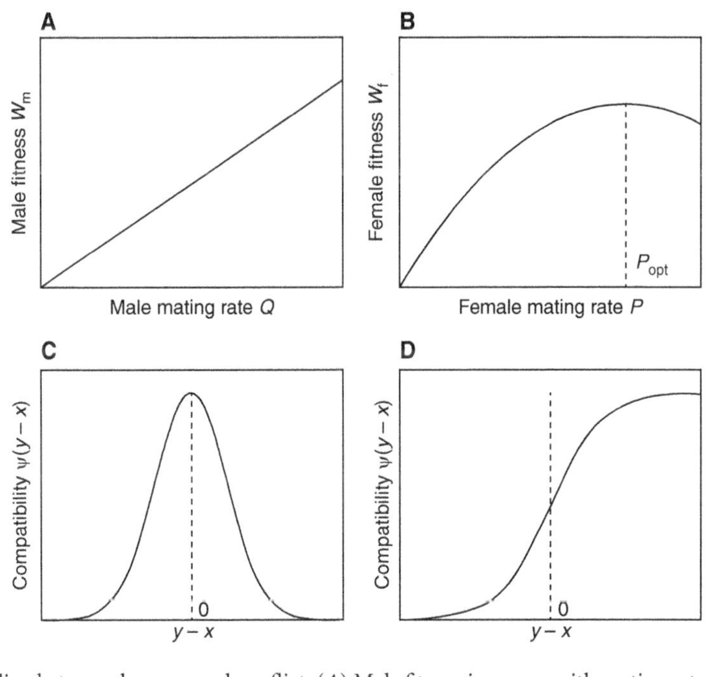

Figure 1. Modeling between-locus sexual conflict. (*A*) Male fitness increases with mating rate. (*B*) For females, there is an optimum mating rate P_{opt}. (*C*) Successful mating requires complementarity of male and female traits ($y = x$). (*D*) Successful mating requires male "offense" trait y to overcome female "defense" trait x.

sexual conflict: continuous coevolutionary chase between the sexes (Gavrilets 2000; Gavrilets et al. 2001; Gavrilets and Waxman 2002; Rowe et al. 2003; Gavrilets and Hayashi 2006), evolution toward an equilibrium (Kondoh and Higashi 2000; Gavrilets et al. 2001; Kimura and Ihara 2009), evolution toward a line of equilibria with subsequent random drift along this line (Gavrilets 2000; Gavrilets and Hayashi 2005; Hayashi et al. 2007), cyclic evolution (Gavrilets et al. 2001; Haygood 2004), Buridan's ass regime involving extensive diversification in female alleles without comparable diversification in male alleles (Frank 2000; Gavrilets and Waxman 2002; Gavrilets and Hayashi 2005; Hayashi et al. 2007), and extensive diversification in both male and female alleles (Frank 2000; Gavrilets and Hayashi 2005; Härdling and Bergsten 2006; Hayashi et al. 2007; Härdling and Karlsson 2009), including a possibility of sympatric speciation (Gavrilets and Waxman 2002; Hayashi et al. 2007). Mathematical models (Gavrilets and Hayashi 2005; Hayashi et al. 2007) also show that different dynamic regimes can be observed with the same set of parameter values but under different initial conditions; populations can switch from one regime to another as a result of stochastic perturbations owing to, say, random genetic drift; and different sets of loci controlling mating and fertilization in the same population can follow different dynamic regimes. In short, the behavior of models of sexual conflict is extremely varied and complex (Rowe et al. 2003; Gavrilets and Hayashi 2005). Frequency-dependent selection is well recognized for its ability to result in complex evolutionary dynamics (Altenberg 1991; Gavrilets and Hastings 1995). In the case of between-locus sexual conflict, selection is not only frequency dependent, but also sex differential, whereas between-sexes interactions include both conflict and cooperation. Therefore, the complexity of resulting evolutionary dynamics is probably not that unexpected.

What does all this tell us about the possibility of speciation by sexual conflict? Out of the six regimes listed above, two regimes are the most intriguing in this regard. The first regime (i.e., continuous coevolutionary chase between the sexes) will eventually lead to allopatric speciation as a by-product if diverging populations keep evolving in different directions and/or at different rates. The sixth regime (i.e., extensive diversification in both male and female alleles) can lead to sympatric speciation if the diversification is in the form of distinct cluster formation. The four other regimes will typically lead to increased genetic variation in the population but not to speciation per se.

The question of which outcome—speciation or increased genetic variation without speciation—is more general can hardly be answered. The answer will depend on a number of specific biological factors and details, which will most likely vary between different organisms or populations. Mathematical models, however, do allow us to get a better understanding of the effects of these factors. However, models often make predictions in terms of more abstract model parameters; the challenge, then, is to link them with empirical data.

Lessells (2006) offered an insightful analysis of evolutionary outcomes of sexual conflict in terms of three sets of variables: (1) benefits of male "offense" and female "defense"; (2) costs of "offense" and "defense"; and (3) feasibility of "offense." Obviously, continuous coevolutionary chase is only possible if it is not constrained by natural selection on the alleles and traits underlying reproduction (so that the costs of "offense" and "defense" are low, whereas feasibility of "offense" is high). Parker and Partridge (1998) argued that the cost of resistance for females can be lower than those for males of imposing a mating. At the same time, the opportunity costs for males are expected to be low (Lessells 2006). If mating requires matching or complementarity of male and female traits (as assumed in Fig. 1C), then male "offense" and female "defense" are more likely to be realized without a systematic change in the underlying traits. In this case, costs may be lower and the opportunity for substantial divergence and speciation higher. If mating is more of a contest (as assumed in Fig. 1C), in which success requires a systematic increase in the efficiency of underlying mating traits, then increasing costs are more likely. These costs will prevent perpetual evolutionary change and constrain speciation (Lessells 2006).

However, the absence of constraints does not guarantee that coevolutionary chase will be observed. Modeling shows (Hayashi et al. 2007) that it is also necessary that the population has relatively low levels of genetic variation and that the plausibility of coevolutionary chase is strongly affected by the genetics of between-sexes interactions. In particular, coevolutionary chase is more plausible if these interactions are mediated by additive quantitative characters than if multiple dominant alleles are involved. In other words, genetic details are also important.

The absence of clear-cut conclusions can make a disappointed reader question the utility of mathematical modeling sexual conflict, but such modeling did clarify the logic of verbal arguments, make them more precise, identify the most important parameters and relevant timescales, and stimulate novel empirical work. Moreover, mathematical modeling has resulted in several novel conclusions that were not included in earlier discussions of sexual conflict. One example of this is that females can reduce the burden of sexual conflict by diversifying genetically rather than by "running" away from males (Buridan's ass regime) (Gavrilets and Waxman 2002). Genetic variation in females is maintained because sexual conflict generates negative-frequency-dependent selection (i.e., rare female types do better) because the males will tend to optimize their mating rate with whichever female type is most abundant. This situation stands in contrast to models of sexual selection in which the interests of males and females typically align. The diversification in females can be answered by diversification in males, potentially leading to sympatric speciation. The extent of diversification and whether or not it results in the formation of distinct genetic clusters within the population depend on a number of different factors including the strength of sexual conflict, the strength of natural selection, the population size, and the genetics of the traits involved. Another conclusion is that genetic variation in female alleles/traits can be larger than that in males. The level of genetic variation in both sexes is also predicted to increase with the strength of sexual conflict. Yet another prediction is that the evolved mating rate is often intermediate between the one optimum for females and the one optimum for males. Therefore, in this case, sexual conflict is not won by one sex but rather there is a (dynamic or static) evolutionary compromise between the sexes.

At the same time, the existing models of sexual conflict are far from being comprehensive (Gavrilets and Hayashi 2005; Lessells 2006). For example, competition between males is not explicitly modeled, spatial models are still lacking, between-sexes interactions are modeled in a rather abstract way despite there being much more currently known about the underlying mechanisms of, say, sperm–egg interactions, the ecological divergence required for coexistence of different species is not considered, and only a single study (Gavrilets and Hayashi 2006) has attempted to model three-way interactions (between male offense, male defense, and female defense) important in Rice's arguments.

EMPIRICAL PATTERNS

Experimental evidence both that high mating rates can be quite detrimental to female fitness and that sexual conflict can drive rapid genetic changes in experimental populations are quite solid (e.g., Rice 1996; Holland and Rice 1999; Pitnick et al. 2001a,b; Rice et al. 2005). How easily these genetic changes result in reproductive isolation and speciation is much less clear-cut. There are three sets of relevant empirical approaches: experimental evolution, analyses of fertilization traits and proteins, and comparative studies of species richness.

Experimental Evolution of Reproductive Isolation

The first study of the evolution of reproductive isolation by sexual conflict under laboratory conditions was that by Martin and Hosken (2003) using populations of the dung fly *Sepsis cynipsea*. In this species, males' armored genitalia often injure females internally during copula. Injuries are visible as sclerotized scars in the female ovipositor, and their occurrence increases with mating frequency. Presumably owing to these injuries, mated females show higher mortality (Blanckenhorn et al. 2002). Females typically attempt to dislodge mounted males by harassing them with vigorous shaking. Martin and Hosken (2003) manipulated population densities and then measured copulation success between a male and a female chosen from populations subject to different or the same treatments for 35 generations. They showed that copulation frequency was greater in pairings within conflict populations than those between populations in which conflict was removed, and this effect was stronger in high-density than in low-density populations. Evidence of female (but not male) resistance to flies from other conflict populations implies that sexual conflict drives divergence (Martin and Hosken 2003; Hosken et al. 2009).

However, similar subsequent experiments have largely failed to provide evidence for the evolution of reproductive isolation. These include experiments with *D. melanogaster* (Wigby and Chapman 2006, but see Ghosh and Joshi 2012), *Drosophila pseudoobscura* (Bacigalupe et al. 2007), bruchid beetle populations *Callosobruchus maculatus* (Gay et al. 2009), water strider *Gerris gillettei* (Gagnon and Turgeon 2011), and the bulb mite *Rhizoglyphus robini* (Plesnar-Bielak et al. 2013).

Several factors can explain the negative results: The experiments may be too short to cause a noticeable effect, the populations could have evolved in a similar direction at a similar rate, or the dynamics imposed by the experimental conditions were different from the "coevolutionary chase."

Fertilization Traits and Proteins

A very large amount of experimental data coming from studies of sperm or pollen competition between closely related species (e.g., Arnold et al. 1993; Wade et al. 1994; Rieseberg 1995; Howard et al. 1998; Howard 1999), as well as from molecular studies of fertilization proteins (e.g., Aguade et al. 1992; Lee and Vacquire 1992; Vacquier and Lee 1993; Metz and Palumbi 1996; Palumbi 1998; Howard 1999; Swanson and Vacquier 2002a,b; Galindo et al. 2003; Landry et al. 2003; Swanson et al. 2003), indicates that in many diverging taxa, traits and proteins related to fertilization evolve extremely rapidly. At least in the case of marine organisms, there now appears to be a consensus that these data can be explained in large part by selection for avoidance of polyspermy (Levitan et al. 2007; Palumbi 2009; Lessios 2011; Vacquier and Swanson 2011). Sperm is under strong selection for increased efficiency in fertilizing eggs, but if more than one sperm penetrates the egg, the resulting zygote in usually inviable. Polyspermy thus represents an important example of sexual conflict over mating rate (Howard et al. 1998). A variety of mathematical models of the effects of polyspermy all converge on the prediction of a rapid coevolutionary chase between the traits of sperm and eggs with sperm evolving to increase their efficiency and eggs evolving molecular "defense" against sperm (Frank 2000; Gavrilets 2000; Gavrilets and Waxman 2002; Haygood 2004; Tomaiuolo and Levitan 2006; Hayashi et al. 2007). Sperm–egg interactions are more likely to require complementarity at the gene-product level than contest (Fig.1C,D) (Rice 1998; Lessells 2006), which, as discussed above, makes continuous "evolutionary chase" more plausible.

Mathematical models of sexual conflict also provide explanations for certain empirical patterns that were previously rather puzzling. These include apparent deceleration in the evolution of sperm lysin genes in abalone (Yang et al. 2000) and complex genetic clustering in sperm protein bindin in sea urchins (Palumbi 1999), in sperm lysin genes in abalone (Swanson et al. 2001), and in sea urchin gamete recognition genes (Vacquier and Moy 1997).

The data also indicate a potential role for reinforcement, because, in some cases, the rate of evolution and the extent of polymorphism correlate with the number of sympatric species (Vacquier and Swanson 2011). However, rapid evolution of fertilization proteins is not universal. For example, for some allopatric species of sea urchins that diverged more than 4–9 million years, there is no evidence of positive selection in bindin alleles that evolve very rapidly in other species (Vacquier and Swanson 2011).

Selection for increased female resistance to mating can also potentially explain rapid evolution of genitalia, which has been observed in many organisms (Eberhard 1985). One can then predict that the differences in genitalic form among congeneric species in which females are protected from male harassment should be less common than differences among congeneric species in which females are vulnerable to harassment by males. This prediction, however, has failed for a large sample of insects and spiders (Eberhard 2004). At the same time, recent comparative studies of water strider populations provide strong evidence for a persistent coevolutionary chase between male and female sexual traits affecting reproductive isolation (Ronn et al. 2007; Perry and Rowe 2012).

Species Richness

The results of comparative studies of the effects of sexual conflict and sexual selection on species richness are equivocal (Parker 2006). Some data show positive correlation between species richness (i.e., the number of species in a clade) and some measures that can be interpreted as proxies for a potential sexual conflict. Arnqvist et al. (2000) were the first to use comparative data to analyze the effects of sexual conflict on large-scale diversification. These investigators contrasted pairs of related groups of insect species differing in the opportunity for sexual conflict. The latter was characterized by the number of matings per female. Arnqvist et al. (2000) showed that groups in which females mate with many males exhibited speciation rates four times as high as in related groups in which females mate only once. In the hoverflies, higher levels of sexual selection are associated with higher species richness (Katzourakis et al. 2001). Barraclough et al. (1995) showed that species richness correlates with the proportion of sexually dichromatic species within taxa of passerine birds. In birds, taxa with promiscuous mating systems tend to be more species-rich than their nonpromiscuous sister taxa (Mitra et al. 1996). Data on *Amphibia* show that sexual size dimorphism is associated with increased species richness, diversification rate, and reduced extinction risk (Lisle and Rowe 2014). However, the prediction about the positive relationship between the number of species in a clade and a potential for sexual conflict does not hold in some mammals, butterflies, spiders (Gage et al. 2002), birds (Morrow et al. 2003), and fish (Ritchie et al. 2005). There are some issues with the interpretation of the results of comparative studies attributed to the difficulties for controlling for phylogenetic effects, sexual selection, and extinction (Panhuis et al. 2001; Coyne and Orr 2004).

Female Polymorphism

As was emphasized above, mathematical theory predicts that sexual conflict can lead to speciation only under specific conditions. If these conditions are not satisfied, populations are expected to diversify. In particular, an interesting prediction emerging from theoretical studies is that females can sometimes get the upper hand in sexual conflict and decrease male harassment by diversifying genetically (Franke et al. 2002; Gavrilets and Waxman 2002; Gavrilets and Hayashi 2005; Härdling and Bergsten 2006; Hayashi et al. 2007). Overall, female sexual polymorphisms appear to be overlooked in the past as most studies of sexual selection focused on male–male competition (Svensson et al. 2009). However, by now there are a number of well-studied cases in which female sexual polymorphism has been shown and linked to sexual conflict. These include higher genetic variation in female reproductive traits in carabid beetles (Sota et al. 2000) and apparent polymorphisms in female genitalia in a spider species (Huber and González, 2001), an African bat bug (Reinhardt et al. 2007), the guppy (Evans et al. 2013), black scavenger flies (Puniamoorthy et al. 2010), butterflies (Cook et al. 1994), diving beetles (Härdling and Bergsten 2006), and damselflies (Van Gossum et al. 2008; Svensson et al. 2009). A study in damselflies explicitly looked at frequency- and density-dependent fecundity patterns (Iserbyt et al. 2013). The results are consistent with the hypothesis that the polymorphism is driven by a conflict between sexes over mating rate, with costly male sexual harassment driving negative-frequency-dependent selection on morph fecundity.

CONCLUDING REMARKS

Earlier experimental, comparative, and mathematical work has provided strong support for the idea that sexually antagonistic selection is a powerful engine of speciation. However, we now much better understand the constraints and limitations on the evolution of reproductive isolation as driven by sexual conflict. Modeling work shows that speciation is only one of several possible evolutionary outcomes of sexual conflict and can occur only under specific conditions. In a similar vein, both experimental evolution studies and comparative analyses of fertilization proteins and of species richness show that sexual conflict leads to, or is associated with, reproductive isolation and speciation in some cases but not in others. Sexual conflict is an engine of speciation. But as with other engines of speciation and diversification—ecological selection, sexual selection, and accumulation of incompatibilities between diverging lineages by mutation and drift—it is most efficient when operating under optimal conditions and can stall otherwise.

ACKNOWLEDGMENTS

I thank W. R. Rice, K. Rooker, L. Rowe, and G. S. van Doorn for comments and suggestions. This work is supported by the National Institute for Mathematical and Biological Synthesis through National Science Foundation Award #EF-0830858, with additional support from The University of Tennessee, Knoxville.

REFERENCES

Aguade M, Miyashita N, Langley CH. 1992. Polymorphism and divergence in the *Mst26A* male accessory gland gene region in *Drosophila*. *Genetics* **132:** 755–770.

Alonzo SH, Warner RR. 2000. Dynamic games and field experiments examining intra- and intersexual conflict: Explaining counterintuitive mating behavior in a Mediterranean wrasse, *Symphodus ocellatus*. *Behav Ecol* **11:** 56–70.

Altenberg L. 1991. Chaos from linear frequency-dependent selection. *Am Nat* **138:** 51–68.

Andres JA, Morrow EH. 2003. The origin of interlocus sexual conflict: Is sex-linkage important? *J Evol Biol* **16:** 219–223.

Arnold ML, Hamrick JL, Bennett BD. 1993. Interspecific pollen competition and reproductive isolation in *Iris*. *J Hered* **84:** 13–16.

Arnqvist G, Rowe L. 2005. *Sexual conflict*. Princeton University Press, Princeton, NJ.

Arnqvist G, Edvardsson M, Friberg U, Nilsson T. 2000. Sexual conflict promotes speciation in insects. *Proc Natl Acad Sci* **97:** 10460–10464.

Bacigalupe LD, Crudgington HS, Hunter F, Moore AJ, Snook RR. 2007. Sexual conflict does not drive reproductive isolation in experimental populations of *Drosophila pseudoobscura*. *J Evol Biol* **20:** 1763–1771.

Ball MA, Parker GA. 2003. Sperm competition games: Sperm selection by females. *J Theor Biol* **224:** 27–42.

Barraclough TG, Harvey PH, Nee S. 1995. Sexual selection and taxonomic diversity in passerine birds. *Proc Biol Sci* **259:** 211–215.

Barta ZN, Houston AI, McNamara JM, Szekely T. 2002. Sexual conflict about parental care: The role of reserves. *Am Nat* **159:** 687–705.

Bateman AJ. 1948. Intrasexual selection in *Drosophila*. *Heredity* **2:** 349–368.

Blanckenhorn W, Hosken D, Martin O, Reim C, Teuschl Y, Ward P. 2002. The costs of copulating in the dung fly *Sepsis cynipsea*. *Behav Ecol* **13:** 353–358.

Bodmer WF. 1965. Differential fertility in population genetics models. *Genetics* **51:** 411–424.

Bonduriansky R. 2011. Sexual selection and conflict as engines of ecological diversification. *Am Nat* **178:** 729–745.

Bonduriansky R, Chenoweth SF. 2009. Intralocus sexual conflict. *Trends Ecol Evol* **24:** 280–288.

Butlin R. 1987. Speciation by reinforcement. *Trends Ecol Evol* **2:** 8–13.

Chapman T, Partridge L. 1996. Female fitness in *Drosophila melanogaster*: An interaction between the effect of nutrition and of encounter rate with males. *Proc Biol Sci* **263:** 755–759.

Ciani AC, Cermelli P, Zanzotto G. 2008. Sexually antagonistic selection in human male homosexuality. *PLoS ONE* **3:** e2282.

Connallon T, Clark AG. 2010. Sex linkage, sex-specific selection, and the role of recombination in the evolution of sexually dimorphic gene expression. *Evolution* **64:** 3417–3442.

Connallon T, Clark AG. 2012. A general population genetic framework for antagonistic selection that accounts for demography and recurrent mutation. *Genetics* **190:** 1477–1489.

Cook S, Vernon J, Bateson M, Guilford T. 1994. Mate choice in the polymorphic African swallowtail butterfly, *Papilio dardanus*: Male-like females may avoid sexual harassment. *Anim Behav* **47:** 389–397.

Coyne J, Orr HA. 2004. *Speciation*. Sinauer Associates, Sunderland, MA.

Crespi B, Nosil P. 2013. Conflictual speciation: Species formation via genomic conflict. *Trends Ecol Evol* **28:** 48–57.

Dawkins R. 1976. *The selfish gene*. Oxford University Press, Oxford.

Dieckmann U, Doebeli M, Metz JAJ, Tautz D. 2004. *Adaptive speciation*. Cambridge University Press, Cambridge.

Dobzhansky TG. 1940. Speciation as a stage in evolutionary divergence. *Am Nat* **74:** 312–321.

Eberhard WG. 1985. *Sexual selection and animal genitalia*. Harvard University Press, Cambridge, MA.

Eberhard WG. 2004. Male–female conflict and genitalia: Failure to confirm predictions in insect and spiders. *Biol Rev Camb Philos Soc* **79:** 121–186.

Evans JP, van Lieshout E, Gasparini C. 2013. Quantitative genetic insights into the coevolutionary dynamics of male and female genitalia. *Proc Biol Sci* **280:** 20130749.

Frank SA. 2000. Sperm competition and female avoidance of polyspermy mediated by sperm-egg biochemistry. *Evol Ecol Res* **2:** 613–625.

Franke ES, Styan CA, Babcock R. 2002. Sexual conflict and polyspermy under sperm-limited conditions: In situ evidence from field simulations with the free-spawning marine echinoid *Evechinus chloroticus*. *Am Nat* **160:** 485–496.

Gage MJG, Parker GA, Nylin S, Wiklund C. 2002. Sexual selection and speciation in mammals, butterflies and spiders. *Proc Biol Sci* **269:** 2309–2316.

Gagnon MC, Turgeon J. 2011. Sexual conflict in *Gerris gillettei* (Insecta: Hemiptera): Intraspecific intersexual correlated morphology and experimental assessment of behaviour and fitness. *J Evol Biol* **24:** 1505–1516.

Galindo BE, Vacquier VD, Swanson WJ. 2003. Positive selection in the egg receptor for abalone sperm lysin. *Proc Natl Acad Sci* **100:** 4639–4643.

Gavrilets S. 1998. One-locus two-allele models with maternal (parental) selection. *Genetics* **149:** 1147–1152.

Gavrilets S. 2000. Rapid evolution of reproductive isolation driven by sexual conflict. *Nature* **403:** 886–889.

Gavrilets S. 2004. *Fitness landscapes and the origin of species*. Princeton University Press, Princeton, NJ.

Gavrilets S, Hastings A. 1995. Intermittency and transient chaos from simple frequency-dependent selection. *Proc Biol Sci* **261:** 233–238.

Gavrilets S, Hayashi TI. 2005. Speciation and sexual conflict. *Evol Ecol* **19:** 167–198.

Gavrilets S, Hayashi TI. 2006. The dynamics of two- and three-way sexual conflicts over mating. *Phil Trans R Soc B* **261:** 345–354.

Gavrilets S, Rice WR. 2006. Genetic models of homosexuality: Generating testable predictions. *Proc Biol Sci* **273:** 3031–3038.

Gavrilets S, Waxman D. 2002. Sympatric speciation by sexual conflict. *Proc Natl Acad Sci* **99:** 10533–10538.

Gavrilets S, Arnqvist G, Friberg U. 2001. The evolution of female mate choice by sexual conflict. *Proc Biol Sci* **268:** 531–539.

Gay L, Eady PE, Vasudev R, Hosken DJ, Tregenza T. 2009. Does reproductive isolation evolve faster in larger populations via sexually antagonistic coevolution? *Biol Lett* **5:** 693–696.

Ghosh SM, Joshi A. 2012. Evolution of reproductive isolation as a by-product of divergent life-history evolution in laboratory populations of *Drosophila melanogaster*. *Ecol Evol* **2:** 3214–3226.

Haig D. 2000. The kinship theory of genomic imprinting. *Annu Rev Ecol Syst* **31:** 9–32.

Härdling R, Bergsten J. 2006. Nonrandom mating preserves intrasexual polymorphism and stops population differentiation in sexual conflict. *Am Nat* **167:** 401–409.

Härdling R, Karlsson K. 2009. The dynamics of sexually antagonistic coevolution and the complex influences of mating system and genetic correlation. *J Theor Biol* **260:** 276–282.

Härdling R, Jormalainen V, Tuomi J. 1999. Fighting costs stabilize aggressive behavior in intersexual conflicts. *Evol Ecol* **13:** 245–265.

Härdling R, Smith HG, Jormalainen V, Tuomi J. 2001. Resolution of evolutionary conflicts: Costly behaviours enforce the evolution of cost-free competition. *Evol Ecol Res* **3:** 829–844.

Hayashi TI, Vose M, Gavrilets S. 2007. Genetic differentiation by sexual conflict. *Evolution* **61:** 516–529.

Haygood R. 2004. Sexual conflict and protein polymorphism. *Evolution* **58:** 1414–1423.

Holland B, Rice WR. 1998. Chase-away sexual selection: Antagonistic seduction versus resistance. *Evolution* **52:** 1–7.

Holland B, Rice WR. 1999. Experimental removal of sexual selection reverses intersexual antagonistic coevolution and removes a reproductive load. *Proc Natl Acad Sci* **96:** 5083–5088.

Hosken DJ, Martin OY, Hy S, Chapman T, Hodgson DJ. 2009. Sexual conflict and reproductive isolation in flies. *Biol Lett* **5:** 697–699.

Howard DJ. 1993. Reinforcement: Origin, dynamics, and fate of an evolutionary hypothesis. In *Hybrid zones and the evolutionary process* (ed. Harrison RG), pp. 46–69. Oxford University Press, New York.

Howard DJ. 1999. Conspecific sperm and pollen precedence and speciation. *Annu Rev Ecol Syst* **30:** 109–132.

Howard DJ, Berlocher SH. 1998. *Endless forms: Species and speciation*. Oxford University Press, New York.

Howard DJ, Reece M, Gregory PG, Chu J, Cain ML. 1998. The evolution of barriers to fertilization between closely related organisms. In *Endless forms: Species and speciation* (ed. Howard DJ, Berlocher SH), pp. 279–288. Oxford University Press, New York.

Huber BA, González AP. 2001. Female genital dimorphism in a spider (Araneae: Pholcidae). *J Zoology* **255:** 301–304.

Iserbyt A, Bots J, Van Gossum H, Sherratt TN. 2013. Negative frequency-dependent selection or alternative reproductive tactics: Maintenance of female polymorphism in natural populations. *BMC Evol Biol* **13**.

Jormalainen V, Tuomi J, Yamamura N. 1994. Intersexual conflict over precopula duration in mate guarding Crustacea. *Behav Proc* **32:** 265–283.

Karlin S. 1972. Some mathematical models of population genetics. *Am Math Monthly* **79:** 699–739.

Katzourakis A, Purvis A, Azmeh S, Rotheray G, Gilbert F. 2001. Macroevolution of hoverflies (Diptera:Syrphidae): The effect of using higher-level taxa in studies of biodiversity, and correlates of species richness. *J Evol Biol* **14:** 219–227.

Kidwell J, Clegg M, Swewart F, Prout T. 1977. Regions of stable equilibria for models of differential selection in 2 sexes under random mating. *Genetics* **85:** 171–183.

Kimura M, Ihara Y. 2009. Replicator-dynamics models of sexual conflict. *J Theor Biol* **260:** 90–97.

Kondoh M, Higashi M. 2000. Reproductive isolation mechanism resulting from resolution of intragenomic conflict. *Am Nat* **156:** 511–518.

Lande R. 1981. Models of speciation by sexual selection on polygenic characters. *Proc Natl Acad Sci* **78:** 3721–3725.

Landry C, Geyer LB, Arakaki Y, Uehara T, Palumbi SR. 2003. Recent speciation in the Indo-West Pacific: Rapid evolution of gamete recognition and sperm morphology in cryptic species of sea urchin. *Proc Biol Sci* **270:** 1839–1847.

Lee YH, Vacquire VD. 1992. The divergence of species-specific abalone sperm lysins is promoted by positive Darwinian selection. *Biol Bull* **182:** 97–104.

Lessells C. 2006. The evolutionary outcome of sexual conflict. *Phil Trans R Soc B* **361:** 301–317.

Lessios H. 2011. Speciation genes in free-spawning marine invertebrates. *Integr Comp Biol* **51:** 456–465.

Levitan DR, terHorst C, Fogarty N. 2007. The risk of polyspermy in three congeneric sea urchins ad its implications for gametic incompatibility and reproductive isolation. *Evolution* **61:** 2009–2016.

Martin O, Hosken D. 2003. The evolution of reproductive isolation through sexual conflict. *Nature* **423:** 979–982.

Metz E, Palumbi S. 1996. Positive selection and sequence rearrangements generate extensive polymorphism in the gamete recognition protein bindin. *Mol Biol Evol* **13:** 397–406.

Michalak P, Noor M. 2003. Genome-wide patterns of expression in Drosophila pure species and hybrid males. *Mol Biol Evol* **20:** 1070–1076.

Mitra S, Landel H, Pruett-Jones S. 1996. Species richness covaries with mating system in birds. *Auk* **113:** 544–551.

Morrow E, Pitcher TE, Arnqvist G. 2003. No evidence that sexual selection is an "engine of speciation" in birds. *Ecol Lett* **6:** 228–234.

Mylius SD. 1999. What pair formation can do to the battle of the sexes: Towards more realistic game dynamics. *J Theor Biol* **197:** 469–485.

Owen A. 1953. A genetical system admitting of 2 distinct stable equilibria under natural selection. *Heredity* **7:** 97–102.

Palumbi SR. 1998. Species formation and the evolution of gamete recognition loci. In *Endless forms: Species and speciation* (ed. Howard DJ, Berlocher SH), pp. 271–278. Oxford University Press, New York.

Palumbi SR. 1999. All males are not created equal: Fertility differences depend on gamete recognition polymorphisms in sea urchins. *Proc Natl Acad Sci* **967:** 12632–12637.

Palumbi S. 2009. Speciation and the evolution of gamete recognition genes: Pattern and process. *Heredity* **102:** 66–76.

Panhuis TM, Butlin R, Zuk M, Tregenza T. 2001. Sexual selection and speciation. *Trends Ecol Evol* **16:** 364–371.

Parker GA. 1979. Sexual selection and sexual conflict. In *Sexual selection and reproductive competition in insects* (ed. Blum MS, Blum NA), pp. 123–166. Academic, New York.

Parker GA. 2006. Sexual conflict over mating and fertilization: An overview. *Phil Trans R Soc B* **361:** 235–259.

Parker GA, Partridge L. 1998. Sexual conflict and speciation. *Phil Trans R Soc B* **353:** 261–274.

Perry JC, Rowe L. 2012. Sexual conflict and antagonistic coevolution across water strider populations. *Evolution* **66:** 544–557.

Pitnick S, Brown WD, Miller GT. 2001a. Evolution of female remating behavior following experimental removal of sexual selection. *Proc Biol Sci* **268:** 557–563.

Pitnick S, Miller GT, Reagan J, Holland B. 2001b. Males' evolutionary responses to experimental removal of sexual selection. *Proc Biol Sci* **268:** 1071–1080.

Plesnar-Bielak A, Skrzynecka AM, Prokop ZM, Kolasa M, Dzialo M, Radwan J. 2013. No evidence for reproductive isolation through sexual conflict in the bulb mite *Rhizoglyphus robini*. *PLoS ONE* **8:** e74971.

Puniamoorthy N, Kotrba M, Meier R. 2010. Unlocking the "Black box": Internal female genitalia in Sepsidae (Diptera) evolve fast and are species-specific. *BMC Evol Biol* **10:** 275.

Reinhardt K, Harney E, Naylor R, Gorb S, Siva-Jothy MT. 2007. Female-limited polymorphism in the copulatory organ of a traumatically inseminating insect. *Am Nat* **170:** 931–935.

Reuter M, Keller L. 2001. Sex ratio conflict and worker production in eusocial *Hymenotera*. *Am Nat* **158:** 166–177.

Rice WR. 1993. Sexually antagonistic genes: Experimental evidence. *Science* **256:** 1436–1439.

Rice WR. 1996. Sexually antagonistic male adaptation triggered by experimental arrest of female evolution. *Nature* **381:** 232–234.

Rice WR. 1998. Intergenomic conflict, interlocus antagonistic coevolution, and the evolution of reproductive isolation. In *Endless forms: Species and speciation* (ed. Howard DJ, Berlocher SH), pp. 261–270. Oxford University Press, New York.

Rice WR, Holland B. 1997. The enemies within: Intergenomic conflict, interlocus contest evolution (ICE), and intraspecific Red Queen. *Behav Ecol Sociobiol* **41:** 1–10.

Rice W, Linder J, Friberg U, Lew T, Morrow E, Stewart A. 2005. Inter-locus antagonistic coevolution as an engine of speciation: Assessment with hemiclonal analysis. *Proc Natl Acad Sci* **102:** 6527–6534.

Rice W, Friberg U, Gavrilets S. 2012. Homosexuality as a consequence of epigenetically canalized sexual development. *Q Rev Biol* **87:** 343–346.

Rieseberg LH. 1995. The role of hybridization in evolution: Old wine in new skins. *Am J Bot* **82:** 944–953.

Ritchie M, Webb S, Graves J, Magurran A, Garcia C. 2005. Patterns of speciation in endemic Mexican Goodeid fish: Sexual conflict or early radiation? *J Evol Biol* **18:** 922–929.

Ronn J, Katvala M, Arnqvist G. 2007. Coevolution between harmful male genitalia and female resistance in seed beetles. *Proc Natl Acad Sci* **104:** 10921–10925.

Rowe L, Cameron E, Day T. 2003. Detecting sexually antagonistic coevolution with population crosses. *Proc Biol Sci* **270:** 2009–2016.

Schluter D. 2000. *The ecology of adaptive radiation*. Oxford University Press, Oxford.

Smith HG, Härdling R. 2000. Clutch size evolution under sexual conflict enhances the stability of mating systems. *Proc Biol Sci* **267:** 2163–2170.

Sota T, Kusumoto F, Kubota K. 2000. Consequences of hybridization between *Ohomopterus insulicola* and *O. arrowianus* (Coleoptera, Carabidae) in a segmented river basin: Parallel formation of hybrid swarms. *Biol J Linn Soc* **71:** 297–313.

Svensson EI, Abbott JK, Gosden TP, Coreau A. 2009. Female polymorphisms, sexual conflict and limits to speciation processes in animals. *Evol Ecol* **23:** 93–108.

Swanson WJ, Vacquier VD. 2002a. The rapid evolution of reproductive proteins. *Nat Rev Genet* **3:** 137–144.

Swanson WJ, Vacquier VD. 2002b. Reproductive protein evolution. *Annu Rev Ecol Syst* **33:** 161–179.

Swanson WJ, Aquadro CF, Vacquier VD. 2001. Polymorphism in abalone fertilization proteins is consistent with the neutral evolution of the egg's receptor for lysin (VERL) and positive Darwinian selection of sperm lysin. *Mol Biol Evol* **18:** 376–383.

Swanson WJ, Nielsen R, Yang QF. 2003. Pervasive adaptive evolution in mammalian fertilization proteins. *Mol Biol Evol* **20:** 18–20.

Tomaiuolo M, Levitan D. 2006. Modeling how reproductive ecology can drive protein diversification and result in linkage disequilibrium between sperm and egg proteins. *Am Nat* **176:** 14–25.

Trivers RL. 1972. Parental investment and sexual selection. In *Sexual selection and the descent of man 1871-1971* (ed. Campbell B), pp. 136–179. Aldine, Chicago.

Vacquier VD, Lee YH. 1993. Abalone sperm lysin: Unusual mode of evolution of a gamete recognition protein. *Zygote* **1:** 181–196.

Vacquier VD, Moy GW. 1997. The fucose sulfate polymer of egg jelly binds to sperm REJ and is the inducer of the sea urchin sperm acrosome reaction. *Dev Biol* **192:** 125–135.

Vacquier VD, Swanson WJ. 2011. Selection in the rapid evolution of gamete recognition proteins in marine invertebrates. *Cold Spring Harb Perspect Biol* **3:** a002931.

van Doorn GS. 2009. Intralocus sexual conflict. *Ann NY Acad Sci* 1168: 52–71.

van Doorn GS, Edelaar P, Weissing FJ. 2009. On the origin of species by natural and sexual selection. *Science* **326:** 1704–1707.

Van Gossum H, Sherratt TN, Cordero-Rivera A. 2008. The evolution of sex-limited colour polymorphism. In *Dragonflies and damselflies: Model organisms for ecological and evolutionary research* (ed. Cordoba-Aguilar A), pp. 219–229. Oxford University Press, Oxford.

Wade MJ, Patterson H, Chang NW, Johnson NA. 1994. Postcopulatory, prezygotic isolation in flour beetles. *Heredity* **72:** 163–167.

Wigby S, Chapman T. 2006. No evidence that experimental manipulation of sexual conflict drives premating reproductive isolation in *Drosophila melanogaster*. *J Evol Biol* **19:** 1033–1039.

Yang ZH, Swanson WJ, Vacquier VD. 2000. Maximum-likelihood analysis of molecular adaptation in abalone sperm lysin reveals variable selective pressures among lineages and sites. *Mol Biol Evol* **17:** 1446–1455.

Sexual Cannibalism as a Manifestation of Sexual Conflict

Jutta M. Schneider

Zoological Institute & Museum, Biocentre Grindel, University of Hamburg, Hamburg, Germany

Correspondence: Jutta.schneider@uni-hamburg.de

Sexual cannibalism is a well-known example for sexual conflict and has many facets that determine the costs and benefits for the cannibal and the victim. Here, I focus on species in which sexual cannibalism is a general component of a mating system in which males invest maximally in mating with a single (monogyny) or two (bigyny) females. Sexual cannibalism can be a male strategy to maximize paternity and a female strategy to prevent paternity monopolization by any or a particular male. Considerable variation exists between species (1) in the potential of males to monopolize females, and (2) in the success of females in preventing monopolization by males. This opens up exciting future possibilities to investigate sexually antagonistic coevolution in a largely unstudied mating system.

Sexual cannibalism, the killing and consumption of potential or actual mating partners in a mating context, has been termed a "pinnacle of sexual conflict" because of the dramatic ending of the act for one mating partner, mostly the male (Elgar and Schneider 2004). This contradiction of traditional sex roles may be one reason why the phenomenon of sexual cannibalism has intrigued naturalists for a long time. In the context of sexual conflict, sexually cannibalistic behavior of females is a harmful trait, and antagonistic traits are expected to evolve in males, which can be considered the reverse of most other examples in which females respond to male harm (see Perry and Rowe 2014). I will discuss potential antagonistic traits to sexual cannibalism in males but will also show that the above view is too simplistic when it comes to spider mating systems characterized by very low male mating rates.

It is important to note that there are different kinds of sexual cannibalism based on very different evolutionary scenarios (Elgar and Schneider 2004; Prenter et al. 2006; Wilder et al. 2009). The most extreme divide exists between cannibalism before sperm transfer, which can only benefit the cannibal, and sexual cannibalism during or after sperm transfer (from here on termed postinsemination sexual cannibalism), which can benefit the cannibal and the victim (Elgar and Schneider 2004). Despite a longer history of research on preinsemination sexual cannibalism, the evolutionary causes and consequences of postinsemination sexual cannibalism are generally less debated.

There are reports (often anecdotal) on the occurrence of sexual cannibalism from diverse invertebrate taxa (Elgar 1992) and it may well occur in all predatory invertebrates that are potentially cannibalistic (Polis 1981). It is beyond

Cite this article as *Cold Spring Harb Perspect Biol* doi: 10.1101/cshperspect.a017731

the scope of this brief review to list and evaluate all reported occurrences. Rather, I will start with a brief account of the generally discussed causes and consequences of sexual cannibalism and will then concentrate on the conflicting interests of the sexes regarding postinsemination sexual cannibalism in mating systems that are characterized by very low male mating rates.

Studies that investigate sexual cannibalism experimentally are mostly concerned with (1) nutritional aspects, (2) the importance of sexual size dimorphism and sexual selection, and, increasingly, (3) behavioral syndromes. The aggressive spillover hypothesis suggests that preinsemination sexual cannibalism is part of a behavioral syndrome in which aggression against mating partners spills over from a foraging context (Arnqvist and Henriksson 1997). There is mixed support for this idea in the few species that have been looked at. In several spider species, females consistently differ in their aggressiveness and these differences affect sexual cannibalism (for a recent debate about the evidence for this hypothesis, see Johnson 2013; Kralj-Fišer et al. 2013b; Pruitt and Keiser 2013).

A majority of studies have taken a unilateral view and have been concerned with the "motivation" of the cannibal; because sexual cannibalism generally occurs in predators, hunger is a well-supported motivation (Wilder et al. 2009). Many predators are food-limited, and, assuming a trade-off between foraging and mating, the balance may tilt toward foraging under particular circumstances (modeled by Newman and Elgar 1991). Food and mate availability will influence the costs and benefits of sexual cannibalism for females and have been one focus of a recent review on sexual cannibalism (Wilder et al. 2009).

In all predatory and cannibalistic animals, mating partners impose selection on each other's abilities to avoid or resist aggression. This selection pressure is asymmetrical if one sex is physically dominant. Indeed, the differences in size between females and males often determine the frequency of sexual cannibalism, perhaps because the potential to resist a cannibalistic attack is size-dependent (Elgar 1992; Wilder and Rypstra 2008). Usually, males are the victims and females are the cannibals. Yet, reversed sexual cannibalism has also been reported and appears to be associated with the reversed pattern in sexual size dimorphism. Examples are the water spider, *Arygoneta aquatica* (Schutz and Taborsky 2005, 2011) and role-reversed wolf spiders (Aisenberg et al. 2011). In the gnaphosid spider, *Micaria sociabilis*, large, young males cannibalize old and relatively smaller females (Sentenska and Pekar 2013). These examples further support the notion that the relative size differences of a mating pair play a part in determining the likelihood of sexual cannibalism. Patterns can be found both on a between-species comparative scale and on a within-species scale (Wilder and Rypstra 2008; Wilder et al. 2009), and they are also reported as an underlying pattern in cannibalism outside a mating context (Bleakley et al. 2013). Furthermore, there is anecdotal evidence for the same pattern in hermaphrodites (e.g., Goto and Yoshida 1985; Michiels et al. 2003), which may constitute a particularly interesting case to study, as the power asymmetries are less obviously related to the male or female role.

In asymmetric encounters, the costs and risks of aggressive behavior toward potential mating partners are low for the dominant partner. Toward smaller males, females could use aggressiveness as a means of partner choice. Indeed, many studies suggest that sexual selection in addition to gaining a meal may be the adaptive value of sexual cannibalism (Prenter et al. 2006). From the female perspective, aggressive behavior directed toward males may serve as a general screening of partner quality, a mechanism often described as indirect mate choice (Elgar and Nash 1988; Prenter et al. 2006; Kralj-Fišer et al. 2012). A screening method implies that females attack every male, and suitors that cannot withstand and persist an attack will be killed and consumed; alternatively, females may differentiate between males and attack and consume only those males that do not meet certain quality criteria (reviewed in Prenter et al. 2006). The latter has been found in wolf spiders (Wilgers and Hebets 2012). The latter mechanism of direct choice is more complex than the indirect one as it requires perception and assessment of quality cues, and large enough benefits of choosiness are expected to match the costs. Mate rejection via sexual cannibalism is considered a particularly extreme case of sexual conflict mostly because rejection can lead to death. Although this may be true for the individual male that loses all future reproductive success, frequencies of preinsemination sexual cannibalism might be rather low (Kralj-Fišer et al. 2013b). Please note that in almost every species, a certain proportion of individuals will be excluded from the mating market and will have no mating success. The claim that prevention of mating success via sexual cannibalism results in more intense sexual conflict than exclusion from mating with less drastic measures has, to my knowledge, never been tested. Because of the scarcity of data on natural frequencies of preinsemination cannibalism, a meta-analysis would not reveal a realistic picture at this stage. Hence, to date, it is not feasible to compare the relative strength of selection imposed by a cannibalistic mate choice strategy against a strategy with less drastic consequences of mate rejection. More studies are needed to unravel the exact nature of sexual selection under the threat of ending as a meal. Below, I will briefly sketch possible responses to selection imposed by sexually cannibalistic females before or during insemination.

RESPONSES TO THE THREAT OF SEXUAL CANNIBALISM

Sexually cannibalistic animals are generally invertebrate predators equipped with sensory and motor skills that facilitate hunting success. For a male who has to approach a potential mate that is a voracious and often larger predator, attention and communication are vital (Schneider and Andrade 2011; Uhl and Elias 2011). The mate approach of, for instance, many spiders and mantises is indeed described as very cautious. In some species, sexual cannibalism may nevertheless occur before copulation because a courting male is unable to reveal his identity to the female, who will then mistake the male for prey (Gould 1984). Such signaling errors may occur when the channels of communication are disturbed. Vibratory signaling could be impaired because of missing legs and chemical signaling owing to poor condition or old age as suggested by Sentenska and Pekar (2013). Cases of sexual cannibalism resulting from mistaken identity emphasize the importance of signaling and incidences are expected to be rare and accidental. Evidently, this nonadaptive explanation does not address the evolution of the phenomenon of sexual cannibalism.

Sexual cannibalism will select for male abilities to assess the potential risks and benefits of a mating attempt. Indeed, perception of cues that elucidate receptivity and perhaps quality of females are reported for several species of spiders (Foelix 2011 and references therein) and mantids (for praying mantis, see Lelito and Brown 2006; Barry et al. 2010; Watanabe and Yano 2012).

Recent studies have shown that the risk of sexual cannibalism before insemination can vary owing to consistent individual differences in aggressiveness of females within species (Pruitt and Riechert 2012), and therefore male sensory ability to detect this variation in danger is expected to evolve (Berning et al. 2012; Johnson 2013; Kralj-Fišer et al. 2013a,b; Pruitt and Keiser 2013). In the bridge spider, *Larinioides sclopetarius*, individuals of both sexes consistently differ in their aggressiveness (Kralj-Fišer and Schneider 2012). The mating pattern is positively assortative in that nonaggressive males preferentially mate with nonaggressive females (Kralj-Fišer et al. 2013a). It remains to be investigated whether males are able to assess female personality and adjust their behavior accordingly. Hunger will also influence the inclination of a female to attack a male (Moya-Larano et al. 2003; see also Rabaneda-Bueno et al. 2008), and, for example, a male black widow can judge from web cues alone whether a female is starved or well fed, the latter of which they prefer (Johnson et al. 2011). Males can also determine whether females are virgin or mated, and already mated females may impose a relatively larger risk of sexual cannibalism (Johnson 2001; Rabaneda-Bueno et al. 2008). However, there are many other possible explanations of males rejecting mated and hungry females. Rigorous experimental approaches are essential to verify that avoidance of sexual cannibalism is behind male behavior, but to experimentally manipulate the risk of sexual cannibalism is challenging (Herberstein et al. 2011).

The differences in risks imposed by different females may be strong enough to favor the evolution of male mate choice. In praying mantis, Barry et al. (2010) found that males prefer larger and more fecund females in a binary choice situation, but poor-condition females are not rejected in a sequential encounter scenario. Barry and Kokko (2010) showed in a model that even under the threat of sexual cannibalism, male mate choice may not evolve if mating opportunities are sufficiently limited. However, even under such conditions, a compromise between the need to secure a mating and mate choice may evolve as has been found in some spiders. This strategy would be analogous to the trading-up strategy used to explain multiple mating in females (Jennions and Petrie 2000). For instance, males of the highly sexually cannibalistic *Argiope bruennichi* can mate maximally twice and they readily mate with low quality females and with siblings suggesting the absence of choosiness. However, they copulate only very briefly with less preferred females and thereby double their chances of surviving their first copulation. Singly mated males can then walk off to search for another female that ideally will be of higher quality (bigynous mating strategy) (Welke and Schneider 2010; Fromhage and Schneider 2012; Welke et al. 2012). In general, mate choice should evolve if mating costs are very high for males, because quality differences between females, if present, matter a lot for the lifetime reproductive success of a male that has a maximum mating rate of two (modeled by Fromhage and Schneider 2012). However, even in systems with low male mating rates and a theoretical expectation of male mate choice, cognitive demands and constraints may have to be taken into account as well. For example, it may be questionable whether a spider brain easily evolves the capacity to evaluate and compare qualities of several potential mating partners and to make a decision based on information that requires memory and comparisons of quality traits. Nevertheless, cognitive costs of male choice may be modest if, for example, a threshold rule is used as a base for the decision to mate or not (Real 1990).

The threat of sexual cannibalism has likely shaped courtship rituals (Robinson and Robinson 1980; Robinson 1982) and is considered responsible for a multitude of mating strategies and adaptations. Males of some orb-web spiders court on a special mating thread, which either connects the hub of the web to a support structure or is produced within the web after removing part of the capture web (Elgar 1991). Generally, the female will leave her position in the hub of the web or in a retreat and move onto this thread for copulation. The male then has the option to drop quickly without landing in the capture web, should the female attack. Courtship on such a thread is therefore considered safer for males. Elgar (1991) suggested that these differences in the mating position and the associated differences in risks for males select for morphological differences; as, for example, leg length, if males with longer legs are better able to escape sexual cannibalism. Risk-sensitive behavioral strategies are also expected. For example, in cannibalistic Mediterranean tarantulas, males approach and court females outside the hunting periods (Moya-Larano et al. 2004). In many species, males adopt opportunistic strategies and copulate while the female is occupied with feeding or while she is molting to maturity (Robinson and Robinson 1980; Fromhage and Schneider 2005c). A few spiders such as *Pisaura mirabilis* induce such an event by presenting a nuptial gift in the form of a prey item (Prokop and Maxwell 2009). Males wrap the gift in silk, which helps the male to remain in contact with the gift if the female interrupts copulation (Andersen et al. 2008). The dense wrapping can also disguise a worthless gift as for example a dried insect carcass without nutritional value (Albo et al. 2012). Furthermore, courting males have been found to induce catalepsy in females (Singer et al. 2000; Becker et al. 2005) or pacify females through tactile courtship (Stoltz et al. 2008).

There is a large body of suggestive evidence of adaptations in males that function to reduce the risk of sexual cannibalism, but there are very few rigorous experimental studies. Experimental approaches have been adopted in two studies of golden orb-web spiders. In *Nephila pilipes*, males are very small in comparison to females and sometimes get killed and eaten during copulation. During courtship, the tiny male surrounds the female body by repeatedly laying down silk, although the silk cannot truly restrain the female (Robinson and Robinson 1973). This behavior is called "mate binding" and was interpreted as a counter adaptation to the threat of sexual cannibalism (Kuntner et al. 2009b). Suggestive evidence for this comes from staged matings showing that mate binding mostly occurs after the female has aggressively terminated the first copulation. In an elegant experiment, Zhang et al. (2011) silenced chemoreception and/or disabled tactile perception of females, and males were manipulated to be able or unable to lay down silk during courting. Males were less able to calm females if they could not produce silk and if female tactile perception was disabled. The results of this study show that the male behavior reduced the likelihood of females reacting aggressively toward a second mating attempt of the same male. In the congener, *Nephila fenestrata*, males have the option to copulate after the female has captured a prey item. However, after some time, they also mate with females that are not feeding (Fromhage and Schneider 2005c). Experimental staging of copulations with and without prey showed that all males survived copulation when the female was feeding, whereas 33% of the males were captured and consumed in the treatment without prey (Fromhage and Schneider 2005c). Sexual cannibalism after copulation is costly for males in this species because surviving males can successfully prevent loss of paternity by mate guarding (Fromhage and Schneider 2005c).

In summary, evidence is accumulating that sexual cannibalism is a selective agent responsible for a high diversity of behavioral, sensory, and morphological counter adaptations in males. For the rest of this review, I will focus on species where postinsemination sexual cannibalism is a regular occurrence and co-occurs with a particular mating system (see below). In these cases, postinsemination sexual cannibalism may be a male strategy to increase paternity share or a female response to male adaptations in the context of sperm competition. Both options evolved under sexual selection and sexual conflict. In the examples below, it is indeed the killing and much less the feeding that is the motivation behind the aggression. In other words, females will attack to kill, but once a kill is made, the female may just as well consume the body. By restricting myself to such cases, I largely ignore the rich literature on nutritional aspects of feeding on conspecifics, which is a highly interesting field in itself.

MALE MATING STRATEGIES IN MONOGYNOUS OR BIGYNOUS MATINGS SYSTEMS

Superficially, the notion that sexual cannibalism can be part of a male mating strategy seems to be at odds with the claim that sexual cannibalism involves a strong sexual conflict. To fully comprehend the phenomenon, we need to dissect the mating system and measure the costs and benefits of killing and being killed for both sexes.

Monogynous mating systems are rare but occur in diverse taxa such as fish, crustaceans, insects, and spiders. Within the entelegyne spider subgroup, there are at least four independent origins of such a mating system in which males are adapted to mate with a single or maximally two females (monogyny, bigyny), whereas the females are polyandrous (Miller 2007; Schneider and Fromhage 2010; Schwartz et al. 2013). A theoretical model suggests that monogyny can evolve when more males than females mate (male biased effective sex ratio) (Fromhage et al. 2005). Such a bias can occur if the tertiary sex ratio is male biased, for example, because more males than females survive to maturation. Assuming a male bias and polyandry, the average expected paternity success of polygamous males will be smaller than one (Fromhage et al. 2005). It then pays to invest maximally into mating with one female only (monogynous male mating strategy), if this will elevate paternity success above the average.

To achieve above average paternity, monogynous males must possess means to protect their sperm against rivals by outcompeting rival ejaculates or preventing female remating (Fromhage et al. 2008).

Allowing or even inviting sexual cannibalism is one way for a male to achieve this goal. For instance, in two *Latrodectus* species, males perform a characteristic somersault during copulation, which will bring their abdomen right on top of the female chelicerae and which may facilitate cannibalism by the female (Forster 1992; Segoli et al. 2008). Thereby, males prolong their copulation duration and reduce the female propensity to remate (Andrade 1996). However, females have two spermathecae that cannot both be inseminated during a single copulation. Males have paired mating organs too, the pedipalps, which are secondary mating organs of spiders that are charged with sperm, and males can only use one pedipalp at the time (see Fig. 1 for an illustration of male and female sexual organs in selected spider species). Hence, although the male benefits from prolonging his first copulation, he also increases the risk of dying before he was able to use his second pedipalp. Males should make sure they survive for as long as necessary to empty both their pedipalps. Male *Latrodectus hasselti* can do so by pulling their inner organs to the front of the abdomen, for instance, away from the area where the female will insert her mouth parts, before they attempt the first copulation (Andrade et al. 2005). Even though the female may start feeding on the copulating male, he has a good chance of surviving the first somersault and mating. Survival enables males to return to the same female where they finally sacrifice their life during the second insertion and the second somersault (Andrade et al. 2005).

In species of *Tidarren* and *Echinotheridion*, two genera within the same family as Latrodectus (Theridiidae), males mature shortly after hatching at a very small body size. The paired pedipalps are very large compared to the body, and during development, males actively detach one pedipalp (Knoflach and van Harten 2001). On finding a female, the one-palped males will inseminate one of the two spermathecae of a single female and invariably die during their only copulation. The dead body will temporally prevent rivals from using the same genital opening. However, the males cannot protect the still virgin side of the female and hence accept the risk that a second male may inseminate the other spermatheca. Inferring from the above model (Fromhage et al. 2005), such a strategy can be evolutionary stable if the expected reproductive success of a male that fertilizes 50% of the eggs of a female is still higher than the expected reproductive success of an alternative strategy. If the male effective sex ratio bias is larger than two, this condition would be met because the average paternity success of a polygamous male would be <50%. Alternatively, the mating strategy will pay off if many females will not mate multiply. The mating system of these one-palped spiders can be safely considered monogynous, although natural mating rates of females are unknown for all these species (Knoflach and van Harten 2001). In the above examples, sexual cannibalism is a regular element of the mating procedure that is actively solicited by the males, either because they position themselves onto the fangs of the females or because they die during mating. Hence, the act of dying during copulation is not against the male interest, although it may oppose female interests. However, males may not be indifferent to the timing of their death if they have additional mating options, regardless of whether these options are with the same or a different female.

A monogynous male mating strategy is only better than a polygynous alternative if the male can efficiently protect his paternity. Male death alone, however, cannot always prevent females from remating or rivals from threatening paternity. Interestingly, very similar mechanisms of postcopulatory and even postmortal paternity protection have evolved independently in those same spider families that are characterized by monogynous species (Foellmer and Fairbairn 2003; Miller 2007; Schwartz et al. 2013). Males of most *Argiope* species, several theridiids such as *Latrodectus and Tidarren*, and some *Nephilidae* break off parts of their genitalia during copulation, which remain inside the female and hinder further inseminations (Uhl et al. 2010).

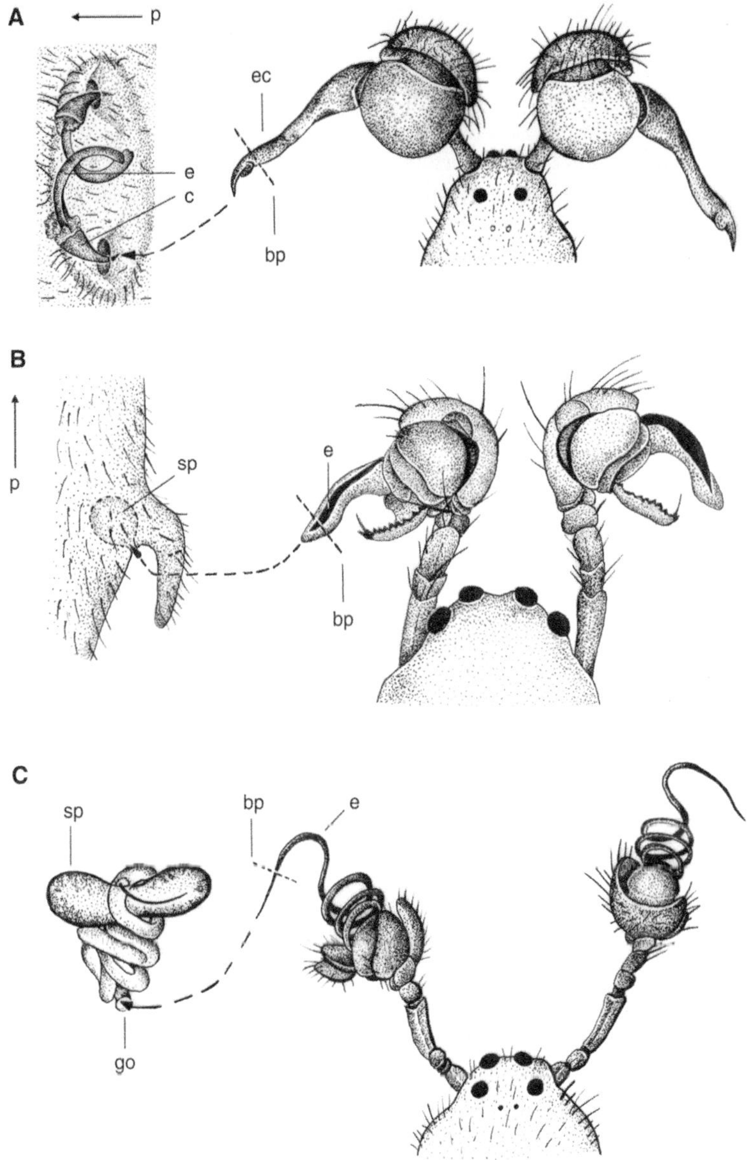

Figure 1. Drawings of male (*right*) and female (*left*) spider genitalia from one representative of three spider families in which mating plugs through genital damage evolved independently. Males possess two pedipalps, which are secondary genitalia that get charged with sperm after maturation; female genitalia possess two copulatory openings each leading to an independent spermatheca. During copulation, males transfer sperm into one spermatheca at a time with one pedipalp. The tip of the intromittent organ breaks and obstructs the copulatory opening (*A*,*B*) or the entrance of the spermatheca (*C*). (*Top*) In *Nephila fenestrata* (Nephilidae), the tip of the male sperm transfer organ (embolic conductor, ec) breaks and remains stuck in the genital opening. Males often use both pedipalps with the same female and obstruct both of her genital openings as depicted. Plugs are visible. (*Middle*) *A. bruennichi* (Araneidae) have very complex pedipalps that unfold during copulation. Some sclerites attach to the protrusion above the female genital openings, the scape. The tip of the embolus (e) breaks and blocks one of two copulatory openings. The spermatheca sits right behind the entrance as depicted connected to the outside by a short duct. The plugs cannot be easily seen from the outside because of the scape. (*Bottom*) In *Latrodectus hasselti* (Theridiidae), males have a coiled embolus (e) that has to be pushed all the way up the equally coiled insemination duct of the female. Here, only the internal insemination duct and one spermatheca are drawn. The tip of the embolus reaches into the lumen of the spermatheca and may break off there. Hence, in this species, the plug is deep inside. p→, point to the anterior end of the female; bp, breakage point; e, embolus; c, conductor; ec, embolic conductor; go, genital opening; sp, spermatheca. *A* redrawn and modified from Fromhage and Schneider (2006), *B* from Uhl et al. (2007), and *C* from Snow et al. (2006). Drawings are not to scale.

Males in some other species plug the opening from the outside; again, others obstruct the connection of the insemination duct to the spermatheca (see Fig. 1 for examples of both patterns). The latter may not prevent further copulations, but a rival will not be able to transfer sperm into the sperm storage organ and will gain no paternity as a result (Snow and Andrade 2004; Snow et al. 2006). Mating plugs also exist in other, less cannibalistic spiders but are then composed of secretions and not of genital fragments at high costs (Uhl et al. 2010).

For the male, the cost of a plugging mechanism based on his genital parts is sterility, as each pedipalp can be used once only caused by the damage. Sterility has no negative implications for reproductive success in a monogynous or bigynous mating system where each pedipalp is needed only once anyway to inseminate and then seal the two spermathecae of a single female. Intriguingly, not every male succeeds in breaking off the tip of the genital or in positioning the piece in the accurate location (e.g., Neumann and Schneider 2011). Another challenge to the goal of protecting ones sperm is that plugs can be removed or fall out. If the probability of overcoming these impediments is summarized as plug effectiveness, there are pronounced differences between species. If rivals can remove or bypass plugs, males can benefit from surviving copulations and defending the female as eunuchs. Males of *N. fenestrata,* for example, often survive copulations but nevertheless leave a plug behind (Fromhage and Schneider 2006). However, males of this species are capable of bypassing plugs of rivals (Fromhage and Schneider 2005c) and can achieve a considerable paternity share with a mated female, thereby reducing the paternity share of the previous male. Therefore, males with damaged genitalia benefit from guarding their female. These eunuchs, which have no reproductive future at stake, will readily escalate a fight and succeed in chasing away much larger opponents (Fromhage and Schneider 2005a; Kralj-Fišer et al. 2011). Males of *Nephilengys malabarensis* copulate by detaching the entire pedipalp, which will continue to transfer sperm into the female independently (Li et al. 2012). Males without pedipalps are better fighters than males with pedipalps (Lee et al. 2012), possibly because the pedipalps are somehow in the way or constitute an extra weight. Hence, although sexual cannibalism often co-occurs with a monogynous mating system, it is not tightly linked.

The above examples show males that invest maximally in protecting their paternity by sacrificing genitalia and their lives, and, in some cases, this involves enforcing monogamy on females. Assuming that females benefit from polyandry, counteradaptations that prevent male monopolization are expected (see Edward et al. 2014). Investigating between species variation and evolutionary correlations of plug effectiveness versus sexual cannibalism may be very revealing.

FEMALE COUNTERSTRATEGIES IN MONOGYNOUS OR BIGYNOUS MATINGS SYSTEMS

Assuming traditional sex roles, females are expected to be selective in their choice of mating partner. In predatory and cannibalistic animals, mate rejection can lead to an extra benefit by ending in a meal (see above). This benefit may be large enough to favor adaptations that facilitate this ending. Natural selection will likely put a limit to such adaptations if they occur before a sufficient sperm load has been secured, as very aggressive females may risk remaining unmated.

Indeed, observations on several spider species suggest that female spiders initially accept any male at least for a first copulation (Wilder and Rypstra 2007). If another mating opportunity arises, females may only remate with a male of higher quality than the previous one (trade-up strategy) (Halliday 1983). Unselective mating of virgins followed by a trade-up strategy is particularly likely if selective females have an increased risk of remaining unmated (Kokko and Mappes 2005). Alternatively, females may mate multiply and mix paternity or they may cryptically favor the sire of highest quality. As Kokko and Mappes (2013) suggest in a recent model, multiple mating might be a better null model for explaining female mating rates than

 Cite this article as *Cold Spring Harb Perspect Biol* doi: 10.1101/cshperspect.a017731

monandry especially if mating is not very costly, a condition that may well apply in sexually cannibalistic spiders. Under conditions in which the number of mates is unpredictable and where it can be very low, females will benefit more from securing a sperm supply than from choosiness. Interestingly, although entelegyne spiders that include all species mentioned here possess two spermathecae (Foelix 2011), the filling of a single one may be enough to ensure fertilization of all eggs (Schneider et al. 2005). It is tempting to speculate that a second spermatheca is a good option to secure the potential of multiple mating in species with a highly efficient plugging mechanism of males.

In polyandrous mating systems, it is widely accepted that transfer of spermatozoa is only one of many functions of copulation (Eberhard 1996) and that sperm competition selects for male strategies that influence sperm use (Kelly and Jennions 2011; Edward et al. 2014) or female remating (Uhl et al. 2009; Alonzo and Pizzari 2013; Kvarnemo and Simmons 2013). In addition to the common costs of copulation, males may cause harm to females (e.g., Chapman et al. 1995; Morrow et al. 2003; Ram and Wolfner 2007; Sirot et al. 2014). If these processes are a function of copulation duration, and male and female interests do not coincide, females will benefit from exerting control over the termination of copulation (Elgar et al. 2000). Sexual cannibalism during mating may be a very efficient and even nutritious means to control the duration of copulation and thereby regulate relative paternity of males. Such a function would imply that sexual cannibalism shortens the duration of copulation (Elgar 1995). However, within species displaying regular sexual cannibalism during copulation, sexual cannibalism has been found to prolong copulation (Andrade 1996; Schneider et al. 2006). Perhaps sexual cannibalism has initially evolved as a female strategy to end copulation, and males have coevolved effective means that nevertheless permit a prolonged attachment of their genitalia. The highpoint of this antagonistic coevolution may be the continuous sperm transfer by a detached pedipalp (remote copulation) (Knoflach and van Harten 2001; Li et al. 2012).

Aggressive behavior during mating may have enabled females to terminate genital contact and control ejaculate transfer. Among other things, this power asymmetry may have triggered a coevolution in male and female genitalia (Kuntner et al. 2009a). Female control depends on the genital coupling mechanism, and the strength of genital attachment will determine whether females can induce detachment rapidly enough to influence sperm transfer. Secure attachment will prevent females from ending copulation as evidenced by cases in which male parts, or whole males, remain stuck in the female genital opening after they have been attacked and even died (Knoflach and van Harten 2001; Schneider et al. 2001; Foellmer and Fairbairn 2003). Spider species vary widely in their genital morphology and genital coupling mechanisms (Eberhard 2004a), and firmness of the genital coupling may be a result of a sexual conflict over copulation duration in concert with sexual selection and natural selection (see Brennan and Prum 2014).

Moreover, in *Argiope lobata*, sexual cannibalism not only prolongs the first copulation, but also significantly enhances plugging success (Nessler et al. 2009). This correlation can be viewed as evidence for male investment into a given female or for female choice. The female choice explanation implies that to disfavor a certain male, females would need to refrain from cannibalism. However, observations suggest that females indiscriminately attack every male during copulation (Nessler et al. 2009). Indeed, female influences may become important after copulation. *A. lobata* females seem to control sperm storage, and how much they store of a given male is independent of copulation duration (Welke and Schneider 2009). Females of the congener *A. bruennichi* also exert control over relative paternity independently of the duration of copulation by favoring males that courted more (Schneider and Lesmono 2009). Although these studies seem straightforward examples of cryptic female choice (Eberhard 2004b), it remains to be seen whether these findings apply to many other species (see Albo et al. 2013). In particular, it is unclear how fitness relevant the option for cryptic female choice will be in nature.

From the male perspective, paternity success may be enhanced not only by engaging in longer copulations, but also by repeated copulations with the same female. Male redback spiders survive the somersault during their first copulation, and their return to inseminate and plug the other side will result in fertilization of the majority of this female's eggs. Females are more likely to cannibalize a male during the first copulation if the male had shown a weak courtship performance. Hence, through sexual cannibalism during the first copulation, polyandrous females restrict the paternity success of a male (Stoltz et al. 2008). Many studies have shown that multiple mating, at least in modest rates, is beneficial for females as, for instance, negative effects of incompatibility are reduced (Zeh and Zeh 2001, 2003; Cornell and Tregenza 2007) and the genetic diversity of their offspring increased (reviewed in Jennions and Petrie 2000). In species in which males have evolved means to protect their paternity, females with two spermathecae that are inseminated via different ducts should have an interest in avoiding repeated copulations with the same male. Few studies have tested this prediction in cannibalistic species, but the available studies found no evidence for females selectively rejecting males with which they previously mated (Fromhage and Schneider 2005b; Tuni and Bilde 2010). There are two explanations for the absence of discrimination. One is that there are no costs or perhaps even benefits of remating with the same male. The other is based on the possible constraint that females do not possess the sensory capacity for discriminating individual males even though they would benefit from avoiding repeated copulations with the same male. This may occur if selection on discrimination would be relatively weak, for example, because occurrences are kept in low frequencies caused by an alternative mechanism such as sexual cannibalism after the first copulation. Thereby, females can prevent a male from returning and monopolizing her. Another prediction from the hypothesis would be that the costs of remating for females have to be larger than the gains for males because otherwise selection on males to remate would be stronger than selection on females to prevent remating.

In monogynous mating systems, males are under particularly strong selection to protect their paternity with a single female, and this may at least in some cases override female interests. If male and female interests are opposed, antagonistic coevolution over monopolization may reveal itself in the mating patterns, for example, in the frequency of sexual cannibalism after the first copulation and in the efficiency of paternity protection. In *A. bruennichi*, 83% of the males that copulate with a virgin female leave a highly effective plug in one of her two genital openings (Nessler et al. 2007). Indeed, females stereotypically attack every male within one second of genital contact, and up to 80% of the males do not survive the first copulation (Fig. 2). A male that survives the first copulation can return and plug the second opening, and he thereby monopolizes paternity with this female with a very high certainty. However, the female does not compromise fertilization success by preventing a second copulation, because a single sperm load contains sufficient sperm to fertilize all of her eggs (Schneider et al. 2005). Males that survive the first copulation by terminating copulation within 10 s still secure a sig-

Figure 2. Sexual cannibalism in *A. bruennichi*. The larger female has wrapped the smaller male in silk after a successful copulation had occurred. (Reprinted, with permission, from Stefan Nessler who holds the copyright.)

nificant proportion of paternity, even if the female remates, by transferring thousands of spermatozoa and setting a plug in this short time. If these patterns evolved under sexual conflict, we would expect a coevolution of female attack timing and frequency, on the one hand, and the speed of sperm transfer, on the other hand. Because of this dramatic copulation scenario, we predicted that females of *A. bruennichi* would benefit from multiple mating. The potential for cryptic female choice that has been shown in the species (Schneider and Lesmono 2009) may be such a benefit. However, so far, other pronounced indirect or direct fitness benefits of multiple mating without the effect of cannibalism have not been found in this species (Welke and Schneider 2012). However, in double mating experiments, female choice has not been included. To date, only a handful of studies exist that identified benefits of polyandry in spiders (e.g., Watson 1998; Uhl et al. 2005; Maklakov and Lubin 2006; Tuni et al. 2013).

In summary, females may not be very selective, although they may apply a general "screening" as mentioned above by testing male resistance to an attack before mating and during mating. Polyandry may be used as a bet-hedging strategy. Mechanisms of cryptic female choice will further enhance the options of such a strategy, as it permits biasing paternity toward a preferred sire. One testable prediction of this hypothesis would be that females of species that do not discriminate males before copulation are more likely to benefit from multiple mating than species that reject males by preventing copulation. In other words, females may rely more strongly on polyandry if their options of precopulatory choice are limited. However, evidence is mixed and more data are required to decide whether or not the hypothesis generally holds.

SEXUAL CANNIBALISM AS PATERNAL INVESTMENT

In some species, males are very successful in monopolizing females, perhaps because females do not resist monogamy or do not actively facilitate polyandry. In such a situation, sexual conflict is modest and male sacrifice could be interpreted as a paternal investment strategy (modeled by Buskirk et al. 1984). However, males are often much smaller than the average prey of the females, and direct fecundity effects of sexual cannibalism were not found in species in which both males and females have low mating rates (Fromhage et al. 2003). Please note that not all species in question have been investigated and that feeding on a male may have nutritional benefits unrelated to his body mass, for example, by providing limiting essential nutrients. Surprisingly, despite this being a long-standing hypothesis, its relevance is still unclear (Wilder et al 2009). A recent experiment provided some suggestive evidence for this idea. Female *A. bruennichi,* which were experimentally fed a male, produced offspring that survived winter conditions better than offspring of females that were experimentally prevented from cannibalism (Welke and Schneider 2012). Consistent with the paternal investment hypothesis, natural mating rates of both sexes in this species are very low and many females mate with only one male (Zimmer et al. 2012). Furthermore, males are very effective in placing mating plugs, leading to a relatively high paternity security in males. Under such circumstances, it may benefit the male to invest in improving his mate's condition, if this enhances offspring fitness, as his sacrifice will unlikely benefit a future rival. Overall, however, the paternal investment hypothesis of sexual cannibalism has received little support, and based on our current knowledge, it is unlikely to contribute generally to a better understanding of the phenomenon. Enhancement of female condition or offspring quality through the male's investment of soma may become relevant once a largely monogamous mating system has evolved and male and female interests coincide (Holland and Rice 1999).

CONCLUDING REMARKS

Sexual conflict and antagonistic coevolution is important in species with sexual cannibalism and the interests of the sexes may diverge over control of copulation duration and remating.

Female predators such as spiders and praying mantis have the potential to kill an unwelcome suitor at low costs, and they may gain a more or less profitable meal to compensate for the energetic costs of the lunge. Thus, we can expect that selection on males is so strong that any trait which reduces the risk of being killed before copulation will rapidly spread. As a result, a proportion of males will still be eaten and killed before insemination, but those are very likely individuals that are in a poor condition or that may have misinterpreted the intentions of the female they approached as being ready to mate, whereas in fact, these were actually on the hunt. However, perhaps as a result of a power asymmetry with dominant females and relatively powerless males, which reduces the costs and risks for the female, sexual cannibalism has become a means in the sexual conflict for control of mating. Despite an easy identification of the victim of the aggression, it is not legitimate to deduce which sex is in control from this observation alone. Indeed, even male victims can benefit from sexual cannibalism.

In this brief essay, I emphasized cases in which sexual cannibalism is more than a side effect of foraging and failed signaling but instead can be considered a regular component of the mating sequence. In monogynous mating systems, males are adapted to invest maximally in fertilizing the eggs of a single female and this investment often includes damage to their genitalia and a loss of their life. Indeed, by accepting or even inviting sexual cannibalism, males can at least sometimes enhance their paternity success by prolonging copulation duration, increasing sperm transfer, facilitating the placement of mating plugs, and perhaps the transfer of manipulative seminal substances. As soon as these mating strategies oppose female interests, sexually antagonistic coevolution will favor female counteradaptations. If females benefit from polyandry, selection will favor adaptations that prevent males from monopolizing fertilization of all her eggs. However, if this female benefit is to retain options for cryptic choice, it is not polyandry per se that is beneficial (Wedell 2013). In contrast, whether or not a particular female benefits from polyandry will depend on the quality and compatibility of her first male. Indeed, there is evidence from redback spiders that females selectively kill disfavored males after the first copulation, although they accept monopolization by preferred males.

In many species, males may apply costly plugs, but they may not always prevent sperm competition. Low plug effectiveness would relax selection on females to kill males during the first copulation. A sexual conflict scenario would be supported if there was a correlation between plug effectiveness and the frequency of premature sexual cannibalism.

Behavioral observations as well as the morphology of the genital apparatus are suggestive of a sexual conflict over control of mating in several spider species. If sexual cannibalism or more accurately the attack behavior of females during copulation is a result of sexual conflict, we can make a number of predictions that still wait to be tested:

1. Sexual cannibalism during the first copulation is particularly common in species with monogynous/bigynous mating systems (active self-sacrifice may have to be excluded here as it cannot be attributed to the female aggressively pushing her interest).
2. Female attacks during mating more often result in male death in species in which male monopolization potential is high.
3. Genital detachment and/or remote copulation occurs when females benefit from control of copulation frequency and duration.

Tests of these comparative predictions would be most revealing if they could be correlated with a phylogenetic hypothesis. The recent advents of molecular methods for reconstructions of phylogenetic trees have been applied to several spider families relevant in this context, and the increasing number of researchers striving to increase our knowledge of spider mating systems allows an optimistic prognosis that the above predictions will be tested in the very near future (Eberhard 2004b). Ideally, the comparative method will be combined with replicated experimental studies (Rönn et al. 2007).

ACKNOWLEDGMENTS

I am grateful to Sergey Gavrilets and Bill Rice for inviting me to contribute to this collection. I thank Mark Elgar, Lutz Fromhage, Jannis Liedtke, Rainer Neumann, Jasmin Ruch, Wiebke Schuett, Gabriele Uhl, and Steffi Zimmer for critically reading the text and two referees for constructive comments. Special thanks to Monika Hänel for drawing the figure and to Claudia Drees for editing.

REFERENCES

*Reference is also in this collection.

Aisenberg A, Costa FG, Gonzalez M. 2011. Male sexual cannibalism in a sand-dwelling wolf spider with sex role reversal. *Biol J Linn Soc Lond* **103:** 68–75.

Albo MJ, Toft S, Bilde T. 2012. Female spiders ignore condition-dependent information from nuptial gift wrapping when choosing mates. *Anim Behav* **84:** 907–912.

Albo MJ, Bilde T, Uhl G. 2013. Sperm storage mediated by cryptic female choice for nuptial gifts. *Proc Biol Sci* **280:** 20131735.

Alonzo SH, Pizzari T. 2013. Selection on female remating interval is influenced by male sperm competition strategies and ejaculate characteristics. *Phil Trans R Soc B* **368:** 20120044.

Andersen T, Bollerup K, Toft S, Bilde T. 2008. Why do males of the spider *Pisaura mirabilis* wrap their nuptial gifts in silk: Female preference or male control? *Ethology* **114:** 775–781.

Andrade MCB. 1996. Sexual selection for male sacrifice in the Australian redback spider. *Science* **271:** 70–72.

Andrade MCB, Gu L, Stoltz JA. 2005. Novel male trait prolongs survival in suicidal mating. *Biol Lett* **1:** 276–279.

Arnqvist G, Henriksson S. 1997. Sexual cannibalism in the fishing spider and a model for the evolution of sexual cannibalism based on genetic constraints. *Evol Ecol* **11:** 255–273.

Barry KL, Kokko H. 2010. Male mate choice: Why sequential choice can make its evolution difficult. *Anim Behav* **80:** 163–169.

Barry KL, Holwell GI, Herberstein ME. 2010. Multimodal mate assessment by male praying mantids in a sexually cannibalistic mating system. *Anim Behav* **79:** 1165–1172.

Becker E, Riechert S, Singer F. 2005. Male induction of female quiescence/catalepsis during courtship in the spider, *Agelenopsis aperta*. *Behaviour* **142:** 57–70.

Berning AW, Gadd RDH, Sweeney K, MacDonald L, Eng RYY, Hess ZL, Pruitt JN. 2012. Sexual cannibalism is associated with female behavioural type, hunger state and increased hatching success. *Anim Behav* **84:** 715–721.

Bleakley BH, Welter SM, McCauley-Cole K, Shuster SM, Moore AJ. 2013. Cannibalism as an interacting phenotype: Precannibalistic aggression is influenced by social partners in the endangered Socorro Isopod (*Thermosphaeroma thermophilum*). *J Evol Biol* **26:** 832–842.

* Brennan PLR, Prum RO. 2014. Mechanisms and evidence of genital coevolution: The roles of natural selection, mate choice, and sexual conflict. *Cold Spring Harb Perspect Biol* doi: 10.1101/cshperspect.a017749.

Buskirk RE, Frohlich C, Ross KG. 1984. The natural selection of sexual cannibalism. *Am Nat* **123:** 612–625.

Chapman T, Liddle LF, Kalb JM, Wolfner MF, Partridge L. 1995. Cost of mating in *Drosophila melanogaster* females is mediated by male accessory-gland products. *Nature* **373:** 241–244.

Cornell SJ, Tregenza T. 2007. A new theory for the evolution of polyandry as a means of inbreeding avoidance. *Proc R Soc B* **274:** 2873–2879.

Eberhard WG. 1996. *Female control: Sexual selection by cryptic female choice.* Princeton University Press, Princeton, NJ.

Eberhard WG. 2004a. Male–female conflict and genitalia: Failure to confirm predictions in insects and spiders. *Biol Rev* **79:** 121–186.

Eberhard WG. 2004b. Why study spider sex: Special traits of spiders facilitate studies of sperm competition and cryptic female choice. *J Arachnol* **32:** 545–556.

* Edward DA, Stockley P, Hosken DJ. 2014. Sexual conflict and sperm competition. *Cold Spring Harb Perspect Biol* doi: 10.1101/cshperspect.a017707.

Elgar MA. 1991. Sexual cannibalism, size dimorphism. and courtship behavior in orb-weaving spiders (Araneidae). *Evolution* **45:** 444–448.

Elgar MA. 1992. Sexual cannibalism in spiders and other invertebrates. In *Cannibalism* (ed. Elgar MA, Crespi BJ), pp. 128–155. Oxford, New York.

Elgar MA. 1995. The duration of copulation in spiders: Comparative patterns. Records of the Western Australian Museum, Perth, Australia, pp. 1–11.

Elgar MA, Nash DR. 1988. Sexual cannibalism in the garden spider *Araneus diadematus*. *Anim Behav* **36:** 1511–1517.

Elgar MA, Schneider JM. 2004. Evolutionary significance of sexual cannibalism. In *Advances in the study of behavior* (ed. Slater JB, et al.), Vol. 34, pp. 135–163. Academic, San Diego.

Elgar MA, Schneider JM, Herberstein ME. 2000. Female control of paternity in the sexually cannibalistic spider *Argiope keyserlingi*. *Proc Biol Sci* **267:** 2439–2443.

Foelix RF. 2011. *Biology of spiders.* Oxford University Press, Oxford.

Foellmer MW, Fairbairn DJ. 2003. Spontaneous male death during copulation in an orb-weaving spider. *Proc Biol Sci* **270:** S183–S185.

Forster LM. 1992. The stereotyped behavior of sexual cannibalism in *Latrodectus hasselti* Thorell (Araneae, Theridiidae), the Australian redback spider. *Aust J Zool* **40:** 1–11.

Fromhage L, Schneider J. 2005a. Virgin doves and mated hawks: Contest behaviour in a spider. *Anim Behav* **70:** 1099–1104.

Fromhage L, Schneider JM. 2005b. No discrimination against previous mates in a sexually cannibalistic spider. *Naturwissenschaften* **92:** 423–426.

Fromhage L, Schneider JM. 2005c. Safer sex with feeding females: Sexual conflict in a cannibalistic spider. *Behav Ecol* **16:** 377–382.

Fromhage L, Schneider JM. 2006. Emasculation to plug up females: The significance of pedipalp damage in *Nephila fenestrata*. *Behav Ecol* **17:** 353–357.

Fromhage L, Schneider JM. 2012. A mate to die for? A model of conditional monogyny in cannibalistic spiders. *Ecol Evol* **2:** 2572–2582.

Fromhage L, Uhl G, Schneider JM. 2003. Fitness consequences of sexual cannibalism in female *Argiope bruennichi*. *Behav Ecol Sociobiol* **55:** 60–64.

Fromhage L, Elgar MA, Schneider JM. 2005. Faithful without care: The evolution of monogyny. *Evolution* **59:** 1400–1405.

Fromhage L, McNamara JM, Houston AI. 2008. A model for the evolutionary maintenance of monogyny in spiders. *J Theor Biol* **250:** 524–531.

Goto T, Yoshida M. 1985. The mating sequence of the benthic arrowworm *Spadella schizoptera*. *Biol Bull* **169:** 328–333.

Gould SJ. 1984. Only his wings remained. *Nat Hist* **93:** 10–18.

Halliday TR. 1983. The study of mate choice. In *Mate choice* (ed. Bateson P), pp. 3–32. Cambridge University Press, Cambridge.

Herberstein ME, Schneider JM, Harmer AMT, Gaskett AC, Robinson K, Shaddick K, Soetkamp D, Wilson PD, Pekar S, Elgar MA. 2011. Sperm storage and copulation duration in a sexually cannibalistic spider. *J Ethol* **29:** 9–15.

Holland B, Rice WR. 1999. Experimental removal of sexual selection reverses intersexual antagonistic coevolution and removes a reproductive load. *Proc Natl Acad Sci* **96:** 5083–5088.

Jennions MD, Petrie M. 2000. Why do females mate multiply? A review of the genetic benefits. *Biol Rev Camb Phil Soc* **75:** 21–64.

Johnson JC. 2001. Sexual cannibalism in fishing spiders (*Dolomedes triton*): An evaluation of two explanations for female aggression towards potential mates. *Anim Behav* **61:** 905–914.

Johnson JC. 2013. Debates: Challenging a recent challenge to the aggressive spillover hypothesis. *Ethology* **119:** 811–813.

Johnson JC, Trubl P, Blackmore V, Miles L. 2011. Male black widows court well-fed females more than starved females: Silken cues indicate sexual cannibalism risk. *Anim Behav* **82:** 383–390.

Kelly CD, Jennions MD. 2011. Sexual selection and sperm quantity: Meta-analyses of strategic ejaculation. *Biol Rev* **86:** 863–884.

Knoflach B, van Harten A. 2001. *Tidarren argo* sp. Nov. (Araneae: Theridiidae) and its exceptional copulatory behaviour: Emasculation, male palpal organ as a mating plug and sexual cannibalism. *J Zool* **254:** 449–459.

Kokko H, Mappes J. 2005. Sexual selection when fertilization is not guaranteed. *Evolution* **59:** 1876–1885.

Kokko H, Mappes J. 2013. Multiple mating by females is a natural outcome of a null model of mate encounters. *Entomol Exp Appl* **146:** 26–37.

Kralj-Fišer S, Schneider JM. 2012. Individual behavioural consistency and plasticity in an urban spider. *Anim Behav* **84:** 197–204.

Kralj-Fišer S, Gregoric M, Zhang SC, Li DQ, Kuntner M. 2011. Eunuchs are better fighters. *Anim Behav* **81:** 933–939.

Kralj-Fišer S, Schneider JM, Justinek Z, Kalin S, Gregoric M, Pekar S, Kuntner M. 2012. Mate quality, not aggressive spillover, explains sexual cannibalism in a size-dimorphic spider. *Behav Ecol Sociobiol* **66:** 145–151.

Kralj-Fišer S, Mostajo GAS, Preik O, Pekar S, Schneider JM. 2013a. Assortative mating by aggressiveness type in orb weaving spiders. *Behav Ecol* **24:** 824–831.

Kralj-Fišer S, Schneider JM, Kuntner M. 2013b. Challenging the aggressive spillover hypothesis: Is pre-copulatory sexual cannibalism a part of a behavioural syndrome? *Ethology* **119:** 615–623.

Kuntner M, Coddington JA, Schneider JM. 2009a. Intersexual arms race? Genital coevolution in *Nephilid spiders* (Araneae, Nephilidae). *Evolution* **63:** 1451–1463.

Kuntner M, Kralj-Fišer S, Schneider JM, Li D. 2009b. Mate plugging via genital mutilation in nephilid spiders: An evolutionary hypothesis. *J Zool* **277:** 257–266.

Kvarnemo C, Simmons LW. 2013. Polyandry as a mediator of sexual selection before and after mating. *Phil Trans R Soc Lond B Biol Sci* **368:** 20120042.

Lee QQ, Oh J, Kralj-Fiser S, Kuntner M, Li DQ. 2012. Emasculation: Gloves-off strategy enhances eunuch spider endurance. *Biol Lett* **8:** 733–735.

Lelito JP, Brown WD. 2006. Natural history miscellany—Complicity or conflict over sexual cannibalism? Male risk taking in the praying mantis *Tenodera aridifolia sinensis*. *Am Nat* **168:** 263–269.

Li DQ, Oh J, Kralj-Fiser S, Kuntner M. 2012. Remote copulation: Male adaptation to female cannibalism. *Biol Lett* **8:** 512–515.

Maklakov AA, Lubin Y. 2006. Indirect genetic benefits of polyandry in a spider with direct costs of mating. *Behav Ecol Sociobiol* **61:** 31–38.

Michiels NK, Raven-Yoo-Heufes A, Brockmann KK. 2003. Sperm trading and sex roles in the hermaphroditic opisthobranch sea slug *Navanax inermis*: Eager females or opportunistic males? *Biol J Linn Soc Lond* **78:** 105–116.

Miller JA. 2007. Repeated evolution of male sacrifice behavior in spiders correlated with genital mutilation. *Evolution* **61:** 1301–1315.

Morrow EH, Arnqvist G, Pitnick S. 2003. Adaptation versus pleiotropy: Why do males harm their mates? *Behav Ecol* **14:** 802–806.

Moya-Larano J, Orta-Ocana JM, Barrientos JA, Bach C, Wise DH. 2003. Intriguing compensation by adult female spiders for food limitation experienced as juveniles. *Oikos* **101:** 539–548.

Moya-Larano J, Pascual J, Wise DH. 2004. Approach strategy by which male Mediterranean tarantulas adjust to the cannibalistic behaviour of females. *Ethology* **110:** 717–724.

Nessler SH, Uhl G, Schneider JM. 2007. Genital damage in the orb-web spider *Argiope bruennichi* (Araneae: Araneidae) increases paternity success. *Behav Ecol* **18:** 174–181.

Nessler S, Uhl G, Schneider J. 2009. Sexual cannibalism facilitates genital damage in *Argiope lobata* (Araneae: Araneidae). *Behav Ecol Sociobiol* **63:** 355–362.

Neumann R, Schneider JM. 2011. Frequent failure of male monopolization strategies as a cost of female choice in the black widow spider *latrodectus tredecimguttatus*. *Ethology* **117:** 1057–1066.

Newman JA, Elgar MA. 1991. Sexual cannibalism in orb-weaving spiders—An economic model. *Am Nat* **138:** 1372–1395.

* Perry JC, Rowe L. 2014. The evolution of sexually antagonistic phenotypes. *Cold Spring Harb Perspect Biol* doi: 10.1101/cshperspect.a017558.

Polis GA. 1981. The evolution and dynamics of intraspecific predation. *Annu Rev Ecol Syst* **12:** 225–251.

Prenter J, MacNeil C, Elwood RW. 2006. Sexual cannibalism and mate choice. *Anim Behav* **71:** 481–490.

Prokop P, Maxwell MR. 2009. Female feeding regime and polyandry in the nuptially feeding nursery web spider, *Pisaura mirabilis. Naturwissenschaften* **96:** 259–265.

Pruitt JN, Keiser CN. 2013. Debates: The aggressive spillover hypothesis: Existing ailments and putative remedies. *Ethology* **119:** 807–810.

Pruitt JN, Riechert SE. 2012. The ecological consequences of temperament in spiders. *Curr Zool* **58:** 589–596.

Rabaneda-Bueno R, Rodriguez-Girones MA, Aguado-De-La-Paz S, Fernandez-Montraveta C, De Mas E, Wise DH, Moya Larano J. 2008. Sexual cannibalism: High incidence in a natural population with benefits to females. *PLoS ONE* **3:** 1.

Ram KR, Wolfner MF. 2007. Seminal influences: *Drosophila* Acps and the molecular interplay between males and females during reproduction. *Integr Comp Biol* **47:** 427–445.

Real L. 1990. Search theory and mate choice: 1. Models of single-sex discrimination. *Am Nat* **136:** 376–405.

Robinson M. 1982. Courtship and mating behavior in spiders. *Annu Rev Entomol* **27:** 1–20.

Robinson MH, Robinson B. 1973. Ecology and behaviour of the giant wood spider *Nephila maculata* (Fabricius) in New Guinea. *Smithson Contrib Zool* **149:** 1–76.

Robinson MH, Robinson B. 1980. Comparative studies of the courtship and mating behaviour of tropical araneid spiders. In *Pacific insects monograph*, 218 pp. Bishop Museum, Honolulu, HI.

Rönn J, Katvala M, Arnqvist G. 2007. Coevolution between harmful male genitalia and female resistance in seed beetles. *Proc Natl Acad Sci* **104:** 10921–10925.

Schneider JM, Andrade MCB. 2011. Mating behaviour and sexual selection, pp. 215–274. Cambridge University Press, Cambridge.

Schneider JM, Fromhage L. 2010. Monogynous mating strategies in spiders. In *Behaviour: Evolution and mechanisms* (ed. Kappeler PI). Springer, Heidelberg, Germany.

Schneider JM, Lesmono K. 2009. Courtship raises male fertilization success through post-mating sexual selection in a spider. *Proc Biol Sci* **276:** 3105–3111.

Schneider JM, Fromhage L, Uhl G. 2005. Extremely short copulations do not affect hatching success in *Argiope bruennichi* (Araneae, Araneidae). *J Arachnol* **33:** 663–669.

Schneider JM, Thomas ML, Elgar MA. 2001. Ectomised conductors in the golden orb-web spider, *Nephila plumipes* (Araneoidea): A male adaptation to sexual conflict? *Behav Ecol Sociobiol* **49:** 410–415.

Schneider JM, Gilberg S, Fromhage L, Uhl G. 2006. Sexual conflict over copulation duration in a cannibalistic spider. *Anim Behav* **71:** 781–788.

Schutz D, Taborsky M. 2005. Mate choice and sexual conflict in the size dimorphic water spider *Argyroneta aquatica* (Araneae, Argyronetidae). *J Arachnol* **33:** 767–775.

Schutz D, Taborsky M. 2011. Sexual selection in the water spider: Female choice and male–male competition. *Ethology* **117:** 1101–1110.

Schwartz SK, Wagner WE, Hebets EA. 2013. Spontaneous male death and monogyny in the dark fishing spider. *Biol Lett* **9:** 20130113.

Segoli M, Arieli R, Sierwald P, Harari AR, Lubin Y. 2008. Sexual cannibalism in the brown widow spider (*Latrodectus geometricus*). *Ethology* **114:** 279–286.

Sentenska L, Pekar S. 2013. Mate with the young, kill the old: Reversed sexual cannibalism and male mate choice in the spider *Micaria sociabilis* (Araneae: Gnaphosidae). *Behav Ecol Sociobiol* **67:** 1131–1139.

Singer F, Riechert SE, Xu HF, Morris AW, Becker E, Hale JA, Noureddine MA. 2000. Analysis of courtship success in the funnel-web spider *Agelenopsis aperta. Behaviour* **137:** 93–117.

* Sirot LK, Wong A, Chapman T, Wolfner MF. 2014. Sexual conflict and seminal fluid proteins. A dynamic landscape of sexual interactions. *Cold Spring Harb Perspect Biol* doi: 10.1101/cshperspect.a017533.

Snow LSE, Andrade MCB. 2004. Pattern of sperm transfer in redback spiders: Implications for sperm competition and male sacrifice. *Behav Ecol* **15:** 785–792.

Snow LSE, Abdel-Mesih A, Andrade MCB. 2006. Broken copulatory organs are low-cost adaptations to sperm competition in redback spiders. *Ethology* **112:** 379–389.

Stoltz JA, Elias DO, Andrade MCB. 2008. Females reward courtship by competing males in a cannibalistic spider. *Behav Ecol Sociobiol* **62:** 689–697.

Tuni C, Bilde T. 2010. No preference for novel mating partners in the polyandrous nuptial-feeding spider *Pisaura mirabilis* (Araneae: Pisauridae). *Anim Behav* **80:** 435–442.

Tuni C, Albo MJ, Bilde T. 2013. Polyandrous females acquire indirect benefits in a nuptial feeding species. *J Evol Biol* **26:** 1307–1316.

Uhl G, Elias DO. 2011. Communication. In *Spider behaviour: Flexibility and versatility* (ed. Herberstein ME), pp. 127–189. Cambridge University Press, Cambridge.

Uhl G, Schmitt S, Schafer M. 2005. Fitness benefits of multiple mating versus female mate choice in the cellar spider (*Pholcus phalangioides*). *Behav Ecol Sociobiol* **59:** 69–76.

Uhl G, Nessler SH, Schneider J. 2007. Copulatory mechanism in a sexually cannibalistic spider with genital mutilation (Araneae: Araneidae: *Argiope bruennichi*). *Zoology* **110:** 398–408.

Uhl G, Nessler S, Schneider J. 2009. Securing paternity in spiders? A review on occurrence and effects of mating plugs and male genital mutilation. *Genetica* **138:** 75–104.

Uhl G, Nessler S, Schneider J. 2010. Securing paternity in spiders? A review on occurrence and effects of mating plugs and male genital mutilation. *Genetica* **138:** 75–104.

Watanabe H, Yano E. 2012. Behavioral response of male mantid *Tenodera aridifolia* (Mantodea: Mantidae) to windy conditions as a female approach strategy. *Entomol Sci* **15:** 384–391.

Watson PJ. 1998. Multi-male mating and female choice increase offspring growth in the spider *Neriene litigiosa* (Linyphiidae). *Anim Behav* **55:** 387–403.

Wedell N. 2013. The dynamic relationship between polyandry and selfish genetic elements. *Phil Trans R Soc Lond B Biol Sci* **368:** 20120049.

Welke KW, Schneider J. 2009. Inbreeding avoidance through cryptic female choice in the cannibalistic orb-web spider *Argiope lobata*. *Behav Ecol* **20:** 1056–1062.

Welke KW, Schneider JM. 2010. Males of the orb-web spider *Argiope bruennichi* sacrifice themselves to unrelated females. *Biol Lett* **6:** 585–588.

Welke KW, Schneider JM. 2012. Sexual cannibalism benefits offspring survival. *Anim Behav* **83:** 201–207.

Welke KW, Zimmer SM, Schneider JM. 2012. Conditional monogyny: Female quality predicts male faithfulness. *Front Zool* **9:** 7.

Wilder SM, Rypstra AL. 2007. Male control of copulation duration in a wolf spider (Araneae, Lycosidae). *Behaviour* **144:** 471–484.

Wilder SM, Rypstra AL. 2008. Sexual size dimorphism predicts the frequency of sexual cannibalism within and among species of spiders. *Am Nat* **172:** 431–440.

Wilder SM, Rypstra AL, Elgar MA. 2009. The importance of ecological and phylogenetic conditions for the occurrence and frequency of sexual cannibalism. *Annu Rev Ecol Evol Syst* **40:** 21–39.

Wilgers DJ, Hebets EA. 2012. Age-related female mating decisions are condition dependent in wolf spiders. *Behav Ecol Sociobiol* **66:** 29–38.

Zeh JA, Zeh DW. 2001. Reproductive mode and the genetic benefits of polyandry. *Anim Behav* **61:** 1051–1063.

Zeh JA, Zeh DW. 2003. Toward a new sexual selection paradigm: Polyandry, conflict and incompatibility. *Ethology* **109:** 929–950.

Zhang SC, Kuntner M, Li DQ. 2011. Mate binding: Male adaptation to sexual conflict in the golden orb-web spider (Nephilidae: *Nephila pilipes*). *Anim Behav* **82:** 1299–1304.

Zimmer SM, Welke KW, Schneider JM. 2012. Determinants of natural mating success in the cannibalistic orb-web spider *Argiope bruennichi*. *PLoS ONE* **7:** e31389.

Mechanisms and Evidence of Genital Coevolution: The Roles of Natural Selection, Mate Choice, and Sexual Conflict

Patricia L.R. Brennan[1,2] and Richard O. Prum[3]

[1]Departments of Psychology and Biology, University of Massachusetts, Amherst, MA 01003

[2]Organismic and Evolutionary Biology Graduate Program, University of Massachusetts, Amherst, MA 01003

[3]Department of Ecology and Evolutionary Biology and Peabody Museum of Natural History, Yale University, New Haven, CT 06520

Correspondence: pbrennan@cns.umass.edu

Genital coevolution between the sexes is expected to be common because of the direct interaction between male and female genitalia during copulation. Here we review the diverse mechanisms of genital coevolution that include natural selection, female mate choice, male–male competition, and how their interactions generate sexual conflict that can lead to sexually antagonistic coevolution. Natural selection on genital morphology will result in size coevolution to allow for copulation to be mechanically possible, even as other features of genitalia may reflect the action of other mechanisms of selection. Genital coevolution is explicitly predicted by at least three mechanisms of genital evolution: lock and key to prevent hybridization, female choice, and sexual conflict. Although some good examples exist in support of each of these mechanisms, more data on quantitative female genital variation and studies of functional morphology during copulation are needed to understand more general patterns. A combination of different approaches is required to continue to advance our understanding of genital coevolution. Knowledge of the ecology and behavior of the studied species combined with functional morphology, quantitative morphological tools, experimental manipulation, and experimental evolution have been provided in the best-studied species, all of which are invertebrates. Therefore, attention to vertebrates in any of these areas is badly needed.

Of all the evolutionary interactions between the sexes, the mechanical interaction of genitalia during copulation in species with internal fertilization is perhaps the most direct. For this reason alone, coevolution between genital morphologies of males and females is expected. Morphological and genetic components of male and female genitalia have been shown to covary in many taxa (Sota and Kubota 1998; Ilango and Lane 2000; Arnqvist and Rowe 2002; Brennan et al. 2007; Rönn et al. 2007; Kuntner et al. 2009; Tatarnic and Cassis 2010; Cayetano et al. 2011; Evans et al. 2011, 2013; Simmons and García-González 2011; Yassin and Orgogozo 2013; and see examples in Table 1).

Cite this article as *Cold Spring Harb Perspect Biol* doi: 10.1101/cshperspect.a017749

Table 1. Examples of coevolution found in a variety of taxa with emphasis on more recent literature and model systems

Taxa	Male structures	Female structures	Evidence	Likely mechanism	References
Mollusks					
Land snails (Xerocrassa)	Spermatophore-producing organs	Spermatophore-receiving organs	Comparative among species	SAC or female choice	Sauder and Hausdorf 2009
Satsuma	Penis length	Vagina length	Character displacement	Lock and key	Kameda et al. 2009
Arthropods					
Arachnids (Nephilid spiders)	Multiple	Multiple	Comparative among species	SAC	Kuntner et al. 2009
Pholcidae spiders	Cheliceral apophysis	Epigynal pockets	Comparative (no phylogenetic analysis)	Female choice	Huber 1999
Harvestmen (Opiliones)	Hardened penes and loss of nuptial gifts	Sclerotized pregenital barriers	Comparative among species	SAC	Burns et al. 2013
Millipedes					
Parafontaria tonominea	Gonopod size	Genital segment size	Comparative in species complex	Mechanical incompatibility resulting from Intersexual selection	Sota and Tanabe 2010
Antichiropus variabilis	Gonopod shape and size	Accesory lobe of the vulva and distal projection	Functional copulatory morphology	Lock and key	Wojcieszek and Simmons 2012
Crustacean					
Fiddler crabs, Uca	Gonopode	Vulva, vagina, and spermatheca	Two-species comparison, shape correspondence	Natural selection against fluid loss, lock and key, and sexual selection	Lautenschlager et al. 2010
Hexapodes					
Odonates	Clasping appendages	Abdominal shape and sensory hairs	Functional morphology, comparative among species	Lock and key via female sensory system	Robertson and Paterson 1982; McPeek et al. 2009
Insects					
Coleoptera: seed beetles	Spiny aedagus	Thickened walls of copulatory duct	Comparative among species	SAC	Rönn et al. 2007
Callosobruchus: *Callosobruchus maculatus*	Damage inflicted	Susceptibility to damage	Full sib/half sib mating experiments	SAC	Gay et al. 2011
	Reduced spines	No correlated response	Experimental evolution	SAC	Cayetano et al. 2011

Cite this article as *Cold Spring Harb Perspect Biol* doi: 10.1101/cshperspect.a017749

Carabid beetles (Ohomopterus)	Apophysis of the endophallus	Vaginal appendix (pocket attached to the vaginal apophysis)	Cross-species matings	Lock and key	Sota and Kubota 1998; Sasabi et al. 2010
Dung beetle: *Onthophagus taurus*	Shape of the parameres in the aedagus	Size and location of genital pits	Experimental evolution	Female choice	Simmons and García-González 2011
Diptera: *Drosophila santomea* and *D. yakuba*	Sclerotized spikes on the aedagus	Cavities with sclerotized platelets	Cross-species matings	SAC	Kamimura 2012
Drosophila melanogaster species complex	Epandrial posterior lobes	Oviscapt pouches	Comparative among species	SAC or female choice	Yassin and Orgogozo 2013
	Phallic spikes	Oviscapt furrows			
	Cercal teeth, phallic hook, and spines	Uterine, vulval, and vaginal shields			
D. mauritiana and *D. sechelia*	Posterior lobe of the genital arch	Wounding of the female abdomen	Mating with introgressed lines	SAC	Masly and Kamimura 2014
Stalk-eyed flies (Diopsidae)	Genital process	Common spermathecal duct	Comparative among species and morphological	Female choice	Kotrba et al. 2014
Tse-tse flies: *Glossina pallidipes*	Cercal teeth	Female-sensing structures	Experimental copulatory function	Female choice	Briceño and Eberhard 2009a,b
Phelebotomine: sand flies	Aedagal filaments, aedagal sheaths	Spermathecal ducts length, base of the duct	Comparative among species	None specified	Ilango and Lane 2000
Heteroptera: Bed bugs (Cimiciidae)	Piercing genitalia	Spermalege (thickened exosqueleton)	Comparative among species	SAC	Carayon 1966; Morrow and Arnqvist 2003
Plant bugs (Coridromius)	Changes in male genital shape	External female paragenitalia	Comparative among species	SAC	Tatarnic and Cassis 2010
Waterstriders (Gerris sp.)	Grasping appendages	Antigrasping appendages	Comparative among species	SAC	Arnqvist and Rowe 2002
Gerris incognitus	Grasping appendages	Antigrasping appendages	Comparative among populations	SAC	Perry and Rowe 2012

Continued

Table 1. *Continued*

Taxa	Male structures	Female structures	Evidence	Likely mechanism	References
Bee assassins (Apiomerus)	Aedagus	Bursa copulatrix	Comparative among species	None	Forero et al. 2013
Cave insects (Psocodea), Neotrogla	Male genital chamber	Penis-like gynosome	Comparative among species	Female competition (role reversal), coevolution SAC	Yoshizawa et al. 2014
Butterflies (Heliconiinae)	Thickness of spermatophore wall	Signa: Sclerotized structure to break spermatophores	Comparative among species	SAC	Sánchez and Cordero 2014
Fish					
Basking shark: *Cetorhinus maximus*	Clasper claw	Thick vaginal pads	Morphological observation	None	Matthews 1950
Gambusia	Gonopodial tips	Genital papillae within openings	Comparative among species	Strong character displacement	Langerhans 2011
Poecilia reticulata	Gonopodium tip shape	Female gonopore shape	Comparative among populations	SAC	Evans et al. 2011
Reptiles					
Anoles	Hemipene shape	Vagina shape	Shape correspondence, two species	Sexual selection	Köhler et al. 2012
Several species	Hemipene shape	Vagina shape	Shape correspondence	Lock and key, female choice, and SAC	Pope 1941; Böhme and Ziegler 2009; King et al. 2009
Asiatic pit vipers	Spininess in hemipenes	Thickness of vagina wall	Two-species comparison	None	Pope 1941
Garter snake: *Thamnophis sirtalis*	Basal hempene spine	Vaginal muscular control	Experimental manipulation	SAC	Friesen et al. 2014
Birds					
Waterfowl	Penis length	Vaginal elaboration	Comparative among species	SAC	Brennan et al. 2007
Tinamous	Penis length/presence	Vaginal elaboration	Comparative among species	Female choice/natural selection	PLR Brennan, K Zyscowski, and RO Prum, unpubl.

Cite this article as *Cold Spring Harb Perspect Biol* doi: 10.1101/cshperspect.a017749

Mammals					
Marsupials	Bifid penis	Two lateral vaginae	Shape correspondence	None	Renfree 1987
Equidna	Bifid penis with four rosettes	Single vagina splits into two uteri	Shape correspondence	None	Augee et al. 2006; Johnston et al. 2007
Insectivores: Short-tailed shrew: *Blarina brevicauda*	S-shaped curve of the erect penis	Coincident curve in the vagina	Shape correspondence	None	Bedford et al. 2004
Common tenrec: *Tenrec caudatus*	Filiform penis (up to 70% of the male's body length)	Internal circular folds in the vagina	Length correspondence	None	Bedford et al. 2004
Rodents: Cape dune mole: *Bathyergus suillus*	Penis and baculum length	Vaginal length	Allometric relationships within species	None	Kinahan et al. 2007
Australian hopping mice (Notomys)	Spiny penis	Derived distal region in the vagina	Morphological observation and two-species comparison	Copulatory lock	Breed et al. 2013
Pig: *Sus domesticus*	Filiform penis end	Cervical ridges	Artificial insemination	Female choice	Bonet et al. 2013
Primates: *Macaca arctoides*	Long and filamentous glans	Vestibular colliculus (fleshy fold) that partially obstructs the entrance to the vagina	Shape correspondence and comparison with close relatives	None	Fooden 1967

The likely mechanism is that suggested by the authors, and it includes sexually antagonistic coevolution (SAC), natural selection, sexual selection, female choice, or none specified. The evidence provided by the studies can be comparative among species or among populations, experimental evolution, cross-species matings, full-sibling (sib)/half-sib matings, shape, and length correspondence. Shape correspondence is often taken as evidence of coevolution, although it is not as conclusive as other approaches.

Male genitalia are among the most variable structures in nature (Eberhard 1985). In contrast, female genitalia have typically been found not to be as interspecifically variable as male genitalia in several studies that specifically examined and described them (Eberhard 1985, 2010a,b). Female genitalia are not studied as often as male genitalia, perhaps because of a male-biased view of evolutionary processes by researchers (Ah-King et al. 2014). However, studying female genitalia is undeniably challenging. Male genitalia are generally kept inside of the body cavity, but are everted before, or during copulation, so their functional morphology can be more easily studied than the internal genitalia of females. Female genitalia also tend to be softer than male genitalia and thus their morphology may be more difficult to describe, and can more easily be distorted on dissection and preservation. Female adaptations to sense or oppose features of male genitalia can be subtle, requiring careful study. Female genital tracts are under multiple sources of selection: not just mating, but also storing sperm, egg laying, birthing, and often interfacing with the terminal portion of the digestive tract. Therefore, selection balancing multiple functions may further constrain morphological evolution in female genitalia. However, even small morphological changes in female genitalia, for example, increases in vaginal muscle, may change a female's ability to choose or reject a male during mating, or to manage the costs of mating. Thus, the functional consequences to male and female genital morphology are hard to predict unless one knows how genitalia function during intromission. Despite these challenges, recent studies have examined variation of female genitalia and evidence is accumulating that features of female genitalia are variable enough to support coevolutionary processes (Polihronakis 2006; Puniamoorthy et al. 2010; Siegel et al. 2011; Showalter et al. 2013; and see additional references in Ah-King et al. 2014).

In this article, we will discuss different hypotheses of genital evolution that predict coevolution; however, this is not a review of that entire subject (but see Eberhard et al. 2010b; Simmons 2013). Rather, we discuss the various mechanisms of genital coevolution differentiating the potentially independent or overlapping roles of natural selection, female choice, and male–male competition (Fig. 1). This classification allows us to distinguish specifically those mechanisms of genital coevolution that involve sexual conflict (i.e., when the evolutionary interests of individuals of different sexes, particularly over mating, are different). We then highlight examples in different taxa organisms with particular emphasis on those that provide evidence of sexual conflict.

GENITAL COEVOLUTION

Genital coevolution results when evolutionary changes in a genital trait found in one sex selects for an evolutionary change in a genital trait

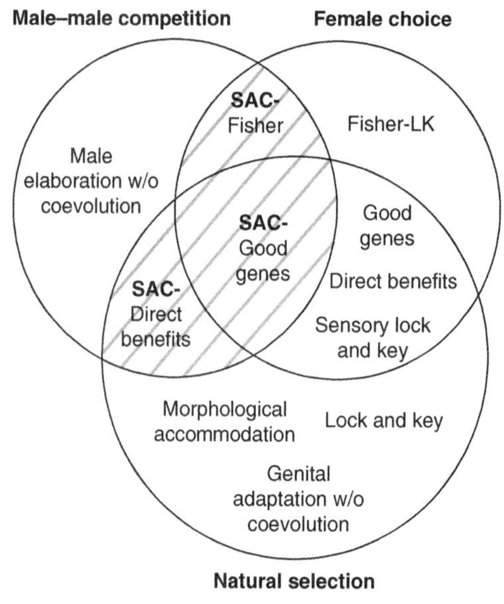

Figure 1. Graphical classification of mechanisms of genital evolution and coevolution. Three circles depict the independent and co-occurring actions of natural selection, female choice, and male–male competition. Different specific versions of genital coevolution can occur depending on which of the three broader evolutionary mechanisms are occurring. Sexual conflict (hatched lines) occurs through the simultaneous action of male–male competition and female choice, or male–male competition and natural selection. SAC, sexually antagonistic coevolution. See text for explanation.

in the other sex. This reciprocal evolutionary change can be shown by both macroevolutionary and microevolutionary research approaches. Comparative studies of species and/or populations can provide evidence of dynamic changes in genital morphology in both sexes. These include investigations of morphological correspondence, compatibility in overall shape, or between morphological structures and corresponding sensing structures. Comparative studies can reveal evidence of arms races in which evolutionary changes in genital morphology of each sex are antagonistic. A microevolutionary approach can include studies of correlation between genital features (both correspondence or antagonistic), experimental manipulation to measure the fitness consequences of changes in genital morphology in both sexes, and most recently, experimental evolutionary approaches that directly inform on the selective pressures that act on genital features, and the presence of coevolutionary changes in both sexes. In Table 1 we summarize examples of systems in which all these diverse types of evidence for coevolution have been reported, and some of them we will discuss in detail.

As our examples outline below, many different genital traits may coevolve, including overall size, shape, functional morphology, size and shape of particular components of the genitalia including but not limited to spines, ridges, flaps, and many other superficial elaborations, nongenital contact structures, and physiology of moveable and erectile and inflatable components of genitalia known in vertebrates and invertebrates.

MECHANISMS OF GENITAL COEVOLUTION

Genital diversification has been hypothesized to result from pleitropy, and natural, sexual (Eberhard 1985; Hosken and Stockley 2004), and social selection (East et al. 1993). Within sexual selection, male–male competition, female choice, and sexual conflict can influence genital morphology. Some of these hypotheses predict genital coevolution between the sexes including natural selection for mechanically feasible copulation, lock and key to prevent hybridizations, female choice of preferred males, and sexually antagonistic coevolution resulting from sexual conflict. Neither pleiotropy, social signaling, or male–male competition explicitly predict coevolution, although changes brought on by these processes may elicit a coevolutionary response on females through a different mechanism.

Here, we discuss mechanisms of genital coevolution in terms of the independent or overlapping forces of natural selection, female choice, and male–male competition (Fig. 1). We identify those examples of sexually antagonistic coevolution between male and female genitalia as the result of the co-occurrence of female choice and male–male competition, or male–male competition and natural selection.

Natural Selection

For Mechanically Feasible Copulation

Natural selection to make copulation functionally possible must always influence genital morphology to achieve the basic function of gamete transfer (Darwin 1871), despite the likely action of other mechanisms in genital elaboration. This process can give rise to a pattern of morphological accommodation (Fig. 1). Genital size is of primary importance in the mechanical interaction during copulation, and therefore genital size covariation indicative of coevolution is likely to be important and widespread. Although size coevolution can also result from sexual selection, natural selection on genitalia to insure successful gamete transfer can help to explain many observed patterns, yet it is rarely discussed (Pereti 2010).

Three lines of evidence suggest that genital size fit is important. First, crossings between species with different genital sizes are almost always impossible or lethal to females, especially when males are much larger (Masly 2012). Second, when size dimorphism is marked, selection can operate to match genital size. In orb-weaving spiders with extreme size dimorphism, tiny males have disproportionally large genitalia (up to 20% of their body mass) to match female genital size (Ramos et al. 2005). This has not been examined when males are larger than females, but females may be predicted to enlarge their genitalia to accommodate large male genitalia (Langerhans 2011). Third, changes in genital size appear to be more constrained than changes in shape. In the *Onthophagus* beetles, the shape of male and female genitalia changes more rapidly than size (Macagno et al. 2011). This pattern was also shown in a study of experimental evolution of male genitalia in *Onthophagus taurus*, in which shape but not size of the aedeagus changed after 19 generations (Simmons et al. 2009), and in an experimental evolution study in mice *Mus musculus*, in which baculum width but not length was shown to change after 27 generations (Simmons and Firman 2014). However, an experimental evolution study of genitalia in *Drosophila* found that both size and shape of the genital arch changed in response to natural and sexual selection (House et al. 2013), so this pattern is not universal. However, male insects tend to show a pattern of negative allometry that suggests that genital size changes little relative to body size compared to other nongenital characters (Eberhard et al. 1998).

For a Lock-and-Key Mechanism to Prevent Hybridization

Defour noted that male genitalia in insects were often species specific and stated that these served to preserve "types" (Defour 1844). The "lock-and-key" hypothesis proposes that variation in genital form coevolves through natural selection against hybridization and to enhance reproductive isolation (Eberhard 1985; Shapiro and Porter 1989). Females avoid the fitness costs of having their eggs fertilized by a male of the wrong species by evolving genitalia that permit only males with the right fit to achieve mechanical coupling leading to fertilization. The male "key" must fit the female "lock." As such, this hypothesis predicts close coevolution in genital shape and mechanical fit during copulation. Shapiro and Porter (1989) provide an excellent review of the history, tests and evidence for and against this hypothesis in the insects. They conclude that although there is no evidence that lock and key explains most instances of genital elaboration in insects, the hypothesis has not really been falsified. A more recent review centers on examining hybridization between species, in which genital morphology seems to play a role in reproductive isolation via mechanical incompatibility (Masly 2012). Lock and key is supported by recent studies of millipede genitalia (e.g., Tanabe and Sota 2008; Wojcieszek and Simmons 2012, 2013).

There are several predictions of the lock-and-key hypothesis that have not been supported (reviewed by Eberhard 1985, 2010a,b; Shapiro and Porter 1989), but they may simply not have been adequately tested (Simmons 2013). According to lock and key, matings between sympatric species should be mechanically difficult, and often this is not true. However, in two closely related sympatric Carabid beetles, interspecific mating are costly for both males and females and result in genital damage owing to mechanical mismatch (Sota and Kubota 1998). Further, genital divergence should be correlated with low hybrid fitness, otherwise selection would be too weak to drive further mechanical isolation; but again there is little evidence for this, an exception being the hybrid zone between the beetles described above, in which females have deformed ovaries (Sota and Kubota 1998). Because lock and key is based on natural selection for reinforcement of species isolation, there should also be character displacement in male genitalia of sister species in zones of sympatry compared to allopatry. Although Eberhard (2010b) argues that this has not been shown, Simmons (2013) details some insect examples in which reinforcement has been shown. Character displacement is involved in genital diversification and coevolution of penis and vagina length in land snails (Kameda et al. 2009), and in male and female genital diversification in fish of the genus *Gambusia* (Langerhans 2011). Finally, the most widespread reason why lock and key is deemed unlikely is the perception that female genitalia are not as variable as male's, that is, the keys are variable, but the locks are not (Eberhard 1985). However, examination of current literature suggests that this conclusion is only weakly supported even in the well-studied insects (Simmons 2013).

Surprisingly, many studies of the lock-and-key mechanism do not examine females at all (e.g., Mutanen et al. 2006; and see references in Ah-King et al. 2014). Some studies also conclude that females are not variable by using either linear estimates of genital size or qualitative examinations of genital shape (Eberhard 2004a,b, 2006, 2010a,b). However, linear estimates do not capture the complexity of genital shape, and size may be selected to be less variable than shape as discussed above. When quantitative measures and ultrastructure analyses of female morphology and shape are conducted, significant genital morphological variation across species has been detected. For example, female genitalia in *Drosophila* were deemed to be fairly invariant across species (Eberhard and Ramirez 2004; Jagadeeshan and Singh 2006), but a recent study showed that in nine species of *Drosophila*, several aspects of female genital morphology covary with male genital traits (Fig. 2) (Yassin and Orgogozo 2013). A geometric morphometric analysis of genital shape among two sister species of water snakes (*Nerodia* sp.) that seemed very similar on visual examination alone, showed that vaginal shape of these two species differed significantly in the degree of bifurcation and the aspect ratio of the vagina (Showalter et al. 2013).

Although there are good reasons to suggest that lock and key is unlikely to explain widespread patterns of genital evolution, quantitative examination of female genitalia should be performed before ruling out the lock-and-key hypothesis on the basis of lack of female genital variation alone. Simmons (2013) has made a similar point.

Although evidence for genital covariation in shape has been used as evidence to support lock-and-key (e.g., Mikkola 1992), mechanically feasible copulation, female choice and sexually antagonistic coevolution can also result in shape coevolution. Therefore, genital coevolution is not an exclusive prediction of lock and key.

Female Choice

Female choice can result in coevolution of different features of genitalia. Just like any other sensory mediated mate choice mechanism, females may be able to use sensory stimuli received during copulation from male genitals as the basis for female choice. This process requires female to receive and evaluate these stimuli. Thus, if males stimulate females with vibrissae, then females must have the ability to contact the vibrissae and detect their movement, shape, or whatever component of male morphology is speculated to stimulate them.

Through sexual selection, mating preferences can coevolve with male traits through either the arbitrary Fisherian mechanism, or by adaptive good genes and direct benefits mechanisms (Fig. 1) (Kirkpatrick and Ryan 1991; Andersson 1994; Prum 2010). Female mate choice can affect male genital evolution when females choose on the basis of interaction with male genitalia

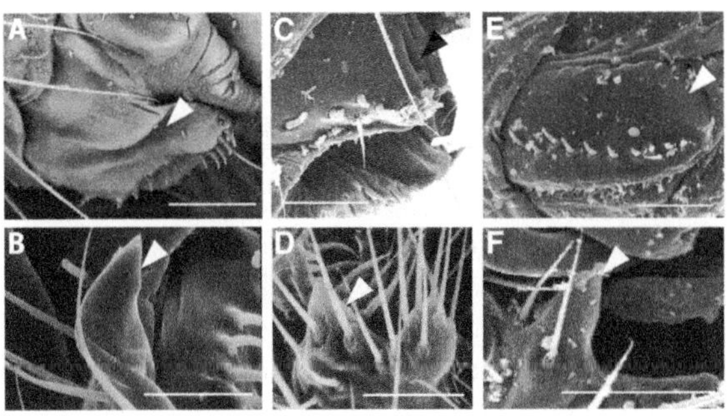

Figure 2. Female genitalia in *Drosophila* were previously thought to be invariant, but recent work has described extensive variation and coevolution in genital morphology. (*A,C,E*) Female oviscapts, and (*B,D,F*) male epandrial posterior lobes. Scale bars, 50 μm. (Reprinted from Yassin and Orgogozo 2013.)

(Eberhard 1996). Thus, male and female genitalia may participate in all the same coevolutionary processes that characterize mating preferences and displays that are mediated by the sensory system during precopulatory interactions. Genital coevolution by female choice requires female promiscuity, so that females can choose among multiple males with whom they mate.

Female mate choice could result in arbitrary "Fisherian," or Lande Kirkpatrick (LK), coevolution in male and female genitalia via either sensory preferences or other anatomical features of female genitalia that allow females to evaluate male features. We discuss this possibility below in birds. As proposed for mate choice in general (Prum 2010), the arbitrary female choice mechanism is actually the null hypothesis of coevolution until evidence of good genes or direct benefits mechanisms can be obtained (Prum 2010). This process can occur when female mating preferences are not under natural selection.

The best support for genitally mediated female mate choice is provided by a series of experiments examining interactions between genital features of males, and female receptors in tsetse fly *Glossina pallidipes*. Modifications of the cercal teeth—the part of the male genitalia that squeezes the female abdomen—resulted in reduction of female sperm storage and ovulation while increasing remating probability. These changes were shown to be dependent on female structures the sense male stimulation (Briceño and Eberhard 2009a,b).

Studies of domestic pig (*Sus domesticus*) provide an example of genital coevolution via female choice in vertebrates. Male pigs have penises that end on a corkscrew filament whereas sows have ridges on their cervix that specifically sense stimulation from the penis during copulation. When sows are artificially inseminated, stimulation of the cervix ridges must occur otherwise the pregnancy rates drop dramatically (Bonet et al. 2013). This is consistent with a role for female choice in shaping pig penis morphology, although no information on male variation of genital features is available to determine its potential correlation with fertilization success. Many other mammals have these filamentous or finger-like endings on their penis (Eberhard 1985), suggesting that female choice may play an important role in shaping this particular class of genital elaboration.

A modified version of mechanical lock and key is known as the sensory lock and key; this hypothesis argues that differences in genital morphology perceived by either sex during copulation can result in premature termination of mating (Eberhard 1985, 2010a,b). Coevolution occurs between components of genitalia associated with sensing "fit," rather than purely mechanical aspects of shape differences that make copulation difficult. However, if mating success is mediated by female sensory perception, then this mechanism is essentially a form of female choice in which female mating preferences are under natural selection against hybridization (Fig. 1). Simmons (2013) argued that sensory lock and key is another type of female choice process that should be viewed as a continuum that enhances species isolation in one end, or that select for male features that increase female fitness on the other. Sensory lock and key to avoid hybridization is supported in some Odonates, in which mechanical isolation via clasping appendages is reported, and coevolution between the male abdominal claspers and female shape, sensory hairs (Robertson and Paterson 1982), and shape of female abdomens (McPeek et al. 2009) prevents heterospecific matings.

Sexually Antagonistic Coevolution Resulting from Sexual Conflict

So far we have discussed cases in which coevolution results in correspondence or fit. However, sexually antagonistic coevolution (SAC) that occurs as a consequence of sexual conflict between males and females can result in coevolutionary processes that reduce fit, or that function to decrease ease of copulation, or to mitigate male damage to the female. SAC occurs when genital adaptations coevolve in each sex to advance control over mating and fertilization initiating a coevolutionary arms race (Chapman et al. 2003; Parker 2006; Brennan and Prum 2012). Both direct and indirect selection are important in shaping evolutionary responses to sexual conflict including conflict over mating, fertilization, and mate identity (Arnqvist and Rowe 2005; Parker 2006; Brennan and Prum 2012). Thus, sexually antagonistic coevolution can proceed through either: (1) a combination of male–male competition for fertilization success and natural selection on female behavior, physiology or morphology to reduce the direct harm of sexual coercion (Arnqvist and Rowe 2005); or (2) a combination of male–male competition and female mate choice to maintain the indirect and direct benefits of choice (Fig. 1) (Parker 2006; Brennan and Prum 2012).

Waterstriders are a well-known example of sexually antagonistic coevolution to reduce direct costs (Arnqvist and Rowe 2002, 2005). Males jump on the back of the female to mate, and they remain attached to her as a form of mate guarding after copulation has taken place. Optimal female mating rate is lower in females than males, as females suffer mating costs from carrying males on their backs. As a result, conflict over mating frequency is strong. Comparative studies have shown that grasping devices in the legs of males that allow them to remain attached to females despite their struggles have coevolved with female antigrasping devices (Arnqvist and Rowe 2002).

Waterfowl are the best-known example of antagonistic genital coevolution in vertebrates. Males have corkscrew penises and penis length is correlated with the frequency of forced extrapair copulations (FEPCs) among species. FEPCs are costly to females, and they detract from otherwise female autonomous precopulatory mate preference. Females have evolved convoluted vaginas that make it difficult for unwanted males to evert their penis fully inside the female when they coerce a mating (Fig. 3A) (Brennan et al. 2007, 2010; Brennan and Prum 2012). Brennan and Prum (2012) hypothesize that female genital resistance morphologies in ducks have evolved through the indirect, genetic benefit of being fertilized by preferred males. Females that can prevent forced fertilizations will have offspring that inherit the male traits that they and other females prefer.

Genital coevolution in waterfowl has been dynamic, both increasing and decreasing in genital length and complexity during evolutionary history (Brennan et al. 2007). FEPCs can be up to 40% of all observed copulations, but only 2%–5% of offspring are sired as a result of unwanted matings. This is likely a combination of the success of female genital elaborations in excluding male penises from everting, as well as strategies for paternity protection pursued by the females' mate, who usually mates with the female immediately after FEPCs have occurred (Brennan and Prum 2012). Ducks also have evolved an explosive eversion mechanism that allows males to quickly inseminate females despite their resistance (Brennan et al. 2010). Female ducks have evolved thickened walls in the proximal end of their vagina presumably to be able to withstand the forces involved with penis eversion (PLR Brennan, K Zyscowski, and RO Prum, unpubl.).

In some cases, coevolved female morphological responses to sexual conflict are easily noticed as in the examples above. However, females can adapt in more subtle ways if behavioral, physiological, or minor morphological modifications are sufficient to allow them to mitigate male inflicted damage, or to assert control over fertilization and mate identity. For example, in red-sided garter snakes (*Thamnophis sirtalis sirtalis*), females exert some physiological control over copulation duration; experimental anesthesia of the female vagina dramatically increased copulation duration (Friesen et al. 2014). Conflict over copulation duration in garter snakes has likely resulted in modification and elaboration of male hemipene spines that allow males to stay attached to females for longer than optimal, as males with clipped spines stayed in copulo for a shorter period of time than intact males (Friesen et al. 2014). Copulation duration results from an interplay between male hemipene spines and female capacity to terminate copulation by contracting her vaginal musculature (Friesen et al. 2014). This example highlights that the lack of overt specialized female structures that function to prevent male access cannot be used as evidence against the role of sexual conflict in genital evolution (contra Eberhard 2004, 2006, 2010b). Female physiological and behavioral re-

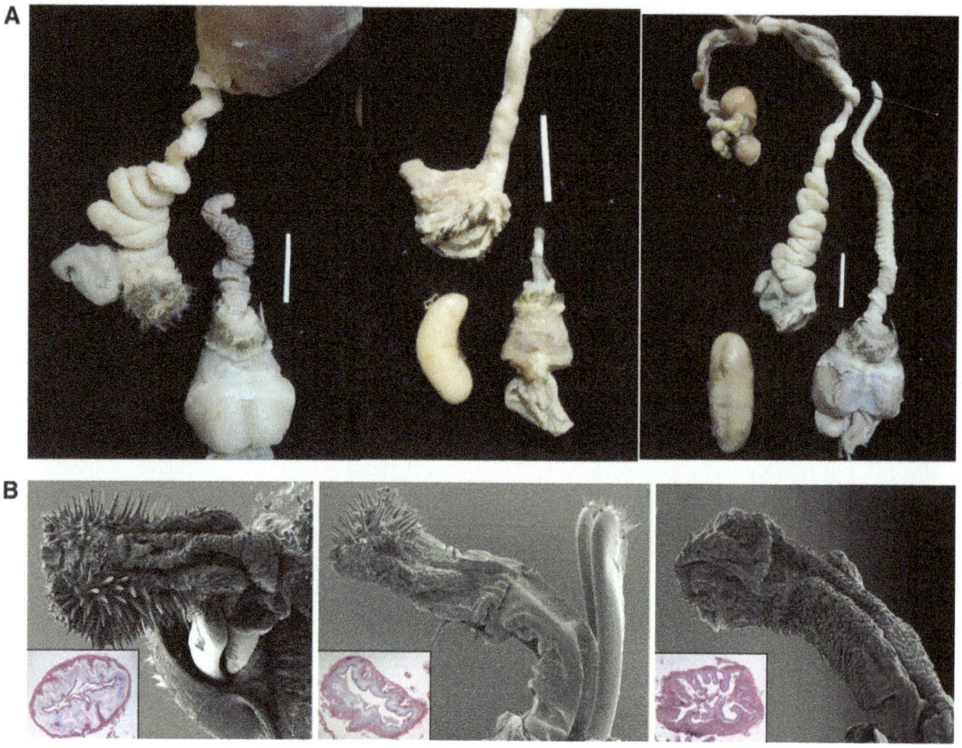

Figure 3. Genital coevolution via sexual conflict. (*A*) Waterfowl, in which the elaboration of the vagina correlates with penis length (vagina on the *left*, penis on the *right* in mallard, harlequin, and long-tailed duck) Scale bars, 2 cm, and (*B*) seed beetles (Callosobruchus), in which the presence and abundance of spines on the aedagus correlates with the thickness of the wall of the female's bursa copulatrix. (From Rönn et al. 2007; reprinted, with permission, from the National Academy of Sciences © 2007.)

sponses to copulation and male genital morphology can be varied, involving behavior, sensory systems, physiology, as well as morphology (also see Simmons 2013 for further discussion). These additional aspects of genital function can coevolve with male genital features. Therefore, studies that integrate copulatory behavior, functional morphology, and careful anatomical examination are needed before ruling out that features of female genitalia have not coevolved with male genitalia.

Distinguishing between Mechanisms of Genital Coevolution

Much effort has been spent trying to generalize predictions that allow distinction between evolutionary hypotheses of genital evolution (Arnqvist 1997; Eberhard 2004a,b, 2006, 2010a,b, 2011; Hosken and Stockley 2004). However, different selective mechanisms of genital coevolution are not necessarily mutually exclusive, and substantially different outcomes are expected through the interactions of different evolutionary mechanisms. They can act simultaneously or sequentially in evolutionary time, and in the same or separate genital traits (Arnqvist and Rowe 2005; Eberhard 2010b; Brennan and Prum 2012; Simmons 2013). In Figure 1, we illustrate how different mechanisms interact and overlap to produce genital evolution and coevolution. Each circle represents one mechanism of genital evolution: female choice, male–male competition, and natural selection.

Natural selection on genitalia can give rise to coevolution via: (1) adaptive size and morphological accommodation (i.e., in spider species in which tiny males have overly large genitalia to match the female), and (2) natural selection against hybridization (mechanical lock and key). Within female choice there are three potential mechanisms: Fisher-LK, good genes, and direct benefits. The Fisher-LK mechanism will occur when there is genetic variation for genital traits and genitally mediated preferences in the absence of natural selection on these preferences. Good genes and direct benefits can occur when female preferences (mediated by genital morphology) are under additional natural selection. This natural selection acts directly on female survival and fecundity, or indirectly on heritable genetic benefits to offspring. Just like any mating preference, all of these specific mechanisms can give rise to preference-trait coevolution, which is in this case nonantagonistic genital coevolution. In addition to good genes and direct benefits processes, the simultaneous action of female choice and natural selection against hybridization can give rise to sensory lock-and-key coevolution.

Male–male competition alone can give rise to genital evolution, but will not, by itself, produce genital coevolution between the sexes.

Sexual conflict occurs in the overlap between (1) male–male competition and natural selection on female survival and fecundity, or (2) male–male competition and female choice (Fig. 1, cross-hatched area) (Parker 2006; Brennan and Prum 2012). Sexual conflict can give rise to sexually antagonistic genital coevolution in various ways. Sexual conflict mediated by direct harm in the absence of female choice will result in SAC based on mitigating direct harm (e.g., in waterstriders). Sexual conflict mediated by female choice can produce SAC when female preferences are based on arbitrary traits (SAC-Fisher), or when they are based on good genes (SAC-good genes), depending on the absence or presence of natural selection on the indirect, genetic benefits of mate choice, that males prevent females from obtaining (Fig. 1). For example, species in which males prevent the autonomous exercise of female choice, such as some waterfowl and Poeciliid fishes, show evidence of SAC. The complexity of the interactions between mechanisms illustrated here makes obvious why distinguishing these hypotheses is so difficult.

An economics approach that fully quantifies costs and benefits to male and female fitness of traits involved in coevolution has been suggested as a possible avenue to distinguish female choice and sexual conflict (Fricke et al. 2009; Simmons 2013). This approach may be possible with some insects in the laboratory, but it is not feasible in most wild, long-lived vertebrates. In addition, measuring current fitness could hide the success of female mitigating strategies against conflict, for example, if females succeed in preventing unwanted males from siring their offspring as waterfowl appear to do (Brennan and Prum 2012). Therefore, experimental manipulation to render putative traits involved in conflict ineffective, as suggested by Eberhard (2011) may be a better alternative, particularly in vertebrates.

Experimental modification of genital traits can be a powerful tool (Eberhard 2011). Experiments with males matched for size but differing only in one feature of genital morphology are needed, although finding enough variation in critical components can be very difficult (but see Okusaki and Sota 2014). Creating males and females with different genital genotypes using experimental evolution can reveal coevolutionary processes underlying genital traits. A recent experiment in seed beetles *Callosobruchus maculatus* generated phenotypic variation via both artificial selection and experimental ablation of spines and elegantly showed that longer spines, are beneficial to male fitness (Hotzy et al. 2012). Similarly, a recent study in *Drosophila* generated variation in male genitalia by creating introgressed lines between *D. mauritiana* and *D. sechelia*. They mated these males with varied morphologies to *D. sechelia* females, and showed that the greater the mismatch between male and female morphology, the greater the harm suffered by females (measured as scarring). This offers strong evidence that the coevolutionary process results from sexual conflict (Masly and Kamimura 2014).

Another possible avenue is to generate phenotypic diversity from phenotypic plasticity. Although previous work in insects showed that development of genitalia are canalized and show little evidence of plasticity (House and Simmons 2007), experiments in barnacles showed that genitalia can respond in a phenotypically plastic manner to wave action (Neufeld and Palmer 2008), and social environment, in which individuals grow shorter penises when colonies are more dense (Hoch 2009; Neufeld 2011). Similarly, we have found that some species of ducks have phenotypically plastic male genitalia in at least some species in response to their social environment (PLR Brennan, K Zyscowski, and RO Prum, unpubl.).

EXAMPLES OF GENITAL COEVOLUTION

Insects

Eberhard (2010b) and Simmons (2013) have provided comprehensive reviews of genital evolution in insects. Both agree that there is ample evidence that sexual selection is responsible for most genital diversity, but that sufficient data to discriminate among mechanisms are lacking in most species. However, Eberhard and Simmons disagree on their interpretation of the evidence in those species for which data exist. Further discussion of the many important points raised by both reviews is beyond the scope of this article but we encourage readers to refer to them for more details. Several examples of coevolution between male and female genital traits are summarized in Table 1 and Simmons (2013).

Experimental evolution and comparative studies combined with knowledge on the fitness effects of mating on females have shown genital coevolution occurs in dung beetles (*Onthophagus* sp.) via female choice (Simmons and García-González 2011) and seed beetles (*Callosobruchus* sp.) via sexual conflict (Cayetano et al. 2011). After 19 generations of sexual selection compared to enforced monogamy, the genital pits of female dung beetles (in which the parameres at the tip of the aedagus will anchor) become smaller and get closer and thus harder to reach, while the shape of their aedagus (external genitalia) changes in males (Simmons and García-González 2011). An interspecific phylogenetic study in dung beetles detected coevolution between the shape of the parameres and the female pygidial flap morphology in which the parameres fit into before intromission (Macagno et al. 2011). Female choice is likely the mechanism by which genitalia have diversified in dung beetles because there are no detectable fitness costs to promiscuity for females, and offspring have higher viability when females mate with multiple males (Simmons 2013).

In contrast, studies in seed beetles have supported a role for sexual conflict in genital evolution. As males in various species have evolved a more spiny aedagus that causes damage to the female reproductive tract, females have coevolved more substantial connective tissue within their copulatory ducts (Fig. 3B) (Rönn et al. 2007). After 21 generations of enforced monogamy in *Callosobruchus macullatus*, males evolved smaller spines on the tip of their aedagus, although no coevolutionary genital response in females was detected (Cayetano et al. 2011). Defensive traits may have a delayed evolutionary response, if they are not selected for reduction until enough reduction in the male damaging trait has been achieved (Cayetano et al. 2011). Alternatively, if the defensive trait is not too costly, there may not be selection to eliminate it. In the absence of other supporting information, the lack of female evolutionary response in seed beetles may be taken as evidence against a genetic correlation between male and female genitalia expected under SAC, so even experimental evolution may not necessarily provide complete answers when examining genital evolution. A genetic covariance exists between male damage and female susceptibility to damage exists in this species as evidenced by a full sib/half sib design (Gay et al. 2011). A recent comparative study in the *Drosophila melanogaster* species complex, revealed previously unknown internal and external female adaptations some of which appear to counteract or prevent the damage cause by male genital structures, therefore likely evolving by SAC, whereas others seeming to facilitate copulation, therefore evolving by female choice (Yassin and Orgogozo 2013).

Arachnids

Spider genitalia are extremely variable and show evidence of mechanical fit during copulation strongly supporting coevolution that may result from female choice for a mechanical fit that would proceed through good genes (Huber 1999; Eberhard and Huber 2010). However, it is unclear how females sense male genital features, when there is no evidence that female spider genitalia have any sensory structures (Eberhard and Huber 2010). There is no evidence to support mechanical lock and key or sexual conflict in this group. A recent study in harvestmen (Opiliones), provides comparative evidence that coevolution of genital structures is dynamic. Female sclerotized pregenital barriers evolve in concordance with the loss of male nuptial gifts to females, and subsequently evolving hardened penes (Burns et al. 2013). Therefore, sexual conflict over male investment and fertilization may be important in this group (Burns et al. 2013). The evolved loss of nuptial gifts appears to be associated with changes in the length of the mating season, providing a direct link to ecology and life-history evolution that is typically lacking in genital evolution studies.

Fish

Poeciliids are becoming an excellent model to study of genital coevolution. Here we summarize some key aspects of their biology from Greven (2005) and Langerhans (2011). Unlike most teleost fishes with external fertilization, male poeciliids have evolved a novel genital intromittent organ or gonopodium modified from anal fin rays. Poeciliids have both elaborate sexually dimorphic coloration patterns and display behaviors that are subject to female mate choice, and males in 38% of species display but also pursue forced copulations. In 58% of poeciliid species, males do not display to females but only coerce copulations. Interspecies comparisons show that species with longer gonopodia exhibit reduced courtship displays (Greven 2005; Martin et al. 2010). The tip of the gonopodium can have structures used for insertion, grasping, stabilizing, and sensing, but whether there is a relationship between gonopodium length and tip elaboration remains unclear (Greven 2005). When hooks are present, males can get stuck inside the female and cause damage to her reproductive tract (Greven 2005).

Evidence of genital coevolution at the population level has been provided in guppies (*Poecilia reticulata*). Male guppies routinely force females to copulate, and gonopodium length is associated with success in forced copulations (Kelly et al. 2000). Forced copulations are costly to females (Magurran and Seghers 1994). An intraspecific study examining different guppy populations in which gonopodium length and shape is known to differ according to predation risk, showed that males with longer gonopodium were more successful at achieving forced copulations, while males with shorter gonopodium tips transferred more sperm into females (Evans et al. 2011). In high predation populations in which males have shorter gonopodium tips, females have wider oviducts than fish from low predation populations, suggesting that coevolution is taking place between male and female genitalia (Evans et al. 2011). Quantitative trait loci of genital traits between these differing populations have provided evidence of genetic covariation in genital traits, and high additive genetic variance, that support a critical prediction of coevolutionary processes under both female choice and sexual conflict (Evans et al. 2013). A recent experiment showed that the claw at the end of the gonopodium, increases sperm transfer only when females are nonreceptive, supporting its potential role in sexual conflict (Kwan et al. 2013). Overall, there is strong evidence of sexual conflict over copulation in guppies, and therefore genital coevolution in this species likely proceeds by sexual conflict.

Squamates

Male squamates have paired intromittent organs known as hemipenes that are often species specific. Female genitalia in snakes have been shown to vary at least among families (e.g., Siegel et al. 2011). Eversion of one hemipenis takes place into the female cloaca and vaginal tract. Not much work has been performed specifically on genital coevolution in this group, although shape correspondence indicative of coevolution seems widespread (Pope 1941; Edgren 1953; Pisani 1976; Böhme and Ziegler 2009; King et al. 2009).

In at least some species, this close shape fit has been used to suggest a coevolutionary process that could result from female choice (Böhme and Ziegler 2009), sexual conflict (King et al. 2009), or natural selection lock-and-key mechanisms (Pope 1941). Our own observations in 12 species of snakes including natricine and non-natricines, and two lizards have also revealed shape correspondence (Fig. 4A) (PLR Brennan, K Zyscowski, and RO Prum, unpubl.). However, shape correspondence is apparently not universal in snakes (Siegel et al. 2011). Similar to the dung beetles, there seems to be correspondence between how spiny the hemipenes are, and the thickness of the vaginal wall in two species of Asiatic pit vipers (Pope 1941). This is likely a widespread phenomenon in squamates as many species have spiny hemipenes that leave imprints inside the female after copulation (Pope 1941).

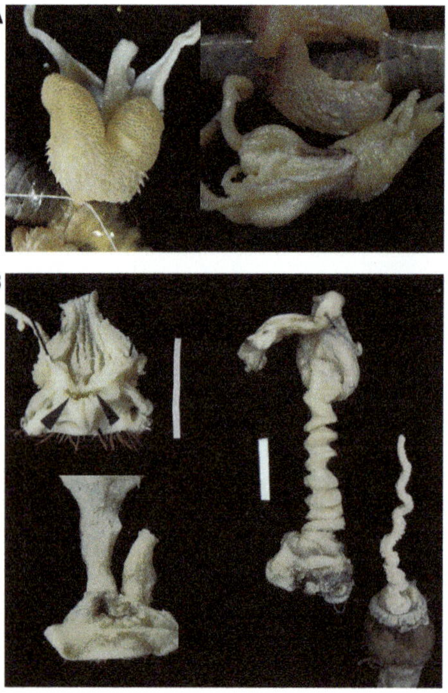

Figure 4. Genital coevolution. (*A*) Snakes, in which the shape of the inflated male hemipenes corresponds with the shape of the female vagina (*Crotalis horridus* and *Coluber constrictor*), and (*B*) tinamous species without a penis have a simple vagina (*Crypturellus undulatus* male cloaca with nonintromittent penis and female on the *left*); the presence of a long penis is associated with an accordion-like vagina (*Tinamus major* on the *right*). Scale bars, 2 cm.

Birds

Genital evolution in birds is certainly more complex than previously appreciated (Brennan 2013), and we recently discovered genital coevolution in tinamous (Family Tinamidae), a basal family of birds, which is within the ratites (Fig. 4B) (PLR Brennan, K Zyscowski, and RO Prum, unpubl.). Some tinamou species have long intromittent penises, whereas others have a reduced nonintromittent penis (i.e., *Tinamus* vs. *Crypturellus*) (Brennan et al. 2008). In species with a nonintromittent penis, female vaginas are a simple tube, whereas in species with an intromittent penis, the female vagina is folded like an accordion (Fig. 4B).

Genital coevolution in tinamous could have evolved by: (1) male–male competition and natural selection, or (2) female choice. Female tinamous are often promiscuous, so there is opportunity for sperm competition and female choice based on genital size and shape. Male tinamous could benefit from having a longer penis to deposit their sperm as close to the sperm storage tubules as possible. If a long penis is favored via male–male competition, then there might be natural selection on females to elongate their genitalia to allow for mechanical fit during copulation. Alternatively, tinamou penis size could also coevolve with vaginal morphology through female choice for greater copulatory stimulation. Female vaginal morphology could provide a morphological mechanism to distinguish among males with different genital lengths. The derived shift to monogamy in *Crypturellus* may have reduced selection on males to insure sperm storage, or eliminated the opportunity for female choice based on male genital features, leading to a reduction in penis length. The potential egg-laying cost to a long folded vagina may be enough to select against female genital elaboration, as tinamous lay some of the largest eggs for any given body size in birds (Davies 2002).

 Cite this article as *Cold Spring Harb Perspect Biol* doi: 10.1101/cshperspect.a017749

Mammals

Among placental mammals, insectivores, rodents, carnivores, bats, and primates generally have a penis bone known as the baculum, or os penis. The baculum is known to be morphologically diverse and subject to sexual selection at least in some groups (Ramm 2007). However, no studies have investigated effects of baculum shape on female genital morphology predicted by several hypotheses of baculum evolution including vaginal friction, induced ovulation, and sperm transfer during prolonged copulation (Patterson and Thaeler 1982; Larivière and Ferguson 2002; Dixson 2012).

In mice, baculum morphology appears to be associated with higher male fitness (Stockley et al. 2013), while experimental evolution studies have shown that mice evolving under high sexual selection (compared to enforced monogamy) change the shape although not the size of their baculum (Simmons and Firman 2014). These changes may be associated with higher mechanical stimulation of females to enhance litter size (Simmons and Firman 2014), although the mechanism by which baculum shape changes affect male fitness remains unknown, and will require an examination of females to determine how they can sense copulatory stimuli.

Primates have highly variable penile morphology. The specialization and characteristics of the primate penis are consistent with sexual selection hypotheses, in that penis length, tip elaboration and the presence of penile spines, are associated with multimale, multifemale groups, and dispersed mating systems in which females are promiscuous (Dixson 2012). In most primates with long bacula, the tip of the bone projects from the tip of the penis to contact the entrance of the cervix, facilitating sperm transport into the uterus (Dixson 2012). This suggests that there should be some coevolution in length between the penis and the vagina. However, unlike the case of penis length, a preliminary analysis of vaginal length found no correlation with mating system (Dixson 2012). A weak relationship seems to exist between the presence of female sexual swellings and long penises. For example, female sexual swellings can increase the distance to the cervical os up to 50% in chimpanzees (Dixson 2012), so a longer penis would be required to contact the female cervix during estrous, and this may explain the long filiform penis of male chimps (Dixson 2012).

In humans, MRI of individuals during coitus in the missionary position showed very close fit in shape, but significantly also that the anterior wall of the vagina lengthened by 1 cm in the preorgasmic phase (Schultz et al. 1999), underscoring the importance of examining copulation itself to understand how male and female genitalia fit together. Even though precopulatory female choice may play a role in penis length in humans (Mautz et al. 2013), whether copulatory female choice for genital features also takes place is unknown. The female vagina is densely innervated, particularly its anterior and posterior walls (Hilleges et al. 1995), and clearly this sensory ability can allow females to sense male penis size. Although previous studies have shown strong variation in genitalia among human ethnic groups (vagina: Pendergrass et al. 2000; penis: data by country compiled in Khan et al. 2012), there has been no rigorous effort to study human genital size coevolution with comparative analyses using standardized methods.

CONCLUSION

Evidence suggests that genital coevolution is widespread, and that both sexual conflict and female choice are commonly supported as the main coevolutionary mechanisms. Investigation of the mechanisms of genital coevolution and sexual conflict requires understanding the independent and co-occurring action of natural selection, female choice, and male–male competition. Genital coevolution can be inferred by comparative studies of closely related species or populations, as well as by microevolutionary studies that experimentally measure the fitness effects of rendering genital features ineffective. Experimental evolution can uncover the selective pressures responsible for genital evolution and establish whether changes in genital features in one sex result in changes in the other sex with fitness consequences. However, lack of evidence for coevolutionary responses in female defensive traits does not necessarily suggest that coevolution has not taken place, but merely that maintenance of female defensive traits may not be too costly. However, lack of understanding of functional morphology of genitalia during copulation, and copulatory behavior, as well as scarcity of quantitative studies that examine female genital variation, hinder further progress in our understanding of the mechanisms responsible for coevolution with few exceptions in model systems.

ACKNOWLEDGMENTS

P.L.R.B. is partly supported by National Science Foundation (NSF) grant IOS 0920344 while some of the work in this article was being conducted. We thank Teri Orr for helpful discussions during the preparation of this manuscript.

REFERENCES

Ah-King M, Barron AB, Herberstein ME. 2014. Genital evolution: Why are females still understudied? *PLoS Biol* **12:** e1001851.

Andersson MB. 1994. *Sexual selection*. Princeton University Press, Princeton, NJ.

Arnqvist G. 1997. The evolution of animal genitalia: Distinguishing between hypothesis by single species studies. *Biol J Linn Soc Lond* **60:** 365–379.

Arnqvist G, Rowe L. 2002. Correlated evolution of male and female morphologies in water striders. *Evolution(NY)* **56:** 936–947.

Arnqvist G, Rowe L. 2005. *Sexual conflict*. Princeton University Press, Princeton, NJ.

Augee ML, Gooden BA, Musser A. 2006. *Echidna: Extraordinary egg-laying mammal.* CSIRO, Collingwood, Australia.

Bedford JM, Mock OB, Goodman SM. 2004. Novelties of conception in insectivorous mammals (Lipotyphla), particularly shrews. *Biol Rev* **79:** 891–909.

Böhme W, Ziegler T. 2009. A review of iguanian and anguimorph lizard genitalia (Squamata: Chamaeleonidae; Varanoidea, Shinisauridae, Xenosauridae, Anguidae) and their phylogenetic significance: Comparisons with molecular data sets. *J Zool Syst Evol Res* **47:** 189–202.

Bonet S, Casas I, Holt WV, Yeste M. 2013. *Boar reproduction: Fundamentals and new biotechnological trends*. Springer, Berlin.

Breed WG, Leigh CM, Speight N. 2013. Coevolution of the male and female reproductive tracts in an old endemic murine rodent of Australia. *J Zool* **289:** 94–100.

Brennan PLR. 2013. Genital evolution: Cock-a-doodle-don't. *Curr Biol* **23:** R523–R525.

Brennan PLR, Prum R. 2012. Sexual conflict in the narrow sense: New insights from waterfowl biology. *Phil Trans R Soc* **367:** 2324–2338.

Brennan PLR, Prum RO, McCracken KG, Sorenson MD, Wilson RE, Birkhead TR. 2007. Coevolution of male and female genital morphology in waterfowl. *PLoS ONE* **25:** e418.

Brennan PLR, Birkhead T, Zyskowski K, Van der Waag J, Prum RO. 2008. Independent evolutionary reductions of the phallus in basal birds. *J Avian Biol* **39:** 487–492.

Brennan PLR, Clark C, Prum R. 2010. Explosive eversion and functional morphology of waterfowl penis supports sexual conflict in genitalia. *Proc Biol Sci* **277:** 1309–1314.

Briceño RD, Eberhard WG. 2009a. Experimental modifications of male genitalia induce cryptic female choice against the male in a tsetse fly. *J Evol Biol* **22:** 1516–1525.

Briceño RD, Eberhard WG. 2009b. Experimental demonstration of possible cryptic female choice on male tsetse fly genitalia. *J Insect Physiol* **55:** 989–996.

Burns MM, Hedin M, Shultz JW. 2013. Comparative analyses of reproductive structures in harvestmen Opiliones reveal multiple transitions from courtship to precopulatory antagonism. *PLoS ONE* **8:** e66767.

Carayon J. 1966. Traumatic insemination and the paragenital system. In *Monograph of Cimicidae (Hemiptera, Heteroptera)*, pp. 81–166. Entomological Society of America, College Park, MD.

Cayetano L, Maklakov AA, Brooks RC, Bonduriansky R. 2011. Evolution of male and female genitalia following release from sexual selection. *Evolution (NY)* **65:** 2171–2183.

Chapman T, Arnqvist G, Bangham J, Rowe L. 2003. Sexual conflict. *Trends Ecol Evol* **181:** 41–47.

Darwin C. 1871. *The descent of man and selection in relation to sex* (reprinted in 1959). Modern Library, New York.

Davies SJJF. 2002. *Ratites and tinamous*, p. 310. Oxford University Press, Oxford.

Defour L. 1844. Anatomie generale des Dipteres [General anatomy of the Dipteridae]. *Ann Sci Nat* **1:** 244–264.

Dixson AF. 2012. *Primate sexuality: Comparative studies of the prosimians, monkeys, apes, and humans.* Oxford University Press, Oxford.

East ML, Hofer H, Wickler W. 1993. The erect "penis" is a flag of submission in a female-dominated society: Greetings in Serengeti spotted hyenas. *Behav Ecol Sociobiol* **336:** 355–370.

Eberhard WG. 1985. *Sexual selection and animal genitalia.* Harvard University Press, Cambridge, MA.

Eberhard WG. 1996. *Female control: Sexual selection by cryptic female choice.* Princeton University Press, Princeton, NJ.

Eberhard WG. 2004a. Male–female conflict and genitalia: Failure to confirm predictions in insects and spiders. *Biol Rev* **79:** 121–186.

Eberhard WG. 2004b. Rapid divergent evolution of sexual morphology: Comparative tests of antagonistic coevolution and traditional female choice. *Evolution (NY)* **58:** 1947–1970.

Eberhard W. 2006. Sexually antagonistic coevolution in insects is associated with only limited morphological diversity. *J Evol Biol* **19:** 657–681.

Eberhard W. 2010a. Evolution of genitalia: Theories, evidence, and new directions. *Genetica* **138:** 5–18.

Eberhard W. 2010b. Rapid divergent evolution of genitalia: Theory and data updated. In *The evolution of primary sexual characters in animals*, pp. 40–78. Oxford University Press, Oxford.

Eberhard W. 2011. Experiments with genitalia: A commentary. *TREE* **26:** 17–21.

Eberhard W, Huber BA. 2010. Spider genitalia: Precise maneuvers with a numb structure in a complex lock. In *The evolution of primary sexual characters in animals*, pp. 249–284. Oxford University Press, Oxford.

Eberhard WG, Ramirez N. 2004. Functional morphology of the male genitalia of four species of *Drosophila*: Failure to confirm both lock and key and male–female conflict. *Annls Entomol Soc Am* **97:** 1007–1017.

Eberhard WG, Huber BA, Rodriguez RLS, Briceño RD, Salas I, Rodriguez V. 1998. One size fits all? Relationships between the size and degree of variation in genitalia and other body parts in twenty species of insects and spiders. *Evolution (NY)* **52:** 415–431.

Edgren RA. 1953. Copulatory adjustment in snakes and its evolutionary implications. *Copeia* **1953:** 162–164.

Evans JP, Gasparini C, Holwell GI, Ramnarine IW, Pitcher TE, Pilastro A. 2011. Intraspecific evidence from guppies for correlated patterns of male and female genital trait diversification. *Proc Biol Sci* **278:** 2611–2620.

Evans JP, van Lieshout E, Gasparini C. 2013. Quantitative genetic insights into the coevolutionary dynamics of male and female genitalia. *Proc Biol Sci* **280:** 1763–1767.

Fooden J. 1967. Complementary specialization of male and female reproductive structures in the bear macaque, *Macaca arctoides*. *Nature* **214:** 939–941.

Forero D, Berniker L, Weirauch C. 2013. Phylogeny and character evolution in the bee-assassins (Insecta: Heteroptera: Reduviidae). *Mol Phylogenet Evol* **66:** 283–302.

Fricke C, Perry J, Chapman T, Rowe L. 2009. The conditional economics of sexual conflict. *Biol Lett* **5:** 671–674.

Friesen C, Uhrig EJ, Squire MJ, Mason R, Brennan PLR. 2014. Sexual conflict over mating in red-sided Garter snakes *Thamnophis sirtalis* as evidenced by experimental manipulation of genitalia. *Proc Biol Sci* **281:** 20132694.

Gay L, Brown E, Tregenza T, Pincheira-Donoso D, Eady PE, Vasudev R, Hunt J, Hosken DJ. 2011. The genetic architecture of sexual conflict: Male harm and female resistance in *Callosobruchus maculatus*. *J Evol Biol* **24:** 449–456.

Greven H. 2005. Structural and behavioral traits associated with sperm transfer in the Poeciliinae. In *Viviparous fishes* (ed. Grier HJ, Uribe MC), pp. 147–165. New Life, Homestead, FL.

Hilleges M, Falconer C, EkmanOrdeberg G, Johanson O. 1995. Innervation of the human vaginal mucosa as revealed by PGP 9.5 immunohistochemistry. *Acta Anat* **153:** 119.

Hoch JM. 2009. Adaptive plasticity of the penis in a simultaneous hermaphrodite. *Evolution (NY)* **63:** 1946–1953.

Hosken DJ, Stockley P. 2004. Sexual selection and genital evolution. *Trend Ecol Evol* **192:** 87–93.

Hotzy C, Polak M, Rönn JL, Arnqvist G. 2012. Phenotypic engineering unveils the function of genital morphology. *Curr Biol* **22:** 2258–2261.

House CM, Simmons LW. 2007. No evidence for condition-dependent expression of male genitalia in the dung beetle *Onthophagus taurus*. *J Evol Biol* **20:** 1322–1332.

House CM, Lewis Z, Hodgson DJ, Wedell N, Sharma MD, Hunt J, Hosken DJ. 2013. Sexual and natural selection both influence male genital evolution. *PloS ONE* **8:** e63807.

Huber BA. 1999. Sexual selection in pholcid spiders (Araneae, Pholcidae): Artful chelicerae and forceful genitalia. *J Arachnol* **27:** 135–141.

Ilango K, Lane RP. 2000. Coadaptation of male aedeagal filaments and female spermathecal ducts of the old world phlebotomine sand flies. *J Med Entomol* **37:** 653–659.

Jagadeeshan S, Singh RS. 2006. A time-sequence functional analysis of mating behaviour and genital coupling in *Drosophila*: Role of cryptic female choice and male sex-drive in the evolution of male genitalia. *J Evol Biol* **19:** 1058–1070.

Johnston SD, Smith B, Pyne M, Stenzel D, Holt WV. 2007. One-sided ejaculation of echidna sperm bundles. *Am Nat* **1706:** E162–E164.

Kameda Y, Kawakita A, Kato M. 2009. Reproductive character displacement in genital morphology in Satsuma land snails. *Am Nat* **173:** 689–697.

Kamimura Y. 2012. Correlated evolutionary changes in *Drosophila* female genitalia reduce the possible infection risk caused by male copulatory wounding. *Behav Ecol Sociobiol* **66:** 1107–1114.

Kelly CD, Godin JG, Abdallah G. 2000. Geographic variation in the male intromittent organ of the Trinidadian guppy *Poecilia reticulata*. *Can J Zool* **78:** 1674–1680.

Khan S, Somani B, Lam W, Donat R. 2012. Establishing a reference range for penile length in Caucasian British men: A prospective study of 609 men. *BJU Int* **1095:** 740–744.

Kinahan AA, Bennett NC, O'Riain MJ, Hart L, Bateman PW. 2007. Size matters: Genital allometry in an African mole-rat (Family: Bathyergidae). *Evol Ecol* **21:** 201–213.

King RB, Jadin RC, Grue M, Walley HD. 2009. Behavioral correlates with hemipenis morphology in New World natricine snakes. *Biol J Linn Soc Lond* **98:** 110–120.

Kirkpatrick M, Ryan MJ. 1991. The evolution of mating preferences and the paradox of the lek. *Nature* **350:** 33–38.

Köhler J, Hahn M, Köhler G. 2012. Divergent evolution of hemipenial morphology in two cryptic species of mainland anoles related to *Anolis polylepis*. *Salamandra* **48:** 1–11.

Kotrba M, Huber J, Feijen HR. 2014. Coevolution of male and female genitalia in stalk-eyed flies (Diptera: Diopsidae). *Org Divers Evol* **14:** 187–201.

Kuntner M, Coddington JA, Schneider JM. 2009. Intersexual arms race? Genital coevolution in Nephilid spiders (Araneae, Nephilidae). *Evolution (NY)* **63:** 1451–1463.

Kwan L, Cheng YY, Rodd FH, Rowe L. 2013. Sexual conflict and the function of genitalic claws in guppies *Poecilia reticulata*. *Biol Lett* **95:** 20130267.

Langerhans RB. 2011. Genital evolution. In *Ecology and evolution of poeciliid fishes*, pp. 228–240. University of Chicago Press, Chicago.

Larivière S, Ferguson SH. 2002. On the evolution of the mammalian baculum: Vaginal friction, prolonged intromission or induced ovulation? *Mamm Rev* **32:** 283–294.

Lautenschlager AD, Brandis D, Storch V. 2010. Morphology and function of the reproductive system of representatives of the genus Uca. *J Morphology* **271:** 1281–1299.

Macagno AL, Pizzo A, Parzer HF, Palestrini C, Rolando A, Moczek AP. 2011. Shape-but not size-codivergence between male and female copulatory structures in onthophagus beetles. *PloS ONE* **6:** e28893.

Magurran AE, Seghers BH. 1994. A cost of sexual harassment in the guppy, *Poecilia reticulata*. *Proc R Soc B* **258:** 89–92.

Martin SB, Albert JS, Leberg PL. 2010. The evolution of the poeciliid gonopodium: Integrating morphological and behavioral traits. In *Viviparous fishes II* (ed. Uribe M, Grier H). New Life, Homestead, FL.

Masly JP. 2012. 170 years of "lock-and-key": Genital morphology and reproductive isolation. *Int J Evol Biol* **2012:** 247352.

Masly JP, Kamimura Y. 2014. Asymmetric mismatch in strain-specific genital morphology causes increased harm to *Drosophila* females. *Evolution (NY)* doi: 10.1111/evo.12436.

Matthews LH. 1950. Reproduction in the basking shark, *Cetorhinus maximus* (Gunner). *Phil Trans R Soc B* **234:** 247–316.

Mautz BS, Wong BBM, Peters RA, Jennions MD. 2013. Penis size interacts with body shape and height to influence male attractiveness. *Proc Natl Acad Sci* **110:** 6925–6930.

McPeek MA, Shen L, Farid H. 2009. The correlated evolution of three-dimensional reproductive structures between male and female damselflies. *Evolution (NY)* **63:** 73–83.

Mikkola K. 1992. Evidence for lock-and-key mechanisms in the internal genitalia of the Apamea moths (Lepidoptera, Noctuidae). *Syst Entomol*, **17:** 145–153.

Morrow EH, Arnqvist G. 2003. Costly traumatic insemination and a female counter-adaptation in bed bugs. *Proc Biol Sci* **270:** 2377–2381.

Mutanen M, Kaitala A, Mönkkönen M. 2006. Genital variation within and between three closely related Euxoa moth species: Testing the lock-and-key hypothesis. *J Zool* **2682:** 109–119.

Neufeld CJ. 2011. Modular phenotypic plasticity: Divergent responses of barnacle penis and feeding leg form to variation in density and wave-exposure. *J Exp Zool B Mol Dev Evol* **316:** 254–262.

Neufeld CJ, Palmer AR. 2008. Precisely proportioned: Intertidal barnacles alter penis form to suit coastal wave action. *Proc Bio Sci* **275:** 1081–1087.

Okuzaki Y, Sota T. 2014. How the length of genital parts affects copulation performance in a carabid beetle: Implications for correlated genital evolution between the sexes. *J Evol Biol* **27:** 565–574.

Parker GA. 2006. Sexual conflict over mating and fertilization: An overview. *Phil Trans R Soc B* **361:** 235–259.

Patterson BD, Thaeler CS Jr. 1982. The mammalian baculum: Hypotheses on the nature of bacular variability. *J Mammal* 1–15.

Pendergrass PB, Reeves CA, Belovicz MW, Molter DJ, White JH. 2000. Comparison of vaginal shapes in Afro-American, Caucasian and Hispanic women as seen with vinyl polysiloxane casting. *Gynecol Obstet Invest* **501:** 54–59.

Pereti AV. 2010. An ancient indirect sex model: Single and mixed patterns in the evolution of scorpion genitalia. In *The evolution of primary sexual characters in animals* (ed. Leonard JL, Córdoba-Aguilar A). Oxford University Press, Oxford.

Perry JC, Rowe L. 2012. Sexual conflict and antagonistic coevolution across water strider populations. *Evolution (NY)* **66:** 544–557.

Pisani GR. 1976. Comments on the courtship and mating mechanics of *Thamnophis* (Reptilia, Serpentes, Colubridae). *J Herpetol* **10:** 139–142.

Polihronakis M. 2006. Morphometric analysis of intraspecific shape variation in male and female genitalia of *Phyllophaga hirticula* (Coleoptera: Scarabaeidae: Melolonthinae). *Ann Entomol Soc Am* **991:** 144–150.

Pope CH. 1941. Copulatory adjustment in snakes. *Field Mus Nat Hist, Zool Ser* **24:** 249–252.

Prum RO. 2010. The Lande–Kirkpatrick mechanism is the null model of evolution by intersexual selection: Implications for meaning, honesty, and design in intersexual signals. *Evolution (NY)* **64:** 3085–3100.

Puniamoorthy N, Kotrba M, Meier R. 2010. Unlocking the "Black box": Internal female genitalia in Sepsidae Diptera evolve fast and are species-specific. *BMC Evol Biol* **101:** 275.

Ramm SA. 2007. Sexual selection and genital evolution in mammals: A phylogenetic analysis of baculum length. *Am Nat* **169:** 360–369.

Ramos M, Coddingtion JA, Christenson TE, Irschick DJ. 2005. Have male and female genitalia coevolved? A phylogenetic analysis of genitalic morphology and sexual size dimorphism in web-building spiders (Araneae: Araneoidea). *Evolution (NY)* **59:** 1989–1999.

Renfree M. 1987. *Reproductive physiology of marsupials*. Cambridge University Press, Cambridge.

Robertson HM, Paterson HE. 1982. Mate recognition and mechanical isolation in Enallagma damselflies (Odonata: Coenagrionidae). *Evolution (NY)* 243–250.

Rönn J, Katvala M, Arnqvist G. 2007. Coevolution between harmful male genitalia and female resistance in seed beetles. *Proc Natl Acad Sci* **104:** 10921–10925.

Sánchez V, Cordero C. 2014. Sexual coevolution of spermatophore envelopes and female genital traits in butterflies: Evidence of male coercion? *PeerJ* **2:** e247.

Sasabe M, Takami Y, Sota T. 2010. QTL for the species-specific male and female genital morphologies in *Ohomopterus* ground beetles. *Mol Ecol* **19:** 5231–5239.

Sauer J, Hausdorf B. 2009. Sexual selection is involved in speciation in a land snail radiation on Crete. *Evolution* **63:** 2535–2546.

Schultz WW, Van Andel P, Sabelis I, Mooyaart E. 1999. Magnetic resonance imaging of male and female genitals during coitus and female sexual arousal. *BMJ* **319:** 1596–1600.

Shapiro AM, Porter AH. 1989. The lock-and-key hypothesis: Evolutionary and biosystematic interpretation of insect genitalia. *Ann Rev Entomol* **34:** 231–245.

Showalter I, Todd B, Brennan PLR. 2013. Intraspecific and interspecific variation of the vagina in two species of water snakes. *Biol J Linn Soc Lond* **111:** 183–191.

Siegel DS, Miralles A, Trauth SE, Aldridge RD. 2011. The phylogenetic distribution and morphological variation of the "pouch" in female snakes. *Acta Zool* **93:** 400–408.

Simmons LW. 2013. Sexual selection and genital evolution. *Aust J Entomol* **53:** 1–17.

Simmons LW, Firman RC. 2014. Experimental evidence for the evolution of the mammalian baculum by sexual selection. *Evolution (NY)* **68:** 276–283.

Simmons LW, García-González F. 2011. Experimental coevolution of male and female genital morphology. *Nat Commun* **2:** 374–377.

Simmons LW, House CM, Hunt J, García-González F. 2009. Evolutionary response to sexual selection in male genital morphology. *Curr Biol* **1917:** 1442–1446.

Sota T, Kubota K. 1998. Genital lock-and-key as a selective agent against hybridization. *Evolution (NY)* **52:** 1507–1513.

Sota T, Tanabe T. 2010. Multiple speciation events in an arthropod with divergent evolution in sexual morphology. *Proc R Soc B* **277:** 689.

Stockley P, Ramm SA, Sherborne AL, Thom MDF, Paterson S, Hurst JL. 2013. Baculum morphology predicts reproductive success of male house mice under sexual selection. *BMC Biol* **11:** 66.

Tanabe T, Sota T. 2008. Complex copulatory behavior and the proximate effect of genital and body size differences on mechanical reproductive isolation in the millipede genus *Parafontaria*. *Am Nat* **1715:** 692–699.

Tatarnic NJ, Cassis G. 2010. Sexual coevolution in the traumatically inseminating plant bug genus *Coridromius*. *J Evol Biol* **23:** 1321–1326.

Wojcieszek JM, Simmons LW. 2012. Micro-CT scanning provides insight into the functional morphology of millipede genitalia. *J Zool* **287:** 91–95.

Wojcieszek JM, Simmons LW. 2013. Divergence in genital morphology may contribute to mechanical reproductive isolation in a millipede. *Ecol Evol* **32:** 334–343.

Yassin A, Orgogozo V. 2013. Coevolution between male and female genitalia in the *Drosophila* melanogaster species subgroup. *PloS ONE* **82:** e57158.

Yoshizawa K, Ferreira RL, Kamimura Y, Lienhard C. 2014. Female penis, male vagina, and their correlated evolution in a cave insect. *Curr Biol* **24:** 1006–1010.

Index

H

I

P

R

S

T

U

V